AF332953

Multiscale Modelling
of Materials

MATERIALS RESEARCH SOCIETY
SYMPOSIUM PROCEEDINGS VOLUME 538

Multiscale Modelling of Materials

Symposium held November 30–December 3, 1998, Boston, Massachusetts, U.S.A.

EDITORS:

Vasily V. Bulatov
Lawrence Livermore National Laboratory
Livermore, California, U.S.A.

Tomas Diaz de la Rubia
Lawrence Livermore National Laboratory
Livermore, California, U.S.A.

Rob Phillips
Brown University
Providence, Rhode Island, U.S.A.

Efthimios Kaxiras
Harvard University
Cambridge, Massachusetts, U.S.A.

Nasr Ghoniem
University of California
Los Angeles, California, U.S.A.

Materials Research Society
Warrendale, Pennsylvania

Single article reprints from this publication are available through
University Microfilms Inc., 300 North Zeeb Road, Ann Arbor, Michigan 48106

CODEN: MRSPDH

Published by:

Materials Research Society
506 Keystone Drive
Warrendale, PA   15086
Telephone (724) 779-3003
Fax (724) 779-8313
Website: http://www.mrs.org/

Library of Congress Cataloging in Publication Data

Multiscale modelling of materials : symposium held November 30–December 3,
    1998, Boston, Massachusetts, U.S.A. / editors, Vasily V. Bulatov, Tomas Diaz
de la Rubia, Rob Phillips, Efthimios Kaxiras, Nasr Ghoniem
        p.cm.—(Materials Research Society symposium proceedings ;
        ISSN 0272-9172 ; v. 538)
        Includes bibliographical references and index.
        ISBN 1-55899-444-0
        1. Materials—Computer simulation—Congresses. 2. Continuum
    mechanics—Computer simulation—Congresses. 3. Length measurement—
    Congresses. 4. Molecular dynamics—Computer simulation—Congresses.
    I. Bulatov, Vasily V. II. Diaz de la Rubia, Tomas III. Phillips, Rob
    IV. Kaxiras, Efthimios V. Ghoniem, Nasr VI. Series: Materials Research Society
    symposium proceedings ; v. 538.
TA405.M887 1999   99-13247
620.1'1'0113—dc21   CIP

Manufactured in the United States of America

PART III: <u>CRYSTAL DEFECTS AND INTERFACES</u>

*Invited Paper

PART IV: <u>NOVEL METHODS FOR MATERIALS MODELLING</u>

*Invited Paper

PREFACE

An emerging theme in computational materials science is that of multiscale modelling. While the definition of "multiscale modelling" itself is developing as new ideas and applications appear, a broad interpretation of this field includes efforts to exploit insights arising either from distinct methodologies, or from the attempt to incorporate multiple mechanisms into the same modelling paradigm. Though multiple scale models are not new (their origins can be traced to early attempts to explain macroscopic phenomena on the basis of molecular interactions within a physical system, or between a system and its boundaries), the topic has recently taken on a new sense of urgency. This is in large part due to the recognition that brute force computational approaches often fall short of allowing for direct simulation of both the characteristic structures and temporal processes found in real materials. As a result, a number of hybrid approaches are now finding favor in which ideas borrowed from distinct disciplines or modelling paradigms are unified to produce more powerful techniques.

Symposium J, "Multiscale Modelling of Materials," at the 1998 MRS Fall Meeting in Boston, Massachusetts, attempted to serve as a forum for taking stock of recent developments in multiscale modelling. The symposium was characterized by high attendance, lively discussion, a series of successful invited talks, and important contributed talks and posters. In addition, this symposium participated in three separate joint sessions with other symposia:

(i)   Symposium K, "Computation of Rates of Activated Processes"

This joint session was engendered in part on the recognition that multiscale efforts are needed not only in order to treat the multiple length scales that arise in materials, but also to treat the multitude of time scales arising in many processes. Indeed, much of the discussion in this joint session was aimed at determining how to connect the exceedingly short time scales characterizing atomic motion to the time scales of interest in physical phenomena, ranging from seconds upward.

(ii)   Symposium M, "Fracture and Ductile vs. Brittle Behavior—Theory, Modelling and Experiment"

Developments that were widely discussed in this symposium were primarily methodological, and concerned the general question of how best to exploit atomic-scale calculations without having to resort to such calculations throughout the medium of interest. A common theme in these models is the use of continuum ideas in one region, and atomic-scale calculations only in the vicinity of what one might loosely call the "process zone." A number of different schemes were presented in which the continuum regions were treated with finite element descriptions, while the regions of full atomic resolution were treated using conventional atomistic arguments. The key difficulty that must be faced in these types of methods is the "handshaking" between the different regions.

(iii) Symposium N, "Microstructural Processes in Irradiated Materials"

A joint session with Symposium N focused on the application of molecular dynamics, kinetic Monte Carlo and rate theory methods to describing microstructure changes in irradiated materials. Presentations ranged from descriptions of the atomistic process of defect production in irradiated materials (a picosecond and nanometer-scale problem) to rate equation-based simulations of damage accumulation and microstructure evolution in irradiated steels (a macroscopic length and time-scale problem). Furthermore, several talks described the coupling of these simulation methods to provide a unified simulation approach across all relevant length and time scales. Applications to embrittlement of reactor pressure vessel steels provided an excellent example of how multiscale modelling can be used to model and understand problems of actual industrial interest in real materials.

One of the specific problem areas in which multiscale modelling has seen increasing success is the investigation of thin-film growth and processing. These ideas were illustrated in this symposium via a variety of models involving a few key elements, the first being a microscopic parameterization of the key atomic-scale processes that take place during the growth process. Once these processes have been parameterized, a kinetic Monte Carlo (KMC) scheme can be implemented in order to simulate the process of interest, in which all relevant microscopic processes are taken into account with appropriate rates. Finally, KMC results can be incorporated into large-scale theoretical models, which serve to describe the film morphology at macroscopic scales. In addition to the modelling of growth processes, a number of efforts addressed different classes of defects in materials including point defects, surfaces and grain boundaries. The role of defects as a key structural component in materials is well known. The fact that such defects are often associated with length scales intermediate between atomic and microstructural features makes them a continuing challenge in the context of materials modelling.

This symposium illustrated the important role of multiscale modelling as a key component in the current efforts to construct a viable field of computational materials science. A number of important problems and challenges remain to be addressed, some major unresolved themes being the seamless coupling of different regions described by disparate methodologies, and the issue of multiple time scales. Nevertheless, the symposium showed that a legitimate foundation for future work has been laid.

Vasily V. Bulatov
Tomas Diaz de la Rubia
Rob Phillips
Efthimios Kaxiras
Nasr Ghoniem

January 1999

MATERIALS RESEARCH SOCIETY SYMPOSIUM PROCEEDINGS

MATERIALS RESEARCH SOCIETY SYMPOSIUM PROCEEDINGS

Volume 532— Silicon Front-End Technology—Materials Processing and Modelling,
N.E.B. Cowern, D.C. Jacobson, P.B. Griffin, P.A. Packan, R.P. Webb,
1998, ISBN: 1-55899-438-6

Volume 533— Epitaxy and Applications of Si-Based Heterostructures, E.A. Fitzgerald,
D.C. Houghton, P.M. Mooney, 1998, ISBN: 1-55899-439-4

Volume 535— III-V and IV-IV Materials and Processing Challenges for Highly Integrated
Microelectonics and Optoelectronics, S.A. Ringel, E.A. Fitzgerald, I. Adesida,
D. Houghton, 1999, ISBN: 1-55899-441-6

Volume 536— Microcrystalline and Nanocrystalline Semiconductors—1998, L.T. Canham,
M.J. Sailor, K. Tanaka, C-C. Tsai, 1999, ISBN: 1-55899-442-4

Volume 537— GaN and Related Alloys, S.J. Pearton, C. Kuo, T. Uenoyama, A.F. Wright, 1999,
ISBN: 1-55899-443-2

Volume 538— Multiscale Modelling of Materials, V.V. Bulatov, T. Diaz de la Rubia, R. Phillips,
E. Kaxiras, N. Ghoniem, 1999, ISBN: 1-55899-444-0

Volume 539— Fracture and Ductile vs. Brittle Behavior—Theory, Modelling and
Experiment, G.E. Beltz, R.L. Blumberg Selinger, K-S. Kim, M.P. Marder, 1999,
ISBN: 1-55899-445-9

Volume 540— Microstructural Processes in Irradiated Materials, S.J. Zinkle, G. Lucas,
R. Ewing, J. Williams, 1999, ISBN: 1-55899-446-7

Volume 541— Ferroelectric Thin Films VII, R.E. Jones, R.W. Schwartz, S. Summerfelt, I.K. Yoo,
1999, ISBN: 1-55899-447-5

Volume 542— Solid Freeform and Additive Fabrication, D. Dimos, S.C. Danforth, M.J. Cima,
1999, ISBN: 1-55899-448-3

Volume 543— Dynamics in Small Confining Systems IV, J.M. Drake, G.S. Grest, J. Klafter,
R. Kopelman, 1999, ISBN: 1-55899-449-1

Volume 544— Plasma Deposition and Treatment of Polymers, W.W. Lee, R. d'Agostino,
M.R. Wertheimer, B.D. Ratner, 1999, ISBN: 1-55899-450-5

Volume 545— Thermoelectric Materials 1998—The Next Generation Materials for Small-Scale
Refrigeration and Power Generation Applications, T.M. Tritt, M.G. Kanatzidis,
G.D. Mahan, H.B. Lyon, Jr., 1999, ISBN: 1-55899-451-3

Volume 546— Materials Science of Microelectromechanical Systems (MEMS) Devices,
A.H. Heuer, S.J. Jacobs, 1999, ISBN: 1-55899-452-1

Volume 547— Solid-State Chemistry of Inorganic Materials II, S.M. Kauzlarich,
E.M. McCarron III, A.W. Sleight, H-C. zur Loye, 1999, ISBN: 1-55899-453-X

Volume 548— Solid-State Ionics V, G-A. Nazri, C. Julien, A. Rougier, 1999,
ISBN: 1-55899-454-8

Volume 549— Advanced Catalytic Materials—1998, P.W. Lednor, D.A. Nagaki, L.T. Thompson,
1999, ISBN: 1-55899-455-6

Volume 550— Biomedical Materials—Drug Delivery, Implants and Tissue Engineering,
T. Neenan, M. Marcolongo, R.F. Valentini, 1999, ISBN: 1-55899-456-4

Volume 551— Materials in Space—Science, Technology and Exploration, A.F. Hepp,
J.M. Prahl, T.G. Keith, S.G. Bailey, J.R. Fowler, 1999, ISBN: 1-55899-457-2

Volume 552— High-Temperature Ordered Intermetallic Alloys VIII, E.P. George, M. Yamaguchi,
M.J. Mills, 1999, ISBN: 1-55899-458-0

Volume 553— Quasicrystals, J-M. Dubois, P.A. Thiel, A-P. Tsai, K. Urban, 1999,
ISBN: 1-55899-459-9

Volume 554— Bulk Metallic Glasses, W.L. Johnson, C.T. Liu, A. Inoue, 1999,
ISBN: 1-55899-460-2

Volume 555— Properties and Processing of Vapor-Deposited Coatings, M. Pickering,
B.W. Sheldon, W.Y. Lee, R.N. Johnson, 1999, ISBN: 1-55899-461-0

Volume 556— Scientific Basis for Nuclear Waste Management XXII, D.J. Wronkiewicz,
J.H. Lee, 1999, ISBN: 1-55899-462-9

Prior Materials Research Society Symposium Proceedings available by contacting Materials Research Society

Part I

Modelling Dislocation Properties and Behavior

DISLOCATIONS AND PLASTICITY IN SILICON CRYSTALS BY 3-D MESOSCOPIC SIMULATIONS

L.P. KUBIN *, A. MOULIN * and P. PIROUZ **
* LEM, CNRS-ONERA, 92322 F-Châtillon Cedex, kubin@onera.fr, amoulin@zig.onera.fr
** Dept. MSE, CWRU, Cleveland, OH 44106-7204, USA, pirouz@cwmsd.mse.cwru.edu

ABSTRACT

Several problems related to the dynamics of dislocation sources and the plasticity of silicon crystals are investigated with the help of a mesoscopic simulation. The questions successively examined are the dynamics of a source of perfect dislocations and the conditions under which perfect or partial dislocations are emitted by a source. This leads to a discussion of the initial steps of the model proposed by Pirouz for mechanical twinning and, further, to the suggestion that a relation may exist between several transitions experimentally observed at low temperatures in elemental or compound semi-conductors: a change in the slope of the yield stress vs. temperature curves, a brittle-to-ductile transition and a change in the nature of the mobile dislocations. Finally, simulations are presented of the yield point phenomenon that is a well-known feature of Si and Ge crystals. The results are discussed in terms of evolutionary laws for the total dislocation density during straining.

INTRODUCTION

Silicon is the model material where dislocation velocity is the best characterized and for which a large amount of experimental data exists in the literature (see e.g. [1] for a review). Dislocation glide in the diamond cubic (dc) structure occurs by the breaking and reconstruction of covalent bonds, which leads to the occurrence of high Peierls forces. Hence, in a large range of temperatures (typically below about 1200 K), the energy involved in the movement of a dislocation from one Peierls valley to the next is larger than the self-energy of the dislocation. As a consequence, the dislocation loops tend to be straight, often forming hexagonal loops with the sides of the hexagon parallel to the three <110> close-packed directions of a {111} slip plane. Dislocation motion then occurs by the formation of kink pairs over the primary Peierls valleys and the migration of the kinks over the secondary Peierls valleys. These processes are usually modeled within the frame of the diffusing kink model [2].

However, to understand plastic flow it is also necessary to model other processes which are less easy to approach experimentally, for instance the mechanisms of dislocation multiplication. In this context, the recent development of fully three-dimensional mesoscopic simulations of dislocation dynamics allows to check the basic assumptions of several existing models. The present paper is thus concerned with several applications of such mesoscopic simulations to the case of silicon crystals.

The main principles of this simulation are first presented. Two types of applications are then discussed. One is concerned with the dynamics of a single Frank-Read source, with a source segment that is either a perfect dislocation or a dislocation dissociated into two Shockley partials. In the latter case, it is shown that the source may in certain conditions emit a partial dislocation, as has been predicted by Pirouz [3]. The existence of a transition between the emission of partial versus perfect dislocations may have several possible consquences on the mechanical properties that are discussed with respect to the available experimental data. Finally, such simulations can also be a powerful tool for the understanding of particular features of the stress-strain curves, as examplified by a study of the yield point phenomenon in relation to the multiplication mechanisms.

MESOSCOPIC SIMULATIONS

Method

The simulation code used in the present work derives from an existing mesoscopic simulation dedicated to the simulation of dislocation dynamics in FCC crystals [4]. By "mesoscopic" is meant

Mat. Res. Soc. Symp. Proc. Vol. 538 © 1999 Materials Research Society

here that dislocations are treated as linear discontinuities in an elastic isotropic continuum. Space is discretized, the smallest distance considered in the continuum being the critical annihilation distance of edge dipoles, $y_e \sim 6b$, where b is the magnitude of Burgers vector of the perfect dislocations [5]. This distance marks the border-line between the domain of dislocation core properties and the domain of elastic interactions, the former being accounted for by phenomenological local rules. As a consequence, the simulation is based on a discrete fcc lattice, mimicking the diamond cubic lattice in the present case, whose parameter is 2.71 nm. The resolved shear stress applied on each dislocation segment includes the applied stress, the line-tension stress and the interaction stress corresponding to long range elastic interactions. More detail about the discretization of the dislocation lines and the implementation of the dynamics and of the local rules can be found in [6, 7].

The modifications introduced were intended to account for the main consequences of the high Peierls forces at the mesoscopic scale, viz. the hexagonal shapes of the gliding dislocation loops and the particular stress vs. velocity laws associated with the kink pair mechanism. As discussed in [8], this involves the discretization of the segment orientations into three elementary characters, screw and mixed $\pm$ 60°, in place of the initial edge/screw description. In addition, a stress and temperature-dependent velocity rule was introduced for each segment. Within the kink diffusion model of Hirth and Lothe [2], the prefactor of the thermally activated velocity depends on whether the length L of the considered segment is smaller or larger than a temperature-dependent critical length L_c. The two corresponding expressions are reproduced here without demonstration.

For $L < L_c$, only one kink pair propagates at any time along the dislocation line and the mean free path of the kink pair is L. The velocity of the dislocation segment is then given by:

$$v = L\frac{\tau^* bh^2}{kT} \upsilon_D \exp\left(-\frac{2E_{kf} + E_{km}}{kT}\right)\exp\left(\frac{2S}{k}\right), \tag{1}$$

where h is kink height, υ_d is the Debye frequency, k the Boltzmann constant, τ^* the resolved shear stress, and T the absolute temperature. E_{kf} and E_{km} denote respectively the formation and migration energies of a single kink. S is a correcting entropy term introduced by Marklund [9]. In this regime, which corresponds to large stresses and low temperatures, the velocity of a segment is proportional to its length.

For $L > L_c$, the mean free path of kinks is smaller than the segment length. The stress velocity law is then of the form:

$$v = \frac{2a_0\tau^* bh^2}{kT} \upsilon_D \exp\left(-\frac{E_{kf} + E_{km}}{kT}\right)\exp\left(\frac{S}{k}\right), \tag{2}$$

where a_o is the lattice parameter. This law applies at high temperatures and low stresses and has been the object of numerous experimental checks [1, 10]. In this regime, the velocity law can be rewritten in the form

$$v = \frac{dx}{dt} = \alpha\tau^* f(T) \tag{3}$$

where x denotes the position of the segment, f(T) includes the temperature dependence and α is a constant. Hence, a change of temperature is strictly equivalent to a change of time scale. Considering for instance the configuration of dislocation loops emitted by a Frank-Read (FR) source, a temperature change modifies the dynamics but not the geometry. The transition between the two velocity regimes of eqs. 1 and 2 has been experimentally observed by Louchet [11] and Hirsch et al. [12].

Validation

Two algorithmic parameters govern the dynamics [8]. One is the elementary time step dt; this parameter determines the frequency of the resolved stress computations, of the dislocation displacements and of the treatment of particular events such as dislocation annihilation, dislocations running out from the free surfaces of the simulated crystal, cross-slip, etc. The other parameter is a discretization parameter, L_d, governing the creation of new dislocation segments at the pinning points of dislocation sources. Thanks to this rather simple set of rules, the present simulation can reproduce the complete operation of a FR source from a segment pinned at its two ends (cf. fig. 1).

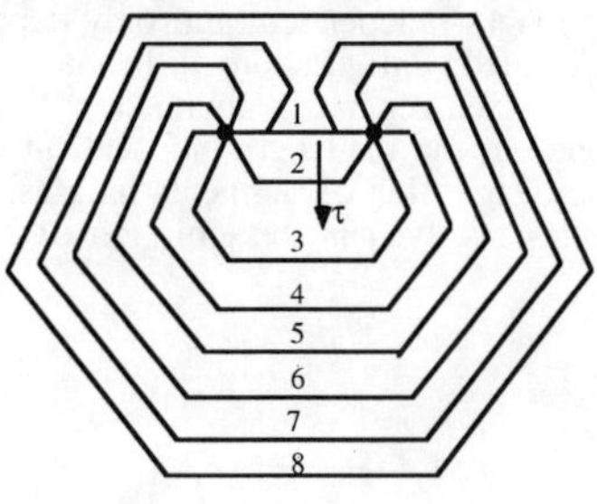

Fig.1. Successive configurations of a dislocation loop emitted by a Frank-Read source in a (1$\bar{1}$1) slip plane (T = 1000 K, τ = 35 MPa, L = 0.81 µm).

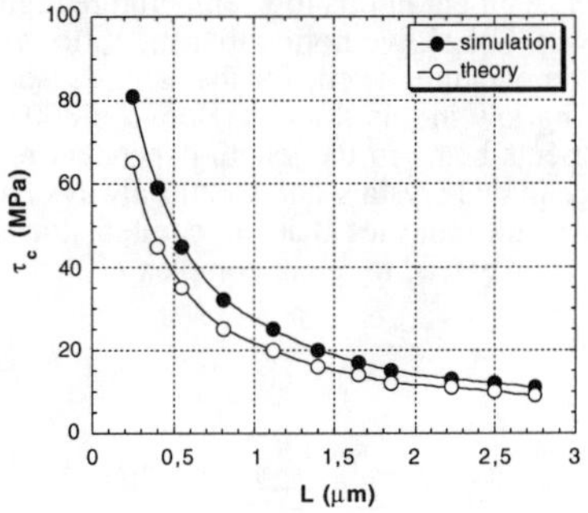

Fig. 2. Critical stress τ_c as a function of source length L (T = 1000 K). The theoretical values refer to circular loops [13].

The two parameters dt and L_d may induce artefacts if their values are not correctly defined. For instance, the time step dt must be kept small enough to ensure the convergence of the equations of motion of individal dislocation segments. It has also to be optimized in order to reduce the computation time, which is the main limitation of the simulation when large dislocation densities are involved. The optimal ranges for these parameters were defined by considering the critical bowing-out stress of an initially pinned segment, τ_c, which is defined as the lower applied stress for which a dislocation loop can be formed. τ_c is a static and temperature-independent quantity, which stems from a static equilibrium between the line tension stress and the applied stress. Fig. 2 shows the variations of τ_c versus the source length at T = 1000 K. The computed values are 20% larger than the ones predicted by Foreman [13] for circular loops, which is due to the fact that we are dealing here with hexagonal loops of larger self-energy.

By checking the stability of the critical stress values with respect to changes of the two parameters, one can define optimum values for these parameters for each simulation condition, as defined by the temperature, the applied stress and the initial source length. As a consequence, all the results presented in what follows are independent of the algorithmic construction of the simulation. It is worth noting that the time steps used in the case of silicon (typically dt = 5×10^{-2} s) are much larger than those used for fcc crystals (typically 10^{-9} s or less). This is due to the low dislocation mobilities induced by the lattice friction.

STUDY OF A SINGLE DISLOCATION SOURCE

The interest of such a simulation is that, in contrast to the available models, it does not make any phenomenological assumption regarding the conditions for dislocation emission by the source. From an investigation of the dynamics of a FR source as a function of stress and temperature, it appears that the critical parameters are the size of the innermost loop and the time it takes to form it (i.e., to reach step 8 of fig. 1). Fig. 3 shows the evolution of the (equivalent) radius of this loop as a function of temperature under a constant applied stress 1.3 times larger than the critical stress and for an initial length of the source segment of 0.8 µm. The double plateau obtained in fig. 3 directly illustrates the influence of the two velocity regimes for the dislocations. According to the theoretical value of the critical transition length L_c, all the segments are in the length-independent regime part of the velocity law (eq. 2) above T = 1000 K. In this domain, which corresponds to the high temperature plateau in fig. 3, a change in temperature can be accommodated by a change in the time scale of the source emission (eq. 3) without any modification in the geometry of the emitted loop. For T < 800 K, all the segments are in the length-dependent regime of the velocity law, which corresponds to the low temperature plateau of fig. 3. Considering the velocities of a given segment in this low temperature regime and at two different temperatures, $v = \alpha f(T)L\tau^*$ and $v' = \alpha f(T')L\tau^*$, we notice that the ratio $v/v' = f(T)/f(T')$ is length-independent and only depends on temperature. Then, for the same reason as before, the emitted configurations follow a simple scaling law in this domain. Between 800 K and 1000 K a mixed regime is obtained, the small segments being in the length-dependent regime and the long ones in the length-independent one. Since in real crystals one should always have a rather wide distribution of segments lengths, this first result indicates that the combination of the simple rules involved in the emission of a FR source may lead to rather complex effects.

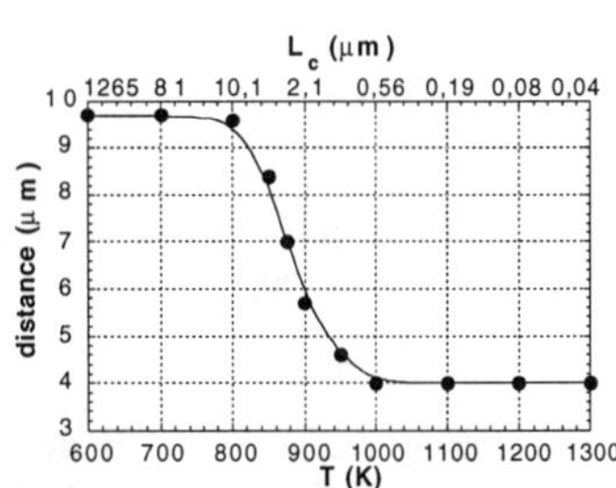

Fig. 3. Radius of the first complete loop vs. temperature (τ = 35 MPa, L = 0.81 μm).

Fig.4. Typical configuration of dislocation loops emitted by a source (same conditions as fig. 1). The crystal size is 550 μm.

In a next step the concentric loops emitted by a single source (cf. fig. 4) are considered. These configurations seem to present a uniform spacing between the successive loops, in agreement with an assumption commonly made in the modeling of source emission. However, closer examination reveals that the spacings can deviate from average by about 10%, the deviations being governed by the evolution of the backstress on the source segment [14,15]. A quantitative study was performed of the effective stresses acting on the innermost dislocation loop in its glide plane and in its cross-slip plane, respectively τ_G^* and τ_{CS}^*. The ratio τ_{CS}^*/τ_G^* is found to increase with the number of emitted loops. This is illustrated by fig. 5 where it is plotted for several values of (stress, temperature) pairs corresponding to the upper yield point of silicon. These values were computed in the steady state or saturation regime, where a new loop is formed by the source every time a loop disappears at the free surface of the simulated crystal. As the ratio τ_{CS}^*/τ_G^* increases monotonically with temperature, the occurrence of cross-slip events is favoured in a purely mechanical way.

To illustrate the complex combination of phenomena involved in the source emission, fig. 6 shows a plot of the dislocation density produced in a slip plane by a single source operating in steady state conditions, as a function of applied stress and temperature. The (stress, temperature) pairs shown in fig. 5 are reported on this plot. The interesting fact is that points 1 and 5 correspond to the same saturation density, while according to fig. 5 the cross-slip probability should be quite different. In fact, point 1 corresponds to high stress/low temperature conditions, where the dislocation density is essentially driven by the length-dependent regime of the velocity law. Due to the high applied stress, the interaction stresses represent a small relative fraction of the effective stress and they cannot influence cross-slip. In contrast, point 5 corresponds to low stress/high temperature conditions, where the dislocation density is limited by the low applied stress. In this case, the elastic interactions between dislocations are significant compared to the applied stress, and this leads to high values of the ratio $\tau_{CS}^{*}/\tau_{G}^{*}$. Hence, configurations 1 and 5 are very similar from a geometrical point of view but are intrinsically governed by quite different mechanisms.

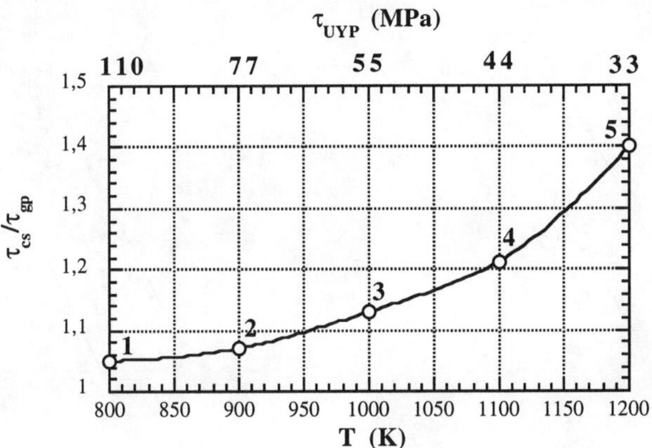

Fig. 5. Simulated values of the ratio τ_{CS}/τ_{G} for couples of stress and temperature values corresponding to the upper yield stress of silicon.

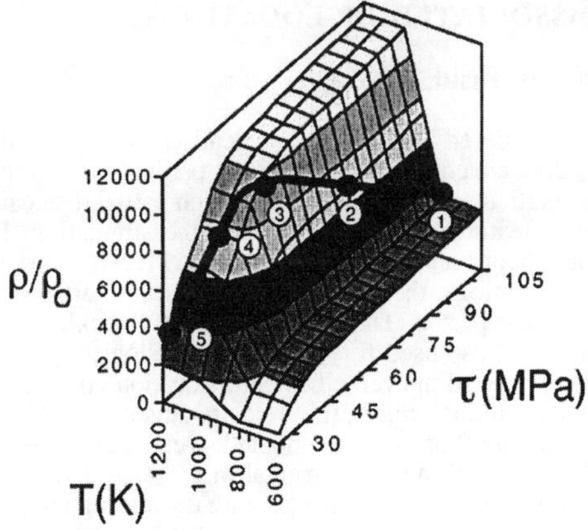

Fig. 6. Saturation density (scaled by the initial density ρ_o) in the glide plane of a source as a function of temperature and applied stress. The source length is 0.81 µm and the 2-D crystal size is 550 µm.

This study can easily be extended to the case where the dislocation mobilities depend on the character of the moving dislocation, provided that the anisotropy of dislocation mobility is known theoretically or experimentally. As an illustration, fig. 7 shows the changes in source configuration obtained in the case where screw and 60° dislocations have different mobilities (the values are those quoted in [1] for germanium and their ratio is denoted by g) and in the case where dislocation segments of opposite sign have different mobilities (the values are typical ones quoted in [1] for III-V compounds).

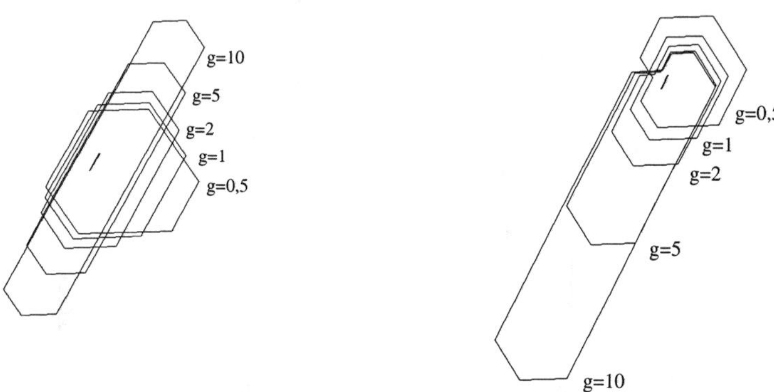

Fig. 7. Influence of the velocity anisotropy ratio g on the geometry of the first formed loop (same conditions as in fig. 1).- (a) between screw and 60° dislocations (case of Ge, $g = v_{60°} / v_{screw}$). - (b) between α and screw, or α and β dislocations (case of III-V compounds, $g = v_{\alpha} / v_{screw, \beta}$).

SOURCES WITH DISSOCIATED DISLOCATIONS

Perfect and Partial Frank-Read Sources

Up to now we have considered only perfect dislocations. The dynamics of dislocation sources made up of dislocations dissociated into two Shockley partial dislocations was investigated with the aim of identifying the particular conditions under which partial dislocations can be emitted [16]. For this purpose, two new features were introduced in the simulation. If γ is the stacking fault energy ($\gamma = 69$ mJ/m^{-2} in silicon) and $b_p = |\mathbf{b}_p| = |1/6<112>|$ is the modulus of the Burgers vector of the Shockley dislocations, the stacking fault induces a drag stress $-\gamma/b_p$ on the leading partial and $+\gamma/b_p$ on the trailing partial. The velocities of the Shockley partials were expressed by the same functional forms as those used for perfect dislocations, as justified by Yasutake et al. [17]. In addition, the dependence of the velocities on dislocation character was introduced. Due to the lack of experimental data on this subject, this velocity anisotropy of the partials was assumed to be temperature-independent and the values adopted were those determined by Wessel and Alexander at 800 K [18]. Although no experimental measurement is available to date, it is well established, however, that this anisotropy increases with decreasing temperature [19].

The initial configuration of a dissociated source segment is shown in fig. 8(a). To characterize the type of dislocation emitted by the source, two quantities are of interest which are defined in fig. 8(b). At the moment that the first loop (made up by the leading partial) closes up, the distance d_1 measures the equivalent of a dissociation width, and d_2 measures the advance of the trailing partial with respect to the initial leading segment. By convention, a partial dislocation is considered as being emitted when a faulted loop is formed ($d_1 >> d_2$) with $d_2 < 200$ nm.

A factor that strongly influences the behavior of the partial dislocations is what might be called the "Escaig effect". For a fixed orientation of the applied stress, the resolved stresses on the Shockley partials are not identical and tend to either constrict or expand the stacking fault ribbon.

The latter case is met with a [100] stress orientation which is the one used here. For straight infinite dislocations, a simplified calculation due to Alexander and Wessel [18] estimates a critical stress of 300 MPa for an infinite extension of the stacking fault ribbon. Under a stress value fixed at 350 MPa in a first step, a clear transition in source emission is observed. This is illustrated in fig. 9 which shows the variations of the two geometrical parameters d_1 and d_2 within a wide range of temperature. Two characteristic source configurations are shown in addition. As expected, perfect dislocations are emitted at high temperatures (1100K), whereas source emission at 800K leads to the formation of a faulted loop. This transition can be interpreted in terms of the coexistence of the two velocity regimes (eqs. 1 and 2) and of their influence on the dynamics of source emission [16].

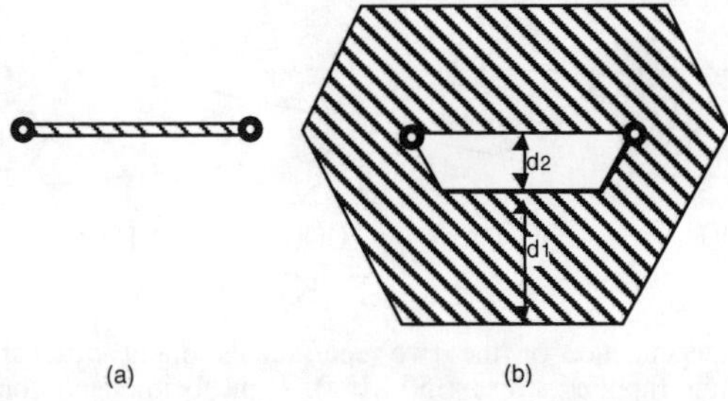

Fig. 8. Initial source configuration (a) and configuration obtained after the formation of the first loop (b). The striped area represents the faulted region.

Consequences

It must be noticed that once a faulted loop has been formed, the source shuts off and cannot re-emit a second partial dislocation. Indeed, this would involve the dissociation of the source dislocation within a faulted plane, thus producing a "forbidden" stacking fault of high energy. This locked configuration actually constitutes the first step of the model proposed by Pirouz [3, 19] for mechanical twinning in semiconductors. Further steps of this model involve the recombination and cross-slip of the source segment, where the order of the partials has been exchanged, and the production of faulted loops in successive slip planes.

A systematic investigation of the transition between perfect and partial emission by dislocation sources was performed as a function of stress and temperature. The results are summarized in fig. 10. This figure shows the values of d_2, the distance between the trailing and leading partials once the first loop has closed up, as a function of stress and temperature. Using the criterion mentioned above to define partial emission, the latter occurs in the domain shown in white in fig. 10. Although this estimate is based on several assumptions and is carried out with a favourable orientation of the applied stress, it shows that the first step of Pirouz model is likely to occur at low temperatures and under rather high stresses.

To fix the ideas, the computed critical stress at 950 K is about 400 MPa, a value much larger than the yield stress of silicon at such a temperature. In silicon, the Brittle to Ductile Transition (BDT) occurs at about 850 K (in non-predeformed crystals, this value is sensitive to the applied strain rate), under stresses typically of the order of a few tens of MPa. Thus, very high stress concentrations are needed to obtain partial emission, and possibly mechanical twinning, in the ductile regime. If, however, one considers compression tests on predeformed specimens and/or under a superimposed hydrostatic pressure, or indentation tests, stress levels can be reached that are typically those predicted by fig. 10 (see e.g. [20, 21]). Hence, the qualitative estimate provided by the simulation indicates that mechanical twinning, or at least its first step according to the model of Pirouz, is more likely to occur below the BDT than above. For the sake of completeness, it must be mentioned that the partial dislocations of a twin can also be nucleated at a surface heterogeneity [19, 22, 23], a situation that cannot be investigated with the present simulation.

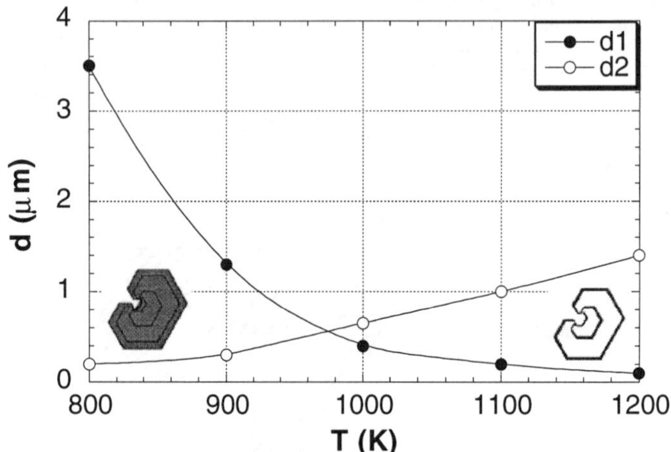

Fig. 9. Temperature dependence of the two coefficients d_1 and d_2 characterizing the source geometry vs. temperature (applied stress: 350 MPa). Typical simulated configurations of perfect and dissociated sources, for large and small values of d_2 respectively, are inset.

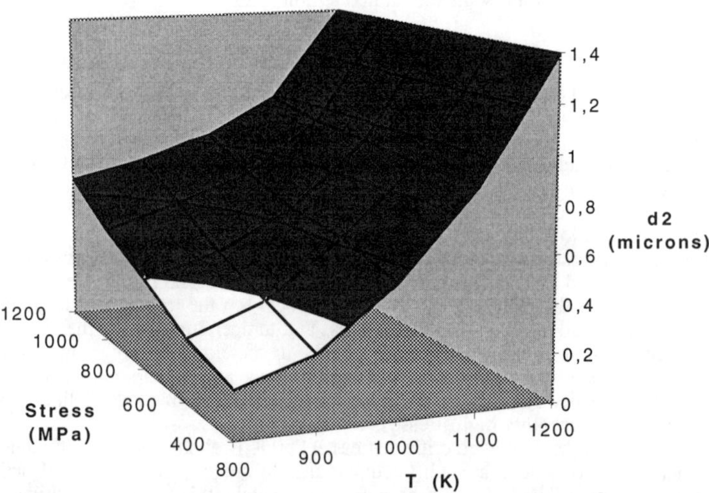

Fig. 10. Temperature and stress dependence of the geometrical parameter d_2. The source is considered as having emitted a partial dislocation when $d_2 < 200$ nm. A partial dislocation is emitted in the domain shown in white and perfect dislocations in the domain shown in black.

The fact that the transition in the mode of operation of a dislocation source occurs approximatively in the same temperature range as the BDT suggests that the two may be related. This possibility has been investigated by Pirouz et al. [24] who suggested a new model correlating these two transitions with a third one occurring in semiconductors in the same temperature range. This last transition is a break that occurs, roughly at the same temperature as the BDT, when the yield stress of a number of semiconductors is plotted as a function of temperature. The experimental results in such materials as Si [20,21], InP, GaAs, InSe [25] and SiC [24], distinctly show this transition in, for instance, the slope of the ln (τ) vs. 1/T plot.

According to the results described above, a FR source operating at high stress and low temperature can only emit one leading partial before being blocked. If the operation of a surface source is considered, it is known that the formation energy of the leading partial is smaller than that of the trailing partial [26]. Under such conditions, and for initially dislocation-free specimens, only mechanical twins or a few leading partials can be generated and the crystal should undergo brittle failure under uniaxial stress. With increasing temperature, a rather sharp transition towards ductile behavior should be observed when the generation of perfect dislocations becomes possible. In a crystal previously predeformed above the BDT, the behaviour of the dislocations introduced by prestrain should depend on the orientation of the applied stress (via the "Escaig effect") and on the strong anisotropy of the partials velocities that develops at low temperature and high stress. The microstructure after a low temperature deformation e.g., under hydrostatic pressure, should then consist of a mixture of constricted or extended dislocations, of leading partials and twins, in relative amounts that depend on the experimental conditions. Such features correspond well to the existing experimental observations (cf. [20] and [21] for the case of silicon).

The abrupt transition expected in the deformation microstructures and behavior of dislocation-free crystals in uniaxial deformation may also provide a new explanation for the particularly sharp character of the BDT observed in dislocation-free silicon crystals [27]. Whether dislocations are generated at the crack tip or produced by FR sources at a short distance from it, the transition towards partial nucleation when the temperature is decreased prevents the formation of a ductile zone and induces brittleness by lack of dislocation generation. Finally, the break in the $\ln(\tau)$ vs. 1/T plots yielded by many experimental observations close to the BDT temperature would, within this hypothesis, simply reflect a change in the nature of the mobile dislocations. The proposed model cannot be made fully quantitative by lack of experimental data, particularly about the low temperature velocities of the partial dislocations, but it appears to be in agreement with a wealth of experimental results.

THE YIELD POINT BEHAVIOR OF SILICON CRYSTALS

A yield point phenomenon is commonly observed in dc semiconducting materials containing a small initial dislocation density, as a result of the low stress sensitivity of the velocities (cf eqs. 1 and 2). Under deformation with a constant applied strain rate, the density of mobile dislocations has to increase by several orders of magnitude before the plastic strain matches the imposed strain rate. This induces a strong increase of the yield stress followed by a yield transient, until both dislocation velocities and densities are stabilized. This phenomenon has been the object of many experimental studies [28] and of a very successfull model proposed by Alexander and Haasen [29]. However, some questions still remain open regarding a few assumptions involved in this model. They have been examined by simulating the behavior of a 3-D crystal containing a random distribution of sources of perfect dislocations sources, identical to the ones discussed above.

The initial configuration of FR sources is characterized by the average length of the screw segments, the range of their length distribution and the total initial dislocation density. All the sources have same Burgers vector ($1/2[\bar{1}01]$, glissile on (111)), so that the simulation operates in easy glide conditions, in a crystal of cubic size and of volume $V=(80\ \mu m)^3$. The stress was applied along a $[\bar{1}23]$ axis and constant strain rate conditions were imposed, with strain rate values in the range 3×10^{-5}-3×10^{-4} s^{-1}. Under the typical conditions where a yield point is recorded experimentally, the latter is qualitatively well reproduced by the simulation as shown in fig. 11. Figure 12 shows the corresponding dislocation microstructure at the upper yield point and under the same conditions. Further, in the range of validity of the simulation, i.e., up to 1200 K, the temperature and strain rate dependence of the upper and lower yield stresses (cf. fig. 13) are in good agreement with those observed experimentally [28] or theoretically predicted from the Alexander and Haasen model [29]. This allowed to fix the parameters describing the initial microstructure at the values that yield the best quantitative agreement with experiment.

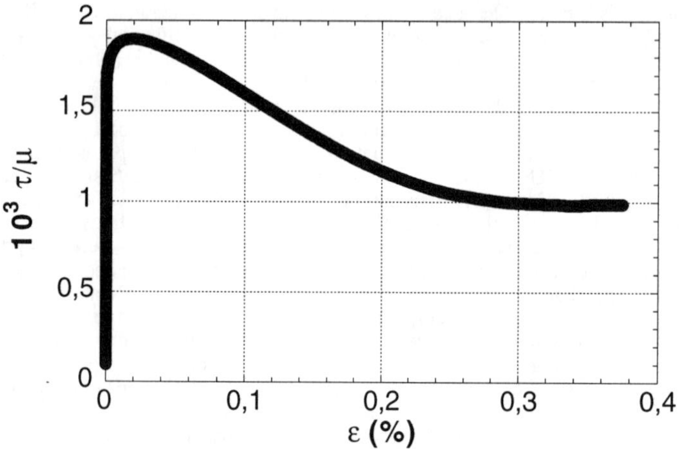

Fig. 11. Simulated stress vs. plastic strain curve exibiting a yield point (T = 1000 K, strain rate: 3×10^{-5} s^{-1}, initial dislocation density: 7.5×10^{7} m^{-2}).

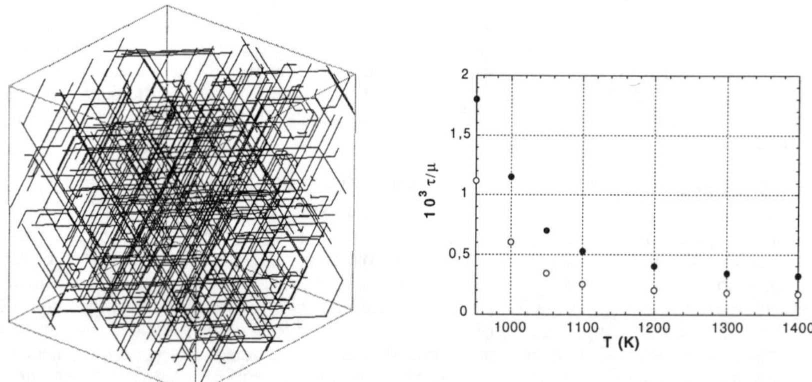

Fig. 12. Typical dislocation configuration at the upper yield point (cf. fig. 11). The size of the simulation box is (80 μm)^3.

Fig. 13. Simulated temperature dependence of the upper (solid circles) and lower (open circles) yield stresses in silicon (strain rate: 3×10^{-5} s^{-1}).

The simulation data make it possible to check the multiplication law proposed by Alexander and Haasen which is of the form

$$d\rho/dt = K\rho v \tau^{*},\qquad(4)$$

where ρ is the total dislocation density (all dislocations are assumed to be mobile), K is a constant, and τ^* is the average effective stress. Up to now, no physical explanation has been proposed for this semi-empirical law which semi-quantitatively fits dislocation density measurements performed by the etch-pit method [29]. Most of the checks of the model have been performed at the lower yield point, but it can be shown (see proof in ref. [14]) that at this point the model is insensitive to the type of multiplication law adopted. Actually, the simulations of the stress-strain curves show that eq. 4 does not accurately predict the position or the amplitude of the multiplication peak. A much better agreement is achieved with a different multiplication law:

$$d\rho/dt = Kv\sqrt{\rho\tau^*}\exp(-\rho/\rho_0), \tag{5}$$

which stems from a simple model describing the activity of fixed FR sources. As shown above, it is possible to model an expanding source reasonably accurately by assuming that the sucessive loops are equidistant. Under such conditions, one can easily show that for a constant stress, the rate of increase of dislocation density is given by an expression identical to eq. 5 except that the exponential factor is omitted. This last term takes statistical account of the fact that once the external loop has reached the surface, a source functions in steady-state ($d\rho/dt = 0$). Then, the dislocation density saturates at a certain value, ρ_0 which increases as the square of the crystal size. Eq. 5 is found to reproduce well the simulated behaviour of the total dislocation density as a function of strain, strain rate, temperature and crystal size.

Thus, it may well be that in dc semiconductors, or more generally in materials with large friction stresses, the laws for dislocation multiplication differ significantly from the ones currently used in other materials such as fcc metals. Indeed, in the latter case, the rate of increase of mobile dislocations is often assumed not to depend explicitly on the effective stress.

Although the multiplication law discussed above yields satisfactory results, it is based on the assumption of a fixed density of potential sources in the crystal. For a full description of multiplication processes, however, it is necessary to know by which mechanism fresh dislocation sources are formed during plastic deformation. Indeed, slip band widening has been often observed experimentally [30] and has been attributed without any real proof to double cross-slip in the bulk or cross-slip assisted by surface effects. As discussed in [14], numerical simulations of cross-slip by interaction of two dislocation pile-ups emitted by FR sources on neighbouring slip planes show that this mechanism is not very effective. This leads to the conclusion that new sources may be nucleated by cross-slip at the surface, at least as long as the dislocation density is below a certain threshold.

CONCLUSION

This study of dislocation dynamics in silicon crystal has focused on the dynamics of Frank-Read sources and on its consequences on dislocation multiplication and mechanical properties. On the one hand, it allowed a check of known properties such as the yield point phenomenon, thus illustrating the validity and the potential of the mesoscopic approach. On the other hand, it provided qualitative or quantitative insight into several mechanisms that are not yet fully understood. Several possible consequences were drawn from the fact that at low temperature and high stress, dissociation effects can strongly affect the generation rate of dislocations. With the recent development of atomistic simulations of the mobility of partial dislocations in silicon, it is hoped that some speculative aspects of the models discussed in the present work will be critically assessed. At a more macroscopic scale, the present study illustrates the strong influence of multiplication mechanisms on the stress-strain behaviour and the limits of our present knowledge regarding the competition between surface and bulk mechanisms for dislocation generation.

REFERENCES

1.   A. George and J. Rabier, Rev. Phys. Appl. **22**, p. 941 (1987).
2.   J. Hirth and J. Lothe, *Theory of Dislocations*, (J. Wiley & Sons, New-York, 1982), p. 544
3.   P. Pirouz, Scripta metall. **23**, p. 401 (1987).
4.   L. P. Kubin, Canova G., Condat, M., Devincre, B., Pontikis, V. and Bréchet Y., Solid State Phenomena, 22 & 23, p. 455 (1992).
5.   J. Lépinoux and L.P. Kubin, Scripta metall. **21**, p. 833 (1987).

6.   B. Devincre in *Computer Simulations in Materials Science*, edited by H.O.K. Kirchner, L.P. Kubin and V. Pontikis (NATO-ASI, vol. E 308 , Kluwer, NL-Dordrecht 1996), p. 309-323.

7.   B. Devincre and L.P. Kubin, Mat. Sci. Eng. **A234-236**, p. 8 (1997).

8.   A. Moulin, M. Condat and L.P. Kubin, Acta mater. **45**,p. 2339 (1997).

9.   S. Marklund, Solid State Com. **54**, p. 555 (1985).

10. M. Imai and K. Sumino, Phil. Mag A **47**,p. 599 (1983).

11. F. Louchet, Phil. Mag A **43**, p. 1289 (1981).

12. P.B. Hirsch, A. Ourmazd and P. Pirouz, Inst. Phys. Conf. Ser. **60**, p. 29 (1981).

13. A.J.E. Foreman, Phil. Mag. A **15**, p. 1011 (1967).

14. A. Moulin, *Doctoral Thesis*, Ecole Centrale Paris (1992).

15. P. Pirouz, Inst. Phys. Conf. Ser. **104**, p. 49 (1989).

16   A. Moulin, M. Condat and L.P. Kubin, Phil Mag A, accepted.

17. K. Yasutake, S. Shimizu, M. Umeno and H. Kawabe, J. Appl. Phys. **61**, p. 940 (1987).

18. K. Wessel and H. Alexander, Phil. Mag. A **35**, p. 1523 (1977).

19. P. Pirouz, in *Twinning in Advanced Materials,* edited by M.H. Yoo and M. Wuttig (The Minerals, Metals & Materials Society, Pittsburgh PA 1994), p. 275-295.

20. J. Castaing, P. Veyssière, L.P. Kubin and J. Rabier, Phil. Mag. A **44**, p. 1407 (1981).

21. J.L. Demenet, *Doctoral Thesis*, University of Poitiers (1987).

22. P. Pirouz and X.J. Ning, Inst. Phys. Conf. Ser. **146**, p. 69 (1995).

23. X.J. Ning and N. Huvey, Phil. Mag. Let. **74**, p. 241 (1996).

24. P. Pirouz, A.V. Samant, M.H. Hong, A. Moulin and L.P. Kubin, submitted.

25. T. Suzuki, T. Yasutomi, T. Tokuoka and I. Yonenaga, Phys. Stat. Sol. (a), in press.

26. S. Brochard, N. Junqua and J. Grilhé, Phil. Mag. A **77**, p. 911 (1988).

27. P.B. Hirsch and S.G. Roberts, Phil. Mag. A **64**, p. 55 (1991).

28. A. George and J. Rabier, Rev. Phys. Appl. **22**, p. 1327 (1987).

29. H. Alexander and P. Haasen, Solid State Physics, **22**, p. 27 (1968).

30. S. Mahajan, D. Brasen and P. Haasen, Acta metall. **27**, p. 1165 (1987).

ON DISLOCATION REACTIONS AND HARDENING MECHANISMS IN 3D DISLOCATION DYNAMICS

H.M. ZBIB*, M. RHEE**, J.P. HIRTH*, T. D. DE LA RUBIA**
*School of MME, Washington State University, Pullman, WA 99164-2920.
**Lawrence Livermore National Laboratory, USA

ABSTRACT

A 3D dislocation dynamics (DD) model for plastic deformation, which connects the macroscopic mechanical properties to the basic physical laws that govern dislocation mobility and related interaction mechanisms, has been under development. In this model there is a set of critical reactions that, basically, determine the overall results of the simulations. These reactions are annihilation, formation of jogs, junctions, dipoles, and cross-slip. In this paper we discuss these reactions and the manner in which they influence the predicted stress-strain behavior in bcc metals. In particular, the formation (zipping) and strength of dipoles and junctions, and effect of jogs, using the dislocation dynamics model are examined. The strengths (unzipping) of these reactions for various configurations are determined by direct evaluation of the elastic interaction forces.

INTRODUCTION

There have been significant advances in understanding the relationship between mechanical behavior of metals and associated dislocation structures. Such structures could be highly heterogeneous and include dislocation cells, slip bands, microshear bands, persistent slip bands and dislocation tangles, all of which are critical to material properties [1-7]. Understanding how these structures form and evolve and how they affect work hardening is, perhaps, one of the most difficult tasks which is still rife with controversy. The main difficulty has been in dealing with large numbers of dislocations. For example, Kuhlmann-Wilsdorf [7] has proposed that the structures can be understood as a progression of low energy thermodynamic states with something like a conventional phase transition taking place between carpet structures and 3D cell structures at the end of Stage II. However, Holt [8] proposed that dislocation structure evolution can be viewed in the spirit of spinodal decomposition by the introduction of local densities of diffusing dislocation populations. This approach has been adopted in later models. In particular, Walgraef and Aifantis [9] developed a model of dislocation patterning to describe the evolution of ordered structures in chemically reacting systems. In this model, several types of dislocation populations are introduced as density functions of position in space, with the evolution determined by diffusion and reaction terms. Although these reaction-diffusion schemes have been successful in modeling 2D dislocation patterns, they have a number of difficulties associated with determination of model parameters and they do not yet address realistic 3D dislocation configurations.

By viewing the dislocation structures problem as a dynamical system, one can establish a number of discrete models to understand the origin of dislocation structures in deformed crystals. Although the method was initiated over a decade ago, most of the original models were two-dimensional and consisted of periodic cells each with dislocations of infinite length. These 2D models have provided some understanding of rules of interaction and glide mechanisms of

Mat. Res. Soc. Symp. Proc. Vol. 538 © 1999 Materials Research Society

dislocations. However, since these 2D models are based on the idealization of infinite dislocation lines, a number of important mechanisms and dislocation interactions are either not included into these models or accounted for in an implicit manner at best. These mechanisms include, cross-slip, junctions, jogs, multiplication by Frank-Read sources, and line tension associated with self-energy. These difficulties have been addressed in a pioneering three-dimensional dislocation model which was developed by Kubin and his co-workers [10] and Canova, et al. [11]. Their model was based on the discretization of dislocation curves into succession of *pure edge* and *pure* screw dislocation segments of fundamental length, corresponding to the *discretized lattice*. More recently, a new approach for 3D dislocation dynamics has been established by Zbib, Hirth and Rhee [12-14]. In this approach arbitrarily curved dislocations are decomposed into piecewise continuous arrays of *mixed straight segments* in a *continuum crystal* and long range interactions are treated using super-dislocations, allowing for the treatment of large dislocation densities.

The 3D discrete dislocation model *(micr3d)* developed at WSU simulates the dynamical behavior of large numbers of dislocations of arbitrary shapes, interaction among groups of 3D dislocations and, therefore, the behavior of prescribed cell walls. In the model there are a number of rules and models for short-range reactions which, in turn, have a decisive effect on the predicted hardening and evolution of the structure. These issues have been, recently, addressed by Rhee et al. [13] who developed numerical rules for implementation into DD models to treat short-range interactions, by introducing a critical force criterion. In this paper, we investigate these reactions and corresponding mechanisms that contribute critically to work hardening during deformation using the dislocation dynamics model. The main issues we address are junction formation and strength through the process of "zipping" and "unzipping", jog formation and strength, and dipole strength. It is shown that the full dynamics of these interactions can be explicitly captured via dislocation dynamics simulations, where the dynamics of these configurations is determined by direct calculation of driving forces and internal stresses.

BASIC EQUATIONS

The complete description of the 3D dislocation model (DD) can be found in [12-14]. Here we give a brief outline of the main features of the model and basic governing equations. The main problem is that of nonlinear interaction of large number of curved dislocations of arbitrary shapes. The crystal is treated as a *continuum* with dislocations restricted to move on crystallographic slip systems. The model has been developed for both fcc and bcc single crystals. For bcc single crystals we consider the {110}<111> and {112}<111> slip systems which are the most close packed slip systems and both are active at low temperatures. The {123}<111> slip systems are less close packed and become active at high temperatures and will be included into the model at a later stage. For fcc single crystals, {111}< 011> slip systems, and hence glissile dislocation lines can only lie on {111} planes.

Each plane contains a number of dislocation curves and loops whose configurations are approximated by a series of straight line segments of *mixed character* as illustrated in Figure 1a. There are a number of advantages in using mixed segments of arbitrary lengths and orientation:

1. Segment length depends upon the local curvature. Dislocation curves with small curvature are segmented with long segments, while those with large curvatures are meshed with shorter segments (typical segment size could vary from 50 b to a few hundred b, where b is the Burgers vector).

2. With mixed segments of arbitrary length and orientations, dislocation interactions such as junctions with any arbitrary configuration can be formed.

3. Mobility and strength of dislocation reactions, such as jog motion through dislocation bow-out at the jog and junction destruction by the process of "unzipping" can be readily captured.

4. The stress field of arbitrary mixed straight dislocation segment in an isotropic medium is given in a closed form.

Long Range Interaction and Self-Force

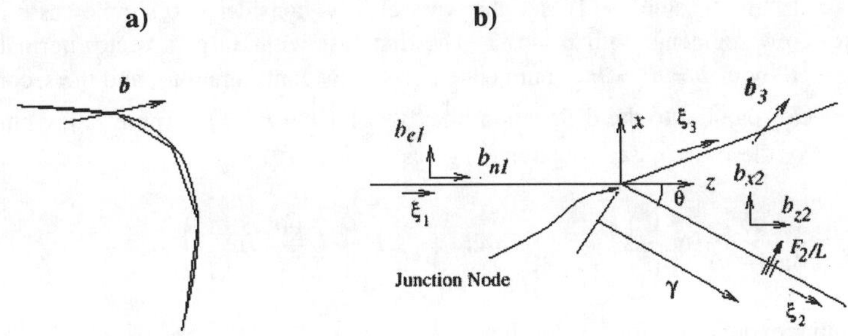

Figure 1. a) Discretization of a dislocation curve, b) Dislocation bend or junction.

The stress field of each segment is given by Hirth and Lothe [15] and de Wit [16]. The interaction force per unit length (Peach-Koehler force "*PK force*") a given segment exerts on a remote segment is evaluated at the center of the remote segment. This approximation is valid for two segments that are far apart from each other since the variation of the interaction force along the segment length is very small. For adjacent segments forming a bend the interaction force and *self-force* varies significantly along the dislocation line; it is singular at the bend and decays as *1/r*. Therefore, we treat this case in a more rigorous way as described by Zbib et al. [12]. The solution for the total interaction force between the two adjacent segments shown in Figure 2b is developed along the same lines described by Hirth and Lothe [15] for a dislocation bent with the same Burgers vector. The two segments could belong to the same dislocation line with the same Burgers vector or they could be at a *junction node* with two different Burgers vector. The result is an average glide force per unit length given by

$$F_2 = \frac{\mu}{4\pi}\ell n\left(\frac{L}{\rho}\right)\Big[\ b_{z1}b_{z2}\frac{\cos\theta-1}{\sin\theta} - b_{z2}^2\frac{v\sin\theta\cos\theta}{1-v} + b_{x1}b_{z2}\frac{v}{1-v} + b_{x2}b_{z2}\frac{v}{1-v}\left(2\sin^2\theta-1\right)$$

$$+ b_{x1}b_{x2}\frac{\cos\theta-1}{(1-v)\sin\theta} + b_{x2}^2\frac{v\sin\theta\cos\theta}{1-v} + b_{y1}b_{y2}\frac{\cos\theta-1}{(1-v)\sin\theta}\ \Big]$$

$$(1)$$

The normal force component out of page is given by

$$F_n = \frac{\mu}{4\pi(1-v)}\ell n\left(\frac{L}{\rho}\right)\left[\; b_{y1}b_{x2}\frac{v\sin\theta + \cos\theta - 1}{\sin\theta} + b_{x1}b_{y2}\frac{\cos\theta - \cos^2\theta - v\sin^2\theta}{\sin\theta} \right.$$

$$+ b_{z1}b_{y2}(1-v)(1-\cos\theta) + b_{y1}b_{z2}v(2-\cos\theta) \qquad (2)$$

$$\left. - b_{z2}b_{y2}v\cos\theta + b_{x2}b_{y2}\cos\theta + b_{x2}b_{y2}v\sin\theta \;\right],$$

where μ is the elastic shear modulus, v is the Poisson's ration, L is the dislocation segment and ρ is the core parameter. For the case of general bend of two segments only (ignore segment "3: in Figure 2a) $b_1 = b_2$ and equations (1) and (2) reduce to the expressions given in [15].

In passing we emphasize that equation (1) accounts explicitly for the *self-force* which gives rise to the so call "line tension". To see this very clearly, consider two simple cases of pure edge and pure screw segments with $\theta = \pi/2$. The first case with Burgers vector normal to the dislocation line "1" with $b = b_{x1} = b_{x2}$ (pure edge-pure screw configuration), and the second case with Burgers vector parallel to the dislocation line "1" with $b = b_{z1} = b_{z2}$ (pure screw-pure edge configuration). For these two cases equation (1) gives

$$F_g = \frac{-\mu b^2}{4\pi(1-v)}\ell n\left(\frac{L}{\rho}\right), \qquad \textit{and} \qquad F_g = \frac{-\mu b^2}{4\pi}\ell n\left(\frac{L}{\rho}\right). \qquad (3)$$

These are familiar expressions for line tension.

Dislocation Mobility

The motion of each dislocation segment is determined by first evaluating the total *PK force* which arises from all other dislocation stress fields and the applied stress, such that

$$F_i = \left(\sum_{\substack{j=1 \\ j\neq i, \\ j\neq i+1 \\ j\neq i-1}}^{N}\left(\sigma_j^D + \sigma^a\right)\cdot b_i\right)\times\xi_i, + F_{i,i+1} + F_{i,i-1}, \qquad (4)$$

where N is the total number of dislocation segments, σ_j^D is the stress tensor from a remote segment j, σ^a is the applied stress tensor, ξ_i is the line sense vector, and $F_{i,i+1}$ and $F_{i,i-1}$ are the interaction forces between segments i and $i+1$, and i and i-1, respectively, as computed from equations (1) and (2). The effective shear stress τ_{ei} on segment i is given by

$$\tau_{ei} = \frac{f_{gi}}{b} - \sigma_f \qquad (5)$$

where $f_{gi} = |F_i \cdot \hat{v}_i|$ is the magnitude of the glide force per unit length with $\hat{v}_i$ being a unit vector in the direction of slip, and σ_f is the friction stress arising from lattice damping effects. The velocity vector of segment i is given by $v_i = v_{gi} \hat{v}_i$ where the basic relation for v_{gi} is discussed below.

The relationship between the glide velocity and the glide force per unit length (or effective shear stress) is temperature-dependent. At high temperatures, dislocations (pure edge, pure screw, and mixed) move by the phonon drag mechanism and the driving effective force is athermal. There are a number of relations for the dislocation glide velocity v_g, including relations of power law forms and forms with an activation term in an exponential or as the argument of a sinh form. Often, the simple power law form is adopted for expedience, e. g. $v_g = v_s (\tau_e / \tau_s)^m$. In a number of cases of pure phonon/electron damping control or of glide over the Peierls barrier a linear form of equation (1), $m = 1$, predicts the results very well. The linear form has been theoretically predicted for a number of cases as reviewed by Hirth and Lothe [15] leading to

$$m^* \dot{v}_{gi} + v_{gi} / M_{gi} = F_g; \quad i = e \text{ (for edge/mixed dislocation),} \\ \text{or } s \text{ (for screw dislocation)} \tag{6}$$

where m^* is the *effective mass* and M_{gi} is the dislocation mobility and, in general, it dependence on the character of the dislocation. Generally, M_{gi} could be, among other things, a fuction of the angle between the Burgers vector and the dislocation line sense, especially at low temperatures. In bcc single crystals, at low temperatures a pure screw dislocation has a rather complex three-dimensional core structure, resulting in a high Peierls stress which is overcome by stress-assisted thermal activation [12]. This leads to a relatively low mobility for screw dislocations while the mobility of mixed dislocations is very high [17].

In a more general treatment, the inertia term should be included in equation (6). However, recent studies have shown that the rise time for a dislocation to reach a steady state is on the order of 10^{-10} s for a dislocation velocity below 0.4 times the shear wave velocity [18]. A typical time step in computer simulation is of the same order, making it possible to neglect the inertia term, especially for low mobility dislocations in bcc metals. This inertia term, however, plays an important role for fast moving dislocations in fcc metals where the mobility is a few orders of magnitude higher than that in bcc.

The result of the above formulation is a set of nonlinear first order differential equations governing the motion of the dislocation segments. The motion of the dislocations give rise to plastic strain D^P and spin W^P, which are calculated from

$$\mathbf{D}^P = \text{Sym} \frac{-1}{V} \sum_{\substack{\text{all segments}}}^{N} (\mathbf{v}_{gi} \times \xi_i) \otimes \mathbf{b}_i \ell_i, \quad \text{and} \quad \mathbf{W}^P = \text{Skew} \frac{-1}{V} \sum_{\substack{\text{all segments}}}^{N} (\mathbf{v}_{gi} \times \xi_i) \otimes \mathbf{b}_i \ell_i, \tag{7}$$

where ℓ_i is the segment length, $\mathbf{v}_{gi}$ is the segment, $\mathbf{b}_i$ is the Burgers vector, ξ_i is the sense vector, and V is the volume of the simulated crystal.

<u>**Short Range Interactions**</u>

For short-range interactions basic theories, which describe the underlying physical mechanisms at the core level, may be rigorously included into the 3DD model. Although this would be important when investigating the local interaction between two dislocations over small distances (close to the core), it would not be desirable numerically or even important when dealing with relatively large numbers of dislocations on a large scale (10's of μm). It would be more numerically efficient to develop rules based on a rigorous investigation of the two-dislocation interaction problem which can then be implemented directly into the 3DD model for large-scale simulations. Therefore, in [13] we developed numerical rules for implementation into the 3DD model to treat short-range interactions. These rules are:

Rule 1: Critical force criterion "$F \geq F^c$" for short-range interaction .
Rule 2: Critical force criterion for annihilation.
Rule 3: Critical-angle criterion for junction formation; $\theta_{AB} \leq \theta_{jn}^c$.

Rule 4: Critical-angle criterion for jog formation; $\theta_{AB} \geq \theta_{jg}^c$.

Rule 5: Critical-angle criterion for jog strength; $\theta_{jg} \leq \theta_{jgs}^c$.

Here θ_{AB} is the angle between two dislocation segments, θ_{jg} is the bow-out angle at a jog, and $\theta_{jn}^c, \theta_{jg}^c$, and θ_{jgs}^c are the *critical* values for junction formation, jog formation and jog strength, respectively. Each of these rules involves a critical value that determines the interaction. We suggest that numerical values for θ_{jn}^c and θ_{jg}^c should be determined from a rigorous investigation of the two-dislocation interaction problem, similar to those performed by Huang et al [19] for the dipole problem. Values for the critical force, jog strength, as well as dislocation mobility, could be quantified explicitly by means of simulations at the core level, such as MD simulations [20]. These simulations are expected to be very extensive since one would have to examine a number of possible combinations of Burgers vectors and slip planes.

BASIC REACTIONS AND STRENGTHENING MECHANISMS

(In the following discussion we consider the deformation of a single crystal Ta at room temperature for which $b = 2.86 \times 10^{-10} m$, $\mu = 70.7 GPa$, $v = 0.339$, $\sigma_f = 3 \times 10^{-5} \mu$).

<u>**Junction Formation – Zipping**</u>

The process of dislocation junction formation has been addressed schematically by many authors, but no quantitative analysis has been performed except by the recent work of Bulatov et al. [21]. The DD dislocation dynamics model is used to examine this process. The process begins when two attractive dislocations gliding on intersecting planes (Figure 2a) meet at the line of intersection and combine if the reaction is favorable (Figure 2b). In the DD simulations this is captured through explicit evaluation of the system dynamics, i.e. there is no need to check if Frank's rule (i.e. $|b_1 + b_2|^2 < |b_1|^2 + |b_2|^2$)is met before combining the segments. If the two segments are attractive (Rule 1) and their energy would be at a minimum by forming a junction, they would approach each other, align themselves into a parallel configuration (Rule 3), react and form a junction. Once initiated, the junction extends along the line of intersection by the process of "zipping" as anticipated by the Friedel-Saada model [22] (Figure 2c). The configurations shown in Figures 2a-h and 3a-b are typical examples of junctions formed using DD simulations.

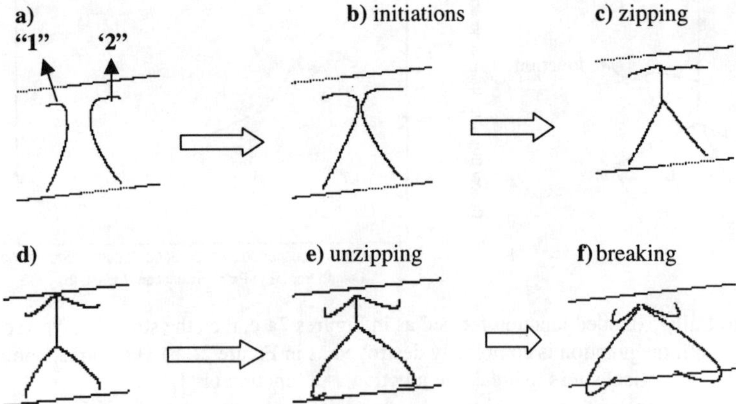

Figure 2. a-b) Junction formation "zipping" with stress of 20 MPa applied in the [001] direction. d-f) Junction destruction "unzipping" with reversed stress (-30MPa). Dislocation "1" : $b/\sqrt{3}[\bar{1}11]$ (110) ; dislocation "2": $b/\sqrt{3}[11\bar{1}]$ $(1\bar{1}0)$. Sessile junction is in the [001] direction with Burgers vector $2b/\sqrt{3}[010]$.

Junction Destruction – "unzipping"

The extension or destruction of the junction takes place by the motion of the dislocation node at the triple point along the line of intersection as anticipated by the Freidel-Saada model [22] and the recent MD simulations of Bulatov et al. [21]. This is illustrated by typical DD simulation results shown in Figs. 2d-f and Figs. 3a,b. Thus, as also pointed out by Bulatov et al [21], the strength and fate of the junction is determined by the force driving nodal stacking-fault energy.

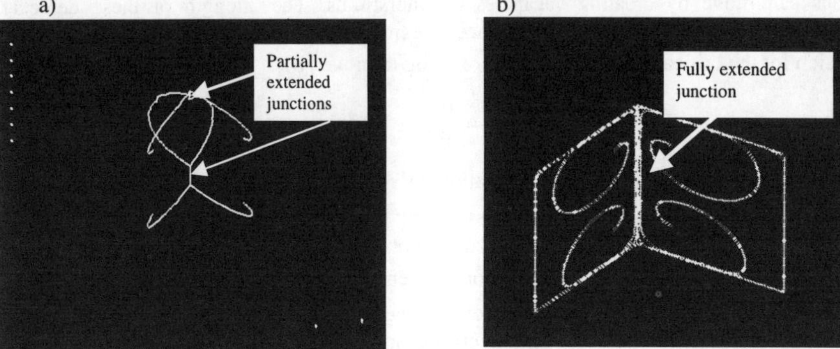

Figure 3. a) Formation of partial junctions and unzipping with stress increased to 50 MPa (not reversed). b) Pileup of dislocations at a fully extended junction.

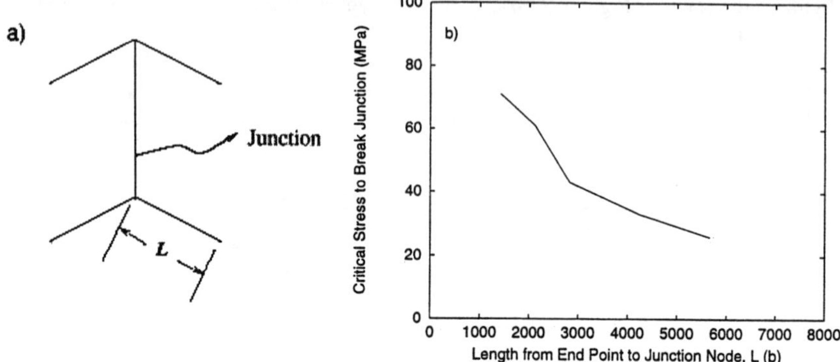

Figure 4. a) Fully extended junction formed as in Figures 2a-c, then the stress is reversed incrementally until the junction is completely destroyed as in Figure 2f. b) The corresponding critical stress to break the junction as a function of "L"

The process of junction formation and destruction is a balance between applied stress, dislocation interaction forces and core energy around the junction node. As the reversed stress is increased to a critical value the forces on the dislocation nodes increase and move the nodes towards each others along the line of intersection. In the simulation, once the junction is formed and reaches a stable configuration as shown in Fig. 4a, the applied stress is reversed incrementally. Due to the line tension effect, a higher stress is required to overcome the elastic interaction energy and core energy to unzip the junction. Figure 4b shows the effect of line tension on the critical stress to unzip and break a junction. It can be deduced from the figure that for a longer dislocation branch from the source (L) less force is required to overcome the elastic energy barrier for the unzipping process to occur. This is due to the fact that the force associated with line tension is larger with larger "*L*", resulting in higher self-force at the junction node.

Jog Strength

Jogs can move by creating vacancies or interstitials. The strength of these defects is balanced by the total line tension of two adjacent segments around a jog, leading to the following expression for the critical bow-out angle θ_{jgs}^{c} for jog to move [14]

$$cos\left(\frac{\theta_{jgs}^{c}}{2}\right) = \frac{W}{\mu b^3} \tag{8}$$

where W is the interstitial or vacancy formation energy. To investigate the effect of the critical jog angle on the corresponding critical stress for jog motion, we carried out DD simulations for the configuration shown in Figure 5. The result is obtained by considering a simple Frank--Read source with a jog located in the middle. For a given θ_{jgs}^{c} the stress is increased incrementally until the jog begins to move. Figure 4a shows a snapshot of the simulation result after the jog has moved from its initial position. As the critical angle decreases, higher stress is required to move the jog as shown in Fig. 5c. However, further decrease in the angle (less than 60 degrees) does not result in a significant increase in the critical stress. This is because once the bow out reaches the unstable configuration as in the Frank--Read source, less stress is required for the dislocation to wrap around the source. In the absence of a jog, the critical stress required to bowout a Frank--Read source of length 6000b or 3000b is 11 MPa or 23 MPa respectively. This value is, indeed, the minimum threshold for the critical stress to bowout the Frank--Read source with a very weak jog, corresponding to a critical jog angle approaching 180 degrees.

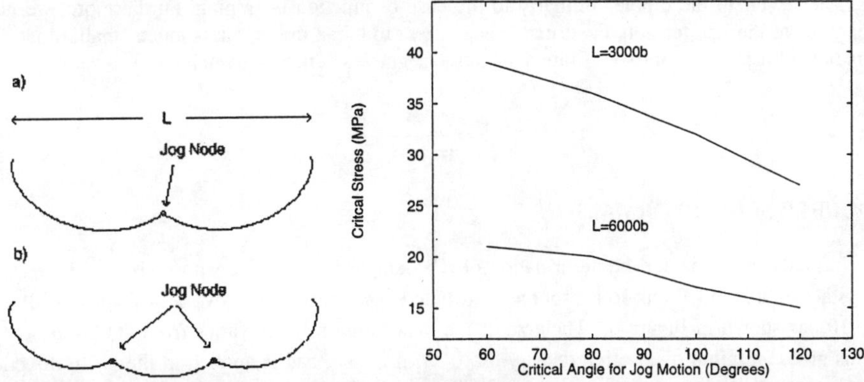

Figure 5. a,b) Pinned dislocations with jogs. c) Critical stress required to bow out and propagate the jogged dislocations as a function of jog critical angle (strength).

Dipole Strength

Although models describing the dipole strength are available, the dipole is assumed to be infinitely long. In our approach, dipole break occurs naturally if the stress required to cause it to unzip apart is sufficient. Consider, for example, the dipole which is formed by the bow-out of dislocations from Frank-Read sources shown in Fig. 6a. The strength of the dipole depends on the normal separation h and the length of the distance from dipole position to the dislocation endpoint d where the dislocations emanate.

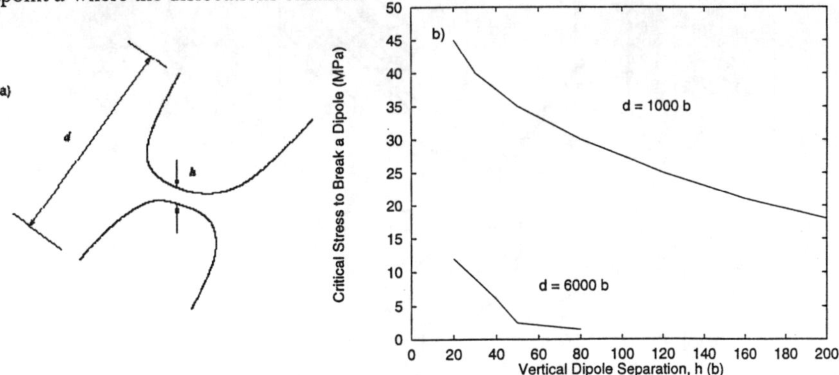

Figure 6. a) Two dislocations with opposite Burgers vectors emanating from two Frank-Read sources on parallel planes separated by distance "h", forming a dipole "lock" at distance $d/2$. b) Critical stress to unzip the dipole.

Figure 6b shows the critical stress to unzip a dipole as a function of the normal separation and the separation between the original sources. For a constant dipole separation h, dipole with a smaller value d requires higher stress to unzip it than the case with larger distance. This is due to the line tension from the dipole to the pinning point. For larger distance of d the curvature of the dislocation at the dipole position is larger. This, in turn, results into a larger line tension, making it easier to unzip the dipole, similarly to the case of junction unzipping. Furthermore, we note that, due to the line tension, the critical shear stress to break the dipole is much smaller that that predicted for the case of two infinite dislocation segments which is given by

$$\tau_c = \frac{\mu b}{8\pi(1-\nu)}\frac{1}{d} \tag{9}$$

DOUBLE SLIP DEFORMATION

Finally, we consider the deformation of a Ta single crystal cube whose size is $10\mu m$. The crystal contains 50 Frank-Read source distributed randomly on the two slip planes (110) and ($\bar{1}$10) as shown in Figure 6. The size of the dislocation sources varies from 5000b to 8000b. This gives an initial dislocation density of 10^{11} / m^2. The load is applied in the [100] direction with a constant strain rate of 10/ s. The resolved shear stresses on both planes are equal, activating both slip systems. The mobility of edge and general dislocations is 10^2 / (Pa. s) and that of screw character is assumed to be smaller by one order of magnitude.

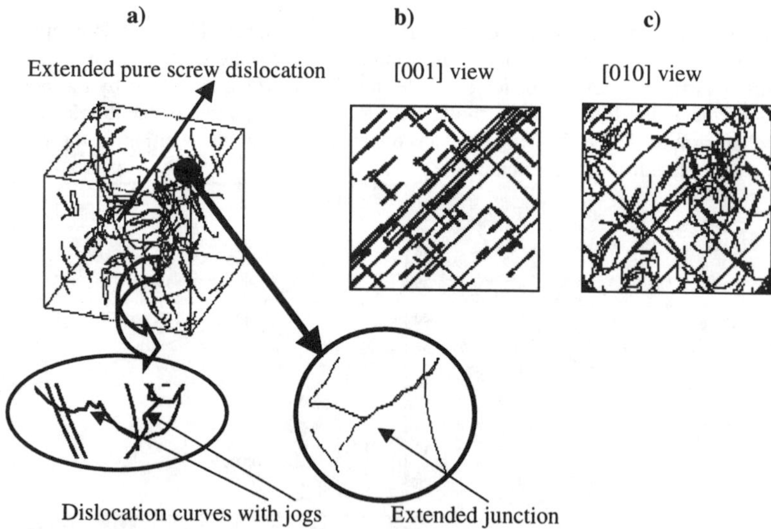

Figure 7. a) DD simulation of double slip deformation, b) [001] view, c) [010] view.

A snapshot of the dislocation structure after 163,000 iterations is shown in Figure 7. As a result of dislocation intersections, many jogs and a few junctions form and their total number increases with further deformation (the total number of jogs in Fig. 7 is 184). The simulation is performed for a jog critical angle (strength) of 120 degrees. The corresponding stress--strain curve is shown in Figure 8. It can be deduced from the figure that significant strain hardening occurs as a result of formation of jogs and functions. The dislocation density in Figure 7 is $4.3 \times 10^{11}/m^2$.

CONCLUSIONS

A number of hardening mechanisms in metals using dislocation dynamics have been investigated. These mechanisms are studied according to DD rules and models that have been developed to treat short-range reactions in bcc and fcc metals. The effect of these mechanisms on the overall stress-strain behavior of metals has been illustrated. The result suggests that the DD simulation model provides a very valuable tool for investigating critical dislocation phenomena that control plastic deformation and hardening in metals. This type of microscopic modeling of deformation provides a natural transition from the atomic scale to the continuum crystal scale. While MD simulations can provide needed rules of dislocation dynamics and interaction based on fundamental principles, the massive DD simulations provides the most rigorous means for transmitting the influence of these critical mechanisms from the dislocation length scale to the macroscopic scale.

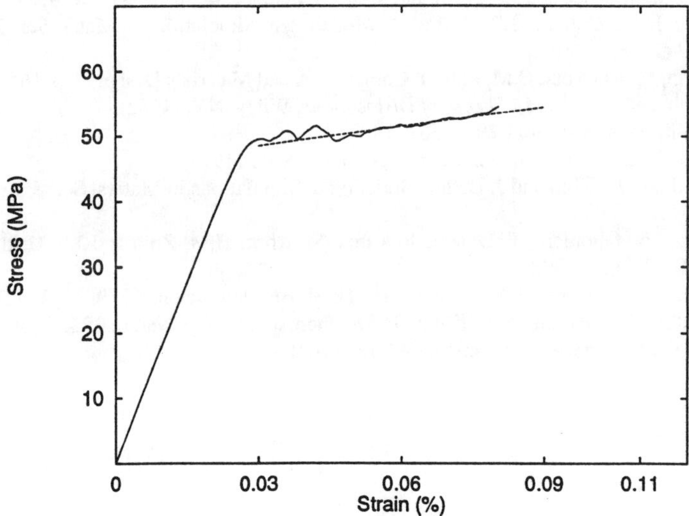

Figure 8. Stress-strain curve as predicted by the DD simulation of double slip deformation, resulting into the formation of jog and junctions and yielding strain hardening.

ACKNOWLEDGMENTS

The support of the Lawrence Livermore National Laboratory under contract No. W-7405-ENG-48 with the DOE, is gratefully acknowledged.

REFERENCES

1. S. Mader, A. Seeger and H.M. Thieringer, J. Appl. Phys. **34** , p. 3376—3386 (1963).
2. H. Mughrabi, Mater.Sci.Engng, **85**, p. 15—31 (1987).
3. M.S. Whelanin, in: *The Physics of Metals,* edited by P.B. Hirsch, Part 2: Defects, Cambridge Univ. Press, p. 98 (1975).
4. U. Essmann and H. Mughrabi, Phil. Mag. A, **40**, No. 6, p.731—756 (1979).
5. P.J. Woods, Phil. Mag. **28**, p. 155—191 (1973).
6. N., Hansen and D. Kuhlmann-Wilsdorf, Matl. Sci. Eng., **81** p 141 (1986).
7. D., Kuhlmann-Wilsdorf, Mat. Res. Innovat, **1**, p. 265 (1989).
8. D. Holt, *J. Appl. Phys.,* **41**, 3197—3201 (1970).
9. E.C. Aifantis, *Sol. Stat. Phenom.,* **3 & 4**, 397—406 (1988).
10. L. P. Kubin, *Phys. Stat. Sol (a),* **135**, 433—443 (1993).
11. G. R. Canova Y. Brechet and L.P. Kubin, in: *Modelling of Plastic Deformation and Its Engineering Applications,* edited by S.I. Anderson et al, Riso National Laboratory, Roskilde, Denmark. (1992).
12. H. M. Zbib, M. Rhee and J.P. Hirth, Int. J. Mech. Science, **40**, p. 113-127 (1998).
13. M. Rhee, H.M. Zbib and J.P. Hirth, J.P., Modeling & Simulations in Mater. Sci. & Engr, **6**, p. 467-492 (1998).
14. J.P. Hirth, M. Rhee and H.M. Zbib, J. Computer-Aided Materials Design, **3**, p. 164-166, (1996).
15. J.P. Hirth and J. Lothe, J., *Theory of Dislocations*, Wiley, NY. (1982).
16. R. de Wit, Phys. Stat. Sol., **20**, p. 567 (1967).
17. N. Urabe N. and J. Weertman, *Mater. Sci. Eng.*, **18**, 41 (1975).
18. J.P. Hirth, H.M. Zbib and J. Lothe, Modeling & Simulations in Maters. Sci. & Enger. **6**, 165-169 (1998).
19. H. Huang, N. Ghoniem, T. Diaz de la Rubia, M. Rhee, H.M. Zbib and J.P. Hirth, to appear in ASME-JEMT, 1999.
20. S.J. Zhou, D.L. Preston, P.S. Lomdahl and D. M. Beazley, Science **279**, p. 1525 (1998).
21. V. Bulatov, F. F. Abraham, L. Kubin, B. Devincre and S. Yip, Nature **392**, p.669 (1998).
22. J. Friedel, *Dislocations,* Pergamon, Oxford (1964).

Simulation of Dislocation Dynamics in the Continuum Limit

K. W. SCHWARZ, IBM Research, P.O. Box 218, Yorktown Heights, NY 10598

Abstract

Peach-Koehler theory is implemented to simulate the motion of three-dimensionally interacting dislocations, located on various glide planes and having any allowed Burgers vector. The self-interaction is regularized by a modified Brown procedure, which remains stable and loses accuracy in a well-controlled manner as atomic dimensions are approached. The method is illustrated by applying it to several problems involving interacting dislocations in an fcc slip system. The strong interaction of two dislocations on intersecting glide planes is investigated with a view towards developing a set of rules to describe the outcome of such interactions. The effect of Frank-Read sources in relaxing a strained layer are illustrated, both for sources on parallel and on intersecting glide planes. Other possible applications of the method are discussed.

Introduction

The numerical simulation of dislocation dynamics on the mesoscopic scale is currently being pursued by a number of groups [1]–[8] using a variety of approaches. Such simulations are not only expected to be useful in exploring many of the classic issues in metallurgy, plasticity, and fracture: they can also elucidate problems which dislocations pose in various aspects of semiconductor manufacturing, problems such as thin-film relaxation, the nucleation and growth of dislocations in the presence of stress concentrators such as impurities or edge fields, and the growth of dislocations from defects generated during ion implantation. Our particular work [9]–[12] has focussed on this latter area. Hence, in addition to being designed for doing very large scale simulations involving many interacting dislocation lines, our computational methods are suitable for doing simulations down to the very small scales approaching the dimensions at which atomistic effects become important. As such, the methods described here should prove useful in connecting the continuum approach to the atomistic calculations which comprise an essential part of the multiscale approach. At the other extreme, they should be useful in implementing the local interaction "rules" that are expected to be used in the large-scale simulations, and indeed, in going beyond this relatively crude approach when the physical situation demands it. In this brief review, we first describe our meth·ds for simulating dislocation motions. We then give a simple example of a calculation which establishes an interaction rule. Finally we illustrate the application of the method to the relaxation of a strained epitaxial $Si_{1-x}Ge_x$ layer on a Si substrate.

Computational Methods

Dislocations can do many complicated things, each of which may be of interest in a particular situation. At the present stage of development, however, mesoscopic simulations are limited to the case of perfect dislocations undergoing glide in an isotropic medium. Such a dislocation feels a force in the glide plane [13]

$$\vec{f}_g = (b_i \sigma_{ij} n_j)\, \hat{n} \times \vec{dl}\,, \tag{1}$$

acting on each line element $\vec{dl}$. Here $\vec{b}$ is the Burgers vector, σ is the stress tensor at the line element, and $\hat{n}$ is the normal to the glide plane. For the situations considered in the present paper, the dislocation motion is assumed to be determined entirely by the instantaneous local force acting on it. Effects such as point-defect absorption, inertia, or sound emission are not considered. In this approximation, we need only to calculate the stress tensor and to formulate some kind of response model to completely determine the dynamics of the dislocations.

The stress tensor includes stresses due to the applied strains, stresses generated by the presence of the dislocations themselves, and additional corrections arising from the presence of boundaries (such as free surfaces). In many problems of interest, the applied strain is specified to be merely some constant value. Sometimes, however, such as when working with semiconductor heterostructures, the determination of the applied stress/strain field can in itself pose a difficult problem. When such fields are required, they must be determined by a separate calculation, and the results stored in look-up tables that allow the program to find the applied stress tensor at every point of interest. Once the applied stresses are specified, the main computational challenge is to determine the effect that the dislocations have on themselves and on each other.

In an infinite, isotropic, and homogeneous elastic medium, the contributions made by the dislocation lines to the stress tensor at a point $\vec{r}$ are given by the Peach-Koehler expression [13, 14], which essentially has the form of a line integral taken over all of the dislocations present. The source term can conveniently be written as

$$\frac{d\sigma_{ij}}{\sigma_0} = -\frac{(1-\nu)}{R^3}[(\vec{b}' \times \vec{R})_i dl'_j + (\vec{b}' \times \vec{R})_j dl'_i]$$

$$- \frac{1}{R^5}(\vec{dl}' \times \vec{b}')_\gamma [R^2(\delta_{i\gamma}R_j + \delta_{j\gamma}R_i - \delta_{ij}R_\gamma) - 3R_i R_j R_\gamma] \tag{2}$$

where the index γ is summed over. Here $\sigma_0 = \mu/4\pi(1-\nu)$, where μ is the shear modulus and ν is Poisson's ratio, and $\vec{R} = \vec{r}' - \vec{r}$. Equation (2) may be thought of as the contribution to σ_{ij} at the point $\vec{r}$ from the element $\vec{dl}'$ at $\vec{r}'$, although it must be remembered that it includes unphysical fields which cancel when the elements are laid end to end to make a closed curve.

Equation (2) allows computation of the stress generated by an arbitrary dislocation, simply by stringing together small line elements. Practically speaking, such elements are not true differentials, and in fact there are many cases in which one would prefer to use somewhat longer straight-line segments. If one defines an arbitrary straight line $\vec{r}' = \vec{s}_0 + \hat{s}s$, and chooses $\vec{s}_0$ such that $\vec{s}_0 \cdot \hat{s} = 0$, (see Fig. 1a) then the indefinite integral of (2) over this line takes the form

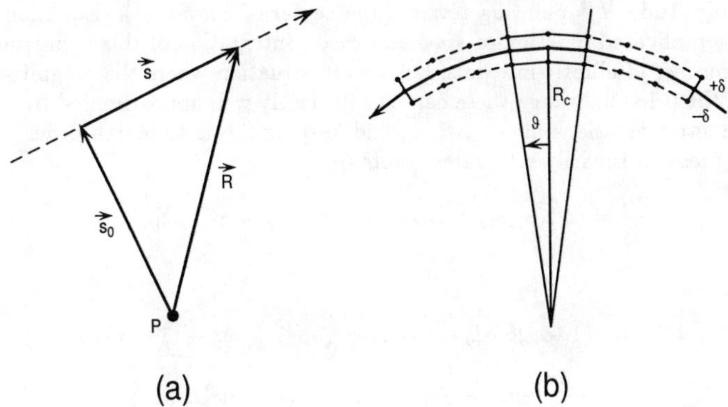

(a)   (b)

Fig. 1: Constructs for the evaluation of the stress tensor at a point P.

$$\frac{\sigma_{ij}}{\sigma_0} = -(1-\nu)[(\vec{b}' \times \vec{F})_i s_j + (\vec{b}' \times \vec{F})_j s_i] + (\vec{b}' \times \vec{s})_\gamma (\delta_{i\gamma} F_j + \delta_{j\gamma} F_i)$$

$$+ G[\delta_{ij} s^2 + s_{0i} s_{0j} (\frac{s^2}{R^2} + \frac{2s^2}{s_0^2}) - (s_i s_{0j} + s_j s_{0i})\frac{s_0^2}{R^2} + s_i s_j (1 - \frac{s_0^2}{R^2})]\,, \qquad (3)$$

where

$$\vec{F} = (\vec{s}_0 s^2 - s_0^2 \vec{s})/s^2 s_0^2 R \,, \qquad (4)$$

and

$$G = (\vec{s} \times \vec{b}')_\gamma s_{0\gamma}/s^2 s_0^2 R \,. \qquad (5)$$

Here $\vec{s} = s\hat{s}$, as indicated in Fig. 1a. The contribution to the stress tensor at P from any segment of this line is just the difference of (3) between the ends of the segment.

In the calculations, the dislocations are represented as lines of discrete points. These points are assumed to be connected by straight line elements, and the contribution of each element to σ_{ij} at a particular point P on the dislocation is obtained using Eq. (3). Such an approach is not adequate for dealing with that part of the dislocation which actually passes through P, since it leads to a logarithmic divergence. A complete calculation of the self-interaction obviously requires an atomistic treatment of the core structure. Less ambitiously, one can treat a section of the line containing P in terms of some *finite core approximation* that is designed to yield the proper asymptotic behavior but that will leave some of the less important and model-dependent features unspecified.

The particular regularization procedure that we have found most useful is to pass a circular arc through P and its two neighboring points. This is then split into two arcs, passing on both sides of P at a distance δ as shown in Fig. 1b. Thus we require the stress tensor contribution at P from the displaced circular arcs shown in Fig. 1b, where δ can

29

be positive or negative. To integrate Eq. (2) over such an arc, we first associate with the point P a local coordinate system $\hat{s}_1 = \vec{s}\,'/|\vec{s}\,'|$, $\hat{s}_2 = \vec{s}\,''/|\vec{s}\,''|$, and $\hat{s}_3 = \hat{s}_1 \times \hat{s}_2$, where $\vec{s}\,'$ is the local tangent of the arc passing through P and $\vec{s}\,''$ is the local curvature (i.e. a vector of magnitude R_c^{-1} pointing towards the center of curvature). Equation (2) then becomes a complicated function of $sin\varphi$ and $cos\varphi$. Integration of this expression over φ leads to a number of elliptic integrals, but for our situation where the magnitude of φ is typically in the $0.1 - 0.2$ range these can be sufficiently well approximated by expanding the trigonometric functions about $\varphi = 0$, and keeping terms to fourth order in φ. The resulting expression is easily integrated to obtain

$$\frac{\sigma_{ij}}{\sigma_0} = [D1(s_{3i}s_{1j} + s_{3j}s_{1i}) + D2(s_{3i}s_{2j} + s_{3j}s_{2i})] \,, \tag{6}$$

where

$$\begin{aligned} D1 &= (1 - \nu)b_1\delta I_0 - \nu b_2\rho I_1 + \frac{1}{2}b_1[(1 + \nu)\rho - (1 - \nu)\delta]I_2 \\ &\quad + \frac{3\nu}{2}b_2\rho I_3 - \frac{b_1}{24}[(1 + 7\nu)\rho - (1 - \nu)\delta]I_4 \,, \end{aligned} \tag{7}$$

$$\begin{aligned} D2 &= b_2\delta I_0 - \nu b_1\delta I_1 + \frac{b_2}{2}[(1 - 2\nu)\rho - \delta]I_2 + \frac{\nu}{6}b_1(3\rho + \delta)I_3 \\ &\quad + \frac{b_2}{24}[(8\nu - 1)\rho + \delta]I_4 \,. \end{aligned} \tag{8}$$

Here $b1 = \vec{b}\,' \cdot \hat{s}_1$, $b2 = \vec{b}\,' \cdot \hat{s}_2$, and $\rho = R_c + \delta$. The integral coefficients are $I_0 = f\varphi/as$, $I_1 = -f/s$, $I_2 = -f\varphi/s + f\ln(s + \varphi)$, $I_3 = f(2a + \varphi^2)/s$, and $I_4 = f\varphi^3/2s - 3aI_2/2$, where $f = R_c^{-3/2}(R_c + \delta)^{-1/2}$, $a = \delta^2/R_c(R_c + \delta)$, and $s = (a + \varphi^2)^{1/2}$. The field due to an arc extending from φ_1 to φ_2 is evaluated by subtracting the value of this expression at φ_1 from that at φ_2. The procedure is carried out for positive and negative δ, and the results averaged to obtain the local contribution to the stress tensor at P. The typical numerical error introduced by truncating the expansion at the fourth order of φ is less than 1% for $\varphi = 0.20$, and less than 5% for $\varphi = 0.50$.

The line-splitting procedure, first suggested by Brown [15], obviously does not reflect the detailed physical nature of the core. Rather, it is an artifice for producing the correct asymptotic behavior as the radius of curvature of the dislocation line becomes large compared to the core size. One may argue about whether this is an optimal procedure, but in fact the use of any heuristic core-regularization procedure introduces relative errors of order $1/ln(R_c/\rho) \sim 0.1 - 0.2$ into *any* nonatomistic model of the dislocation motion. Here, ρ is a core-size parameter, which for our procedure equals $\delta/2$. Clearly, ρ now represents a phenomenological number which must be evaluated by comparison with experiment or from microscopic theory. Fortunately, the core-size parameter enters only in the logarithm, so that its exact value is not crucial.

To do an actua computation, one must define a *local* neighborhood about each point of interest P, to be treated in whatever core approximation one likes. The *nonlocal* contribution from outside this neighborhood is then obtained by summing over straight-line segments. We have found that, in order to avoid artificial instabilities, the finite-core approximation must be applied to a local region which is large compared to the core radius parameter ρ. This can be conveniently achieved within the context of Brown's recipe by refining its application as shown by the dashed lines in Fig. 1b. Here, in addition to splitting the arc passing through P and its nearest neighbors, we have also split the straight-line elements further away from P. The field at P is then determined as usual by averaging over the split portions. The effect is to treat the core as having a finite thickness over a certain length

$$L \gg \rho \tag{9}$$

irrespective of the point spacing Δl used to describe the dislocation line. This *modified Brown procedure* suffers no instability problems, it converges smoothly to the exact result as $\Delta l \to 0$, and it is insensitive to exactly where the local/nonlocal crossover is made. These characteristics permit proper error control and convergence tests. It also enables us to extend dislocation calculations down to nanometer scales, while suffering only an expected gradual loss of accuracy as $ln(R_c/\rho) \to 1$.

Much greater complications are involved if one needs to correct the dislocation fields to account for the presence of of free surfaces, or interfaces between different materials. In principle, the correction due to an infinite plane free surface can be computed element by element, as shown in the paper of Gosling and Willis [16], but their expressions are computationally much too demanding for most purposes, and of course do not deal with more complicated boundaries. For the moment, we are limited to using crude image approximations to handle this problem, although better approximations may soon be available [8]. This is not a major limitation for most applications, for which boundary corrections are either relatively small or else may be treated in an averaged way.

Finally, to complete the dynamical description, it is necessary to specify how the dislocation responds to a force. Again, various kinds of behavior can be found in real systems, but we limit ourselves to the simplest description: the dislocation line is assumed to move everywhere with a velocity proportional to the local force as determined from Eq. (1). This linear-response, diffusive-drift model of dislocation motion [14] assumes negligible inertial effects and the absence of any local pinning mechanisms. Although in the present work the proportionality factor between velocity and force (the mobility) is assumed to be constant, in practice the mobility is typically a strong function of temperature, and can furthermore depend on the position and orientation of the dislocation element.

Parallel Implementation

As a final part of the methodology description, we briefly outline the parallel implementation of the program, which is designed to operate on a scalable, distributed-memory machine running in loosely synchronous mode. The entire collection of dislocations is treated as a single string of points described by a set of one-dimensional arrays. The specification arrays include the three coordinates of each point, linked lists connecting the points (which need not be labelled sequentially), arrays specifying the Burgers vector and the glide plane of each point, an array specifying the nature of each point (start point, end point, free-surface termination, etc.), and other specifiers as needed. The string is created by or read into a master node, which cuts it into sections of nearly equal length, include four overlapping ghost points on each end, and distributes it to the other (slave) nodes. This provides practically perfect load balancing.

Although the present focus is on investigating relatively small systems of interacting dislocations, the program is structured with the aim of eventually handling large tangles and networks. Thus as much work as possible is carried out within the slave nodes. Simple tasks, such as evaluation of the local force contributions, interpolation between points, reconnections to boundaries, and movement of the individual points, can be carried out within each node and only require updating of the ghost points after every time step. Tasks such as the evaluation of the nonlocal forces and the search for reconnections require every point to look at every other point. This is done by using ring-shift calls to move the information in each slave node through all of the others. At present, our program treats the N^2 problem of evaluating the nonlocal forces by brute force. The addition of an $NlnN$ method will be required before large problems can be attacked.

The program has the ability to adaptively interpolate and remove points, and to change the dislocation tangle topology by reconnection. In a single node, such operations are conveniently accomplished by simply changing the linked-list pointers. To do these operations in a parallel environment, however, raises certain difficulties. For example, since the string is subdivided without regard to the spatial position of the points, reconnections can occur between points in different nodes, requiring one to break the string and terminate it by new ghost points within each node. To facilitate this kind of operation, each slave node is provided with an *update ledger* listing the locations of the ghost points that it is responsible for updating and the points within the node responsible for providing the update information. After every time step, all of the necessary update information is concatenated into a single ledger, along with the location for which it is intended. Each node then treats this ledger as a mailbox, looking at it to find out which of its points need to be changed, and to copy the information needed. With this device, it is possible to effectuate reconnection or similar operations while remaining in the parallel environment, simply by changing the local structure of the update ledgers.

Study of a Forest Interaction

Some dislocation encounters evolve in a rather obvious way, the reconnections which occur during the operation of a Frank-Read source being good examples. Not so for the more general case of dislocations colliding at an angle, as in the case of the "forest intersections" leading to work hardening in metals. We here study a simple example of such an interaction, with a view towards the question of whether one can develop a simple rule to describe its outcome. Let us look at the the evolution of two initially straight dislocation lines, one (1) lying on a $(1\bar{1}1)$ plane and the other (2) on an intersecting (111) plane. Dislocation 1 is allowed to have various starting configurations, which we can classify in terms of the angle ϕ wich it makes with the [110] direction and the distance Δ which initially separates it from dislocation 2. Here it is assumed that no applied stresses are acting on the dislocations and that the Burgers vector is chosen to be $\frac{1}{2}[\bar{1}01]$ for both lines. Numerical investigation of even this small family of interactions shows that several different kinds of behavior can occur.

In looking at the interactions shown in Fig. 2, it is instructive to concentrate one's attention on the region near the glide plane edge, since it is only here that the lines can interact strongly. Thus, if dislocation 1 is put at $\phi = 0°$ at a distance $\Delta = 40\,nm$, one can see that the lines move apart there so that the interaction is repulsive (Fig. 2a). If, however, ϕ is decreased only a little, the interaction suddenly switches character as the lines now evolve towards each other. As they do so, they mutually distort each other so that they are more or less antiparallel and aligned with the glide plane edge when they come together (Fig. 2b). Since they attract strongly in this configuration, it does not require an atomistic calculation to conclude that this kind of an *attractive instability* leads to a reconnection. As ϕ is decreased still further, the attractive character of the interaction remains, but its detailed nature changes subtly, until at $\phi = -180°$ lines 1 and

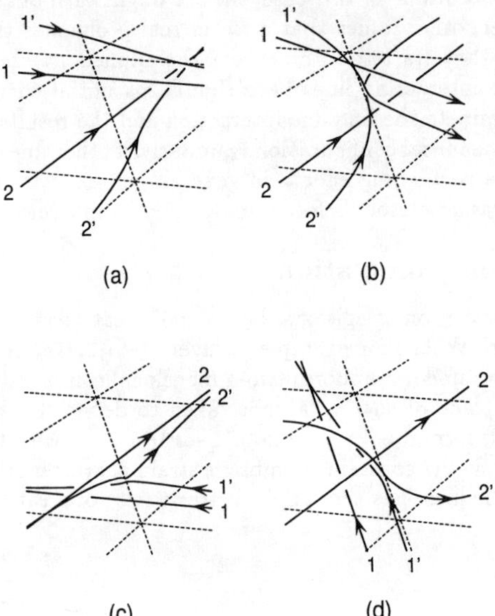

Fig. 2: Outcome of the interaction for $\Delta = 40\,nm$ and several values of ϕ. In each case the primed lines show the final configuration. (a) $\phi = 0°$; (b) $\phi = -15°$; (c) $\phi = -180°$; (d) $\phi = -240°$.

2 are in a crossed configuration (Fig. 2c) even as the cores touch. Now, it is not necessarily obvious how the evolution will proceed to a conclusion, and an atomistic calculations would be useful to sort things out. If the starting configuration of line 1 is rotated even more (Fig. 2d), the lines once again approach each other in an antiparallel fashion, but it is obvious that the distortion symmetry is different from that of Fig. 2b, with the lines now approaching each other from the opposite direction. Finally, at $\phi = -250°$, the interaction again becomes repulsive.

Suppose now that one is interested in treating a more realistic problem, involving many interacting dislocations which run into each other frequently as they evolve. It is extremely inefficient to resolve each such strong interaction fully in the manner of Fig. 2, since very small time steps are required to do so. Hence, one would like to replace this set of calculations by an interaction rule prescribing what happens when such an encounter occurs. Assuming that for equal Burgers vectors an attractive instability always leads to reconnection (an assumption still needing atomistic justification), a simple rule describing the results illustrated in Fig. 2 would then state that if line 1 is oriented within a certain range of angles, it will reconnect; outside of this range, the two lines will repel. Further calculations show that, in fact, the range of angles over which reconnection occurs is quite insensitive to large variations in the initial separation Δ between the lines. Of course, one also needs to consider other directions for dislocation 2 and other possible Burgers vector assignments. Thus, while it does not seem too difficult to establish a complete set of rules for two interacting dislocations in this case, the set might turn out to be rather large.

There are, however, other issues that arise in real problems: the dislocations might not be straight when they start to interact and the applied stresses will not be zero. It is obvious that if two interacting lines have significant initial curvatures, then the line tension forces will dominate the line-line interaction and the results will depend strongly on the particulars of the initial configuration. Similarly, if other lines happen to be nearby, their fields will have a perturbing effect. In general, one expects that lines will have to approach each other rather closely before a reasonably simple set of rules can be applied.

A More Ambitious Simulation

The growth and relaxation of epitaxial layers is of great technical interest in semiconductor manufacturing. When, for example, a layer of $Si_{1-x}Ge_x$ is grown epitaxially on pure Si, the layer grows under a uniform two-dimensional compressive strain of up to 4%. If the normal to the plane of such a layer is taken to define the $\hat{z}$ direction, the stress tensor takes the simple form $\sigma_{xx} = \sigma_{yy} = 2\epsilon_0\mu(1+\nu)/(1-\nu)$, with the other components equal to zero. Here, ϵ_0 is the constant membrane strain characteristic of such a layer, and the medium outside the layer has zero applied stress. Also of great interest is the capped

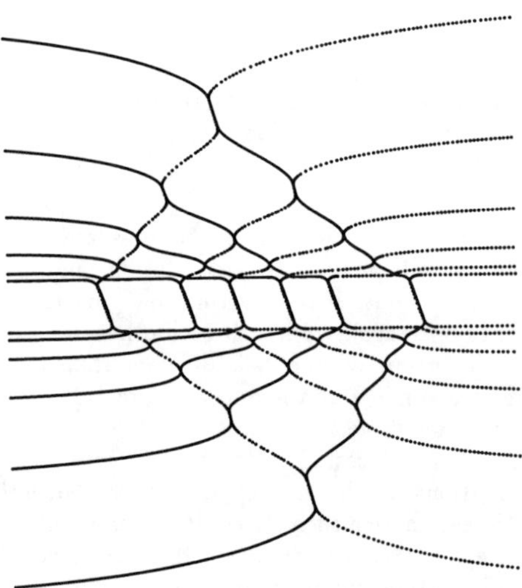

Fig. 3: Dislocation structure built by two Frank-Read sources on parallel glide planes. Threading arms propagate in from both sides of the figure. This particular calculation was done on a $100\,nm$ capped layer, on glide planes eight glide plane spacings apart, and $\epsilon = .0090$. The dislocation on the plane in front is shown as a solid line, the one on the back plane is dotted. The high quality of the computation can be judged by the fact that the dots are the actual mesh points used in the calculation.

layer, which one can think of as a region of uniform two-dimensional applied stress in an infinite medium. Although a capped layer looks like a symmetrized version of an ordinary layer, it is in fact much simpler, since it does not require one to deal with a free surface.

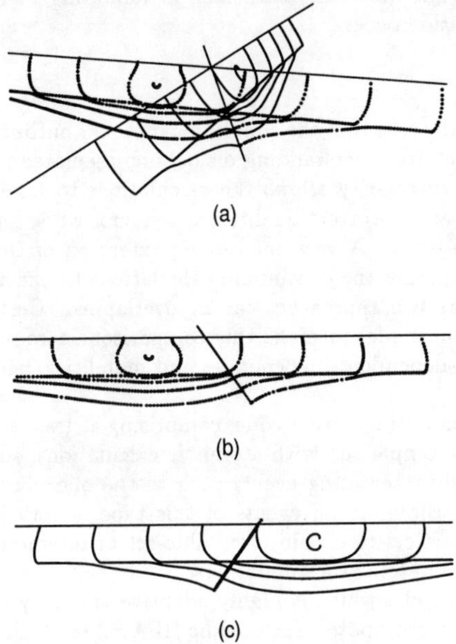

Fig. 4: Fully resolved calculation of the pattern created by two different Frank-Read sources, asymmetrically placed on intersecting glide planes, in a $100\,nm$ layer with a free surface and under a strain of 0.020. One instant in time is shown. Dislocations lying on the (111) plane are shown as solid lines, while the actual computational mesh points are used to depict the lines lying on the $(1\bar{1}1)$ plane. For clarity, the substrate-layer interface is not drawn, but its location is obvious. (a) perspective view; (b) viewed in the $[\bar{1}10]$ direction ; (c) viewed in the $[\bar{1}\bar{1}0]$ direction.

On the assumption that relaxation processes in these layers are driven by the operation of a low density of Frank-Read type sources, one is impelled to study the interaction between the dislocations produced by such sources, both for the case where the interacting sources lie on parallel glide planes, and for the case where they lie on intersecting glide planes. The results of such studies are illustrated in Figs. 3 and 4. It is found that sources operating on proximate parallel glide planes interact strongly to form immobile dipoles which thread the strained layer, and are connected to elaborate dislocation networks extending far out into the unstrained substrates. Sources on intersecting glide planes, on the other hand, can interact more strongly in the form of topology-changing interactions such as reconnection, and jog or junction formation. A relatively simple example, involving only reconnection, is shown in Fig. 4. One finds that even in this case, complicated corner pileup structures are produced in the substrate, involving dislocation which pass from one glide plane to the other. In actual experimental studies, both the network patterns and the corner pileup structures are characteristically observed in the substrates of relaxed layers. Hence in this case, the numerical simulations are useful both in explaining the experimental observations and in validating a simple theoretical picture of how the layer relaxation occurs.

Conclusions

Figures 2 – 4 make the case that the simulation methods outlined here are suitable for a true multiscale approach to understanding dislocation-mediated phenomena. The proper treatment of the core singularity allows the calculations to be extended down towards atomistic scales, maintaining correct qualitative behavior while losing accuracy in a well-controlled way as $\rho/R \to 0$. A very interesting extension of the work presented here, therefore, will be to compare the continuum calculations to exact atomistic calculations in the regime where the two approaches can be overlapped. Certain refinements can be added to the continuum model to make this comparison more meaningful: dissociation into partials, direction-dependent line energies and mobilities, better surface treatments, and lattice anisotropy.

An improved version of the current code, comprising at least some of the above refinements, and verified by comparison with atomistic calculations, should prove most useful in investigating strongly interacting events such as the one illustrated by Fig. 2. The number of different possible strong events of this type is very large, especially in bcc metals, and it is impractical to obtain a suitable set of interaction rules for large-scale simulations directly from atomistic simulations.

Finally, the present code, which is highly adaptive and fully parallelized to run on a distributed-memory supercomputer, such as the IBM SP series, is suitable for developing efficient large-scale simulations. To the extent that rules can be found which adequately predict the outcomes of strong local interactions, these can be utilized by the program to speed up its processing. Simulations such as those shown in Fig. 4, however, have already shown that the rule-based approach is likely to run into trouble as the density of the dislocations becomes large. An advantage of the approach presented here is that, when this occurs, the program can deal with such situations without breaking stride, slowing down and resolving the interaction, down to the atomistic level if necessary, in order to decide how to proceed. This scheme, in which a large-scale parallel simulation applies time-saving short cuts when it can, but resolves difficult situations more fully, seems to offer the optimal combination of efficiency and physical validity.

References

[1] L.P. Kubin, G. Canova, M. Condat, B. DeVincre, V. Pontikis, and Y. Bréchet, Solid State Phenomena **23-24**,455 (1992).

[2] B. DeVincre and L.P. Kubin, Model. Simul. Mater. Sci. Eng. **2**, 559 (1994).

[3] L.P. Kubin, B. DeVincre, G. Canova, and Y. Bréchet, Key Engineering Materials **103**, 217 (1995).

[4] D. Raabe, Z. Metallkd. **87**, 493 (1996).

[5] N.M. Ghoniem and M. Bacaloni, Eng. Rept. UCLA/MATMOD-97-01, Dept. of Mech. and Aerospace Eng., UCLA, Los Angeles, CA (1997).

[6] M. Tang, L.P. Kubin, and G.R. Canova, submitted to Acta Materialia.

[7] H.M. Zbib, M. Rhee, and J.P. Hirth, Int. J. Mech. Sci. **40**, 113 (1998).

[8] L.B. Freund and R. Kukta (private communication).

[9] K.W. Schwarz and J. Tersoff, Appl. Phys. Lett. **69**, 1220 (1996).

[10] K.W. Schwarz, Phys. Rev. Lett. **78**, 4785 (1997).

[11] K.W. Schwarz and F.K. LeGoues, Phys. Rev. Lett. **79**, 1877 (1997).

[12] K.W. Schwarz, J. Appl. Phys. (in press).

[13] M.O. Peach and J.S. Koehler, Phys. Rev. **80**, 436 (1950).

[14] J.P. Hirth and J. Lothe, *Theory of Dislocations*, 2nd Ed. (Wiley, New York, 1982).

[15] L.M. Brown, Philos. Mag. **10**, 441 (1964)

[16] T.J. Gosling and J.R. Willis, J. Mech. Phys. Solids **42**, 1199 (1994).

DISCRETE DISLOCATIONS INTERACTING WITH A MODE I CRACK

H.H.M. CLEVERINGA[1], E. VAN DER GIESSEN[1] and A. NEEDLEMAN[2]
[1]Delft University of Technology, Koiter Institute Delft, Mekelweg 2, 2628 CD Delft, The Netherlands
[2]Brown University, Division of Engineering, Providence, RI 02912, USA

ABSTRACT

Small scale yielding around a plane strain mode I crack is analyzed using discrete dislocation dynamics. The dislocations are all of edge character, and are modeled as line singularities in an elastic material. At each stage of loading, superposition is used to represent the solution in terms of solutions for edge dislocations in a half-space and a complementary solution that enforces the boundary conditions. The latter is non-singular and obtained from a linear elastic, finite element solution. The lattice resistance to dislocation motion, dislocation nucleation, dislocation interaction with obstacles and dislocation annihilation are incorporated into the formulation through a set of constitutive rules. A relation between the opening traction and the displacement jumps across a cohesive surface ahead of the initial crack tip is also specified, so that crack initiation and crack growth emerge naturally. Material parameters representative of aluminum are employed. Two cases are considered that differ in the strength and density of dislocation obstacles. Results are presented for the evolution of the dislocation structure and the near-tip stress field during the early stages of crack growth.

INTRODUCTION

The interaction between plastic flow and the actual process of material separation plays an important role in setting the fracture response of structural materials. It is well known that plastic dissipation in the material around a crack tip results in a macroscopic work of fracture that is much larger then the work required to separate the crack surfaces. On the other hand, the micro or nano scale patterning of plastic flow can lead to dislocation free regions and/or dislocation pile-ups that give rise to stress concentrations that promote fracture.

Analyses using classical continuum plasticity have provided much insight into the interaction between plastic flow and fracture processes. However, when the actual separation process takes place on an atomic scale there is an inherent difficulty in using continuum plasticity. According to classical continuum plasticity, the maximum stress attained at a blunted mode I crack tip is of the order of 3–5 times the material's flow strength. Typically, such stress levels are too low to cause atomic separation. This is of particular significance when cohesive surface models of fracture are used in conjunction with continuum plasticity, as in [3, 4]. To overcome this limitation, a model was proposed by Suo *et al.* [5], where there is an elastic strip inside which the crack propagates. Because the strip remains elastic, stresses become high enough for atomic separation to develop.

There is experimental evidence that plastic flow over small material size scales requires higher stress levels than does plastic flow over larger size scales. Nonlocal plasticity theories have been developed, e.g. [6, 7], to model this effect. Such theories give rise to higher stress levels in the vicinity of a crack tip than do classical plasticity theories, see [7]. In contrast to the model of Suo *et al.* [5], where the high stresses arise from a lack of plastic flow, the

Mat. Res. Soc. Symp. Proc. Vol. 538 © 1999 Materials Research Society

nonlocal theories implicitly attribute the increased stress to concentrations associated with dislocation structures.

In this paper, we carry out full boundary value problem solutions for small scale yielding of a mode I crack in plane strain, with plastic flow arising from the collective motion of large numbers of edge dislocations. The fracture properties of the material are embedded in a cohesive surface constitutive relation so that crack initiation and crack growth emerge as natural outcomes of the boundary value problem solution. Attention is restricted to small strains and the dislocations are modeled as line defects in an isotropic linear elastic solid.

In our analyses, the plastic stress-strain response and the evolution of the dislocation structure are outcomes of the boundary value problem solution. By way of contrast, in a conventional continuum formulation the plastic stress-strain response is an input, while early dislocation analyses have postulated a dislocation structure in the crack tip vicinity, e.g. [8]. More recently, dislocation dynamics studies have been carried out where the dislocation structure evolved in response to the imposed loading; near a stationary mode I crack tip by e.g. [9, 10] and near a mode III growing crack by Zacharopoulos *et al.* [11]. A main focus in these studies is the ductile-brittle transition in initially dislocation-free materials, where dislocations are nucleated from the crack tip. In this study, dislocation nucleation takes place from two dimensional Frank-Read sources distributed in a process window surrounding the crack tip.

THEORY

We analyze the two-dimensional plane strain small-scale yielding problem sketched in Fig. 1, with dislocations restricted to a process window as shown in the figure. Symmetry about the crack plane is assumed. Remote from the crack tip, displacements corresponding to the linear elastic mode I K-field are applied, and crack initiation and growth are modeled using a cohesive surface framework, as in [3].

The origin is at the initial crack tip, with the x_1 and x_2 directions in the crack plane and perpendicular to the crack plane, respectively. Behind the initial crack tip is a traction-free surface. Immediately ahead of the crack tip $T_1 = 0$ and $T_2 = -T_n(u_2)$, with T_n representing a traction separation law that will be specified later. Further ahead of the initial crack tip, there is no cohesive surface and the condition $\dot{u}_2 = 0$ is applied to enforce symmetry.

At each time step, an increment of the mode I stress intensity factor $\dot{K}_I \Delta t$ is prescribed. At the current instant, the stress and strain state of the body is known, so the forces between dislocations can be calculated. Edge dislocations on multiple slip planes which have an angle of $\pm 30°$ with the x_2-axis are considered. On the basis of these forces we update the dislocation structure, which involves the motion of dislocations, the generation of new dislocations, their mutual annihilation, their pinning at obstacles, and their exit into the open crack. After this, the new stress and strain state can be determined.

The method for determining the current state of the body with the new dislocation distribution is similar to that in [1]. The current state of the body in terms of the displacement, strain and stress fields is written as the superposition of two fields ([1, 2, 12]),

$$u = \tilde{u} + \hat{u}, \quad \epsilon = \tilde{\epsilon} + \hat{\epsilon}, \quad \sigma = \tilde{\sigma} + \hat{\sigma}. \tag{1}$$

The (~) fields are the superposition of the singular fields of the individual dislocations in their current configuration. In previous work [1, 2] the individual dislocation fields were taken to be those of an edge dislocation in an infinite medium. Here, solutions that describe the

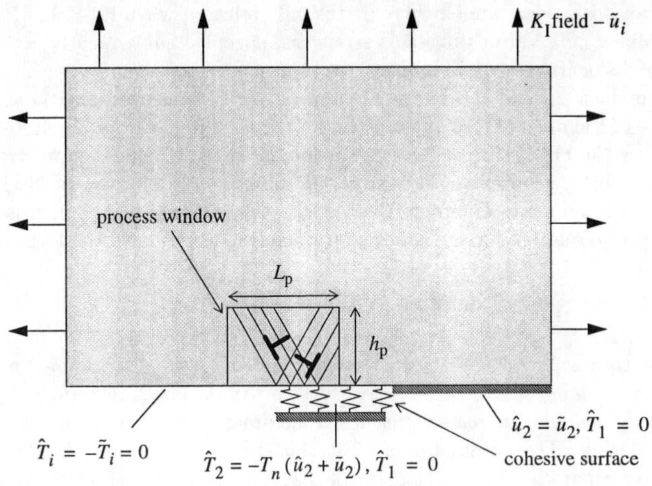

Figure 1: Mode I crack problem with the imposed boundary conditions.

stress and displacement field of an edge dislocation in a half-space are used [13], with the traction-free surface corresponding to the crack plane $x_2 = 0$.

The boundary conditions are imposed through the (^) fields. The sum of the (~) and the (^) fields in (1) gives the solution that satisfies all boundary conditions. Since the (^) fields are smooth in the region of interest, the boundary value problem for them can conveniently be solved using a finite element method. The size of the region analyzed is $1000\mu\mathrm{m} \times 500\mu\mathrm{m}$ and a finite element mesh of 120×100 bilinear quadrilateral elements is used. The process window in Fig. 1 is specified by $L_\mathrm{p} = 10\mu\mathrm{m}$ and $h_\mathrm{p} = 12.5\mu\mathrm{m}$ and in it there is a fine mesh of 80×80 quadrilateral elements.

Dislocation motion is assumed to occur only by glide with no cross slip so that dislocations remain on their slip plane. The Peach-Koehler force f^i acting on the ith dislocation is given by

$$f^i = n^i \cdot \left(\hat{\sigma} + \sum_{j \neq i} \sigma^j\right) \cdot b^i, \tag{2}$$

with n^i the slip plane normal, b^i the Burgers vector of dislocation i and σ^j is the stress field of dislocation j. The direction of this force is in the slip plane and normal to the dislocation line. The magnitude of the glide velocity v^i of dislocation i is taken to be linearly related to the Peach-Koehler force f^i through the drag relation $f^i = Bv^i$. The value for B is taken as $B = 10^{-4}\mathrm{Pa\,s}$, which is a representative value for aluminum.

New dislocations are generated by simulating Frank-Read sources in two dimensions by point sources on a slip plane. The sources generate a dislocation dipole when the magnitude of the Peach-Koehler force exceeds a critical value $\tau_{\mathrm{nuc}}b$ during a period of time t_{nuc}. The source strengths are distributed randomly about a mean value $\tau_{\mathrm{nuc}} = 50\,\mathrm{MPa}$ and $t_{\mathrm{nuc}} = 10\,\mathrm{ns}$. The distance between this dislocation pair is chosen so that their mutual attractive force is equal to $\tau_{\mathrm{nuc}}b$. To model locking of dislocations at small precipitates or at dislocations on other slip planes we introduce obstacles as fixed points on a slip plane. A dislocation that glides against an obstacle is pinned there and is only released when its Peach-Koehler force exceeds the value $\tau_{\text{obs}}b$. Various obstacle strengths, specified subsequently, are considered. Annihilation of dislocations with opposite signed Burgers vector occurs when they are within a material dependent, critical annihilation distance $L_e = 6b$. Dislocations can also glide into the free surface of the open crack. When they do, they disappear from the system but leave a lattice step on the crack surface. Since slip planes are positioned symmetrically about $x_2 = 0$, when a dislocation exits the computational region across the (still closed) plane ahead of the crack, a dislocation enters the computational region on the mirror slip plane.

The constitutive relation for the cohesive surface is taken to have the form ([14])

$$T_n(\Delta_n) = e\sigma_{\text{max}}\frac{\Delta_n}{\delta_n}\exp(-\frac{\Delta_n}{\delta_n})$$ (3)

where Δ_n is the total separation of the cohesive surface (twice the displacement along $x_2 = 0$) and T_n is the traction normal to the cohesive surface. As the cohesive surface separates, the magnitude of the traction increases, reaches a maximum and then approaches zero with increasing separation. The parameters used in this study are $\sigma_{\text{max}} = 1.2\,\text{GPa}$ and $\delta_n = 4b$ giving a work of separation, $\phi_n = e\sigma_{max}\delta_n$ of $3.26\ \text{J/m}^2$. The work of separation can be related to a reference stress intensity factor K_0 defined by

$$K_0 = \sqrt{\frac{E\phi_n}{(1 - \nu^2)}}.$$ (4)

With $E = 70\,\text{MPa}$ and $\nu = 0.33$, representative for aluminum, $K_0 = 0.506\,\text{MPa}\sqrt{\text{m}}$.

RESULTS

The calculations are carried out for a loading rate of $\dot{K}_I = 50\text{GPa}\sqrt{\text{m}}/\text{s}$. This rather high loading rate is used to reduce the computer time needed for crack growth because resolving the dislocation dynamics requires a small time step, $\Delta t = 0.5\text{ns}$. The results to be presented are for a slip plane spacing of $86b$. There are 401 slip planes oriented at $+30°$ and 401 at $-30°$ to the x_2-axis. The Burgers vector is $b = 0.25\text{nm}$, which is appropriate for aluminum. The material is taken to be initially dislocation-free.

Two cases are analyzed. In case I, a random source distribution of $\rho_{\text{src}} = 49.2/\mu\text{m}^2$ and a random obstacle distribution of $\rho_{\text{obs}} = 98.4/\mu\text{m}^2$ is chosen. The obstacle strength is $\tau_{\text{obs}} = 150\text{MPa}$. Case II has the same source distribution and strength as case I, but only half the density of obstacles, $\rho_{\text{obs}} = 49.2/\mu\text{m}^2$. For case II, the obstacle strength is increased to $\tau_{\text{obs}} = 200\text{MPa}$. For both case I and case II, two realizations of the random source and obstacle locations are analyzed.

Figure 2a shows the value of the applied K_I as a function of the amount of crack growth Δa for all four calculations. The value of K_I at the initiation of crack growth depends on the particular realization, with no systematic difference seen between the results for cases I and II. On the other hand, the crack growth resistance, the mean slope of the K_I versus Δa curve, is substantially higher for the two realizations of case II than for the two realizations of case I. Thus, the cases with a lower density of higher strength obstacles have a significantly higher crack growth resistance. In all four calculations, crack growth tends to occur in "spurts;" periods of crack growth occur at more or less constant K_I that are separated by periods where the crack is essentially stationary while K_I increases. A similar mode of crack

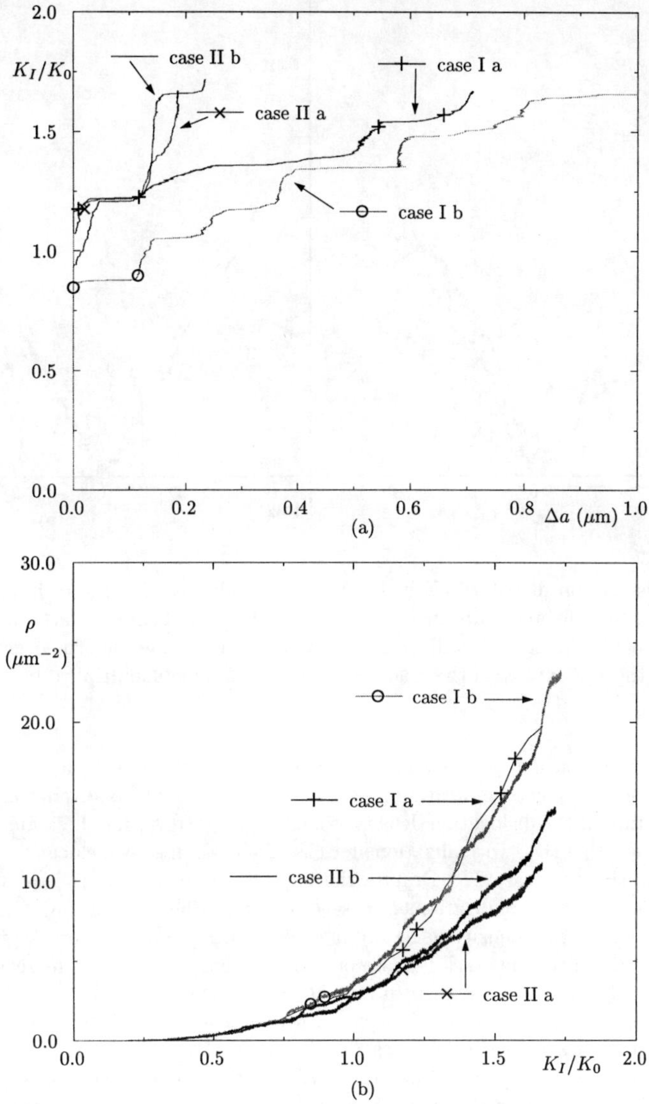

Figure 2: The applied stress intensity factor K_I versus crack growth Δa (a) and the dislocation density versus K_I (b) for the two realizations of case I and case II. The symbols + indicate states that are further explained in Figs. 3a, 4, 6 and 7a; × in Fig. 3b and ○ in Fig. 5.

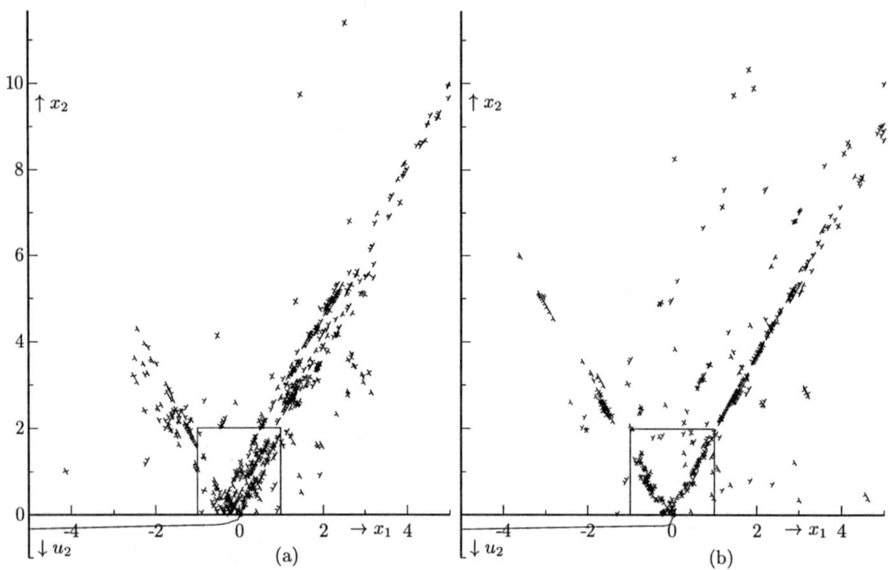

Figure 3: Dislocation distributions in the process window at $K_I/K_0 = 1.18$; (a) case Ia and (b) case IIa. Distances are in μm. The small boxes indicate the area in which more detailed dislocation structures will be shown subsequently. The deformed crack profile is shown below the x_1-axis, with the crack opening displacements multiplied by a factor of 10.

growth was seen by Zacharopoulos *et al.* [11]. During the first "spurt" the crack grows along more than 2 mesh elements, indicating no correlation between crack advance and mesh size.

The evolution of the dislocation density versus the applied K_I for these four calculations is shown in Fig. 2b. The two realizations for case I show a higher dislocation density than the two realizations for case II. Despite the higher dislocation density, the crack growth resistance is less for both realizations of case I than for either realization of case II. Also, note that the dislocation density generally increases during periods of crack growth as well as during periods where the crack is more or less stationary while K_I is increasing.

Dislocation distributions for case Ia and IIa are shown in Fig. 3 at $K_I/K_0 = 1.18$. In case II most dislocation activity takes place on only a few slip planes, whereas in case I many more slip planes are active. Dislocation pile ups are larger for case II due to the higher obstacle strength. In both cases, most dislocation activity takes place on slip planes that cross the crack plane behind the crack tip and that have an angle of $-30°$ with the x_2-axis. The curve below the x_1-axis shows the opening of the crack and the location of the crack tip. Since symmetry about the x_1-axis is assumed, the opening of the lower half of the crack surface is shown for clarity. The opening displacements are magnified by a factor 10 to give a better view. The profile of the crack opening for case I is more rounded than for case II, which is attributed to the fact that more slip planes are active.

Continuum solutions for stationary crack tip fields in single crystals under plane strain

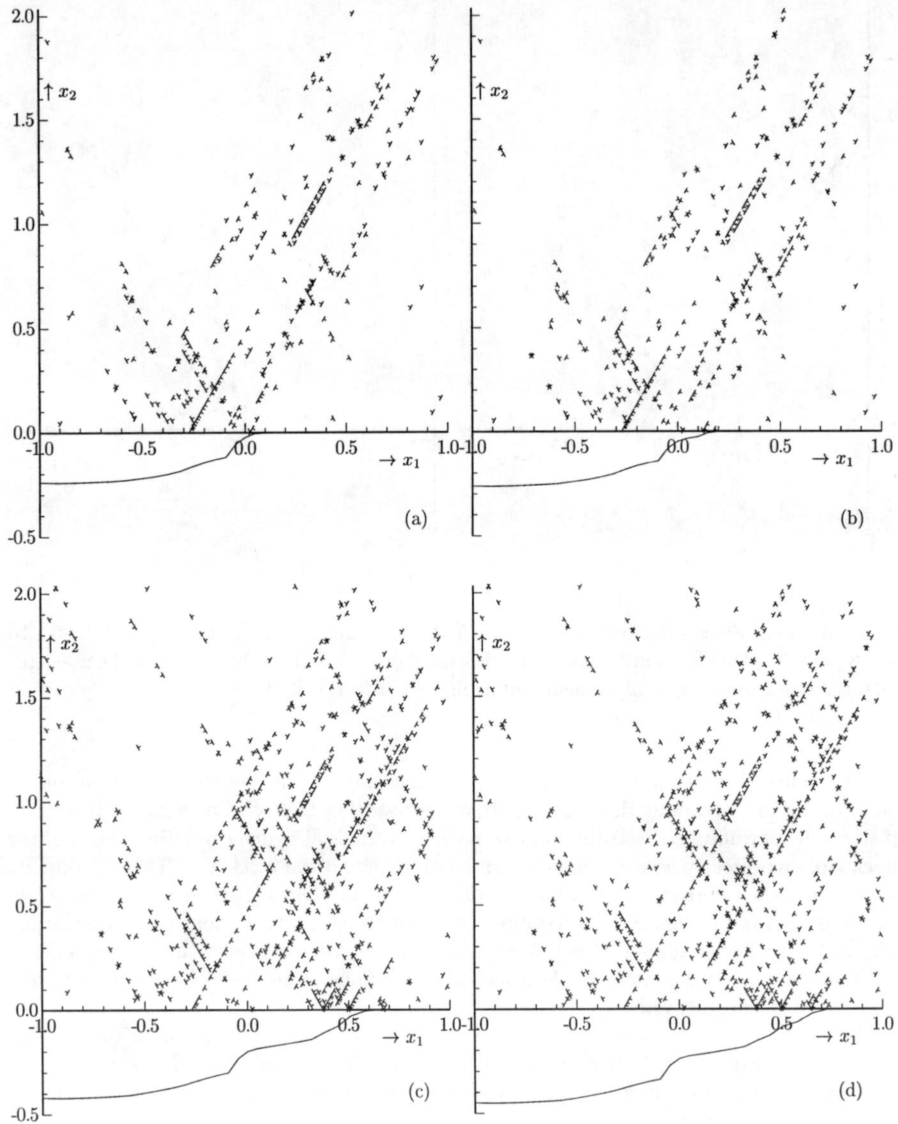

Figure 4: Dislocation distributions for case Ia at four loading stages. (a) $K_I/K_0 = 1.18$, (b) $K_I/K_0 = 1.23$, (c) $K_I/K_0 = 1.52$ and (d) $K_I/K_0 = 1.57$. Distances are in μm. The deformed crack profile is shown below the x_1-axis, with the crack opening displacements multiplied by a factor of 10.

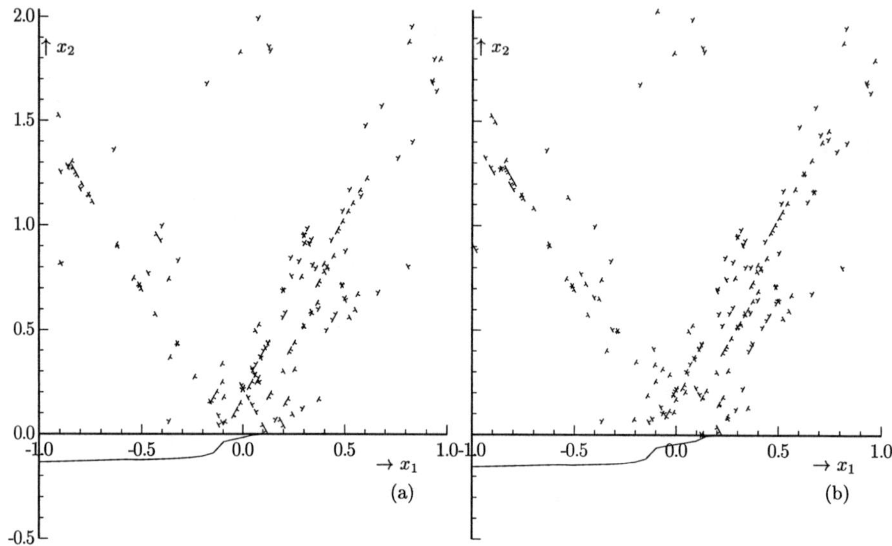

Figure 5: Dislocation distribution for case *Ib* at two loading stages, (a) $K_I/K_0 = 0.85$, (b) $K_I/K_0 = 0.90$. Distances are in μm. The deformed crack profile is shown below the x_1−axis, with the crack opening displacements multiplied by a factor of 10.

conditions have been presented in [15] and [16]. Rice [15] obtained an analytical solution for small strains and assuming the solid is everywhere at yield. This solution exhibits three rays of stress discontinuity representing narrow regions of single slip; one eminating ahead of the crack tip, one directly above it and one emanating behind the crack tip. The one directly above the crack tip gives rise to a kink mode. Cuitiño and Ortiz [16] carried out numerical solutions allowing for finite deformations and found three similar regions of concentration. The dislocation structures in Fig. 3 give evidence of concentrations that correspond to the two shear mode directions in the continuum solutions. The angles here are set by the orientation of the slip systems.

To illustrate what happens closer to the crack tip with the dislocation distribution, Fig. 4 shows dislocation distributions at four stages of loading for case *Ia*. These figures also show the shape of the crack surface. By comparing the various snapshots, it is seen that most dislocation activity takes place on slip planes that cross the crack plane in front of the crack tip. Upon closer observation, one sees that from $K_I/K_0 = 1.18$ to $K_I/K_0 = 1.23$, between Figs. 4(a) and (b), the crack propagates a small distance and then gets locked at a dislocation. From $K_I/K_0 = 1.52$ to $K_I/K_0 = 1.57$, between Figs. 4(c) and (d), a similar event occurs. As the crack grows, new slip planes in front of the current crack tip become active. The dislocation structures show that there are always a few dislocations located close to the crack tip. Apparently, the period during which crack growth occurs is long enough for

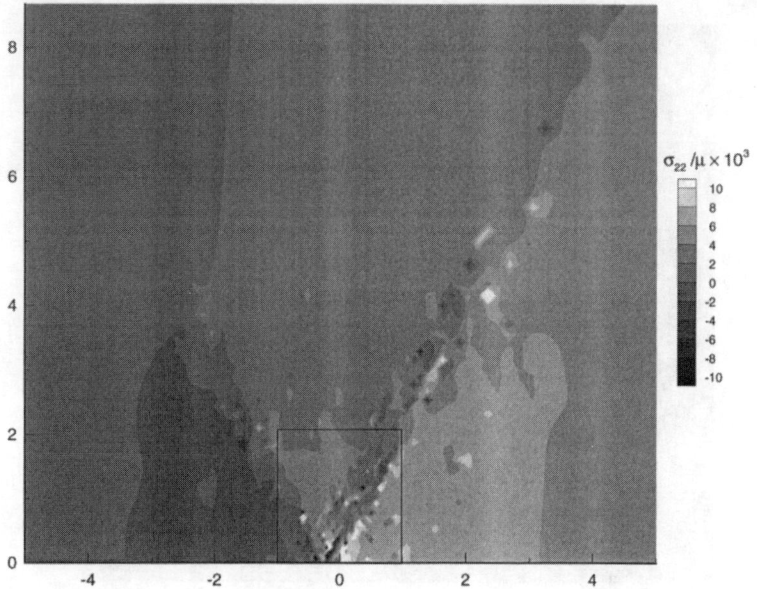

Figure 6: Contours of normal stress σ_{22} in the process window for case Ia at $K_I/K_0 = 1.18$. Distances are in μm.

sources ahead of the crack tip to generate new dislocation pairs. Generally, one or several of these dislocations glide towards the (still closed) crack plane. These dislocations then act to obstruct further crack motion. The plots for the higher K_I values show that the crack surface stays rounded after the crack has grown. At certain times, many dislocations glide out of the material into the crack free surface, causing a relatively large step. This step then remains as the crack propagates.

Figure 5 shows the dislocation distribution for case Ib at $K_I/K_0 = 0.85$ and at $K_I/K_0 = 0.90$. Between fig. 5(a) and (b) the crack propagates over a small distance. In this situation crack growth occurs due to changes in the local dislocation structure. Dislocations ahead of the crack tip move away, the crack starts to propagate and then other dislocations move towards the crack plane, obstructing further crack motion. The value of K_I at initiation for case Ib is much lower than the value of K_I at the beginning of crack growth for case Ia. In this case, there are no dislocations close to the crack tip, which is probably why the stress field at the crack tip is high enough for decohesion.

This can be verified by looking at the local stress fields inside the process zone as computed in (1c). A contour plot of the opening stress σ_{22} at $K_I/K_0 = 1.18$ for case Ia is shown in Fig. 6. The region shown in this figure is the full process window. It gives an indication of how the elastic K_I field is relaxed by the evolved dislocation structure, especially above the active slip planes. In front of the crack tip and below the activated slip planes, high stresses are still present.

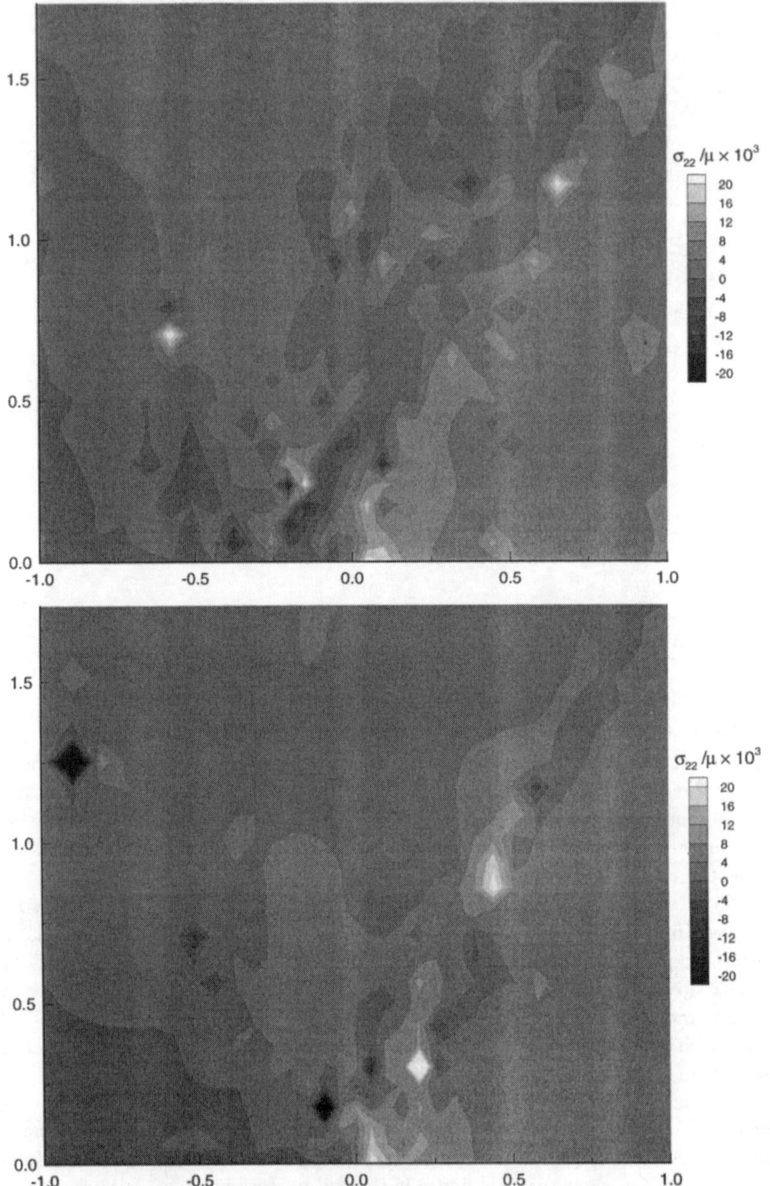

Figure 7: Contours of normal stress σ_{22}; (a) case Ia at $K_I/K_0 = 1.18$ and (b) case Ib at $K_I/K_0 = 0.85$. Distances are in μm.

Figure 7 shows the stress distribution very close to the crack tip for two cases; case Ia at $K_I/K_0 = 1.18$ and case Ib at $K_I/K_0 = 0.85$. At this magnification, the stress peaks of the individual dislocations are clearly visible. According to Fig. 2a, the crack in the plotted state is about to 'jump' forward in both cases. Indeed, the figure shows that the stress level close to the crack tip is more or less the same for both cases.

CONCLUSIONS

Preliminary results of a plane strain analysis of small scale yielding around a mode I crack tip have been presented where plastic flow arises from the motion of large numbers of dislocations. The material is initially dislocation free and dislocation nucleation occurs from Frank Read sources distributed randomly in the material. Special dislocation nucleation or emission from the crack tip is not incorporated. Two obstacle densities were considered with different obstacle strengths. It was found that the cases with the lower density of obstacles but the higher obstacle strength gave rise to the larger crack growth resistance. In all cases, crack growth occurred in "spurts" with the dislocation activity close to the crack tip having a strong influence on the crack growth behavior.

ACKNOWLEDGMENTS

We acknowledge stimulating discussions with R. LeSar and D.J. Srolovitz. The work of H.H.M. Cleveringa is part of the research program of the "Stichting voor Fundamenteel Onderzoek der Materie (FOM)" which is supported financially by the "Nederlandse Organisatie voor Wetenschappelijk Onderzoek (NWO)." A. Needleman and E. Van der Giessen acknowledge support from the Materials Research Science and Engineering Center on *On Micro- and Nano-Mechanics of Materials* at Brown University (NSF Grant DMR-9632524).

REFERENCES

[1] E. Van der Giessen and A. Needleman, *Modeling Simul. Mater. Sci. Eng.* **3**, 689–735 (1995).

[2] H.H.M. Cleveringa, E. Van der Giessen and A. Needleman, *Acta Mat.* **45**, 3163–3179 (1997).

[3] A. Needleman, *J. Mech. Phys. Solids* **38**, 289–324 (1990).

[4] V. Tvergaard and J.W. Hutchinson, *J. Mech. Phys. Solids* **40**, 1377–1397 (1992).

[5] Z. Suo, C.F. Shih and A.G. Varias, *Acat Metall. Mater.* **41**, 1551–1557 (1993).

[6] N.A. Fleck and J.W. Hutchinson, *J. Mech. Phys. Solids* **41**, 1825–1857 (1993).

[7] N.A. Fleck and J.W. Hutchinson, *Adv. Appl. Mech.* **33**, 295–361 (1997).

[8] J. Weertman, I.H. Lin and R. Thomson, *Acta Metall.* **31**, 473–482 (1983).

[9] P.B. Hirsch and S.G. Roberts, *Scr. Metall.* **23** 925–930 (1989).

[10] V.R. Nitzsche and K.J. Hsia, *Mat. Sci. Engng A* **176** 155 (1994).

[11] N. Zacharopoulos, D.J. Srolovitz and R. LeSar, *Acta Mat.* **45**, 3745–3763 (1997).

[12] V. Lubarda, J.A. Blume and A. Needleman, *Acta Metall. Mater.* **41**, 625–642 (1993).

[13] L.B. Freund, *Adv. Appl. Mech.* **30**, 1–66 (1994).

[14] J.H. Rose, J. Ferrante and J.R. Smith, *Phys. Rev. Lett.*, **47**, 675–678 (1981).

[15] J.R. Rice, *Mech. Mat.*, **6**, 317–335 (1987).

[16] A.M. Cuitiño and M. Ortiz, *Model. Simul. Mat. Sci. Engin.*, **1**, 255–263 (1992).

ATOMIC SCALE SIMULATION OF CROSS SLIP AND SCREW DISLOCATION DIPOLE ANNIHILATION

TORBEN RASMUSSEN*

Centre for Atomic–scale Materials Physics and Department of Physics
Technical University of Denmark, DK–2800 Lyngby, Denmark
and
Materials Research Department, Risø National Laboratory, DK–4000 Roskilde, Denmark

ABSTRACT

Atomistic simulations are used to study cross slip of a single screw dislocation as well as screw dislocation dipole annihilation in Cu. A configuration space path technique is applied to determine, without presumptions about the saddle point, the minimum energy path of transition for cross slip. The cross slip process is that proposed by Friedel and Escaig, and the energy of the in–plane constriction initiating cross slip is determined. A minimum stable dipole height much smaller than previously inferred from experimental studies is found. Relaxed screw dislocation dipoles adopt a skew configuration due to the anisotropy of Cu. The path technique is applied to investigate annihilation of stable screw dislocation dipoles, and the energy barrier for annihilation as a function of dipole height is determined for both homogeneous and heterogeneous cross slip leading to the annihilation. The results might be used as quantitative input into meso–/macro–scopical modelling approaches which rely on parameters deduced from either simulation or experiment.

INTRODUCTION

Recently so–called *multiscale modelling* has attracted increased attention in the materials science community [1, these proceedings]. However, while the desire to model across different length and time scales is not new, the renewed interest is primarily a consequence of the dramatic increase in computational capabilities acquired during the 1990s. Naturally the development of new and improvement of older computational methods are of great significance, too.

It is not within the scope of this paper to define, or even discuss, multiscale modelling. Rather, it is the intention to present results obtained by atomistic simulation of cross slip and the closely related annihilation of screw dislocation dipoles in the fcc metal Cu. These basic phenomena need a fully three dimensional atomistic treatment that has only recently become possible. Besides the fundamental interest in such phenomena it is the hope that the results might be used as input into modelling carried out at coarser length scales.

Modelling the collective behaviour of many individual dislocations at the *meso–scale* by dislocation dynamics simulations (e.g. [2, 3]) is now a well–known technique. This approach ultimately aims at connecting the microscopical properties of dislocations and the macroscopical properties of real materials. In doing so, the method has to rely on parameters controlling the behaviour of the individual dislocation and its interactions with other defects. Information about such parameters has to come from either experiment or simula-

---

*Present address: Theoretical Division, Los Alamos National Laboratory, Los Alamos, NM 87545, USA.

Mat. Res. Soc. Symp. Proc. Vol. 538 © 1999 Materials Research Society

tion. The latter could be atomistic simulations that are capable of providing quantitative information about, e.g., dislocation junction strength [4], kink pair formation energy [5], and parameters pertinent to cross slip and screw dislocation dipole annihilation which is the subject of this paper.

CROSS SLIP

Cross slip of screw dislocations is a prominent dislocation process in plastic deformation. In deformed materials the dislocation density is controlled by dislocation generation and dislocation annihilation. At room temperature, cross slip is the only known mechanism for dislocation annihilation. In cyclic deformation experiments the build–up of persistent slip bands is attributed to irreversible cross slip of screw segments resulting in the dipolar structure of edge dislocations, e.g., Ref. [6]. Furthermore, the cell patterning seen in monotonic deformation is also connected to cross slip which can account for the dislocation storage and thus the strain hardening. In Al, where cross slip is relatively easy, dislocation dynamics simulations show no cell patterning in the absence of cross slip [2]. Hence, attempts to model plasticity, either numerically or analytically, must include cross slip properly.

Cross slip is an intricate mechanism involving long range elastic interactions as well as atomistic (core) effects. In this paper the focus will be on the mechanism proposed by Friedel and Escaig (FE) [7, 8], even though the mechanism proposed by Fleischer [9, 10] might operate under certain conditions [11]. In the FE cross–slip mechanism for a dissociated screw dislocation the two Shockley partials must recombine (constrict) in the primary glide plane before the subsequent redissociation in the cross–slip plane. The redissociation creates two non–equivalent twisted constrictions on the dislocation denoted "edgelike" and "screwlike" [12] because of the characters of the partials in the vicinity of the constrictions. The activation energy for the stress free FE cross–slip mechanism, defined as the sum of the energies of a pair of infinitely separated edgelike and screwlike constrictions, is $2.7\,\mathrm{eV}$ in Cu, with the energies of the edgelike and the screwlike constriction being $3.8\,\mathrm{eV}$ and $-1.1\,\mathrm{eV}$, respectively [12]. The constriction energy is defined as the difference in energy between a system containing a screw dislocation with a single constriction (either screwlike or edgelike) and a system with two parallel partials at their equilibrium separation. In principle the constrictions attract even for an infinite separation, but a critical minimum separation between essentially non–interacting constrictions can be found as $50-60\,\mathrm{b}$ [12]. The energy of the initial in–plane constriction has been estimated as $\simeq 1.6\,\mathrm{eV}$ [13]. However, the focus in the earlier simulations [12, 13] was on the transition state of the process, and no attempt to make an accurate determination of the in–plane constriction energy was made. Cross slip has also been addressed theoretically by methods based on elasticity theory [8, 14, 15, 16], and the in–plane constriction has been treated in Refs. [17, 18].

METHODS

Since the activation energy of cross slip in Cu is $2.7\,\mathrm{eV}$, it is unrealistic to simulate a cross–slip event directly by standard molecular dynamics methods. Instead a configuration space path technique, the *nudged elastic band* (NEB) method [19, 20], is applied to determine the minimum energy path (MEP) of transition for a single cross slipping dislocation and the MEPs and corresponding transition states for annihilation of stable screw dislocation dipoles. An MEP is a path in configuration space with zero force perpendicular to the path. It connects relaxed initial and final states of the system, and has the desired property of tracing out the path through the saddle point configuration of the transition. In the NEB method the entire path of transition is relaxed simultaneously. The real continuous path is represented by a finite number of replicas of the system distributed along the path by harmonic spring forces. The dynamics of the path is controlled by forces derived from the atomic potential. The forces are modified in the following manner: Parallel to the path, only spring forces are used, and perpendicular to the path, only forces from the potential are used. This *nudging* procedure ensures the continuity of the path while effectively decoupling the action of the springs from the dynamics of the path controlled by the potential. The relaxed initial and final states of the system, i.e., the first and the last replica in the path, are not acted on by any force. The spring constants can be varied along the path, thus allowing a better resolution of certain parts of configuration space to be obtained.

The atomic interactions are described by a many–body potential derived from the effective–medium theory [21, 22]. The potential is slightly modified [12] so as to produce a reasonable value for the intrinsic stacking–fault energy, γ, e.g., [23]. The values for γ and the $\langle 110 \rangle \{111\}$ shear modulus, μ, are $56 \, \mathrm{mJ \, m^{-2}}$ and $56 \, \mathrm{GPa}$, respectively.

The geometry of the systems is very similar in the two kinds of simulations. Either one screw dislocation with Burgers vector $\boldsymbol{b} = \frac{1}{2}[110]$ or two dislocations $(\pm \boldsymbol{b})$ are located in a parallelepiped shaped computational cell with $(1\bar{1}1)$ and $(\bar{1}11)$ planes as boundaries parallel to the dislocation(s). The length of the perfect Burgers vector is $b = 0.26 \, \mathrm{nm}$ in Cu. In the case of a single dislocation, periodic boundary conditions are applied along the dislocation line, whereas in the dipolar case full periodic boundary conditions are applied. The full periodic boundary conditions in the dipolar case are justified because a dipole is a "defect" with zero Burgers vector. The initial dipole configurations consist of two dissociated screw dislocations in parallel glide planes with a separation between the centres of the dislocations parallel to the glide planes of $\simeq 4 \, \mathrm{nm}$. The perpendicular separation between the two initial glide planes, the dipole height, is denoted h. To determine equilibrium structures the systems are relaxed by a standard energy minimisation scheme [24].

To visualise the different dislocation configurations the *common neighbour analysis* (CNA) method [25] is used. This method provides a means of assigning a label to each atom corresponding to that atom's local crystallographic environment. In its simplest form CNA can distinguish between atoms situated in a local fcc environment or in a local hcp environment or neither of these possibilities. Hence, with the CNA method it is possible to identify dislocation cores (neither fcc nor hcp) and intrinsic stacking faults (hcp) between Shockley partials. In all pictures of dislocation configurations, which will follow, the atoms close to dislocation cores are coloured dark grey, and the atoms belonging to intrinsic stacking faults are coloured light grey. The atoms in a local fcc environment are white.

CROSS SLIP OF A SINGLE SCREW DISLOCATION

Cross slip has been addressed earlier by atomistic simulations [12, 13]. Hence, it is not the purpose to recapitulate earlier findings here, but rather to illustrate, by way of a recent simulation, the versatility of the NEB method. The following discussion is focused on the nucleation of the cross–slip event, i.e., the formation of the in–plane constriction initiating the transfer to the cross–slip plane, rather than the transition state of the process. In NEB simulations of cross slip with equidistantly spaced replicas [13], the resolution of the transition state is very good (for 18 - 34 replicas), whereas the resolution around the point

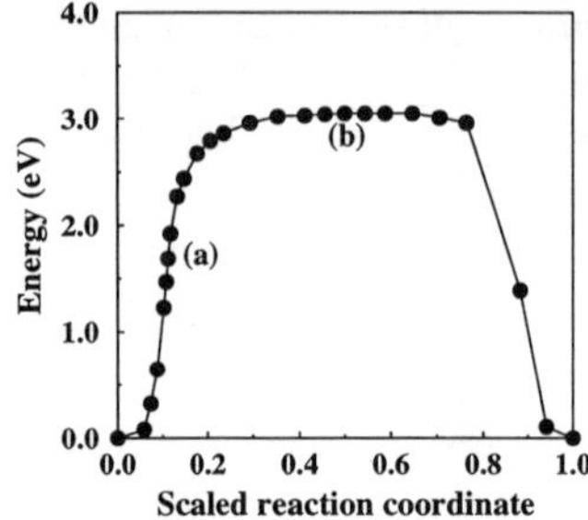

Figure 1: Energy vs. reaction coordinate along the MEP for cross slip of a single dislocation. (a) and (b) are the in–plane configuration and the transition state, respectively, and refer to Fig. 2.

in configuration space corresponding to the in–plane constriction is less satisfactory. In order to improve the resolution around the in–plane constriction, a new simulation with varying spring constants along the path has been performed.

The initial and final states are taken as a relaxed screw dislocation in either of the two possible glide planes. In the case of stress free cross slip, the two glide planes are equivalent, and the cross–slip process is thus "symmetrical" around the transition state (due to the periodic boundary conditions along the dislocation). This fact is used to limit the total number of replicas in the simulation, by focusing only on the first part of the path. The system used in the present work has the same dimensions perpendicular to the dislocation as in Ref. [13], i.e., the width is 9.5 nm, but the length of the dislocation is increased to 120 b (30 nm). An NEB simulation with 26 replicas, each of which consisted of 221 880 atoms, was performed. The spring constants was chosen so that the first part of the path had a high density of replicas, whereas the last part just played the role of connecting the first part to the final state. Figure 1 shows the energy as a function of scaled "reaction coordinate" along the path, with 0 and 1 corresponding to the initial and the final states, respectively. The transition state is that of the FE mechanism with two twisted constrictions separated by exactly half of the length of the dislocation. The activation energy is 3.2 eV, somewhat higher than the 2.7 eV referred to above. However, the difference is easily explained by a small offset of the dislocation away from the centre of the computational cell in the initial and the final states. A correction for the elastic energy associated with the offset can be estimated to $\simeq 0.2\,\text{eV}$, and the remaining difference can be attributed to a finite size effect [13, 26].

In this simulation, special care has been taken to improve the resolution in the part of configuration space corresponding to the in–plane constriction that initiates cross slip. Since the path is not continuous, it is not possible to assert precisely whether a given configuration is exactly what is meant by an in–plane point constriction in the classical dislocation sense. In this work the "definition" will be that the in–plane constriction is the configuration just prior to the configuration where the dislocation starts to develop in the cross–slip plane. With a sufficiently high resolution of the relevant part of configuration space, it is possible to determine the energy and configuration quite accurately. Figure 2 shows the in–plane constriction and the transition state together with projected energies defined as the difference in the local energy of a 1 b thick slice of the computational cell and a 1 b thick slice of a reference system with straight, parallel partials. The projected energies thus show the energy deviation of the actual configuration from a configuration

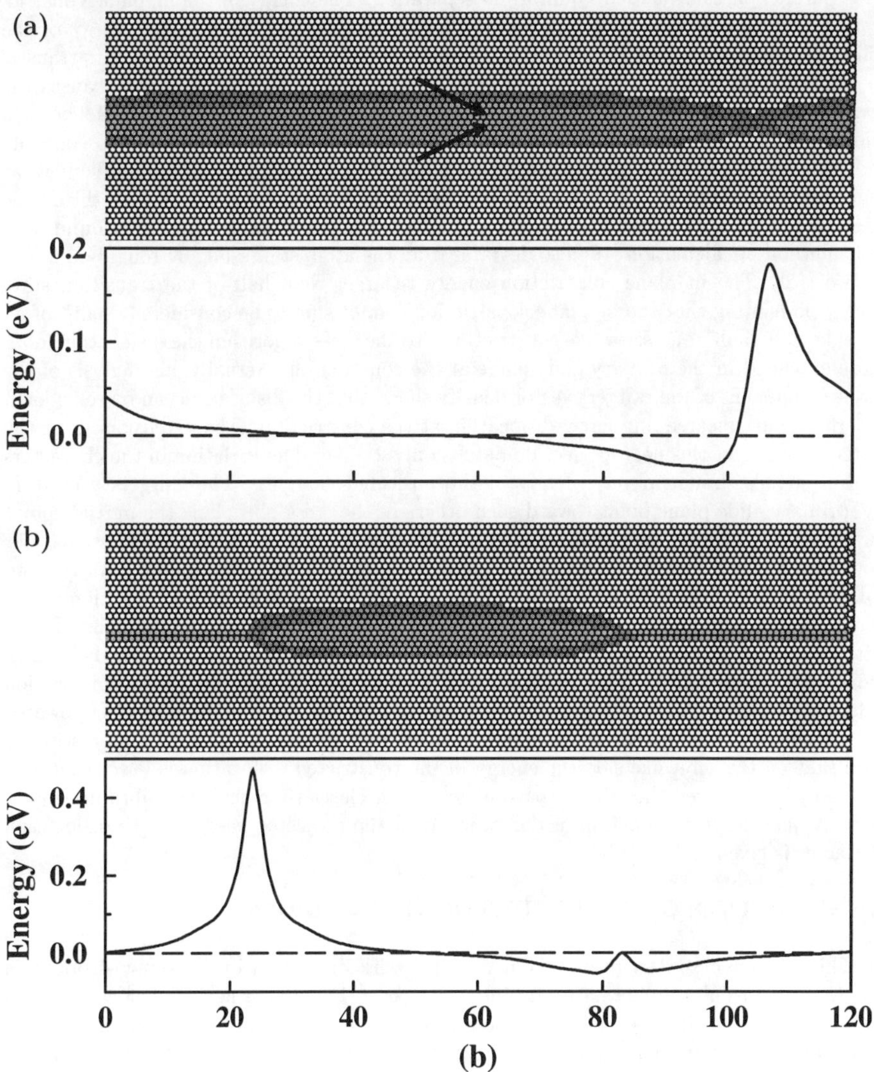

Figure 2: (a) The in–plane constriction in the primary glide plane just prior to cross slip. The partials acquire screwlike (edgelike) character to the left (right) of the constriction as is clearly reflected in the projected energy. (b) The transition state with two twisted constrictions. The constriction to the left (right) is edgelike (screwlike). The strings of light grey atoms on either side of the constrictions belong to the stacking fault between the Shockley partials in the cross–slip plane (not shown). In the primary plane the Burgers vectors toward each other, and in the cross–slip plane they point away from each other.

with parallel partials at their equilibrium separation. The energy of the in–plane constriction, following the "definition", is 1.7 eV after the elastic correction. The energy of the following replica, i.e., the configuration where the dislocation has just initiated its transfer to the cross–slip plane, is 1.9 eV. The finite size effect [26] mentioned earlier, causes the energies of the constrictions to be approximately 0.1 eV higher than they would be in a wider simulation cell. If this effect is taken into account it seems reasonable to conclude that the energy of the in–plane constriction is 1.7 ± 0.1 eV in Cu. This value is somewhat higher than an estimate based on elasticity theory [18] which gives $\simeq 1.1$ eV with the elastic constants of Cu obtained with the EMT potential. The transition state energy found with that method of calculation [16] also deviates from the atomistic result, by roughly a factor of two [12]. The in–plane constriction energy is larger than half of the transition state energy, indicating that the in–plane constriction cannot simply be considered "half" of an edgelike and "half" of a screwlike constriction. To the CNA algorithm the dislocation looks quite extended in the primary glide plane at the constriction. Actually, an analysis of the screw component of the Burgers vector density shows that the dislocation can be considered a perfect, but smeared out screw dislocation at the constriction. The extension along the dislocation line of this near–perfect domain is only $2 - 3$ b. The variation in the characters of the partials shows up very clearly in the projected energies. The Burgers vectors in the primary glide plane point toward each other. In the cross–slip plane the partials point away from each other, thus giving rise to the distinction between the twisted constrictions. For the transition state the peaks in the projected energy can be associated with the two constrictions, and the difference between the edgelike and the screwlike constriction is evident. The projected energy for the in–plane constriction shows the same behaviour. To the left of the constriction, the partials acquire a more screwlike character, and to the right a more edgelike character. Since the energy of the positive edgelike part of the constriction is larger in magnitude than the negative screwlike part, the peak is slightly offset towards the edgelike side compared to the actual location of the constriction. It is interesting to note that on the screwlike side the energy of the constricted dislocation is lower than two parallel partials, even though the partials are much closer than their equilibrium separation. A qualitative explanation of the variation of the projected energies with dislocation character is given in Ref. [12].

ANNIHILATION OF SCREW DISLOCATION DIPOLES

This section describes results from ongoing work on simulation of screw dislocation dipoles. The results divide naturally into two parts: For dipole heights below a certain critical value, the dipoles annihilate spontaneously. For heights larger than the critical value, an energy barrier for annihilation exists. It is the latter, activated part that is the subject of this section.

For dipole heights less than 1.0 nm (i.e., 4 or less $(1\bar{1}1)$ planes between the initial glide planes) the dipoles annihilate without an energy barrier. Beyond that height, stable dipoles are formed. Experimentally [27] stable dipole heights down to ~ 50 nm have been reported, but this number should only be considered an upper limit to the minimum stable dipole height. Due to the anisotropy of Cu, the stable dipoles form skew equilibrium configurations [28], see Fig. 3. Similar skew dipoles have been observed experimentally in Ni, and good agreement with anisotropic elasticity was obtained [29]. The splitting width of the dislocations comprising the dipole is also affected by the dipolar structure. For dipole

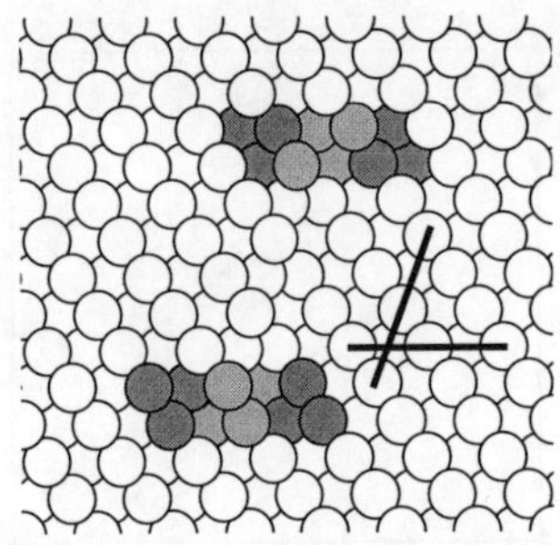 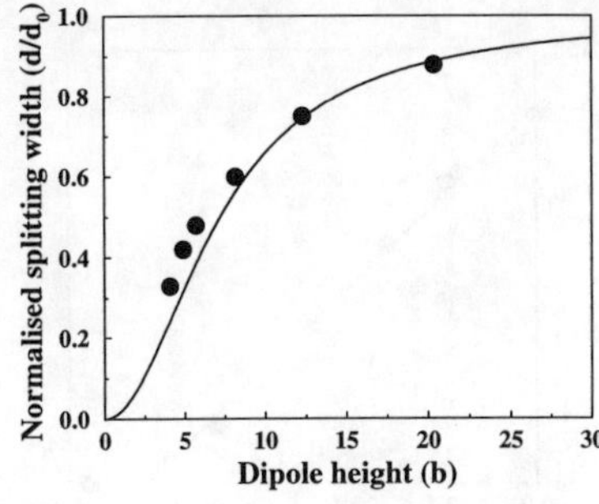

Figure 3: Left: Relaxed dipole configuration for $h = 7$ ($1\bar{1}1$) planes (5.7 b). The dislocations are perpendicular to the paper and have Burgers vectors $\frac{1}{2}[110]$ (upper) and $\frac{1}{2}[\bar{1}\bar{1}0]$ (lower). The two possible glide planes are ($1\bar{1}1$) (horizontal) and ($\bar{1}11$) (inclined) as indicated. Right: Normalised splitting widths of the dislocations of the equilibrium dipole structure vs. dipole height. The splitting width of an isolated screw dislocation is $d_0 = 6.5$ b in Cu.

heights less than ~ 20 b (5 nm) the splitting widths decrease rapidly with decreasing dipole height, as can be seen in the right panel of Fig. 3. The full line is the result of a simple, isotropic model, where the splitting of one of the dislocations in the dipole is modelled as though the other dislocation is a perfect screw dislocation. Despite the simplicity of the model, it still captures the trend quite accurately.

The NEB method is applied to determine the minimum energy paths and the corresponding activation energies of the annihilation processes for varying dipole heights. Two kinds of simulations are performed. In simulations with short (5 b) dislocation segments, the dislocations are kept straight throughout the annihilation process, i.e, the cross slip initiating annihilation takes place homogeneously. In simulations with longer dislocation segments (up to 60 b), the dislocations are flexible, and the cross–slip process becomes heterogeneous. The initial states are relaxed dipole configurations of varying dipole heights and dislocation "lengths", and the final states are perfect fcc crystals. Due to space limitations, only the 3D simulations will be discussed in some detail here.

The energy barrier for annihilation as a function of inverse dipole height is seen in figure 4 for both the 2D and the 3D case. The points corresponding to infinitely high dipoles are taken as the stress free cross–slip activation energy of a single dislocation [12, 13]. The annihilation processes are initiated by cross slip of one of the dislocations. In the 2D case the activation energies show a linear dependence on the inverse dipole height, Fig. 4 left panel. The point for 5 ($1\bar{1}1$) planes that deviates from this trend is special in that the transition state for this dipole height is different from the others [30]. The 2D transition state is a state where the Shockley partials of the cross slipping dislocation are so close that the dislocation might be considered a perfect screw dislocation. Hence, in 2D the transition states are determined by the nucleation of the (homogeneous) cross slip.

In the case of 3D simulations, the computational cells become very large. The side lengths perpendicular to the dislocations are 15 nm, and dislocation lengths up to 60 b (15 nm) are used. The largest systems then consist of 261 360 atoms, and the paths typically consist of 26 replicas. Three dipole heights are examined: $h = 6, 10,$ and 15 ($1\bar{1}1$)

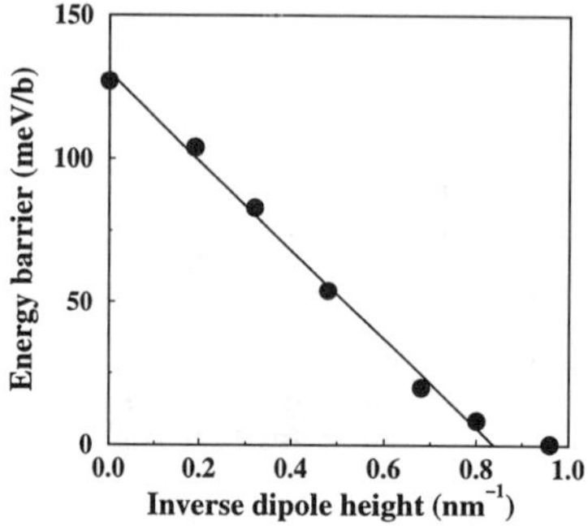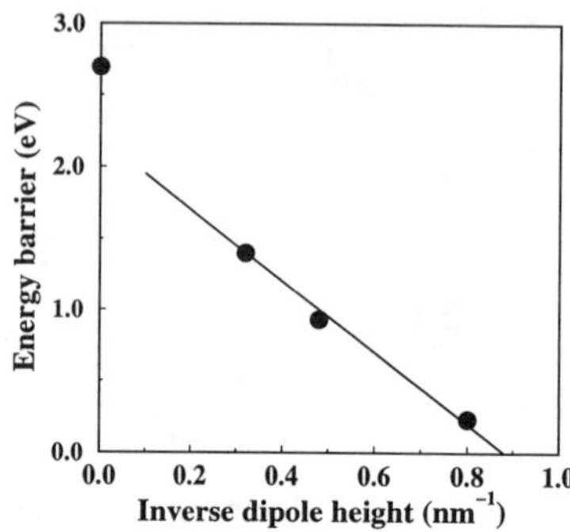

Figure 4: Annihilation energy as a function of inverse dipole height. Left: Dislocation "length" 5 b (homogeneous cross slip). The dipole spacings are (from right to left) 5, 6, 7, 10, 15, and 25 $(1\bar{1}1)$ planes. The point for an infinitely high dipole is for stress free cross slip of a single dislocation (Ref. [13]). Right: Heterogeneous cross slip leading to annihilation. The dipole spacings are (from right to left) 6, 10, and 15 $(1\bar{1}1)$ planes. The point for an infinitely high dipole is for stress free cross slip of a single dislocation (Ref. [12]).

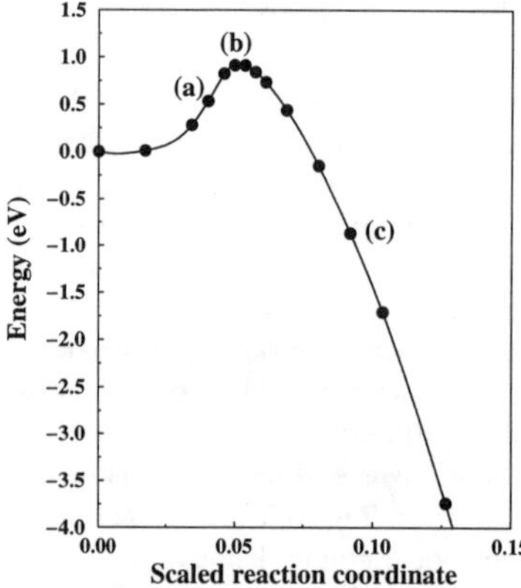

Figure 5: Energy vs. scaled reaction coordinate along the MEP for annihilation of a dipole of height $10\,(1\bar{1}1)$ planes (2.1 nm). (a) - (c) correspond to the top, 2nd, and 3rd row of Fig. 6.

planes. Figure 5 shows the energy vs. scaled reaction coordinate for the annihilation process of a dipole of height $10\,(1\bar{1}1)$ planes and dislocation "length" 60 b. The cross slip is heterogeneous and the barrier is 0.93 eV, i.e., a reduction by a factor of 3 compared to the stress free FE mechanism. The activation energies for $h = 6$ and $15\,(1\bar{1}1)$ planes are 0.23 and 1.4 eV, respectively. The annihilation activation energies for the three dipole heights considered in this work show a linear dependence on the dipole height, Fig. 4 right panel. However, the linearity cannot be extrapolated with confidence into the region of high $(h \geq 10\,\mathrm{nm})$ dipoles, and another (non–linear) relation is perhaps more likely.

The 3D cross–slip mechanism leading to annihilation is qualitatively similar to the

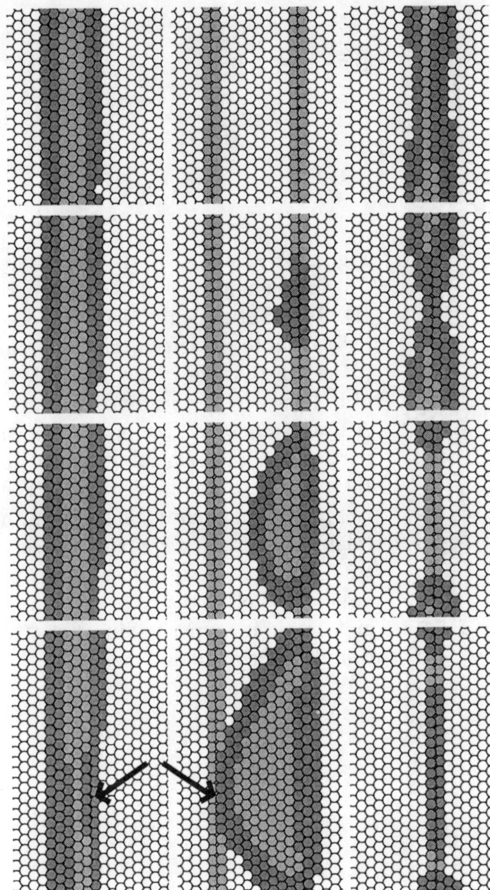

Figure 6: Details of the cross slip process leading to annihilation (part I). The dipole height is $10\,(1\bar{1}1)$ planes, and the dislocation "length" is $60\,b$. For each row the figure shows the initial glide plane for the negative (positive) dislocation in the left (right) column. The centre column shows the cross–slip plane. Top: Formation of a constriction entirely in a single plane. 2nd row: Transition state. The dislocation is slightly extended in the cross–slip plane, but the twisted constrictions have not developed yet. 3rd row: The cross slipping dislocation is attracted to the opposite dislocation. Bottom: Formation of a short segment of stair rod dislocation (indicated by arrows) at the intersection of the cross–slip plane and the glide plane of the non cross slipping dislocation. Notice the highly asymmetrical configuration of the cross slipped loop.

FE mechanism. The transition states are reached after the recombined segment of screw dislocation starts to develop in the cross–slip plane, but for none of the dipole heights considered in this work is it fair to speak of a transition state resembling the FE transition state with two twisted constrictions and intrinsic stacking faults in both glide planes. The transition state for $h = 10\,(1\bar{1}1)$ planes is shown in the second row of Fig. 6.

In the FE mechanism a stress is assumed to act on the opposite edge components of the partials of the cross slipped segment. The transition state is reached when the attractive constriction–constriction interaction is balanced by the driving force stemming from the widening of the cross slipped segment by the applied stress [8]. For dipoles an additional term must be considered. As soon as the cross slipping dislocation develops in the cross–slip plane, and is attracted to the other dislocation, the dislocation–dislocation interaction energy is lowered significantly. Hence, the system consisting of *both* dislocations reaches the transition state for smaller constriction separations on the cross slipping dislocation than in the usual FE mechanism for an isolated dislocation. The change in interaction energy

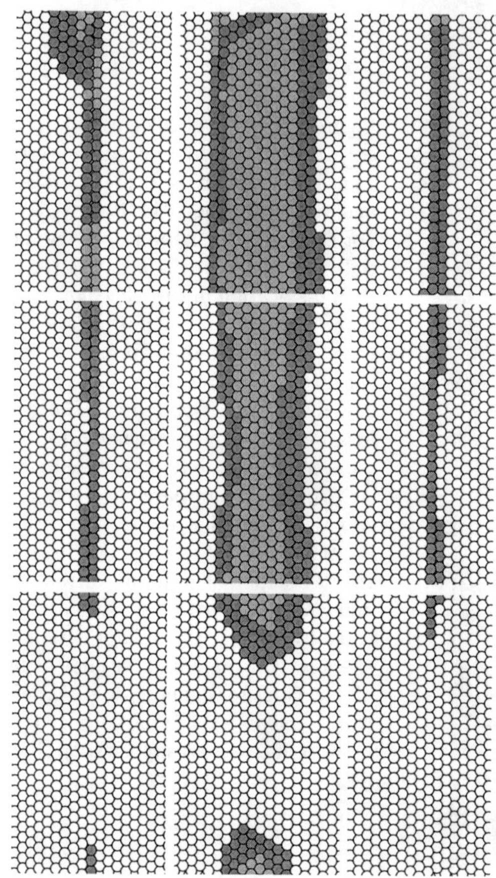

Figure 7: Details of the cross slip process leading to annihilation (part II). Top: Formation of a new Shockley partial to the left in the cross–slip plane (also visible in the glide plane of the negative dislocation). 2nd row: Attraction between the newly formed partial and the remaining partial in the cross–slip plane. 3rd row: The two partials annihilate, thus creating a "slab" of perfect fcc crystal at the centre of system. Two dislocation loops are pinched off in the cross–slip plane, and move in opposite directions to complete the annihilation.

associated with a change of separation between two perfect screw dislocations from r_1 to r_2, is given by $\Delta E = (\mu b^2)/(2\pi)\ln(r_1/r_2)$. A cross slipped segment of $10\,b$ thus lowers the interaction energy by $\sim 0.5\,\text{eV}$ when the separation is decreased from $10\,b$ to $9.5\,b$. Escaig has shown [8] that in the case of an isolated dislocation the effect of the bow–out of the cross slipped segment is negligible compared to the effect of the widening of the cross slipped segment in the cross–slip plane. Here, however, the bow–out cannot be ignored, since it results in a considerable lowering of the dislocation–dislocation interaction energy. For increasing dipole hights a gradual shift towards a more regular FE mechanism can be expected, and the annihilation activation energy becomes that of the cross slip activation energy for a single, isolated screw dislocation for infinitely high dipoles. Unfortunately, an examination of intermediate dipole heights is not within the capabilities of the present atomistic approach.

It is possible to monitor the dislocation reactions leading to the annihilation. The cross slipped $\frac{1}{2}[110]$ dislocation is attracted to the negative dislocation, and the annihilation proceeds by standard dislocation reactions, see Figs. 6 and 7. The $\frac{1}{6}[211]$ partial to the left in the cross–slip plane, is attracted to the $\frac{1}{6}[2\bar{1}1]$ partial to the right in the glide plane of the negative dislocation (Fig. 6, 3rd row), and they react to produce a sessile stair rod dislocation ($\frac{1}{3}[001]$) at the intersection of the two planes, Fig. 6 bottom row. The stair rod attracts the other partial of the non cross slipping dislocation ($\frac{1}{6}[\overline{121}]$), and the two dislocations react to produce a new Shockley partial ($\frac{1}{6}[\overline{121}]$) which is glissile in the cross–slip plane, Fig. 7 upper row. The newly generated Shockley partial and the remaining $\frac{1}{6}[12\bar{1}]$ partial in the cross–slip plane attract and annihilate, leaving part of the crystal defect free. Two dislocation loops are pinched off in the cross–slip plane by these reactions. The two loops move in opposite directions to complete the annihilation. Similar reactions have been observed in atomistic simulations of surface nucleated screw dislocation annihilation [12].

CONCLUSION

The *nudged elastic band* method has earlier been used to determine the actual cross–slip mechanism of a dissociated screw dislocation in an fcc metal. In this work the method is used to determine the energy of the in–plane constriction initiating cross slip in Cu. The method is also successfully applied to the case of screw dislocation dipole annihlation. Activation energies and transition states for annihilation is obtained in 2D as well as in 3D simulations. It is argued that the lowering in dislocation–dislocation interaction energy plays a significant role in determining the activation energy and transition state for the 3D annihilation process. Due to space limitations only a brief report of the results has been given here. A more elaborate account is forthcoming [30].

Cross slip, dislocation dipole annihilation, and other three dimensional dislocation mechanisms are recognised as playing a key role in several aspects of plastic deformation. It is therefore important to obtain quantitative accurate information about such mechanisms in order to provide reliable input into modelling of plasticity carried out at coarser length scales. Atomic scale simulation has proven to be well suited for this purpose, and to complement elastictity theory where its use is questionable. It is the hope that the work presented here, and elsewhere, might be a step in the direction of providing useful results, obtained by atomistic simulation, to be used in meso–/macro–scopical simulation approaches. If such a connection between different simulation approaches is eventually made, it will certainly be fair to speak of *multiscale modelling*.

ACKNOWLEDGMENTS

It is a pleasure to thank Karsten W. Jacobsen, Torben Leffers, Ole B. Pedersen, and Tejs Vegge for their contributions to the work which is presented here in part. Centre for Atomic–scale Materials Physics (CAMP) is sponsored by the Danish National Research Foundation. The present work was in part financed by The Danish Research Councils through grants #9501775, and was done as a collaboration between CAMP and the Engineering Science Centre for Structural Characterization and Modelling of Materials, Risø.

REFERENCES

[1] J. V. Carstensen, T. Leffers, T. Lorentzen, O. B. Pedersen, B. F. Sørensen, and G. Winther, editors, *Modelling of Stucture and Mechanics of Materials from Microscale*

to Product. Risø National Laboratory, Roskilde, Denmark (1998).

[2] B. Devincre and L. P. Kubin, Materials Science and Engineering A **234**, 8 (1997), and references therein.

[3] M. Rhee, H. M. Zbib, J. P. Hirth, H. Huang, and T. de la Rubia, Modelling Simul. Mat. Sci. Eng. **6**, 467 (1998).

[4] D. Rodney and R. Phillips, *Structure and Strength of Dislocation Junctions: An Atomic Level Analysis*, (1998), submitted.

[5] M. S. Duesbery and W. Xu, Scripta Mat. **39**, 283 (1998).

[6] O. B. Pedersen, Philosophical Magazine A **73**, 829 (1996).

[7] J. Friedel, In *Dislocations and Mechanical Properties of Crystals*, eds. J. C. Fisher et al., John Wiley & Sons (1957).

[8] B. Escaig, In *Dislocation Dynamics*, eds. A. R. Rosenfeld, G. T. Hahn, A. L. Bement Jr., and R. I. Jaffee, pages 655–677, McGraw–Hill Series in Materials-Science and Engineering (1968); Jour. de Phys. (France) **29**, 255 (1968).

[9] R. L. Fleischer, Acta Metallurgica **7**, 134 (1959).

[10] M. J. Marcinkowski, K. Sadananda, and N. J. Olson, Cryst. Latt. Def. **5**, 187 (1974).

[11] L. M. Clarebrough and C. T. Forwood, Phys. Stat. Sol. A **32**, K15 (1975).

[12] T. Rasmussen, K. W. Jacobsen, T. Leffers, and O. B. Pedersen, Physical Review B **56**, 2977 (1997).

[13] T. Rasmussen, K. W. Jacobsen, T. Leffers, O. B. Pedersen, S. G. Srinivasan, and H. Jónsson, Physical Review Letters **79**, 3676 (1997).

[14] G. Saada, Materials Science and Engineering A **137**, 177 (1991).

[15] M. S. Duesbery, N. P. Louat, and K. Sadananda, Acta metall. mater. **40**, 149 (1992).

[16] W. Püschl and G. Schoeck, Mat. Sci. and Eng. A **164**, 28 (1993).

[17] A. N. Stroh, Proc. Phys. Soc. B **67**, 427 (1954).

[18] W. Püschl, Phys. Stat. Sol. (B) **162**, 363 (1990).

[19] G. Mills, H. Jónsson, and G. Schenter, Surface Science **324**, 305 (1995).

[20] H. Jónsson, G. Mills, and K. W. Jacobsen, In *Classical and Quantum Dynamics in Condensed Phase Simulations*, eds. B. J. Berne, G. Ciccotti, and D. F. Coker, World Scientific (1998).

[21] K. W. Jacobsen, J. K. Nørskov, and M. J. Puska, Physical Review B **35**, 7423 (1987).

[22] K. W. Jacobsen, P. Stoltze, and J. K. Nørskov, Surface Science **366**, 394 (1996).

[23] N. M. Rosengaard and H. L. Skriver, Physical Review B **47**, 12865 (1992).

[24] P. Stoltze, *Simulation methods in atomic–scale materials physics*, Polyteknisk Forlag, Lyngby (1997).

[25] D. Faken and H. Jónsson, Computational Materials Science **2**, 279 (1994).

[26] T. Rasmussen, *Atomic Scale Simulation of Dislocation Reactions*, PhD thesis, Technical University of Denmark (1998).

[27] U. Essmann and H. Mughrabi, Philosophical Magazine A **40**, 731 (1979).

[28] A. J. Morton and C. T. Forwood, Cryst. Latt. Def. **4**, 165 (1973).

[29] C. T. Forwood and P. Humble, Aust. J. Phys. **23**, 697 (1970).

[30] T. Rasmussen, K. W. Jacobsen, T. Leffers, O. B. Pedersen, and T. Vegge, *Simulation of structure and annihilation of screw dislocation dipoles*, (1999), to be submitted.

COUPLED MESO-MACRO SIMULATIONS OF PLASTICITY: VALIDATION TESTS

C. Lemarchand*, B. Devincre**, L.P. Kubin** and J.L. Chaboche*
* DMSE, ONERA, BP 72, 92322 Chatillon Cedex, France
** LEM, CNRS-ONERA, BP 72, 92322 Chatillon Cedex, France

ABSTRACT

A new numerical model has been developed in order to study the plastic deformation of mesoscopic crystalline samples under complex boundary conditions. The model is based on the coupling of two types of simulations, a *dislocation dynamics* and a *finite element* code. The former accounts for the dislocation-based plastic properties of the material, thus replacing the usual constitutive form, while the latter treats the boundary value problem and cares of the mechanical equilibrium. In order to test the hybrid simulation and examine its accuracy, the self-stress fields of straight dislocations have been computed in a single crystal of finite size and compared with the predictions of the isotropic elasticity theory. The excellent agreement obtained emphasizes the enormous potential of such hybrid methods for a rigorous treatment of meso-macro problems in plasticity.

INTRODUCTION

In the last decade, studies on the plastic deformation of crystalline materials have been revived by the development of dislocations dynamics (DD) simulations [1, 2, 3, 4, 5, 6]. Within the frame of these computer models, the motion and interaction of dislocation lines in an elastic continuum is treated in a discretized manner, leading to a new approach of the collective properties of large dislocation densities at the mesoscopic scale, *i.e.*, at the scale of the dislocation microstructure [7, 8]. The ultimate objective of such studies is to build up plasticity models valid at the macroscopic scale but based on physically justified constitutive forms taking into account dislocation properties.

Eventhough they are extremely promising, the DD simulations have important limitations which can be summarized as follows:

i) In many materials, there is a lack of accurate input regarding several properties associated with the dislocation core behaviour.

ii) The amplitude of the plastic strain accessible to the simulations is restricted by the computing task stemming from the calculation of dislocation-dislocation elastic interactions.

iii) The boundary conditions currently used in the DD simulations are rather crude and may introduce size effects or artefacts due to the fact that stress equilibrium conditions are not cared of.

In order to remedy to the last two weaknesses and to study strongly heterogeneous materials submitted to complex loadings, such as thin epitaxial layers or alloys containing large volumic fractions of precipitates (Ni-base superalloys), we have recently proposed to couple a DD simulation with a finite elements (FE) code [9]. The main advantage of such a hybrid code is to provide a rigorous treatment of the boundary value problem as well as an efficient computation of the temporal and spatial behaviour of the local stress field. Indeed, with this new computer model, the dislocation dynamics can be solved without explicitly introducing the dislocation-dislocation elastic interactions at the mesoscale.

The approach used to couple the two simulations codes is only briefly sketched in section 2, a detailed presentation of the algorithm being postponed to a further publication. Section 3 is dedicated to the presentation and discussion of the first validation tests on the hybrid code, *i.e.*, to the computation of the self-stress fields of screw and edge straight dislocations in a single crystal of small finite size.

NUMERICAL METHOD

The principles of the coupling of the DD and FE simulations can be understood in simple terms. The FE part functions in a standard manner, except that the prediction for the local plastic strain and stress during a loading step is no longer given by a material constitutive form but by the output of the DD simulation.

The simulation box is made up of an elastic continuum containing discreet dislocation lines. Superimposed to the DD discretization lattice, which accounts for the crystallographic properties of dislocation glide, is a FE mesh of cubic elements containing either 8 or 20 nodes. All the computations reported in the next section were performed with a $13 \cdot 13 \cdot 13$ mesh corresponding to a cubic volume element of side $15 \mu m$. The simulation length scale is then no longer arbitrary, as it is fixed by the dislocations Burgers vector.

The boundary value problem is treated by the FE code which accounts for the loading conditions and different types of possible conditions at the external surfaces of the simulation box. In the present study, a simple shear is applied along two external faces of the simulated crystal and free surface conditions are used for the four other faces. For the latter, the condition imposed on the external nodes of the FE mesh is written: $\sigma \cdot \vec{n} = \vec{0}$, where $\vec{n}$ is the surface normal. At the beginning of each step of the FE simulation, of duration Δt, the nodal displacements are calculated from a balance force criterion and the total strain increments, $\Delta \epsilon$, are deduced at each Gauss point of the mesh.

At this stage, a rescaling of the time step has to be introduced. Indeed, dislocation mobilities are quite large in the case of f.c.c. crystals that is implicitly considered here, hence small time steps on the order of $\delta t = 10^{-9} s$ are required to properly integrate the dislocation dynamics [8]. Such a value is substantially smaller than the time increment $\Delta t \approx 1000 \cdot \delta t$ used by the FE code. A small strain increment, suitable for the DD simulation, $\delta \epsilon = \Delta \epsilon \cdot (\delta t / \Delta t)$ is then prescribed at the Gauss points. Then, from the strain increment at step i the stress increment is computed at each Gauss point according to $\delta \sigma_i = \mathbf{K}(\delta \epsilon_i - \delta \epsilon_{i-1}^P)$, where $\mathbf{K}$ is the matrix of elastic coefficients and ϵ_{i-1}^P is the plastic strain produced by the DD part during the previous time step. From the definition of the stresses at the Gauss points, the local stresses can be interpolated at any position in the crystal, particularly at the dislocations coordinates, thus defining the resolved effective forces acting on the mobile dislocation lines.

At this point, it is necessary to emphasize that the discreet nature of the plastic shear associated with dislocation glide has to be be homogenized before it can be dealt with by the FE code. For this reason, the plastic strain measured at the Gauss points is an average value carried out by summing up all the elementary dislocation displacements over all the active slip systems in a reference volume centered at the Gauss points. The size and the shape of this homogenization volume must be initially adjusted to avoid too strong gradients between the Gauss points (see ref. [9] for more detail)

The DD part of the hybrid code is used to produce plastic flow by moving the dislocations during the time interval Δt, after what the internal forces at the Gauss points are calculated by the FE code. The global equilibrium of the FE mesh is then tested. If the residual force is smaller than a prescribed value, a new loading increment is applied. If not, the step is iterated again during a time interval Δt and new nodal displacements are tested until the equilibrium condition is realized.

It is important to notice that within the present frame, the local stresses are computed only at a fixed number of points, the Gauss points, in contrast with the situation met with DD simulations where it has to be evaluated for each dislocation segment. Therefore, the CPU cost stemming from the calculation of the internal forces on the dislocations does not increase with the number of dislocation segments needed to model the total dislocation density.

DISLOCATION STRESS FIELDS

Within the hybrid code, the elastic fields of the dislocations present in the simulation box, as well as the elastic interaction forces that result from them, are computed as part of the global boundary value problem. The simplest way to validate the output of the DD-FE code consists, therefore, of computing the stress fields of simple dislocation configurations and of checking the result with respect to classical results from the elastic theory of dislocations. For the sake of simplicity, isotropic elasticity is used here. However, the hybrid code can be set to handle anisotropic elasticity without any additional difficulty, by just modifying the matrix of elastic constants.

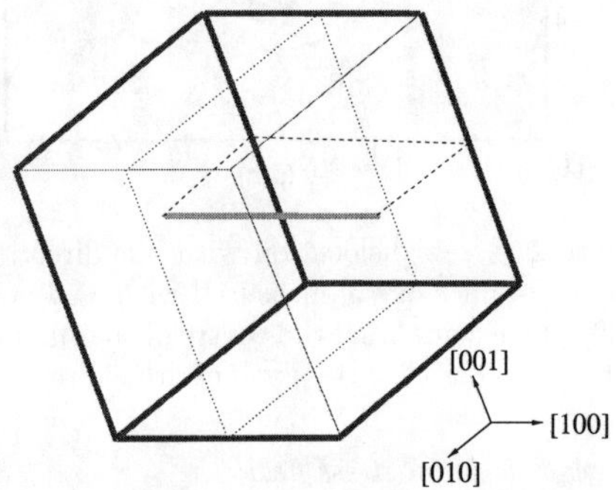

Figure 1: Scheme of the simulation configuration used to test dislocation self-stress fields. The simulation box boundaries are drawn in bold lines. Free surface conditions are used. The [100] dislocation line (in gray) is moved from the (010) surface to the box centre. The Burgers vectors are [100] and [010], respectively, for screw and edge dislocations. The reference plane (dotted lines) where stresses are measured is normal to the dislocation line.

Figure 1 shows the configuration adopted for the measurement of the self-stress field of screw and edge dislocations. A dislocation is initially introduced at a free surface and moved inside the simulation box by the DD code under an appropriate applied shear stress. To simplify the geometry, the line direction is set parallel to [100] and its Burgers vector is either [100] or [010], for the screw and edge dislocations respectively. When the dislocation is exactly at the centre of the cube, the DD simulation is stopped and the stresses are computed in a (100) plane normal to the dislocation line and going through the centre of the simulation box. For comparison, we make use of the classical stress fields predicted by the elastic theory in an infinite crystal.

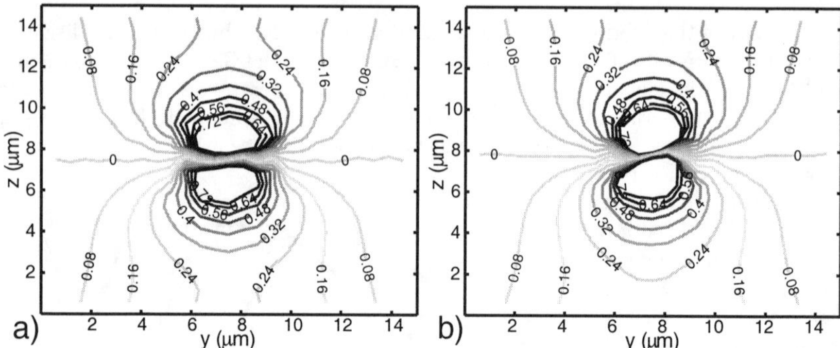

Figure 2: Screw dislocation with line direction and Burgers vector parallel to [100]. Computed isovalues of σ_{12} measured with a) 8 nodes or b) 20 nodes per mesh element, along a (100) plane containing the centre of the simulation box. Free surface conditions are imposed at the boundaries. Stress values are expressed in MPa.

Screw dislocation stress field

In agreement with expectation, only two components of the stress tensor of the screw dislocation are found to be non-zero, σ_{12} and σ_{13} [10]. The isovalues of these two shear stresses are related to each other by a rotation of $\pi/2$ around the dislocation line axis. For this reason, only σ_{12} is plotted here. In fig. 2 are shown the numerical calculations performed with 8 and 20 nodes. The corresponding theoretical solution is plotted in figure 3. In these figures and in the following ones a cut-off has been introduced on the stress maximum amplitude for the sake of clarity.

A detailed examination of these three figures suggests the following remarks.

i) A good quantitative agreement is found between the computed and theoretical isovalues at a sufficient distance from both the dislocation core and the external boundary of the simulation box.

ii) The influence of the free surfaces is well reproduced by the simulation. For instance, in figure 2, one can check that the σ_{12} isovalues vanish along the two vertical z boundaries which correspond to the intersection of the plot plane with the (010) traction-free surfaces of the simulation box. Hence, the effects of the so-called image forces on the self-stress appear to be correctly included into the calculation of the Peach-Koehler

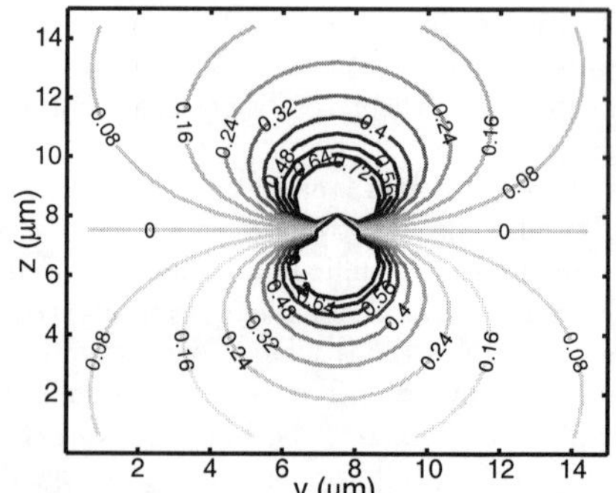

Figure 3: Theoretical isovalues of σ_{12} for a screw dislocation in an infinite isotropic elastic medium. Stress values are expressed in MPa.

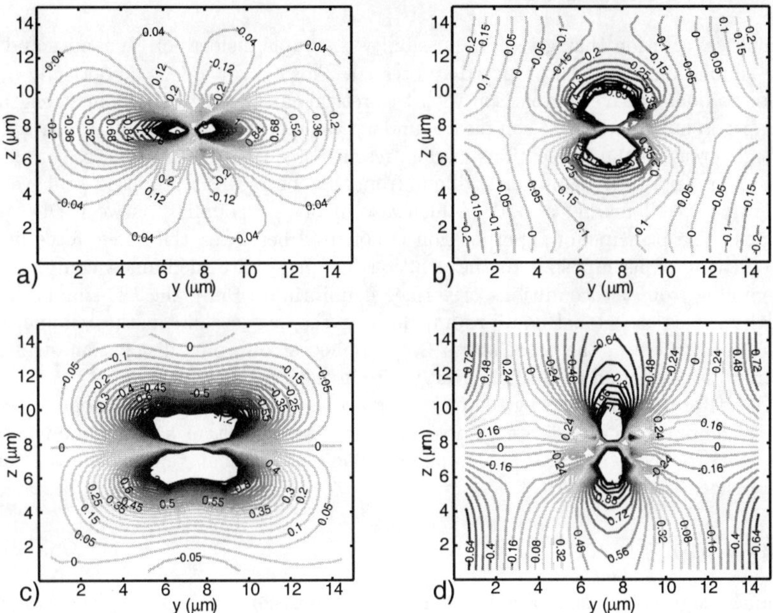

Figure 4: Isovalues of a) σ_{23}, b) σ_{11}, c) σ_{22} and d) σ_{33} for an edge dislocation, computed with 20 nodes per element mesh. Free surface conditions are imposed at the boundaries. Stress are in MPa units.

force on the dislocation. As mentioned above, the latter is derived from the stress solution at the nodes of the FE mesh.

iii) Close to the dislocation core, the stress values are more accurate when the computations are performed with 20 nodes per element mesh. This result is not surprising, since the 20 nodes elements are well known to better account for large strain gradients like the ones met close to stress divergency at the geometric center of dislocations.

Edge dislocation stress field

The stress field around an edge dislocation has been computed only with 20 nodes elements. The line direction is still [100], as in the case of the screw dislocation, and the Burgers vector is [010]. The simulated results are plotted in figure 4, which reproduces the iso-stress values for the non-zero components of the stress tensor, σ_{23}, σ_{11}, σ_{22} and σ_{33}, respectively. Like in the previous case of the screw dislocation, a comparison with the theoretical isovalues (cf. ref. [10] p. 77) can be performed, leading to the same conclusions. The contribution of the free surfaces to the dislocation stress field is correctly accounted for. For instance, the influence of the image forces manifests itself by the fact that σ_{22} vanishes at the trace of the (010) surfaces and the sum of σ_{23} and σ_{33} vanishes at the trace of the (001) surfaces.

CONCLUDING REMARKS

A numerical solution that gives the possibility to couple dislocation dynamics and finite elements simulations has been proposed and tested. This hybrid code is numerically efficient and supports quantitative comparison with theoretical calculations. Its main strength is to provide a rigorous treatment of complex boundary value problems, allowing in particular to treat image forces on the dislocations in the presence of surfaces or interfaces.

The present approach significantly differs from the one recently proposed by Fivel et al. [11], following Needleman et al. [12], which also involves a coupling between DD and FE simulations. The principle of superposition is not used here, i.e., the image force problem is not solved by superimposing to the solution for an infinite medium a complementary term stemming from the conditions of surface equilibrium. Here, the FE simulation part numerically solves the global equilibrium, both at the surface and in the volume, of the simulation box. Hence, the total stress (i.e., applied stress, plus dislocation stress, plus image stress) is computed as a whole everywhere in the elastic continuum.

The $O(N^2)$ problem of treating the pair interactions of N dislocation segments is reduced to a stress computation on a finite number of nodes. An important additional advantage is that the local rotations can also be computed, which is a critical step for treating cell and subgrain dislocation structures.

ACKNOWLEDGMENTS

The authors gratefully acknowledge stimulating discussions with Dr. A. El-Azab (PNL).

REFERENCES

[1] L.P. Kubin, G. Canova, M. Condat, B. Devincre, V. Pontikis, and Y. Brechet. *Solid State Phenomena*, 23-24:455, 1992.

[2] B. Devincre and L. P. Kubin. *Modelling Simul. Mater. Sci. Eng.*, 2:559–570, 1994.

[3] M. C. Fivel, T. J. Gosling, and G. R. Canova. *Modelling Simul. Mater. Sci. Eng.*, 4:581, 1996.

[4] K.W. Schwartz and F. K. LeGoues. *Phys. Rev. Lett.*, 79:1877, 1997.

[5] H. M. Zbib, M. Rhee, and J. P. Hirth. *Int. J. Mech. Sci.*, 40:113, 1998.

[6] M. Tang, L. P. Kubin, and G. R. Canova. *Acta mater.*, 46:3221, 1998.

[7] B. Devincre and M. Condat. *Acta metall. mater.*, 40:2629, 1992.

[8] B. Devincre. In H. O. Kirchner, V. Pontikis, and L.P. Kubin, editors, *Computer Simulation in Materials Science*, pages 309–323. Amsterdam:North-Holland, 1996.

[9] C. Lemarchand, Chaboche J. L., B. Devincre, and L. P. kubin. In *3rd European Mechanics of Materials Conference*, 1998.

[10] J.P. Hirth and J. Lothe. *Theory of dislocations*. MacGraw-Hill: New York, 1982.

[11] M. Fivel. PhD thesis, INP Grenoble FRANCE, 1997.

[12] V. Lubarda, J. A. Blum, and A. Needleman. *Acta metall. mater.*, 41:625, 1993.

DYNAMICS OF DISSOCIATED DISLOCATIONS IN SI : A MICRO-MESO SIMULATION METHODOLOGY

W. Cai*, V.V. Bulatov**, J.F. Justo***,S. Yip* and A.S. Argon****
* Department of Nuclear Engineering, MIT, Cambridge, MA 02139
** Lawrence Livermore National Laboratory, Livermore, CA 94550
*** Instituto de Fisica da Universidade de Sao Paulo, Sao Paulo, Brazil
**** Department of Mechanical Engineering, MIT, Cambridge, MA 02139

ABSTRACT

The theory of dislocation motion in materials with high Peierls stress relates dislocation mobility to the underlying kink mechanisms. While one has been able to describe certain qualitative features of dislocation behavior, important details of the atomic core mechanisms are lacking. We present a hybrid micro-meso approach to modeling the mobility of a single dislocation in Si in which the energetics of defect cores and kink mechanisms are treated by atomistic simulation, while dislocation motion under applied stress and at finite temperature is described through kinetic Monte Carlo. Three important aspects pertaining to treating the details of local structure and dynamics of kinks are incorporated in our approach: (1) Realistic complexity of (multiple) kink mechanisms in the dislocation core. (2) Full Peach-Koehler formalism for treatment of curved dislocation. (3) Detailed investigation of interaction between partials. This simulation methodology is used to calculate micron-scale dislocation mobility, with no adjustable parameters; specifically we obtain temperature and stress dependent velocity results that can be compared with experimental measurements.

INTRODUCTION

Experimental data on dislocation mobility in semiconductors, notably in Si, is often fit to a simple empirical equation relating dislocation velocity v to temperature T and applied stress τ

$$v(\tau, T) \sim \frac{\tau^m}{k_B T} \exp(-\frac{Q}{k_B T}), \tag{1}$$

where m is a stress sensitivity exponent, Q is an activation energy, and k_B is the Boltzmann's constant. The kink diffusion model [1] considers dislocation motion as a net result of thermally activated nucleation, migration and recombination of kinks, under the action of applied stress. This simple model connects Q and m in equation 1 to the energetics and mobility of dislocation kinks

$$\begin{aligned} Q &= E_{form} + E_{mig}, \\ m &= 1, \end{aligned} \tag{2}$$

where E_{form} is the single kink formation energy and E_{mig} is the kink migration barrier.

For Si, parameters E_{form} and E_{mig} were first obtained by atomistic calculations based on the empirical inter-atomic potentials [2]. These values were then plugged in equations 1 and 3 to predict the temperature dependence of dislocation velocity, ostensibly in agreement with experiment. More recently, in order to remove the remaining discrepancies between theory and experiment, parameters E_{form} and E_{mig} were re-calculated using more accurate atomistic approaches, including Tight Binding [3] and the density functional theory calculations [4]. However, despite impressive recent developments in this area, some experiments and atomistic simulations suggest that the kink diffusion model, while capturing some essential physics, is not an adequate starting point for a first principles theory of dislocation motion by kink mechanisms. In particular, the model does not account for the existence of multiple kink species in the dislocation core [5, 6] and for the extended nature of dislocations in semiconductors [7].

We propose a two-scale computational approach which combines atomistic simulations of dislocation kink mechanisms with mesoscopic (continuum) modeling of large scale dislocation behavior in Si. First, atomistic calculations based on inter-atomic potentials are used to obtain formation and migration energies of various kinks and core defects involved in dislocation motion. Then, these energetics are used to parameterize the rates of kink formation and migration events

Mat. Res. Soc. Symp. Proc. Vol. 538 ©1999 Materials Research Society

- key parameters for the meso-scale Kinetic Monte Carlo (KMC) simulations. The resulting atomistic-mesoscopic model is used to model dislocation motion on the micron scale, in agreement with existing experimental data, providing a computationally efficient framework for parameter-free theory of dislocation motion in Si.

TWO-SCALE MODEL

Atomistic Scale

Similar to FCC materials, dislocation in Si have $1/2\langle 110 \rangle$ Burgers vectors and glide on $\{111\}$ planes. Owing to a very high Peierls potential, dislocations tend to stretch along the primary Peierls valleys, i.e. $\langle 110 \rangle$ directions. Full dislocations of screw and 60° (mixed) characters are known to split into two Shockley partials glued together by a ribbon of stacking fault (SF), 30 to 70 Å wide. The partials are of either 30° or 90° character with respect to the preferred $\langle 110 \rangle$ directions. Since full dislocations can move only through the the nucleation of double kinks, followed by kink migration and recombination on each partial, the focus has been on kink mechanisms in the two partial dislocations.

Earlier atomistic simulations [5, 8] showed that multiple species of kinks exist in both 30° and 90° partials. Recently, we proposed a general approach to classification of kinks and other dislocation core defects into topological classes, according to the symmetry and symmetry breaking in the dislocation core [9]. Using this simple *a priory* analysis as a guide, we obtained a complete list of all possible kinks in both partials. Then, we employed a newly developed inter-atomic potentials for Si [10] to examine the atomic pathways and energetics of nucleation and migration for all relevant kink species. The results are listed in Table 1.

Table 1: The formation energy and migration barrier of kinks and Anti-Phase-Defect (APD) on 90° and 30° partials in Si, by atomistic simulations using EDIP

		E_{form}(eV)	E_{mig}(eV)			E_{form}(eV)	E_{mig}(eV)
	APD	0.65	0.39		APD	0.49	0.65
	LR	0.70	0.62		LK	0.65	1.46
90°	RL	0.70	0.62	30°	RK	0.72	0.56
	LL	0.84	0.55		LC	0.90	1.04
	RR	1.24	0.71		LK′	1.49	0.62
					RK′	0.39	0.89

Mesoscopic Scale

The data presented in the table is sufficient to develop a mesoscale model of dislocation motion in Si incorporating the atomistic details of kink mechanisms [11]. In the model, relevant degrees of freedom are no longer the atomic positions in and around the core, but rather the identities and positions of the core defects. Coupling to the atomistic energetics is established through the activation parameters of kink mechanisms taken directly from Table 1 and plugged in the rate equations of the standard Transition State Theory [12],

$$ w = w_0 \exp(-\frac{E}{k_B T}), \tag{3} $$

where w_0 is the frequency factor, E includes $2E_{form}$ for the rate of double kink nucleation and is equal to E_{mig} for the rate of kink migration.

At finite temperature and stress, the activation parameter E in equation 3 should be replaced by a stress dependent free energy $F(\tau)$, or free enthalpy, taking into account both the activation entropy contribution and the coupling to applied stress. The latter coupling is introduced here in the form of work produced by external stress on displacing the atoms from their initial positions to the barrier configuration, for each kink mechanism considered,

$$F(\tau) = F(0) - A^*b\tau. \tag{4}$$

Here $F(0)$ is the stress free activation barrier, b is the Burgers vector of the partial, A^* is the area swept by the partial in transforming from its original state to the saddle point for a given kink mechanism, and $A^*b\tau$ is the amount of work done by the external stress. Concerning the entropy terms, Marklund argued that these can be very significant for the kink mechanisms in Si [13]. However, since no reliable data on kink activation entropies is available, we disregard these potentially large contributions to the activation rates. At present, work is underway in our group to evaluate the missing entropy terms using exact methods of adiabatic switching [14] and thermodynamic integration [15].

Another important aspect of dislocation motion in Si, ignored in the earlier version of the mesoscopic model [11], is the interaction between moving partials. Möller recognized that this interaction can be a source of unusual mobility behavior observed in Si and Ge at low stress [7]. In particular, two regimes of double-kink nucleation are possible; one, in which nucleation takes place on each partial independently (uncorrelated regime) and another, in which two double-kinks form on two partials simultaneously (strongly correlated regime). The latter behavior is typical for low stress conditions, but the correlations become progressively weaker with increasing stress. Möller's theory contained several adjustable parameters that were fitted to a limited experimental data on dislocation mobility and then used to calculate the overall shape of the velocity-vs-temperature and velocity-vs-stress functions. The predicted kinetics showed several realistic features, including existence of an apparent starting stress and a super-linear dependence of dislocation velocity on stress.

Following Möller's ideas we developed our mesoscopic model for the case in which two interacting partial dislocations move together under stress. In the present model, both the stacking fault forces and the elastic terms of the partial-partial interactions are fully accounted for. In particular, the stress acting on the various segments of a moving dislocation is evaluated using the full Peach-Koehler formalism [1]. The simulation proceeds as follows. First, the rates $\{w_i, i = 1 \cdots n\}$ of all possible kink events (nucleation, motion, recombination) are calculated using equations 3,4. In doing so, the stress distribution along the line is evaluated for a given dislocation configuration. Second, a single event is randomly chosen from the current list with a probability proportional to this event's share in the total transition rate, according to the usual Kinetic Monte Carlo (KMC) procedure [16]. Then, the stress field along the dislocation line is updated and the whole cycle is repeated to simulate a stochastic sequence of kink nucleation, migration and recombination events. To imitate an infinitely long dislocation line, periodic boundary conditions are applied where the kink leaves and reenters the simulation length. So far, in order to make numerical implementation easier, we have introduced only a single kink species in the model. Formation and migration energies of this generic kink are chosen as averages over the whole kink family listed in Table 1. Extension to a more realistic case of the multiple kink mechanisms is rather straightforward.

Initial trial runs showed that generation of double-kinks can be very ineffective, especially in the case of low stress. This is because kinks need to migrate for a sufficiently long distance and recombine with kinks from the other nucleation sites to contribute to the overall dislocation motion. However, since the kinks from a newly formed pair are typically very close to each other, the probability of their mutual recombination immediately after nucleation is exceedingly high. Consequently, lots of the KMC steps are wasted in sampling the nucleation of such "non-sustainable" kink pairs. To eliminate this waste of effort, we developed a special procedure to calculate the rate of production of sustainable kink pairs, under the assumption that the local stress remains unchanged during the whole process of pair unraveling. Now, each sustainable kink pair is generated in the KMC simulation in one "giant" step.

RESULTS

Dislocation Motion

Figure 1 shows the output of a typical KMC simulation in which a screw dislocation is moving at a constant temperature (900K) in response to a constant stress (10MPa). The two solid lines indicate the positions of the two partials as functions of time. After a short transient, the dislocation reaches a steady state of motion in which the separation between the partials remains close to the equilibrium value of 30$\mathring{A}$. By fitting the displacement curve to a straight line, we obtain the steady dislocation velocity, at $v = 4 \times 10^{-8}$cm/s in this case. To reduce the statistical errors, we ran several KMC simulations under the same conditions.

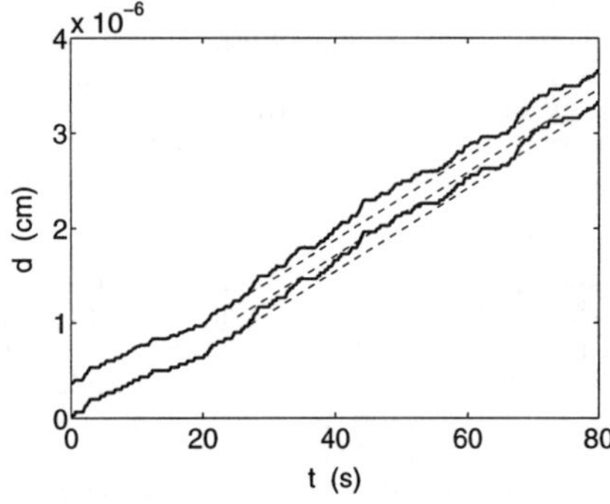

Figure 1

Position of the two 30° partials of a screw dislocation as a function of time, at $T = 900$K, stress $\tau = 10$MPa. Steady dislocation motion is observed after certain relaxation time. Velocity extracted from straight line fitting is $v = 4 \times 10^{-8}$cm/s

$v - T$ **Dependence**

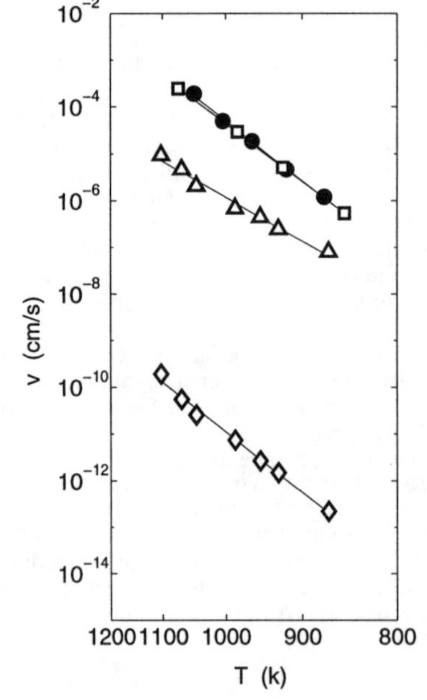

Figure 2

Screw dislocation velocities at different temperatures under stress $\tau = 10$MPa

●   : experiments by Imai and Sumino
□   : experiments by George at slightly
△   : KMC results using EDIP
◇   : KMC results using TB

Two series of KMC simulations were performed to examine the temperature dependence of dislocation velocity. In Figure 2 the simulations are compared with experiment: the filled circles are the experimental measurements reported by Imai and Sumino [17] and the squares are the experimental data points from George [18]. The two sets of simulation results were obtained using different atomistic input for the kink energetics: the triangles are for the EDIP model and the diamonds are for a different set of atomistic parameters extracted from the Tight Binding calculations of Nunes et. al. [3].

The slope of each data set is the activation energy for dislocation mobility. These slopes are identical for both sets of experimental data, at $Q = 2.20$ eV, and are reasonably close to the predicted values, at $Q = 1.67$ (EDIP) and $Q = 2.31$ (TB). At the same time, the absolute values of the predicted velocities differ more significantly from the experiments. This discrepancy can be, at least in part, attributed to the missing entropy terms in the rates of kink nucleation and migration. We estimate that, in order to make quantitative comparison meaningful, we need to know the activation parameters of kink mechanisms to within 0.1 eV. Such an accuracy is within reach of the parameter-free first principle approaches, but only for the ground state, zero temperature energetics. On the other hand, finite temperature calculations required to examine the entropy contributions to dislocation mobility from first principles remain prohibitive at present.

$v - \tau$ Dependence

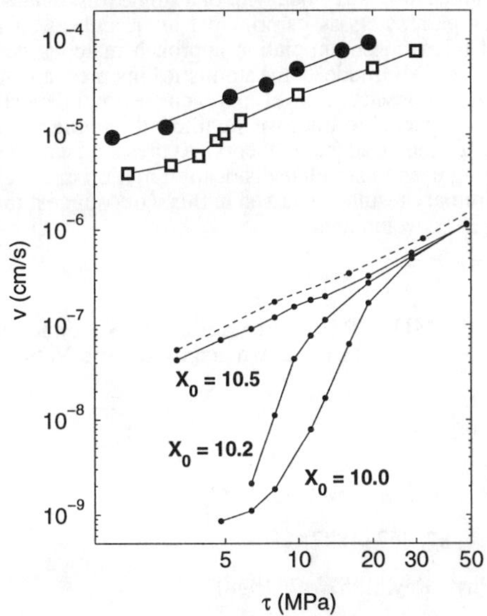

Figure 3

Screw dislocation velocities under different stress at temperature $T = 1000$K

● : experiments by Imai and Sumino
□ : experiments by George (at a slightly lower temperature, $T = 983$K)
— : KMC results with different X_0
- - : KMC results with paritals decoupled

Figure 3 shows the stress dependence of dislocation velocity. The same two sets of experimental data are plotted as in Figure 2. In this case, however, the experimental data points contradict each other at low stress, below 8 MPa. The filled circles from Imai and Sumino follow a simple linear dependence, while the squares from George indicate a low stress anomaly, often referred to as a starting stress behavior.

Small dots connected by the solid lines are the KMC results obtained using slightly different sets of input parameters. In fact, the only difference between the three simulated curves was in the value of the stacking fault energy (SFE) which determines the equilibrium separation between two partials at rest, X_0. The simulations show that a small change in X_0 can have a dramatic effect on dislocation mobility at low stress, changing from a very prominent starting stress behavior at $X_0 = 10.0$ to a complete lack of thereof at $X_0 = 10.5$. X_0 is determined by the balance between the linear elastic repulsion between the two partials, their attraction due to the stacking fault, and the interaction with external stress, if any. In particular, in the case of a screw dislocation dissociated in two 30° partials, the formula for X_0 is

$$X_0 = \frac{\gamma_{SF} - \tau_e \frac{b}{2\sqrt{3}}}{\mu b^2 (\frac{1}{4} - \frac{1}{12(1-\nu)})}, \tag{5}$$

73

where γ_{SF} is the stacking fault energy, μ is the shear modulus, ν is the Poisson's ratio and τ_e is the so-called Escaig stress, which is the edge component of the stress resolved in the glide plane that pushes the partials together.

Figure 3 shows that both behaviors observed experimentally are reproduced in our KMC simulations. When X_0 is expressed in the units of lattice spacing in the direction of dislocation motion, a strong low stress anomaly is observed if X_0 is an integer. On the other hand, if X_0 is a half integer, the stress dependence remains linear even in the low stress range. Shown for comparison (dashed line) is another simulation in which the interaction between two partials is artificially removed. We conclude that the starting stress behavior is a result of coupling between two moving partials and that, if X_0 happens to be half integer, this coupling is effectively suppressed.

CONCLUSIONS

The simulation procedure presented in this work is an adaptation of the Dislocation Dynamics approach pioneered by Kubin and Canova [19] for modeling collective behavior of large dislocation ensembles. Here the same approach is used to simulate large scale behavior of a single (dissociated) dislocation. The model uses atomistic activation parameters as its principal input and, as such, contains no adjustable parameters. Our hybrid micro-meso simulation approach matches very accurately the dislocation behavior in the mesoscopic KMC model to the atomistic kink mechanisms in the dislocation core. Predictive capabilities of the proposed two-level approach are limited mostly by the quality of its atomistic input. The latter is expected to improve significantly owing to the rapid development of the parameter-free methods of condensed matter theory. At present the model, with its current set of atomistic parameters, is being used to simulate dislocation motion in a wide range of temperature and stress conditions. Preliminary results discussed in this work suggest that the approach has a considerable potential for further development.

ACKNOWLEDGEMENT

This work has been supported supported by the MIT MRSEC Program of the National Science Foundation under award number DMR 94-00334 and by the Lawrence Livermore National Laboratory.

REFERENCES

[1] J.P.Hirth and J.Lothe, *Theory of Dislocations*, (Wiley, New York, 1982).

[2] M.Heggie and R.Jones, Inst. Phys. Conf. Ser., **87**, 367 (1897).

[3] R.W.Nunes, J.Bennetto and D.Vanderbilt, Phys. Rev. B, **57**, 17 (1998).

[4] A.Valladares, J.A.White, and A.P.Sutton, Phys. Rev. Lett., **81**, 4903 (1998).

[5] V.V.Bulatov, S.Yip and A.S.Argon, Philos. Mag. A, **72**, 453 (1995).

[6] W.Xu and J.A.Moriaty, Comp. Mater. Sci., **9**, 348 (1998).

[7] H.J.Möller, Acta. Metall., **26**, p. 963 (1978).

[8] R.W.Nunes, J.Bennetto and D.Vanderbilt, Phys. Rev. Lett., **77**, 1516 (1996).

[9] V.V.Bulatov, J.F.Justo, W.Cai, and S.Yip, Phys. Rev. Lett., **79**, 5042 (1997).

[10] V.V.Bulatov, J.F.Justo and S.Yip, submitted to Philos. Mag. A (to be published).

[11] V.V.Bulatov in *Modelling of Structure and Mechanics of Materials from Microscale to Product*, edited by J.V.Carstensen, T.Leffers, T.Lorentzen, O.B.Pedersen, B.F.Sørensen, and G.Winther (19th Risø International Symposium on Matrials Science 1998), p. 39-60.

[12] G.W.Halsey, H.J.White and H.Eyring, Text res. J., **15**, 295 (1945).

[13] S.Marklund, Solid State Commun. **54**, 555 (1985).

[14] M.de Koning and A.Antonelli, Phys. Rev. B, **55**, 735 (1997).

[15] D.Frenkel and B.Smit, *Understanding Molecular Simulation From Algorithm to Application*, (Academic, San Diego, 1996).

[16] A.B.Bortz, M.H.Kalos and J.L.Lebowitz, J. Comput. Phys., **17**, 10 (1975).

[17] M.Imai and K.Sumino, Philso. Mag. A, **47**, 599 (1983).

[18] A.George, J. Phys. (Paris), **C6**, 40 (1979).

[19] L.P.Kubin, G.Canova, M.Condat, B.Devincre, V.Pontikis, and Y.Brechet, Solid State Phenomena, **23-24**, 455 (1992)

ATOMISTIC SIMULATIONS OF CROSS-SLIP PROCESSES IN MODEL FCC STRUCTURES AND L1$_0$ TiAl

S.Rao*

Materials Directorate, Wright Laboratory, WL/MLLM, Wright-Patterson AFB, OH 45433
* UES Inc, 4401 Dayton-Xenia Rd, Dayton, OH 45432

ABSTRACT

Three dimensional (3D) cross-slipped core structures of a/2[110] screw dislocations in model FCC structures are simulated using lattice statics within the Embedded Atom Method (EAM) formalism using potentials fitted to the elastic and structural properties of FCC Ni as well as L1$_0$ TiAl. 2 and 3-D Green's function (GF) techniques are used to relax the boundary forces in the simulations. The core structure of the constrictions are diffuse. At large separation distances between Shockley partials in the unconstricted configuration, the two constrictions formed by cross-slip onto a cross {111} plane have significantly different energy profiles suggesting that self-stress forces dominate the energetics of the cross-slip process. The variation in cross-slip energy with stacking-fault energy is in reasonable agreement with continuum predictions, excepting at high fault energy values as in L1$_0$ TiAl. Cross-slip energies estimated for Cu, Ni and γ-TiAl from these calculations show reasonable agreement with experimental data. The cross-slip energy shows a significantly weaker dependence on Escaig stress as compared to elasticity calculations, with the activation volume for the cross-slip process being approximately 20b^3 at an applied Escaig stress of 10$^{-3}\mu$ in Cu.

1. INTRODUCTION

Cross slip plays an important role in the plastic deformation of metallic materials. It is considered to be the mechanism through which screw-dislocation segments annihilate each other and form low-energy structures in deformation processes like work hardening, fatigue and creep[1]. Cross-slip of screw dislocations between different glide planes is thought to result in the formation of glide-plane obstacles in hexagonal close-packed metals, L1$_2$, B2 and L1$_0$ based intermetallic alloys[2-8]. Cross-slip can also act to weaken materials by allowing the bypass of obstacles in dispersion-hardened structures such that the Orowan limit is not achieved[9].

Several different mechanisms have been proposed for cross-slip in FCC and HCP materials[10-12]. Some of these mechanisms have been described recently[13,14]. The Shoeck-Seeger mechanism[10] assumes that the original dislocation, dissociated into Shockley partials in the glide plane, has to recombine over a significant portion of its length and then bow out to a critical configuration in the cross slip plane (Figs.1a and 1b). Friedel[15], Schoeck and Seeger[10] suggested that the energy of the bowed out loop significantly decreases if the dislocation is allowed to redissociate in the cross-slip plane. As soon as redissociation enters into the picture (Fig.1c), the cross-slip process can be purely thermally activated without any applied stresses. According to the Friedel-Escaig mechanism of cross-slip, as soon as a Stroh type constriction[16] (Fig.1d) forms on the glide plane due to thermal activation, the dislocation at the constriction point can immediately redissociate in the cross-slip plane and expand. Activation enthalpies and volumes calculated using the Friedel-Escaig mechanism seem to be in fair agreement with experimental data in FCC materials[14,17].

Early theoretical treatments of cross-slip in FCC and HCP structures were mostly performed using continuum elasticity theory[10,13,15-19]. The effect of an applied Escaig stress on the cross-slip plane (glide stress on the edge components of the Shockley partials in the cross-slip plane) for the structure and energetics of the cross-slipped configurations were also considered. The treatment of the constriction was done within the Stroh approximation of a point constriction[16]. The classical treatment of the point constriction by Stroh was recently re-evaluated by Saada[20]. Saada[20] generalized Stroh's treatment by removing the point constriction from the analysis and introducing a finite distance of b-3b as the recombination distance for the Shockley partials. The constriction energy is significantly reduced by this modification when the initial separation distance between

Shockley partials is small. As a result of the various approximations in the continuum calculations[20], the published cross-slip energies for an identical initial separation distance between Shockley partials in the unconstricted configuration can vary almost by a factor of 4.

Considering the influence of cross-slip processes on material properties, and the significant difficulties that one encounters in a continuum treatment of the problem, more accurate treatments of this fundamental problem are needed. Recent 2 and 3-D simulations of dislocation dynamics on the mesoscopic scale[21-25] are completely dependent upon reliable input data or 'rules' for cross-slip conditions. Calculations of the cross-slip process at the atomistic level should be very useful and may give new insights into the problem. Proper treatment of the Friedel-Escaig process of cross slip in FCC structures requires atomistic simulations of 3D configurations of dislocations[14,26-27]. Very few 3D atomistic simulations of dislocation configurations have been published to date[14,26-33]. With the advent of faster computers, new methods such as more reliable interatomic potentials based on the embedded atom method (EAM) for FCC materials[34-35] and the 3D lattice GF method to establish the cell boundary conditions[32], these simulations become feasible.

Fig.1. Schematics of cross-slip according to different proposed mechanisms:-(a and b) Seeger-Schoeck mechanism, (c) Friedel mechanism and (d) Stroh type constriction.

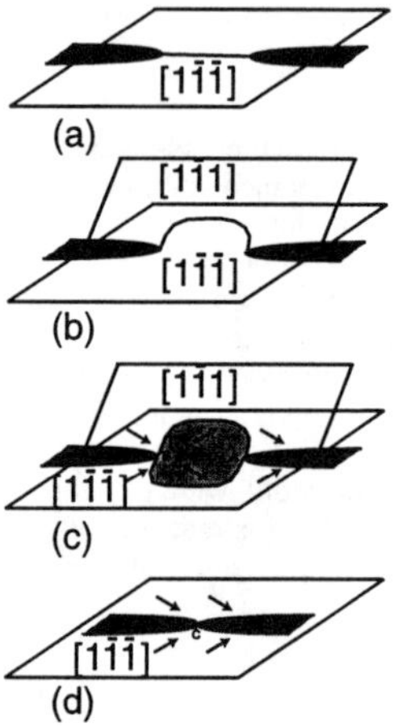

In this manuscript, a 3D atomistic simulation technique combining molecular statics and GF methods is used to investigate cross-slip processes in model FCC structures as well as $L1_0$ TiAl. The boundary conditions of the simulations are such that the Friedel-Escaig mechanistic description of cross-slip is obtained after relaxation. The constrictions occurring in the cross-slip process are simulated individually as well as in pairs. This allows a determination of the formation energies of the constrictions, as well as their interaction energies. The empirical embedded atom method (EAM) potential developed for FCC Ni [**Ni(1)**] by Voter and Chen[36] is used in most of the calculations. In addition, another EAM potential for FCC Ni [**Ni(2)**], which gives a higher stacking-fault energy as well as an EAM potential for $L1_0$ TiAl[37-38] [**NTA02**] are used to study the variation of cross-slip energy with stacking or complex stacking-fault energy. For the calculations with an applied stress, a general state of stress is applied such that there is no resolved Escaig stress on the cross-slip plane but an Escaig stress extending the Shockley partials on the glide plane exists. The results of the atomistic simulations are compared with continuum calculations of cross-slip in FCC-based structures as well as experimental data.

2. POTENTIALS

A comparison of the properties of Ni given by **Ni(1)** and **Ni(2)** shown in Table 1 indicates that the EAM potentials give almost identical properties, excepting for the stacking-fault energy which changes from $58 mJ/m^2$ for the **Ni(1)** potential to $119 mJ/m^2$ for the **Ni(2)** potential. The properties of γ-TiAl given by **NTA02** is summarized in Table 1[38], with the complex stacking fault energy, γ_{csf}, involved in the dissociation of a/2<110> screw dislocations into Shockley partials in $L1_0$ structures having a value of 320 mJ/m^2. Elasticity calculations, as well as current atomistic simulations, predict a dissociation of the a/2[110] screw dislocation into two Shockley partials[39]. The equilibrium distance of this spreading are given in Table 2 for the Ni and $L1_0$ TiAl potentials. The scaled separation distance between Shockley partials in the unconstricted configuration, is d/b

= 6.0-7.5, d/b = 3.0-5.0 for **Ni(1)** and **Ni(2)**, respectively and d/b = 1.0-2.0 for L1$_0$ TiAl. Note the deviation of the continuum computation, particularly for the **Ni(2)** potential, which has a high γ_{sf} and **NTA02**, which has a high γ_{csf}.

3.1. SIMULATION OF DISLOCATION CORE STRUCTURES

For simulations of infinite a/2[110] screw dislocations, a cylindrical crystal was constructed that was 7.5nm in radius and one periodic unit along the dislocation line direction. The dislocation line direction was [110], and the length of the periodic unit in the simulations was $a_0/\sqrt{2}$ in the Ni simulations and $\sqrt{2}\,a_0$ in the L1$_0$ TiAl simulations. The x and y axes in the simulation cell were parallel to [1$\bar{1}$2] and [1$\bar{1}\bar{1}$], respectively. Atomic displacements calculated using anisotropic elasticity theory[40] was used to introduce the a/2[110] screw dislocation into the simulation cell. The origin for the displacement field was chosen to be in-between atomic positions. A flexible-boundary technique based on the GF method was used in the simulations to obtain a true

Table 1: Various structural and elastic properties given by the EAM potentials, Ni(1) and Ni(2), for FCC Ni and NTA02 for L1$_0$ TiAl. Fault energies are in mJ/m2, elastic constants are in units of 10^{12} dynes/cm^2, structural energy and structural energy differences are in eV/atom.

Property	Ni(1)	Ni(1)	NTA02
a (nm)	0.352	0.352	0.4033
E_c	-4.45	-4.45	-4.87
C_{11}	2.44	2.44	2.02
C_{12}	1.49	1.49	0.95
C_{44}	1.25	1.19	0.83
E_v	1.60	1.83	1.88
E_{sf}	58	119	200
E_{bcc} - E_c	—	0.091	—
E_{hcp} - E_c	—	0.010	—
C_{13}	—	—	1.24
C_{33}	—	—	2.37
C_{66}	—	—	0.54
E_{csf}	—	—	320
E_{apb} (111)	—	—	550
E_{apb} (001)	—	—	91
c/a	—	—	0.9912

Table 2. Spacing between Shockley partials in units of b, for the Ni potentials, Ni(1) and Ni(2) and the L1$_0$ TiAl potential, NTA02.

	Atomistic (d/b)	Elastic (d/b)
Ni(1)	7.8	6.1
Ni(2)	5.2	3.0
NTA02	0.0-2.0	1.5

Table 3. Applied 3-D strain (a) and stress (b) tensors in the glide plane coordinate system, 112, 111 and 110. Stress tensor is in units of μ. The applied stress tensor resolved on to the cross-slip plane coordinate system, 112, 111 and 110 is given in (c).

-0.0007408	0.0016000	0.0000000
0.0000000	0.0004076	0.0000000
0.0000000	0.0000000	0.0003331

(a)

-0.0012400	0.0010000	0.0000000
0.0010000	0.0012400	0.0000000
0.0000000	0.0000000	0.0000000

(b)

-0.0015300	-0.0000020	0.0000000
-0.0000020	0.0015300	0.0000000
0.0000000	0.0000000	0.0000000

(c)

local minimum energy structure with little residual boundary forces[14,32]. The a/2[110] screw dislocation was made to dissociate on different {111} planes by varying the origin for the initial displacement field.

3.2. CROSS-SLIP SIMULATIONS

The cross-slipped configuration of the a/2<110> screw dislocation involves two constrictions: a constriction of the dislocation which continuosly transitions the dislocation from the octahedral glide plane to the cross-slip plane, and a constriction which continuosly transitions the dislocation from the cross-slip to the glide plane. The structure and energetics of the two individual constrictions are expected to be different from each other[14,26-27]. Therefore, the atomistic simulation of cross slip is divided naturally into two separate simulations: simulation of a positive constriction and simulation of a negative constriction. The procedure used to obtain the initial conditions for simulations of the constrictions are described in detail by Rao et.al.[14]. The total length of the simualtion cell was 45b, where b is the lattice periodicity along the [110] direction in a fcc or an L1$_0$ structure. The radius of the simulation cell was 4.4nm for FCC Ni and 4.0nm for L1$_0$ TiAl simulations, and 3D GF BC's were employed in all the simulations[14,32].

Interaction energies between the constrictions at four separation distances, 3, 5, 11 and 21b, were evaluated using **Ni(2)**. The methodology used to obtain the interaction energies between the constrictions from the fully-relaxed atomic positions corresponding to the positive and negative constrictions are described in detail by Rao et.al.[14]. For evaluating the effect of an Escaig stress on cross-slip energy, Escaig stresses were applied on the atomistically relaxed cross-slip pair configurations. An Escaig stress was applied on the glide plane, with the corresponding applied Escaig stress on the cross-slip plane zero. The applied strain tensor, in the glide plane coordinate system for a given Escaig stress is given in Table 3 along with the corresponding stress tensors in the glide and cross-slip plane coordinate systems. The constriction pair configurations were atomistically relaxed, with the central layers of the two constrictions fixed. The residual forces that develop at the boundary of the simulation cell are relaxed using 3D GF BC's.

Fig.2. Differential displacement plots of the screw component of the core of a/2[110] screw dislocations, obtained using three different EAM potentials, NTA02, Ni(1) and Ni(2).

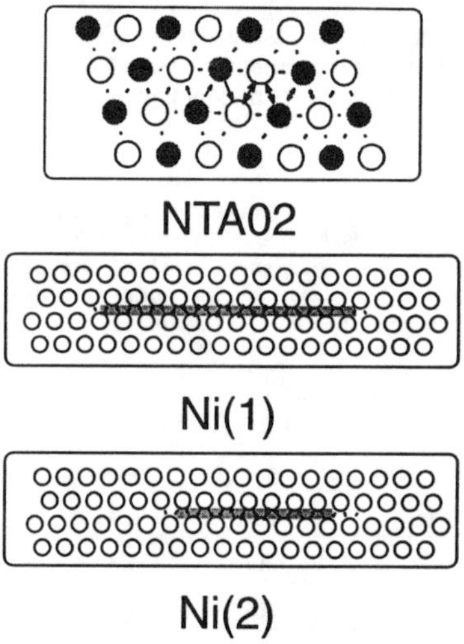

NTA02

Ni(1)

Ni(2)

4. RESULTS AND DISCUSSION

4.1. CORE STRUCTURE OF a/2[110] SCREW DISLOCATIONS

Figure.2 depicts the core structures obtained for the a/2[110] screw dislocation with **Ni(1)**, **Ni(2)** and **NTA02** using the differential displacement field method[32,42]. As expected the core becomes constricted with increasing stacking or complex stacking-fault energy. Figures 3a and b show differential displacement-field plots of the screw components of the atomistically relaxed core configuration of a/2[110] screw dislocations in L1$_0$ TiAl at the positive and negative constrictions involved in the cross-slip process. The differential displacement-field plots are plotted for different perpendicular slices along the dislocation line. Thin cylindrical slices perpendicular to the dislocation axis are taken of the original unperturbed lattice with the thickness of the slice equal to a[110]. The relative displacements of all the atoms within the slice at both the positive and negative constrictions are used to obtain the differential displacement-field plot for that particular slice in figures 3a and 3b. At the central plane of both the constrictions, the differential displacements are more or less equally spread over both the octahedral planes and the core structures are diffuse, as opposed to the classical point constriction[16]. However, within a few slices away from the central plane , the core displacements at the a/2[110] screw dislocation are predominantly contained within one of the two octahedral planes, the $(1\bar{1}\bar{1})$ glide plane on the positive side and the $(1\bar{1}1)$ cross slip plane on the negative side for the positive constriction, and vice versa for the negative constriction.

4.2. ENERGETICS OF CROSS-SLIP OF a/2[110] SCREW DISLOCATIONS - UNSTRESSED

The differential energy, δE, between the constricted dislocation in the cross-slip event and the original unconstricted configuration, for each cylindrical periodic unit along the dislocation line is plotted in fig.4 for $L1_0$ TiAl. The thickness of each cylindrical slice is taken to be a[110] and the position of the shell is taken to be its mean position along the dislocation line in the unperturbed lattice, 2nb. For each shell, the differential energy δE was obtained by summing up the differences in energy between identical atoms in the constricted and unconstricted configurations within a cylindrical radius of 2.75nm. This radius was chosen to avoid any surface effects on the differential-energy plots. The differential-energy plot is shown for both the positive and negative constrictions. The integral of the differential-energy curve over the length of the dislocation line gives the formation energy of the constrictions at zero applied stress. The total energy of the positive and negative constrictions are identical and equal to 0.25eV. The

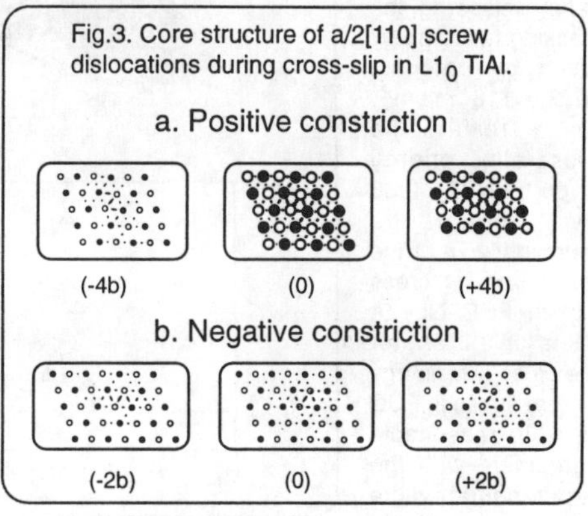

sum of the formation energies of the two constrictions is found to be approximately 0.50 eV for **NTA02**. With **Ni(1)**, the total energy for the cross-slip event is found to be 4.85eV, with the positive constriction having a formation energy of 6.30eV and the negative constriction, -1.45eV[14]. Similarly, with **Ni(2)**, the cross slip energy is 2.30eV, with the positive constriction having a formation energy of 3.20eV and the negative constriction, -0.90eV[14].

There are four contributions to the differential energy δE: i) change in energy of interaction between the two Shockley partials; ii) an increase in dislocation length at the constrictions; iii) a decrease in area of the stacking or complex stacking fault between the two Shockley partials at the constrictions and iv) a change in character of the Shockley partials at the constrictions. The first two factors contribute to an increase in energy of the dislocation at the constrictions. The third contribution is expected to be negative. The energy change due to a change in character of Shockley partials at the constrictions is expected to be positive at the positive constriction, and negative at the negative constriction. The atomistic simulation results suggest that the energy changes due to change in character of the Shockley partials, the fourth contribution, dominates the energetics of the two constrictions when d/b is high. Clearly, the cross-slip energy is decreasing with increasing SF or CSF energy or decreasing d/b ratio. Continuum treatments of the cross-slip process predict a decreasing cross-slip energy with decreasing d/b ratio as[16]

$$E_c \propto (d/b)\sqrt{\ln(d/b)} \tag{1}$$

Present atomistic simulation results on cross-slip energies as a function of d/b ratio is collected together in fig.5 along with the expected continuum scaling according to eq.(1). Figure 5 shows that the scaling of cross-slip energy according to eq.(1) breaks down at extremely small Shockley partial spacings, $d/b \sim 1$-2, as in L1$_0$ TiAl.

4.3. COMPARISON OF RESULTS ON CROSS-SLIP ENERGIES WITH EXPERIMENT

The experimental value for the stacking fault energy in FCC Ni[40,42] is approximately 125 mJ/m^2, closer to the stacking fault energy given by potential **Ni(2)**, 118 mJ/m^2. As a result, the cross-slip energy given by **Ni(2)**, 2.35 eV, should be representative of the real value for cross slip in FCC Ni. A value of 2.35eV for the cross-slip energy at zero stress in FCC Ni is in reasonable agreement with the experimental value for the energy of cross slip in FCC Ni[43]. Similarly, the cross-slip energy in FCC Cu can be estimated from these atomistic simulations since it scales according to eq.1. The d/b value given by **Ni(2)** for the unconstricted a/2[110] screw dislocation (Table 2), is in reasonable agreement with the d/b value observed in FCC Cu[44] As a result, the cross-slip energy in FCC Cu can be obtained from the atomistic result for cross-slip energy in **Ni(2)** by simply scaling by the shear modulus of Cu relative to the shear modulus of Ni. This gives a value of 1.07-1.28eV for the cross-slip energy in FCC Cu. This is in reasonable agreement with the experimental value of 1.15eV deduced for the cross-slip energy in Cu[45].

The yield stress anomaly in γ-TiAl, for certain orientations of the applied stress, has been attributed to the formation of a high density of pinning points or jogs on a/2[110] screw dislocations resulting from double cross-slip[8]. Assuming that the elementary cross-slip process leads to the formation of pinning points on a/2[110] screw dislocations, for a cross-slip energy of 0.5eV, the average spacing between pinning points[39] is calculated to be 46nm at 773K, which is in reasonable agreement with the experimentally observed densities[6,8].

Fig.4. Differential energy, δE, as a function of position along the dislocation line at the positive and negative constrictions (NTA02).

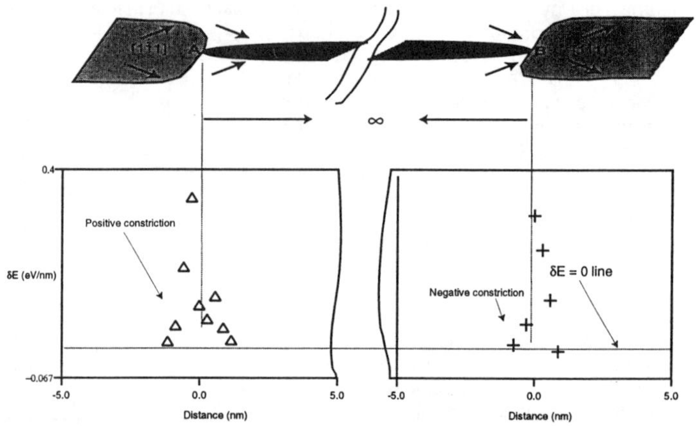

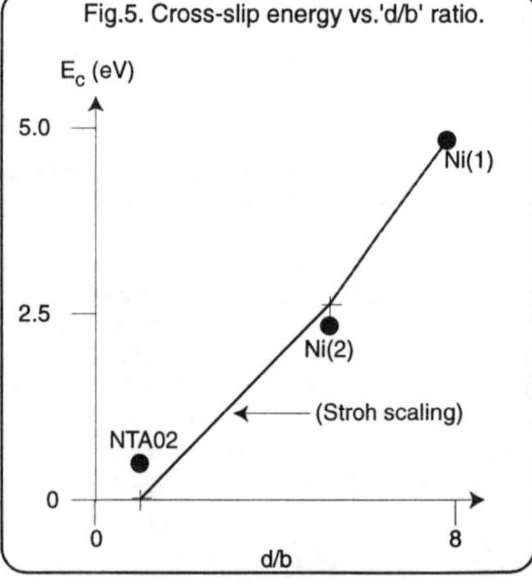

Fig.5. Cross-slip energy vs.'d/b' ratio.

4.4. CONSTRICTION INTERACTION ENERGIES AND ESCAIG STRESS EFFECTS

The atomistic results for the energy of the cross-slipped configuration as a function of the separation distance 'l' between the two constrictions, E(l), scaled relative to the energy at infinite separation distance, E_c, obtained with **Ni(2)**, is shown in Fig.6. Superposed on the atomistic results is a least squares fit of the form

$$E(\lambda) / E_c = 1 - (1 / \lambda) K_{int} \tag{2}$$

where K_{int} characterizes the strength of interaction between the positive and negative constrictions. Figure 6 also compares the atomistic results with continuum calculations by Puschl and Schoeck[19]. The change in character of Shockley partials toward more edge-like at the positive constriction and toward more screw-like at the negative constriction is explicitly taken into account in the atomistic calculations. This results in a weaker interaction between the two constrictions, since parallel edge and screw dislocations are expected to have very little interaction with each other[46].

The energy of the cross-slipped configurations at separation distances of 3,5,11 and 21b were also evaluated with Ni(2) at applied Escaig stresses of 0.00045, 0.0009 and 0.0045µ on the glide plane, with no Escaig stress on the cross-slip plane. Differences in energies between the relaxed cross-slipped configurations and the infinite dislocation confined to the glide plane with an identical applied Escaig stress was determined. The applied Escaig stresses were such that they tended to increase the equilibrium Shockley partial splitting of the a/2[110] screw dislocation on the glide plane, resulting in a lower energy core structure relative to the unstressed one. Atomistic results at each applied Escaig stress were fitted to an equation of the form similar to eq.(2a), with K_{int} a function of the applied Escaig stress, τ. Considering the work done term as simply, $E_w =$ -τb$_e$d$_0$l, where b$_e$ is the magnitude of the edge component of the Shockley partials, a$_0 \sqrt{6}$/12, and maximising the energy with respect to l, the activation energy for cross-slip as a function of the applied Escaig stress 'τ', $E_1(\tau)$, was determined. Such a plot is shown in Fig 7. The atomistic results on the activation energy for cross-slip with applied Escaig stress shows a much weaker dependence as compared to previous continuum calculations[13,17,20]. This results in an activation volume for cross slip , -dE/dτ, to be approximately 20b^3 at 10^{-3}µ in FCC Cu, which is an order of magnitude smaller than the measurements made by Bonneville and Escaig[46] as well as continuum calculations[13].

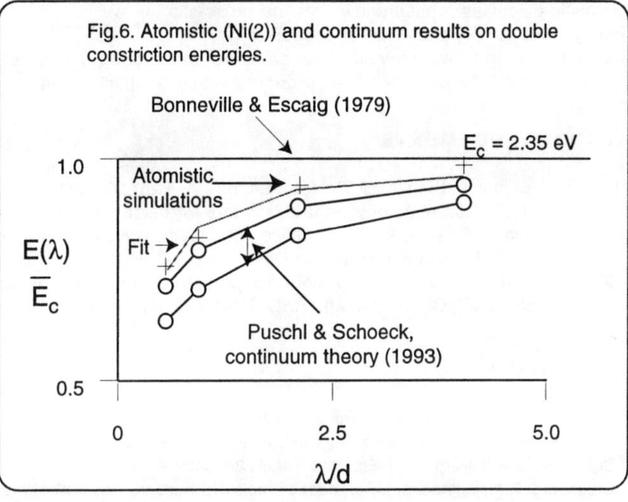

Fig.6. Atomistic (Ni(2)) and continuum results on double constriction energies.

5. SUMMARY

3D atomistic simulations using EAM potentials along with 2 and 3D GFBC techniques have been used successfully to study cross-slip of a/2[110] screw dislocations in model FCC structures as well as L1$_0$ TiAl. Differential displacement-field plots of the core structure of a/2[110] screw dislocations near the constrictions occurring in the cross-slip process reveal that the core structures are diffuse. At high d/b ratios, the energetics of the two constrictions are significantly different from each other, with the positive constriction having a positive formation energy and the negative constriction having a negative formation energy. This suggests that self-stress forces dominate the energetics of the constrictions. Atomistic results on the scaling of cross-slip energy with d/b ratio is in reasonable agreement with continuum predictions, being $\propto (d/b)\sqrt{\ln(d/b)}$ excepting at low d/b ratios. The cross-slip energies are determined to be 2.35, 1.07-1.28 and 0.5eV in FCC Ni, Cu, and $L1_0$ TiAl, compared to a measured value of 1.15 eV in Cu and ~ 0.5eV in $L1_0$ TiAl. Atomistic calculations also show that the cross-slip energy is weakly dependent on applied Escaig

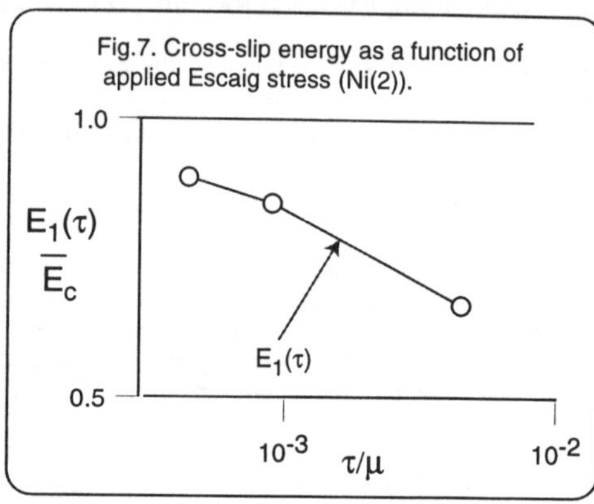

Fig.7. Cross-slip energy as a function of applied Escaig stress (Ni(2)).

stresses, with an activation volume for cross slip of the order of $20b^3$ at $10^{-3}\mu$ in FCC Cu. This result on the activation volume for cross slip is an order of magnitude smaller than previous predictions based on continuum theory.

ACKNOWLEDGEMENTS

The author wishes to thank Drs. P. Hazzledine, T.A. Parthasarathy and C. Woodward of UES Inc. and Drs. D. Dimiduk and J.P. Simmons of Wright Laboratory for helpful discussions during the course of this work. This work was supported by the Air Force Office of Scientific Research and was performed while S.I. Rao worked at the USAF Wright Laboratory, Materials Directorate, WL/MLLM under contract Nos. F33615-91-C-5663 and F-33615-96-5258. This work was supported in part by a grant of HPC time from the CEWES DOD HPC centers: CEWES CRAY-YMP, CRAY-C90 and ASC CRAY C90 supercomputers.

REFERENCES

[1] Jackson, P.J., 1985, Prog.Mater.Sci., **29**, 139.

[2] Paidar, V., Pope, D.P., and Vitek, V., 1984, Acta metall., **30**, 435.

[3] Couret, A., and Caillard, D., 1988, Acta.metall., **36**, 215.

[4] Greenberg, B.A., Antonova, O.V., Indenbaum, V.N., Karkina, L.I., Notkin, A.B., Ponomarev, M.V., and Smirnov, L.V., 1991, Acta.metall.mater., **39**, 233.

[5] Dimiduk, D.M., 1991, J.Physique III, **1**, 1025.

[6] Louchet, F., and Viguier, B., 1995, Phil.Mag. A, **71**, 1313.

[7] Shi, X., Pollock, T., Mahajan, S., and Arunachalam, V.S., 1997, Mat.Res.Soc.Symp.Proc., **460**, 493.

[8] Sriram, S., Dimiduk, D.M., Hazzledine, P.M., and Vasudevan, K.,1997, Phil.Mag.A, **76**, 965.

[9] Humphreys, F.J., and Hirsch, P.B., 1970, Proc.Royal.Soc.A, **318**, 73.

[10] Schoeck, G., and Seeger, A, 1955, Defects in crystalline solids, Phys.Soc.London, 340.

[11] Fleischer, R.L., 1959, Acta.metall., **7**, 134.

[12] Escaig, B., 1968, Proc.Battelle.Colloq. Dislocation Dynamics, edited by A.R. Rosenfield, G.T. Hahn, A.L. Bement Jr and R.I. Jaffee (New York: McGraw Hill), p655.

[13] Duesbery, M., Louat, N.P. and Sadananda, K., 1991, Acta.metall.mater., **40**, 149.

[14] Rao, S.I., Parthasarathy, T.A., and Woodward, C., 1998b, accepted for publication in Phil.Mag.A.

[15] Friedel, J., 1957, Dislocations and Mechanical Properties of Crystals (Wiley: New York), p330.

[16] Stroh, A.N., 1954, Proc.Phys.Soc.B, **67**, 427.

[17] Bonneville, J., and Escaig, B., 1979, Acta metall., **27**, 1477.

[18] Puschl, W., 1990, Phys.stat.sol.b, **162**, 363.

[19] Puschl, W., and Schoeck, G., 1993, Mat.Sci. and Eng.A, **164**, 286.

[20] Saada, G., 1991, Mat.Sci. and Eng.A, **137**, 177.

[21] Kubin, L.P., Canova, G., Condat, M., Devincre, B., Pontikis, V., and Brechet, Y., 1992, Solid state phenomenon, **23-24**, 455.

[22] Mills, M.J., and Chrzan, D.C., 1992, Acta metall., **40**, 3051.

[23] Devincre, B., Veyssiere, P., Kubin, L.P., and Saada, G., 1997, Phil.Mag. A, **75**, 1263.

[24] Devincre, B., and Kubin, L.P., 1994, Modelling Simul.Mater.Sci.Eng., **2**, 559.

[25] Tang, M., Kubin, L.P., and Canova, G.R., 1998, Acta Mater., **46**, 3221.

[26] Rasmussen, T., Jacobsen, K.W., Leffers, T., and Pedersen, O.B., 1997, Phys.Rev.B, **56**(6), 2977.

[27] Rasmussen, T., Jacobsen, K.W., Leffers, T., and Pedersen, O.B., 1997, Mat.Sci and Eng. A, **234-236**, 544.

[28] Duesbery, M.S., 1983, Acta metall., **31**, No.10, 1747.

[29] Parthasarathy, T.A., Dimiduk, D., and Saada, G., 1993, Mat.Res.Soc.Symp.Proc., **288**, 311.

[30] Bulatov, V.V., Yip, S., and Argon, A.S., 1995, Phil.Mag.A, **72**, 1995.

[31] Parthasarathy, T.A., and Dimiduk, D, 1996, Acta Mater., **44**, 2237.

[32] Rao, S.I., Hernandez, C., Simmons, J.P., Parthasarathy, T.A., and Woodward, C., 1998a, Phil.Mag.A, **77**, No.1, 231.

[33] Simmons, J.P., Rao, S.I., and Dimiduk, D, 1998, accepted for publication in Phil.Mag.letters.

[34] Daw, M.S.,and Baskes, M.I., 1984, Phys.Rev. B.,**29**, 6443.

[35] Finnis, M.W., and Sinclair, J.E., 1984, Phil.Mag.A, **50**, 45.

[36] Voter, A.F., and Chen,S.P., 1987, Mater.Res.Soc.Symp.Proc., **82**, 175.

[37] Rao, S.I., Parthasarathy, T.A., and Woodward, C., 1991, Mat.Res.Soc.Symp.Proc., **213**, 125.

[38] Simmons, J.P., Rao, S.I., and Dimiduk, D, 1997, Phil.Mag.A, **75**, 1299.

[39] Hirth, J.P. , and Lothe, J., 1982, Theory of dislocations, second edition (New York: John Wiley and Sons).

[40] Stroh, A.N., 1958, Phil.Mag, **3**, 625.

[41] Vitek, V., 1974, Cryst.Lattice.Defects, **5**, 1.

[42] Carter, C.B., and Holmes, S.M., 1977, Phil.Mag.A, **35**, 1161.

[43] Clement, P.N., and Coulomb, P., 1974, Phil.Mag., **30**, 363.

[44] Stobbs, W.M., and Sworn, C.H., 1971, Phil.Mag., **24**, No.2, 1365.

[45] Bonneville, J., Escaig, B., and Martin, J.L., 1988, Acta.Metall., **36**, 1989.

[46] Hull, D., and Bacon, D.J., 1984, Introduction to dislocations, third edition (Pergamon).

EFFECT OF NON-GLIDE COMPONENTS OF THE STRESS TENSOR ON DEFORMATION BEHAVIOR OF BCC TRANSITION METALS

K. ITO* and V.VITEK**
*Department of Materials Science and Engineering, Kyoto University, Sakyo-ku, Kyoto, 606-8501, Japan.
**Department of Materials Science and Engineering, University of Pennsylvania, Philadelphia, PA 19104-6272.

ABSTRACT

In this paper we demonstrate by atomic computer simulation that the non-Schmid slip behavior in bcc metals is a direct consequence of the non-planar core structure of 1/2<111> screw dislocations and their response to the applied stress tensor. The analysis has been carried out in detail for tantalum using the Finnis-Sinclair type central force many-body potentials. Two distinct non-Schmid effects have been discerned. The first is twinning-antitwinning slip asymmetry on {112} planes. This is an intrinsic property of the bcc structure and depends on the sense of the applied glide stress. The second non-Schmid effect is extrinsic and is controlled by the non-glide shear stresses perpendicular to the total Burgers vector on {110} planes into which the stress-free core of screw dislocations spread.

INTRODUCTION

The Schmid law [1] which states that the plastic flow begins when the resolved shear stress on a slip system reaches a critical value, embodies two distinct assertions. First, that this critical stress does not depend on the sense of shearing and second that it is not affected by any other components of the applied stress tensor. While both these rules apply in fcc metals, neither of them is satisfied in bcc metals which, in general, exhibit a strong dependence of the critical resolved shear stress (CRSS) on orientation of the applied load together with a large variability of the slip geometry. These features of the plastic deformation, though differing in detail from material to material, are common to all bcc metals, ranging from alkali to transition metals (for reviews see [2-8]), and have been found even in more exotic materials with bcc structure, such as the molecular crystal hexamine [9] and solid He^3 [10]. This suggests that they are intrinsic to the bcc structure and it has now been well established that they are governed by the special features of the cores of a/2<111> screw dislocations. These cores are sessile, spread into several fractional dislocations that are confined to different {110} planes of the zone of the corresponding <111> vector [2-4].

Duesbery and Vitek [11] proposed recently that bcc metals exhibit two types of non-Schmid behavior [11]. First is the dependence of the slip on the sense of shearing. This is manifested by the ubiquitous twinning-antitwinning asymmetry of slip on {112} planes. This is an intrinsic property of the bcc structure related to the fact that {111} planes, which are perpendicular to the slip direction, are not mirror planes in this structure. The second is the effect of non-glide components of the stress tensor upon the glide process. This extrinsic effect arises because the core structure of a/2<111> screw dislocations can be altered by the effect of the shear components of the applied stress tensor that are perpendicular to the Burgers vector owing to the force they exert upon the edge components of the fractional dislocations. This influence of the non-glide shears is analogous to the Escaig effect [12, 13] which plays an important role in the cross-slip in fcc crystals and arises because the Shockley partials forming a/2<110> screw dislocations possess edge components.

In this study we first demonstrate by atomistic computer simulations of the motion of 1/2<111> screw dislocations that these two types of non-Schmid behavior are, indeed, always present. The most important result of this investigation is then determination of those non-glide stress component that are most influential in the non-Schmid behavior. The calculations have been carried out for tantalum and molybdenum but in this paper we only discuss tantalum.

Mat. Res. Soc. Symp. Proc. Vol. 538 © 1999 Materials Research Society

METHOD OF CALCULATION

In order to investigate the effect of applied stresses on deformation behavior of bcc transition metals we have performed molecular statics simulations of $1/2<111>$ screw dislocations. The atomic interactions were described by the Finnis-Sinclair type [14] central force many-body potentials. In the present calculations, the potential for tantalum constructed by Ackland and Thetford [15] was used. Periodic boundary conditions were applied in the direction of the dislocation line, [111], chosen as direction of the z axis. The block of atoms in the form of a rectangular parallelepiped (45a x 45a) is divided into an inner part in which the atoms are fully relaxed and an outer part in which the atoms are assigned fixed displacements evaluated in accordance with the elastic anisotropic displacement field due to the dislocation studied. Similarly, the applied stress has been imposed via the corresponding elastic displacement field evaluated using anisotropic elasticity. The following three types of externally applied stresses have been employed in the present study: Pure shear stresses parallel to the Burgers vector with various maximum resolved shear stress (mrss) planes, tensile and compressive stresses with several differently oriented tension/compression axes, and judiciously chosen combinations of glide and non-glide shear stresses. However, we only discuss in more detailed the latter two cases.

EFFECT OF APPLIED STRESSES ON DEFORMATION BEHAVIOR

Tension-compression tests

Six tensile and compressive axes, [001], [012], $[\bar{2}38]$, [011], $[\bar{1}22]$ and $[\bar{1}11]$, have been selected to demonstrate the effect of corresponding applied stresses on deformation behavior in tantalum. The mrss plane for the [001] axis is $(\bar{1}\bar{1}2)$, for the [012] and $[\bar{2}38]$ axes $(\bar{1}01)$ and for the last three orientations $(\bar{2}11)$. It is a common practice to introduce angle χ as the angle between $(\bar{1}01)$ and the mrss plane; this angle is then $-30°$, $0°$ and $+30°$ for the above three types of mrss planes [2, 3]. The angle between the Burgers vector and the direction of tensile and/or compressive axes is marked λ and it determines the non-glide shear stress components.

Figure 1 shows the critical resolved shear stress (CRSS) on the mrss plane in tension-compression tests for various orientations. For comparison, critical shear stresses on the mrss plane when loading in pure shear are also shown. The CRSS in tension is lower than in compression for [001] axis when the mrss plane is $(\bar{1}\bar{1}2)$; the opposite is true for the orientations, [011], $[\bar{1}22]$ and $[\bar{1}11]$, when the mrss plane is $(\bar{2}11)$. This asymmetry results from the twinning-antitwinning slip asymmetry on $\{112\}$ planes. However, this is not the only asymmetry observed. For orientations with the $(\bar{1}01)$ mrss plane, the CRSS in tension is always lower than in compression and both are lower than in pure shear. In fact, the CRSS in tension-compression tests always differs from that in pure shear. When a $\{112\}$ plane is the mrss plane this difference is large when the shearing is in the antitwinning sense, but small when the shearing is in the twinning sense. For the case of antitwinning sense of shearing on a $\{112\}$ plane dislocation motion was not observed for compression along [001] and for tension the CRSS is a strong function of λ, decreasing significantly with increasing λ. For orientations with the $(\bar{1}01)$ mrss plane, the CRSS also decreases with increasing λ but only slightly and even for slip in the twinning sense, the CRSS is lowest in compression along $[\bar{1}11]$ which corresponds to the largest λ. These results suggest that the CRSS depends not only on the glide shear stress but also on the non-glide shear stress components.

Investigation of the effect of the applied stress on the core structure and eventual glide of $1/2[111]$ screw dislocations reveals significant differences between pure shear, tension or compression. Figure 2 shows several cases when the dislocation core structure changes differently in pure shear than in tension-compression tests with the same mrss plane; this is then reflected in different slip behavior. In this figure the core structure is depicted using the usual differential displacements maps [3, 4] in the [111] projection. The circles represent atoms within one $a/2[111]$ period that are positioned in three different (111) layers. The screw components of the relative displacements of pairs of atoms due to the dislocation are depicted as arrows positioned between atoms. Long dashed arrows mark the slip plane and the direction of the dislocation motion.

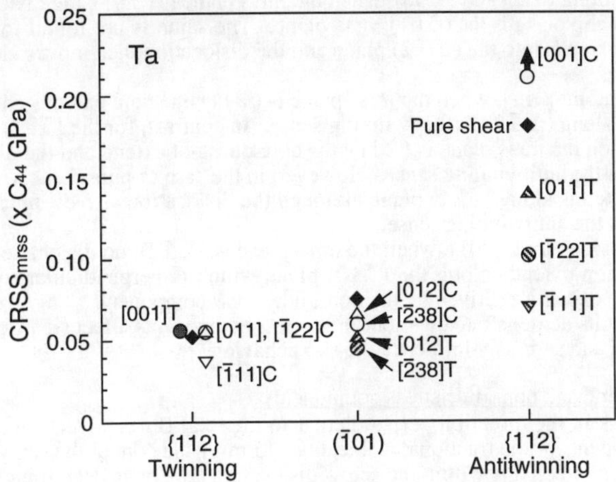

Fig.1. CRSS on the mrss plane in tension-compression tests with various orientations of the tensile/compressive axes. T and C mark tension and compression, respectively. The values of the CRSS on the mrss plane for the case of pure shear are also shown.

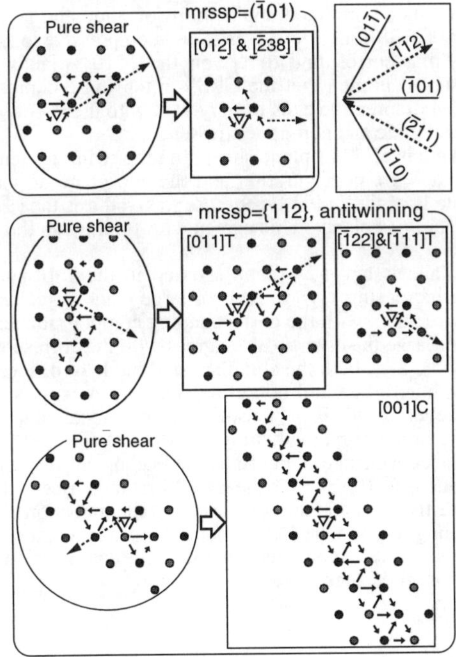

Fig.2. Effect of the applied stress on dislocation core and choice of the slip plane.

For tension along [012] and [$\bar{2}$38], when the mrss plane is ($\bar{1}$01), the dislocation core does not extend and the slip plane is the ($\bar{1}$01) mrss plane. The same is not found for pure shear. In this case the core extends into the ($\bar{1}\,\bar{1}$2) plane and the dislocation then moves along this plane in the twinning sense.

For tension along [011], when the mrss plane is ($\bar{2}$11), the core extends and the dislocation eventually moves along ($\bar{1}\,\bar{1}$2) in the twinning sense. In contrast, for the [$\bar{1}$22] and [$\bar{1}$11] axes (increasing λ), when the mrss plane is ($\bar{2}$11), the core does not extend and the dislocation moves along the ($\bar{2}$11) in the antitwinning sense. However, in the case of pure shear with the same mrss plane, the core extends along (1$\bar{2}$1) plane although the dislocation also eventually moves along the ($\bar{2}$11) plane in the antitwinning sense.

For compression along [001], when the mrss plane is ($\bar{1}\,\bar{1}$2), no dislocation motion occurs since the dislocation extends along the ($\bar{1}$10) plane which is perpendicular to the mrss plane ($\bar{1}\,\bar{1}$2); this core extension is driven by the non-glide stress component.

All these results demonstrate that non-glide stress components affect significantly the core of screw dislocations and seriously influence the slip behavior.

<u>Combination of glide and non-glide stress components</u>

Shear stress in the direction perpendicular to the total Burgers vector (τ_{12}) can act upon small edge components of the fractional dislocations forming the core of the screw dislocation. In tantalum the edge components within the screw dislocation core have been found to spread along three {110} planes. Hence, the τ_{12} on {110} planes is likely to be the non-glide shear stress that affects the CRSS. In order to examine this assertion, we have studied combinations of glide and non-glide stresses chosen such that the situations encountered in tension-compression tests can be reproduced. For instance, for tension-compression tests for which the mrss plane is ($\bar{1}$01), there is zero non-glide shear stress in the ($\bar{1}$01) plane while in the ($\bar{1}$10) and (0$\bar{1}$1) planes there are non-glide shear stresses with opposite signs.

Figure 3 shows the dependence of the CRSS, σ_{23}, on the mrss plane in the direction of the Burgers vector on τ_{12} for such combined applications of glide and non-glide stresses. In this figure we also display the CRSS for various tensile and compressive loadings. These are plotted for those values of τ_{12} that correspond to τ_{12} on the {110} planes for the corresponding tension/compression tests. It is seen that the CRSS in tension-compression tests with various orientations of the tensile/compressive axes fit very well onto the σ_{23}-τ_{12} dependencies found in calculations with combined glide and non-glide stresses.

When the mrss plane is a {112} plane sheared in the twinning sense, the CRSS is almost independent of τ_{12} but the core configuration and the slip plane do vary with τ_{12}. For large negative τ_{12} the slip plane is of the {110} type; close to τ_{12}=0 it is the ($\bar{2}$11) mrss plane and for positive τ_{12} it is the ($\bar{1}\,\bar{1}$2) mrss plane. This agrees with findings in the corresponding tension-compression tests.

When the mrss plane is a {112} plane sheared in the antitwinning sense, the CRSS decreases strongly with increasing τ_{12}. In fact for large negative values of τ_{12} no dislocation motion takes place, in agreement with the corresponding compression test with [001] axis. The rapid fall of the CRSS for large positive τ_{12} also agrees with the corresponding tensile tests. For the [$\bar{1}$11] and [$\bar{1}$22] tensile axes both the slip plane and mode of dislocation motion agree with the calculations for combined glide and non-glide stresses; the slip is along the ($\bar{2}$11) in the antitwinning sense. However, for the [011] tensile axis the dislocation moves along the ($\bar{1}\,\bar{1}$2) in the twinning sense while in the combined study it moves along the ($\bar{2}$11) in the antitwinning sense; at the same time the core changes induced by the loading are very similar in both cases.

When the mrss plane is ($\bar{1}$01) (χ=0°), the CRSS first increases with increasing τ_{12}, reaches a maximum at a small positive τ_{12} and then decreases. At the transition the slip transfers from the ($\bar{1}\,\bar{1}$2) plane in the twinning sense to the ($\bar{1}$01) plane. This again agrees with the corresponding tension tests. Similarly, dislocation core changes found for compression agree with those found for combined glide and non-glide stresses.

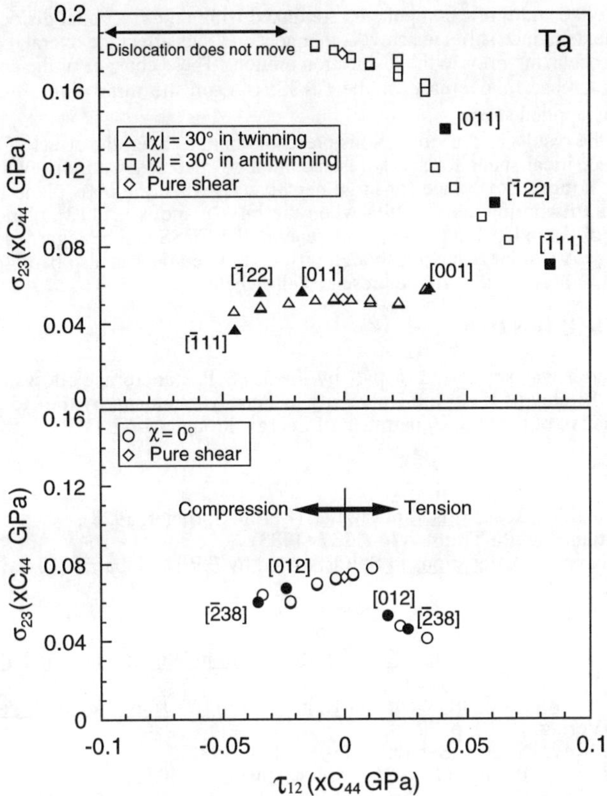

Fig.3. The dependence of the critical shear stress on the mrss plane, σ_{23}, on τ_{12} found in calculations with combined glide and non-glide stresses (open symbols). The filled symbols represent the CRSS obtained in tension-compression tests with various orientations. The negative and positive τ_{12} correspond to the situations encountered in compression and tension tests, respectively.

DISCUSSION

The results of computer simulations presented in this paper illustrate that tantalum exhibits two distinct types of non-Schmid behavior. The first is related to the twinning-antitwinning asymmetry of slip on {112} planes. Based on the simulations of pure shear and tension-compression tests, this asymmetry is such that the CRSS is always lower when the mrss plane is closer to a {112} plane sheared in the twinning sense than to a {112} plane sheared in the antitwinning sense. As mentioned earlier, this is an intrinsic effect related to the symmetry of the bcc lattice [11].

The second is an extrinsic effect of non-glide stress components. The calculations demonstrate that the non-Schmid effect arising from the influence of non-glide stress components is controlled by the shear stress in the direction perpendicular to the total Burgers vector on the {110} planes into which the stress-free core spreads. These shear stress components affect the core structure by exerting a force on the edge components of the fractional dislocations forming the core of a/2<111> screw dislocations. While the non-glide stress components cannot move the dislocation, the edge and screw components are linked within the fractional dislocations and shifts in edge components induce shifts in screw component and thus affect the overall core structure and transformations occurring prior to the dislocation motion. These changes of the core structure may lead to either increase or decrease of the CRSS, σ_{23}, on the mrss plane, depending on the orientation of the applied stress.

Based on the results of the simulations presented in this paper, the effect of the τ_{12} on $\{110\}$ planes upon the critical shear stress, σ_{23}, is such that σ_{23} decreases rapidly with increasing τ_{12}. No dislocation motion takes place for large negative τ_{12} when the mrss plane is a $\{112\}$ plane sheared in the antitwinning sense while when the mrss plane is a $\{112\}$ plane sheared in the twinning sense σ_{23} is independent of τ_{12}. In general, the CRSS can be regarded as a function of the ratio, τ_{12}/σ_{23}. A similar behavior, though different in details, has also been found in the case of molybdenum. These results will be presented elsewhere.

ACKNOWLEDGMENTS

This research was supported in part by the JSPS Postdoctoral Fellowship for Research Abroad during the stay of KI at the University of Pennsylvania and by the Advanced Strategic Computing Initiative of the U.S. Department of Energy through LLNL (VV).

REFERENCES

1. E. Schmid and W. Boas, Kristallplastizität, (Berlin: Springer, 1928).
2. J.W. Christian, Metall. Trans. A**14**, 1237 (1983).
3. M.S. Duesbery, in Dislocations in Solids, edited by F.R.N. Nabarro, (Elsevier: Amsterdam, 1989), p. 67.
4. V. Vitek, Prog. Mater. Sci., **36**, 1 (1992).
5. G. Taylor, Prog. Mater. Sci., **36**, 29 (1992).
6. J.W. Christian, in Proc. 3rd Int. Conf. on Reinstoffe in Wissenschaft und Technik, (Berlin, Akademie-Verlag, 1970), p. 263.
7. B. Sestak, in Proc. 3rd Int. Conf. on Reinstoffe in Wissenschaft und Technik, (Berlin, Akademie-Verlag, 1970), p. 221.
8. L.P. Kubin, Rev. Deform. Behav. Mater., **4**, 181 (1982).
9. J. Dipersio and B. Escaig, Phys. Stat. Sol. (a), **40**, 393 (1977).
10. A. Sakai, Y. Nishioka and H. Suzuki, J. Phys. Soc. Japan, **46**, 881 (1979).
11. M.S. Duesbery and V. Vitek, Acta Mater. **46**, 1481 (1998).
12. B. Escaig, J. Phys. Paris, **29**, 225 (1968).
13. B. Escaig, J. Phys. France, **35**, C7 (1974).
14. M.W. Finnis and J.E. Sinclair, Phil. Mag., A**50**, 45 (1984).
15. G.J. Ackland, and R. Thetford, Phil. Mag. A**56**, 15 (1987).

TEMPERATURE EFFECTS AND FAST-MOVING SCREW DISLOCATIONS AT HIGH STRAIN RATE DEFORMATIONS

A. ROOS, E.D. METSELAAR, J.TH.M. DE HOSSON, H.H.M. CLEVERINGA*, E. VAN DER GIESSEN*
Laboratory of Applied Physics, Materials Science Center, Netherlands Institute of Metals Research, University of Groningen, Nijenborgh 4, 9747 AG Groningen, The Netherlands. e-mail: hossonj@phys.rug.nl.
*Department of Mechanical Engineering, Micromechanics of Materials Group, Netherlands Institute for Metals Research, Delft University of Technology, Mekelweg 2, 2628 CD Delft, The Netherlands.

ABSTRACT

In this paper, shear deformation at high strain rates is modeled within the framework of discrete dislocation plasticity. The method of discrete dislocation plasticity is extended to incorporate the temperature rise induced by moving dislocations. Also, the stress and displacement fields of a screw dislocation on inclined planes in a periodic structure are developed. The influence on the temperature rise on various micro-mechanical processes is discussed.

INTRODUCTION

Plastic deformation at high strain rates, e.g. in the case of perforation, may lead to strain localization in shear bands. The strain rate is accommodated by generating a large density of dislocations, and by high velocities of the dislocations themselves. At present, detailed investigations of the generation of local stress peaks on the length scale of dislocations are scarcely available.

During the deformation process mechanical energy is converted into heat. The heat remains localized due to the small time lapse of the process and the time for heat conduction decreases upon increasing strain rate. As a matter of course the actual thermal diffusion length depends also on the thermal diffusivity, e.g. the thermal conductivity and thermal diffusivity of Cu is two orders of magnitude larger than of Ti. Consequently, compared to Cu the thermal diffusion length in Ti is one order of magnitude smaller. It has been observed experimentally that the temperature rise may reach values as high as the melting temperature of Titanium (~1660 °C)[2]. The temperature rise may trigger thermal softening and shear instability, which will occur at smaller strain with increasing strain rate. Depending on the thermo-physical properties like the thermal diffusion length the flow stress may decrease rapidly with respect to neighboring regions, i.e. localizing the deformation even more.

This work extends a computer simulation methodology based on the framework of discrete dislocation plasticity [3,4,5]. The extension is twofold. Firstly, stress and displacement fields of screw dislocations moving on inclined slip planes are developed. Secondly, the motion of strings of dislocations moving on horizontal slip planes is directly coupled to the temperature field.

COMPUTATIONAL METHOD

1. Computational cell, discrete dislocation plasticity and the screw dislocation

The commonly used stress and displacement fields in the literature [6,7] describe dislocations in an infinite elastic body. In the case of a finite body or sample, these fields will not comply with the boundary conditions (in terms of prescribed displacements or tractions as a function of time).

Mat. Res. Soc. Symp. Proc. Vol. 538 © 1999 Materials Research Society

Therefore, a correction is necessary, which traditionally is interpreted in terms of "image dislocations", but which [3] proposed to handle by a finite element procedure. This leads to a decomposition of the problem in two fields. Referring to [3,4,5] for details, the final fields of stress and displacement are the sum of the dislocation fields ($\sim$) and the correction fields ($\wedge$):

$$u_\alpha = \tilde{u}_\alpha + \hat{u}_\alpha, \quad \tilde{u}_\alpha = \sum_i \tilde{u}_\alpha^i; \quad \sigma_{\alpha\beta} = \tilde{\sigma}_{\alpha\beta} + \hat{\sigma}_{\alpha\beta}, \quad \tilde{\sigma}_{\alpha\beta} = \sum_i \tilde{\sigma}_{\alpha\beta}^i \quad \text{with } \alpha, \beta \in \{1,2\}. \tag{1}$$

Throughout this paper the calculations are performed in a Cartesian (x_1, x_2, x_3) laboratory reference frame. A two-dimensional unit cell which is periodic in the x_1-direction and of width×height = $2w \times 2h$ is used. The shear is applied at the top and bottom surfaces through the kinematic boundary conditions $u_{0,1}(t) = \pm h \dot{\gamma} t$ and $u_{0,2}(t) = 0$ along $x_2 = \pm h$, where $\dot{\gamma}$ denotes the strain rate. The overall stress needed to sustain the shear strain rate is the average shear stress $\sigma_{12}(x_1, \pm h)$ over the top and bottom surfaces. The solution $\hat{\sigma}_{\alpha\beta}$ of the linear elastic boundary value problem is found using a finite-element technique with quadrilateral elements. The calculation of the total dislocation stress $\tilde{\sigma}_{\alpha\beta}$ has to take into account not only the stress fields of all dislocations i, but also of all their replicas n in the periodic boxes. Previously, the case of a fast-moving edge dislocation moving in the x_1-direction, the (straight) dislocation line aligned parallel to the x_3-axis and its Burgers vector of magnitude b_i in the positive or negative x_1-direction was treated [5]. In this paper we consider a straight screw dislocation i moving at a constant velocity v_i in the (x_2, x_3)-plane, its dislocation line and its Burgers vector b_i aligned parallel to the x_1-axis. The slip plane is inclined to the horizontal (x_1, x_3)-plane at an angle θ_i. A second coordinate system $(u_i, v_i) \equiv (x_3 \cdot \cos(\theta_i) + x_2 \cdot \sin(\theta_i), -x_3 \cdot \sin(\theta_i) + x_2 \cdot \cos(\theta_i))$ can be identified, which rotates the x_2 and x_3 axes around the x_1-axis by the angle θ_i.

The coordinates of dislocation i at time t are denoted by $(x_2^i(t), x_3^i(t))$. When very high strain-rates are applied, the dislocations achieve velocities for which the stress and displacement fields become velocity-dependent, as discussed in [6,7], for instance. For screw dislocations, the relevant limiting velocity is the shear wave velocity a_2, defined by $a_2^2 = \mu / \rho$ (with μ the shear modulus and ρ the material density). Analytical summation over all replicas (each at a distance $2w$ apart) in the x_3-direction yields the displacement field of the string of dislocations consisting of the current dislocation i and all its replicas on the slip plane. The following definitions will be used below:

$$\beta_{2,i}^2 = 1 - v_i^2 / a_2^2; \quad \Delta x_\alpha^i = x_\alpha - x_\alpha^i(t), \ \Delta \xi_\alpha^i = \Delta x_\alpha^i / w \text{ with } \alpha \in \{2, 3\}$$

$$\Delta \xi_2^{\prime i} = \frac{\Delta \xi_2^i}{(1 + \beta_{2,i}^2) + (1 - \beta_{2,i}^2)\cos(2\theta_i)}; \quad \delta^i(x) = \begin{cases} -1 \\ +1 \end{cases} \text{if} \quad \begin{matrix} x \in (-3,-1) \\ x \in (+1,+3) \end{matrix}; \tag{2}$$

$$\Gamma^i = k \text{ if } \left(1 - \beta_{2,i}^2\right)\sin(2\theta_i)\Delta \xi_2^{\prime i} - 2\arctan\left(\frac{\tanh\left(\pi \beta_{2,i} \sin(2\theta_i)\Delta \xi_2^{\prime i}\right)}{\tan(\theta_i)}\right) < -\pi \Delta \xi_3^i + 2k\pi < -\frac{\pi \Delta \xi_2^i}{\tan(\theta_i)}$$

Noting that $\tilde{u}_2^i = \tilde{u}_3^i = 0$ for a screw dislocation, the high-velocity displacement fields $\tilde{u}_1^i$ seen at a distance $(\Delta x_2^i, \Delta x_3^i)$ of dislocation i are given by (following the procedure as outlined in [3] and using the dislocation fields as given in [6])

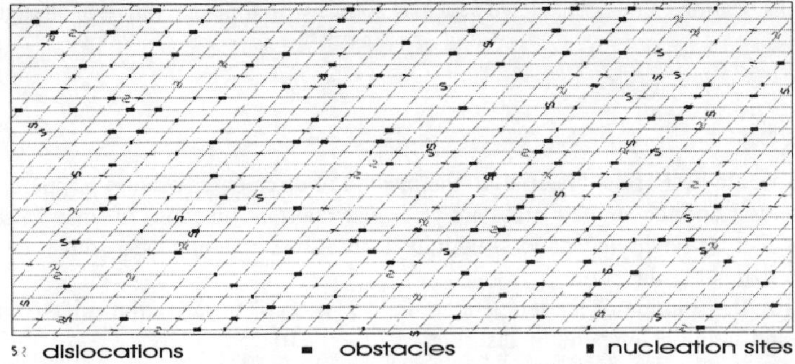

Figure 1: *Typical configuration during a simulation with* $\dot{\gamma} = 10^4\ s^{-1}$. *The computational cell has* $2w\times2h=2\times2\ \mu m$ *and material parameters are taken from Al.*

$$\tilde{u}_1^i\left(\Delta\xi_2^i,\Delta\xi_3^i,\theta_i\right)= \frac{b_i}{2\pi}\arctan\left\{\frac{\tan\!\left(\frac{\pi}{2}\Delta\xi_3^i-\frac{\pi}{2}\left(1-\beta_{2,i}^2\right)\sin(2\theta_i)\Delta\xi_2^{\prime i}\right)+\tanh\!\left(\pi\beta_{2,i}\Delta\xi_2^{\prime i}\right)}{\tanh\!\left(\pi\beta_{2,i}\Delta\xi_2^{\prime i}\right)-\tan\!\left(\frac{\pi}{2}\Delta\xi_3^i-\frac{\pi}{2}\left(1-\beta_{2,i}^2\right)\sin(2\theta_i)\Delta\xi_2^{\prime i}\right)\tan(\theta_i)}\right\}$$
$$-\frac{b_i}{2}\delta^i\left(\tfrac{\pi}{2}\Delta\xi_3^i-\tfrac{\pi}{2}\left(1-\beta_{2,i}^2\right)\Delta\xi_2^{\prime i}\right)-b_i\Gamma^i\left(\Delta\xi_3^i,\Delta\xi_2^i\right) \tag{3}$$

The shear stress component along the u_i-axis is given by

$$\tilde{\sigma}_{xu}^i\left(\Delta\xi_2^i,\Delta\xi_3^i,\theta_i\right)= \frac{2\mu b_i}{w}\frac{\cos^2(\theta_i)}{3+4\cos(2\theta_i)+\cos(4\theta_i)+2\beta_{2,i}^2\sin^2(2\theta_i)}\times$$
$$\left\{\frac{\beta_{2,i}\sin(\theta_i)\sin\!\left(\pi\Delta\xi_3^i+\pi\left(1-\beta_{2,i}^2\right)\sin(2\theta_i)\Delta\xi_2^{\prime i}\right)+\cos(\theta_i)\sinh\!\left(2\pi\beta_{2,i}\Delta\xi_2^{\prime i}\right)}{\cos\!\left(\pi\Delta\xi_3^i+\pi\left(1-\beta_{2,i}^2\right)\sin(2\theta_i)\Delta\xi_2^{\prime i}\right)-\cosh\!\left(2\pi\beta_{2,i}\Delta\xi_2^{\prime i}\right)}\right\} \tag{4}$$

whereas the shear stress component along the v_i-axis reads

$$\tilde{\sigma}_{xv}^i\left(\Delta\xi_2^i,\Delta\xi_3^i,\theta_i\right)= \frac{2\mu b_i\beta_{2,i}}{w}\frac{\cos^2(\theta_i)}{3+4\cos(2\theta_i)+\cos(4\theta_i)+2\beta_{2,i}^2\sin^2(2\theta_i)}\times$$
$$\left\{\frac{\cos(\theta_i)\sin\!\left(\pi\Delta\xi_3^i+\pi\left(1-\beta_{2,i}^2\right)\sin(2\theta_i)\Delta\xi_2^{\prime i}\right)+\beta_{2,i}\sin(\theta_i)\sinh\!\left(2\pi\beta_{2,i}\Delta\xi_2^{\prime i}\right)}{\cos\!\left(\pi\Delta\xi_3^i+\pi\left(1-\beta_{2,i}^2\right)\sin(2\theta_i)\Delta\xi_2^{\prime i}\right)-\cosh\!\left(2\pi\beta_{2,i}\Delta\xi_2^{\prime i}\right)}\right\} \tag{5}$$

A typical configuration during a simulation is shown in figure 1.

2. Dislocations as heat sources and temperature field

Consider the dislocation i with Burgers vector of length b_i to be a line source of heat moving at velocity v_i due to an applied resolved shear stress $\sigma_{12}^{Applied}$. In general it is assumed that a fraction $\gamma \sim 0.8$ of the plastic work done on a dislocation is converted into heat [8]. Then, each second a quantity of heat per unit line length is emitted at a rate of [9]

$$\dot{q}_i = \gamma b_i \sigma_{12}^{Applied} v_i \quad W\,m^{-1}, \tag{6}$$

so that the dislocation can be considered as a heat source with time derivative of strength

$$\dot{Q}_i = \frac{\gamma b_i \sigma_{12}^{Applied} v_i}{\rho c} \quad K\, m^2\, s^{-1}.$$ (7)

with c the specific heat of the material. In the method of discrete dislocation plasticity the dislocation velocity v_i and the applied stress are piecewise constant functions of time. In the following, we consider a straight dislocation i with its dislocation line along the x_3-direction and moving in the horizontal x_1-direction

From [10], the solution of the temperature field $T(x_1, x_2, t)$ for an initial surface temperature $T_s(x_1, x_2, t)$ and an initial temperature distribution $T_i(x_1, x_2, 0)$

$$T(x_1,x_2,t)= \iint u_{t'=0}\, T_i(x_1',x_2',0)\,dx_1'dx_2' + \int_0^{t-\varepsilon}\left[-\kappa\int T_s(x_1',x_2',t')\frac{\partial u}{\partial n_i}ds + \iint \dot{Q}u\,dx_1'dx_2'\right]dt',$$ (8)

where $u(x_1, x_1', x_2, x_2', t, t')$ denotes the temperature at (x_1, x_2) at the time t due to an instantaneous line source of strength unity generated at (x_1', x_2') at the time t', the solid being initially at zero temperature, and the surface being kept at zero temperature.

Since the shear banding takes place in an adiabatic fashion, there is no heat flux across the upper and lower surface of the slab, so the temperature at these surfaces is taken to be constant (in this case: zero) throughout the simulation. In the present approach, there are no left and right surfaces, since the periodicity in the horizontal direction is already taken into account by the string of replicas of each dislocation. This can be expressed as $T_s(x_1, x_2, t) = 0$, where the s denotes the top and bottom surfaces only. This leaves the surface integration equal to zero, and we are left with

$$T(x_1,x_2,t)= \iint u_{t'=0}\, T_i(x_1',x_2',0)\,dx_1'dx_2' + \int_0^{t-\varepsilon}\iint \dot{Q}\,u\,dx_1'dx_2'dt'.$$ (9)

During the simulations, the time increases in discrete time steps. The first term accounts for the further evolution of the temperature distribution generated in previous time steps whereas the second term takes into account the heat being generated in the current time step. From this point of view, $t = 0$ s denotes the start of a discrete time step. The second term is developed below.

Consider the region bounded by two parallel planes $x_2 = -h$ and $x_2 = h$ with an *instantaneous* line source of unit strength at (x_1^i, x_2^i) at time $t = 0$. If the surface planes are kept at zero temperature, the temperature distribution for $t > 0$ is given by [11]

$$u(x_1,x_1^i,x_2,x_2^i,t,0)= \sum_{m=1}^{\infty} \frac{e^{-\frac{(\Delta x_1^i)^2}{4\kappa t}}\, e^{-\frac{\kappa m^2\pi^2 t}{4h^2}}}{h\sqrt{4\pi\kappa t}}\sin\left[\frac{m\pi}{2}\left(\frac{x_2}{h}+1\right)\right]\sin\left[\frac{m\pi}{2}\left(\frac{x_2^i}{h}+1\right)\right].$$ (10)

The computational cell contains dislocations $i = 1\ldots k$ (k integer) at $(x_1^i(t), x_2^i(t))$, moving at a velocity $v_i(t)$ in the positive x_1-direction. In the computations, the dislocation velocities are constant during each time step. Defining the start of the time step as $t = 0$, the dislocation position can be written as $(x_1^i(0)+v_i\cdot t, x_2^i(0)) \equiv (x_1^i+v_i\cdot t, x_2^i)$. Dislocation i represents a *continuous* heat source generating a quantity of $\dot{q}_i$ units of heat per second. Each dislocation has associated with it an infinite string of replicas n, located at $(x_1^i + 2nw + v_i t, x_2^i)$ (n integer). Following [11], consider the quantity $\dot{q}_i\, dt'$ liberated at the infinitesimal time interval dt' at t'. The temperature at t-ε $\geq t' \geq 0$ due to the heat emitted from 0 to t is (the second term on the right-hand side of eq. 9)

$$T(x_1,x_2,t) = \int_0^{t-\varepsilon} dt' \sum_{i=1}^{k} \sum_{n=-\infty}^{\infty} \sum_{m=1}^{\infty} \frac{\dot{q}_i\, e^{-\frac{\left(\Delta x_1^i - 2nw - v_i t'\right)^2}{4\kappa(t-t')}}\, e^{-\frac{\kappa m^2 \pi^2 (t-t')}{4h^2}}}{\rho c h \sqrt{4\pi\kappa(t-t')}} \sin\left[\frac{m\pi}{2}\left(\frac{x_2}{h}+1\right)\right]\sin\left[\frac{m\pi}{2}\left(\frac{x_2^i}{h}+1\right)\right]. \quad (11)$$

Due to v_i and $\dot{q}_i$ being constant during the time step, the integration can be taken inside the summation. Performing the integration and taking the limit $\varepsilon \to 0$ yields

$$T(x_1,x_2,t) = \sum_{i=1}^{k} \sum_{n=-\infty}^{\infty} \sum_{m=1}^{\infty} \frac{\gamma\left|b_i \sigma_{12}^{Applied}\right|\sin\left[\frac{m\pi}{2}\left(\frac{x_2}{h}+1\right)\right]\sin\left[\frac{m\pi}{2}\left(\frac{x_2^i}{h}+1\right)\right]}{2h\rho c\sqrt{1+\left(\pi\kappa m/h v_i\right)^2}}\, e^{-v_i \frac{\Delta x_1^i - 2nw}{2\kappa}} \times$$

$$\times \left\{ e^{-|v_i|\frac{\left|\Delta x_1^i - 2nw\right|\sqrt{1+(\pi\kappa m/h v_i)^2}}{2\kappa}}\, \mathrm{erfc}\left(\frac{\left|\Delta x_1^i - 2nw\right| - \sqrt{1+\left(\pi\kappa m/h v_i\right)^2}\,|v_i|t}{\sqrt{4\kappa t}}\right)\right.$$

$$\left. - e^{|v_i|\frac{\left|\Delta x_1^i - 2nw\right|\sqrt{1+(\pi\kappa m/h v_i)^2}}{2\kappa}}\, \mathrm{erfc}\left(\frac{\left|\Delta x_1^i - 2nw\right| + \sqrt{1+\left(\pi\kappa m/h v_i\right)^2}\,|v_i|t}{\sqrt{4\kappa t}}\right)\right\}. \quad (12)$$

An example of temperature fields in Cu and Ti is shown in figure 2.

DISCUSSION

In this paper, the method of discrete dislocation plasticity has been extended in two ways: the stress and displacement fields of an infinite string of screw dislocations moving on inclined slip planes have been developed, as well as the temperature field generated by an infinite string of dislocations moving on a horizontal plane. In literature, there have been discussions about the temperature rise cause by moving dislocations ([9] and references therein). The temperature field used in [9] is valid only for a steady state at $t \to \infty$. In the present case, the loading conditions are

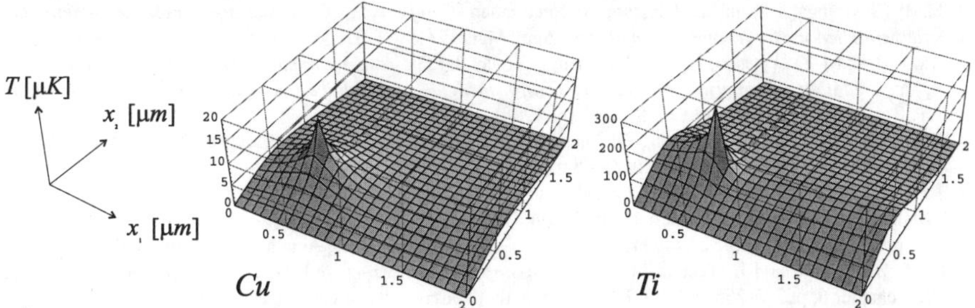

Figure 2: *Temperature rise (μK) generated by an edge dislocation moving during 10^{-7} s at a constant velocity in the x_1-direction in a periodic cell of $2h \times 2w = 2 \times 2$ μm. Both dislocations are driven by an applied shear stress of $\sigma_{12}^{Applied}/\mu_{Cu} = 10^{-4}$. These pictures are calculated using eq. (12) with $-8 \leq n \leq 8$ and $m=1..25$.*

such that it is no longer possible to consider a steady state. Inserting the temperature field of equation (12) into equation (9), it is now possible to directly study the temperature rise of dislocations moving at a piecewise constant velocity and at a finite time.

The temperature rise and the possibly associated localization of heat may effect the dislocation dynamics in various ways. At low strain rate and low stress values the dislocation motion is thermally activated and the effective obstacle strength may be lowered substantially. At high strain rates and high stress values the phonon drag will be affected, whereas the viscosity increases even further because of relativistic effects. In future work we will include these aspects on the cross slip events so as to study the influence of a temperature rise on the dislocation dynamics when going from an obstacle controlled plasticity regime into a fully phonon drag controlled plasticity regime.

It should be realized that the above treatment applies to plasticity in a single crystal rather than in polycrystalline materials. Nevertheless, dislocation pattern formation in the three stages of work hardening can be studied in more detail. In particular the formation of dislocation grids (sheets) in stage II and thermally activated rearrangements of dislocations in stage III can be investigated. This is in particular relevant when realizing that correlations between the flow stress and the average dislocation density may not be physically significant, since the flow stress will most likely be determined by local interactions between non-uniformly distributed dislocations [12]. Consequently, the modeling approach presented here is a more appropriate method than trying to interpret the statistical effects of the various potential contributions to work hardening.

ACKNOWLEDGEMENTS

The work described in this paper is supported by IOP-Metals under project number C94.703.RG.TF, the Netherlands Institute for Metals Research under project number MS97006 and the "Stichting voor Fundamenteel Onderzoek der Materie (FOM)" which is financially supported by the "Nederlandse Organisatie voor Wetenschappelijk Onderzoek (NWO)".

REFERENCES

1. For instance in M.A. Meyers, L.E. Murr, K.P. Staudhammer (editors), *Shock-wave and high-strain-rate phenomena in materials (Explomet '90)*, Marcel Dekker New York (1992)
2. W.H. Holt, W. Mock Jr., W.G. Soper, C.S. Coffey, *Reverse-ballistic impact study of shear plug formation and displacement in Ti_6Al_4V alloy*, J. Appl. Phys. **73** (1993), 3753.
3. E. van der Giessen, A. Needleman, *Discrete dislocation plasticity: a simple planar model*, Modelling Simul. Mater. Sci. Eng. **3** (1995), 689-735.
4. H.H.M. Cleveringa, E. van der Giessen, A. Needleman, *Comparison of discrete dislocation and continuum plasticity predictions for a composite material*, Acta. Mat. **45** (1997), 3163.
5. A. Roos, J.Th.M. De Hosson, H.H.M. Cleveringa, E. van der Giessen, *A computer simulation of metal perforation*, in *Simulation of materials processing: theory, methods and applications (Numiform '98)* (J. Huétink, F.P.T. Baaijens eds.), A.A. Balkema Rotterdam (1998), 303.
6. J. Weertman, J.R. Weertman, *Moving Dislocations*, in *Dislocations in Solids* (ed. F.R.N. Nabarro), North Holland Publ. Comp., (1980) volume 3, chapter 8, pp. 1-59.
7. J.P. Hirth, J. Lothe, *Theory of Dislocations* (1st ed.), Wiley New York (1968), chapter 7.
8. J.J. Manson, A.J. Rosakis, G. Ravichandran, Mech. Of Materials, **17**, 135 (1994).
9. J.D. Eshelby, P.L. Pratt, *Note on the heating effect of moving dislocations*, Acta Met. **4** (1956), 560.
10. J.Th.M. De Hosson and E. Teeuw, in : *Lasers in surface engineering*, (ed. N. Dahotre), ASM International, (1998), chapter 6, pp205-255 and A.J. Kloosterman, thesis, university of Groningen, 1998.
11. H.S. Carslaw, J.C. Jaeger, *Conduction of heat in solids*, Oxford (1959).
12. J.Th.M. De Hosson, O. Kanert, A.W. Sleeswijk, *Dislocations in solids investigated by nuclear magnetic resonance*, in *Dislocations in Solids* (ed. F.R.N. Nabarro), North Holland Publ. Comp., (1983) volume 6, chapter 32, pp 441-553.

THREE-DIMENSIONAL NUMERICAL SIMULATION OF INTERACTING DISLOCATIONS IN A STRAINED EPITAXIAL SURFACE LAYER

R. V. KUKTA[†] and L. B. FREUND[‡]

[†]Div. of Engineering and Appl. Science, California Inst. of Tech., Pasadena, CA 91125
[‡]Division of Engineering, Brown University, Providence, RI 02912

ABSTRACT

In the context of elastic dislocation theory, the glide of arbitrarily shaped dislocations near a free surface is simulated numerically. The simulation is made possible by utilizing a compact line-integral representation of the elastic field produced by a piecewise planar dislocation loop near a free surface. The interaction between a threading dislocation and a misfit dislocation in its path is simulated and the blocking effect is shown to be much more significant than found in simulations which neglect free surface effects.

INTRODUCTION

A few methods have been developed for evaluating the elastic field of a curved disloca-tion near a free surface or a bi-material interface. All of these methods are computationally intensive and most are impractical for all but the simplest problems. Until very recently, all of the methods required the evaluation of a surface integral. Gosling and Willis[1] obtained a line-integral expression for evaluating the elastic fields of a dislocation near a free surface. Although this was a substantial improvement over the surface integral methods, it is still computationally intensive and requires a large effort to implement. Here, a much simpler line-integral is introduced which can in principle be used for any bounded solid provided the elastic Green's function is known. The result has been used to simulate dislocation glide near a free surface, as well as near the interface of an elastic solid and a viscoelastic fluid.[2]

A numerical scheme for simulating the glide of dislocation loops in piecewise planar configurations is outlined. The scheme is used to the model a fundamental interaction between dislocations in strained epitaxial surface layers, that is, the blocking of a threading dislocation by an interface misfit dislocation in its path. The blocking effect is studied in detail for the case in which the misfit dislocation is held fixed. The case for which both dislocations are permitted to glide freely is also addressed.

A KINETIC RELATION FOR DISLOCATION GLIDE

To simulate dislocation glide, a relationship between the force on a dislocation and its velocity is needed. For simplicity, the following linear kinetic relation is chosen:

$$\mathbf{v} = M\mathbf{f} \tag{1}$$

where $\mathbf{v}$ is the local glide velocity, $\mathbf{f}$ is the normal glide force on the dislocation, and M is mobility. Due to the presence of an applied stress $\boldsymbol{\sigma}$, the force is the Peach-Koehler force $\mathbf{f} = -(\mathbf{b} \cdot \boldsymbol{\sigma}\mathbf{n})\mathbf{s}$ where $\mathbf{b}$ is the Burgers vector, $\mathbf{n}$ is the glide plane unit normal, and $\mathbf{s}$ is the local unit normal to the dislocation line in the glide plane. Here the slipped region is assumed to lie in the $-\mathbf{s}$-direction and the Burgers vector is defined as $\mathbf{b} = \mathbf{u}^- - \mathbf{u}^+$

Mat. Res. Soc. Symp. Proc. Vol. 538 © 1999 Materials Research Society

where $\mathbf{u}^-$ and $\mathbf{u}^+$ are respectively the displacements as a point on the slipped surface is approached from the $-\mathbf{n}$ and $+\mathbf{n}$ directions.

The self force, that is, the force resulting from the stress induced by the dislocation itself, is generally unbounded in the linear theory unless some regularization is adopted. Several regularizations exist in the literature, including the Periels-Nabarro model and core cut-off models. The simpliest regularization is that of Brown[3] which suggests using the stress $\sigma^* = [\sigma(+\rho\mathbf{s}) + \sigma(-\rho\mathbf{s})]/2$ in the Peach-Koehler fomula; the singularity is removed by averaging the stress taken at two points removed a distance ρ from the dislocation line. Although this regularization is used here, and commonly in numerical simulations of dislocations , a more suitable regularization is perhaps that of Gavazza and Barnett.[4]

DISCRETIZATION OF THE KINETIC RELATION

The dislocation line L is discretized such that its motion is described by nodal velocities. The velocity of a node A in the i direction is V_{iA}. Pointwise along L, the velocity is

$$v_i(l) = \sum_{A=1}^{n} V_{iA} N_A(l) \tag{2}$$

where n is the number of nodes, $N_A(l)$ is the shape function at node A, and l is arclength along the dislocation line. The kinetic relation (1) is cast into weak form by taking its inner product with respect to a virtual velocity $\delta\mathbf{v}(l)$ which has the same admissible form as $\mathbf{v}$. By argueing that the weak form holds for all admissible $\delta\mathbf{v}(l)$, the following discrete kinetic relation is obtained:

$$F_{iB} = \sum_{A=1}^{n} V_{iA} K_{AB} \tag{3}$$

where

$$F_{iB} = \oint_L f_i(l) N_B(l)\, dl \quad \text{and} \quad K_{AB} = M^{-1} \oint_L N_A(l) N_B(l)\, dl. \tag{4}$$

Here, F_{iB} is the nodal driving force on node B in the i-direction and K_{AB} is the global velocity "stiffness" matrix. Assuming that the force $f_i(l)$ can be determined, the motion of the dislocation is determined by time stepping according to the velocities V_{iA} which are obtained by inverting (3). In most cases, adaptive repositioning of the nodal positions is needed to maintain accuracy in time.

A SIMPLE LINE-INTEGRAL FOR THE ELASTIC FIELD

A line-integral representation of the displacement due to a dislocation loop can be obtained quite easily for a dislocation which is assumed to lie in a piecewise planar configuration. Such a configuration can be constructed through superposition of planar loops, hence only the planar case is addressed. The displacement field due to a planar dislocation in an isotropic elastic solid is given by[5]

$$u_m(\mathbf{z}) = -\mu b_k n_l \int_A \frac{\partial}{\partial x_l} G_{km}(\mathbf{x}, \mathbf{z})\, da - \mu n_k \int_A b_l \frac{\partial}{\partial x_l} G_{km}(\mathbf{x}, \mathbf{z})\, da \tag{5}$$

where A is the slipped surface and μ is shear modulus. The summation over repeated indices is implied. The elastic Green's function $G_{km}(\mathbf{x}, \mathbf{z})$ is the k-displacement at the point $\mathbf{x}$ due to a unit point force at $\mathbf{z}$ in the m-direction. Clearly the second term in (5)

Figure 1. Illustration of slip plane coordinates x_2' and x_3' used in obtaining a line integral expression for the displacement due to a dislocation loop L.

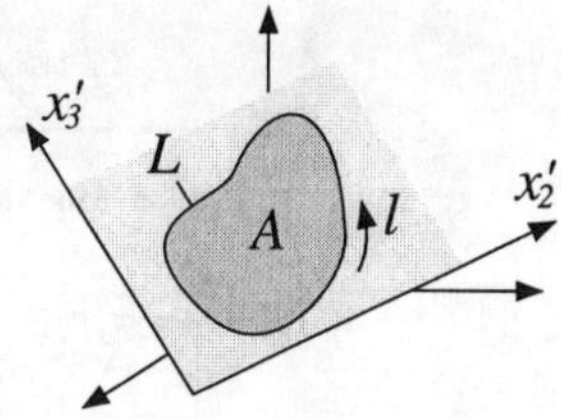

can be partially integrated since $b_l \partial/\partial x_l$ is a directional derivative in the plane on the loop. By introducing the Green's function in terms of the Papkovitch-Neuber potentials φ_{ij} and φ_{0j}, that is, as

$$G_{ij} = \varphi_{ij} - \frac{1}{4(1-\nu)} \frac{\partial}{\partial x_i} \left(r_k \varphi_{kj} + \varphi_{0j} \right) \tag{6}$$

where $r_k = x_k - z_k$, a second in-plane directional derivative can be identified and a similar partial integration can be done. The result is

$$u_m(\mathbf{z}) = -\mu b_k n_l \int_A \frac{\partial \varphi_{km}}{\partial x_l} \, da - \mu n_k \oint_L b_i s_i H_{km} \, dl \tag{7}$$

where $H_{km} = 2G_{km} - \varphi_{km}$. No simple manipulations can be made on the first term in (7) to transform it into a line integral, but, for relatively simple cases the term can be partially integrated using brute force. With respect to the orthogonal slip plane coordinates shown in Figure 1, the result is

$$u_m(\mathbf{z}) = -\mu b_k n_l \oint_L \frac{\partial L_{km}}{\partial x_l} \, dx_3' - \mu n_k \oint_L b_i s_i H_{km} \, dl \tag{8}$$

where

$$L_{km} = \int_y^{x_2'} \varphi_{km} \, dx_2'. \tag{9}$$

Any convenient value can be assigned to the lower limit y in (9). In the case of an elastic half space or two joined elastic half spaces, the potential φ_{km} can be integrated with little difficulty. Furthermore, for these cases, x_2' can be chosen to be parallel to the interface in order to exploit the rotational symmetry in the Green's function. The Green's function in terms of the Papkovitch-Neuber potentials was found by Mindlin[6] for an elastic half space and by Rongved[7] for two joined elastic half spaces.

The stress due to the dislocation loop can be calculated from (8) using the isotropic stress-strain relationship with strain $\boldsymbol{\epsilon} = \text{sym}(\mathbf{u})$. The interaction force between loops is calculated using the Peach-Koehler formula.

CONSIDERATION OF A LOOP EMERGING AT A FREE SURFACE

Special consideration is required for a dislocation loop which terminates at a free surface. The glide force per unit length along a straight dislocation which intersects a free surface is[8]

$$f(d) = \frac{1}{d} \left[E(\beta) \cot\theta - \frac{\partial E(\beta)}{\partial \theta} \right] \tag{10}$$

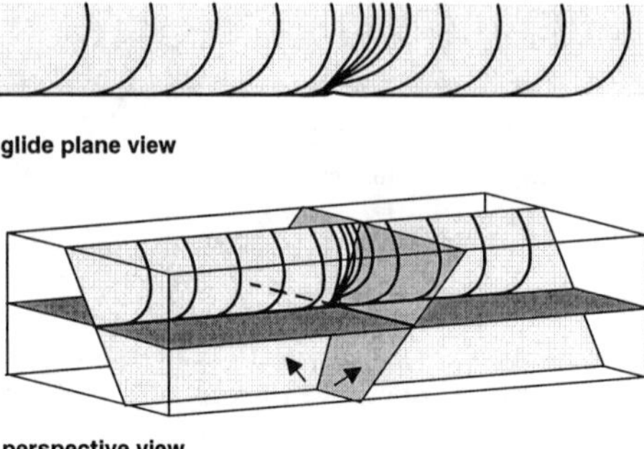

Figure 2. The calculated evolution of a threading dislocation as it passes (from left to right) over an interface misfit dislocation in the strained surface layer. Each line represents the dislocation line at successive time intervals of equal size. The interface misfit dislocation is held fixed. The film thickness is $h/b = 160$ and the mismatch strain is $\epsilon_0 = 0.35\%$. Arrows denote the directions of the Burgers vectors.

where θ is the angle between the dislocation line and the surface, d is the distance along the dislocation line as measured from the surface, $E(\beta)$ is the pre-logarithmic energy factor, and β is the angle between the dislocation line and the Burgers vector. It is clear that the force is generally singular at the surface. Further consideration of (10) reveals that for a given Burgers vector there exists at least one orientation angle for which the force is non-singular at the interface and a single angle θ_{eq} exists at which the dislocation is in stable equilibrium. Hence, (10) provides the angle with which the dislocation must approach the surface according to the elastic theory. Unless some regularization is invoked this boundary condition must be imposed. This is done in the numerical simulation.

INTERACTION OF A THREADING DISLOCATION WITH A MISFIT SEGMENT

A crystalline film deposited epitaxially onto a lattice mismatched substrate will be strained an amount ϵ_0 which is dependent upon the degree of mismatch. This mismatch strain provides the driving force for the nucleation and propagation of dislocations in the film. It is well known that a film of thickness larger than some critical value h_{cr} is unstable with respect to the formation of misfit dislocations which lie along the film/substrate interface. A mechanism by which misfit dislocations form is through the propagation of threading dislocations through the film. A threading dislocation is a segment which crosses the thickness of the film as it connects a misfit dislocation to the free surface. If the film is thicker than h_{cr}, a threading segment will propagate in a forward manner, laying down a misfit segment of increasing length along the interface. Otherwise such a dislocation is not stable and, if introduced, would would recede thereby drawing the misfit segment out of the system. The numerical scheme introduced earlier was tested against an analytic solution for the critical thickness,[9] and agreement was found to be excellent.

Interactions between threading dislocations and misfit dislocations become inevitable as the misfit dislocation density is increased. One such interaction is a threading dislocation traveling towards a misfit dislocation on an intersecting glide plane. Using analytical techniques, this interaction was approximated by Freund[10] who demonstrated that for a given strain ϵ_0, there exists a definite film thickness above which the threading dislocation

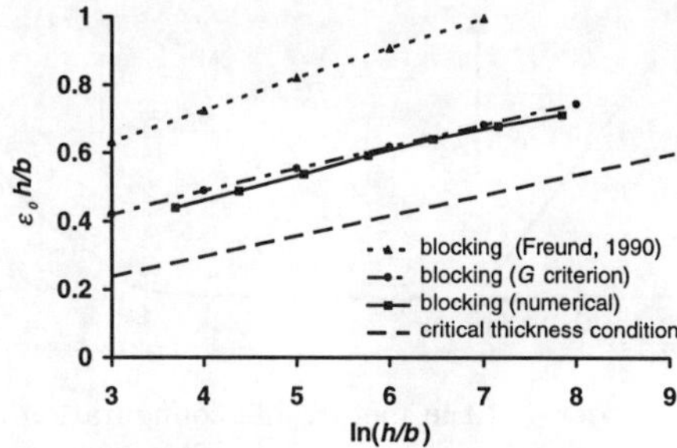

Figure 3. Plot of the numerically determined blocking criterion for the case outlined in Figure 2 as compared with two blocking estimates and the critical thickness condition. Points below each blocking line represent the combination of ϵ_0 and h for which a threading dislocation is predicted to be blocked. Points below (above) the critical thickness line represent systems below (above) the critical thickness.

will pass over the misfit segment and below which it is blocked. The force exerted by the misfit dislocation on the threading dislocation opposes the effect of mismatch stress. This interaction was treated numerically for the case of an embedded strained layer by Schwarz and Tersoff.[11] They found that in the absence of free surface effects, the blocking effect is not as substantial as Freund's model suggests and, in fact, in films only slightly thicker than h_{cr} a threading dislocation will pass over an intersecting misfit dislocation. Here, the numerical simulation is used to evaluate the blocking effect in a film having a free surface.

Plotted in Figure 2 is a sample calculation in which the threading dislocation is not blocked. The bold lines plot the evolution of the threading dislocation at equal time intervals and the shaded planes denote the two glide planes and the film/substrate interface. The film surface is a (001) plane and the Burgers vectors of the threading dislocation and the misfit dislocation lie along $[\bar{1}01]$ and $[101]$, respectively, each with length $b = 3.84$Å. The glide plane of the threading dislocation has the normal $[1\bar{1}1]$. In this case the misfit dislocation was held rigidly in a fixed position. The threading dislocation is approximated by piecewise linear segments. The elastic constants of the film and the substrate were taken to be equivalent with $\mu = 43$ GPa and $\nu = 1/4$. The film (thickness $h = 160b$) is biaxially strained an amount $\epsilon_0 = 0.35\%$ due to lattice mismatch and hence the membrane stress in the plane of the interface is $\sigma_{11}^0 = \sigma_{22}^0 = 2\mu\epsilon_0(1 + \nu)/(1 - \nu)$ due to lattice mismatch.

Although the dislocation is not blocked, it clearly hesitates while passing over the misfit segment. The speed of the dislocation decreases by as much as 90%. In a film having the slightly smaller thicknesses of $h = 150b$, the threading dislocation was found to be blocked by the misfit dislocation. This is compared with the critical thickness $h_{cr} \approx 100b$ for the same system. The blocking effect in this case is substantial.

For a mismatch strain of $\epsilon_0 = 0.35\%$, the critical blocking thickness lies somewhere between $h/b = 150$ and $h/b = 160$. Determining the blocking criterion by iteratively running the simulation is a tedious process of narrowing in on the highest value of mismatch strain at which the dislocation is blocked for a given film thickness. This was done for a range of film thicknesses for the case of addressed above. The result is presented in Figure 3 along with the critical thickness condition, Freund's blocking criterion, and a modified version of Freund's criterion[2] based upon the generalized force G on the dislocation (G

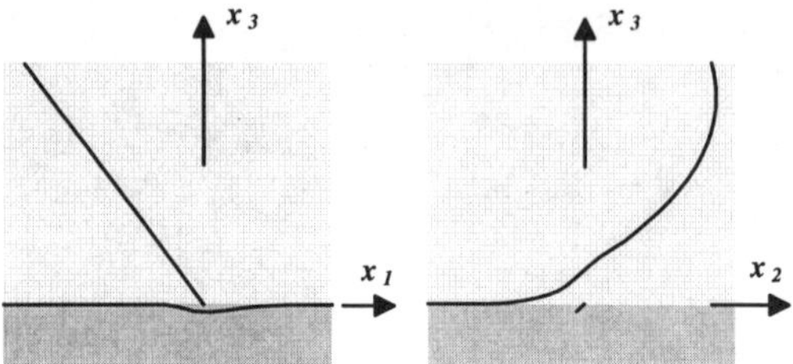

Figure 4. Two planar views of the metastable configuration resulting from a simulation in which the misfit dislocation is free to glide. The case is the same as that shown in Figure 2 but with a film thickness $h/b = 150$ which is quite close to the critical blocking condition with the misfit dislocation held fixed.

criterion). Numerical values of the critical thickness condition obtained from the simulation are indistinguishable from the analytical line plot here.

Schwarz and Tersoff[11] obtained a similar plot for the embedded layer case. The critical thickness for a given mismatch in an embedded layer is $2h_{cr}$ where h_{cr} is the critical thickness for a surface layer. It is therefore fitting to compare the result of Schwarz and Tersoff for a layer thickness of $2h$ with the current result for a thickness of h. For a comparison of the value of strain below which blocking occurs, the film thickness is taken as $h/b = 160$. Since the slip systems used by Schwarz and Tersoff differ from those of Figure 3, the calculation was repeated with their system to form a direct comparison. The comparison is made in terms of the ratio of blocking strain to the strain of the critical thickness condition ϵ_0/ϵ_{cr}. According to Figure 3, blocking occurs in the surface layer at $\epsilon_0/\epsilon_{cr} \approx 1.48$. For the embedded layer $\epsilon_0/\epsilon_{cr} \approx 1.13$ and, using the same slip systems, the results is $\epsilon_0/\epsilon_{cr} \approx 1.35$ for the surface layer. The two systems are not equivalent, and the results are quite different. The blocking effect is apparently less significant in an embedded layer than in a surface layer.

As the threading dislocation approaches the misfit dislocation, the misfit dislocation will either bow down into the substrate or bow up towards the threading dislocation. The interaction depends on the particular slip systems. In either case, the increase in self-energy associated with the bowing (the line tension) tends to keep the dislocation relatively straight. If the dislocation bows into the substrate, then the force due to the mismatch is removed and the free surface tends to pull the misfit dislocation back towards the film/substrate interface. Although, the dislocation may bow due to its interaction with the threading dislocation, there are restoring forces which limit the extent of the bowing. To examine if the blocking effect is characteristically different when the misfit dislocation is allowed to glide freely, the case treated in Figure 2 is reconsidered but with a film thickness $h/b = 150$. With the misfit dislocation held fixed, this thickness is slightly below the critical thickness for blocking. Yet, with the misfit dislocation free to glide, it is found that the threading dislocation is still blocked. Figure 4 shows the calculated metastable configuration resulting from the simulation. At least for this case (at this film thickness), freeing up the misfit dislocation does not have a significant effect on the blocking criterion. If the misfit dislocation was on a different glide plane or if cross slip was permitted, the same conclusion may not be reached

ACKNOWLEDGEMENTS

This work was supported in part by the Office of Naval Research under Contract N00014-95-1-0239, and in part by the MRSEC program of the National Science Foundation under Award DMR-9632524. This support is gratefully acknowledged.

REFERENCES

1. T.J. Gosling and J.R. Willis, *J. Mech. Phys. Solids* **42**, 1199 (1994).
2. R.V. Kukta, PhD thesis, Brown University, 1998.
3. L.M. Brown, *Phil. Mag.* **10**, 441 (1964).
4. S.D. Gavazza and D.M. Barnett, *J. Mech. Phys. Solids* **24**, 171 (1976).
5. J.P. Hirth and J. Lothe, J., *Theory of Dislocations*, 2nd ed. (John Wiley & Sons, Inc., New York, 1982), p. 99.
6. R.D. Mindlin, *Physics* **7**, 195 (1936).
7. L. Rongved, in *Proceedings*, (Second Midwestern Conference on Solid Mechanics, 1955), pp. 1-13.
8. J. Lothe, in *Fundamental Aspects of Dislocation Theory*, edited by J.A. Simmons, R. deWitt and R. Bullough, (Spec. Pub. 317, **I**, Nat. Bur. Stand., U.S., 1970), pp. 11-22.
9. L.B. Freund, *J. Mech. Phys. Solids* **38**(5), 657 (1990).
10. L.B. Freund, *J. Appl. Phys.* **68**(5), 2073 (1990).
11. K. Schwarz, K. and J. Tersoff, *Appl. Phys. Lett.* **69**(9), 1220 (1996).

IMAGE FORCES ON 3D DISLOCATION STRUCTURES IN CRYSTALS OF FINITE VOLUME

A. El-Azab

Pacific Northwest National Laboratory, MSIN k5-26, 902 Battelle Boulevard, P.O. Box 999, Richland, WA 99352, Anter.El-Azab@pnl.gov

ABSTRACT

The present work aims at studying the image stress and image Peach-Koehler force fields for three-dimensional dislocation configurations in a single crystal of finite volume. It is shown that the image stress field is significant within the entire crystal volume, and that the image Peach-Koehler force can be of the same order of magnitude as the direct interaction force calculated from the infinite domain solution. The results demonstrate that image stress gives rise to long-range interaction forces that are important in meso-scale dynamics of dislocation structures.

INTRODUCTION

Simulation of meso-scale plastic deformation is based on tracking the spatio-temporal behavior of a discrete dislocation ensemble in a finite crystal space. These simulations are currently used to compute the stress-strain law for single crystals [1-3], study the relaxation of stresses in thin film and analyze nano-indentation experimental results [4]. The dynamics of dislocation ensembles can only be studied in crystals of finite size. Therefore, in order to conduct the simulations in a self-consistent manner, it is necessary that the total stress field in the simulated crystal space satisfy proper traction boundary condition. Under the assumption that linear elasticity theory applies, the total stress field responsible for the evolution of the discrete dislocation ensemble consists of two superposed contributions; an externally-imposed stress and the long-range stress arising due to dislocations themselves. The applied stress is determined from the applied boundary traction. The stress field due to dislocations must satisfy either: a) consistency traction boundary condition if the simulated crystal space is a part of a much larger (essentially infinite) single crystal, or (2) traction-free boundary condition if the simulated crystal is of finite size. To enforce traction-free boundary condition in the second case, the singular stress field is modified by what the so-called "image stress." It has been shown that, for the former case, the "consistency stress" field is the same as the image stress field under the assumption that the simulated crystal volume is statistically representative of the macroscopic crystal [5].

The present work focuses on computation of the image stress field for arbitrary elementary dislocation structures in bounded crystals. As mentioned above, this stress field is the same as the consistency stress field for simulated bulk crystal volume. A crystal volume of cubic shape is considered here. Glide dislocation loops are arbitrarily located inside this crystal volume, and the image stress field and Peach-Koehler force are computed. The latter are compared with the direct interaction force computed from the singular stress field contribution; recall that the singular stress field is computed in a crystal of infinitely extended size.

Following this introduction, the traction boundary value problem for the image stress field is stated. Here, the Finite Element Method (FEM) is used to solve this problem. The image Peach-Koehler force is calculated from the image stress field. Specific results are discussed, with emphasis on the magnitude of the image forces relative to the direct interaction force.

Mat. Res. Soc. Symp. Proc. Vol. 538 © 1999 Materials Research Society

THE BOUNDARY VALUE PROBLEM

Figure 1 illustrates the boundary value problem for the image stress field. Denote the finite crystal volume by Ω and its boundary by $d\Omega$. Let L refer to the total dislocation line included within the volume Ω. The total stress field $\boldsymbol{\sigma}$ of an arbitrary dislocation configuration within the finite crystal volume Ω is given as follows

$$\boldsymbol{\sigma}(\mathbf{x}) = \boldsymbol{\sigma}^a(\mathbf{x}) + \boldsymbol{\sigma}^s(\mathbf{x}) + \boldsymbol{\sigma}^i(\mathbf{x}) \ ; \quad \mathbf{x} \in \Omega, \tag{1}$$

where $\boldsymbol{\sigma}^a$ is the applied stress, $\boldsymbol{\sigma}^s$ the singular stress, and $\boldsymbol{\sigma}^i$ the image stress field. Considering an isotropic elastic crystal, the singular stress field is given by [6]

$$\boldsymbol{\sigma}^s(\mathbf{x}) = \frac{\mu}{4\pi} \int_L (\mathbf{b} \times \nabla') \frac{1}{R} \otimes d\mathbf{L}' + \frac{\mu}{4\pi} \int_L d\mathbf{L}' \otimes (\mathbf{b} \times \nabla') \frac{1}{R}$$

$$- \frac{\mu}{4\pi(1-v)} \int_L \nabla' \cdot (\mathbf{b} \times d\mathbf{L}')(\nabla \otimes \nabla - I\nabla^2)R \quad ; \ R = |\mathbf{x} - \mathbf{x}'|; \ \mathbf{x}, \mathbf{x}' \in \Omega, \tag{2}$$

in which $\mathbf{b}$ is the Burgers vector, $\nabla = \partial / \partial x_i \mathbf{e}_i$, the primed variables (operator) refer to values (operation performed) at $\mathbf{x}'$, $\mathbf{I}$ is the unit dyadic, and μ and v are the shear modulus and Poisson's ratio, respectively. Let $\mathbf{t}$ be the traction left by the entire dislocation configuration on the boundary $d\Omega$. The image stress field is then given by the solution to the following boundary value problem

$$\nabla \cdot \boldsymbol{\sigma}^i(\mathbf{x}) = \mathbf{0}; \qquad\qquad \mathbf{x} \in \Omega$$

$$\mathbf{n} \cdot \boldsymbol{\sigma}^i(\mathbf{x}) = -\mathbf{t}(\mathbf{x}); \qquad \mathbf{x} \in d\Omega, \tag{3}$$

where $\mathbf{n}$ is the normal to the boundary. The singular stress field can be evaluated by discretizing the total dislocation line, and adding the (analytical) contributions from all line elements [1]. The solution of the boundary value problem (3) can only be performed numerically in a three-dimensional domain; the FEM method is used here for this purpose.

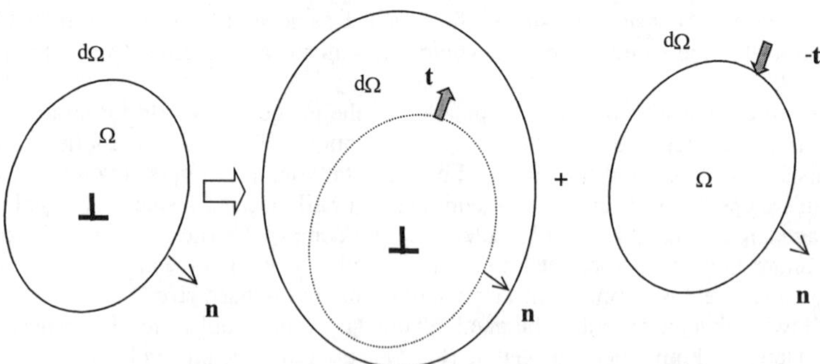

Figure 1: Schematic of boundary value problem for the image stress field.

THE IMAGE STRESS

For the purpose of computing the image stress field, a crystal of cubic shape with a side of 1000b is used. Poisson's ratio of 0.3 is considered. The global axes (x,y,z) of the cubic volume are chosen to coincide with the crystallographic directions [100], [010] and [001], respectively; see figure 2. The crystal is considered to be BCC, in which the primary slip systems are of $<111>\{110\}$ type, with Burgers vector $a<111>/2$, where a is the lattice constant. The linear dimensions of dislocation loops are chosen in units of the magnitude of the Burgers vector.

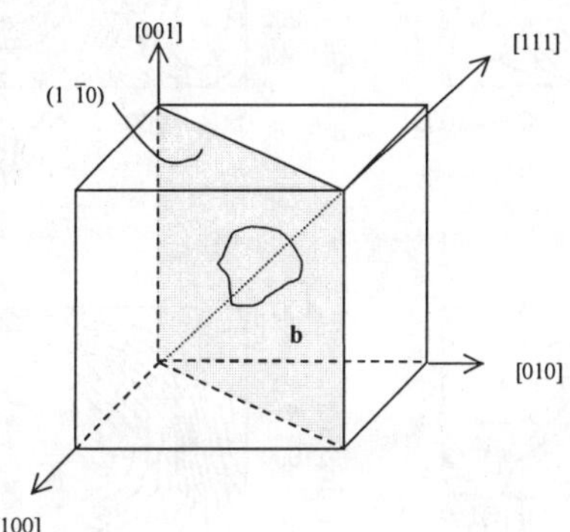

Figure 2: A dislocation loop on the slip system [111](1$\bar{1}$0) in a BCC crystal.

The image stress field in a crystal of bounded volume shows strong dependence on the overall geometry of the dislocation configuration and the crystal boundary. Figure 3 shows contour plots of the non-zero image stress components due to a glide loop of radius 250b, on the slip system $[111](1\bar{1}0)$. The plane of the loop passes through the origin and the loop is positioned at the center of the crystal. It is worth recalling that the image stress field in a semi-infinite crystal decays as 1/r from the free surface, and that these fields are significant only if the defect is in the close vicinity of the free surface. In the present case, the image stress field exhibits maximum values away from the crystal boundaries; see figure 3, even if the dislocation structure giving rise to these stresses is not closely positioned near the boundary. Consequently, available analytical solutions for half spaces can not be used to approximate the image fields of dislocation elements near the surface. The asymmetry exhibited by the non-zero image stress components shown in figure 3 is a result of the choice of the dislocation loop configuration relative to the computational volume.

The significance of the image stresses in the bounded crystal can be demonstrated by recalling Albenga's theorem [7]; that is the volume average of the stress field arising in a crystal of finite volume due to crystal defects is identically zero. Therefore

$$< \sigma^s + \sigma^i >_\Omega = \frac{1}{\Omega} \int_\Omega (\sigma^s + \sigma^i)\, d\Omega = 0 \quad \Rightarrow \quad < \sigma^i >_\Omega = - < \sigma^s >_\Omega . \tag{4}$$

This yields a quantitative estimate of the long-range stress contribution that arises due to the image stress.

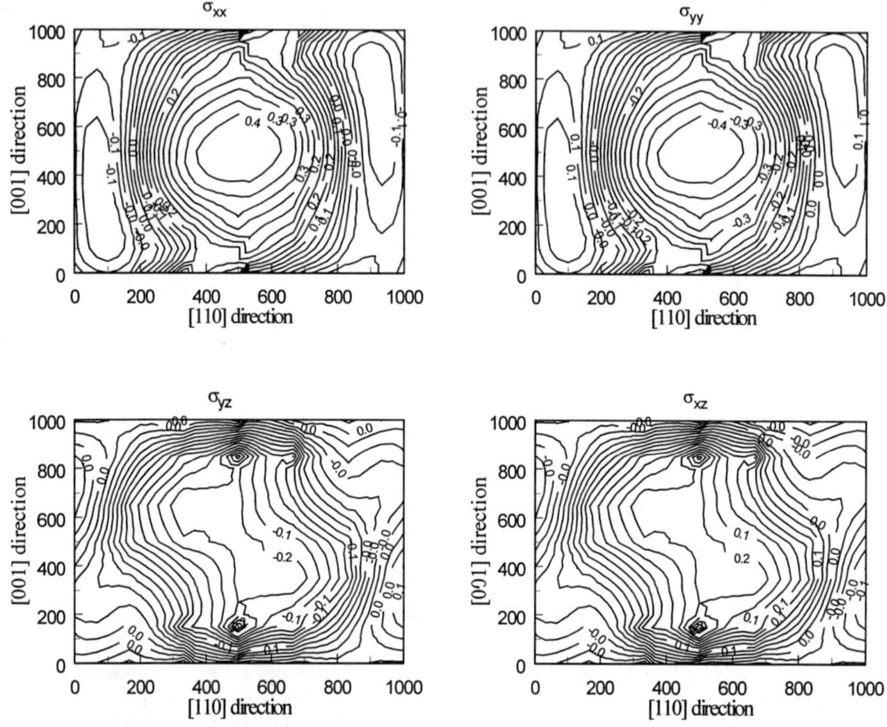

Figure 3: Contour plots of the non-zero image stress components for a loop of radius 250b on the (1 1 0) slip plane with Burgers vector $a[1\,1\,1]/2$. The loop's center is the same as the center of the cube. . The [110] axis is scaled by $\sqrt{2}$.

PEACH-KOEHLER FORCE

Consider a dislocation element of Burgers vector $\mathbf{b}$ and a unit tangent vector ξ. If the local stress field interacting with the dislocation element is σ, the glide Peach-Koehler force is defined by $\mathbf{F} = \mathbf{b} \cdot \sigma \times \xi = (\mathbf{b} \cdot \sigma \cdot \mathbf{n}^{gp})\eta$, in which $\mathbf{n}^{gp}$ is the normal to the glide plane and $\eta = \mathbf{n}^{gp} \times \xi$ is a unit vector along the glide direction, i.e., normal to the direction of dislocation line. In figure 4, the Peach-Koehler force field ($\mathbf{F} = \mathbf{b} \cdot \sigma^{i} \cdot \mathbf{n} = b\tau_{rss}$) induced by a dislocation loop of radius 250b on its glide plane $(1\,\bar{1}\,0)$ is shown for two the coplanar Burgers vectors $a[111]/2$, figure 4(a), and $a[11\,\bar{1}]/2$, figure 4(b). A dislocation element lying on this glide plane would experience a glide force whose magnitude is shown by the contour levels. It is clear that the image force field exhibits larger values close to the center. The sign changes indicate that the direction of the glide force changes from one region to another.

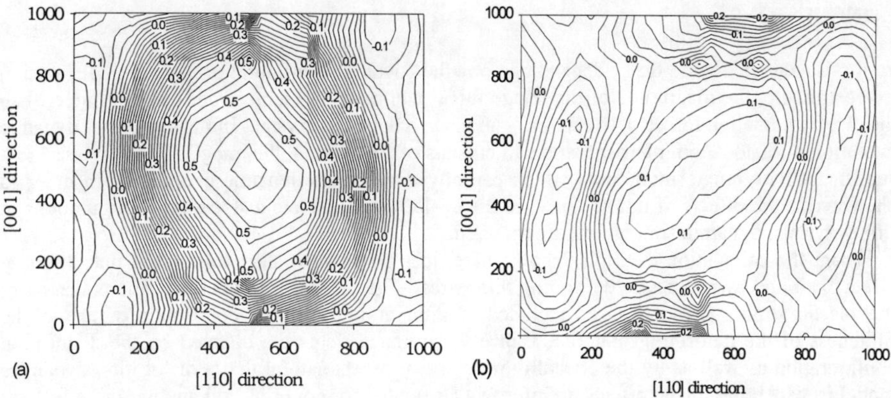

Figure 4: Contour plots of the magnitude of the image Peach-Koehler force due to a loop of radius 250b on the $(1\bar{1}0)$ slip plane with Burgers vector $a[111]/2$. The loop is at the center of the cube. The [110] axis is scaled by $\sqrt{2}$.

An example of comparing the image and self-forces for single dislocation loops is shown in figure 5. Again, the plane of the loop passes through the origin; see figure 2. For those shown in figure 5(a), each loop is located at the cube center. When the loop size becomes larger, the image and self-forces become comparable; figure 5(a). The variation in self-force around the loop is due to the fact that the character of the dislocation line changes between pure edge, mixed and pure screw. As the loop becomes larger, its curvature and, in turn, the self-force become smaller. When the loop is shifted towards one of the cube faces, the image force become stronger on the part of the loop closer to the boundary. Yet, on the rest of the loop, the image force still can not be neglected.

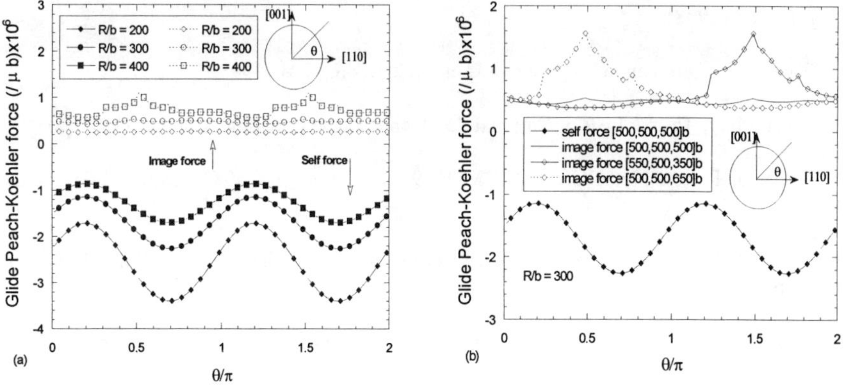

Figure 5: Image and direct Peach-Koehler forces for glide loops on the $(1\bar{1}0)$ slip plane with Burgers vector of $a[111]/2$ type. The effect of loop size is shown in (a), while (b) shows effect of loop position.

CONCLUSIONS

It is shown here that, depending on the overall configuration of the dislocation microstructure, the magnitude of the image forces can be of the same order or even higher than the direct interaction forces between dislocations in a bounded crystal. In the context of dynamic evolution of dislocation microstructures in crystals of finite sizes, the image forces give rise to a significant long-range interaction, which partially dictates the dynamic evolution. In order to characterize the effects of image fields on the evolution of discrete dislocation configurations in finite crystals, dynamic simulations are required.

From the calculation examples shown here, it is possible to shed some light onto the size effect exhibited by plastically deforming thin wires. The dynamics of the dislocations that carry the plastic strains in such cases included significant image force contributions due to the finiteness of the deforming material. Since these image fields are dictated by the dislocation configuration as well as by the crystal dimensions, it seems natural that a dislocation dynamics model in which the image effects are properly accounted for could indeed capture the effects of size on plastic deformation in very small crystals.

ACKNOWLEDGMENT

This work was supported by LLNL under award number B339033 at the University of California, Los Angeles. The support received at PNNL during preparation of the manuscript is greatly appreciated.

REFERENCES

1. L.P. Kubin, G. Canova, M. Condat, B. Devincre, V. Pontikis and Y. Bréchet, Solid State Phenomena, **23&24**, 455 (1992).

2. M. Rhee, H.M. Zbib, J.P. Hirth, H. Huang and T. de la Rubia, Modelling Simul. Mater. Sci Eng. **6**, 467 (1998).

3. M. Tang, L.P. Kubin and G.R. Canova, Acta Metall. **46**, 3221 (1998).

4. M. Fivel, Ph.D. Thesis, Institut National Polytechnique de Grenoble, France, 1997.

5. A. El-Azab, Report No. UCLA-MATMOD-97-02, University of California, Los Angeles, 1997.

6. J.P. Hirth and J. Lothe, *Theory of dislocations*, 2nd Ed. (John Wiley & Sons, New York, 1982) p. 106.

7. V.L. Indenbom and J. Lothe, *Elastic Strain Fields and Dislocation Mobility* (North-Holland, 1992) p. 14.

A PEIERLS-NABARRO MODEL FOR KINK-PAIR GENERATION IN SCREW DISLOCATIONS AT LOW TEMPERATURES

A.H.W. NGAN
Department of Mechanical Engineering, The University of Hong Kong, Pokfulam Road, Hong Kong, P.R. China

ABSTRACT

Kink-pair generation in three-fold screw dislocations in the bcc lattice is investigated within the framework of the generalised Peierls-Nabarro model. Using a piece-wise plane strain approximation, the apparent activation energy is predicted to vary with stress in a parabolic manner which is in good agreement with experimental findings.

INTRODUCTION

Most of the current understanding of the behaviour of screw dislocation cores in the bcc lattice is by and large derived from atomistic simulation studies performed in the last three decades or so. These atomistic investigations lead to the current perception that i) $\frac{1}{2}<111>$ screw dislocation cores are in general three-fold dissociated with intrinsically high Peierls stress of the order of $10^{-2}\mu$, μ being the shear modulus, ii) their slip behaviour violates Schmid's law, iii) their motion at zero-temperature may follow a path along $\{101\}$ or a zig-zag path averaging along a twinning or anti-twinning $\{112\}$ plane, and iv) their motion can be affected by non-glide stress components. At finite temperatures, motion of screw dislocations does not happen in a plane-strain manner but is effected by the generation and nucleation of kink-pairs, and at low deformation temperatures, it is the nucleation of these kink-pairs that forms the rate determining step for slip. The application of atomistic simulation to the investigation of kink-pair generation, however, has so far been very limited, as kink problems are essentially 3-D problems in which the number of atoms to be handled may be too large. In this respect, the most frequently used research tool is still the continuum approach involving the line tension approximation and the phenomenological concept of Peierls potential [1]. The Peierls potential, being the energy potential of the dislocation with respect to its "position", is purely phenomenological because the constraint that must be applied to keep the dislocation at a general non-equilibrium position cannot be clearly defined. Also, the introduction of an inner cut-off radius in the line tension disconnects the whole approach from the vast body of literature on the core behaviour of the dislocation.

In the light of these difficulties, the present work aims at developing a semi-continuum model which combines certain elements of the atomistic and continuum picture of the dislocation kink-pair problem. The model is essentially an extension to the Peierls-Nabarro model, in which the dislocation core is envisaged as comprising of linear elastic regions connected to one another along certain surfaces over which a non-linear misfit force law acts. The misfit force law is readily established as the γ-surface, which can be calculated using atomistic means for specific materials. With the introduction of the misfit surfaces, the elastic field of the dislocation no longer carries any singularities and so the introduction of the dubious inner cut-off is not necessary.

THEORY

The bcc lattice obeys three-fold rotational symmetry about $<111>$ and in the generalised P-N model [2,3], a screw dislocation along this direction is considered to be composed of three 120° elastic wedges with the wedge boundaries along three $\{101\}$ planes as shown in Fig. 1. The wedges are strained into an anti-plane strain manner so that when they are welded together at the far field, the long-range field of the screw dislocation is established there. At the centre of the dislocation, the total Burgers vector content is partitioned amongst the misfit displacements of the adjacent wedge faces as shown in Fig. 1. The interaction between adjacent wedge faces is characterised by a non-linear force law $\gamma[\Phi]$, where Φ is the misfit displacement. The total energy per unit length E_{tot} of the dislocation is therefore composed of three parts: i) the strain energies of

Mat. Res. Soc. Symp. Proc. Vol. 538 © 1999 Materials Research Society

the wedges, ii) the misfit energy of the three cuts, and iii) the work done by the applied stress τ_a. Each of these energy terms is a functional of the boundary displacement functions $u_{i\pm}$ defined for each wedge relative to the position of the wedge tip. E_{tot} can therefore be expressed as [4]:

$$E_{tot} = \frac{3\mu}{4\pi} \sum_{i=1}^{3} \int_0^\infty \int_0^\infty \{ \frac{\eta^{1/2}}{\xi^{3/2} + \eta^{3/2}} [u_{i-}'(\xi)u_{i+}(\eta) + u_{i+}'(\xi)u_{i-}(\eta)] - \frac{\eta^{1/2}}{\xi^{3/2} - \eta^{3/2}} [u_{i+}'(\xi)u_{i+}(\eta) + u_{i-}'(\xi)u_{i-}(\eta)] \} d\eta d\xi$$

$$+ \int_0^\infty \gamma \left[\Phi = \Delta_{1|2}b - u_{1-}(r) - u_{2+}(r) \right] dr + \int_0^\infty \gamma \left[\Phi = (1 - \Delta_{1|2} - \Delta_{3|1})b - u_{2-}(r) - u_{3+}(r) \right] dr$$

$$+ \int_0^\infty \gamma \left[\Phi = \Delta_{3|1}b - u_{3-}(r) - u_{1+}(r) \right] dr + \tau_a \cos(120° - \chi) \int_0^\infty \left[\Delta_{1|2}b - u_{1-}(r) - u_{2+}(r) \right] dr$$

$$+ \tau_a \cos\chi \int_0^\infty \left[(1 - \Delta_{1|2} - \Delta_{3|1})b - u_{2-}(r) - u_{3+}(r) \right] dr + \tau_a \cos(120° + \chi) \int_0^\infty \left[\Delta_{3|1}b - u_{3-}(r) - u_{1+}(r) \right] dr, \tag{1}$$

in which the double integral terms represent the strain energy, the γ terms the misfit energy and the τ_a terms the work done against the applied stress τ_a. In this equation, b is the Burgers vector, $\Delta_{i|j}$ the fractional Burgers vector content of the $i|j$ cut, and χ the angle between the maximum resolved shear stress plane and the 2|3 cut in Fig. 1. At equilibrium without an applied stress, the core should adopt the symmetrical three-fold configuration for which $\Delta_{1|2} = \Delta_{3|1} = {}^1/_3$, and when the core becomes planar along, say, the 2|3 cut under thermal agitation or a large enough τ_a, $\Delta_{1|2} = \Delta_{3|1} = 0$. The atomic force law $\gamma[\Phi]$ can be calculated for specific materials using atomistic simulation [5]. The static core configuration under any predefined stress level can then be obtained by minimising E_{tot} in eqn. (1) with respect to $u_{i|j}$ and $\Delta_{i|j}$ using a variational technique. For example, the Peierls stress, i.e. the stress required to constrict the three-fold core to the planar configuration, estimated by a Frenkel force law is $\sim 0.04\ \mu$ when $\chi = 0°$ [6], which is in good agreement with typical values calculated by atomistic simulation or measured by experiments.

At finite temperatures, the dislocation will not move as a rigid line but will have to go through a saddle configuration in the form of a kink-pair illustrated schematically in Fig. 2. In this configuration, the dislocation core changes continuously from the fully dissociated state into the fully recombined state through a kink, and then back to the fully dissociated state through another

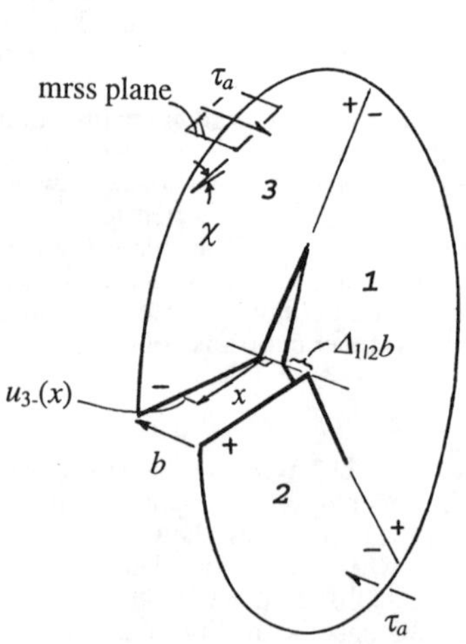

Fig. 1 - Three-fold Screw Dislocation Core

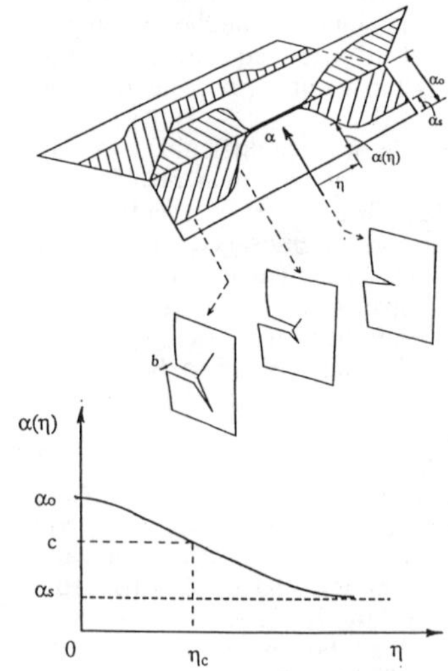

Fig. 2 - Double Kink Configuration

kink with opposite sign as shown in Fig. 2. The geometry represents a 3-D elasticity problem and the mathematics can be greatly simplified by invoking the piecewise anti-plane strain approximation, in which we assume that for each section perpendicular to the dislocation line, the local degree of recombination can be characterised by a generalised coordinate α, whose value varies from $\alpha_s \approx 0$ for the fully dissociated state to $\alpha_o \approx 1$ for the fully recombined state. The kink shape is then marked by the function $\alpha(\eta)$, where η is the spatial coordinate along the dislocation line. The section at η is assumed to have a screw direction displacement field u^α in cylindrical coordinates given by

$$u^\alpha(r,\theta,\eta) = u^\Delta(r,\theta) + \alpha(\eta)\, f(r,\theta), \tag{2}$$

where $u^\Delta(r,\theta)$ is the screw direction displacement of the stress free, 3-fold symmetrical anti-plane strain state, and $f(r,\theta)$ is the displacement change on going from the $u^\Delta(r,\theta)$ reference state to the fully recombined planar state. In eqn. (2) are implied two approximations. The first is a linearisation approximation, in which the saddle path the core configuration moves along is approximated by the "straight line" joining the starting three-fold and ending two-fold states. The second is the piecewise anti-plane strain approximation as discussed above, which is valid only at the high stress (or low temperature) regime during which the rate-determining step involves kink-pair nucleation. If kink-pair separation is instead the rate-determining step as in the case of the low stress (or high temperature) regime, the two opposite kinks are well-formed and sharp. In this situation, the long-range elastic interaction between the two opposite kinks will dominate and the classical treatment using Volterra fields will be more valid [7].

With the assumptions spelt out in eqn. (2), the different energy contributions of the double kink configuration can be calculated as follows.

i) Self energy E_{self} - Using the stress-free, symmetrical three-fold dissociated state as the reference, the energy change of a partially recombined core given by eqn. (1) may be expressed as $\delta E_{tot} = \delta E_{self} - \tau_a V^*$, where δE_{self} comprises of the strain and misfit energy changes, and the τ_a term in which V^* is the activation volume represents the work done change. It has been shown beforehand [2,3] that the energy per unit length of the equilibrium planar state is higher than the three-fold dissociated state by an amount $\Delta E \sim 0.03\ \mu b^2$. ΔE constitutes the energy barrier that must be overcome before the ground state three-fold configuration can be transformed into the mobile planar state, and in the absence of thermal agitation, this is solely overcome by an applied stress equal to the zero-temperature Peierls stress. Thus in the double kink configuration shown in Fig. 2, a slab situated within the fully recombined planar segment (with $\alpha \sim 1$) would have energy per unit length ΔE higher than one situated well within a ground state segment (with $\alpha \sim 0$), and so it seems reasonable to associate the self-energy δE_{self} of a partially recombined slab with the corresponding α value by a piecewise parabolic relationship:

$$\delta E_{self}(\alpha) \approx \begin{cases} \Delta E \alpha^2 / c & \text{for } 0 \leqslant \alpha \leqslant c \\[6pt] \Delta E - \Delta E(\alpha-1)^2/(1-c) & \text{for } c \leqslant \alpha \leqslant 1 \end{cases} \tag{3}$$

where c is a constant between 0 and 1. Detailed consideration of the strain and misfit energy terms in eqn. (1) using a Rayleigh-Ritz approach to be published elsewhere [4] shows that the lowest energy path can be well approximated by eqn. (3) with c set to 0.2 and ΔE to 0.03 μb^2. With these prescriptions, the total self-energy of the double kink configuration E_{self} is then given by

$$E_{self} = \int_0^\infty \delta E_{self}(\alpha)\, \mathrm{d}\eta \ . \tag{4}$$

ii) Interaction energy E_{int} - Provided that $\alpha(\eta)$ is slowly varying, the interaction energy E_{int} between the slabs is given by

$$E_{int} = \sum_{i=1}^{3} \frac{E}{2} \int_0^\infty d\eta \iint_{wedge\ i} \left[\frac{\partial}{\partial \eta} \delta u_\eta^i(r,\theta) \right]^2 r\, dr\, d\theta \tag{5}$$

where E is the Young's modulus, and $\delta u_\eta{}^i(r,\theta)$ is the displacement change in polar coordinates of the interior of wedge i when the core goes from the initial three-fold configuration to a general configuration specified by α under the anti-plane strain condition. From eqn. (2), $\delta u_\eta{}^i(r,\theta) = \alpha(\eta) f_i(r,\theta)$, so that eqn. (5) can be rewritten as

$$E_{int} = \int_0^\infty \frac{[\alpha'(\eta)]^2}{2} EP\, d\eta \tag{5a}$$

where
$$P = \sum_{i=1}^{3} \iint_{wedge\ i} [f_i(r,\theta)]^2 r\, dr\, d\theta. \tag{6}$$

P can be calculated numerically from eqn. (6) [4] but physically, $\sqrt{P}$ is simply the root-mean-squared change in the displacement on going from the three-fold to the planar configuration multiplied by an effective size of the core, and so $\sqrt{P}$ is of the order of b^2.

iii) Work-done E_w by the applied stress τ_a - The work done term in eqn. (1) as the symmetrical three-fold core recombines into the planar state, within Eshelby's superposition spirit, is equal to the summation over the cuts ilj of the products of the stress component acted on each cut and the area change of the cut δA_{ilj} during the recombination. The work done for a partially recombined section $d\eta$, within the linearisation approximation, is thus given by:

$$\tau_a dV^* = \tau_a \cos(120° - \chi)\alpha(\eta)\delta A_{1l2}d\eta + \tau_a \cos\chi\,\alpha(\eta)\delta A_{2l3}d\eta + \tau_a \cos(120° + \chi)\alpha(\eta)\delta A_{3l1}d\eta,$$

but because $\delta A_{1l2} \simeq \delta A_{3l1} < 0 \ (= -A_2,\ \text{say})$ and $\delta A_{2l3} > 0 \ (= A_1,\ \text{say})$, this equals to

$$\tau_a dV^* = \alpha(\eta)\tau_a \cos\chi (\delta A_{2l3} + \delta A_{1l2})d\eta$$
$$= \alpha(\eta)\tau_a \cos\chi\, A^* d\eta$$

where $A^* = A_1 + A_2$ is approximately constant with respect to orientation. Obviously, A^* is of the order of a few b^2. For one half of the kink-pair configuration, the total work done is therefore

$$E_w = \tau_a A^* \cos\chi \int_0^\infty \alpha(\eta)\, d\eta \quad . \tag{7}$$

The total energy H of the whole kink-pair is

$$H = 2 \times \int_0^\infty \{[\delta E_{tot}(\alpha) - \delta E_{tot}(\alpha_s)] - \tau_a A^* \cos\chi\,(\alpha - \alpha_s) + EP\alpha'^2/2\}\, d\eta \tag{8}$$

where α_s marks the static anti-plane strain configuration under τ_a and is given by $\alpha_s = \tau_a A^* \cos\chi\, c/(2\Delta E)$.

PREDICTIONS

Orientation Dependence of Peierls Stress

The zero-temperature Peierls stress τ_p is the value of τ_a when $\alpha_s = c$ in the anti-plane strain condition, and is given by

$$\tau_p = \frac{2\Delta E}{A^* \cos \chi} \ . \tag{9}$$

Since $\Delta E \sim 0.03 \ \mu b^2$ and A^* is of the order of a few b^2, within the allowable range $\pm 30°$ for χ, $\tau_p \sim 10^{-2} \ \mu$, which is the order of magnitude observed experimentally. From eqn. (9), τ_p is orientation dependent, showing a minimum at $\chi = 0°$ and maximum at $\chi = \pm 30°$. Such a τ_p-χ relationship is indeed what is observed in Fe and Mo [8]. The magnitude of variation in τ_p from $\chi = 0°$ to $\pm 30°$ estimated by eqn. (9) is $\sim 10^{-3} \ \mu$, which is again in very good agreement with experimental results for Fe and Mo. Generally speaking, for bcc metals which exhibit this type of τ_p-χ relationship, the slip plane would always be $\{101\}$ irrespective of the stress orientation χ, and this is indeed what is assumed to happen in the present model. In the atomistic simulation performed by Vitek and co-workers [9], this corresponds to the behaviour of the J_2 potential which yields the widest dissociation on $\{101\}$ planes. If dissociation is of limited extent on $\{101\}$ planes, there is a general tendency for the core to glide on a $\{112\}$ plane, and the selection of twinning or anti-twinning $\{112\}$ is dependent on the stress orientation. The associated τ_p-χ relationship would exhibit a maximum at the χ value at which the slip plane changes from twinning to anti-twinning $\{112\}$, and in general, τ_p for twinning slip is lower than that for anti-twinning slip. This twinning/anti-twinning asymmetry in the slip behaviour of bcc materials like Ta, Li-Mg, AgMg and β-CuZn is not predicted by the present model simply because the detailed atomic arrangement along the screw direction is not modelled here.

Critical Kink Shape

At finite temperatures, the critical kink shape can be obtained by extremising H in eqn. (8) with respect to $\alpha(\eta)$. The predicted critical kink height, defined here as $\alpha_o = \alpha(0)$, is found to be

$$\alpha_o = (\ \sqrt{1\text{-}c} \ + 1 \ ) - t \ [\ \sqrt{1\text{-}c} \ + (1\text{-}c) \ ], \tag{10}$$

where $t = \tau_a/\tau_p$ is a normalised stress. The critical kink height therefore decreases as the applied stress t increases, and this is in qualitative agreement with atomistic calculations [7]. The kink separation may be defined to be the separation $2 \times \eta_c$ between the two points $\eta = \pm \eta_c$ at which $\alpha(\pm \eta_c) = c$ (see Fig. 2), and

$$\eta_c = \ \sqrt{(1-c)EP/(2\Delta E)} \ \tan^{-1}\left[-\sqrt{c/(1-c)}\right]. \tag{11}$$

The critical kink separation is therefore stress independent, again in qualitative agreement with atomistic simulation [7].

Activation Energy

The minimised value of H from eqn. (8) is found to be

$$H = (1 - t)^2 \ \left[\sqrt{2(\Delta E)cEP} + 2\Delta E\eta_c/(1\text{-}c)\right] \ . \tag{12}$$

The apparent activation energy is therefore predicted to vary parabolically with the applied shear stress t. Fig. 3 shows the experimental data of activation energy vs stress for potassium and high purity iron. It can be seen that the experimental data for both Fe and K fall on the same relationship $\sqrt{(H/\mu b^3)} \approx (0.30 \pm 0.03) \times (1\text{-}t)$, where $t = \ \tau_a/\tau_p$. Furthermore, for the case of Fe, this relationship also holds for two different tensile orientations as indicated. The Table shows the experimental values of $\sqrt{(H(\tau_a{=}0)/\mu b^3)}$ for a few other bcc metals, which indicate clearly that the parameter falls in a narrow range around 0.30 ± 0.03. The constancy of the relationship $\sqrt{(H/\mu b^3)} \approx 0.3 \times (1\text{-}t)$ is a strong verification for the present model, which predicts the same relation as expressed in eqn. (12) if one takes the expected values $\Delta E \sim 0.03 \ \mu b^2$, $c \sim 0.2$ and $EP \sim 0.1 \ \mu b^4$.

Fig. 3 also shows the atomistic simulation results for K by Duesbury [7], which significantly overestimate the kink energy at the intermediate to large stress regime.

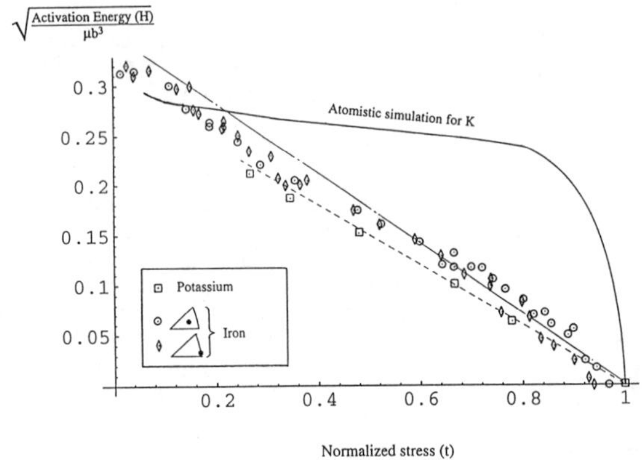

Fig. 3 - Activation Energy vs Normalised Applied Stress

[Ref. for K: Basinski, Duesbury and Murty (1981); for Fe: Aono, Kuramoto and Kitajima (1981).]

Table - Experimental Values of Zero-stress Activation Energies

	Cr	Ta	V	Mo	W	Nb
$\sqrt{(H(\tau_a=0)/\mu b^3)}$	0.25-0.30	0.28-0.31	0.34	0.30	0.28	0.30
References	← experimental data quoted in Dorn and Rajnak (1964) →					Conrad and Hayes (1963)

CONCLUSIONS

The movement of 1/2<111> screw dislocations in Fe-like bcc metals can be modelled within the framework of the generalised Peierls-Nabarro model. The apparent activation energy is predicted to vary with stress in a parabolic manner, in good agreement with experimental findings for iron and potassium.

ACKNOWLEDGMENT

This research was supported by an earmarked research grant (# HKU 572/96E) from the Research Grants Council, Hong Kong.

REFERENCES

1. See, e.g. T. Suzuki, H. Koizumi and H.O.K. Kirchner, Acta Metall. Mater., **43**, 2177 (1995) and references therein.
2. A.H.W. Ngan, Phil. Mag. Lett., **72**, p. 207 (1995).
3. A.H.W. Ngan, J. Mech. Phys. Solids, **45**, p. 903 (1997).
4. A.H.W. Ngan, to appear in Phil. Mag. A.
5. M.S. Duesbury, and V. Vitek, Acta Mater., **46**, 1481 (1998).
6. A.H.W. Ngan in *Strength Theory: Application, Development & Prospects for 21st Century*, edited by M.H. Yu and S.C. Fan (Science Press, Beijing China 1998), p. 151-156.
7. M.S. Duesbury, Acta Metall., **31**, 1747 (1983).
8. T. Suzuki, S. Takeuchi and H.Yoshinaga, *Dislocation Dynamics and Plasticity*, Springer-Verlag Berlin Heidelberg, 1991, Chp. 6.
9. V. Vitek, Crystal Lattice Defects, **5**, 1 (1974).

KINETIC MONTE CARLO SIMULATION OF DISLOCATION DYNAMICS

KARIN LIN[*,**], D. C. CHRZAN[**,***]
* Department of Physics, University of California, Berkeley, CA 94720
** Division of Materials Sciences, Lawrence Berkeley National Laboratories, Berkeley, CA 94720
*** Department of Materials Science and Mineral Engineering, University of California, Berkeley, CA 94720

ABSTRACT

A kinetic Monte Carlo simulation of dislocation motion is introduced. The dislocations are assumed to be composed of pure edge and screw segments only, and are assumed to be confined to a fixed lattice. The stress and temperature dependence of the dislocation velocity is studied. It is also noted that the simulated dislocations display kinetic roughening.

INTRODUCTION

Dislocation motion is the primary means of plastic deformation in most materials. Thus, any attempt to develop models of plastic deformation must include, at a minimum, a rudimentary description of dislocation dynamics. Development of a truly predictive description of dislocation dynamics is nontrivial, and requires the solution of a complicated many-body problem. Central to solving this problem is the understanding of the motion of a single dislocation propagating under an applied stress.

Much work has gone into the development of the current understanding of dislocation dynamics. For refractory body-centered-cubic materials, in which one expects a large Peierls stress (arising from the dislocation core structure), dislocations are thought to move through a corrugated potential surface via the nucleation of double-kink pairs and the lateral motion of those kinks. Rapid progress towards understanding these aspects of dislocation dynamics in specific materials has been made both experimentally [1] and theoretically [2, 3, 4, 5, 6].

One expects that certain features of the dislocation dynamics will be present in a broad class of materials. For example, it is expected that double-kink nucleation will be a thermally activated process, and that motion of the kinks may also be thermally activated[6]. Further, it is expected that inclusion of long-ranged elastic forces is important to the development of a quantitative model of dislocation dynamics. It is worthwhile, therefore, to explore the large scale behavior of model systems retaining these generic features. Study of such model systems should aid in the identification of relevant atomic scale calculations, as well as provide a means to study the effects of the "many-body" nature of the problem.

In the current paper, results from the study of one such model are presented. The model is based on isotropic elasticity theory, and reflects the stochastic and "many-body" aspects of the dynamics of dislocation motion. An analogy is drawn between the model and rudimentary theories of crystal growth. This analogy suggests that the dislocations display kinetic roughening.

Mat. Res. Soc. Symp. Proc. Vol. 538 © 1999 Materials Research Society

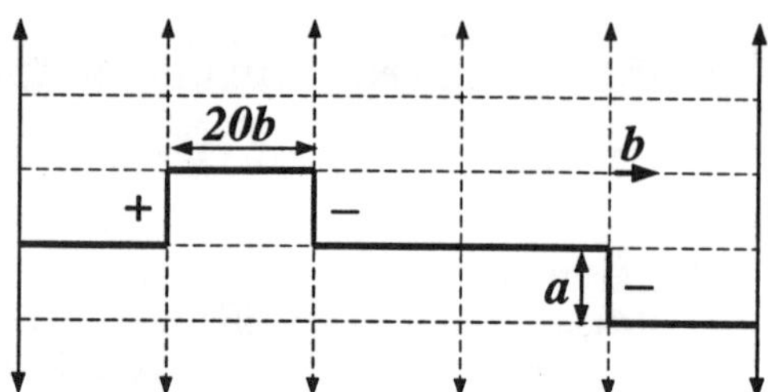

Figure 1: Discrete model of a dislocation. Screw segments lie in Peierls minima and edge segments are present only as kinks. The + and − signs indicate positive and negative kinks, respectively.

SIMULATION

Representation of the Dislocation

The dislocation is modeled as a system of pure screw and pure edge segments on a rectangular net that is infinite in the direction of motion and finite in the direction perpendicular to motion (see Fig. 1). Screw segments are assumed to lie in Peierls minima. The dislocation moves via the processes of double-kink nucleation and lateral kink motion, which are brought about by thermal fluctuations and an optional applied stress. Parameters are chosen to be consistent with the (110) slip plane of tantalum. In the direction perpendicular to motion, the minimum possible discretization distance is the Burgers vector b. However, it is found that kink pairs separated by a distance of b are unstable at normal stresses and temperatures. The kink pair separation distance is thus chosen to be $20b$, which gives stable kink pairs at most stresses and temperatures and improves computational efficiency by reducing the number of nucleation and kink motion events for a dislocation of a given length.

Dislocation Evolution

Evolution of the dislocation is controlled by standard kinetic Monte Carlo procedures[7]. The initial state of the dislocation is a perfectly straight, single segment in screw orientation. The allowed events at each stage of the simulation include double-kink nucleation, annihilation of an existing kink pair, and lateral motion of a single kink. In the presence of an applied stress, both events that occur in the direction of motion favored by the stress and events that occur against the stress are allowed, since thermal fluctuations can overcome the additional energy bias caused by the stress. The rates of each tabulated event are then calculated according to

$$R = \omega_o \exp(-\Delta E_{barr}/k_B T) \tag{1}$$

where ω_o is an attempt frequency and ΔE_{barr} is the energy barrier associated with moving between two configurations.

The calculation of ΔE_{barr} includes several terms. For double-kink nucleation and annihilation, the following energies are calculated:

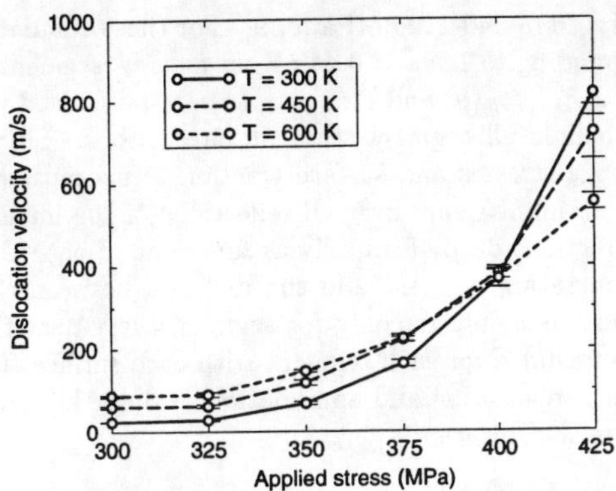

Figure 2: Dislocation velocity as a function of stress for three temperatures.

$$\Delta E_{mid} = \Delta E_s(a/2) + \Delta E_P \tag{2}$$

$$\Delta E_{final} = \Delta E_s(a) \tag{3}$$

where ΔE_P is the Peierls energy and

$$\Delta E_s(y) = \Delta E_{s,elastic}(y) \pm \sigma b(20b)y \tag{4}$$

is the elastic energy and work done by or against an applied stress associated with moving a screw segment a distance y. Once ΔE_{mid} and ΔE_{final} are determined,

$$\Delta E_{barr} = \begin{cases} \Delta E_{mid} & \Delta E_{mid} \geq \Delta E_{final} \\ \Delta E_{final}/2 & \text{otherwise} \end{cases} . \tag{5}$$

This scheme allows for a more accurate choice of energy barrier in the case where the difference in the elastic and stress energy required to move a screw segment to its final position compared to the midpoint exceeds the Peierls barrier. The factor of 2 ensures that detailed balance is maintained. For lateral kink motion,

$$\Delta E_{barr} = \Delta E_e(20b)/2 + \Delta_{mig} \tag{6}$$

where Δ_{mig} is a kink migration energy and

$$\Delta E_e(x) = \Delta E_{e,elastic}(x) \pm \sigma bax \tag{7}$$

is the elastic energy and work done by or against an applied stress associated with moving an edge segment a distance x. Again, the factor of 2 maintains detailed balance.

The quantities ω_o, ΔE_P, $\Delta E_{s,elastic}$, $\Delta E_{e,elastic}$, and ΔE_{mig} are determined as follows. The attempt frequency ω_o is chosen by taking the fundamental mode of the Granato-Lücke vibrating string model [6], $\omega_o = C_t \pi / L$. Using a length $L = 20b$ and the speed of sound in tantalum, $C_t = 2060$ m/s, one finds $\omega_o = 1.23 \times 10^{12}$. Measurements of the Peierls stress in tantalum vary widely, from 340-1100 MPa [8, 9]. For these simulations, σ_P was chosen to be 750 MPa, corresponding to $\Delta E_P = 0.44$ eV for a screw segment of length $20b$.

The calculations of $\Delta E_{s,elastic}(y)$ and $\Delta E_{e,elastic}(x)$ are performed within isotropic elasticity theory [6] and include all segment-segment interactions. Since the dislocation is in a finite medium, image stresses and surface traction terms must be included [10, 11]. Properly, this requires an infinite sum over all reflections of the images on both surfaces; however, since the interactions drop off rapidly as functions of both the distance from the surface of the stress-generating segment and the distance between the segment and the point of evaluation, and are significant only for segments very near the surface, only the first term in the sum is included for each segment with each surface. In order to minimize the surface effects further, nucleation and annihilation of single kinks are allowed to occur at a distance of $10b$ from the surface.

RESULTS

Dislocation Velocity

Fig. 2 displays the steady state velocity of the dislocation as a function of applied stress for three temperatures, using the full simulation including detailed calculation of elastic energies. As expected, for any given temperature the dislocation velocity increases with applied stress. The point at which the curves cross, *i.e.* ~ 385 MPa, corresponds to the zero-temperature yield stress *in the limit of an infinite-length dislocation.* (The effects of finite size will be discussed elsewhere.)

The crossing of the curves can be understood as follows. There are essentially six processes that govern the motion of the dislocation: kink pair nucleation with/against an applied stress, kink pair annihilation with/against the stress, and lateral kink motion with/against the stress. One may suppose that the velocity of the dislocation is determined primarily by the complementary processes of kink pair nucleation with the stress and kink pair annihilation against the stress. This is not unreasonable, since kink pair nucleation against the stress occurs rarely and lateral kink motion can occur only after kinks have been nucleated. To explore the relative importance of these two primary processes, the energy barrier associated with each process was plotted as a function of stress, assuming an isolated kink pair in the middle of an otherwise flat dislocation. Since the energy barrier for kink pair nucleation with the stress decreases with increasing stress and the energy barrier for kink pair annihilation against the stress increases with increasing stress, the curves must eventually cross, and do so around 410 MPa. The ratio of the two rates, assumed to be related to the velocity, is given by

$$R_{nuc}/R_{ann} = exp(-(E_{nuc} - E_{ann})/kT) \qquad (8)$$

where E_{nuc} and E_{ann} are the energy barriers associated with kink pair nucleation and annihilation, respectively. At low stresses, $E_{nuc} > E_{ann}$ and the exponent in Eq. 8 is negative. Thus, as T increases, the ratio of the rates (and thus the dislocation velocity) also increases. This produces the expected behavior. However, at high stresses, the exponent in Eq. 8 is positive, indicating that it is energetically favorable to nucleate double-kinks. As T increases, then, the exponent approaches zero and the velocity decreases.

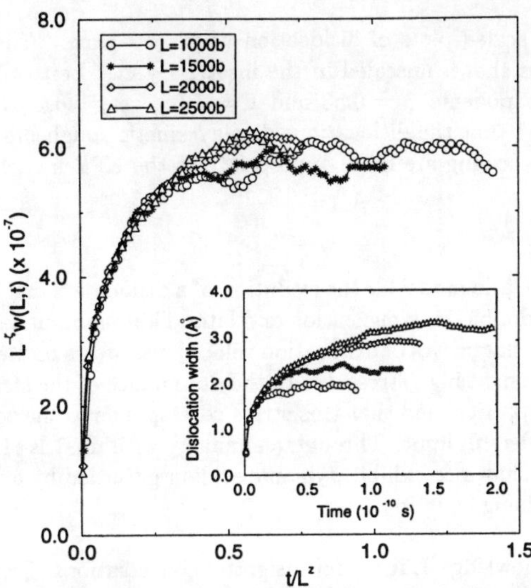

Figure 3: Dislocation width vs. time for four dislocation lengths, scaled according to Eq. (9) using exponents of $\zeta = 0.53, z = 1.55$ (unscaled data are shown in the inset). Data are for an applied stress of 300 MPa and temperature 450 K.

Dynamic Scaling of Dislocation Width

As a way of characterizing the dislocation dynamics, an analogy between dislocation propagation through the corrugated Peierls potential and the growth of thin films on a one-dimensional substrate is insightful [6]. In the analogy, double-kink nucleation/annihilation corresponds to adatom deposition/evaporation on the flat substrate, and kink motion corresponds to addition/removal of an atom at a ledge.

It is a characteristic of crystal growth models that the surfaces display dynamic scaling behavior (for a nice review, see [12]). This scaling behavior is revealed in several structural aspects of the growing surfaces. A commonly studied aspect is the dynamic scaling of the interface width, defined for a dislocation as

$$w(L,t) = L^{-1/2} \left\{ \sum_{i=1}^{N} (h_i(t) - \bar{h}(t))^2 \right\}^{1/2} \tag{9}$$

where $h_i(t)$ is the position of screw segment i in the dislocation at time t (analogous to the height of a growing thin film), $\bar{h}(t)$ is the average position of the dislocation at time t, and N is the number of segments (*i.e* the length of the dislocation is $L = 20bN$).

For simple models of growth it is known that the interface width obeys the dynamic scaling form [13]

$$w(L,t) = L^{\zeta} f(t/L^z) \tag{10}$$

where ζ and z are scaling exponents and $f(x)$ is a scaling function. The exponents often observed are those expected from the analysis of the Kardar-Parisi-Zhang (KPZ) equation [14], which gives $\zeta = 1/2$ and $z = 3/2$.

Fig. 3 contains scaled plots of dislocation width vs. time. (The data, averaged over at least 200 trials, is shown unscaled in the inset.) By eye, best collapse of the data was achieved with the exponents $\zeta = 0.53$ and $z = 1.55$, very close to the KPZ exponents. These results suggest that the dislocations display kinetic roughening, and that the exponents governing this scaling are those appropriate to the KPZ model of thin film growth.

CONCLUSIONS

A kinetic Monte Carlo model for the evolution of a dislocation in a finite medium moving via the processes of double-kink nucleation and lateral kink motion has been developed and studied. It is shown that curves of dislocation velocity vs. stress plotted for several different temperatures will cross at high stress due to the dominance of the stress term in the calculation of the energy barrier, and that this stress corresponds to the zero-temperature yield stress in the infinite-length limit. Through an analogy with models of thin film growth, the dislocation width is shown to exhibit dynamic scaling governed by exponents predicted by the Kardar-Parisi-Zhang equation.

The authors acknowledge J. Hirth for insightful conversations. Computer time provided by the National Energy Supercomputer Research Center is gratefully acknowledged. This work is supported by the Department of Energy, Office of Basic Energy Sciences, Division of Materials Science under contract DE-AC03-76SF00098.

REFERENCES

1. H. R. Kolar, J. C. H. Spence, and H. Alexander, Phys. Rev. Lett. **77**, 4031 (1996).

2. J. Bennetto, R. W. Nunes, and D. Vanderbilt, Phys. Rev. Lett. **79**, 245 (1997).

3. W. Xu and J. A. Moriarty, Comp. Mat. Sci. **9**, 348 (1998).

4. 4. Y. M. Huang, J. C. H. Spence, and O. F. Sankey, Phys. Rev. Lett. **74**, 3392 (1995).

5. A. Seeger and P. Schiller, in *Physical Acoustics*, edited by W. P. Mason (Academic, New York, 1966), Vol. 3A.

6. J. P. Hirth and J. Lothe, *Theory of Dislocations*, 2nd ed. (Krieger Publishing Company, Malabar, Florida, 1992).

7. P. A. Maksym, Semicond. Sci. Technol. **3**, 594 (1988).

8. T. Suzuki, K. H, and H. O. K. Kirchner, Acta mater. **43**, 2177 (1995).

9. S. Nemat-Nasser and J. B. Isaacs, Acta mater. **45**, 1227 (1997).

10. Y. Maurissen and L. Capella, Philos. Mag. **29**, 1227 (1974).

11. Y. Maurissen and L. Capella, Philos. Mag. **30**, 679 (1974).

12. J. Krug and H. Spohn, in *Solids Far From Equilibrium*, edited by C. Godrèche (Cambridge University Press, Cambridge, 1992).

13. F. Family and T. Vicsek, J. Phys. A **18**, 75 (1985).

14. M. Kardar, G. Parisi, and Y. C. Zhang, Phys. Rev. Lett. **56**, 889 (1986).

Part II

Defect Dynamics and Microstructural Evolution

COMPUTER SIMULATION OF DEFECT PRODUCTION AND BEHAVIOUR IN DISPLACEMENT CASCADES IN METALS

D.J. BACON, F. GAO, A.V. BARASHEV AND YU.N. OSETSKY
Materials Science and Engineering, Department of Engineering
The University of Liverpool, Brownlow Hill, Liverpool L69 3GH, U.K.

ABSTRACT

Recent research using molecular dynamics to simulate radiation damage due to displacement cascades in metals is reviewed. It includes results dealing with the effect on defect formation of primary knock-on atom energy and irradiation temperature. Clear dependencies and trends have emerged in these areas. In terms of the development of models to describe the evolution of radiation damage microstructure, the important parameters are not only the total number of Frenkel defects but also the distribution of their population in clusters and the form and mobility of these clusters. Results on these aspects are reviewed and it is shown that computer simulation is providing detailed information that paves the way for successful development of models of the evolution of damage beyond the stage of the cascade process.

INTRODUCTION

Displacement cascades are formed by primary knock-on atoms (PKAs) with a kinetic energy of more than a few hundred eV and are the primary source of damage during heavy-ion and fast-neutron irradiation of metals. Here we review knowledge that has been gained by computer modelling on the nature and properties of the defects they create. The cascade process occurs with length and time scales of the order of nm and ps, respectively, and is ideally suited to study by atomic-scale computer simulation using the method of molecular dynamics (MD). (See reviews [1-4] and references cited therein.) The subsequent evolution of the defects created in cascades gives rise to important changes that affect the performance of metals, and understanding these is crucial for assessing material suitability for service where radiation damage occurs. The evolution can occur over long periods of time with a characteristic length scale consistent with the evolving microstructure, and so is well-suited to continuum and/or stochastic Monte Carlo modelling. These approaches are addressed by other authors in these proceedings. It should be noted, however, that the input parameters for such treatments require quantitative knowledge of not only the number of the defects formed in the primary state but also their distribution and properties. This is where atomic-scale computer simulation is required.

The initial 'collision' phase of a cascade lasts a few tenths of a ps and creates a hot disordered core with some liquid-like characterisitics surrounded by regions of crystal displaced outwards. It is followed by the 'thermal-spike' phase lasting several ps, during which the majority of the displaced atoms in the outer regions quickly return by athermal relaxation to their lattice sites, but due to the strong disorder that persists in the core for a longer time (several ps), some atoms cannot regain lattice sites and so become self-interstitial atoms (SIAs) at the periphery of the core. The vacant sites are finally formed within the core when it crystallises. A few SIAs may created by individual replacement-collision sequences ejected in the collision phase, but these represent only a minor part of the population of defects produced by cascades.

Mat. Res. Soc. Symp. Proc. Vol. 538 © 1999 Materials Research Society

THE EFFECT OF PKA ENERGY FOR DIFFERENT METALS

We shall now consider the relationship between the number, N_F, of SIA-vacancy pairs, i.e. Frenkel defects, created by a cascade and the kinetic energy, E_p, of the PKA. The industry standard for this, which leads directly to an estimate of the displacements per atom (DPA) in irradiated metals, is the NRT formula [5,6], expressed as

$$N_{NRT} = 0.8E_{dam} / 2\overline{E}_d , \tag{1}$$

where N_{NRT} is the value of N_F in the NRT formulation, $\overline{E}_d$ is the value of the threshold displacement energy averaged over all crystallographic directions and E_{dam} is the damage energy available for elastic collisions, i.e. E_p with inelastic losses subtracted. (As noted in [2], since electronic losses are not been included in most simulations, the use of E_{dam} for E_p in eq. (1) is appropriate for comparing the MD data with the NRT prediction.)

The binary collision model on which the NRT formula is based does not accurately describe atomic interactions in the thermal spike and seriously overestimates the number of defects produced (see below). All the MD simulations to date show that defect production by displacement cascades in metals is not as efficient as indicated by the NRT formula. In fact, N_F is typically only 20 to 40% of N_{NRT} for a given cascade energy when E_p is larger than about 1-2 keV. By consideration of MD results for PKA energy up to either 5 or 10 keV for several metals, it was shown in [2] that the following empirical relationship between N_F and E_{dam} gives a good fit to the simulation data:

$$N_F = A (E_p)^m , \tag{2}$$

where A and m are constants which are weakly-dependent on the material and temperature. The wider range of applicability of eq. (2) is shown in fig. 1 by using more recent data for N_F for E_p up to 10 keV for Cu [7], α-Ti [8] and Ni_3Al [9], 20 keV for α-Zr [8] and 40 keV for α-Fe [10,11]. Each data point is the mean N_F value for at least four cascades at that energy and all the data were obtained for simulations of model crystals at a temperature of 100 K. We have added the value of N_F reported for 25 keV cascades in Cu at 10 K [12]. Almazouzi et al. [13] have recently simulated cascades in Al and Ni at 10 K, with E_p up to 15 and 30 keV, respectively, and their data for N_F is included in the figure.

Eq. (2) provides a good fit to the data and appears to have validity over a wide energy range. A line showing the NRT estimate for N_F with an $\overline{E}_d$ value of 40 eV is included in fig. 1 and demonstrates the reduced efficiency with which simulated cascades produce Frenkel defects in comparison with the NRT prediction. This reduction occurs because most SIAs are produced at the periphery of the disordered core and their close proximity to the vacancies and the high kinetic energy of the core during the thermal spike assist SIA-vacancy recombination. The MD data are consistent with experiments [14-16] which imply that visible defect production efficiency in pure metals and alloys under cascade-producing irradiation with ions and neutrons is approximately one quarter of the NRT value.

One conclusion to be drawn from the MD studies is that, with the potentials currently available, the value of N_F has no discernible dependence on crystal structure, for Al, Cu and Ni are FCC, α-Fe is BCC, α-Ti and α-Zr are HCP and Ni_3Al has the ordered $L1_2$ structure. This implies that any generic differences observed experimentally between the damage microstructure in pure BCC, FCC and HCP metals are not due to differences in defect production efficiency in

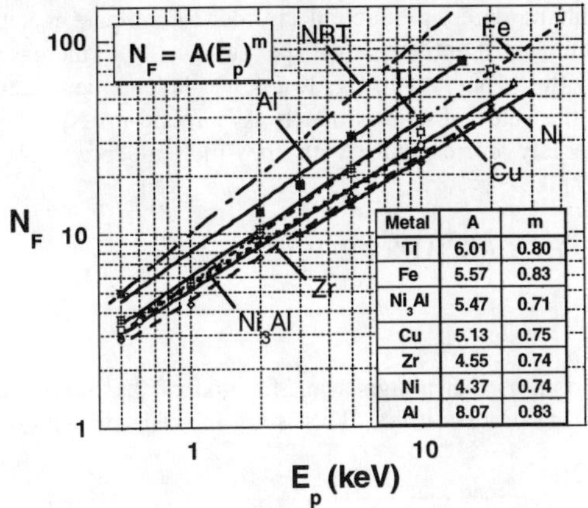

Metal	A	m
Ti	6.01	0.80
Fe	5.57	0.83
Ni$_3$Al	5.47	0.71
Cu	5.13	0.75
Zr	4.55	0.74
Ni	4.37	0.74
Al	8.07	0.83

Fig. 1.   Log-log plots of N_F vs E_p for Ni$_3$Al and the four pure metals Cu, Fe, Ti and Zr at 100K, and Al and Ni at 10K, demonstrating the power-law dependence of eq. (2) in the text. The inset table shows the values of m and A (with E_p in keV) obtained with the best-fit lines shown in the Figure. The data for Al and Ni were kindly supplied by Almazouzi et al. [13].

the cascade process itself. Even Ni$_3$Al exhibits a similar dependence of Frenkel defect number on cascade energy to the pure metals.

Another outcome is that the differences in N_F that do exist between the metals appear to be due mainly to the atomic mass, for N_F is highest in Al, followed by Ti and Fe. As can be seen from the Table inserted in fig. 1, as the atomic mass increases, parameter A in eq. (2) decreases, suggesting either an enhancement of recombination due to thermal spike effects or a scaling law arising from an increase in $\overline{E}_d$. Furthermore, the exponent m in the power-law relationship also reflects a dependence on the atomic mass of the metal, for it is approximately 0.8 for Al, Ti and Fe, and nearer 0.75 for the other metals. As suggested in [11] and [13], for example, the higher value obtained by fitting eq. (1) across the whole range of E_p for the lighter elements may arise because as cascade energy increases, there is an increased tendency for cascades to break up into sub-cascades. Since one cascade produces fewer defects on average than two separate cascades of the same total energy, sub-cascade formation leads to an increase in m. This transition occurs at lower energy in the lighter metals, as can be discerned for Al and Fe.

DEFECT PRODUCTION IN ALLOYS

A few studies have considered the effects of alloying elements in solution. Cascades of up to 2 keV in energy were simulated in Cu containing up to 15 at% Au in solution in [17]. The presence of Au decreased the length of focused displacement events in the ballistic phase and thereby enhanced the intensity and lifetime of the disorder of the thermal spike. However, it was found that N_F did not depend on the alloy composition, at least for the E_p range considered. Cascades of 1-20 keV energy in Fe-Cu alloys have also been investigated by MD to see if N_F is affected by the presence of up to 1% Cu in solution and if cascades change either the cluster distribution of solute or produce clusters of Cu atoms in association with point defects [18]. The number of Frenkel pairs formed in the alloy was found to be similar to that in pure Fe and no evidence was found for significant Cu-Cu clustering during the cascade process. The Cu-Cu binding energy is too small and the cascade lifetime too short.

Most computer modelling of cascades in alloys has been concerned with ordered alloys. The number of Frenkel pairs produced is comparable with that for pure metals (see fig. 1 for Ni$_3$Al). However, additional damage effects can occur, either because disorder is created when sites in the individual sub-lattices become occupied by atoms of the wrong type to form antisite defects or crystalline order is not restored and an amorphous structure is formed. In fact, zones of either disordered crystal or amorphous structure are the dominant product of cascade damage in such materials, e.g. [19-20]. Computer simulations have been used to investigate both aspects, e.g. [9,21-24]. Here, we simply consider antisite defect production in order to compare it with N_F. Simulations [9] of Ni_3Al show that the number of antisite defects per cascade, N_{AS}, increases with increasing E_p, as shown in fig. 2. As with the N_F data in fig. 1, a power law applies again, but with an exponent of 1.25, in contrast to the value of approximately 0.71 found for N_F. The zone of antisites formed by a cascade largely corresponds with the 'molten' core [9,21], demonstrating the importance of the thermal spike.

CLUSTERING OF INTERSTITIALS IN THE CASCADE PROCESS

The clustered fraction

An important outcome of MD simulations is that a significant fraction of the interstitial population is produced in clusters in the cascade process itself. This is not inconsistent with the experimental measurement of diffuse X-ray scattering in copper irradiated with neutrons at 4.6 K [25]. It is of importance because interstitial clusters are thermally stable and, if they can move away from their parent cascades by rapid one-dimensional glide (see below), can be absorbed preferentially at sinks such as dislocations and boundaries. Vacancy clusters, in contrast, are not stable and dissociate into point defects at high enough temperature. This leads to a 'production bias' in favour of vacancies in the supply of point defects from cascades [26].

The probability of clustering and the size of the largest clusters increase with increasing PKA energy, and a higher proportion of SIAs than vacancies form

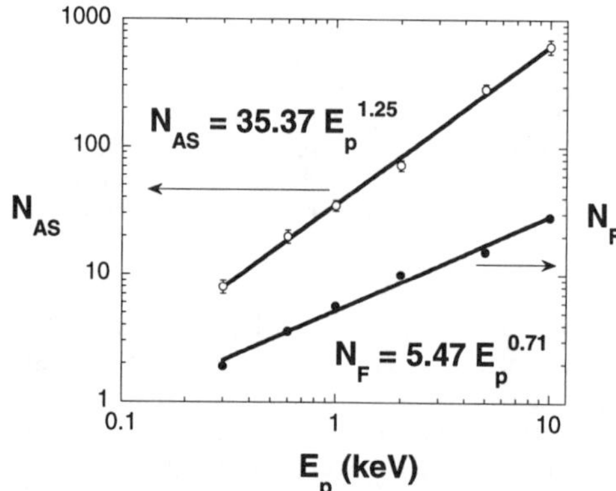

Fig. 2. *The average total number of antisite defects, N_{AS}, and Frenkel pairs, N_F, per cascade in Ni_3Al at 100K as a function of E_p [9].*

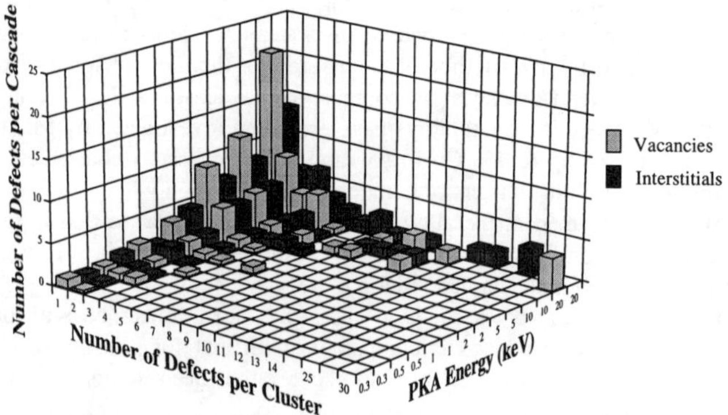

Fig. 3. *Data for the number of SIAs in interstitial clusters and vacancies in vacancy clusters per cascade as a function of cascade energy in α-Zr at 100K [8]. The data were obtained by averaging over all cascades at each energy.*

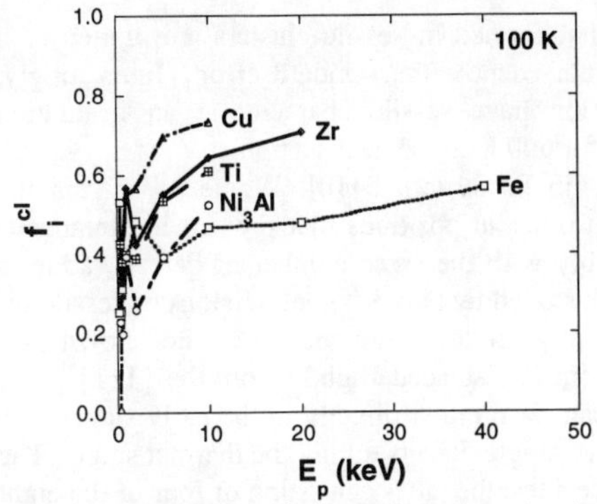

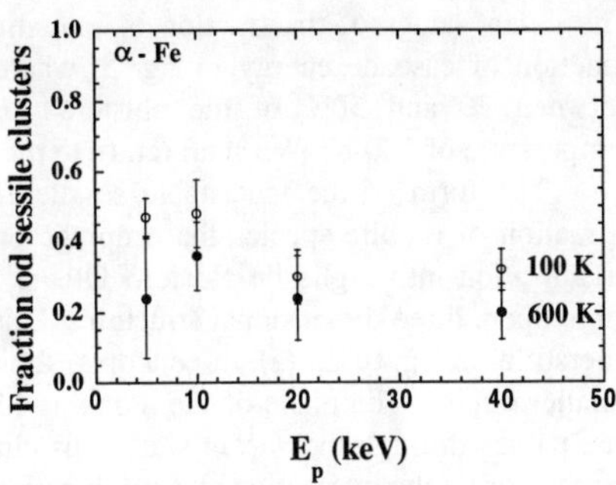

Fig. 4.   Fraction, f_i^{cl}, of SIAs that survive in clusters of two or more in Cu, α-Fe, α-Ti, α-Zr and Ni$_3$Al at 100K.

Fig. 5.   The fraction of SIA clusters formed in a metastable sessile configuration by cascades in α-Fe as a function of PKA energy and irradiation temperature [10].

clusters. These features hold true for all the metals studied to date and are demonstrated by the example of the histogram for cascades of up to 20 keV in energy in α-Zr in fig. 3 [8]. (A cluster is defined such that every defect has at least one other in a nearest-neighbour position.) Although these trends are general, differences in the level of clustering may exist between different metals. For example, fig. 4 shows the fraction, f_i^{cl}, of interstitials found to exist in clusters of size two or larger for MD simulation of cascades in Cu, α-Fe, α-Ti, α-Zr and Ni$_3$Al at 100 K. (The sources of this data are the references cited above for fig. 1.) f_i^{cl} varies from metal to metal, being highest in copper and lowest in iron, and does not seem to be related to the differences in N_F. Since f_i^{cl} is very similar in the two HCP metals Ti and Zr, despite the significant differences in their atomic mass and N_F values, it is tempting to speculate that the differences in f_i^{cl} between the metals may be due to the influence of crystal structure in restricting the ability of SIAs to move and cluster during the thermal spike. Confirmation of this must await studies of a wider range of metals.

The glissile-sessile nature of interstitial clusters

Most of the clustered SIAs formed in cascades simulated by MD are seen to have high mobility in the context of the time-scale of the simulations (see next section). They exhibit a morphology akin to perfect dislocation loops and are therefore 'glissile', i.e. each can glide on the prism defined by its periphery and Burgers vector. Not all clusters have this form, however, for 'sessile' clusters have been reported in Fe [10], Zr [8] and Ti [27]. They consist of metastable arrangements of SIAs that have not managed to reorganise into a stable, loop-like form by the end of the thermal spike. They are not restricted in size and one of 25 SIAs was formed by a 20 keV cascade in Zr. (Note that another class of immobile interstitial cluster can exist, such as that based on faulted Frank loops in the FCC metals. They can be formed from sets of <100> dumbbells [31], are intrinsically sessile and are not considered further here.) Since the metastable sessile defects would not be able to migrate away from their parent cascade without converting first to their stable glissile state, their population and properties, and their lifetime for conversion to the glissile form at different temperatures, are of interest.

These issues have been considered for cascades with energy between 5 and 40 keV in α-Fe at 100 and 600 K [10]. At least four simulations were carried out at each energy and temperature, and were stopped when the cascade zone had cooled to the ambient temperature (after about 20-30 ps). The glissile and sessile components of the clustered fraction of SIAs were then identified.

The mean values of the fraction of interstitials that formed in sessile clusters are plotted as a function of cascade energy in fig. 5, where the bars denote the standard error. Interestingly, between 30 and 50% of the clustered population have sessile character at an irradiation temperature of 100 K. We shall return to the data for 600 K in the next section.

The form of the metastable sessile clusters in Fe is varied [10]. Some arise from the attraction of mobile species that impinge on each other in positions that prevent instantaneous organisation into a glissile cluster. Others, possibly with the same number of defects, adopt a more open, three-dimensional structure. This is illustrated by two different versions of sessile di-interstitial in fig. 6. In (a), three atoms, shown as large circles, share one lattice site, shown by a smaller circle. The plane of the atoms is {111}, but is displaced slightly from the {111} plane that passes through the vacant site. This cluster can form either directly in the early stages of a cascade or by the interaction and combination of two single defects during the thermal spike. The cluster in (b) arises from 12 atoms sharing 10 lattice sites, the latter consisting of four of the eight nearest neighbours and all six second-nearest neighbours of an atom in the BCC structure. (The smaller light grey circles in fig. 6(b) represent sites in this cell that are occupied by atoms.) This symmetric, three-dimensional cluster was observed in about half of the cascades simulated. Higher-order SIA clusters in α-iron exhibit more complicated variants of these defects [10].

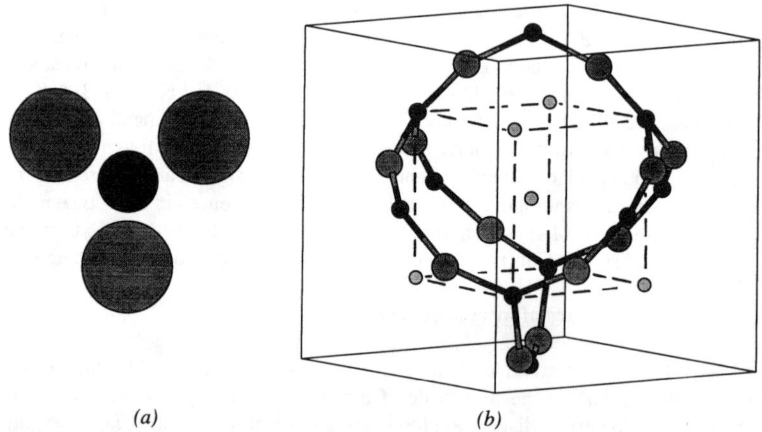

(a)   *(b)*

Fig. 6.   Two versions of metastable di-interstitials created in cascade simulations of α-Fe. The large spheres represent atoms displaced to interstices and the smaller dark spheres represent vacant lattice sites. In (b), the smaller light spheres represent the sites of a BCC unit cell (shown by dashed lines) occupied by atoms. The defect in (a) is three atoms in a {111} plane sharing one lattice site, whereas the one in (b) is 12 atoms sharing 10 sites.

The mobility of glissile interstitial clusters

Detailed investigation of the structure and motion of SIA clusters has been carried out by MD for Cu and Fe using several different interatomic potentials, e.g. [28-32], and a preliminary study for Zr has been reported [33]. In qualitative terms, the properties do not depend on the type of interatomic potential used. Mobile clusters have the form of densely-packed, parallel crowdions, with the crowdion axis usually along the close-packed direction, i.e. <110> in FCC, <111> in BCC and <11$\bar{2}$0> in HCP, and this is the direction of their perfect Burgers vector **b** in the dislocation description. The crowdion form persists even in metals where the stable single SIA is a dumbbell rather than a crowdion. In α-Fe, clusters of more than about 15-20 <100> crowdions describe small, stable, perfect loops with **b** = <100>.

The crowdion clusters exhibit thermally-activated, one-dimensional glide along the crowdion direction and only small clusters (<4) change their glide direction during the period of MD simulation (~5ns). During their motion, they produce displacements of the atoms in the model crystallite and analysis shows that it cannot be described as classical diffusion [32]. Nevertheless, some dynamical characteristics have been estimated by using MD to study the movement of the centre of gravity of the SIAs in moving clusters over periods of the order of ns. An important feature of motion is that the effective correlation factor is greater than 1, i.e. a cluster that has moved by one step in one direction has a high probability of making the next step in the same direction. This effect increases at low temperature and is probably related to the response of crowdion clusters to specific phonons [31].

The jump frequency for clusters containing from 2 to 19 <111> crowdions in α-Fe at different temperatures is presented as an Arrhenius-like plot in fig. 7(a). The estimated 'activation energy' is within the range 0.022-0.026 eV for all clusters and is close to the migration energy estimated for the low-temperature <111> single crowdion mechanism (0.023 eV). A cluster of <100> crowdions behaves in a similar fashion [31] with a slightly higher activation energy (0.028 eV). The jump frequency of a cluster of 37 <110> crowdions in α-Fe is presented for different temperatures in fig. 7(b) [32]. The vacancy cluster data in fig. 7(b) is described in the Discussion section.

The data for α-Fe have been analysed by Barashev et al. [34] in order to compare the average jump frequency of the individual crowdions in a cluster with that of the cluster as a whole. The jump frequency of a cluster is found to be approximately equal to the average jump frequency of an individual crowdion in the cluster divided by the number of crowdions. This is consistent with the interpretation that the movement of the centre of mass of an SIA cluster is the result of near-independent jumps of the individual crowdions. However, it should be stressed that there is more complexity in the mobility of SIA clusters in terms of size than represented in this

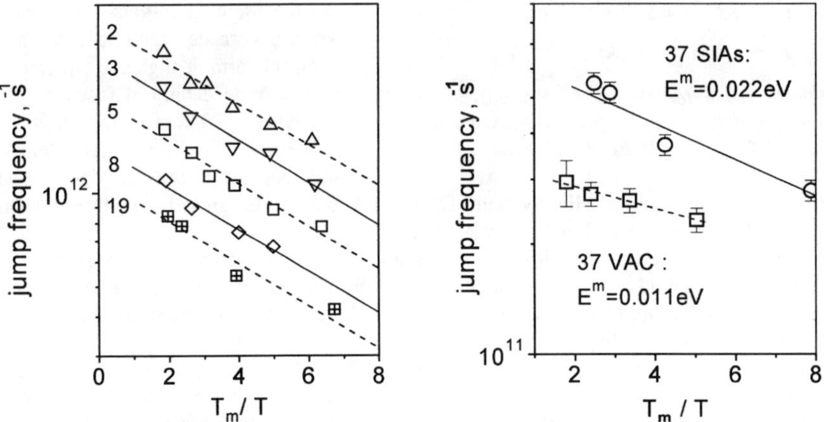

Fig. 7. (a) Jump frequency of the centre of mass of SIA clusters consisting of <111> crowdions in α-Fe at different temperatures T [31,32]. The number at the left of each data line is the number of interstitials in the cluster and Tₘ is the melting temperature (1811K). (b) Jump frequency of the centre of mass of a clusters consisting of either 37 <110> crowdions or 37 vacancies in α-Fe at different temperatures [32].

simple picture. For example, when perfect loops with $\mathbf{b} = 1/2<110>$ in Cu grow beyond about 64 interstitials, they dissociate on their glide prism and this restricts thermally-activated motion [32].

THE EFFECTS OF IRRADIATION TEMPERATURE ON CASCADE DAMAGE

The motion and dissociation of point defects and their clusters are thermally-activated and so temperature has a significant effect on the evolution of radiation damage in metals. In this section, however, we only consider the role of temperature on the production of damage in the cascade process itself, which, as noted earlier, has a lifetime of a few ps at most. In most reported MD simulations of cascades, the PKA was generated in a lattice equilibrated at 100 K or less. The effect of ambient bulk temperature, T_{irr}, on cascade damage in both Cu and Fe was reported in [7,35] but the MD block was treated as an adiabatic system. A more rigorous method to allow for heat to be dissipated into a surrounding medium with the same thermal characteristics as the atomic system has been developed and applied to α-Fe [36] and Ni_3Al [37].

Results for N_F as a function of T_{irr} in Fe and Ni_3Al for $E_p = 10$ keV are plotted in fig. 8, where each point is the mean for at least four cascades and the bars represent the standard error.

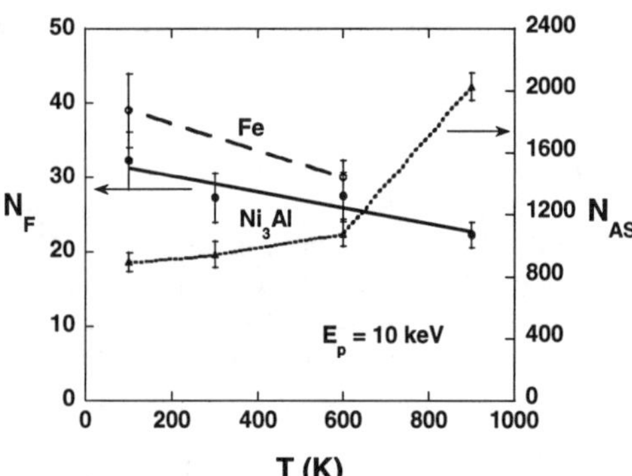

Fig. 8.   Data for N_F per 10 keV cascade as a function of T_{irr} in α-Fe [36] and Ni_3Al [37]. The values of N_{AS} in Ni_3Al are also plotted.

The effect on N_F of increasing T_{irr} is to decrease the number of defects produced, but only by a few percent per hundred degrees. The similar trend in Fe and Ni_3Al suggests that it does not depend on crystal structure and alloy state. It is believed to be due to the increase in the lifetime of the thermal spike [36,37], which allows more defect motion to take place and so leads to more interstitial-vacancy recombination.   Another contributory factor may be the reduction in the interstitial-vacancy separation because cascades tend to have a more compact form at higher temperature due to the shorter length of focused collision sequences.   These same effects also influence the interstitial clustering fraction, f_i^{cl}, discussed earlier, but in contrast to N_F, it increases slightly with T_{irr} [36,37], due to the greater mobility and smaller separation of the SIAs in the longer thermal spike phase.

In the case of the ordered alloy Ni_3Al, the simulations [37] show that the final number, N_{AS}, of antisite defects greatly exceeds N_F at all temperatures and, furthermore, increasing T_{irr} actually enhances the production of antisite defects, particularly at high temperature, as shown in fig. 8. This reflects a very dense form of the damage zone at high temperature.

It was noted above that a sizeable fraction of the SIA clusters formed by cascades in α-Fe at 100 K has a sessile form. This fraction is temperature dependent, however, as can be seen in fig. 5 [10]. The fraction of interstitials in clusters increases between 100 and 600 K, but the fraction of them in sessile clusters decreases, due to the higher probability of a transformation from sessile to glissile form at the higher temperature. Nevertheless, the sessile fraction is still significant, and it may be remarked that the largest cluster (of 18 interstitials) generated in all of the cascades represented in fig. 5 was sessile in form.

DISCUSSION

The good fit of eq. (2) to the MD data for N_F as a function of E_p suggests that this power-law relationship has wide applicability, irrespective of the crystal structure, over a wide range of PKA energy up to high energies where sub-cascades are formed. Thus, an important result to have emerged from MD simulations is that there are no significant variations between different metals in terms of the efficiency of Frenkel pair production in cascades. This shows that differences in the response of metals to radiation damage probably arises in most cases from the properties of the defects involved in evolution after the primary damage state.

Another key result from MD simulations is that SIA clusters form directly in cascades in all metals. Some are large (tens of SIAs) and it is of interest that firm experimental evidence in support of this has now been reported [38]. Many of these clusters seen in MD simulations are mobile and the recent studies referred to here provide insight into the nature of this mobility. The mobility arises from the clustered crowdion form of these extended defects, which allows them to behave as interstitial dislocation loops with **b** equal to 1/2<110> in FCC, 1/2<111> and <100> in BCC and 1/3<11$\bar{2}$0> in HCP. They are therefore perfect dislocations and, although their mobility depends on size, they are <u>intrinsically</u> glissile. In contrast with this, clusters that form as faulted dislocation loops, such as **b** = 1/3<111> in FCC, 1/2<110> in BCC and 1/2<10$\bar{1}$0> in HCP, are <u>intrinsically</u> sessile and cannot exhibit glide motion. This shows that stacking fault energy cannot be ignored in assessing whether interstitial clusters may be mobile or not.

We should also remark that <u>vacancy</u> clusters that collapse to form dislocation loops with a perfect Burgers vector are also intrinsically glissile. In fact, a recent MD study [32] has shown that their mobility in Cu and α-Fe is only slightly lower than that of clusters of the same number of interstitials, as illustrated by the data for the jump frequency of vacancy and SIA loops as a function of temperature in Fe in fig. 7(b). This result demonstrates that the motion of such defects should not be neglected in the modelling of damage evolution. Of course, if vacancy loops have a partial Burgers vector or form stacking fault tetrahedra, they are intrinsically sessile. Also, if vacancy clusters do not collapse to a dislocation structure, as occurs most commonly in MD simulations of cascades, they cannot move by glide, only by diffusion of individual members of the cluster.

As described in the preceding section, a significant fraction of the interstitial clusters created in cascade simulations of α-Fe has a sessile form, even at high temperature, although they are not intrinsically sessile loops. If these defects were to persist over long time-scales, they could have possibly important consequences for damage evolution, and so their stability has been investigated [10] by annealing representative ones individually by MD for long periods ($\geq$ 1ns) to determine the temperature-dependence of their lifetime before transformation to the more stable glissile form of <111> crowdions. The lifetime of several clusters of different size is plotted in fig. 9. Most transform in well under 1ns at temperatures in the range 500–900K, with an activation energy for the transformation of about 0.35-0.55eV. This implies that they mainly affect the annealing of cascades rather than microstructure evolution in the longterm. However, an eight-interstitial cluster based on the defect shown in fig. 6(b) did not transform in the 1ns simulations, even at 1500K [10]. In view of this, the stability of this defect is currently under investigation using different interatomic potentials.

Although the lifetime of most sessile SIA clusters may be short, they can interact with glissile clusters. In fact, simulations reported in [10] provide an example where such an interaction led to a change in glide direction, i.e. **b**, of a moving cluster. Effects such as this may offer a way for SIA clusters which glide one-dimensionally to change the direction of their motion.

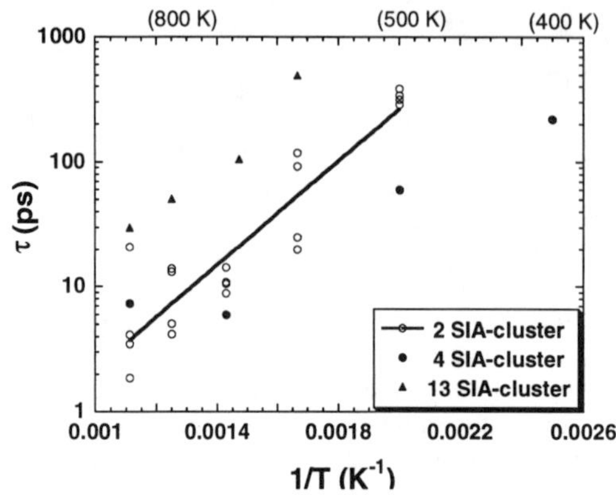

Fig. 9. The lifetime of several different metastable SIA clusters in α-Fe as a function of annealing temperature [10].

The interatomic potentials used in the simulations reviewed here are suitable for the scale of MD modelling required for cascade studies, i.e. MD cells containing ~10^4 to 10^6 atoms, but lack directionality in the atomic bonds. It is not known to what extent a more accurate description of the interatomic forces for, say, the BCC transition metals would affect the production of defects by the cascade process. Most probably, the ballistic phase and the early part of the thermal-spike phase would not be influenced to any significant extent, but later on angular forces that could affect defect behaviour would play a part. It seems unlikely that it would seriously alter the general trends for N_F and f_i^{cl} discussed here, but it could influence the mobility and clustering behaviour of SIAs. This could come about, for instance, if the <110> dumbbell single interstitial in the BCC metals does not actually adopt the <111> crowdion form when in clusters of two or more, which is the configuration found with the potentials currently in use.

The form of the vacancy component of cascades is an important element in damage evolution. However, it should be noted that the vacancy clusters observed in most MD simulations are not collapsed into forms with dislocation character, whereas vacancy dislocation loops and stacking fault tetrahedra (a few nm in size and created in the vacancy-rich core of cascades) produced by cascade collapse have been observed by transmission electron microscopy in ion-irradiated foils of many metals [39,40]. As discussed in [3,41], there is a view, based on both computer simulation and experiment, that the mechanisms leading to the formation of collapsed vacancy clusters in ion-irradiated thin foils do not apply to conditions in the bulk. Also, it is not clear whether the short-range, many-body interatomic potentials used for the vast majority of cascade simulations provide a suitable framework for describing atomic behaviour in a super-saturation of vacancies. Recent simulations [42,43] of vacancy clusters in Cu using short-range, many-body potentials and long-range, non-equilibrium pair potentials have shown how sensitive some of the properties of these clusters are to the form of the interatomic potential. There is undoubtedly a need for the development of potentials that can be used with confidence for all aspects of cascade and defect simulation in metals.

ACKNOWLEDGEMENTS

This research was supported by the Engineering and Physical Sciences Research Council, Magnox Electric plc and the University of Liverpool and an ORS Award of the Committee of Vice-Chancellors and Principles to AVB.

REFERENCES

1. D.J. Bacon and T. Diaz de la Rubia, J. Nucl. Mater. **216**, 275 (1994).

2. D.J. Bacon, A.F. Calder, F. Gao, V.G. Kapinos and S.J. Wooding, Nucl. Instrum. and Meth. B **102**, 37 (1995).

3. D.J. Bacon, A.F. Calder and F. Gao, Rad. Eff. Def. Sol. **141**, 283 (1997).

4. D.J. Bacon, A.F. Calder and F. Gao, J. Nucl. Mater. **251**, 1 (1997).

5. M.J. Norgett, M.T. Robinson, and I.M. Torrens, Nucl. Eng. Design **33**, 50 (1975).

6. Standard E521, *ASTM Annual Book of Standards* (1989).

7. W.J. Phythian, A.J.E. Foreman, R.E. Stoller, D.J. Bacon, A.F. Calder, J. Nucl. Mater. **223**, 245 (1995).

8. S.J. Wooding, L.M. Howe, F. Gao, A.F. Calder and D.J. Bacon and, J. Nucl. Mater. **254**, 191 (1998).

9. F. Gao and D.J. Bacon, Phil. Mag. A **71**, 43 (1995), ibid., **71**, 65 (1995).

10. F. Gao, D.J. Bacon and Yu.N. Osetsky, submitted to J. Nucl. Mater.

11. R.E. Stoller, in *Microstructure of Irradiated Materials*, eds. I.M. Robertson, L.E. Rehn, S.J. Zinkle and W.J. Phythian, Sympos. Proc. vol. 373 (MRS Pittsburgh, 1995) p. 21.

12. T. Diaz de la Rubia and W.J. Phythian, J. Nucl. Mater. **191-194**, 108 (1992).

13. A. Almazouzi, M.J. Caturla, T. Dias de la Rubia and M. Victoria, J. Nucl. Mater. in press.

14. R.S. Averback, R. Benedek and K.L. Merkle, Phys. Rev. B **18**, 4156 (1978).

15. P. Jung, J. Nucl. Mater. **117**, 70 (1983).

16. J.H. Kinney, M.W. Guinan and Z.A. Munir, J. Nucl. Mater. **122-123**, 1028 (1984).

17. H.F. Deng and D.J. Bacon, Phys. Rev. B **53**, 11376 (1996).

18. A.F. Calder and D.J. Bacon, in *Microstructure Evolution During Irradiation*, eds. T. de la Rubia et al., Sympos. Proc. vol. 439 (MRS Pittsburgh, 1997) p. 521.

19. M.L. Jenkins, J. Nucl. Mater. **216**, 124 (1994).

20. N.Q. Lam, P.R. Okamoto and M. Li, J. Nucl. Mater. **251**, 89 (1997).

21. T. Diaz de la Rubia, A. Caro and M. Spaczer, Phys. Rev. B **47**, 11483 (1993).

22. M. Spaczer, A. Caro, M. Victoria and T. Diaz de la Rubia, Phys. Rev. B **50**, 13204 (1994).

23. H. Zhu, R.S. Averback and M. Nastasi, Phil. Mag. A **71**, 735 (1995).

24. M. Spaczer, A. Almazouzi, R. Schublin and M. Victoria, Rad. Eff. Def. Sol. **141**, 349 (1997).

25. R. Rauch, J. Peisl, A. Schmalzbauer and G. Wallner, J. Nucl. Mater. **168**, 101 (1989).

26. C.H. Woo and B.N. Singh, Phil. Mag. A **65**, 889 (1992).

27. S.J. Wooding, D.J. Bacon and W.J. Phythian, Phil. Mag. A **72**, 1261 (1995).

28. B.D. Wirth, G.R. Odette, D. Maroudas and G.E. Lucas, J. Nucl. Mater. **244**, 185 (1997).

29. N. Soneda and T. Diaz de la Rubia, Phil. Mag. A in press.

30. Yu.N. Osetsky, A. Serra, V. Priego, F. Gao and D.J. Bacon, in *Diffusion Mechanisms in Crystalline Solids*, eds. Y. Mishin et al., Symp. Proc. vol. 527 (MRS Pittsburgh, 1998) p. 49.

31. Yu.N. Osetsky, V. Priego and A. Serra, B.N. Singh and S.I. Golubov, Phil. Mag. A, in press; in *Diffusion Mechanisms in Crystalline Solids*, eds. Y. Mishin et al., Sympos. Proc. vol. 527 (MRS Pittsburgh, 1998) p. 59.

32. Yu.N. Osetsky, D.J. Bacon and A. Serra, , these proceedings; Phil. Mag. Lett., submitted.

33. B. Whiting and D.J. Bacon, in *Microstructure Evolution During Irradiation*, eds. T. de la Rubia et al., Sympos. Proc. vol. 439 (MRS Pittsburgh, 1997) p. 389.

34. A.V. Barashev, Yu.N. Osetsky and D.J. Bacon, these proceedings.

35. A.F. Calder and D.J. Bacon, J. Nucl. Mater. **207**, 25 (1993).

36. F. Gao, D.J. Bacon, P.E.J. Flewitt and T.A. Lewis, J. Nucl. Mater. **249**, 77 (1997).

37. F. Gao and D.J. Bacon, submitted to Phil. Mag. A; these proceedings.

38. M.A. Kirk, M.L. Jenkins and H. Fukushima, these proceedings.

39. C.A. English and M.L. Jenkins, Mater. Sci. Forum **15-18**, 1003 (1987).

40. C.A. English, A.J.E. Foreman, W.J. Phythian, D.J. Bacon and M.L. Jenkins, Mater. Sci. Forum **97-99**, 1 (1992).

41. K. Nordlund, R.S. Averback and M. Ghaly, these proceedings.

42. Yu. N. Osetsky, M. Victoria, A. Serra, S.I. Golubov and V. Priego, J. Nucl.Mater. **251**, 34 (1997).

43. Yu. N. Osetsky, A. Serra, M. Victoria, V. Priego and S. I. Golubov, Phil. Mag., A in the press.

SHOCK WAVE RESPONSE OF MATERIALS AT DIFFERENT LENGTH SCALES

Y.M. GUPTA
Institute for Shock Physics and Department of Physics, Washington State University, Pullman, WA 99164-2814, ymgupta@wsu.edu

ABSTRACT

After a brief introduction to shock wave propagation in solids, recent experimental developments to understand the real time response of shocked materials at various length scales are outlined. Results from two representative studies are summarized: x-ray diffraction to examine elastic-plastic deformation in shocked LiF crystals; and picosecond optical absorption to examine structural transformation in shocked CdS crystals. Issues related to examining material processes at different length scales in shock wave experiments are discussed briefly.

INTRODUCTION

Plane shock wave experiments are well suited to examine the time-dependent processes governing deformation induced changes in solids. Some important features of these experiments are: (i) the ability to characterize the macroscopic strain state exactly, (ii) loading times are comparable to time scales associated with molecular/atomic/ionic motions, (iii) deformations can be imparted precisely along different crystal orientations to evaluate anisotropy effects, and (iv) a large range of compression and temperature variations can be produced in condensed materials. Hence, shock wave compression experiments play an important role in materials physics and chemistry. Contemporary problems of interest and the considerable literature in this field can be accessed through proceedings of the biennial topical conferences.[1]

A shock wave is a supersonic disturbance across which large changes in compression, particle velocity, and internal energy occur in a near discontinuous manner.[2] Because of the single event nature of the experiments and the simultaneous variation of several continuum (or thermodynamic) parameters, scientific studies have been dominated by experimental measurements and analyses at the continuum scale.[3] Although shock wave compression can result in profound changes in condensed materials (elastic-plastic deformation, structural phase changes, insulator-metal transitions, molecular dissociation, chemical reactions, etc.),[1] a detailed understanding of these phenomena at different length scales has been limited by the experimental/analytic developments applicable to the loading conditions associated with these experiments.

Before discussing recent experimental advances, which constitute the thrust of this paper, brief comments about computational simulations of shock wave phenomena are in order. As far as the dynamics of processes are concerned (not calculations of final thermodynamic states), the computations have been dominated by continuum modeling of material response and the resulting simulations of measured wave profiles.[4] Advances in computational capabilities have provided improved molecular dynamics (M.D.) computations of shock induced physical and chemical changes.[5,6] More physically realistic M.D. simulations of elastic-plastic deformation, structural changes, and chemical reactions in shocked materials are expected to provide good insight into

Mat. Res. Soc. Symp. Proc. Vol. 538 © 1999 Materials Research Society

the governing processes at different length scales. To fully utilize improved
computational simulations of shock induced changes, there is an important need to
obtain real time measurements at different length scales. The work at Washington
State University has focused on this need[3] and some recent developments are
presented here.[7,8]

In this paper, two types of shock wave induced phenomena are discussed: elastic-
plastic deformation and structural phase changes. The stresses of interest for the
representative examples being considered range between 1-10 GPa. Given the
materials (LiF and CdS) and the modest stresses of interest, the resulting temperature
changes are too small to have an appreciable effect. Hence, the shock wave can be
viewed as a purely mechanical disturbance for the present discussion.

BACKGROUND

The scientific problem of interest can be defined[3] by considering the schematic
representation shown in Figure 1. The top part of the figure represents a half-space
subjected to a rapid impulsive loading resulting in uniaxial strain in the material. Such a
strain state can be produced experimentally through flyer plate impact, use of high
explosives, or radiation deposition. Because of material compressibility, the
compression imparted at the surface of the half space moves through the material as a
step discontinuity or a shock wave. The shock wave velocity depends on the peak
longitudinal stress and the sample material properties.

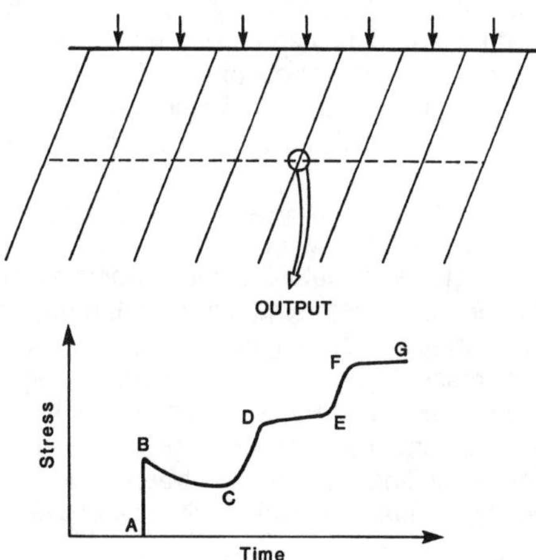

FIGURE 1: Uniaxial strain loading of a material (top); hypothetical wave
structure measured by an in-situ, idealized stress transducer (bottom).

The wave shape or the occurrence of a multiple wave structure, at a particular
location in the sample, as shown in the bottom part of the figure is a consequence of the
material's response to the rapid and large increase in stress. The nonlinear coupling
between the large amplitude shock wave and the resulting material response requires
that proper analyses of the experimental data consider both temporal and spatial

dependence of shock induced changes in material properties.[9] Although the observed multiple wave structure is caused by uniaxial strain induced changes (e.g. elastic-plastic deformation and/or structural transformation) in the material, the lack of such a wave structure does not always mean that significant material changes did not occur. In many cases, the "continuum scale" measurements may be too coarse to resolve the changes at smaller length scales.

An important element of our investigations has been to examine and understand the state of the material through points ABCDEFG. In particular, we want to determine the atomic/molecular structure corresponding to the stress-history shown in Figure 1. We emphasize that determining and understanding the material response at a smaller length scale does not preclude the need for a good understanding at the continuum scale. Since the shock wave represents a macroscopic or continuum deformation, knowledge of the continuum material response is necessary to an understanding at the lower length scales. The situation is somewhat, though not exactly, analogous to understanding dislocation motion and dynamics in quasi-static, uniaxial stress experiments.

To discuss the material phenomena in the next two sections, it is convenient to decompose the longitudinal stress, under uniaxial strain loading, into a mean stress and a deviator:

$$\sigma_x \equiv \frac{\sigma_{mm}}{3} + S_x \tag{1}$$

Increasing the shock wave amplitude or the longitudinal stress results in a larger mean stress and a larger stress deviator in the material. The mean stress increase can cause pressure induced changes (e.g. structural phase transformations) and the deviator stress increase results in elastic-plastic deformation. For these reasons, shock wave experiments are relevant to the fields of static high pressure physics and chemistry, and mechanics of materials. Often, effects due to mean stress increase and deviator stress increase are treated in an uncoupled manner; although incorrect, such an approach is used to simplify the problem.

Finally, it should be noted that the governing equations for uniaxial strain loading do not involve lateral stresses and temperature.[3] Hence, determination of these two continuum variables in shock wave experiments is not common and requires special efforts.[10,11]

ELASTIC-PLASTIC DEFORMATION IN SHOCKED LiF

Figure 2 shows typical stress-volume curves for an elastic and an elastic-plastic solid subjected to shock wave loading. The stress history corresponding to the latter behavior is shown as a two wave structure on the right side. The amplitude of the first wave corresponds to the onset of yielding and is referred to as the Hugoniot Elastic Limit (HEL).[12] The wave profile shown is for a rate-independent response. The more realistic (and interesting) situation of the time-dependent response and related analyses may be seen in the wave profile measurements on shocked LiF crystals.[13]

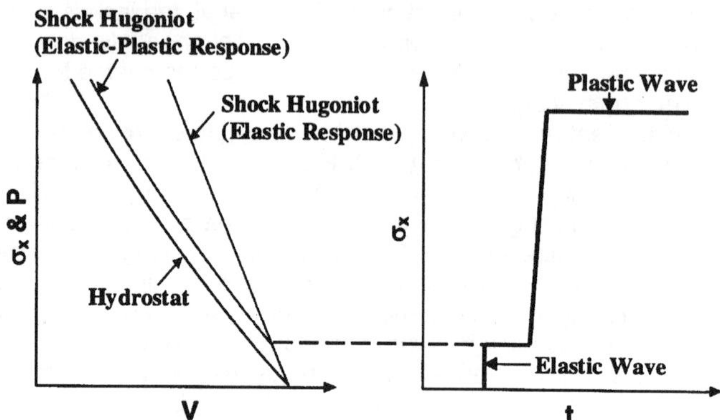

FIGURE 2: Typical stress-volume curves for an elastic and elastic-plastic solid subjected to shock wave loading (left); the stress history corresponding to an elastic-plastic solid (right).

The continuum measurements and analyses on shocked LiF crystals represent likely the most comprehensive examination of time-dependent, elastic-plastic deformation (and its explanation in terms of dislocation dynamics) in a shocked solid.[13-18] Although many important issues like the role of crystal slip systems, effects of point defects, role of impurity clustering, surface preparation, etc. were addressed, the precise dislocation mechanisms governing the extremely rapid stress relaxation in pure LiF crystals has not been established. The difficulty is primarily due to the lack of real time, experimental measurements at microscopic and sub-microscopic scales. Continuum scale measurements, by themselves, cannot be used to determine atomic scale mechanisms. While real time, in-situ examination of individual dislocations in shocked solids is unrealistic, examination of the crystal lattice is expected to be an important step in establishing atomic mechanisms.

The nature of the crystal lattice under shock deformation has been a subject of considerable interest for more than four decades. In 1958, C.S. Smith proposed a hypothesis regarding the lattice structure for elastic, and elastic-perfectly plastic compression.[19] For elastic compression, the shock wave front at any instant, represented a discontinuity separating the uncompressed lattice from the compressed lattice. The latter could be viewed as the crystal lattice subjected to one-dimensional compression at the unit cell level. For plastic compression in a material with negligible strength, the compressed lattice behind the wave front could be viewed as an isotropically compressed lattice. In Smith's model, a large number of dislocations were produced at the plastic wave front to accommodate the lattice mismatch on the two sides.

While Smith's hypothesis or variations of it have been widely used, experimental verification has been difficult. To examine elastic-plastic deformation in shocked solids at the lattice level, we have developed an experimental capability to carry out real time x-ray diffraction measurements in shocked single crystals.[20]

Using this experimental capability, Paulo Rigg has performed experiments to obtain x-ray diffraction measurements in shocked LiF crystals.[8,21] A schematic view of the experimental set up is shown in Figure 3. A pulsed x-ray source (~50 ns duration) is used to obtain diffraction data from LiF crystal planes normal to the shock propagation direction. Plane shock waves are produced in the LiF samples (backed by vitreous carbon) using flyer plates mounted on projectiles. Diffraction peaks are recorded on an appropriately synchronized detector for both the unshocked and the shocked crystal. Effects due to sample translation and tilt, caused by the propagating shock wave, are accounted for.[8]

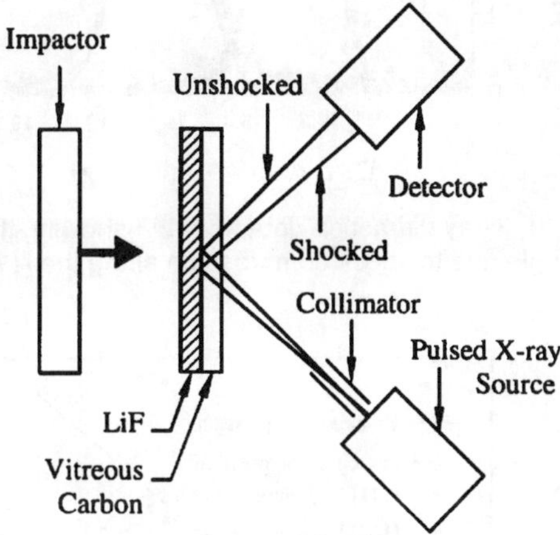

FIGURE 3: Experimental configuration used for x-ray diffraction measurements in shock-compressed LiF.

Experiments have been carried out for shock propagation along both the [111] and the [100] orientations. For the stress range used in the present work (up to 4 GPa), uniaxial compression along [111] results in an elastic response.[16] Uniaxial strain compression along [100] results in an elastic-plastic response.[13] Since the continuum or stress-strain response for these two orientations is known quantitatively, the macroscopic strain state for the two sets of experiments is known.

Figure 4 shows typical x-ray diffraction data; the peaks shown are for shock propagation along [111]. These data provide the changes in the inter-planar spacing through the Bragg relation. Figure 5 summarizes the inter-planar spacing change as a function of macroscopic strain for both orientations.[21] Also included in this figure are the results from two bounding calculations: in one case, the unit cell is assumed to be compressed uniaxially along the direction of shock propagation; in the other case, the unit cell is assumed to be compressed isotropically. With these two assumptions, the inter-planar spacing for the two bounding cases can be determined as a function of macroscopic strain.[21]

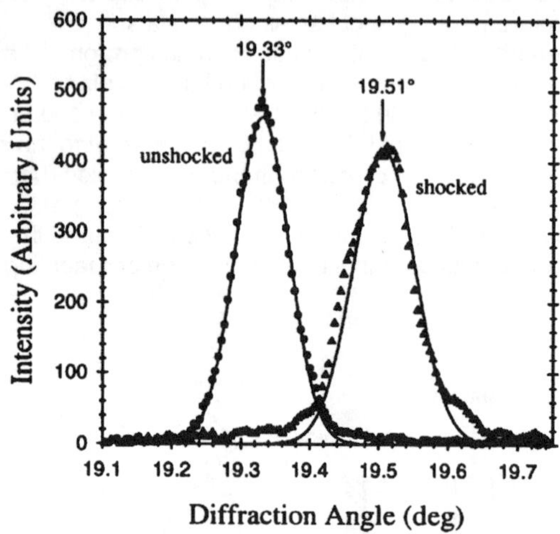

FIGURE 4: X-ray diffraction data from LiF showing the shift in diffraction angle due to shock compression along the [111] direction.

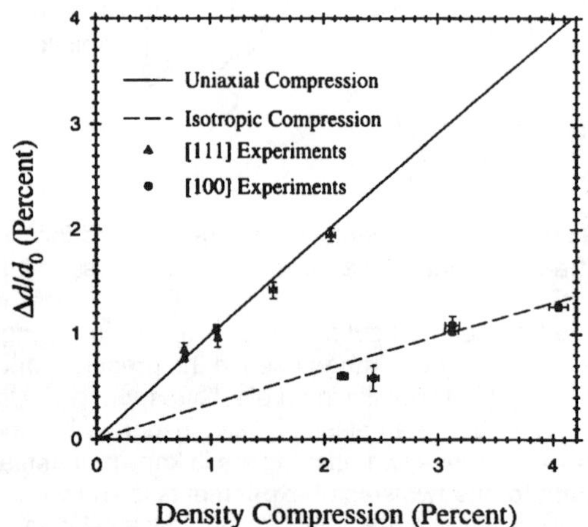

FIGURE 5: Comparison of interplanar spacing changes versus density compression for shock propagation along the [111] and [100] directions. The solid and dashed line represent predictions based on uniaxial and isotropic compression of the unit cell, respectively.

The results in Figure 5 show clearly that elastic compression, as expected, results in uniaxial compression of the unit cell. Elastic-plastic compression of a very soft crystal results in an isotropic compression of the unit cell. It is important to emphasize that the [111] data were valuable in establishing the validity of our experimental approach. Experiments are currently underway to directly determine inter-planar spacing in directions perpendicular to the shock wave propagation direction.[8]

PHASE CHANGE IN SHOCKED CdS

Shock wave experiments provide a convenient approach to study pressure induced structural phase changes. Figure 6 shows a typical pressure – volume plot for a material undergoing a polymorphic phase change accompanied by a significant volume compression at the transformation. Shock compression of such a material results in a two wave structure as shown on the right side of Figure 6. The amplitude of the first wave corresponds to the transformation pressure and the second wave amplitude depends on the peak pressure in the experiment. In this discussion, we have neglected material strength; hence, the distinction between longitudinal stress and pressure can be ignored.

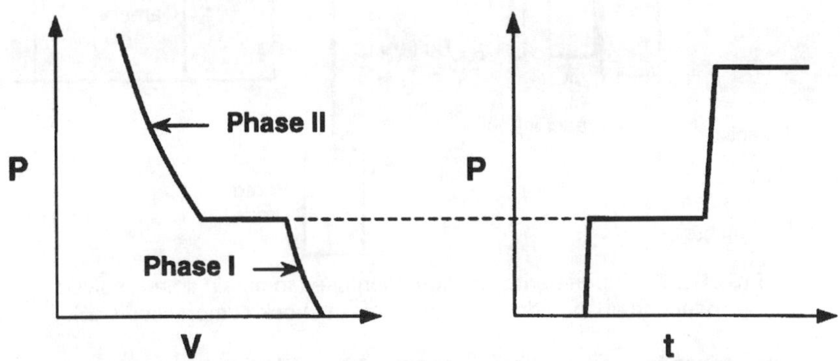

FIGURE 6: Typical pressure-volume curve for a material undergoing a polymorphic phase change (left) resulting in a two-wave structure (right).

The pressure-induced wurtzite to rock salt phase transition in CdS, occurring at about 3 GPa, has been examined under both static high pressure and shock compression.[22-27] Shock wave studies by Gupta and coworkers have examined the time and orientation dependence of the transformation through time-resolved, continuum measurements.[25-27] These findings suggest: (i) the mean stress corresponding to the transformation pressure is independent of shear stress and crystal orientation, (ii) the P-V results from the shock wave experiments are in good agreement with the static data, and (iii) the transformation appears to proceed by a two step process; the time resolution (10-15 ns) of the earlier continuum work could not resolve the initial stages of the transformation.[26-27]

To achieve higher time resolution (~100 ps) and to obtain information about the electronic structure directly in the shocked state, experimental developments were carried out to obtain picosecond, optical absorption measurements in shock compression experiments.[7,28] The phase change in CdS is well suited for examination of the dynamics of the transition because the wurtzite and rock salt phases exhibit a direct and indirect band gap with gap energies of 2.5 and 1.5-1.7 eV, respectively. Thus, fast time resolution absorption measurements permit examination of the initial stages of the transformation and the electronic structure of the crystal directly behind the shock front. Figure 7 shows the experimental configuration used by Marcus Knudson.[8,28]

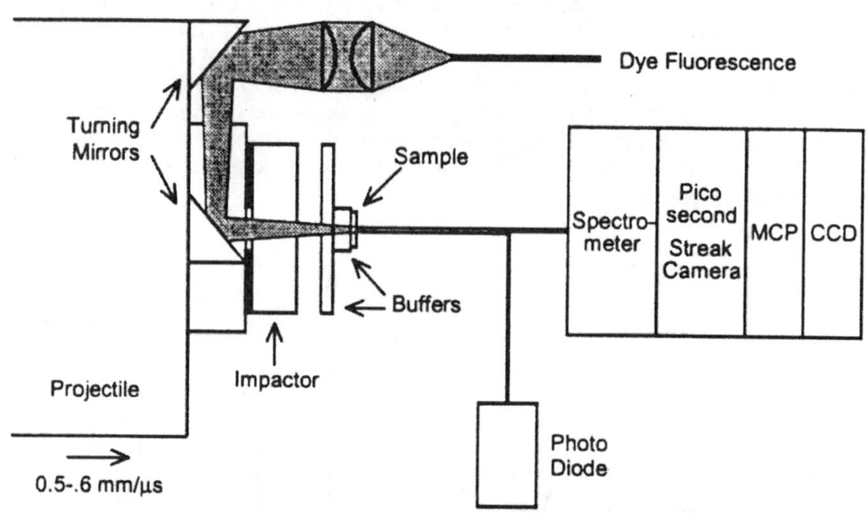

FIGURE 7: Experimental configuration used to obtain time-resolved, picosecond absorption measurements in shock-compressed CdS.

Fluorescence from a broad band source (~12 ns FWHM) was used to cover the wavelength spectrum between 500 and 700 nm. A single pass transmission experiment is carried out using mirrors mounted on the projectile and appropriate optical windows. The transmitted light is directed to a spectrometer (for spectral resolution), picosecond streak camera (for temporal resolution) and then recorded on a CCD as intensity vs. wavelength vs. time.

Shock waves are generated in the thin CdS samples (approx. 5 mm laterally by 0.25 mm thick) by impacting transparent impactors on transparent buffers placed in front of the CdS crystals. Choice of impactors, buffers and impactor velocities provide the stress amplitudes of interest. The total duration of the experiments ranged between 12-25 ns and required synchronization to within 3-4 ns. Further details about the experimental developments may be seen in Refs. 7 and 28.

A representative result[29] is shown in Figure 8. The top half of the figure displays the absorbance vs. wavelength vs. time. Zero time corresponds to the shock wave arrival at the sample. As expected, the sample shows no absorbance between 540-700 nm in the ambient state. Immediately upon the shock wave arrival (within 100 ps), the absorbance increases. The wavelength dependence of the absorbance change is better seen in the bottom half of Figure 8; the linear dependence with time reflects the propagation of the shock wave in the sample.

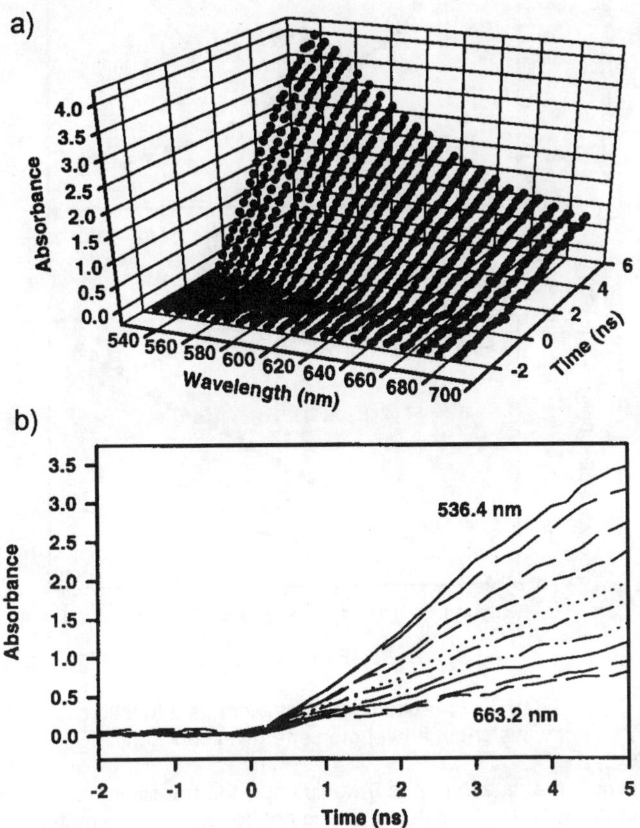

FIGURE 8: Absorption data from CdS showing the temporal and wavelength dependence for shock compression along the c-axis.

Using the shock wave velocity in CdS, the absorbance data can be used to determine the absorption coefficient as a function of the wavelength[29] as shown in Figure 9. In the lower part of Figure 9 the square root of the absorption coefficient is plotted as a function of photon energy. The data in Figure 9 show the occurrence of an indirect band gap material (with a gap energy of 1.52 eV) behind the shock front.

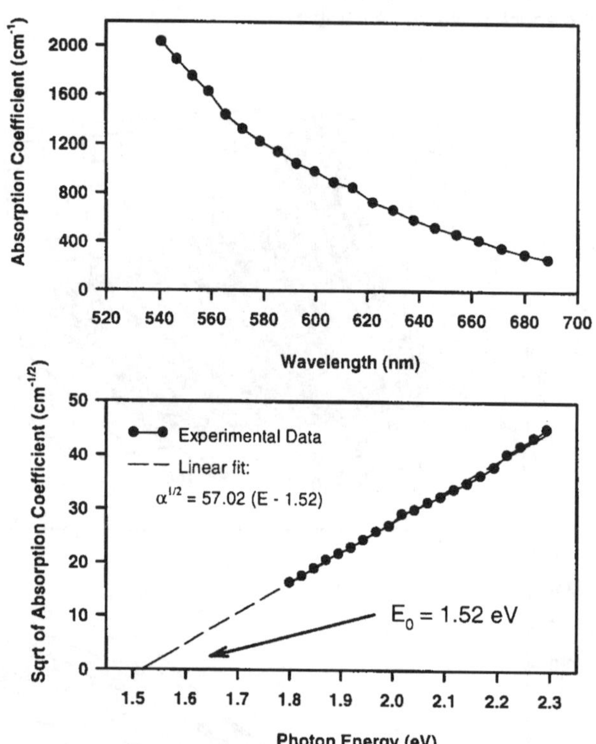

FIGURE 9: Absorption coefficient as a function
of wavelength and photon energy in shocked CdS.

At first glance, the data in Figure 9 would imply the transition from the wurtzite to the rocksalt phase in 100 ps. However, we do not believe that the material has transformed to the rock salt phase for the following reasons. It is unlikely that a 20 percent volume change can occur in 100 ps. Also, the continuum results in shocked CdS suggest a two step process to the final state;[26,27] the second step occurs in 100-200 ns. Hence, based on all available data, we believe that the initial transformation (~100 ps) is to a metastable structure which has an electronic structure similar to the rocksalt phase.[29]

Through a series of ab-initio calculations involving total energy and electronic structure calculations in conjunction with his experimental results, Knudson has made a strong case for the following sequence of events.[7] Within 100 ps, the wurtzite phase transforms to a metastable state identified as a face centered tetragonal (fct) structure. Subsequently, the fct structure relaxes to the rocksalt structure. Because the fct and the rocksalt structure were shown to have nearly identical electronic structures, the optical absorption data cannot be used to examine the second step.

CONCLUSIONS

In this paper we have summarized briefly recent experimental developments that provide insight into the shock wave response of materials at different length scales. Real time measurements at microscopic and sub-microscopic scales in shock wave experiments are inherently difficult. Both x-ray diffraction and optical spectroscopy measurements require high sensitivity detectors, excellent synchronization, and careful determination of the photon energy ranges of interest.

The examples presented here demonstrate that measurements at different length scales are needed to get insight into the atomic mechanisms that govern elastic-plastic deformation and phase changes. Measurements at smaller length scales do not eliminate the need for continuum measurements. Instead, measurements at different length scales should be viewed as complementary approaches to develop a comprehensive picture.

Finally, as pointed out by the CdS study,[30] computations are essential to interpreting the measurements at smaller length scales. While the molecular dynamics computations and experimental measurements do not yet overlap in time resolution and in sample sizes, progress in both computations and experiments is encouraging. A good understanding of the role of extended defects on atomic/molecular processes in shocked solids, and an understanding of material response at the mesoscopic scale remain important challenges.

ACKNOWLEDGEMENTS

The results reported here were taken from the innovative work carried out by Marcus Knudson and Paulo Rigg during the course of their Ph.D. dissertations. They are sincerely thanked for sharing their findings. The author has greatly benefited from his many discussions with Professor George Duvall over the years. Support by the Office of Naval Research, the Army Research Office, and the Department of Energy is gratefully acknowledged.

REFERENCES

1. S.C. Schmidt, D.P. Dandekar, and J.W. Forbes, *Shock Compression of Condensed Matter – 1997* (American Institute of Physics, New York, 1998).

2. Y.M. Gupta, in *Encyclopedia of Physics*, edited by R.M. Besancon (Van Nostrand Reinhold Co., New York, 1985) pp. 1109-1115.

3. Y.M. Gupta, in *Shock Waves in Condensed Matter – 1991*, edited by S.C. Schmidt, R.D. Dick, J.W. Forbes, and D.G. Tasker (North-Holland, Amsterdam, 1992) pp. 15-26.

4. J.N. Johnson and L.M. Barker, J. Appl. Phys. **40**, 4321 (1969).

5. B.L. Holian and P.S. Lomdahl, Science **20**, 2085 (1998).

6. D.W. Brenner, D.H. Robertson, M.L. Elert, and C.T. White, Phys. Rev. Lett. **70**, 2174 (1993).

7.   M.D. Knudson, Ph.D. Thesis, Washington State University, 1998.

8.   P.A. Rigg (Ph.D Thesis work in progress).

9.   J.B. Aidun and Y.M. Gupta, J. Appl. Phys. **69**, 6998 (1991).

10. R. Feng, Y.M. Gupta, and M.K.W. Wong, J. Appl. Phys. **82**, 2845 (1997).

11. G.I. Pangilinan and Y.M. Gupta, J. Appl. Phys. **81**, 6662 (1997).

12. G.E. Duvall, in *Physics of High Energy Density* (Academic Press, New York, 1971).

13. J.R. Asay, et. al., J. Appl. Phys. **43**, 2220 (1972).

14. Y.M. Gupta, G.E. Duval, and G.R. Fowles, J. Appl. Phys. **46**, 532 (1975).

15. Y.M. Gupta, J. Appl. Phys. **46**, 3395 (1975).

16. Y.M. Gupta, J. Appl. Phys. **48**, 5067 (1977).

17. J.E. Vorthman and G.E. Duvall, J. Appl. Phys. **53**, 3607 (1982).

18. K.S. Tunison and Y.M. Gupta, Appl. Phys. Lett. **48**, 1351 (1986).

19. C.S. Smith, Trans. AIME 212, 574 (1958).

20. Y.M. Gupta, K. Zimmerman, E. Zaretsky, P. Rigg, P. Bellamy, D. Savage (unpublished).

21. P.A. Rigg and Y.M. Gupta, Appl. Phys. Lett. **73**, 1655 (1998).

22. A.L. Edwards, et. al., J. Phys. Chem. Solids **11**, 140 (1959).

23. G.A. Samara and A.A. Giardini, Phys. Rev. **140**, 388 (1965).

24. P. Cervantes, et. al., Phys. Rev. B **54**, 17585 (1996).

25. Z.P. Tang and Y.M. Gupta, J. Appl. Phys. **64**, 1827 (1988).

26. Z.P. Tang and Y.M. Gupta, J. Appl. Phys. **81**, 7203 (1997).

27. S.M. Sharma and Y.M. Gupta, Phys. Rev. B **58**, 5964 (1998).

28. M.D. Knudson, K.A. Zimmerman, and Y.M. Gupta, Rev. Sci. Instr. (accepted for publication).

29. M.D. Knudson and Y.M. Gupta, Phys. Rev. Lett. **81**, 2938 (1998).

30. M.D. Knudson, Y.M. Gupta, and A.B. Kunz, Phys. Rev. B. (accepted for publication).

EVOLUTION OF FRACTAL PARTICLES IN SYSTEMS
WITH CONSERVED ORDER PARAMETER

S.V. Kalinin[1*], D.L. Gorbachev[2], K.V. Tomashevitch[1], A.Yu. Borisevich[1*], A.A. Vertegel[1],
A.J. Markworth[3], Yu. D. Tretyakov[1]
[1] Dept. of Chemistry, Moscow State University, 119899, Moscow, RUSSIA
[2] UVD Sakhalinskoi Oblasti, Sakhalin, RUSSIA
[3] Dept. of Mat. Sci. Eng., The Ohio State University, 2041 College Rd., Columbus, Ohio 43210
[*] Now at the Dept. Mat. Sci. Eng., University of Pennsylvania, Philadelphia, PA 19104

ABSTRACT

In the present research, the evolution of fractal aggregates in systems with conserved order parameter is studied. The aggregates are generated by diffusion limited aggregation (DLA). This model describes such important processes as annealing of dendrite inclusions in solids, healing of cracks in ceramics, temperature-induced transformations in composites, relaxation of rough surfaces, aging of colloid particles, etc. It is shown that the evolution in fractal media differs significantly from the evolution from the initially homogeneous state and leads to the different values of the scaling exponent.

INTRODUCTION

It is known that the geometric structure of many artificial and natural objects can be best described in terms of fractal geometry [1-3]. Spatial inhomogeneity pertinent for these systems significantly influences their behavior in various physico-chemical processes. In recent years, certain specific features of diffusion and reaction processes in fractal media have been extensively studied [4-6]. However, little is known about structural changes of fractal aggregates during different physical processes. The relevant physical examples include annealing of dendrite inclusions in solids, healing of cracks in ceramics, temperature-induced transformations in composites, relaxation of rough surfaces, aging of colloid particles, etc. In all these cases, the driving force for temporal evolution is the tendency to decrease the surface or interfacial energy of the system. In the present research, Ostwald-ripening-type evolution of fractal aggregate under conservative conditions is studied. The process is assumed to be limited by volume diffusion.

COMPUTER SIMULATIONS

The fractal aggregates were generated by a diffusion limited aggregation (DLA) mechanism on a square lattice. Particles were launched, one at a time, from a reflecting equal-event boundary and performed a random walk until collision with a growing cluster would take place. Upon collision, the particle either sticks to the cluster or reflects and continues its random walk. The sticking probabilities, P_n , to the site with n neighbors were defined as:

$$P_n = \min\left[e^{E(n-4)}, 1\right] \qquad (1)$$

where E is a constant. Larger E values correspond to increasing probability for the particle to stick to a site having a larger number of neighbors and thus lead to denser clusters. For these simulations, E values were taken to be 0, 0.1, ..., 2. The cluster size was chosen to be $3 \cdot 10^5$ particles. The fractal dimension, D, as determined by the box-counting method, was 1.72 ± 0.01 for all clusters. However, a lower cut-off of fractality R_a ($R_a \sim 2S/P$, where S is the 2D volume of the fractal aggregate, and P is surface) increases with E (Fig. 1).

Mat. Res. Soc. Symp. Proc. Vol. 538 © 1999 Materials Research Society

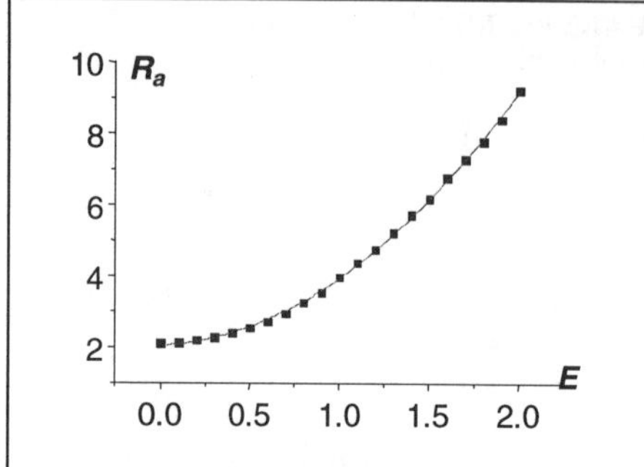
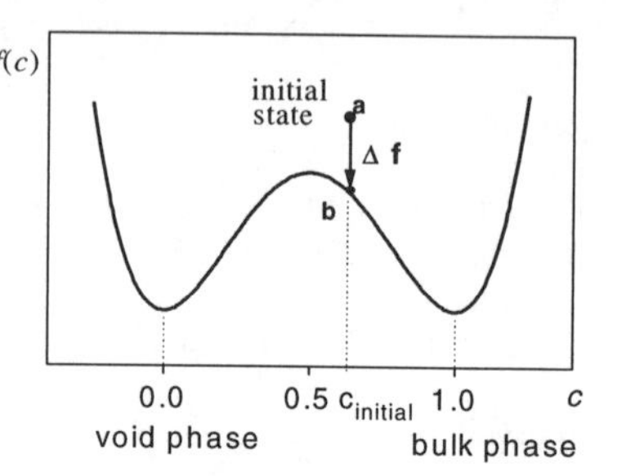

Fig.1. Lower cut-off of fractality R_a for DLA clusters grown with different E values.

Fig. 2. Free energy density $f(c)$ as a function of dimensionless concentration c.

The temporal evolution of a fractal cluster was described by the evolution equation for systems with conserved order parameter (c is a dimensionless concentration, $1 \geq c \geq 0$, M - mobility) [7,8]:

$$\frac{\partial c}{\partial t} = -M\nabla^2 \frac{\delta F}{\delta c}, \tag{2}$$

The free-energy functional was taken to be of the Landau-Ginzburg form:

$$F = \int_V \left[f(c) + \frac{1}{2} k_c (\nabla c)^2 \right] dV, \tag{3}$$

where k_c is a constant related to interfacial energy [9]. The free energy density $f(c)$ was taken as $f(c) = A\,(c\text{-}c^2)^2$ with two symmetric minima corresponding to bulk ($c = 1$) and void ($c = 0$) phases (Fig. 2). Thus, the resulting evolution equation was:

$$\partial c/\partial t = -M\nabla^2[k_c\nabla^2 c\text{-}2A(c\text{-}3c^2+2c^3)]. \tag{4}$$

Equation (4) was solved by the explicit Euler method on a 512x512 grid with a 0.01 timestep. Parameters in Eq. (4) were taken as $M = 1$, $k_c = 1$, $A = 0.25$. A concentration field corresponding to a fractal cluster generated by the modified DLA mechanism was used as the initial condition, $c(\mathbf{r},0)$. Cyclic boundary conditions were used. During the evolution, the correlation function of the object defined as:

$$C(x,t) = \langle c(\mathbf{r}+\mathbf{r}',t), c(\mathbf{r}',t)\rangle, \tag{5}$$

where average is taken over the whole system, was calculated. Bulk and surface fractal dimensions of clusters were also measured by means of the box-counting method.

RESULTS AND DISCUSSION

Depending on the ratio between the gyration radius, R_g, and lower cut-off of fractality, R_a, observed scenarios of evolution can be roughly subdivided into evolution with fragmentation of the initial cluster (Fig. 3) and evolution without fragmentation (Fig. 4).

In the first case, the initial cluster rapidly dissociates, forming many small particles (Fig. 3b). Smaller particles dissolve and larger particles grow as in the classical LSW model [10]. However, due to the inhomogeneous spatial distribution of particles stipulated by the branched structure of the initial aggregate, they have a characteristic elongated form with larger axis directed parallel to the branch of the cluster from which it evolved (Fig. 3c). Subsequent evolution leads to spheroidization of fragments with increase of the separation between them. However, the geometric position of the spherical fragments resembles the structure of the initial cluster (Fig. 3d).

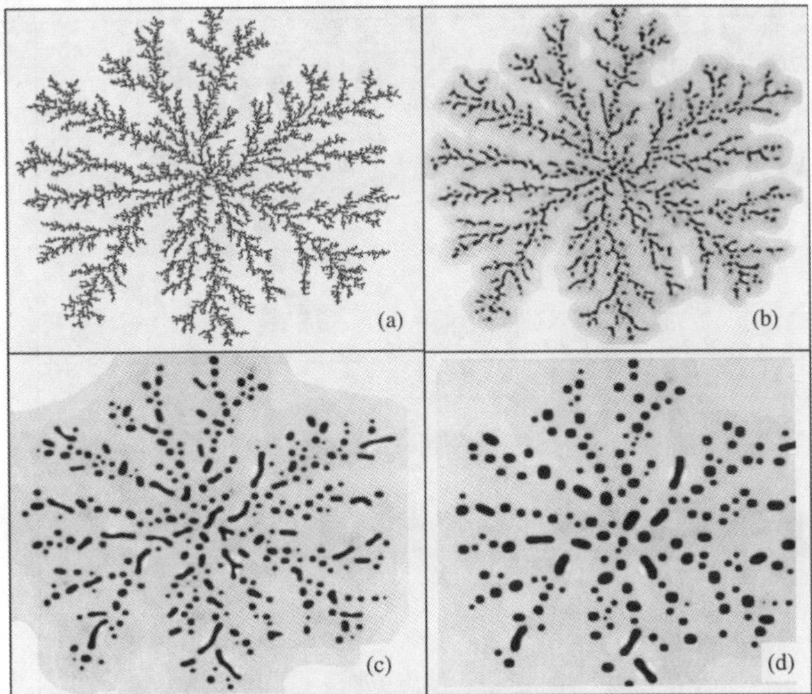

Fig. 3. Evolution of a fractal aggregate ($E = 0$) with fragmentation. (a) – initial cluster, (b) $t = 10$, (c) $t = 100$, (d) $t = 400$.

During evolution by the second scenario, the finer details of the structure of the object are smoothed out, while the larger structural features remain constant. The characteristic size of smoothed regions increases with time. However, the overall shape of the aggregate resembles that of the initial cluster. Evidently, both scenarios will eventually lead to the formation of a single spherical particle, but that would require an extremely long computational time.

The structural changes that occur during the evolution of fractal aggregates were investigated by the analysis of correlation functions of resulting objects. It was shown that for large x the correlation function, $C(x,t)$, for time t is almost the same as that for the initial cluster, $C(x,0)$. The effective correlation function of the system, $\check{C}(x,t) = C(x,t) - C(x,0)$, has a characteristic shape (Fig. 5). Due to the increase of the characteristic size of the particles, the value of $C(x,t)$ grows in comparison with $C(x,0)$, thus leading to positive values of $\check{C}(x,t)$ for small x. At the same time, for relatively large x the neighborhood is depleted, thus leading to the characteristic minimum on $\check{C}(x,t)$. For large x $C(x,t)$ does not change compared to $C(x,0)$ and thus $\check{C}(x,t) \approx 0$.

The first zero of $\check{C}(x,t)$ roughly corresponds to the mean size of single-phase domains. The temporal dependence of $x_c = x_{\check{C}(x)=0}$ shown on Fig.6 can be well approximated by the equation

$$x_c = A\,(t+B)^{\alpha}, \tag{6}$$

where A and B are constants and α is a scaling exponent. The constant B takes into account the early-stage evolution of the system [11,12].

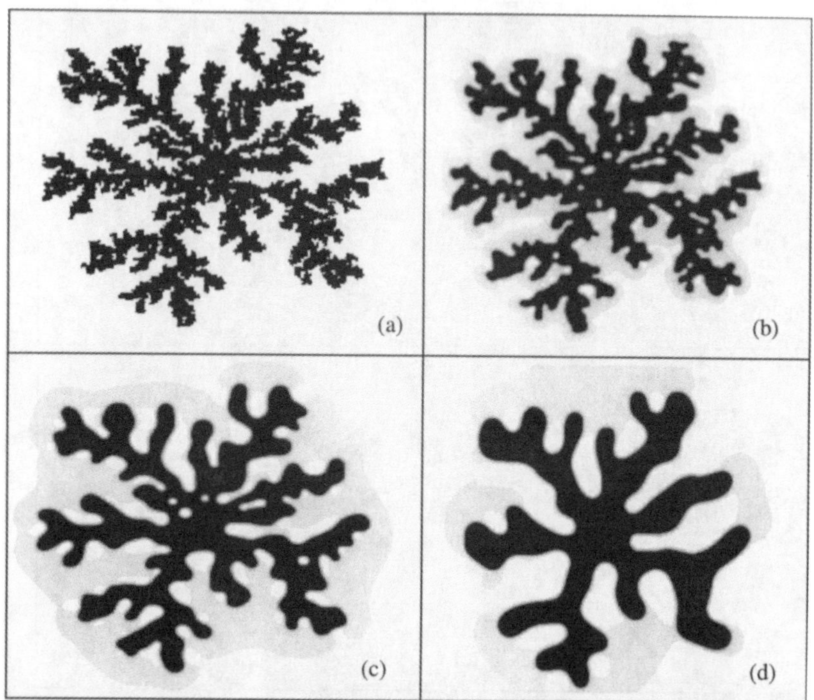

Fig. 4. Evolution of a fractal aggregate ($E = 1.5$) without fragmentation. (a) – initial cluster, (b) $t = 10$, (c) $t = 100$, (d) $t = 400$.

Structural changes occurring during evolution were also studied by box-counting analysis of the surface. The cluster was covered by a square grid with characteristic size l, and the number of squares, N, containing at least one point belonging to the cluster surface was plotted as a function of l in log-log coordinates. For relatively small l, the slope of the corresponding curve is 1 (the surface is one-dimensional), whereas for larger l, the slope is equal to the fractal dimension of the object $D = 1.71$. The intersection of these two linear segments defines the characteristic length R_a, also proportional to the characteristic lengthscale on which the fractal structure is destroyed (Fig.7).

Results of the correlation function analysis, $x_C - t$, for clusters with different E, are shown in Table I, and for the box-counting analysis, $R_a - t$, in Table II. For evolution without fragmentation, values obtained for the scaling exponents, α, agree well. It should be noted that values obtained for B are relatively small ($B \ll$ characteristic time of the evolution), i.e., the characteristic size of the single phase domain grows by a power law, $L(t) \sim t^{\alpha}$.

For evolution with fragmentation, values obtained for the scaling exponents, α, are larger than in the first case. Numerical values of scaling exponents determined from correlation function analysis and box counting analysis are significantly different, probably due to the problems with the box-counting analysis for the system having relatively small isolated particles. Interestingly, for later stages of evolution, transition to $\alpha = 1/3$ takes place when the initial fractal structure of the cluster is completely destroyed, and evolution proceeds as in the initially homogeneous system.

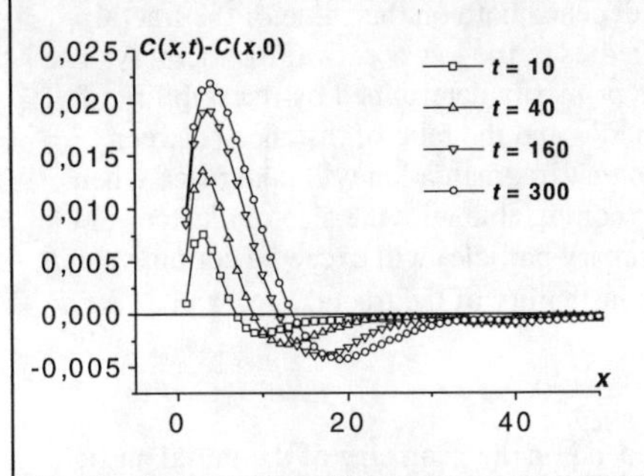

Fig. 5. The effective correlation function $\check{C}(x,t)$ of fractal cluster with E=2 for different times of evolution.

Fig.6. First zero of effective correlation function as a function of time for fractal clusters with different E.

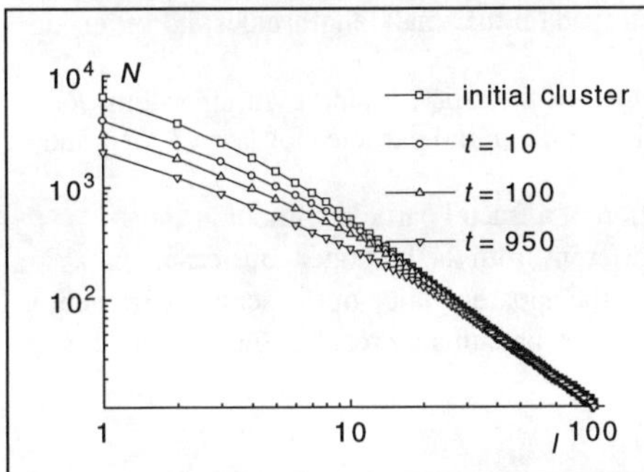

Fig. 7. Box counting analysis of fractal cluster with E=2 for different times of evolution.

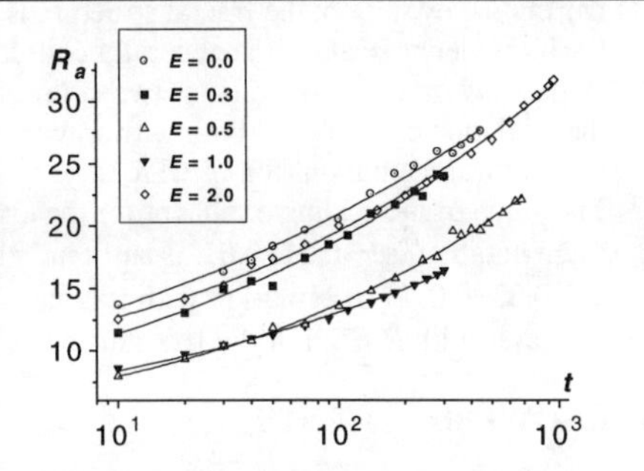

Fig.8. Lower cut-off of fractality as a function of time for fractal clusters with different E.

Table I. Parameters A, B and α in Eq. (6) obtained from the analysis of correlation functions.

E	A	B	α
0.0	2.32 ± 0.05	2.2 ± 0.9	$\mathbf{0.249 \pm 0.003}$
0.3	2.44 ± 0.07	5.7 ± 1.3	$\mathbf{0.262 \pm 0.005}$
0.5	2.73 ± 0.08	7.0 ± 1.4	$\mathbf{0.252 \pm 0.005}$
1.0	3.92 ± 0.03	3.2 ± 0.3	$\mathbf{0.216 \pm 0.001}$
1.5	4.40 ± 0.05	13.0 ± 1.0	$\mathbf{0.205 \pm 0.002}$
2.0	5.73 ± 0.06	-1.0 ± 0.4	$\mathbf{0.213 \pm 0.002}$

Table II. Parameters A. B and α in Eq. (6) obtained from the box-counting analysis of surface.

E	A	B	α
0.0	8.7 ± 0.5	0.0 ± 3.0	$\mathbf{0.19 \pm 0.01}$
0.3	6.6 ± 0.6	-1.2 ± 3.6	$\mathbf{0.23 \pm 0.02}$
0.5	4.2 ± 0.3	2.0 ± 3.0	$\mathbf{0.25 \pm 0.01}$
1.0	4.7 ± 0.2	6.6 ± 1.9	$\mathbf{0.22 \pm 0.01}$
1.5	5.0 ± 0.5	32 ± 11	$\mathbf{0.21 \pm 0.01}$
2.0	7.3 ± 0.3	3.7 ± 2.7	$\mathbf{0.21 \pm 0.01}$

Evidently, the observed scenario of evolution depends both on the value of the fractal dimension, D, of the cluster and the ratio of gyration radius to the lower cut-off of fractality. The stability of the cluster with respect to fragmentation is primarily determined by the stability of tree-like structure of branches, i.e. by the number of nodes and the ratio of distance between nodes to the width of a single branch. We thus assume that fragmentation will take place when the ratio between the characteristic size of the branch (comparable with the size of cluster) and its effective width (comparable with the size of the primary particle) will exceed a certain value, λ_{inst} , which, in turn, may serve as the criterion for the instability of the fractal aggregate.

CONCLUSIONS

The evolution path of the system was shown to depend on the geometry of the initial particle as described by the lower cut-off of fractality, R_a, and gyration radius, R_g , and the value of the fractal dimension, D. The following main features of the evolution were determined:

1. During the evolution, the fractal structure is destroyed on the small lengthscales; however, on the larger lengthscales, it is generally conserved.
2. Depending on the initial ratio between lower cut-off of fractality, R_a and gyration radius, R_g, the evolution can proceed both with fragmentation of the initial particle (for large R_a/R_g) and without fragmentation (for small R_a/R_g).
3. The value of the scaling exponent for the evolution of a fractal particle without fragmentation, $\alpha = 0.21\pm0.01$, is substantially different from the homogeneous case, $\alpha = 0.33\pm0.01$. For evolution with fragmentation, the apparent value of the scaling exponent increases with R_a/R_g. For the late stages of evolution α presumably reaches the value of 1/3.

ACKNOWLEDGEMENTS

This work was partially supported by the RFBR grant No. 96-03-33122a, Universities of Russia program Grant No. 98-06-14-5650, NATO Linkage grant No. 970608, and ISSEP grants a98-1724, s98-1716, and s98-1348.

REFERENCES

1. B.B. Mandelbrot, <u>The Fractal Geometry of Nature</u>, San Francisco, Freeman (1982).
2. <u>Fractals and Disordered Systems</u>, Eds. A. Bunde, S. Havlin, Springer-Verlag (1992).
3. <u>The Fractal Approach to Heterogeneous Chemistry</u>, Ed. D. Avnir, John Wiley (1988).
4. J.S. Havlin, D. Ben-Avraham, Adv. Phys., **36**, 695-798 (1987).
5. M. Sahimi, J. Chem. Eng., **64**, 21-44 (1996).
6. M. Giona, W.A. Schwalm, M.K. Schwalm, A. Adrover, Chem. Eng. Sci., **51**, 4717-4729 (1996), **51**, 4731-4744 (1996), **51**, 5065-5076 (1996).
7. Y. Wang, L.Q. Chen, A.G. Khachaturyan, in H.O. Kirchner *et al.* (eds.) <u>Computer Simulations in Materials Science</u>, 325-371, Kluwer (1996).
8. A.J. Bray, Adv. Phys., **43**, 357-459 (1994).
9. J.W. Cahn, J.E. Hilliard, J. Chem. Phys., **28**, 258-267 (1958).
10. I.M. Lifshitz, V.V. Slyozov, Zh. Exp. Teor. Phys, **35**, 479-488 (1958) (in Russian).
11. G. Ramirez-Santiago, A.E. Gonzalez, Physica A, **236**, 75-84 (1997).
12. J.D. Gunton, M. San Miguel, P.S. Sahni, in <u>Phase Transitions</u>, eds. C.Domb, J.E. Lebowitz, 267-466, Academic Press, London (1983).

MULTISCALE MODELING OF RECRYSTALLIZATION

M.A.MIODOWNIK[*], E.A.HOLM[*], A.W.GODFREY[**], D.A. HUGHES[**], R. LESAR[+].
[*]Sandia National Laboratories, Albuquerque, New Mexico.
[**]Sandia National Laboratories, Livermore, California.
[+]Los Alamos National Laboratory, Los Alamos, New Mexico.

ABSTRACT

We propose a multi length scale approach to modeling recrystallization which links a dislocation model, a cell growth model and a macroscopic model. Although this methodology and linking framework will be applied to recrystallization, it is also applicable to other types of phase transformations in bulk and layered materials. Critical processes such as the dislocation structure evolution, nucleation, the evolution of crystal orientations into a preferred texture, and grain size evolution all operate at different length scales. In this paper we focus on incorporating experimental measurements of dislocation substructures, misorientation measurements of dislocation boundaries, and dislocation simulations into a mesoscopic model of cell growth. In particular, we show how feeding information from the dislocation model into the cell growth model can create realistic initial microstructures.

INTRODUCTION

The role of recrystallization models is to predict the influence of external variables such as strain, strain rate, and temperature on microstructure and texture. Macroscopic models characterize the microstructure in terms of state variables, such as dislocation content, cell size and texture [1]. These are related to the external variables via constitutive equations of recrystallization and recovery derived either analytically or empirically by fitting to experiment. Macroscopic models have achieved considerable success when applied systematically to specific systems of interest, such as multi-pass rolling of steels [2,3]. However they do not include any details about the microstructure of the alloy, except in terms of average properties. It is becoming increasing clear that heterogenities present in the microstructure greatly influence the nucleation process [1,4].

Deformation microstructures consist of cellular patterns of dislocation boundaries. In so much that this is similar to a grain structure, it might seem that the Monte Carlo Potts model would be well suited to simulate evolution of such a structure. The Potts model is extremely successful in modeling annealing processes such as grain growth [5], and secondary recrystallization [6]. It has been applied to primary recrystallization by a number of workers, but in all cases nucleation has been imposed on the simulation either by directly inputing nuclei [7] or by imposing nucleation criteria [8]. In both cases the model is employed as a macroscopic model to investigate the effect of heterogeneous nucleation on the final microstructure and texture. These studies do not reveal information about the nucleation mechanism of recrystallization.

The central question is how do high angle recrystallization nuclei emerge from an evolving structure of dislocation boundaries? There is currently a great deal of experimental work characterizing dislocation substructures. Since it is not feasible to map out the deformation structure of a material, structural laws need to be developed through statistical sampling of small volumes using the TEM. This approach has been pioneered by Hughes et al.[9] who have proposed a scaling law for dislocation boundaries in medium to high stacking fault energy fcc metals. Such work is important because these laws can allow models of cell growth to incorporate realistic starting microstructures. Experimental work to elucidate the kinetic law for the motion of dislocation boundaries is also a vital piece of the jigsaw. Many workers have assumed that dislocation boundaries will migrate under curvature driven growth. It is not clear that this is valid; in particular one would expect dislocation climb and glide to greatly influence boundary migration.

Mat. Res. Soc. Symp. Proc. Vol. 538 ©1999 Materials Research Society

We model recrystallization using a multiscale approach. A dislocation model is employed to examine the dynamics of dislocation cell structures and in particular to develop a kinetic law for dislocation boundary motion. This is fed into a mesoscopic simulation in which we model boundaries instead of dislocations, allowing us to model 3D microstructures and simulate a large number of cells. The initial boundary misorientation distribution (BMD) is compared to the experimental BMDs measured by Hughes et al [9] . Parameters such as nucleation rates and nuclei orientation are fed from the mesoscopic model into a macroscopic model of recrystallization.

In this short paper we describe briefly the dislocation model and the cell growth model, and show how feeding information between them enables us to develop realistic starting microstructures which compare well with experiment.

DISLOCATION MODEL

A 2D Monte Carlo dislocation model was used to create experimentally realistic dislocation cell structures. The simulation volume was discretized into a triangular lattice with periodic boundary conditions and Burger vectors along each principle direction. The initial configuration consisted of N randomly placed edge dislocations. The energy of the system was defined in terms of the sum of the potential between pairs of dislocations *(i, j)* in the system.   Using the stress tensor of a periodic array of dislocation *(j)*, the force on another dislocation *(i)* can be calculated using the Peach-Kohler equation. The interaction potential between the dislocations is then given by [10]:

$$\Phi_{ij} = \int \left(b_{i,x}\sigma_{j,xx} + b_{i,y}\sigma_{j,xy} \right) dy \qquad (1)$$

where $\sigma_{j,xx}$ and $\sigma_{j,xy}$ are components of the stress tensor of dislocation *(j)* and $b_{i,x}$ and $b_{i,y}$ are components of the burgers vector of dislocation *(i)* in the x and y directions. The energy, E, of the system is the sum of the interaction potentials. The system is relaxed using a Monte Carlo method. A dislocation is chosen at random and the change in energy, ΔE, associated with a proposed move of unit distance along its slip direction is calculated. This move is accepted if $\Delta E < 0$. At present the simulation does not include dislocation climb, dislocation reactions or annihilations, though these will be introduced imminently. It should be noted we are not using the model to simulate plastic deformation but to investigate the lattice misorientations present in dislocation cell structures.

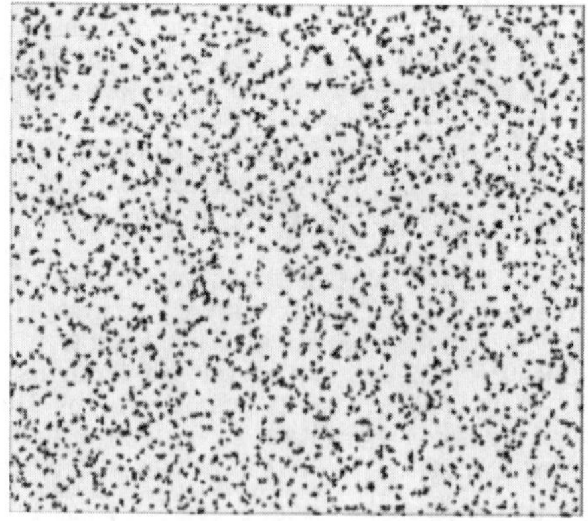

Figure 1. A relaxed dislocation structure with 5000 dislocations.

Figure 1 shows the relaxed dislocation structure for a system with 5000 dislocations. There is little appearance of cell formation, though a large number of dislocation dipoles are formed.

Figure 2. The skew-symmetric orientation map of the relaxed dislocation structure of figure 1.

A quantitative measure of the orientation of the lattice is given by the skew-symmetric rotation matrix:

$$\theta_{x,y} = \frac{1}{2}\left(\frac{\partial u_y}{\partial x} - \frac{\partial u_x}{\partial y}\right) \qquad (2)$$

where u_y and u_x are the displacement components of each site *(x,y)*.

These orientations are plotted in figure 2. A degree of cell formation is evident, although it should be noted that the relaxed structure and random configuration (not shown) share the same general features. The rotation angle, $\Delta\theta$, of each cell is defined as, $\theta_{x,y}$-θ_{ave}, where θ_{ave} is the average orientation (very close to zero degrees in this simulation). The probability function of $\Delta\theta$ is shown in figure 3, showing that the relaxed dislocation structures have a Gaussian distribution of misorientations. We feed this result into the Cell Growth Model to generate initial microstructures.

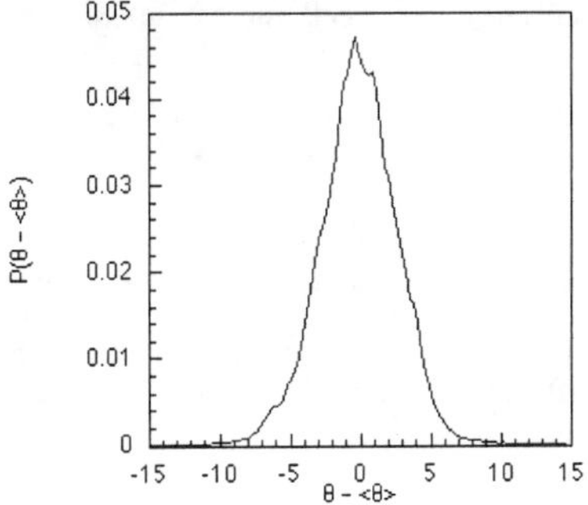

Figure 3. The distribution of rotation angles in the dislocation model

CELL GROWTH MODEL

We use a modified Potts model to simulate cell growth of dislocation substructures. A simulation volume is discretized into a simple cubic lattice of $N_x \times N_y \times N_z$ sites. An equiaxed cell structure is used as a starting configuration, see figure 4. Each site has a spin number i, where $i= 1....N_c$ cells in the system. A set of rotation angles m_i is defined which has a Gaussian distribution, and a set of axes $[uvw]_i$ are defined which are randomly distributed with a uniform probability on a unit sphere. The transformation matrix R_i of a rotation of m_i about $[uvw]_i$ is calculated and the orientation of each cell, O_i is then given by:

$$O_i = R_i O_{ref} \qquad (3)$$

This procedure reproduces the distribution of orientations exhibited by the dislocation model, i.e. a Gaussian distribution with a mean of zero. Misorientations across cell boundaries are used to measure the BMD of the system which is shown in figure 5. The standard deviation of the Gaussian rotation angles, m_i, was found to determine the average cell boundary misorientation but did not influence the shape of the BMD which scales with average misorientation angle. At present no dynamical simulations have been carried out. Boundary migration will be implemented using a Monte Carlo Metropolis algorithm incorporating the kinetic law derived from the dislocation model.

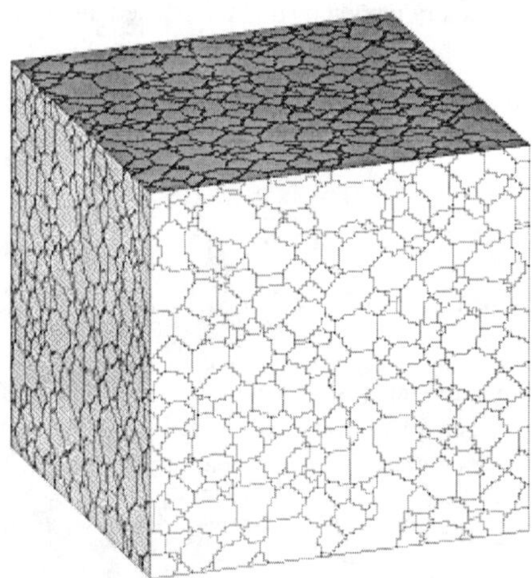

Figure 4. The equiaxed cell microstructure

COMPARISON WITH EXPERIMENT

Dislocations developed during monotonic plastic deformation of medium to high stacking fault energy fcc metals organize into boundaries that subdivide an original grain at two length scales [9]. The larger scale is defined by long, continuous dislocation boundaries called geometrically necessary boundaries (GNBs). Within the volumes defined by these boundaries lower angle incidental dislocation boundaries (IDBs) are formed. The BMDs of GNBs and IDBs each evolve differently as a function of strain.

To develop nucleation criteria based upon local heterogeneities in the deformation microstructure, it is necessary to construct starting microstructures that contain both these boundary types (GNBs and IDBs), and where each type has the experimentally observed misoirentation distribution. Hughes et al. have shown that for a wide range of fcc metals, the misorientation distributions of the IDBs exhibit a scaling property, with the varying distributions collapsing to a single curve when the misorientation probability density distribution is scaled by the average misorientation [9] according to

$$p(\theta, \theta_{av}) = \theta_{av} f\left(\frac{\theta}{\theta_{av}}\right) \qquad (4)$$

with

$$f(x) = \frac{\alpha^{\alpha}}{\Gamma(\alpha)} x^{\alpha-1} \exp(-\alpha x) \qquad (5)$$

where $\Gamma(\alpha)$ is the gamma function evaluated at argument α (the fitting parameter), with $\alpha = 3$. Figure 5 shows the comparison between this curve and that constructed as a starting microstructure in the Cell Growth Model. The good correlation between the two is encouraging, however it should be noted that GNBs have not been included in the model and are likely to have an important influence on the development of recrystallization nuclei. This is the focus of our current efforts.

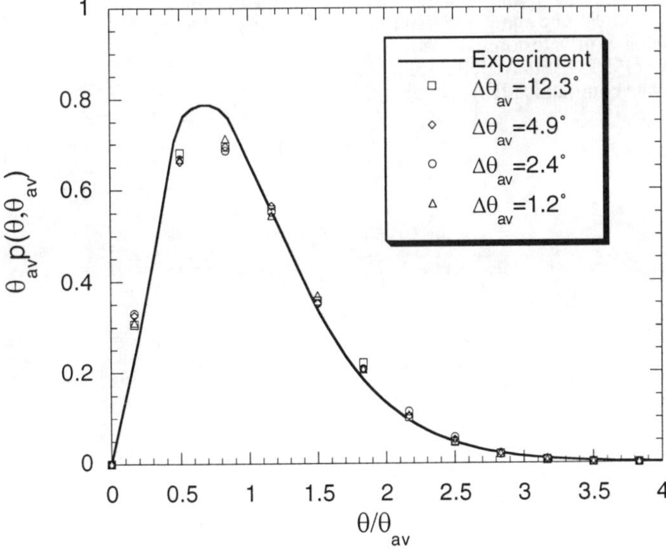

Figure 5. A Comparison of the BMD of the Cell Growth model with the experimentally fitted BMD

CONCLUSIONS

We have modeled the formation of dislocation sub structures by linking a dislocation model to a cell growth model. It should be noted we are not using the dislocation model to model plastic deformation, but to investigate the lattice misorientations present in dislocation cell structures. The Gaussian spread of rotation angles which comes out of the dislocation model when applied the cell model produces a BMD that is very close to the experimental BMD of IDB boundaries. The next stage will be to incorporate dynamics into the dislocation model to measure a kinetic equation for dislocation boundaries. This will then be implemented into the Cell Growth model so that the behavior of large systems cells may be studied.

ACKNOWLEDGMENTS

This work was performed at Sandia National Laboratories, supported by the U.S. Department of Energy under contract number DE-AC04-94AL85000.

REFERENCES

1   F.J.Humphreys & M.Hatherly, *Recrystallization and Related Annealing Phenomena*, (Pergamon Press, Oxford, 1996)
2.   C.M.Sellars, (1990) Mats. Sci. and Tech. **6**, p.1072.
3.   C.M.Sellars & J.A.Whiteman (1979) Met. Sci. **13**, p.325.
4.   J. Hjelen, R. Ørsund & E.Nes (1991) Acta Metall. **39**, p.1377.
5.   Ling S. and Anderson M.P. (1992) JOM, Metals and Materials Society, TMS, Sept, p.30.
6.   Srolovitz D.J., Grest G.S. and Anderson M.P. (1985), Acta Metall., **33**, p.2233.
7.   Srolovitz D.J., Grest G.S. and Anderson M.P. (1986), Acta Metall., **34**, p.1833.
8.   Holm and Battaile (private communication)
9.   Hughes et al. (1997), Acta Metall., **45**, p.105.
10.   R.Lesar (private communication)

COMPUTATION OF MICROSEGREGATION AND MICROSTRUCTURE IN SOLIDIFICATION WITH FLUID CONVECTION

D. JURIC
George W. Woodruff School of Mechanical Engineering
Georgia Institute of Technology, Atlanta, GA 30332-0405

ABSTRACT

The effects of fluid flow on the dendritic solidification morphology of pure materials is studied using a computational methodology based on a two-dimensional front tracking/finite difference method. A general single-field formulation is presented for the full coupling of phase change, fluid flow, and heat transport in both the solid and liquid phases. This formulation accounts for interfacial rejection/absorption of latent heat, interfacial anisotropies, discontinuities in material properties between the liquid and solid phases, shrinkage/expansion upon solidification and motion and deformation of the solid. Numerical results are presented for the dendritic solidification of pure succinonitrile in a shear flow. Comparison with solidification into a quiescent liquid indicates that fluid convection increases the overall rate of solidification while the growth rate of the leading dendrite tip is unchanged.

INTRODUCTION

Nearly all materials of engineering interest have, at some point, solidified from a liquid state. The exact nature of the solidification process determines the microstructure and thus the physical properties of the solid material. In most solidification situations some degree of fluid motion exists, whether by buoyant natural convection or forced convection due to electromagnetic stirring or some mechanical motion. Fluid motion may also be induced by the shrinkage or expansion of the solid upon solidification. An understanding of how the fluid motion affects the dynamics of the solidification process and the resulting microstructure is crucial to many manufacturing technologies.

A variety of numerical methods have been successful in capturing the essential features of microscale solidification dynamics and morphologies of both pure materials and alloys. Some recent examples include finite element based [1–3], phase-field [4–8] and front tracking [9,10] methods. However, until recently these methods have only been applied to the problem without fluid convection. A notable extension of the phase-field method to include convection has recently been developed by Diepers et al. [11] and was applied by Tong et al. [12] to the study of a freely growing dendrite in a flowing melt. A front-tracking method for phase change problems including convection has also recently been developed for calculations of boiling flows [13]. As will be described below, this front-tracking method can be directly applied to the solidification of pure materials. With the addition of a solute transport equation the method can be easily extended to binary alloy solidification. However in this paper we will restrict our attention to pure materials. A report of our work on flow and solidification of alloy grains is available [14]. Here we describe the extension and application of the front tracking method developed for dendritic growth in a quiescent melt [9] to direct simulations of solidification microstructures in pure materials in the presence of fluid convection. The formulation accounts for interfacial rejection/absorption of latent heat, interfacial

Mat. Res. Soc. Symp. Proc. Vol. 538 © **1999 Materials Research Society**

anisotropies, discontinuities in material properties between the liquid and solid phases and shrinkage/expansion upon solidification. In addition, the method allows for arbitrarily complex deformations of the solid-liquid interface, topology changes and solid body motion and deformation.

MATHEMATICAL FORMULATION

We write a single set of conservation equations for the transport of mass, momentum and energy for the entire domain including both solid and liquid phases. The phase boundary is treated as an imbedded interface by adding to these conservation laws the appropriate source terms for surface tension, interphase mass transfer, jumps in material properties and rejection/absorption of latent heat. These source terms are in the form of delta functions localized at the interface and are selected in such a way as to satisfy the correct jump conditions across the phase boundary.

We begin by specifying the material properties which are considered to be constant but not generally equal for each phase. As a consequence, the bulk phases are incompressible but we allow for volume expansion or shrinkage at the phase interface due to the density change upon solidification. Equations for the material property fields can be written for the entire domain using an indicator function, $I(\mathbf{x}, t)$, which has the value 1 in the solid phase and 0 in the liquid phase. I is similar to the phase-field variable in the phase field method, however here, as will be shown below, we determine I from the known position of the tracked interface rather than use it to determine the position of the interface. The values of the material property fields at every location are given by

$$b(\mathbf{x}, t) = b_L + (b_S - b_L)\, I(\mathbf{x}, t)\,, \tag{1}$$

where the subscripts S and L refer to the solid and liquid phases respectively. b represents density, ρ, viscosity, μ, specific heat, c or thermal conductivity, k. Since I is constant except in a local region near the interface, we can express the gradient of I as a local surface integral

$$\nabla I = \int_{\Gamma(t)} \mathbf{n}\, \delta\, (\mathbf{x} - \mathbf{x}_i)\, ds\,, \tag{2}$$

where $\mathbf{n}$ is the unit normal to the interface, defined to point into the solid phase and $\mathbf{x}_i = \mathbf{x}(s, t)$ is a parameterization of the phase interface, $\Gamma(t)$. $\delta\,(\mathbf{x} - \mathbf{x}_i)$ is a three-dimensional delta function that is non-zero only where $\mathbf{x} = \mathbf{x}_i$. Taking the divergence of Eq. (2) results in

$$\nabla^2 I = \nabla \cdot \int_{\Gamma(t)} \mathbf{n}\, \delta\, (\mathbf{x} - \mathbf{x}_i)\, ds\,. \tag{3}$$

Thus we find $I(\mathbf{x}, t)$ by solving the above Poisson equation where the right hand side is a function only of the known interface position at time t. The interface is advected in a Lagrangian fashion by integrating

$$(d\mathbf{x}_i/dt) \cdot \mathbf{n} = \mathbf{V} \cdot \mathbf{n} \tag{4}$$

where $\mathbf{V}$ is the interface velocity vector. We assume that the tangential components of the interface velocity, $\mathbf{V}$, and material velocity, $\mathbf{u}$, at the interface are equal, *i.e.* no slip at the interface.

With the aid of Eq. (2) and the definition of ρ, Eq. (1), the conservation of mass for the entire domain can be expressed as

$$\nabla \cdot \rho\mathbf{u} = -\frac{\partial \rho}{\partial t} = \mathbf{V} \cdot \nabla\rho = \int_{\Gamma(t)} (\rho_S - \rho_L)\,\mathbf{V} \cdot \mathbf{n}\,\delta\,(\mathbf{x} - \mathbf{x}_i)\,ds\ . \tag{5}$$

By adding and subtracting the quantity $\mathbf{u} \cdot \nabla\rho$ to the right side of Eq. (5), we can write the equivalent statement

$$\nabla \cdot \rho\mathbf{u} = \mathbf{u} \cdot \nabla\rho + \int_{\Gamma(t)} (\rho_S - \rho_L)\,(\mathbf{V} - \mathbf{u}) \cdot \mathbf{n}\,\delta\,(\mathbf{x} - \mathbf{x}_i)\,ds\ . \tag{6}$$

The integral term in the above equation can now be more intuitively viewed as a local interfacial mass transfer source/sink due to expansion/shrinkage upon solidification. If $\rho_S = \rho_L$ or if there is locally no phase change $(\mathbf{V} \cdot \mathbf{n} = \mathbf{u} \cdot \mathbf{n})$ then Eq. (6) reduces to the customary incompressible constraint, $\nabla \cdot \mathbf{u} = 0$. We further note that the numerical implementation of Eq. (6) is also more robust since the surface integral represents a delta function *perturbation* to an existing smooth field $(\mathbf{u} \cdot \nabla\rho)$ whereas the surface integral in Eq. (5) attempts to satisfy mass conservation with delta functions *directly*.

The momentum equation, in conservative form, is

$$\frac{\partial \rho\mathbf{u}}{\partial t} + \nabla \cdot \rho\mathbf{u}\mathbf{u} = -\nabla P - \rho\mathbf{g} + \nabla \cdot \boldsymbol{\tau} + \int_{\Gamma(t)} \gamma_o\kappa\mathbf{n}\,\delta\,(\mathbf{x} - \mathbf{x}_i)\,ds\ , \tag{7}$$

where P is the pressure, $\mathbf{g}$ is the gravitational force, γ is the surface tension coefficient (assumed constant in this equation), κ is twice the mean interface curvature and $\boldsymbol{\tau}$ is the deviatoric stress tensor for a Newtonian fluid $\boldsymbol{\tau} = \mu\left(\nabla\mathbf{u} + \nabla\mathbf{u}^T\right)$. Note that we use this same constitutive relation for both the solid and liquid and, in the computations, make the assumption that the solid is a very viscous liquid. In this way we obtain a crude constitutive material model for the solid which allows solid motion and deformation.

The thermal energy equation with an interfacial source term to account for liberation or absorption of latent heat, L, is

$$\frac{\partial \rho c T}{\partial t} + \nabla \cdot \rho\mathbf{u}cT = \nabla \cdot k\nabla T + \int_{\Gamma(t)} \dot{m}\,[L + (c_L - c_S)\,T_{\text{sat}}]\,\delta\,(\mathbf{x} - \mathbf{x}_i)\,ds \tag{8}$$

where we have neglected the viscous dissipation. T is the temperature, T_{sat} is the equilibrium freezing temperature and $\dot{m}$ is the interfacial mass flux, $\dot{m} = \rho_L\,(\mathbf{V} - \mathbf{u}_L) \cdot \mathbf{n} = \rho_S\,(\mathbf{V} - \mathbf{u}_S) \cdot \mathbf{n}$.

It is important to recognize that away from the interface the single field formulation, Eqs. (6), (7) and (8), reduces to the customary mass, momentum and thermal energy equations for each of the bulk phases while integration of these equations across the interface reveals that the formulation naturally incorporates the correct mass, momentum and energy balances across the interface.

To complete our formulation, a form of the Gibbs-Thomson condition for the interface temperature, $T_i = T(\mathbf{x}_i)$, must be satisfied at the phase boundary. Here we use

$$T_i - T_{\text{sat}} + \frac{\gamma(\mathbf{n})T_{\text{sat}}\kappa}{\rho_S L} + \varphi(\mathbf{n})\dot{m} = 0 \tag{9}$$

where $\gamma(\mathbf{n})$ and $\varphi(\mathbf{n})$ are the anisotropic surface tension and inverse kinetic mobility coefficients, respectively.

Equations (3), (4), (6), (7), (8) along with the interface temperature condition, Eq. (9) are solved using the front tracking/finite difference method for phase change described in detail by Juric and Tryggvason [13]. Briefly, these equations are solved iteratively for the correct normal interface velocity, $\mathbf{V} \cdot \mathbf{n}$, that will satisfy the interface temperature condition, Eq. (9). For the spatial discretization, we use the MAC method of Harlow and Welch [15] along with a phase change projection algorithm, similar to that of Chorin's [16] for the time integration. In explicit front tracking, the phase interface is represented discretely by Lagrangian markers connected to form a front which lies within and moves through the stationary Eulerian grid. As the front moves and deforms, interface points are added, deleted and reconnected as necessary throughout the calculation. Thus the interface can exhibit arbitrarily complex interface deformations and topology changes.

DENDRITIC SOLIDIFICATION IN A SHEAR FLOW

We simulated the dendritic growth of a pure material with properties approximately corresponding to that of succinonitrile [17]. The simulations were performed in a 2D, horizontally periodic box of dimensions 11.1×11.1 μm with an undercooling of -16 K and a 300×300 grid resolution. The solid, attached to the bottom wall, grows upward from an initially perturbed interface. Four-fold anisotropic surface tension provides preferred growth in the horizontal and vertical directions initially.

The left column of figure (1) shows the growth of dendrites into a quiescent liquid at four times: $t = 0$, 0.538, 1.076 and 1.614 μs. The same calculation but now with flow is shown to the right in Fig. (1) at the same four times. The flow is induced by the movement of the bottom boundary to the right with a constant velocity of $U_w = 411 cm/s$. For the situation with flow in Fig. (1), the density of the solid is no longer equal to the liquid but is roughly 5% greater which results in shrinkage upon solidification and a resulting inflow of liquid from the top of the domain. The solid phase was modeled as a liquid of 100 times greater viscosity than the liquid phase.

Fluid convection dramatically alters the growth morphology. The presence of flow results in a fuller sidebranch structure and suppresses the emergence of distinct primary dendrite arms. The shear flow causes the solid to form and fill the domain more rapidly. At 1 μs, 25% more solid has formed due to the presence of fluid convection. The dendrite tip velocities in the case with flow are initially greater than the case without flow. However at later times and throughout most of the rest of the calculation, the tip velocities are roughly equal with and without the presence of liquid motion. These gross effects can be attributed to the influence of convection on the thermal envelope of the growing solid. Convection transports cooler liquid into the interstices of the dendritic structure thereby increasing the temperature gradients at the solid/liquid interface which in turn results in larger interfacial growth velocities. Fluid forces also act to bend and deform the dendrites since the solid is modeled as a very viscous liquid.

CONCLUSIONS

We have described a general formulation and method for the direct numerical simulation of dendritic solidification microstructures in pure materials. The method was applied to the two-dimensional solidification of succinonitrile in a shear flow. The effect of fluid flow was seen to have an important influence on the morphology and growth dynamics. While these

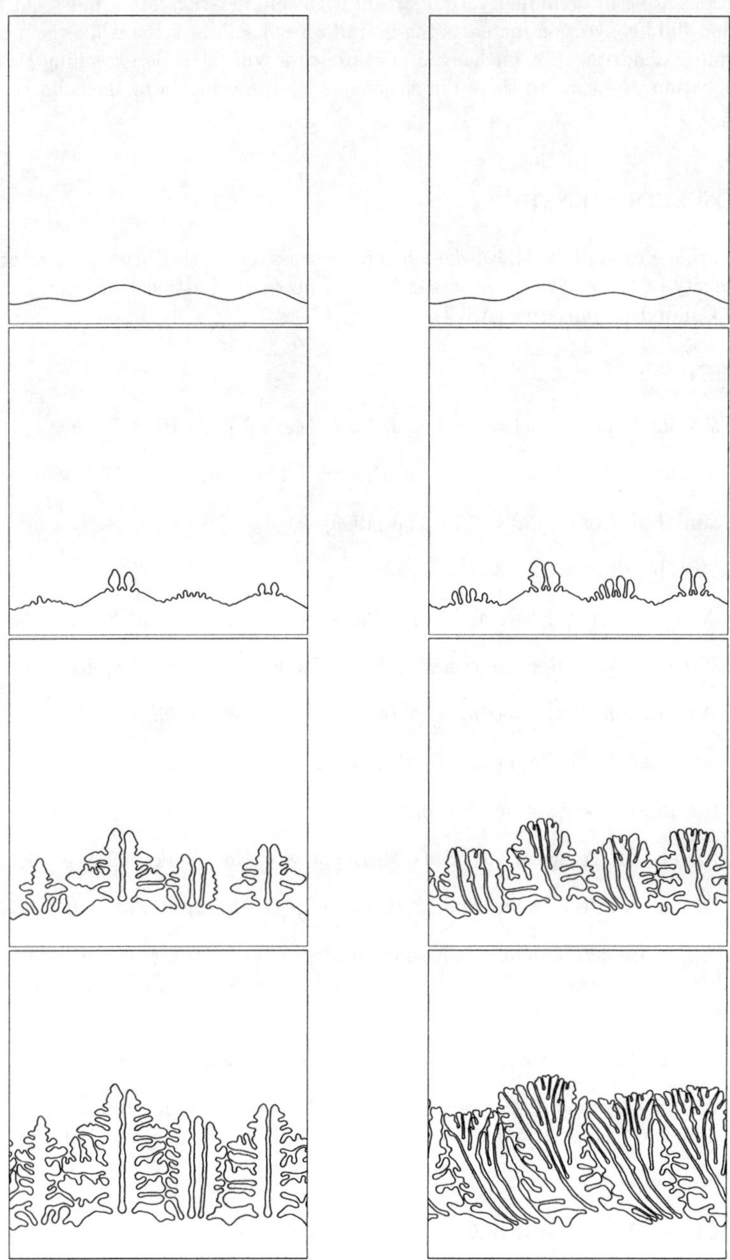

Figure 1: Comparison of dendritic growth of a pure material with no fluid flow (left column) and with fluid flow induced by pulling the solid to the right (right column).

computations are still preliminary, comparison with solidification into a quiescent liquid indicates that fluid convection increases the overall rate of solidification while the growth rate of the leading dendrite tip is unchanged. Future work will focus on extending this method for solidification problems to three dimensions and improvements to the solid constitutive modeling.

ACKNOWLEDGMENTS

The author gratefully acknowledges helpful discussions with Christoph Beckermann of the University of Iowa. This work was supported by the Department of Energy/Advanced Strategic Computing Initiative (ASCI).

REFERENCES

[1] K. Tsiveriotis and R. A. Brown, Int. J. Num. Meth. Fluids,**16**, 827 (1993).

[2] L.H. Ungar, N. Ramprasad and R. A. Brown, J. Sci. Comput.,**3**, 77 (1988).

[3] A. Schmidt, J. Comp. Phys., **125**, 293 (1996).

[4] R. Kobyashi, Physica D, **63**, 410 (1993).

[5] B.T. Murray, A.A. Wheeler and M.E. Glicksman, J. Cryst. Growth, **154**, 386 (1995).

[6] A.A. Wheeler, W.J. Boettinger and G.B. McFadden, Phys. Rev. A, **45**, 7424 (1992).

[7] J.A. Warren and W.J. Boettinger, Acta Metall., **43**, 689 (1995).

[8] A. Karma and W.-J. Rappel, Phys. Rev. Letters, **77**, 4050 (1996).

[9] D. Juric and G. Tryggvason, J. Comp. Phys, **123**, 127 (1996).

[10] D. Juric and G. Tryggvason, (ASME FED-Vol. 234, New York, 1995), p. 141.

[11] H.J. Diepers, C. Beckermann and I. Steinbach, in *Solidification Processing 1997*, p. 426.

[12] X. Tong, C. Beckermann and A. Karma, in *Modeling of Casting, Welding and Advanced Solidification Processes VIII*, edited by B.G. Thomas and C. Beckermann (TMS, Warrendale, PA, 1998) pp. 613-620.

[13] D. Juric and G. Tryggvason, Int. J. Multiphase Flow, **24**, 387 (1998).

[14] D. Juric, in *Modeling of Casting, Welding and Advanced Solidification Processes VIII*, edited by B.G. Thomas and C. Beckermann (TMS, Warrendale, PA, 1998) pp. 605-612.

[15] F.H. Harlow and J.E. Welch, Physics of Fluids, **8**, 2182 (1965).

[16] A.J. Chorin, Mathematics of Computation, **22**, 745 (1968).

[17] M.E. Glicksman, R.J. Schaefer and J.D. Ayers, Metall. Trans. A, **7**, 1747 (1976).

MODELING OF SEGREGATION AND MICROSTRUCTURAL EVOLUTION NEAR GRAIN BOUNDARY IN FE-CR-NI ALLOY UNDER IRRADIATION

N. Sakaguchi*, S. Watanabe** and H. Takahashi*

* Center for Advanced Research of Energy Technology, Hokkaido University,
  Kita-13, Nishi-8, Kita-ku, Sapporo 060 8628, Japan
** Department of Materials Science, Faculty of Engineering, Hokkaido University,
  Kita-13, Nishi-8, Kita-ku, Sapporo 060 8628, Japan

ABSTRACT

We have investigated the solute segregation and simultaneous evolution of extended defects in an Fe-Cr-Ni alloy during irradiation by computer simulation. It sheds a light on the accomplishment of performing "the combined total calculation" or "the muliscale modeling" which deals with both radiation-induced segregation and various kinds of internal sink evolution. The formation of dislocation-free zone (DLFZ) was predicted in the vicinity of a grain boundary. It indicated that DLFZ formation is controlled by solute diffusional process via point defects diffusion near the grain boundary and the activation energy obtained by the width of DLFZ corresponds to the half of the value of the radiation-enhanced solute diffusivity.

INTRODUCTION

The austenitic stainless steel has been considered as a first wall material in near term fusion experimental reactors [1,2], primarily because there exists a large database of irradiated properties for this materials. Although irradiation with energetic particles generates numerous microscopic and non-equilibrium phenomena which influence the macroscopic material properties [3,4], a prediction of the response of the materials to the fusion first reactor conditions is not yet completed even with the use of the most recent experiments and computational models.

Important radiation-induced phenomena include radiation-induced segregation (RIS), radiation-enhanced diffusion, radiation-induced precipitation, and radiation-induced amorphization as well as microstructural changes. These radiation-induced phenomena are known to be associated with the point defects introduced by collision between high energy particles and atoms in an alloy. Among those phenomena, the RIS near grain boundaries is known to cause a significant degradation in physical, chemical and mechanical properties [5,6] in austenitic stainless steels. It is thus of great importance to investigate the retardation of radiation-induced solute redistribution [7-9], viz. segregation, and also the behavior of simultaneous microstructural evolution under irradiation.

In the present study, we investigated the solute segregation and simultaneous evolution of various extended defects such as faulted dislocation loops and network dislocations in an Fe-Cr-Ni alloy during irradiation by a numerical calculation. It sheds a light on the accomplishment of performing "the combined total calculation" which deals with both radiation-induced phenomena and various kinds of internal sink evolution. The present calculation was conducted by solving coupled rate equations for the solute, point defect concentrations and the evolution of the faulted dislocation loops and network dislocations at or near a grain boundary. The relationship between the solute segregation and the heterogeneous formation of dislocations near the grain boundary was investigated.

THEORY AND MODELS

The calculation was performed to evaluate the rate equations for solvent atoms, solute atoms and point defects (vacancies and interstitial atoms) under various irradiation conditions. We also treated the nucleation process of the faulted dislocation loops formed by clustering of interstitial atoms and the evolution of the loops and network dislocations [10-12] on the basis of rate theory. Although di-interstitial atoms may be mobile, we assumed in the present study that the di-inter-

Mat. Res. Soc. Symp. Proc. Vol. 538 © 1999 Materials Research Society

stitial was a nucleus of a dislocation loop because the diffusion parameters on di-interstitial atoms are still unknown thoroughly.

Modeling of Radiation-Induced Segregation

The approach is to solve the coupled diffusion equations for vacancies, interstitial atoms, solvent and solute atoms to provide the terms which couple defect and atom fluxes [13]. The continuity equations used are,

$$\frac{\partial C_k}{\partial t} = -\nabla \cdot J_k \, , \qquad (k \text{ denotes solvent Fe, solute Cr and Ni}) \tag{1}$$

$$\frac{\partial C_v}{\partial t} = -\nabla \cdot J_v + K_0 - K_{vi}C_vC_i - S_vD_v(C_v - C_v^{eq}) \, , \tag{2}$$

$$\frac{\partial C_i}{\partial t} = -\nabla \cdot J_i + K_0 - K_{vi}C_vC_i - S_iD_iC_i - 2K_{ii}C_i^2 \, , \tag{3}$$

where C_k is the concentration of k-atoms, C_v and C_i are the concentration of vacancies and interstitial atoms, K_0 is the generation rate of point defects by irradiation, K_{vi} is the recombination rate of point defects, D_v and D_i are the diffusivities of vacancies and interstitial atoms, and K_{ii} is the clustering rate of single-interstitial to di-interstitial which is the nucleus of dislocation loops, respectively. The flux equations for k-atoms (J_k), vacancies (J_v) and interstitial atoms (J_i) are described as following:

$$J_k = -D_k\nabla C_k + d_k^v C_k\nabla C_v - d_k^i C_k\nabla C_i \, , \tag{4}$$

$$J_v = -D_v\nabla C_v + C_v\sum_k d_k^v\nabla C_k \, , \tag{5}$$

$$J_i = -D_i\nabla C_i - C_i\sum_k d_k^i\nabla C_k \, , \tag{6}$$

where d_k^v is the partial diffusion coefficient of k-atom via vacancies, and d_k^i is the partial diffusion coefficient of k-atom via interstitial atoms, respectively.

The internal sink strength, $S_{v,i}$, for vacancy and interstitial is given by following equation.

$$S_{v,i} = 2\pi\bar{r}_l Z_{v,i}^l C_l^{tot} / \Omega + Z_{v,i}^n \rho_n \, , \tag{7}$$

where $\bar{r}_l$ is the mean radius of faulted dislocation loops, C_l^{tot} is the total concentration of the loops, Ω is the atomic volume, ρ_n is the density of network dislocation, $Z_{v,i}^l$ is the bias factor of faulted dislocation loops and $Z_{v,i}^n$ is the bias factors of network dislocations for vacancies and interstitial atoms, respectively. In this calculation, we treated the simultaneous evolution of both faulted dislocation loops and network dislocations and we indicate the model for each dislocation.

Modeling of Faulted Dislocation Loop

In the present model, we treated the evolution of the faulted dislocation loops by treating a multisize-class description of the loop population [10,11]. The loop size is divided up into a number of size classes, and the time constant for the growth of size class between radii $r_{l(j)}$ and $r_{l(j+1)}$ is

$$\frac{1}{K_{l(j)}} = \int_{r_{l(j)}}^{r_{l(j+1)}} \left\{ \frac{2\pi}{\ln(r_o / r_c)}[Z_i^l D_i C_i - Z_v^l D_v \{C_v - C_v^l(r)\}] \right\}^{-1} dr \ , \tag{8}$$

where the integration corresponds to the loop radial growth velocity, r_o is the dislocation cell radius which is a half of mean dislocation space, r_c is the capture radius of the loop for point defects, respectively. C_v^l is the equilibrium concentration of vacancy near the loop described as follows:

$$C_v^l(r) = C_v^{eq}\exp\left[-\frac{\Omega}{kT}\left\{\frac{\mu b_l}{4\pi(1-\nu)r}\ln\frac{4r}{b_l} + \frac{\gamma_{sf}}{b_l}\right\}\right] , \tag{9}$$

where C_v^{eq} is the thermal equilibrium concentration of vacancy, μ is the bulk modulus, ν is the poisson's ratio, b_l is the Burgers vector of the loop, and γ_{sf} is the stacking fault energy, respectively.

The concentration of loops in the j-th size class ($C_{l(j)}$) is then given by the following equation:

$$\frac{\partial C_{l(j)}}{\partial t} = K_{l(j-1)}\{1 - P_{unf(j-1)}\}C_{l(j-1)} - K_{l(j)}C_{l(j)} \ , \tag{10}$$

where $P_{unf(j)}$ is the fraction of unfaulting from the faulted loop to the network dislocation while it grows with radii $r_{l(j)}$ to $r_{l(j+1)}$, which is determined by the following expression:

$$P_{unf(j)} = \frac{2r_{l(j)}(r_{l(j+1)} - r_{l(j)}) + (r_{l(j+1)} - r_{l(j)})^2}{(\pi r_o / 2)^2 - r_{l(j)}^2} \ . \tag{11}$$

The size distribution of loops between the nucleus size and the maximum radius is represented on a histogram that preserves an essential features of the size distribution [14].

Modeling of Network Dislocation

The model for the evolution of the dislocation structure includes three components, one of them is solely related to the irradiation and others are thermal ones. The irradiation component is the generation term caused by unfaulting from the faulted loops to the network dislocations. The thermal components are a high-temperature climb source term (Bardeen-Herring sources) and a thermal annihilation term caused by the stress-assisted directional diffusion of vacancies. Models of this type [15,16] have been developed for the study of creep processes. Network dislocations can be recovered by climb and glide processes leading to annihilation. In present model, it was assumed that the climbing is a rate controlling process. The rate equation describing the evolution of the dislocation networks is,

$$\frac{\partial \rho_n}{\partial t} = \frac{2\pi}{\Omega}\sum_{j=1} r_{l(j)}P_{unf(j)}K_{l(j)}C_{l(j)} + K_{BH}\rho_n - A\rho_n^2 \ , \tag{12}$$

where the first term represents the unfaulting mechanism, K_{BH} is the generation rate by Bardeen-Herring sources, and A is the annihilation rate by recovery process [16].

Initial and Boundary Conditions

The initial conditions to thermodynamical equilibrium state of the alloys are as follows: For point defects,

Table 1 The main parameters used in the present study.

Parameter	Notation	Value
Vacancy jump frequency via Fe	$\upsilon_{v\text{-}Fe}$	$2.5 \times 10^{13} s^{-1}$
Vacancy jump frequency via Fe	$\upsilon_{v\text{-}Cr}$	$3.5 \times 10^{13} s^{-1}$
Vacancy jump frequency via Fe	$\upsilon_{v\text{-}Ni}$	$1.5 \times 10^{13} s^{-1}$
Vacancy migration energy	$E_m{}^v$	$1.05 eV$
Interstitial jump frequency via Fe	υ_i	$2.5 \times 10^{12} s^{-1}$
Fe-interstitial migration energy	$E_{Fe\text{-}i}{}^m$	$0.3 eV$
Cr-interstitial migration energy	$E_{Cr\text{-}i}{}^m$	$0.3 eV$
Ni-interstitial migration energy	$E_{Ni\text{-}i}{}^m$	$0.9 eV$
Ni-interstitial binding energy	$E_{Ni\text{-}i}{}^b$	$0.75 eV$
Vacancy formation enthalpy	H_v^f	$1.9 eV$
Vacancy formation entropy	S_v^f	$5 k_B$
Lattice constant	a_0	$0.358 nm$
Faulted loop bias for vacancy	$Z_v{}^l$	1.0
Faulted loop bias for interstitial	$Z_i{}^l$	1.5
Network dislocation bias for vacancy	$Z_v{}^n$	1.0
Network dislocation bias for interstitial	$Z_i{}^n$	1.25
Initial dislocation density	ρ_0	$1 \times 10^{-10} m^{-2}$

$$C_v^{eq}\Big|_{t=0} = \exp\,(S_v^f/k)\,\exp\,(-H_v^f/kT)\,, \tag{13}$$

$$C_i^{eq}\Big|_{t=0} = \exp\,(S_i^f/k)\,\exp\,(-H_i^f/kT)\,, \tag{14}$$

where S_v^f and S_i^f, and H_v^f and H_i^f are effective formation entropy and enthalpy for vacancy and interstitial, respectively. For solvent and solute atoms,

$$C_k\Big|_{t=0} = C_k^0\,, \tag{15}$$

where C_k^0 represents the initial bulk concentration of k-atom. For dislocations,

$$C_{l(j)}\Big|_{t=0} = 0\,, \tag{16}$$

$$\rho_n\Big|_{t=0} = \rho_n^0\,, \tag{17}$$

where ρ_n^0 is the initial dislocation density.

In the present work, it is assumed that a grain boundary acts as perfect sink site for point defects. At the grain boundary, we have thus,

$$\frac{\partial C_v}{\partial t}\Big|_{x=0} = 0\,, \tag{18}$$

$$\frac{\partial C_i}{\partial t}\Big|_{x=0} = 0\,. \tag{19}$$

Model calculations were performed for a ternary Fe-15wt.%-20wt.%Ni alloy as a model alloy for an austenitic 316 stainless steel. The main physical parameters used in these calculations has been given previous work [17].

RESULTS AND DISCUSSION

The present model can predict both RIS and dislocation evolution simultaneously. The interaction of these processes is of interest because the local composition changed by RIS influence the defect clustering, whilst the internal sink strength of the faulted dislocation loops and of the network dislocations is an important factor for RIS. In this section, we discuss the interrelationship between RIS and the formation of heterogeneous dislocation near a grain boundary.

In the present calculation, we considered a spatial resolution for the energy dispersive spectroscopy (EDS) analysis with a finite-size probe. The profile of calculation concentration was convoluted by the normalized two dimensional Gaussian distribution for the beam intensity profile and for one dimensional concentration profile. The averaged probe radius was assumed to be 10nm. The averaged technique for calculation is necessary for simulating the actual profile data obtained by TEM-EDS analysis [17].

Radiation-Induced Segregation

Fig.1 showed the typical theoretical and experimental results on RIS near a grain boundary in an Fe-15wt.%Cr-20wt.%Ni alloy after electron irradiation by 1MeV HVEM to 1dpa at a damage rate of 1×10^{-3}dpa/sec at 573K [18]. Due to the irradiation, Ni segregated at the grain boundary, but Cr moved away from there. It is recognized that good consistent between theoretical and experimental results. It is also shown that local minimal concentration of Ni and maximal one of Cr were formed in the vicinity of the grain boundary. This is due to diffusion of Ni atom towards the grain boundary or Cr atom diffusion away from the grain boundary.

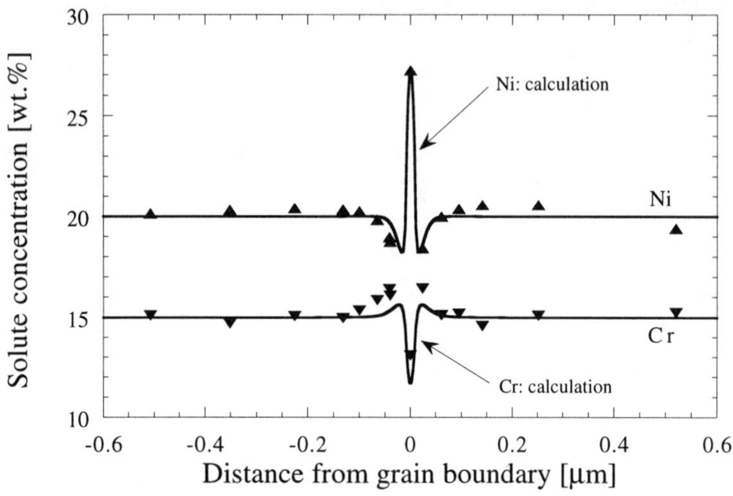

Fig.1 Solute concentration profiles near a grain boundary in Fe-15wt.%Cr-20wt.%Ni alloy after electron irradiation to 1dpa at 573K. Calculation results are shown by solid curves.

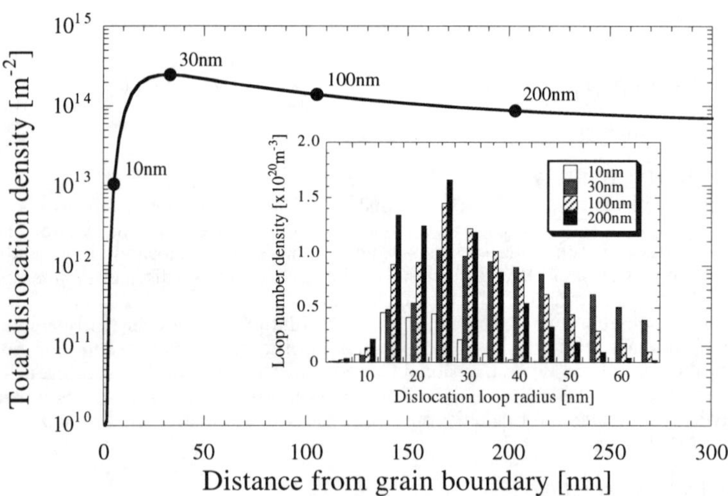

Fig.2 Calculated spatial distribution of dislocation density near a grain boundary after irradiation to 1dpa at 573K. Inset figure shows the loop size distribution at each positions from the grain boundary.

Heterogeneous Dislocation Formation

Fig.2 showed a spatial distribution of faulted dislocation loops and network dislocations near a grain boundary obtained by the calculation after irradiation to 1dpa at 1×10^{-3}dpa/sec and at 573K. The inset shows the loop size distribution at several positions from the grain boundary. It can be seen that a peak density of dislocation loop exists at a distance from grain boundary and there is a region of less dislocations with the width about 30nm in the vicinity of the grain boundary. This is corresponding to a denuded zone for dislocations. We defined this region as dislocation-free zone (DLFZ) near a grain boundary [19]. We have actually observed DLFZ formation as shown in Fig.3. The width of it was about 30nm at both sides of a grain boundary during electron irradiation at the same condition of the above calculation. Frank type faulted interstitial loops with Burgers vector of (a/3)<111> as shown by the black dots in the photograph were observed. It is not clearly observed in Fig.3, but the contrast of void-like was often observed in the interior of the grain.

Interrelationship between RIS and DLFZ formation

The DLFZ is formed as a result of the diffusion of point defects towards a grain boundary sink. That is, the generation of supersaturated point defects near the grain boundary is suppressed and thus DLFZ formed without the defect clustering appears. Also, the diffusion of point defects induces the solute composition changes near the grain boundary due to the inverse-Kirkendall. Therefore, it is essential to understand the relationship between RIS and DLFZ generated near the grain boundary. We considered the characteristic widths of RIS and DLFZ formed due to irradiation. The widths of RIS (W_{RIS}) and DLFZ (W_{DLFZ}) were defined as shown in Fig.4. TheW_{RIS} indicates the nodal point of concentration profiles. The local minima of Ni concentration, W_{RIS} corresponds to W_{DLFZ} which is related to the diffusional behavior of the point defects. Fig.5 showed the relation between W_{RIS} and W_{DLFZ} after irradiation to 1dpa at 473K up to 673K. The widths of RIS and DLFZ are enlarged with increasing of temperature. It is clearly shown that there is a linear relationship between W_{RIS} and W_{DLFZ}.

The W_{DLFZ} corresponds to the critical capture radius of the grain boundary for point defects, r_c^{GB} obtained by Eqs.(1-6) assuming steady state as following expression:

$$W_{DLFZ} \propto r_c^{GB} \propto \sqrt{D_i C_i} . \tag{20}$$

W_{RIS} is the characteristic diffusion length of solute atom towards or away from the grain boundary and is given,

$$W_{RIS} \propto \sqrt{D^{irr} t} , \tag{21}$$

$$D^{irr} = D_v C_v + D_i C_i , \tag{22}$$

where D^{irr} is the radiation-enhanced solute diffusivity.

At steady state,

$$D_v C_v \cong D_i C_i . \tag{23}$$

At a given time, W_{RIS} is proportional to the square root of D^{irr}. Thus a linear relationship between W_{RIS} and W_{DLFZ} is obtained.

Fig.6 showed the Arrhenius plot of both W_{DLFZ} and D^{irr} after irradiation to 1dpa. In the figure, the circles denote experimental results from HVEM electron irradiation [18] at the same conditions as the calculation. It is obtained that the value of the activation energy obtained from the relation of width of DLFZ is a half value of the activation energy for solute diffusion under irradiation. This indicates that DLFZ is corresponding to solute diffusional process corresponding to the point defects diffusion. In the vicinity of a grain boundary, nucleation process of the

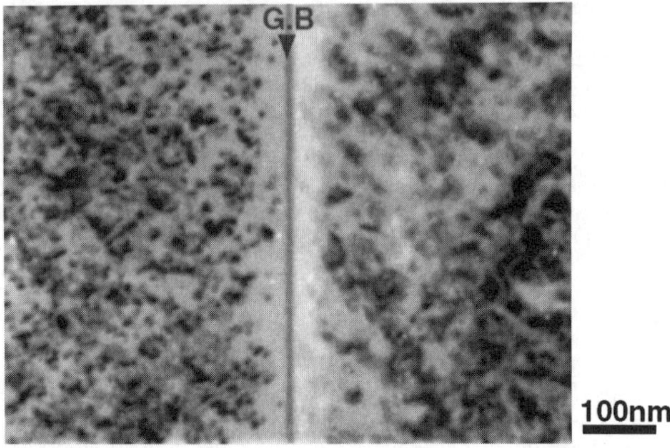

Fig.3 Formation of DLFZ near a grain boundary in Fe-15wt.%Cr-20wt.%Ni alloy by means of TEM observation after electron irradiation to 1dpa at 573K.

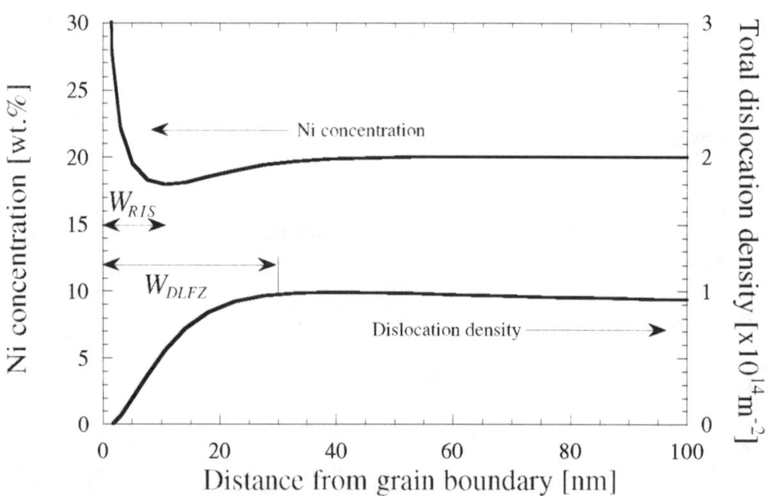

Fig.4 Determination of width of RIS (W_{RIS}) and width of DLFZ (W_{DLFZ}). W_{RIS} indicates the nodal points of the solute concentration profiles.

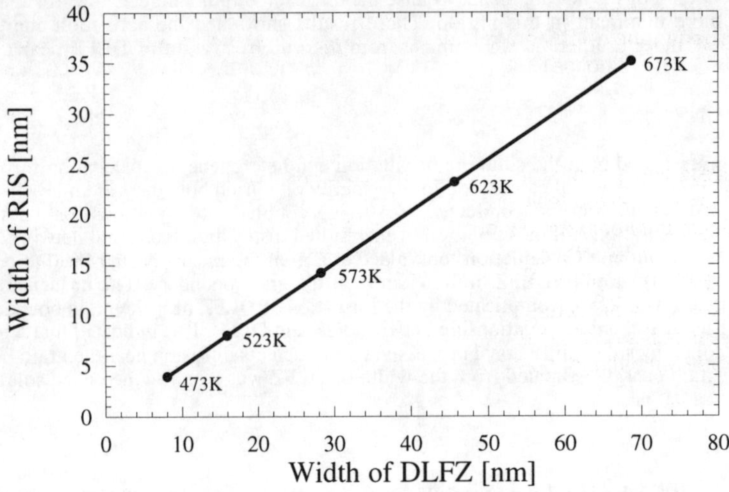

Fig.5 Theoretically predicted dependence of W_{RIS} and W_{DLFZ} after irradiation to 1dpa at each temperature. A linear relationship between W_{RIS} and W_{DLFZ} is shown.

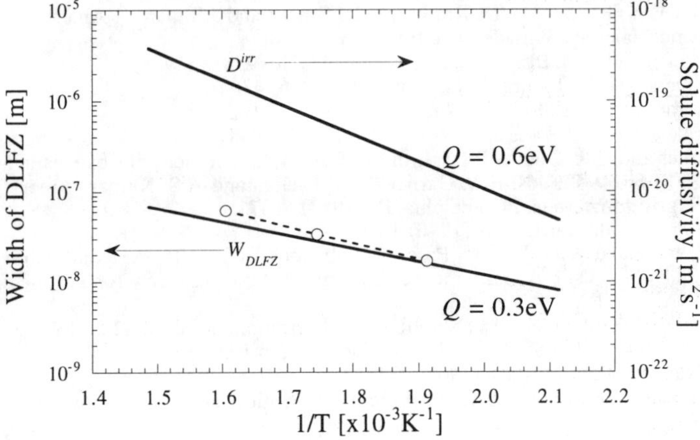

Fig.6 Arrhenius plot of W_{DLFZ} and D^{irr} after irradiation to 1dpa. Circles show experimental results using HVEM at the same conditions with simulation.

faulted dislocation loops is not dominant because the diffusion of point defects towards the grain boundary is more dominant in this region. These results show that the activation energy for solute diffusion under irradiation can estimate from *in-situ* observation of DLFZ near a grain boundary.

CONCLUSIONS

We have investigated both the solute redistribution and heterogeneous dislocation formation during irradiation by numerical calculation in the vicinity of a grain boundary in an Fe-Cr-Ni alloy. The present calculation was conducted by solving coupled rate equations for the solute, point defects concentration and the evolution of the faulted dislocation loops and network dislocations. Ni enrichment and Cr depletion took place at a grain boundary during irradiation. We have clarified the formation of DLFZ in the vicinity of the grain boundary. The influence of internal sink evolution on RIS is complicated by the formation of DLFZ near the grain boundary. It was also clarified that there is a relationship between RIS and DLFZ. It is indicated that width of DLFZ is controlled by solute diffusional process via point defects diffusion near the grain boundary. The activation energy estimated from the width of DLFZ was a half value of the solute diffusion under irradiation.

REFERENCES

1. H. Gross, H.P. Fuchs, H.J. Lippert and W. Dambeitz, Nucl. Eng. Des. **108** (1988) 433.
2. H. Hanien and I. Aho-Mantila, Proc. 3rd Int. Symp. on Environmental Degradation of Materials in Nuclear Power System - Water Reactors, eds. G.J. Thsis and J.R. Weeks (The Metallurgical Society of AIME, Warrendal, PA, 1988) p. 77.
3. R.E. Stoller, J. Nucl. Mater. **174** (1990) 289.
4. R.E. Stoller, P.J. Maziasz, A.F. Rowcliffe and M.P. Tanaka, J. Nucl. Mater. **155** (1988) 1328.
5. J.F. Bates, R.W. Powell and E.R. Gilbert, Proc. 10th Conf. on Effects of Radiation on Materials, ASTM STP 725, eds. D. Kramer, H.R. Brager and J.S. perrin (American Society for Testing and Materials, Philadelphia, PA 1980) p. 713.
6. E.P Simonen and S.M. Bruemmer, Corrosion '93, No. 93615 (1993).
7. R.A. Johnson and N.Q. Lam, Phys. Rev. **B 13** (1976) 4364.
8. P.R. Okamoto and L.E. Rehn, J. Nucl. Mater. **83** (1979) 2.
9. S. Watanabe and H. Takahashi, J. Nucl. Mater. **208** (1994) 191.
10. R.E. Stoller and G.R. Odette, Proc. 13th Conf. of Radiation Induced Changes in Microstructure, ASTM STP 955, eds. F.A. Garner, N.H. Packan and A.S. Kumar (American Society for Testing and Materials, Philadelphia, PA 1987) p. 371.
11. R.E.Stoller, Metal. Trans. **A 21** (1990) 1829.
12. B.B. Glasgow and W.G. Wolfer, Proc. of 12th Conf. of Effects on Radiation on Materials, ASTM STP 870, eds. F.A Garner and J.S. Perrin (American Society for Testing and Materials, Philadelphia, PA 1985) p. 453.
13. A.M Yacout, N.Q. Lam and J.F. Stubbins, Nucl. Instr. and Meth. **B 42** (1989) 49.
14. Y. Katoh, Doctoral Thesis, Tokyo University, 1995, in Japanese.
15. F.R.N. Nabarro, Phil. Mag. **16** (1967) 231.
16. J. Gittus, Viscoelasticity and Creep Fracture in Solids, eds. John Wiley and Sons, New York, 1975.
17. S. Watanabe, N. Sakaguchi, N. Hashimoto and H. Takahashi, J. Nucl. Mater. **224** (1995) 158.
18. N. Wake, Thesis, Hokkaido University, 1997, in Japanese (unpublished).
19. S. Watanabe, N. Sakaguchi, S. Mochizuki and H. Takahashi, Phil. Mag. Lett. **74** (1996) 351.

CHARACTERIZING AND MODELING PLASTIC STRAIN INHOMOGENEITY IN THIN METALLIC SHEETS

X. LI and W. TONG
Department of Mechanical Engineering, Yale University, New Haven, CT 06520-8284
wei.tong@yale.edu

ABSTRACT

Using a newly developed plastic strain measurement technique based on digital image correlation, surface plastic deformation of polycrystalline aluminum alloys in a thin sheet form has been experimentally characterized at a length scale comparable to that of the thickness of the aluminum sheets but much larger than the average size of individual grains. Both static and dynamic local straining patterns in these aluminum alloys have been observed and these strain patterns can not be simulated using the conventional plasticity models. The texture clustering of grains may contribute to the static local plastic strain patterns detected in an Al-Mg alloy. Two distinctive dynamic straining behaviors resulted from the dynamic strain aging of dislocations due to the alloying elements have been experimentally established for 5XXX and 6XXX alloys, respectively. First observation of dynamic strain inhomogeneity is also made in a sheet metal specimen deforming predominately in a plane strain state.

INTRODUCTION

It is well known that the microstructure of polycrystalline metals and alloys is highly heterogeneous. Macroscopically uniform plastic deformation of a sheet metal becomes increasingly non-homogenous when the field of view decreases to the dimension below the thickness of the sheet metal and approaching the size of individual crystalline grains. Such plastic strain inhomogeneity can be spatially distributed in a static or dynamic manner and plays a major role in affecting the surface finish and formability of sheet metals.

However, the spatial variation of the material microstructure and hence the mechanical properties at these length scales is not considered directly in conventional plasticity theories. Predictive capabilities of many existing plasticity analyses of sheet metal forming are rather limited, especially in developing new alloys and new manufacturing processes. One of the obstacles to the incorporation of an effective statistical description of the spatial heterogeneity of polycrystalline materials in a plasticity theory is the lack of understanding of the characteristics of such heterogeneity and its effect on the local plastic deformation behavior over the range of length scales.

This paper presents a novel experimental approach in characterizing plastic strain inhomogeneity in thin metallic sheets. A high accuracy and robust surface plastic strain mapping technique is used to monitor the growth and development of strain inhomogeneity across the wide span of microscopic to macroscopic scales in flat sheet specimens up to necking failure under both uniaxial and plane strain extension conditions. In the following, plasticity experiments of sheet metals using such a new technique will be describe first. Results on both static and dynamic strain patterns in several aluminum alloys will then be presented. The origins of the observed strain patterns will be briefly discussed. Finally, conclusions of our investigation on sheet metal plasticity will be summarized.

EXPERIMENTS

Whole-Field Deformation Mapping

Whole-field measurements of plastic deformation over a range of relevant length scales are extremely valuable to the study of inhomogeneity of plastic deformation in sheet metals. A robust technique using digital image processing has been developed in the past several years [1-5]. This technique has since been further improved and enhanced so plastic strains of sheet metal specimens can be mapped with high accuracy and reliability [6-11]. As shown in Fig.1, the principle of the technique is to extract the local displacement field point-by-point by comparing a pair of digital images of the specimen surface before and after a plastic strain increment.

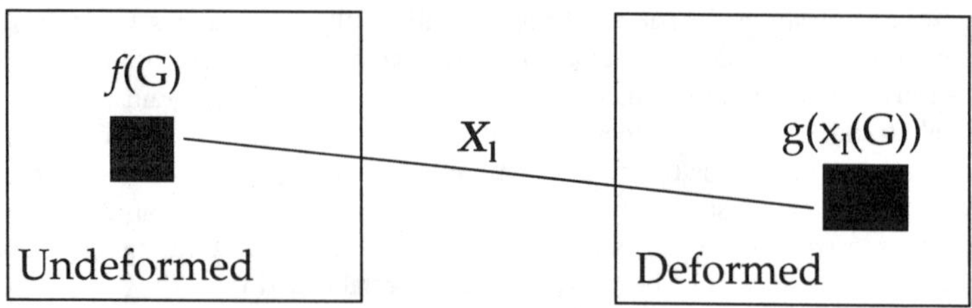

Fig. 1. Principle of digital image based plastic strain technique.

A lot of efforts have been directed to ensure the measurement accuracy and reliability of strain distributions at both small and large deformation regimes [6-11]. Currently, errors in local strain measurements can be routinely limited to 100-150 microstrains [6,7]. Using a newly developed imaging system calibration method, such errors can be further reduced to about 20-40 microstrains [10,11].

Two Types of Plastic Strain Maps

A series of digital images are acquired during a typical experiment of a sheet metal. Both cumulative and incremental strain maps can be easily obtained by the digital image based measurement method, see Fig.2. The incremental strain map or strain rate map is computed by comparing the adjacent image pairs of the deforming specimen and is in particular a highly sensitive means to investigate the dynamic characteristics of a plastic deformation process [8].

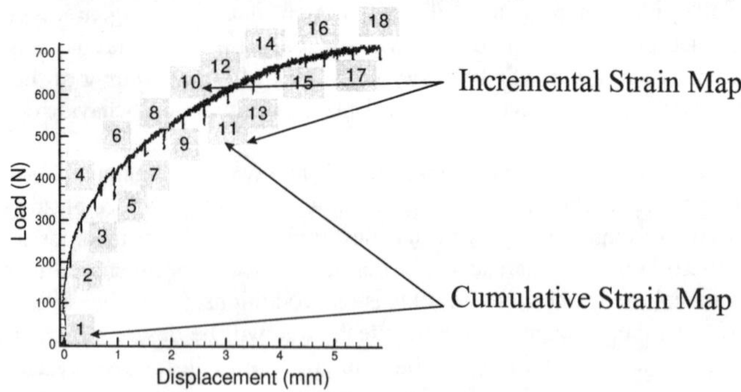

Fig.2 Schematic of computing cumulative and incremental strain maps.

RESULTS

Both uniaxial and plane strain tensile tests were carried out on thin aluminum sheets with a nominal thickness of 1 mm. Results of uniaxial tensile tests will be presented first in the following. Fig.3 shows the geometry of a uniaxial tensile specimen. The contrast pattern on the specimen surface was created by inkjet printing to facilitate the digital image processing.

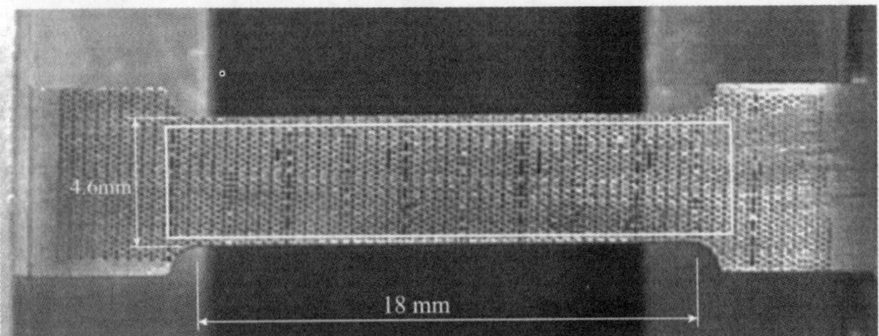

Fig. 3 Image of a uniaxial tensile specimen with a printed contrast pattern. The tensile loading direction is horizontal. The thickness of the specimen is about 1 mm.

In-plane cumulative plastic strains over a 3.6mm-by-3.6 mm region of the specimen gage section were measured and the results of one of strain components are given in Fig. 4. The specimen was made of a recrystalized Al-5%Mg alloy. The static strain inhomogeneity is evident at a length scale comparable to the thickness of the sheet, which is much larger than the grain size of the material (70 microns). Such strain inhomogeneity may be related to the local texture clustering of grains at the same length scale [12].

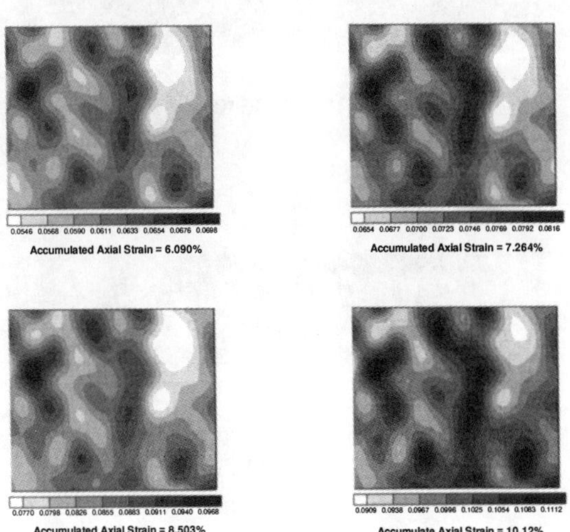

Fig.4 Cumulative axial strain maps of recrystalized Al-5%Mg.

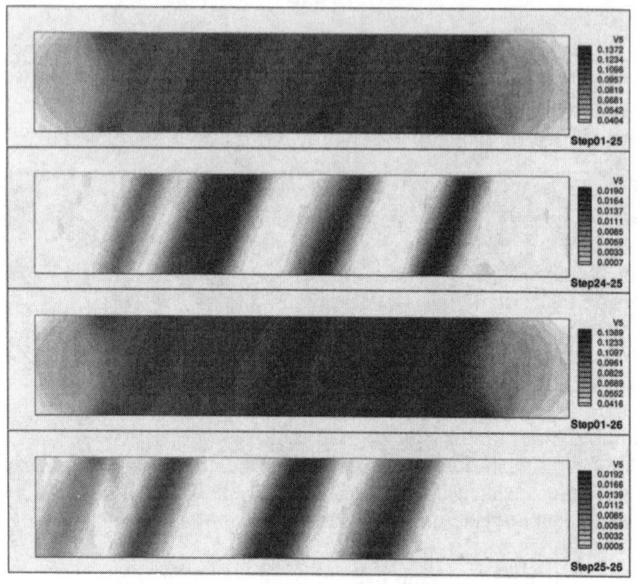

Fig. 5 Cumulative and incremental axial strain maps of AA5182-O alloy at two selected loading steps during a uniaxial tensile test.

Fig. 6 Cumulative and incremental axial strain maps of AA6111-T4 alloy at two selected loading steps during a uniaxial tensile test.

Both cumulative and incremental axial strain maps are shown in Fig. 5 and Fig. 6 respectively for two commercial aluminum alloys, AA5182-O and AA6111-T4. The specimen region of the strain maps in both figures is outlined by a white rectangle shown in Fig. 3. After a few percent of plastic deformation initially, non-steady, localized deformation across the specimen gage section was observed in both alloys. Such a dynamic straining pattern is particularly evident in terms of incremental strain maps. Unlike the Luders band commonly observed in mild steel, these deformation patterns evolved and persist throughout the entire deformation process up to necking and final failure of the flat tensile specimens. The band-like deformation regions in AA5182-O are inclined to the loading axis with an angle about 50 degrees and multiple bands appeared across the entire gage section (see Fig. 5). On the other hand, the localized deformation regions in AA6111-T4 eventually developed into a single band and the macroscopic orientation of the band is almost perpendicular to the loading axis. These two distinctive dynamic behaviors of local plastic deformation may be related to the two different strengthening mechanisms in these two alloys, namely, solid solution and precipitation strengthening mechanisms.

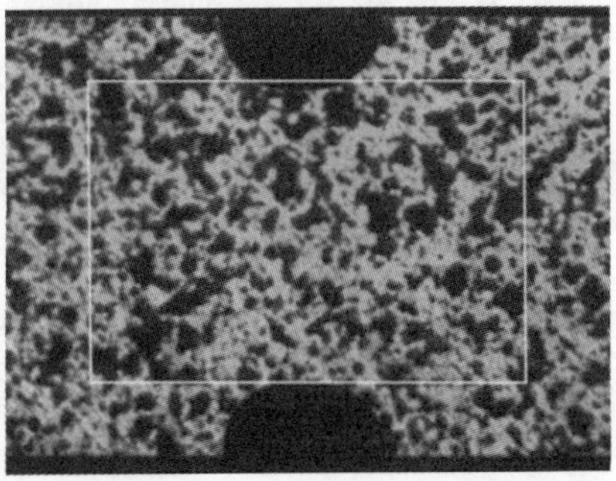

Fig. 7 Image of a plane strain tensile specimen with a decorated surface pattern.
The field of view of the image is about 12.5mm-by-9.4 mm.

The geometry of a compact plane strain tensile specimen is shown in Fig. 7. The surface of the specimen was decorated by a spray of black-and-white Latex paint. The tensile loading direction is again horizontal and the total length of the gage section is much smaller than the width of the gage section. The strain state at the center region of the gage section was found to be nearly a plane strain state according to a finite element analysis [13]. Both cumulative and incremental strain maps of an AA6111-T4 specimen are shown in Fig. 8. The region of the strain maps is outlined by a white rectangle shown in Fig. 7. While the cumulative strain patterns are similar to those predicted by a conventional plasticity theory, the incremental strain patterns as shown in Fig. 8 are non-steady and irregular, totally different from those given by a conventional plasticity theory [13].

CONCLUSIONS

Plastic strain inhomogeneity in sheet metals and its evolution with increasing deformation has been quantitatively characterized at length scales from 100 microns up to 100 mm. An incremental whole-field strain mapping technique has been successfully applied to detect non-

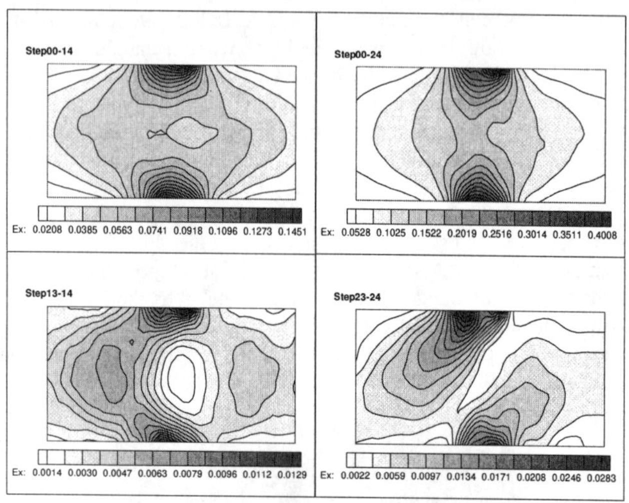

Fig. 8 Cumulative and incremental axial strain maps of AA6111-T4 alloy
at two selected loading steps during a tensile test.

steady deformation patterns in aluminum alloys under quasistatic loading conditions. Incorporation of the mesoscale inhomogeneity of the material microstructure and alloying effects is being made to model the evolution of the observed plastic strain patterns in thin metallic sheets over such a range of length scales.

ACKNOWLEDGMENTS

The research reported here have been supported by NSF (Dr. S. Saigal), Lawrence Livermore National Labs (Dr. W.E. King) and Alcoa Technical Center (Dr. O Richmond). Their financial and technical assistance is gratefully acknowledged.

REFERENCES

1. H.A. Bruck, S.R. McNeill, M.A. Sutton, and W.H. Peters, Exp. Mech. **29**, p361 (1989).
2. M.R. James, W. L. Morris, B.N. Cox, Exp. Mech. **30**, p60 (1990).
3. E.A. Franke, D.J. Wenzel, and D. L. Davison, Rev. Sci. Intrum. **62**(5), p.1270 (1991).
4. D.J. Chen, F.P. Chiang, Y.S. Tan, and H.S. Don, Appl. Opt. **32**, p.1839 (1993).
5. G. Vendroux and W. G. Knauss, Exp. Mech. **38**(3), p.182 (1998).
6. W. Tong, Exp. Mech. **37**(4), p.168 (1997).
7. B.W. Smith, X. Li, and W. Tong, Exp. Tech. **22**(4), p19 (1998).
8. W. Tong, J. Mech. Phys. Solids (in press, 1998).
9. W. Tong and X. Li, Meas. Sci. Tech. (submitted, 1998).
10. X. Li and W. Tong, Opt. Eng. (submitted, 1998).
11. W. Tong and X. Li, Exp. Mech. (submitted, 1998).
12. H. Weiland, Alcoa Technical Center, private communication (1998).
13. X. Li, unpublished research, Yale University, New Haven, CT (1998).

CHARACTERIZATION OF CAST METALS
WITH PROBABILITY DISTRIBUTION FUNCTIONS

M. STEINZIG, F. H. HARLOW*
Group T-3, Los Alamos National Laboratory, Los Alamos, NM 87545

ABSTRACT

Characterization of microstructure using a probability distribution function (PDF) provides a means for extracting useful information about material properties. In the extension of classical PDF methods developed in our research, material characteristics are evolved by propagating an initial PDF through time, using growth laws derived from consideration of heat flow and species diffusion, constrained by the Gibbs-Thomson law. A model is described here that allows for nucleation, followed by growth of nominally spherical grains according to a stable or unstable growth law. Results are presented for the final average grain size as a function of cooling rate for various nucleation parameters. In particular we show that the model describes linear variation of final grain size with the inverse cube root of cooling rate. Within a subset of casting parameters, the stable-to-unstable manifests itself as a bimodal distribution of final grain size. Calculations with the model are described for the liquid to epsilon phase transition in a plutonium 1 weight percent gallium alloy.

INTRODUCTION

Many of the physical processes that govern solidification are understood, and direct numerical simulations (DNS) such as those of Rappaz and Gandin [1] or Juric and Tryggvason [2] are being used to generate accurate characterizations of interesting features such as dendritic growth and coring. These DNS calculations are somewhat limited by the magnitude of scales being covered, which range from the sub-micron scale at nucleation to final grain sizes that might be as large as a centimeter. The computing power necessary for such a simulation is prohibitively large for most problems of practical interest (such as metal casting) and will probably remain so for the foreseeable future. In addition, the extraction of bulk material properties from the results of a DNS simulation are difficult. The use of a probability distribution function (PDF) to describe the stochastics of phase change may prove to be more practical in terms of today's computer hardware, and eventually be more useful for extracting such bulk material properties as average elastic modulus, thereby allowing prediction of the material response to a variety of insults. In this paper, we discuss the formulation of a simple PDF model for the nucleation of grains in a liquid, and use the results of the nucleation to infer the variations in final grain size.

We are interested in the solidification of plutonium, which can undergo a series of phase transitions while cooling to room temperature. The α phase is often undesirable, and so small amounts of gallium are used to alloy the material and limit the transition, as can be seen in Fig. 1. In this work, we present simulations of the liquid to ε transition, in an attempt to determine the final size of ε grains.

FORMULATION

The example described here assumes a problem homogeneous in space, and the only variable considered is R, the grain radius. The PDF is defined such that

$$P(R)dR \qquad (1)$$

Mat. Res. Soc. Symp. Proc. Vol. 538 © 1999 Materials Research Society

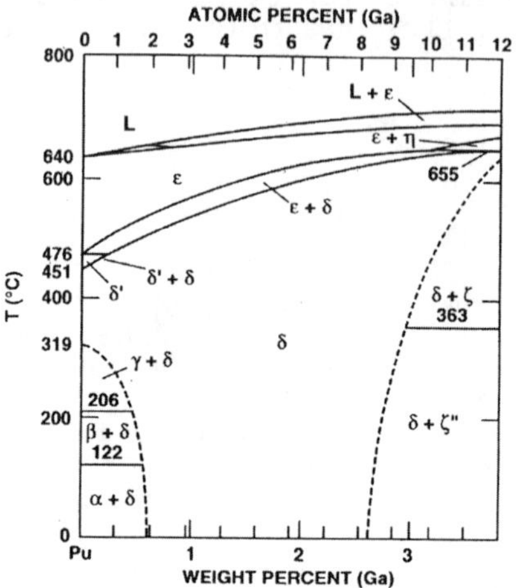

Figure 1. Plutonium-gallium phase diagram for low gallium concentrations [3]

is the probable number of grains per unit volume of size R in an interval of size space dR. The PDF is evolved using a Liouville equation, such that the total number of grains in any arbitrary interval of R space is conserved.

$$\frac{\partial P}{\partial t} + \frac{\partial P \dot{R}}{\partial R} = 0 \tag{2}$$

In addition to the Liouville equation, we have an energy equation

$$\frac{\partial}{\partial t} T_b = \dot{T} + \frac{L}{\rho c_p} \frac{\partial f_s}{\partial t} , \tag{3}$$

in which ρ is density, c_p is specific heat, T_b is the bulk temperature, L is the latent heat, f_s is the fraction solidified and $\dot{T}$ is a specified term representing heat diffusion. The probable number of grains multiplied by the volumetric change rate can be integrated over all size space to give the rate of change of fraction solidified, as

$$\frac{df_s}{dt} = \int_0^\infty 4\pi R^2 \dot{R} P dR . \tag{4}$$

We have used two growth rate equations

$$\dot{R}_s = \frac{K_1 K_2}{R} \left(\frac{1}{R_c} - \frac{1}{R} \right) \tag{5}$$

and

$$\dot{R}_u = \beta \frac{K_1 K_2}{4R_c^2} , \tag{6}$$

with constants defined as

$$K_1 = \frac{k}{L} , \quad K_2 = \frac{2\sigma T_s}{L} ,$$

in which σ is the surface tension, k is the thermal conductivity of the liquid, T_s is the equilibrium solidification temperature, and R_c is the critical radius defined by the Gibbs-Thompson equation, so that for a sphere

$$R_c = \frac{K_2}{\Delta T} \ .$$

The undercooling is defined as

$$\Delta T = T_s - T_b + m_\ell C_+ \ ,$$

where m_ℓ is the slope of the liquidus line and C_+ is the concentration of gallium just outside of the grains. Equation (6) is used for grains that exceed the Mullins and Sekerka [4] stability criterion of $R/R_c \geq 7$, while Eq. (5) is used for those grains with $R/R_c \leq 7$. The β parameter in the unstable growth rate equation is the ratio of the equivalent grain radius to the distance from the center of a grain to the longest tip. This formulation is meant to account for the dendritic growth phase, where growth is strongly tip-led, and the growing grain is no longer spherical, so that R represents an equivalent radius, i.e., the spherical radius with the same volume as the actual grain. No experimental results about the shape of growing ε grains has been located, and we have chosen β as 0.8 for these simulations.

Because of the large gallium diffusivity in both liquid and ε-phase plutonium, the solidification is thermally controlled, and the lever rule is invoked for relating gallium concentration on the liquid side of the grain boundary to f_s. The liquidus line is fit with a constant-slope approximation for use in the Gibbs-Thomson equation.

The derivation of these equations is covered in more detail by Steinzig [5]. For this exercise, the equations are solved using a Lagrangian finite-difference code. The initial PDF represents a flat distribution, P_0, of nucleation sites, from the smallest at $R_{\min}$ to the largest at R_m.

Each simulation begins with liquid plutonium at a bulk temperature equal to the critical temperature of the largest nucleation site in the melt,

$$T_b = T_s - \frac{K_2}{R_m} + m_\ell C_+ \ .$$

As heat is extracted, (through the $\dot{T}$ term in the global energy equation) the temperature drops and progressively smaller sites are nucleated from the initial distribution in a manner similar to that described by Thevoz et al. [6]. Those nucleation sites that are activated become growing grains, and release latent heat that is included in the energy equation and works to counter the effect of $\dot{T}$. In some cases, T_b reaches a minimum and begins to increase toward T_s. This recalescence is shown in the curves plotted in Fig. 2 for various cooling rates; the higher the cooling rate, the more nucleation sites become activated.

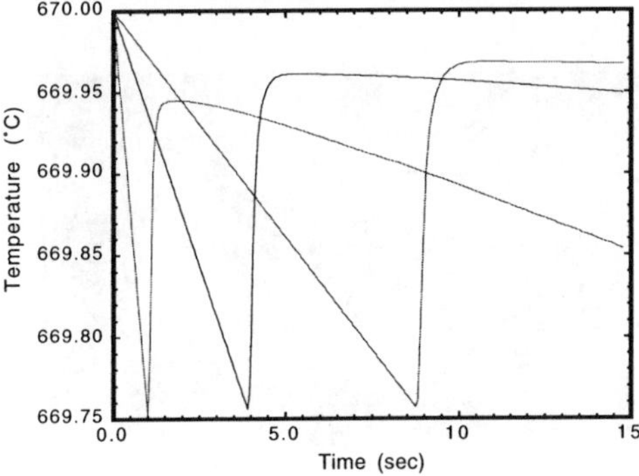

Figure 2. Calculated cooling curves for, from left to right, $\dot{T} = 0.249$, 0.062, and 0.028°C/sec.

It is expected that the number of nucleation sites activated (N_a), has a direct impact on the final grain size. Experimental evidence of this effect is given by the plots of Gardner [7], who shows a linear relationship when final grain size for δ-phase plutonium is plotted as a function of the inverse of the square root of cooling rate, as shown in Fig. 3. For lack of conclusive evidence regarding the structure of the ε-phase grains, we postulate a close correlation with δ-phase grain size when the cooling rate is low and the number of nucleation sites for the $\varepsilon \rightarrow \delta$ transition is likely to be controlled by the geometry of the ε-phase grains. At high cooling rates (e.g., quenching) the $\varepsilon \rightarrow \delta$ nucleation occurs at numerous sites throughout the ε-phase grains, resulting in patchworks of elongate δ-phase grains with directionality controlled by a relationship similar to that described by Kurdjumov and Sachs for ferrite growing into austenite. This structure, as described by Johnson [8], is shown is Fig. 4, which allows tentative inference regarding the ϵ-phase grain sizes, but more extensive data will be required for incisive comparison with our calculated results. Thus our calculations for this paper focus on two issues:

— The variations of final ε-phase grain size with cooling rate for various initial nucleation densities.

— The consequences of grain-growth rate switch from stable, Eq. (5), to unstable, Eq. (6).

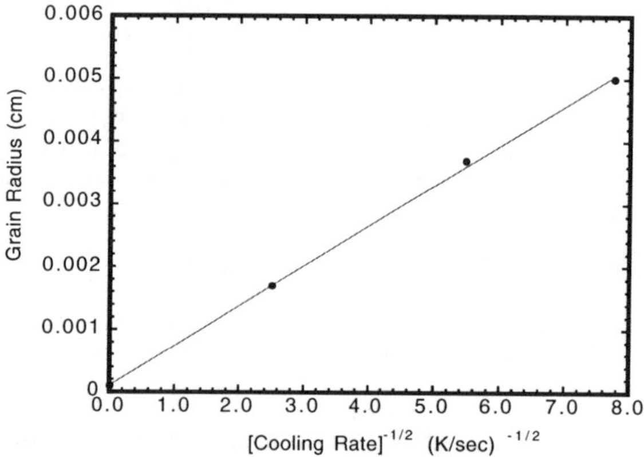

Figure 3. Variations of δ-phase plutonium grain size with cooling rate [7].

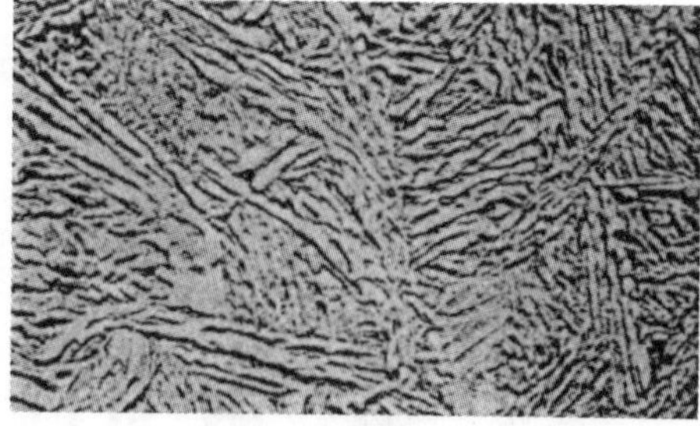

Figure 4. Patchwork structure of δ-phase grains [8]. Width of picture represents approximately 250 μm of actual surface.

Figure 5 shows the variations in ε-phase grain size plotted as a function of the inverse cube root of the cooling rate, $\dot{T}^{-1/3}$. With $P_0 = 10^8\,\mathrm{cm}^{-4}$ (top curve), the grain growth is predominantly unstable, $(R > 7\,R_c)$, the switch occurring quite soon after nucleation. With $P_0 = 10^{10}\,\mathrm{cm}^{-4}$ (bottom curve), the growth becomes unstable $(R > 7\,R_c)$ for a brief period in the early stages of growth, and then reverts to stable growth $(R < 7\,R_c)$ for the rest of the solidification. To show the contrast between stable and unstable growth, we used values of P_0 that are too small to produce the very small ϵ grain size (~ 0.005 cm) inferred from the δ-grain patchwork signature data for ε-phase grain size. Despite the simple structure of the initial PDF, a linear variation of final grain size with $\dot{T}^{-1/3}$ is well described even for the rather narrow parameter range in which the switch from stable to unstable growth occurs during the middle stages of the growth process. In this middle range, with $P_0 = 10^9\,\mathrm{cm}^{-4}$, some of the grains grow unstably while others remain stable throughout their growth. The result is a bimodal distribution of final grain size, shown in Fig. 5 by the two middle curves.

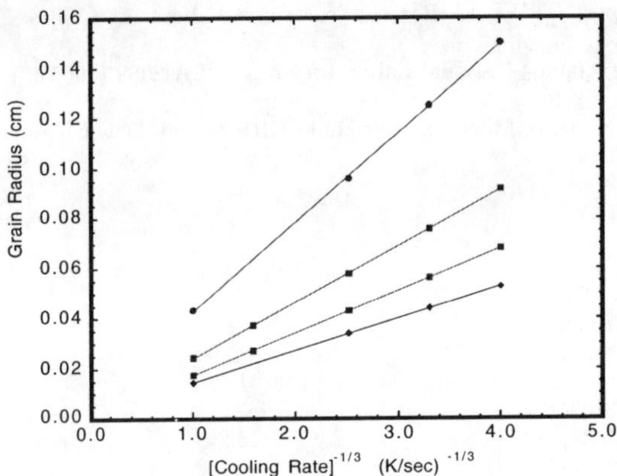

Figure 5. Variations of ε-phase plutonium grain size with cooling rate.

CONCLUSIONS AND DISCUSSION

The variation of final grain size with cooling rate is calculated to be linear when plotted as a function of $\dot{T}^{-1/3}$. Three datum points given by Ferrera, et al. [9] for δ-phase plutonium are consistent with this same $\dot{T}^{-1/3}$ variation, whereas the four datum points given by Gardner suggest $\dot{T}^{-1/2}$. The effects of changing from unstable to stable grain growth are manifested principally by the development of a bimodal distribution of final sizes over a small range of nucleation-site parameters. The PDF approach appears to be a useful and efficient way to characterize gain-size distributions in numerical metal-casting codes.

ACKNOWLEDGEMENT

We are pleased to acknowledge valuable discussions with Thomas Zocco, Frank Gibbs, Damir Juric, Douglas Kothe, Dana Knoll, Christian Charbon, Professor C. Beckermann, and Professor M. Rappaz. This work was performed under the auspices of the United States Department of Energy.

REFERENCES

1. M. Rappaz and Ch.-A. Gandin, Acta metall. mater. Vol 41, #2, p. 345, (1993)

2. D. Juric and G. Tryggvason, J. of Computational Physics **123**, 127 (1996).

3. F. H. Ellinger, C. C. Land, and V. O. Struebing, J. Nucl. Mater. **12**, 228 (1964).

4. W. W. Mullins and R. F. Sekerka, Journal of Applied Physics **34** 323–329 (1965).

5. M. Steinzig, Doctoral Dissertation, New Mexico State University, expected in May 1999.

6. Ph. Thevoz, J. L. Desbiolles, M. Rappaz, Metalurgical Transactions A **20A**, 311 (1989).

7. H. R. Gardner, Report BNWL-13, Battelle Northwest Laboratory (1965).

8. K. A. Johnson, Los Alamos National Laboratory report LA-2989 (1964).

9. D. W. Fererra, J. H. Doyle, M. R. Harvey, Report RFP-1800, Dow Chemical USA, (1972).

LARGE SCALE MOLECULAR DYNAMICS SIMULATIONS
OF A LIQUID CRYSTALLINE DROPLET
WITH FAST MULTIPOLE IMPLEMENTATIONS

ZHIQIANG WANG, JAMES LUPO, SOUMYA S. PATNAIK, ALAN MCKENNEY, RUTH PACHTER
Air Force Research Laboratory, Materials & Manufacturing Directorate, AFRL/MLPJ, Wright-Patterson AFB, OH 45433-7702

ABSTRACT

The Fast Multipole Method (FMM) offers an efficient way (order $O(N)$) to handle long range electrostatic interactions, thus enabling more realistic molecular dynamics simulations of large molecular systems. The performance of the fast molecular dynamics (FMD) code, a parallel MD code being developed in our group, using the three-dimensional fast multipole method, shows a good speedup. The application to the full atomic-scale molecular dynamics simulation of a liquid crystalline droplet of 4-n-pentyl-4'-cyanobiphenyl (5CB) molecules, of size 35,872 atoms, shows strong surface effects on various orientational order parameters.

INTRODUCTION

Atomic level simulations are important in understanding the structure-property relations of materials. For example, the electro-optics of a polymer dispersed liquid crystal (PDLC) can be tuned by varying its constituents. The understanding of the surface anchoring transition from homeotropic (normal) to homogeneous (in plane) of LCs on different polymer walls in the nematic temperature region will certainly help the design of such materials[1]. Because of the size of the LC droplet(typical in microns), very large scale molecular dynamics (MD) simulations of such systems are required.

The most costly part of the large scale MD simulation is the handling of long-range interactions, specifically the *electrostatic* non-bonded forces. An obvious direct calculation requires $O(N^2)$ operations per timestep to compute these forces. In many cases, the simple cutoff method is inappropriate. The method of Ewald summation improves the scaling to $O(N^{(3/2)})$. However, the fast multipole method[2], which uses a multiscale hierarchy of partitions of the volume and a divide-and-conquer strategy to compute the power series, allows all the forces to be computed to any specified accuracy in $O(N)$ operations. Still, the cost of computing all electrostatic interactions remains high. In order to further improve the computing efficiency, we developed a new code named FMM3D (an implementation to FMM in 3-dimensional space), which contains a variety of schemes for computing multipole translations. This code has been incorporated into the FMD (fast molecular dynamics) program we developed which is aimed at very large scale parallel computing on the DoD high performance computer platforms.

In this paper, we first briefly describe the FMM algorithm and the computational details, and then present the results of the FMD simulated structure and various orientational orders in a liquid crystalline system of 944 4-n-pentyl-4'-cyanobiphenyl (5CB) molecules with a free boundary. To our knowledge, this is the largest size simulation so far for a 5CB cluster with all atomic interactions considered. This work is intended to be a first step for the full atomic scale study of 5CB droplets in a polymer dispersed liquid crystal.

METHODS AND COMPUTATIONAL DETAILS

In the Fast Multipole Method(FMM), the computational cube enclosing the whole system of interest is divided into cells in a hierachy of an octal tree structure. Each level-l cell is devided into 8 cubes, forming level-$(l+1)$ cells, or children, and so on until some chosen finest level L of 8^L cubes is reached. The force on a particle in a finest cell is divided into two terms according to the distance. The first is the direct Coulomb interactions due to the nearest and second nearest cells. The rest of the interactions with distant cells are calculated by a field expressed by a local Taylor expansion about the center of the cell where the particle is in. This power series is calculated by first calculating the multipole moments in all the finest level cells, then translating them to both multipole and local expansions at different levels of cells through a multiscale approach of linear transformations between the coefficients of the power series at different expansion origins. These multipole translations usually have a cost of $O(8^L * p^4)$, where p is the truncated term of multipole expansion. The Greengard and Rokhlin's fast translation scheme [3], used in our FMM3D code, reduces the cost to $O(8^L * p^2)$. FMM3D is implemented in SPMD (single program multiple data) style. The detailed descriptions of FMM and FMM3D can be found in our previous paper [4] and references therein.

The molecular dynamics code FMD is derived, in part, from NAMD version 1.4[5]. Changes include: (1) substituting FMM3D for DPMTA (distributed parallel multipole tree algorithm); (2) porting the parallel system from PVM to MPI; (3) adding support for NCSA HDF (Hierarchical Data Format) formatted input/output files; and (4) adopting NDGM (Network Distributed Global Memory) and the DICE (Distributed Interactive Computing Environment) package[6] as the visualization and control interface. The core of FMD retains the object-oriented C++ basis of NAMD. FMD is portable on a wide variety of platforms, ranging from stand-alone Intel cpu based Linux workstations, to massively parallel processors such as the Cray T3E, IBM SP2, and SGI Origin 2000.

FMD was benchmarked with a set of five models ranging in size from 892 to 85,750 atoms. Fig. 1 shows the parallel scaling achieved on up to 32 SGI Origin 2000 nodes for 1,000 MD timesteps. The models are BPTI and Bactetiorhodopsin (BR) proteins and 5CB clusters with the number of molecules ranging from 18 (marked as 5CB×1) to 2124 (5CB×118). The scaling performance improves as the number of atoms increases. For the largest model, the scaling with number of processors, P, increases as $P^{0.78}$. Similar scaling has been seen on up to 48 IBM SP2 and 224 CRAY T3E nodes. Processing speeds, such as seen on the T3E, achieved 5.06×10^{-8} cpu-seconds per atom per node per timestep.

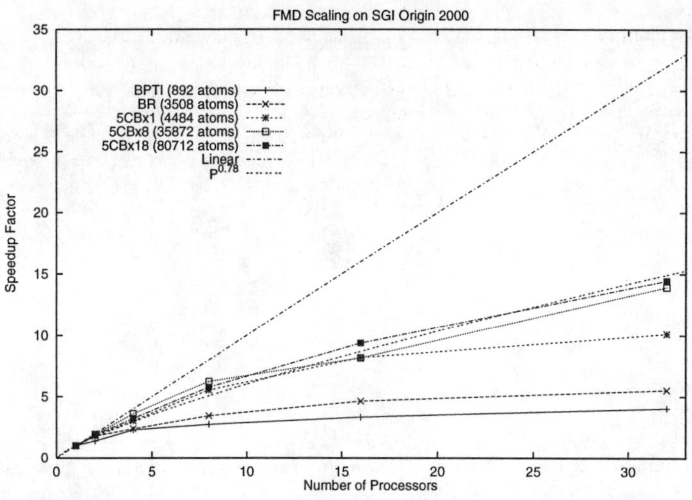

Figure 1: FMD execution time scalings

RESULTS AND DISCUSSION

For the results reported in this section, the initial configuration of the simulation is a cluster of 944 5CB molecules or 35,872 atoms in a cube of size $81.8 \times 68.8 \times 68.8$ Å^3, consistent with the experimentally measured density at its nematic phase. The molecules are oriented along the longer dimension of the box with nematic order parameter of ca. 0.6. The initial temperature is 300K, within the known range of its bulk nematic phase(295 to 308 K). Velocity was rescaled to stablize the temperature. The MD time step is taken as 1 fs. The long-range interactions are updated in a cycle of 10 time steps. The number of multipole expansion terms is typically 6 and FMM finest cell level, L, is 4.

Four orientational order parameters are examined, which are defined as the second rank Legendre polynormial $P_2(x) = (3x^2 - 1)/2$:

$$O = \frac{1}{N} \sum_{i=1}^{N} P_2(\vec{d_i} \cdot \vec{n_i}) \tag{1}$$

where $\vec{d}$ is the director of the liquid crystal molecule, calculated as its shortest principal inertia axis, and $\vec{n}$ determines the type of order parameter as follows [7]: 1) *nematic*: $\vec{n}$ is the director of other molecules (The summation should be over all pairs (both i and j) with the argument in P_2 as $(\vec{d_i} \cdot \vec{n_j})$); 2) *bipolar*: $\vec{n}$ is the local meridian that lies on the plane defined by the droplet axis and the radial vector $\vec{r_i}$ of the center of mass of the molecule while being perpendicular to $\vec{r_i}$ itself; 3) *radial*: $\vec{n}$ is the radial vector $\vec{r_i}$; and 4) *field*: $\vec{n}$ is the external field or some specified direction.

The order parameter is one if the molecular director $\vec{d_i}$ is in perfect alignment with the order vector $\vec{n_i}$; while 0 if $\vec{d_i}$ is random, and -0.5 if $\vec{d_i}$ is perpendicular to the order vector.

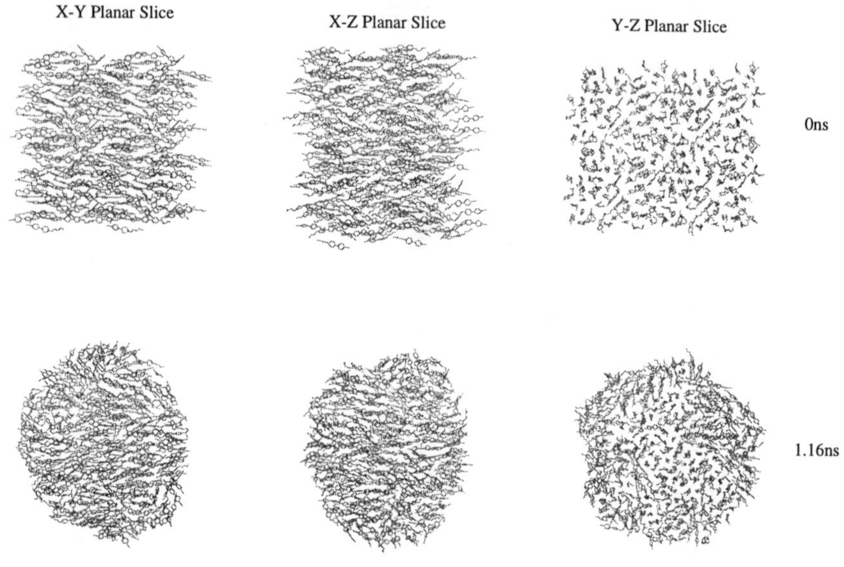

Figure 2: Slice views of the 5CB cluster at time 0 and 1.16 ns. The slices are passing the spheric center with thickness of 20 Å.

Therefore, a radial order parameter of -0.5 indicates that the molecules are tangential to the sphere.

Fig. 2 shows slice views of the 5CB system at the initial time and 1.16 ns. The change of the system shape and molecular orientations on the surface are obvious. In fact, the cluster transformed from a cube to a sphere in about 500 picoseconds. However, the reconstruction and orientation order changes continue throughout the simulations.

Fig. 3 shows the local density and nematic order parameters of the 5CB droplet from the center to the surface at different MD timesteps. The solid curve is for the initial system. The numbers marked are the MD time in ps, thus the data are for MD runs ranging from 0 to about 1 ns. For Fig. 3, the 90% density radius of the final sphere should be about 38Å. The increase of the density near the surface is partly due to the migration of molecules from the corner edge. It is interesting to note that the density on the surface increases to slightly larger than the bulk density at long time.

The nematic order shown in Fig. 3 decreases significantly in the 25 to 35 Å radial region. Even near the center, the nematic order is continuously weakend up to 1 ns. This shows that the effect of the free surface on the orientation order is deep for this droplet size, although caution should be taken that the order decrease near the center may be caused by a less nematic order even in the bulk phase. On the surface, the nematic order

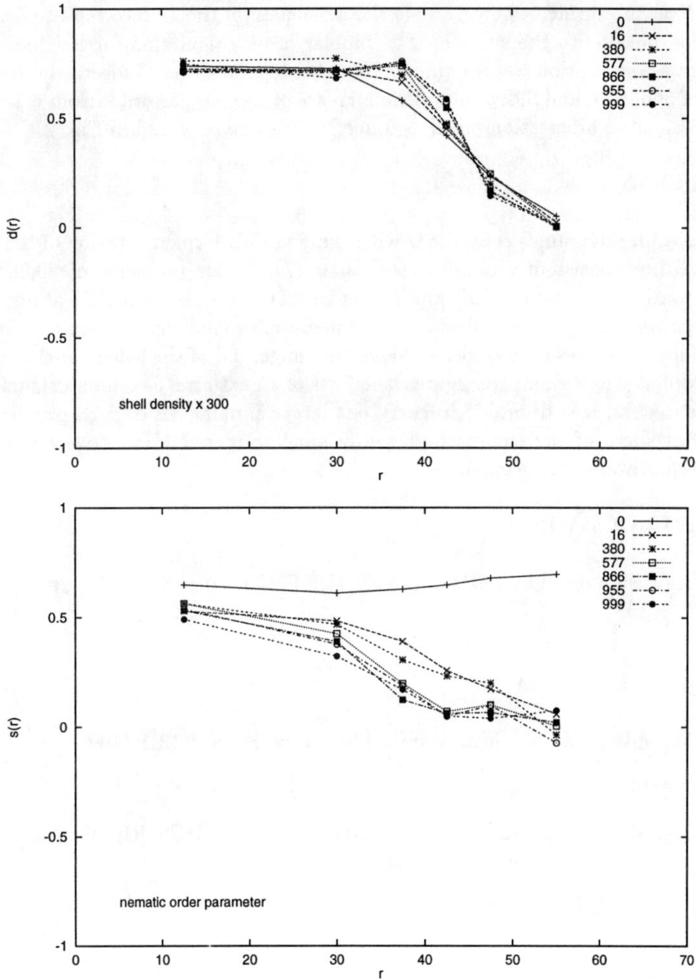

Figure 3: The local density and nematic order parameters of a 5CB droplet at different time steps (marked numbers in *ps*), *vs.* the distance from the center (in Å).

is totally broken. Similar strong size effects are also observed for the field order parameter, which shows the order along the initial orientation. Clearly, a larger size is desirable for the simulation of such a material.

Does the strong surface effect create some other surface related order? The calculated radial order parameter shows the decrease of the value from about 0 near the center to about -0.5 on the surface, showing that the molecules on the surface tend to orient along the direction parallel to the surface. The bipolar order calculation further indicates that this tangential orientation has a tendency towards bipolar order. Indeed, the results with MD time of 866, 955, and 999 *ps* show the increase of the bipolar order from 0.2 to 0.5 and 0.6. The tangential orientation on the surface can also be observed in Fig. 2.

CONCLUSION

The molecular dynamics code FMD with the FMM3D implementation of the fast multipole algorithm consistently scales better than $1/P^{3/4}$, for problems of 30,000+ atoms with full atomic interactions. The application to a 5CB droplet of 35,872 atoms with free boundary shows deep surface effects on the nematic orientational order at temperature 300K. Surface molecules tend to orient along the tangential of the sphere with some trends of bipolar order. To further investigate the effects of an external field and a structured wall of polymer matrix, it is desirable to carry out larger simulations of such problems. More speed improvement of the current high performance targeted MD software is needed, in addition to hardware enhancements.

ACKNOWLEDGEMENT

This work is supported by DOD Common High Performance Software Support Initiative (CHSSI).

References

[1] K. R. Amundson and M. Srinivasarao, Phys. Rev. E **58**, R1211 (1998).

[2] L. Greengard and V. Rokhlin, J. Comp. Phys. **73**, 325 (1987).

[3] L. Greengard and V. Rokhlin, Research Report YALEU/DCS/RR-602, Yale University Dept. of Computer Science (unpublished).

[4] A. McKenney, R. Pachter, S. Patnaik, and W. Adams, Mat. Res. Soc. Symp. Proc **408**, 99 (1996).

[5] M. Nelson, W. Humphrey, A. Gursoy, A. Dalke, L. Kalé, R. Skeel, K. Schulten, and R. Kufrin, International Journal of Supercomputer Applications and High Performance Computing **10**, 251 (1996).

[6] J. A. Clarke, IEEE Computational Science & Engineering 55 (1997).

[7] E. Berggren, C. Zannoni, C. Chiccoli, P. Pasini, and F. Semeria, Phys. Rev. E **50**, 2929 (1994).

COMPUTATIONAL MODELING OF MULTIPLE DOMAIN PATTERN FORMATION

Tao Huang, Tomohiro Tsuji, Alejandro D. Rey and Musa R. Kamal,
Department of Chemical Engineering, McGill University,
3610 University Street, Montreal, Quebec, Canada, H3A 2B2

ABSTRACT

We propose a physical model of multiple domain nucleation and growth in order to perform the cross-scale simulation on envelope profiles of the domain and the internal structure. The model utilizes two coupled field: a nonconserved envelope vector field of the local macroscopic growth domains and a vector field of the local mesoscopic lamellae directors. The envelope evolution equation includes an envelope profile function, the domain growth and impingement kinetics, the relaxation of the envelope by a surface tension, a lowest-order nonlinear term, a noise term, and a surface diffusion term. The splay, twist and bend deformation and the interaction between lamellae and noise field are considered in the free energy density equation of the lamellar director field. Results of large-scale computer simulation are reported which are compared with experiments.

INTRODUCTION

During the last decade, major research efforts have been aimed for developing a fundamental understanding of a wide range of interfacial pattern formation phenomena[1] , such as, dendrite, dense-branching morphology (DBM), and fractal morphology. KPZ equation [2] is now widely accepted as describing lateral growth processes such as the Eden model process and growth by ballistic deposition. These studies are also related to material engineering and device manufacture. However, multiple domain pattern formation poses an unsolved problem as to how polycrystalline entities, such as, spherulites, are formed with crystallographically equivalent radii. Spherulites are commonly formed from the melts which are of relatively high viscosity and the crystallization is slow. Spherulites have been recognized in minerals, organic compounds, polymers, and protein crystal growth. It is well known that spherulite is the most common morphology observed in polymer field.

Despite the wide diversity of physical and chemical properties among spherulite-forming melts, the growth and subsequent morphology are surprisingly universal. An explanation of the existence of spherulitic growth should be independent of the fine details of each individual system, and should be based only on those features which are shared by all spherulite-forming melts. Phenomenologically, spherulites generally develop from a single crystal precursor via repetitive branching and splaying of the direction of the fastest growth which thus eventually becomes radial within a spherical envelope. The initially puzzling transverse orientation of molecular chains was understood once it was appreciated that chain-folded lamellar crystals grew radially outward in spherulites. However, the nature of the splaying entities and the mutual organization of lamellae in spherulites could not be ascertained with certainty. On the other hand, it is well-known that the multiple domain pattern results from impingement of growing nuclei. The multiple domain pattern depends on the spatio-temporal distribution of nucleation sites and on the growth velocities.

Mat. Res. Soc. Symp. Proc. Vol. 538 © 1999 Materials Research Society

In this paper, we propose a physical model to perform the cross-scale simulation on the envelope profiles of the domain and the internal structure. This physical model of multiple domain pattern formation was developed based on experimental [3] and theoretical studies [4]. It consists of three parts: postnucleation model, domain growth model and lamellae director model. A nucleation equation, an envelope equation of domain growth, and a lamellae equation utilize two coupled field: a non-conserved envelope vector field of the local macroscopic growth domains and a vector field of the local mesoscopic lamellae directors. The computational modeling have been performed based on this model to predict and analyze the lamellae organization based on the mesoscopic lamellae director and domain patterns of multiple spherulitic structures.

GOVERNING EQUATION AND COMPUTATION

To avoid the physical details of structural changes of molecular conformation in initial nucleation stage and analytical complexity of the coupling due to external force effects, we simply treat the heat transfer, pressure and stress, flow as external source to the thermal change, and assume that the thermal field varies on time-scales much longer than those of phase order parameters. We consider a vector,

$$\Psi(\vec{X}, t) = \{T(\vec{X}, t), P(\vec{X}, t), \sigma(\vec{X}, t), ...\} \tag{1}$$

with components corresponding to individual parameters of the state, such as, temperature, $T(\vec{X}, t)$, pressure, $P(\vec{X}, t)$, stress, $\sigma(\vec{X}, t)$, etc., where $\vec{X}$ is a position vector), as a variety of time-dependent external fields to describe crystallization conditions. For the large system in a real industrial process, $\Psi(\vec{X}, t)$ is understood as an effective local parameter, rather than boundary value. Non-uniform distribution of $\Psi(\vec{X}, t)$ in the large systems could be solved by process modeling, which is desired from the dynamic equations with appropriate boundary conditions.

1. Postnucleation Model

For systems of relatively high viscosity, many collective length scales are very large and hence a phenomenological (quasiuniversal) description is much more appropriate. With respect to kinetic phenomena at first-order phase transitions, the characteristic time scale of relaxation and diffusion of polymeric macromolecules is much slower for small molecular materials or metal systems. The kinetics of an initially undercooled solid-liquid melt can be studied by means of a generalized phase-field model, which describes the dynamics of an ordering non-conserved field $\phi(\vec{X}, t)$ (e.g., solid-melt order parameter) coupled to a conserved field, a variety of time-dependent external conditions, $\Psi(\vec{X}, t)$. It is assumed that it evolves according to the phenomenological Langevin equation [5] appropriate for a non-conserved order parameter given by

$$\frac{\partial \phi(\vec{X}, t)}{\partial t} = -\Gamma_\phi \frac{\delta G[\phi(\vec{X}, t), \Psi(\vec{X}, t)]}{\delta \phi(\vec{X}, t)} + \zeta(\vec{X}, t) \tag{2}$$

where Γ_ϕ is a rate parameter that sets the time scale for phase evolution, $G[\phi(\vec{X}, t), \Psi(\vec{X}, t)]$ is Gibbs potential to be specified. The $\zeta(\vec{X}, t)$ is the thermal noise from a heat bath, which we assume to obey Gaussian and white correlated stochastic process with the properties,

$$< \zeta(\vec{X}, t) >= 0, < \zeta(\vec{X}, t)\zeta(\vec{X}, t') >= 2T\eta\delta(t - t') \tag{3}$$

The energy balance requires that the latent heat released at the transition equates the temperature change of the melt multiplied the specific heat, i.e:

$$\frac{\partial \Phi(\vec{X}, t)}{\partial t} = D \, \nabla^2 \, \Phi(\vec{X}, t) - \frac{1}{\Delta\phi} \frac{\delta\phi(\vec{X}, t)}{\delta t} \tag{4}$$

In computational performance, the critical nucleation radius is depends on the combination of the capillarity length (the material properties) and crystallization kinetics by Eq.(2). The in-situ spatio-temporal evolution of the nucleation process can be simulated by coupling Eq.(2) with the domain growth equation given in the following section.

2. Multiple Domain Growth Envelope Equation

Based on generalized Langevin Equation [5], the dynamical variable(s) of a system can be described by a systematic force, a force due to systems potential, and a fluctuating force. The balance of the forces is established for the dynamical evolution of the system. We propose an envelope evolution equation that represents the multiple domain growth and impingement kinetics, as follows:

$$\frac{\partial \varphi_i(\vec{u}_i, t)}{\partial t} = \vec{v}(t) \cdot \vec{u}(t) \prod [1 - \delta\{A(t)\}] + D_{u\perp} \frac{\partial^2 \varphi}{\partial u^2} + \gamma \Delta^2 \varphi_i + \frac{\lambda}{2} (\Delta\varphi_i)^2 + \varepsilon\xi \tag{5}$$

Here, $\varphi_i(\vec{u}, t)$ is an envelope profile function, $\vec{v}$ is domain growth velocity vector, and vector $\vec{u} = (\cos\theta, \sin\theta, 0)^T$ represents a unit vector along one spherulite center coordinate on an envelope, and θ is the angle. The term, $\prod [1 - \delta\{A(t)\}]$, describes the impingement dynamics of nearest neighbor domains, $\delta\{A(t)\}$ is the Dirac delta function, the parameter A is compensation factor. In 2D cases, $A = \{[x_{ix} + \cos\theta\phi_i(\theta, t) - x_{jx} + \cos\theta_j\phi_j(\theta_j, t)]^2 + [x_{iy} + \sin\theta\phi_i(\theta, t) - x_{jy} + \sin\theta_j\phi_j(\theta_j, t)]^2\}^{1/2}$. The delta function signifies that, if two envelope positions are at the same place, the grain will stop growing. Therefore, the first term represents the domain growth and impingement kinetics. The second term represents the effect of the diffusion, where $D_{u\perp}$ is a diffusion coefficient.

The underlined terms is combination of surface tension effect, nonlinear growth term, and a noise term, where γ is surface tension, ε is thermal fluctuation strength, and ξ is the thermal noise. It is interesting to find out that this underline terms is similar to KPZ equation. The KPZ equation was introduced and studied by Kardar, Parisi, and Zhang [2], to describe the height profile above a basal substrate in a co-moving coordinate system, which is responsible for the lateral growth of the Eden model process and growth by ballistic deposition. It is very important to note that the proposed domain envelope equation is a basic equation for kinetics of multiple domain envelope profiles, which represents the physical nature of the kinetics of the multiple domain envelopes evolution.

3. Lamellar Director Model

To describe the internal structure of the polycrystalline entities, we assume that the mesoscopic crystalline lamellar director is a unit vector of the major axis direction and changes along the length of a mono-lamellae, where $\vec{l}$ is presented in terms of lamellar direction θ_l. There are three possible configurations: (1) splay, which represents the tilting a lamellae toward another; (2) bend, which represents the angle between two adjacent lamellae; and (3) twist, which arises by counter-rotation at the lamellae ends.

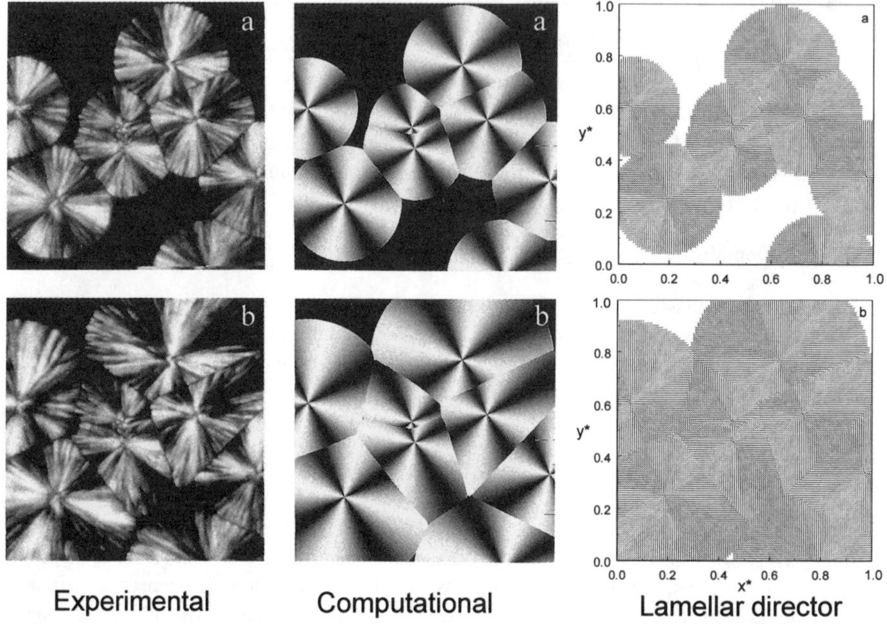

Figure 1: The results of computational modeling of multiple domain pattern and the internal lamellae director field compared with the experimental observation of polypropylene spherulite growth pattern.

The curvature energy of the lamellae director field is similar to Frank energy equation in liquid crystals [6]:

$$F_l = \frac{C_1}{2}(\nabla \cdot \vec{l})^2 + \frac{C_2}{2}(\vec{l} \cdot \nabla \times \vec{l})^2 + \frac{C_3}{2}|\vec{l} \times \nabla \times \vec{l}|^2 \tag{6}$$

where C_1, C_2 and C_3 are elastic coefficients. For two dimensional case, $\vec{l} = (\cos\theta_l, \sin\theta_l, 0)$, and the splay and the bend energies are the follows:

$$F^*_{splay} = \frac{1}{2}\left[\cos\theta_l \frac{\partial\theta_l}{\partial y^*} - \sin\theta_l \frac{\partial\theta_l}{\partial x^*}\right]^2 \tag{7}$$

$$F^*_{bend} = \frac{1}{2}\left[\sin\theta_l \frac{\partial\theta_l}{\partial y^*} + \cos\theta_l \frac{\partial\theta_l}{\partial x^*}\right]^2 \tag{8}$$

We employ the light intensity equation as $I = I_0 \sin^2 2\theta_l$, where θ_l is the local orientation angle of the lamellar. By combining with the envelope equation, the simulations are able to reproduce the main optical features of the spherulites under polarized light.

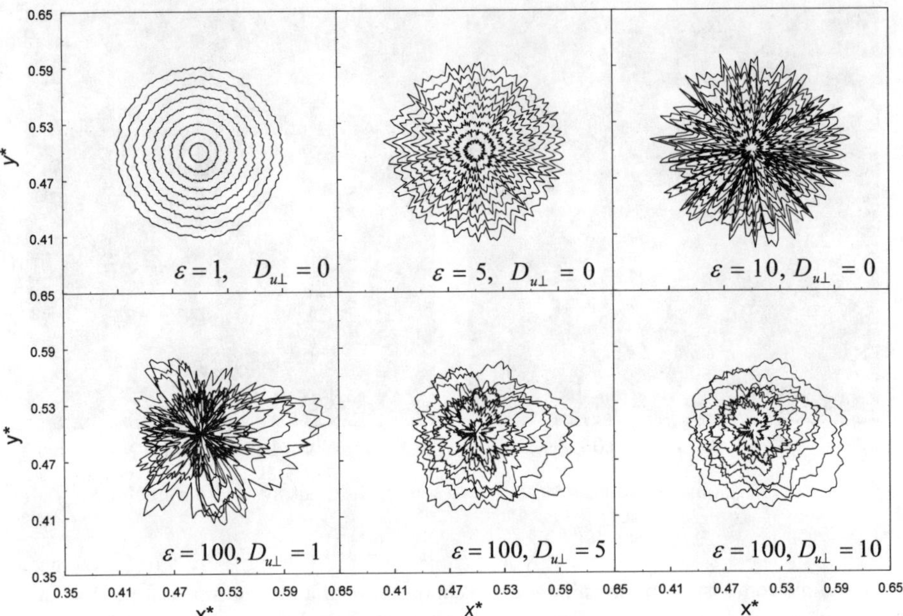

Figure 2: Computational results of the envelope profile evolution of isolated domain growth with different system parameters.

RESULTS AND DISCUSSIONS

Computational modeling has been performed, based on the coupling of the above equations to predict and analyze the lamellar organization based on the mesoscopic lamellar director, and domain patterns of spherulitic structures. The simulation results and the corresponding scientific visualization show the spherulitic textures as Maltese cross gray images. Figure 1 shows a typical result of computational modeling of multiple domain pattern and the internal lamellae director field compared with the experimental observation of polypropylene spherulite growth pattern. Within the spherulite, the computational results indicate that the characteristic of lamellae is the orientation deviations caused strongly by the large splay distortion, and by the bend distortion in some regions. It is clearly shown that both energies at the spherulite grain boundaries are higher than within the spherulite grains. The energy profiles agree exactly with the corresponding vector plot of the mesoscopic lamellar director. The higher energy at grain boundaries is due to the mis-matching of the orientation of the mesoscopic lamellar directors at the grain boundary between two neighboring spherulites. Figure 2 presents the computational results of the envelope profile evolution of isolated domain growth with different system parameters. Figure 3 shows the computational domain patterns of two sets of different system parameters. These results show the influence of the system parameters on the multiple domain pattern formation.

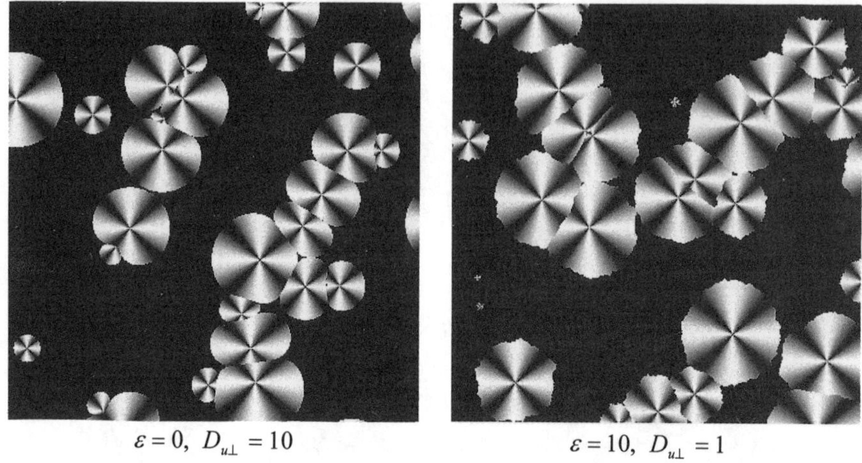

Figure 3: Computational results of two sets of different system parameters.

Dynamic comparison between large scale computational and experimental results have also been accomplished by quantitatively matching the simulated growth patterns to the experimental images. There is very good agreement between the computational results and experimental observations. Furthermore, this model is being extended to modeling the structure development in real polymer processing.

CONCLUSIONS

Computational modeling based on the proposed envelope evolution equation associated with a postnucleation model and the curvature energy of the lamellae director field predicts the nucleation-growth processes and domain patterns of spherulitic structures. Simulations and visualizations represent the spherulitic textures as the Maltese cross. It provides not only the functional data on envelope profiles of the spherulites but also the spherulite internal structure. It has major implications to understand spherulitic multiple domain pattern formation.

REFERENCES

1. J.S. Langer, Phys. Today, Oct., 24, (1992).

2. M. Kardar, G. Parisi and Y.C. Zhang, Phys. Rev. Lett. **56**, 889 (1986).

3. Tao Huang, A.D. Rey and M.R. Kamal, Macromolecules, **31**, 7791 (1998).

4. Tao Huang, Tomohiro Tsuji, A.D. Rey and M.R. Kamal, Phys. Rev. E, **58**, 789 (1998).

5. K. Sekimoto, Prog. Theor. Phys. Suppl., **130**, 17 (1998).

6. F.C. Frank, Disc. Faraday Soc., **25**, 19 (1958).

From Molecular Dynamics to Kinetic Rate Theory:
A Simple Example of Multiscale Modeling

ROGER E. STOLLER[*] AND LAWRENCE R. GREENWOOD[**]

[*] Metals and Ceramics Division, Oak Ridge National Laboratory, P.O. Box 2008, Oak Ridge, TN 37831-6376, USA

[**] Environmental Technology Division, Pacific Northwest National Laboratory, P.O. Box 999, Richland, WA 99352, USA

ABSTRACT

Radiation damage formation in iron has been investigated using the method of molecular dynamics simulation. The MD simulations have been used to determine primary defect production parameters for cascade energies up to 50 keV at temperatures from 100 to 900K. The energy dependence of these parameters has been used to determine appropriate neutron-energy-spectrum averaged damage production cross sections for various irradiation environments. Two applications of these effective cross sections are discussed. The first is an evaluation of neutron energy spectrum effects in commercial fission reactor pressure vessels. The second example deals with the use of these cross sections in the source term of a kinetic model used to predict void swelling and microstructural evolution. The simulation of the primary damage event by MD involves times less than 100 ps and a size scale of a few tens of nm, while the kinetic simulation encompasses several years and macroscopic sizes. This use of the MD results to develop an improved source term for rate theory modeling provides a simple example of multiscale modeling.

INTRODUCTION

Models based on reaction rate theory have been broadly and successfully applied to simulate radiation-induced microstructural evolution.[1-4] Typically, the use of these models involves the simultaneous solution of a modest number of differential equations to predict phenomena such as void swelling, irradiation creep, or embrittlement. The time scale of interest for these processes is determined by atomic diffusion rates and the desired in-service lifetime of irradiated components. The rate theory is well suited to span this time range from seconds to years, and a size scale from micro-meters to macroscopic dimensions. However, the source term in the rate equations is dictated by atomic displacement cascades, events that occur on a time scale of a few tens of pico-seconds and a few tens of nano-meters in space. The method of molecular dynamics (MD) is an appropriate tool for simulating these displacement cascades.

MD has seen extensive application in simulating displacement cascades in irradiated materials,[5-12] and recent advances in computing have enabled researchers to examine higher energy events with their requirement for larger blocks of atoms. These same computing resources have permitted cascade simulations to be completed in sufficient numbers to allow for statistically meaningful trends to be determined as a function of simulation energy and irradiation temperature. For example, the comparison of iron and copper contained in Ref. 9 is based on a database of over 600 cascades with energies from 60 eV to 10 keV and irradiation temperatures of 100, 600, and 900 K. The results used in this work represent an extension of the collaborative study presented in Ref. 9, with the peak cascade energy increased to 50 keV.[13,14]

The analysis of displacement cascade simulations provides useful information on point defect formation at specific cascade energies. However, integral damage parameters that account for the energy dependence of the primary knockon atom (PKA) spectra are needed for more

203

Mat. Res. Soc. Symp. Proc. Vol. 538 © 1999 Materials Research Society

general applications in radiation damage modeling. Since the energy range accessible by MD simulations now includes the domain of cascades produced by high energy neutrons, it is appropriate to use the MD results to develop these integral parameters for different neutron irradiation environments. The integral parameters can be used to compare the effects of neutron energy spectrum as well as providing improved estimates for the average atomic displacement rate in these environments. The use of the MD-based parameters in kinetic radiation damage models represents one simple example of multiscale modeling.

MD SIMULATION METHOD

The details of the MOLDY code and the interatomic potential used are discussed in detail elsewhere.[7,9,15,16] The interatomic potential used was a many-body type, originally developed by Finnis and Sinclair for α-iron[16] and subsequently modified by Calder and Bacon[7] in order to permit its use in dynamic simulations. The original version of this potential was fitted to a number of physical parameters; the primary modification of Calder and Bacon was to the repulsive pair term in order to fit the pressure-volume relationship and to reproduce the displacement threshold surface.[7] The simulations were carried out using an atom block with periodic boundary conditions and at constant pressure. Since the boundary atoms were not damped to remove heat, the initiation of a cascade event led to some lattice heating. The primary impact of the temperature increase (up to 300 K at 1 ps) is to extend the lifetime of the thermal spike; it has no impact on the ballistic phase of the cascade. In addition, the temperature increase does not appear to influence defect survival since this has been shown to exhibit very little temperature dependence.[7,9]

Prior to conducting the cascade simulations, a block of atoms is thermally equilibrated for about 10 ps to establish thermal equilibrium. This equilibrated atom block is then used as the starting point for the subsequent cascade simulations, with the largest atom block used in this work being 2.25 million atoms for the 50 keV cascades. The simulations are initiated by giving one of the lattice atoms a defined amount of kinetic energy and a specific direction. Multiple cascades are differentiated by changing the location of the initial PKA or further equilibrating the block. As discussed in Ref. 9, the initial kinetic energy is analogous to the damage energy in the standard NRT displacement model.[17] In the terminology of the NRT model, a 50 keV cascade simulation energy (or damage energy) corresponds to a PKA energy of 78.7 keV.[9,17] This is the average PKA energy in an elastic collision between a 2.3 MeV neutron and an iron atom. The difference between the PKA and damage energy is the kinetic energy lost to electronic excitation.

RESULTS OBTAINED FROM MD CASCADE SIMULATIONS

Building on the work published in Refs. 7 and 9, iron cascade simulations up to 50 keV have been completed and used in this analysis. Three parameters have been extracted from the analysis because of their relevance to radiation damage simulation: the total number of surviving point defects, the total fraction of surviving interstitials in clusters, and the fraction of interstitials in clusters containing 10 or more interstitials. At each energy, an average value was obtained from several cascades (typically 8 to 10). In order to provide a standard basis of comparison, the averages were divided by the number of displacements calculated from the NRT model.[17] These normalized values are shown in Figs. 1 (a-c), where the standard deviation about the average is included on the higher energy points. Results are shown for all three temperatures in Figs. 1 (a) and (b), but only for 100K in Fig. 1(c). The 10 and 20 keV points have been slightly displaced along the axis to differentiate the three temperatures.

There does not appear to be a systematic dependence on irradiation temperature in the results shown in Fig. 1; therefore, a decision was made to treat all the "data" as a single database for this analysis. The solid lines shown in Fig. 1 were obtained from a least-squares fit to the data,

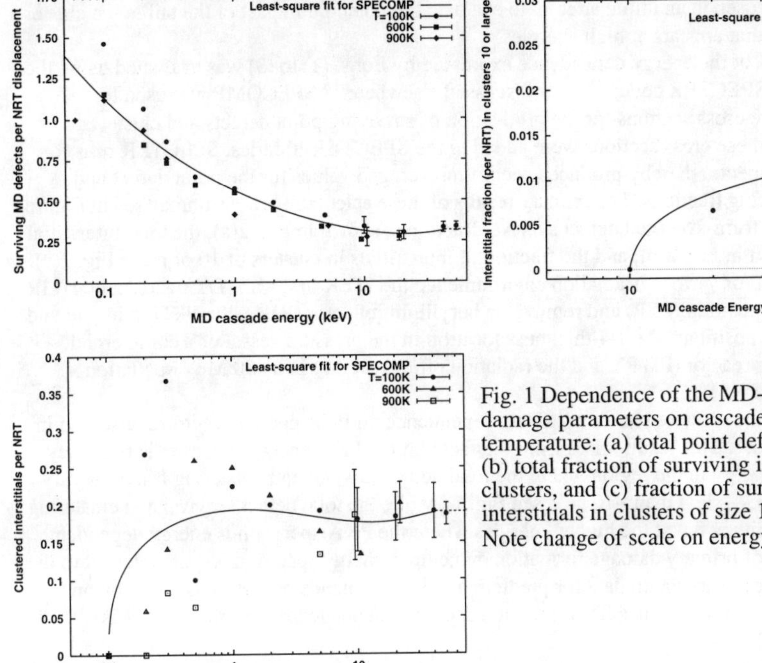

Fig. 1 Dependence of the MD-derived primary damage parameters on cascade energy and temperature: (a) total point defect survival, (b) total fraction of surviving interstitials in clusters, and (c) fraction of surviving interstitials in clusters of size 10 or larger. Note change of scale on energy axis in (c).

yielding the following three functions for the cascade survival efficiency (η), the total interstitial clustering fraction (f_{icl}^{T}), and the interstitial fraction in clusters of 10 or larger (f_{icl}^{10}):

$$\eta = 0.5608 \cdot E_{MD}^{-0.3029} + 3.227x10^{-3} \cdot E_{MD} \tag{1}$$

$$f_{icl}^{T} = [0.097 \cdot ln(E_{MD} + 0.9)]^{0.3859} - 7x10^{-6} \cdot E_{MD}^{2.5} \tag{2}$$

$$f_{icl}^{10} = 4.40376x10^{-3} \cdot ln(E_{MD} - 0.9)^{0.7453} \tag{3}$$

The line is shown dashed above 40 keV in Figs. 1 (a) and (b), because this was the highest energy that had been completed at the time the fitting was done. The value at 40 keV was used for all higher energy damage energies in the analysis discussed below. This was based on the nearly asymptotic behavior seen in Figs. 1 (a) and (b), and the subsequent 50 keV cascades provided limited confirmation for this assumption. As discussed elsewhere,[11,18] subcascade formation appears to be responsible for this behavior. Because there were no clusters of size 10 or larger below 10 keV, Eqn. (3) was not obtained until recently when the 50 keV cascades were completed. There are relatively few such clusters even at the higher energies, and the standard deviations are greater than the absolute value of the averages on the points shown in Fig. 1(c).

However, the large clusters could be a significant nucleation site for extended defects, so this parameter was chosen in an initial attempt to evaluate the potential impact of the tail in the cluster size distribution that appears at high energies.

The impact of the energy dependence expressed by Eqns. (1) to (3) was evaluated using the SPECOMP and SPECTER codes,[19,20] as discussed elsewhere.[18] SPECOMP was used to calculate effective cross sections for the production of surviving point defects and clustered interstitials, and these cross sections were added to the SPECTER libraries. SPECTER runs for various neutron spectra thereby produced spectrum-averaged values for the point defect and interstitial clustering fractions. The primary results of these calculations are summarized in Figure 2. The PKA-spectrum-averaged defect survival fraction is shown in Fig. 2(a), the total interstitial clustering fraction in Fig. 2(b), and the fraction of interstitials in clusters of 10 or more Fig. 2(c). Values are shown for various irradiation environments: the ITER first wall (DT fusion), the HFIR peripheral target position (PTP) and removable beryllium reflector (RB*), the FFTF midcore and below-core (BC) positions, the 1/4-thickness location in the pressure vessel of a commercial pressurized water reactor (PWR), and the radiation effects facility in a 450 MeV spallation neutron source (IPNS).

As pointed out previously, there is little dependence on PKA energy spectrum observed in Figs. 2 (a) and (b).[18] This result arises from the fact that the PKA energy spectrum is relatively independent of energy in the region where the total defect survival and clustering fractions vary with energy (i.e. E_{MD} less than ~10 keV, see Fig. 1), while the total defect survival and clustering fractions are relatively flat at the higher energies where the PKA spectrum is energy dependent. The dependence of primary damage formation on neutron energy spectrum is quite relevant to the use of fission reactors to obtain data for predicting the performance of materials in fusion and spallation neutron environments. Therefore, the spectrum dependence seen in Fig. 2(c) is

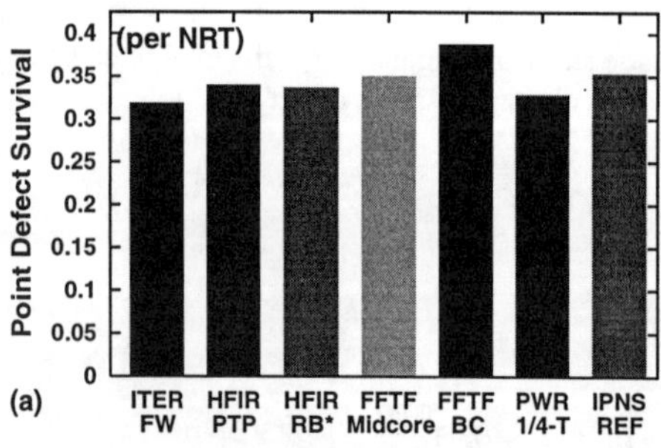

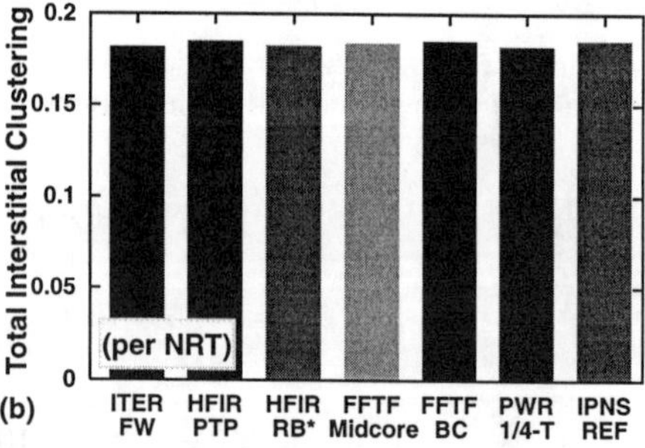

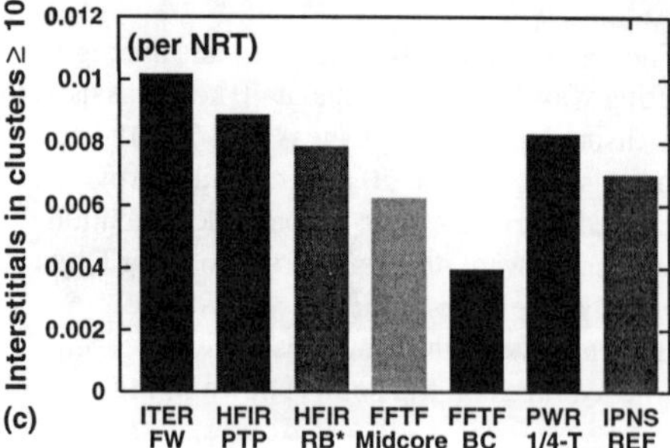

Fig. 2 Spectrum-averaged defect production cross sections (per NRT dpa) for various irradiation environments: (a) total point survival, (b) total interstitial clustering fraction, and (c) interstitial fraction in clusters of size 10 or more.

noteworthy. The significance of the interstitial cluster size distributions need further investigation, but this is the first of the primary damage parameters that has shown a high level of dependence on the neutron (or PKA) energy spectrum.

APPLICATION OF MD-BASED PARAMETERS

Two relatively simple examples can be used to illustrate the application of the MD-based primary damage parameters. The first is relevant to light water reactor pressure vessel (RPV) steels, and the second to predictions of void swelling under fast neutron irradiation.

One difference between commercial pressurized water reactors (PWR) and boiling water reactors (BWR), is that the distance between the outer fuel elements and the RPV is greater in the BWR than in the PWR. This space is filled with water, leading to both a lower neutron flux and a softer neutron spectrum impinging on the BWR RPV. In light of the energy dependence of defect survival illustrated by Fig. 1(a), concerns have been raised about the impact of the softer neutron spectrum on radiation-induced embrittlement in BWRs. Specifically, the concern is whether or not the PKA spectra are sufficiently different between the two types of reactors that the embrittlement per unit fluence might be higher for the BWR. In addition, the neutron energy spectrum shifts as a result of attenuation, and this could produce different responses in material irradiated in surveillance capsules than in the RPV at the 1/4 or 3/4 thickness locations.

The NRT-normalized, spectrum-averaged defect production cross sections for typical BWR and PWR neutron spectra are shown in Fig. 3 for four locations: the last transport node in the water before entering the RPV, the 1/4-T and 3/4-T RPV positions, and the first transport node in the cavity beyond the RPV. Note that the values for interstitials in large clusters are referenced to the right-hand scale in the figure. It is clear that for both total defect survival and total interstitial clustering, there is essentially no difference between the PWR and BWR parameters, nor is there any significant effect of the spectrum shift that arises from attenuation. The total interstitial clustering values for the two reactors are indistinguishable at each location. However, as might be expected from Fig. 2, a much more significant effect is observed for the fraction of interstitials in clusters of 10 or more. Since only a small fraction of the interstitials produced are in such clusters, more analysis is required to assess the significance of this spectrum effect.

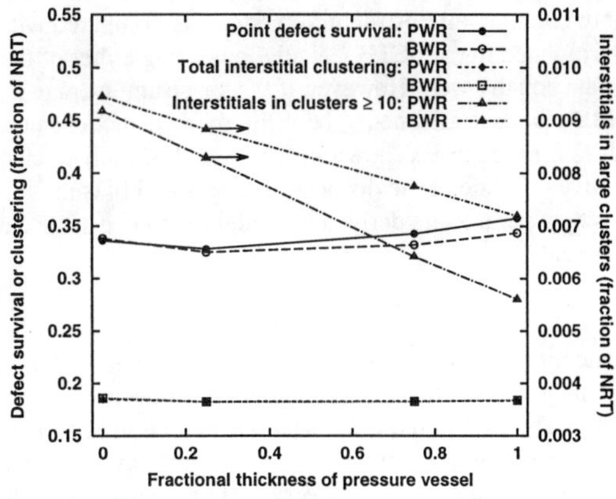

Fig. 3 Comparison of primary damage parameters as a function of location through through pressurized (PWR) and boiling (BWR) water reactor pressure vessels.

Void swelling has been the subject of extensive theoretical and experimental investigation. Models which simulate the kinetics of point defect processes using the reaction rate theory provide generally good agreement with the data.[3] However, even the most detailed models do not

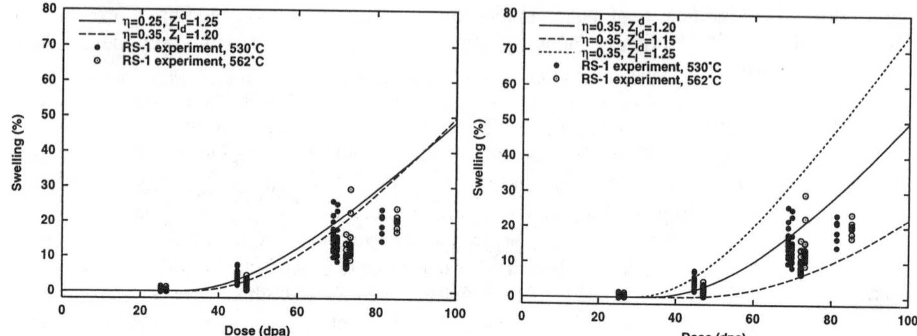

Fig. 4 Evaluation of dislocation bias based on spectrum-averaged defect production cross section and swelling data from FFTF: (a) varying both bias and point defect survival, and (b) varying bias with calculated point defect survival.

include a complete description of the relevant physics, and all models rely on material and irradiation parameters that are not always well known. Because of the uncertainty about input parameters, data are used both to help calibrate the models and to evaluate their predictive capability. One limitation in this process is that multiparameter models may yield nearly equivalent predictions with different parameter sets. This limitation can easily be demonstrated by a simple calculation of the void swelling rate under steady state conditions.

If the point defect kinetics are controlled by absorption at dislocations with sink strength S_d, the swelling rate is proportional to the product of the cascade survival efficiency η and the net dislocation-interstitial bias (Z_i^d-1):

$$\frac{dV}{dt} = \frac{\eta G_{dpa}}{S_d}(Z_i^d - 1) \tag{4}$$

In this case, data fitting can not be used to obtain a unique value for either η or Z_i^d. This is shown in Fig. 4(a), where the predictions of a more complex microstructural model[3] are compared with swelling data for AISI 316 stainless steel irradiated in the FFTF. Calculations using either of the parameter sets shown in the figure fit the data equally well. However, if the spectrum-averaged FFTF value of η=0.35 obtained from the MD results is assumed to be valid, then the model can be used to discriminate between possible values for the bias as shown in Fig. 4(b). It should be clear that this example does not provide a definitive estimate of the dislocation-interstitial bias in stainless steels. Rather, it simply illustrates how parameters derived in modeling on one scale can be used to help improve models on another scale.

SUMMARY

The results obtained from MD cascade simulations up to 50 keV have been parameterized and neutron-energy-spectrum-averaged, damage production cross sections have been calculated for several neutron irradiation environments. The MD-based cross sections can be used to make direct comparisons of the various irradiation environments, and to provide an improved estimate for the primary radiation damage source term in kinetic models. Very little dependence on the neutron energy spectrum is seen in two of the three parameters that have been analyzed. A third parameter, the fraction of surviving interstitials in clusters of 10 or larger exhibits a much higher degree of spectrum dependence. More work is required to determine the significance of this parameter in radiation-induced microstructural evolution or mechanical property changes.

Multiscale modeling can be conducted in several ways. The development of hybrid models that bridge different time and length scales is one approach. An alternate approach that involves passing selected information between different models is illustrated by the example of using MD-based parameters in a kinetic microstructural model. These two approaches could be referred to as handshaking and handing-off, respectively.

ACKNOWLEGEMENTS

Research sponsored by the Office of Nuclear Regulatory Research, U.S. Nuclear Regulatory Commission under inter-agency agreement DOE 1886-N695-3W with the U.S. Department of Energy and by the Division of Materials Sciences and the Office of Fusion Energy Sciences, U.S. Department of Energy under contract DE-AC05-96OR22464 with Lockheed Martin Energy Research Corp., and the Office of Fusion Energy Sciences, U.S. Department of Energy at the Pacific Northwest National Laboratory.

REFERENCES

1. A. D. Brailsford and R. Bullough, Philosophical Transactions of the Royal Society of London **302**, 87 (1981).
2. L. K. Mansur, Nuclear Technology **40**, 5 (1978).
3. R. E. Stoller and G. R. Odette, "A Composite Model of Microstructural Evolution in Austenitic Stainless Steel Under Fast Neutron Irradiation," *Radiation-Induced Changes in Microstructure,* ASTM STP 955, F. A. Garner, N. H. Packan and A. S. Kumar, Eds., American Society of Testing and Materials, Philadelphia, 1987, pp. 371-392.
4. R. E. Stoller, "Pressure Vessel Embrittlement Predictions Based on a Composite Model of Copper Precipitation and Point Defect Clustering," *Effects of Radiation on Materials,* ASTM STP 1270, D. S. Gelles, R. K. Nanstad, A. S. Kumar, and E. A. Little, Eds., American Society of Testing and Materials, Philadelphia, 1996, pp. 25-59.
5. R. S. Averback, T. Diaz de la Rubia, and R. Benedek, Nucl. Inst. and Meth. **B33**, 693 (1988).
6. A. J. E. Foreman, W. J. Phythian, and C. A. English, Phil. Mag. **A66**, 571 (1992).
7. A. F. Calder and D. J. Bacon, J. Nucl. Mater. **207**, 25 (1993).
8. D. J. Bacon and T. Diaz de la Rubia, J. Nucl. Mater. **216**, 275 (1994).
9. W. J. Phythian, R. E. Stoller, A. J. E. Foreman, A. F. Calder, and D. J. Bacon, J. Nucl. Mater. **223**, 245 (1995).
10. D. J. Bacon, A. F. Calder, F. Gao, V. G. Kapinos, and S. J. Wooding, Nucl. Inst. and Meth. **B102**, 37 (1995).
11. R. E. Stoller, G. R. Odette, and B. D. Wirth, J. Nucl. Mater. **251**, 49 (1997).
12. R. E. Stoller, JOM (formerly Journal of Metals) **48**, 23 (1996).
13. R. E. Stoller, "Point Defect Cluster Formation in Iron Displacement Cascades up to 50 keV," these proceedings.
14. R. E. Stoller, "Primary Defect Formation in bcc Iron: The Role of Cascade Energy and Temperature and Pre-existing Defects," presented at the International Workshop: Basic Aspects of Differences in Irradiation Effects Between fcc, bcc, and hcp Metals and Alloys, Cangas de Onis, Spain, 15-20 October 1998, to be submitted to J. Nucl. Mater.
15. M. W. Finnis, "MOLDY6-A Molecular Dynamics Program for Simulation of Pure Metals," AERE R-13182, UK AEA Harwell Laboratory (1988).
16. M. W. Finnis and J. E. Sinclair, Phil. Mag. **A50**, 45 (1984) and Erratum, Phil. Mag. **A53**, 161 (1986).
17. M. J. Norgett, M. T. Robinson, and I. M. Torrens, Nucl. Eng. and Des. **33**, 50 (1975).

18. R. E. Stoller and L. R. Greenwood, "Subcascade Formation in Displacement Cascade Simulations: Implications for Fusion Reactor Materials," accepted for publication in Journal of Nuclear Materials (1998).

19. L. R. Greenwood, "SPECOMP Calculations of Radiation Damage in Compounds," *Reactor Dosimetry: Methods, Applications, and Standardization,* ASTM STP 1001, H. Farrar IV and E. P. Lippincott, Eds., American Society of Testing and Materials, West Conshohocken, PA, 1989, pp. 598-602.

20. L. R. Greenwood and R. K. Smither, "SPECTER: Neutron Damage Calculations for Materials Irradiations," ANL/FPP/TM-197, Argonne National Laboratory, Argonne, IL, January 1985.

KINETIC LATTICE MONTE CARLO SIMULATIONS OF CASCADE AGING IN IRON AND DILUTE IRON-COPPER ALLOYS

B. D. WIRTH and G. R. ODETTE
Dept. of Mechanical and Environmental Engineering, University of California, Santa Barbara, CA 93106

ABSTRACT

Neutron hardening and embrittlement of pressure vessel steels is due to a high density of nm scale features, including copper-manganese-nickel rich precipitates and what are generally believed to be defect cluster-solute complexes. It has been postulated that the sub nanometer defect cluster-solute complexes form directly in displacement cascades. Cluster-complexes that are thermally unstable mediate the effect of flux on embrittlement kinetics. Larger cluster-complexes, that are relatively thermally stable for irradiation times up to 1 Gs, cause embrittlement in low copper steels. Robust characterization of these two types of so-called matrix defects has been an elusive goal. In this work, Kinetic Lattice Monte Carlo (KLMC) simulations of the long term evolution of the vacancy-rich cascade core regions were carried out for both pure iron and dilute iron-copper alloys at the nominal irradiation temperature of 563°K up to times when the vacancy clusters completely dissolve. Energetics were based on lattice embedded atom method potentials. Special time scaling and pulse annealing techniques were used to deal with the enormous range of inherent time scales involved, viz., rapid free vacancy jumps to slow emission from large complexes. Three-dimensional clusters rapidly form, containing a wide range of vacancies, as well as copper atoms in alloys. Small complexes are very mobile and growth takes place primarily by coalescence. The vacancy clusters ultimately dissolve at times from less than 0.1 to more than 100 MS. These simulations support the hypotheses that cascade cluster-complexes constitute both thermally stable and unstable matrix defect features.

INTRODUCTION

Displacement cascades initially develop and cool over ps time scales, resulting in a shell of self-interstitial atom (SIA) and SIA clusters, surrounding a vacancy-rich core [1-4]. This spatially correlated defect production initiates a long chain of events in reactor pressure vessel steels that produces irradiation embrittlement. It is well established that embrittlement is caused by the formation of a high number density ($> 10^{23}$ /m^3) of ultra finescale features, consisting of nanometer-sized copper-manganese-nickel precipitates and so-called matrix features. The latter are postulated to be bound sub-nanometer defect-solute cluster-complexes [5,6]. The objective in this work was to simulate the long-term rearrangement (aging) of defects in displacement cascades in both pure iron and simple iron-copper alloys in order to address the key question: *Are cluster-complexes produced in cascades and, if so, what are their characteristics and properties?*

While not the focus of this paper, one important issue is the fate of the SIA clusters. Single, di and tri SIA are very mobile and undergo limited cascade recombination with vacancies while they rapidly migrate to sinks. However, a significant fraction of SIA form larger clusters in the cascade expansion and initial cooling phases. Thus the behavior and fate of the SIA clusters is a key issue. Our previous atomistic modeling in pure iron showed that SIA clusters form perfect (unfaulted) a/2<111>, hexagonal dislocation loops with formation energies consistent with continuum elasticity predictions [7,8,9]. However, the loops are more extended than a normal edge dislocation, and can be viewed as being highly kinked. The SIA loops undergo very easy, one-dimensional diffusional-type migration, as a result of the intrinsic presence and easy propagation of the kinks.

If they were not mobile, SIA clusters would undergo bias driven growth producing larger loops that, notably, are <u>not</u> observed in RPV steels for typical irradiation conditions

Mat. Res. Soc. Symp. Proc. Vol. 538 © 1999 Materials Research Society

[10]. Unresolved issues associated with SIA loops include: interactions with solutes and trapping, motion in drift fields and interactions with and ultimate annihilation at sinks. However, it is reasonable to conclude that the high intrinsic mobility of SIA clusters explains the absence of larger loops. A corollary is that the one-dimensional diffusional properties of the loops limit in-cascade recombination, leaving vacancy rich cores that evolve over much longer time scales.

SIMULATION METHODOLOGY

The initial configuration for the KLMC cascade aging simulations consisted of vacancy rich core configurations from MD cascade simulations performed by Stoller [3], with the SIA and SIA clusters removed. This approach gives an upper bound on the number of cascade vacancies; however, a similar number would be generated in higher energy cascades. Simulations were performed in iron and dilute iron-copper alloys using embedded atom-method type unrelaxed lattice interatomic (LEAM) potentials. Iron and copper potentials were from Finnis and Sinclair [11] and Ackland et. al. [12], respectively; and the iron-copper potential was developed by fitting the dilute heat of solution for copper in iron, the copper-vacancy binding energy and the iron-copper {110} interface energy [13]. All simulations reported here are for a nominal temperature of 563°K.

The eight jump probabilities for each cascade vacancy (i) jumping to position (j) that results in a change in energy of ΔE_{ij}, depending on the local environment, were computed using a Boltzmann factor and summed to find the total probability of a vacancy jump as,

$$P_i = \Sigma_{1,8} \exp\{-(\Delta E_{ij} + E_m)/kT\} \tag{1}$$

where the saddle point migration E_m was taken as 0.9 eV (obtained from MD simulations) minus $\Delta E_{ij}/2$. Modifying E_m by the $\Delta E_{ij}/2$ term represents a rather ad-hoc attempt to model the effect of the local environment on jump frequencies; detailed molecular statics calculations suggest that this probably represents an upper-bound influence of any such effect [13]. The total jump probabilities for each cascade vacancy (much less than 1) are normalized to the highest total probability. In combination with an effective attempt frequency (v, taken as 10^{14} s^{-1}, including entropic effects), the probability of this vacancy exchange ($P_{m,t}$) sets the effective 'time-step' for a 'sweep' of all vacancies with a lower probability of jumping, as $\Delta t_{LMC} = (v P_{m,t})^{-1}$. The specific jump of the maximum probability vacancy(ies) and that of other lower probability vacancies, if any, are based on a standard random number Metropolis test. Thus, at least one, and in some cases many more than one, vacancy jump occurs in each KLMC sweep.

In iron-copper alloys, vacancy jumps also produce copper diffusion, hence, clustering and complex formation. Clustering is promoted by the strong binding between copper (or other solutes) both with each other, as well as with vacancies ($\approx$ 0.14 eV [14]) and interfaces. As described below, small vacancy clusters are very mobile and impart mobility to any copper associated with a complex. Thus the highly correlated initial vacancy arrangements in cascades continue to undergo correlated evolution over long periods of time; this can be self-consistently described with reasonable accuracy by the KLMC algorithm. Approximations in the simulation are discussed below.

The initial vacancy fate is to either cluster or diffuse to the cell boundary, which is assumed to be a defect sink. Clusters can grow by absorbing single vacancies, which may be emitted from other clusters, or by coalescence with mobile clusters. Emitted vacancies may also annihilate at the cell boundary. The jump rate and sequence depends on the local environment; thus the time increment changes as the cascade evolves. The increments start out short, for unbound, isolated vacancies, but increase enormously as larger bound cluster-complexes develop. As the cluster-complexes grow, the cluster mobility drops and non-emission jumps that change the shape, but not the cluster-complex position or size, become far more probable; hence, the efficiency of the KLMC algorithm decreases dramatically. To circumvent this problem, a 'pulse annealing' technique is used to accelerate the computation of vacancy emission from the clusters.

After twenty 'inefficient' sweeps, the temperature is increased by 160°K in order to identify the least strongly bound vacancy, and to determine its most probable emission trajectory. This vacancy is then 'required' to make three additional jumps along this pre-determined trajectory at the actual simulation temperature. Other vacancies in the simulation are also allowed to jump (or not) during this period. The effective time increment adjustment factor for 'pulse annealing' is the ratio of the total jump probability ($P_{i,t}$) to the product of the probabilities for the pre-selected series of jumps ($P_{i,j}$), $\Delta t_{PA} = \Delta t_{LMC} \Pi_{j=1,3}(P_{i,t}/P_{i,j})$. Emitted vacancies can be captured by the emitting or another cluster or be annihilated at the cell boundary.

As a partial test of the pulse annealing method, multiple KLMC simulations, consisting of fifteen and five simulations for the 20 and 10-member vacancy cluster, respectively, were performed in pure iron at 563 K to obtain the average total dissolution times. These average times were compared to rate theory [15] predictions, assuming self-diffusion, interface energy and atomic density values of 5.4×10^{-31} m^2/s, 1.7 J/m^2 and 8.5×10^{28} /m^3 respectively, consistent with the EAM iron potential. Agreement is excellent for the smaller clusters (2.4 vs. 2.0 x 10^3 s for KLMC vs. rate theory). However, for the larger clusters the KLMC average lifetime is 2.0×10^6 s vs. the 2.2×10^5 s predicted by rate theory. These differences are partly due to the $\Delta E_{ij}/2$ term in the migration energy; and are equivalent to small uncertainties in the interface energy (e.g., the 20 vacancy cluster rate theory dissolution time is 1.0×10^6 s for an interface energy of 1.5 J/m^2).

In addition to the pulse anneal method, there are other approximations in the current KLMC simulation method that should be noted. First, and foremost, the quantitative accuracy of the simulations depends on the validity of the interatomic potentials, that are approximate at best. Further, the effect of each jump on subsequent ones should be evaluated immediately rather than after a complete sweep. Finally, a more rigorous residence time algorithm would provide a slightly better estimate of the time scale. However, our simpler approach is believed to be a good semi-quantitative, and certainly qualitative, approximation to cluster-complex evolution during cascade aging [6]. In this regard, it should be noted that the physics dictating the evolution and ultimate fate of cascade vacancies primarily derives from their high degree of geometric correlation, rather than the exact kinetic sequence-of-events. Overall the KLMC simulations are believed to accurately simulate the size and character of the complexes, with dissolution time estimates good to at least an order of magnitude for the LEAM potential used.

RESULTS AND DISCUSSION

Pure Iron: Starting from a core rich in isolated (mostly) vacancies the KLMC simulations showed a very rapid formation of a population of small, three-dimensional vacancy clusters, on average requiring only a few vacancy jumps. Subsequently, the clusters migrated, coarsened and dissolved. Reduction of the net residual vacancy concentration in the cascade core by emission-migration-annihilation events was continuous, but progressively slowed with cluster growth. Clusters grew initially primarily by the coalescence of smaller mobile clusters, and later by emission-absorption processes. A single or few large clusters finally dissolve over a wide range of times mediated by their size and stochastic effects. The size of the largest cluster was controlled by both the initial number and density of vacancies and once again stochastic effects. Multiple simulations, using three different initial cascade configurations, resulted in a wide range of dissolution times geometrically averaged to obtain the mean cluster lifetimes. The cascades included (energy-MD simulation time): 20 keV-200 ps; 20 keV-10 ps; and 40 keV-10 ps. The sequence of events was similar in all cases.

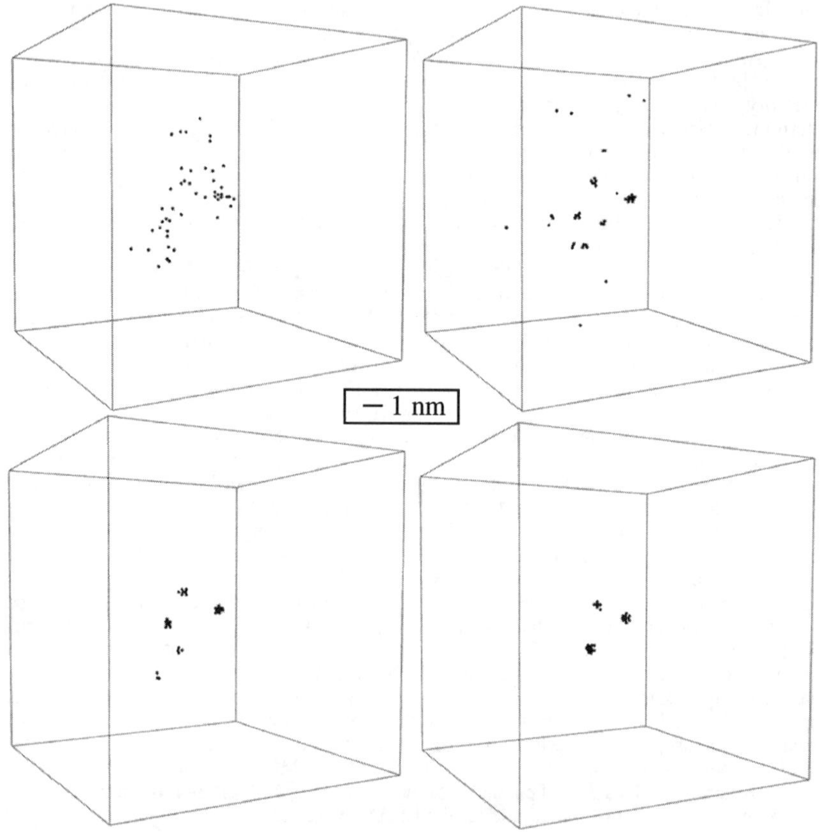

Figure 1 --   KLMC snapshots of the vacancy distribution during cascade aging in pure iron a) initially, b) 70 μs, c) 460 μs and d) 7100 s.

To illustrate the cascade aging evolution consider a 20 keV-200 ps case starting with 47 vacancies in the cascade core, shown in Figure 1a. Within 70 μs a number of three-dimensional clusters formed, including one containing 12 vacancies and a number of small and highly mobile di, tri and tetra-vacancy clusters, shown in Figure 1b. Since the elapsed time is still quite short, only one vacancy had been annihilated and nine vacancies had yet to cluster. By 460 μs, 13 vacancies had annihilated at the boundary and cluster diffusion-coalescence-dissolution produced 5 clusters containing 3, 4, 7, 8 and 12 vacancies, shown in Figure 1c. Again, the smaller clusters (3, 4) were very mobile and coalesced by 930 μs. After 2 vacancies were emitted from the 8-member cluster, the resulting 6-member cluster began to rapidly diffuse, coalescing with the 7-member cluster at 7100 s; as shown in Figure 1d, three clusters remain at this time, containing 6 (after a single emission from a 7 vacancy cluster) , 12 and 13 vacancies. Vacancy emission dissolution continued as the 6-member cluster completely dissolved and the 13-member shrank until it became mobile.

By 1×10^5 sec, the two remaining clusters (containing 12 and 5 vacancies) coalesced to form a 17 vacancy cluster. The single, large cluster began to shrink, emitting vacancies until it reached about 6 vacancies, when it began to diffuse before ultimately dissolving. All vacancies were annihilated by 9.3×10^6 sec.

The range of cluster lifetimes observed in the KLMC simulations varied between about 3.3×10^4 and 1.1×10^7 s. Dissolution times are influenced by: both the maximum cluster size (here, from 9 to 27 vacancies, contributing about a factor of 80 variation in lifetime) and inherent stochastic scatter (contributing about a factor of 40 difference in lifetime). The 20 keV-10 ps cascade had a mean lifetime of 2.3×10^5 vs. 9.1×10^5 s for the 20 keV-200 ps case. Notably the 20 keV-10 ps cascade, with the shorter mean lifetime, initially had a larger number vacancies (78 vs. 47), but was noticeably less dense. The overall geometric mean lifetime of the 20 keV cascades of 4.3×10^5 s is close to the experimental value of about 3×10^5 s obtained by Odette and co-workers [6,16] for thermally unstable matrix defects. The mean lifetime for the 20 keV cascades is about five time less than for the 40 keV cascades (2.1×10^6 s).

Iron-Copper Alloys: Similar KLMC cascade aging simulations were also performed in dilute iron-copper alloys starting by superimposing a random 0.3% copper distribution on the initial cascade vacancy distribution. The sequence-of-events was similar to that observed in pure iron. The key difference was the formation of a population of copper-vacancy cluster-complexes. Small complexes, consisting of less than a total of ten vacancies and copper atoms, were quite mobile and cluster-complex growth occurred by coalescence. Clusters-complexes ranged in size from two to three vacancies with six or more copper atoms to twenty or more vacancies with two to three copper atoms. In general, smaller vacancy clusters were associated with larger numbers of copper atoms. This is partly a result of the high mobility of smaller vacancy complexes, resulting in more local gettering copper atoms. The larger vacancy cluster-complexes were typically associated with fewer copper atoms, indicative of their relatively low mobility. By the time of complete vacancy dissolution, residual copper clusters ranged in size from 4 to 15 atoms.

There were initially a larger number of smaller complexes in the iron-copper alloy, compared to pure iron; but the maximum cluster sizes that eventually developed were similar in the two cases. However, stabilization of small complexes by copper markedly increased the overall vacancy cluster lifetime. The mean lifetime for the 20 keV-200 ps cascade was about 10^7 s, or roughly a factor of ten higher than in the case of pure iron; the corresponding lifetime for the 40 keV-10 ps cascade was about 7×10^7 s, or a 35 fold increase over the pure iron estimate.

Effects Of Long Range Diffusion Interactions And Other Solutes: Given the large range of cluster-complex lifetimes, long range diffusion of both defects generated in other cascades and solutes initially outside the cascade region would interact with and modify the isolated cascade process which are the focus of the present study. For example: a) the larger complexes could serve as preferred sites for bias driven nanovoid growth; b) copper can be considered as a surrogate for other chemically active solutes, like manganese and nickel -- however, the lower self-interaction energies for these elements would result in somewhat more dilute solute aggregation than in the case of strongly bound copper; c) combinations of vacancy cluster dissolution, weak solute-solute binding/solute cluster dissolution, and long range solute diffusion may be the source of dilute solute atmospheres reported in some atom-probe studies; d) alternately, and perhaps less frequently, cascade solute clustering could promote the formation of manganese-nickel (a strong interaction) rich precipitates which have low thermodynamically driven homogeneous nucleation rates. These phenomena will be evaluated in future research.

Summary and Significance of the Simulation Results: Although the quantitative results cannot be taken too seriously, the qualitative, physical significance of the simulations are profound. First, there can be no question that vacancy cluster-solute complexes form in the core of high energy cascades, albeit at times greater than 100 ps. Further, the spectrum of sizes, chemistries and vacancy cluster lifetimes strongly supports the hypotheses that these features are a source of both unstable and stable matrix defects. Further, cascade induced clusters may serve as favored nucleation sites for precipitates, growing nanovoids and dilute solute atmospheres.

SUMMARY AND CONCLUSIONS

KLMC simulations were used to study the long term evolution of the vacancy-rich core regions formed in displacement cascades in iron and dilute iron-copper alloys. The simulations revealed a number of new phenomena, including: a) rapid formation of three-dimensional cluster-complexes containing a wide range of vacancies and copper atoms; b) a high mobility of small complexes, with growth occurring primarily by coalescence and copper scavenging; c) a wide range of vacancy cluster dissolution times from less than 0.1 to more than 100 MS, thus constituting both thermally stable and unstable matrix defect features; e) cascade clusters that may be nucleation sites for growing nanovoids and precipitates, that may include manganese-nickel phases even at low copper levels; and f) the residues of complexes that have first grown by long range diffusion, but then partially dissolve (dissolution may be complete for the vacancy component), constituting a possible source of solute enriched regions.

ACKNOWLEDGMENTS

Helpful discussions with G. Lucas and D. Maroudas (UCSB) and support by the US Nuclear Regulatory Commission (NRC-04-94-049) are gratefully acknowledged.

REFERENCES

1. A. Seeger, *Proc. 2nd UN Int. Conf. on Peaceful Uses of Atomic Energy*, Geneva, Vol. 6, United Nations, New York (1958) 20
2. M. T. Robinson, *J. Nucl. Mat.* **216** (1994) 1.
3. R. E. Stoller, *MRS Soc. Symp. Proc.* **373** (1995) 21.
4. A. F. Calder and D. J. Bacon, *J. Nucl. Mat.* **207** (1993) 25.
5. G. R. Odette, MRS Soc. Symp. Proc., **373** (1995) 137.
6. G. R. Odette and B. D. Wirth, *J. Nucl. Mater.* **251** (1997) 157.
7. B. D. Wirth, G. R. Odette, D. Maroudas, and G. E. Lucas, *J. Nucl. Mat.* **244** (1997) 185.
8. B. D. Wirth, G. R. Odette, D. Maroudas, and G. E. Lucas, *MRS Soc. Symp. Proc.* **504** (1998), in press.
9. B. D. Wirth, *On the Character of Nano-scale Features in Reactor Pressure Vessel Steels Under Neutron Irradiation*, Ph.D. Dissertation, University of California, Santa Barbara (1998).
10. M. L. Jenkins, M. A. Kirk and W. J. Phythian, J. Nucl. Matrl. **205** (1993) 16.
11. M. W. Finnis and J. E. Sinclair, *Phil. Mag.* **A 50 (1)** (1984) 45.
12. G. J. Ackland, G. I. Tichy, V. Vitek, and M. W. Finnis, *Phil. Mag.* **A 56** (1987) 735.
13. B. D. Wirth and G. R. Odette, *MRS Soc. Symp. Proc.* **481** (1998), in press.
14. A. Moslang, E. Albert, E. Recknagel and A. Weidinger, Hyperfine Interactions, 15/16 (1983) 409.
15. G. R. Odette, *Neutron Irradiation Effects in Reactor Pressure Vessel Steels and Weldments*, IAEA IWG-LMNPP-**98/3**, International Atomic Energy Agency, Vienna (1998) 438.
16. E. V. Mader, *Kinetics of Irradiation Embrittlement and the Post-Irradiation Annealing of Nuclear Reactor Pressure Vessel Steels*, Ph.D. Dissertation, University of California Santa Barbara (1995).

KINETIC MONTE CARLO SIMULATIONS OF FECU ALLOYS

C. DOMAIN*, C.S. BECQUART**, J.C. VAN DUYSEN*,**
* EDF - DER Département EMA, Les renardières, F-77818 Moret sur Loing Cédex, France
christophe.domain@edf.fr, jean-claude.van-duysen@edf.fr
** Laboratoire de Métallurgie Physique et Génie des Matériaux, UMR 8517, Bât. C6,
Université Lille I, F-59655 Villeneuve d'Ascq Cédex, France
charlotte.becquart@univ-lille1.fr

ABSTRACT

The steel vessels of pressurized water reactors are embrittled by neutron irradiation. It is well known that copper atoms play an important role in the embrittlement and that different Cu-containing defects such as Cu-rich clusters (sometimes called atmospheres), Cu precipitates and Cu-vacancy complexes have been identified experimentally. It is still difficult to link the formation of these defects to the primary damage resulting from the neutron inducing displacement cascades. Therefore, we investigate the evolution of the primary damage in FeCu alloys using kinetic Monte Carlo simulations based on a vacancy diffusion mechanism. The calculations rely on adapted, phenomenological, n-body potentials that satisfactorily reproduce properties of FeCu. At room temperature, experimentally identified defects such as Cu-vacancy complexes (one Cu atom bound to three or four vacancies) form in the course of our simulations. Furthermore, it appears that complex defects such as a Cu atom linked to two vacancies are very mobile and are responsible for the formation of small Cu clusters.

INTRODUCTION

The primary damage created by the displacement cascades has been explored by computer simulations for over 40 years, primarily by Molecular Dynamics (MD) simulations. MD is indeed a very powerful technique to study events with nanometer length scales and picosecond time duration. The displacement cascades initiated by the interaction of a neutron and one atom (the Primary Knocked-on Atom) fall within these bounds and a great deal has been learned from MD about the primary damage in terms of the amount of residual point defects and their tendency to form clusters. For reviews on this subject see for instance [1][2]. Unfortunately, the subsequent evolution of the defects which induces the change in mechanical properties observed in irradiated materials cannot be scrutinized with this technique as a result of their excessive duration and one is forced to turn to other methods such as Monte Carlo (MC) simulations or a mean field approach.

The mean field approach, as indicated by its name, cannot deal with fluctuations. In the case of the evolution of primary damage, which is very inhomogeneous in essence (for each cascade the primary damage is concentrated in a confined area a few nanometers wide), mean field approaches are inappropriate and MC techniques are more suitable for the problem. Specifically, the Kinetic Monte Carlo (KMC) method is a clever way of introducing time in a Monte Carlo (MC) simulation and performing one configuration change at each step. We therefore used a KMC technique to assess the room temperature vacancy driven evolution of the primary damage obtained from MD simulations of displacement cascades in low Cu content FeCu alloys (%Cu =0.1 and 0.2). The kinetics of the evolution of interstitials are commonly regarded to be orders

Mat. Res. Soc. Symp. Proc. Vol. 538 © 1999 Materials Research Society

of magnitude faster than that of vacancy diffusion and thus vacancies and interstitial processes can be modeled separately, at least to a first order approximation.

COMPUTATIONAL PROCEDURE

The model used in this work is the residence time algorithm derived by Young et al. [3] for the diffusion of a vacancy in a solid. It is assumed that diffusion proceeds via vacancy jumps towards nearest neighbor atoms and that the vacancy jump is a thermally activated process whose frequency (for vacancy n and jump i) is given by:

$$\Gamma_{n,i} = \nu \exp\left\{-\frac{E_a}{kT}\right\} \tag{1}$$

where E_a is the activation energy and ν is the attempt frequency. For each step, one computes the probabilities (given by $\Gamma_{n,i}$) corresponding to all possible vacancy jumps. One move is chosen at random using a weight corresponding to its probability and then the mean residence time corresponding to the chosen transition is calculated. This is given by the inverse of the sum of probabilities $\Gamma_{n,i}$. The attempt frequency, ν, is supposed to be independent of the alloy configuration and was taken to be $3.65\times10^{15}\text{s}^{-1}$ after [4]. This value is based on a vacancy self diffusion coefficient of $3\times10^{-4}\text{m}^2\text{s}^{-1}$ in Fe [4]. Many heuristics can be employed to estimate the activation energy barrier, E_a. We have chosen an environment-dependent form for E_a that satisfies the detailed balance rule and is given by:

$$E_a = E_{a0} + \frac{E_f - E_i}{2} \tag{2}$$

E_{a0} is the energy barrier, taken to be constant and equal to 0.9 eV giving rise to a temperature dependent scaling factor. This value for the energy barrier is halfway between the vacancy migration energy in pure Fe (0.55 eV [5]) and in Fe with a small amount of C (1.28 eV [6]). The probability of two vacancies switching sites was set to zero. E_f and E_i are the total potential energies of the system before (E_i) and after (E_f) the vacancy jump. In this work, they are calculated using an Embedded Atom Method (EAM) interatomic potential. EAM potentials, because of their many-body terms, are more adapted than pair potentials to the simulation of metals, especially in the presence of defects such as vacancies. They do not require too much computational effort and therefore are nowadays commonly used for the simulation of displacement cascades.The EAM potentials were derived by Ludwig et al. [7] and were adapted (hardened) to the simulations of displacement cascades by Raulot [8]. The potential range takes into account the three first neighbouring shells. It gives: a Partial Molar Energy of Cu at infinite dilution (dilute heat of solution) of 0.49 eV (" experimental " 0.317 [9]; 0.49 [7]; 0.59 [10], the values vary from one author to the other as they are computed from Cu solubility limits which are not very precisely known and themselves very scattered); for pure Fe, a vacancy formation energy of 1.63 eV (experimental E_{vac}^{f} =1.53-2 eV [11][5]) and its migration energy of 0.66 eV (experimental E_{vac}^{mig} =0.55 eV [6]). The relaxed binding energies given by our model are summarized in Table I, which also presents the values obtained with two other published Fe-Cu potentials. To our knowledge, the only experimental value published for the Cu-vacancy binding energy is 0.14 eV [12] which is close to our value. For the Cu atom bound to a divacancy (VV-Cu), the reference state is the divacancy and a noninteracting Cu atom.

Table I Relaxed binding energies (eV) in the α-Fe matrix: Vacancy-Vacancy, Vacancy-Cu, Cu-Cu, divacancy-Cu.

	This work		Osetsky [13]		Ackland [9]	
Neighbors	1st	2nd	1st	2nd	1st	2nd
V-V	0.16	**0.21**	-0.02	**0.20**	0.14	**0.19**
Cu-V	**0.19**	-0.03	**0.14**	0.02	**0.087**	0.04
Cu-Cu	**0.2**	-0.02	**0.20**	0.08	**0.075**	0.035
VV-Cu (the two vacancies are 2nd neighbors to each others and 1st neighbors to the Cu atom)	**0.39**		-		**0.17**	

To minimize computational expense, the simulations were performed on a rigid body centered lattice. In this scheme, the relaxations around vacancies or solute atoms are not taken into account and coherent-incoherent phase transitions cannot be reproduced. However, in the first stages of precipitation in FeCu alloys, the Cu precipitates remain coherent with the matrix below a critical size of 4-5 nm [14]. Therefore as long as the precipitates remain below the critical size, a body centered grid can be used. Furthermore, the relaxed and unrelaxed binding energies for our potential are similar in trends. Working on a rigid array (equivalent to using the unrelaxed atomic positions to determine the activation energies) should not change the results too much, at least from a qualitative point of view.

In a first step, we validate our model by comparing its predicted precipitation kinetics in FeCu alloy with experimental results [15]: one vacancy is introduced in a Fe-1.34%Cu alloy to follow the degree of advancement of the precipitation process.

Then, at 300K, the evolution of some cascade debris (up to 20 keV) was studied with this model. The obtained results are also compared to experimental data obtained from room temperature irradiation experiments [16]. To do this, all residual interstitial atoms remaining at the end of the cascade (after ~15 ps, when further defect recombination is believed to be minimal and therefore neglected) were removed and all other atoms were assigned to the nearest perfect lattice site to generate the input for the KMC run.

Following [17], the boundaries were assumed to be vacancy sinks (to study the evolution of one cascade by itself). In this scheme, the vacancies that do not cluster ultimately annihilate.

PRECIPITATION KINETICS IN FECU ALLOYS

Precipitation kinetics predicted by our model at low temperatures are compared to experimental results obtained on Fe-1.34%Cu alloys [15]. The degree of advancement of the precipitation ξ is given by:

$$\xi(t) = \frac{C_{Cu}(0) - C_{Cu}(t)}{C_{Cu}(0) - C_{Cu}(\infty)} \tag{3}$$

$C_{cu}(t)$ is the copper concentration in the matrix at time t. The advancement factor can be modelled by a Johnson-Mehl-Avrami type law:

$$\xi(t) = 1 - \exp\left(-\left(\frac{t}{\tau}\right)^n\right) \tag{4}$$

From the experimental data [15], we deduced the evolution of τ for temperatures greater than 693K using an exponential law fit. From this law, the kinetic factor τ was extrapolated to the range of temperatures we are interested in (close to room temperature).

The vacancy concentration (one vacancy for 54000 atoms) of the simulation is very large compared to the experimental equilibrium vacancy concentration. Therefore to compare our results with the experimental data, the simulation time was rescaled by the factor $C_v^{simu} / C_v^{exp}(T)$ with an experimental vacancy formation energy of 1.4 eV after [4].

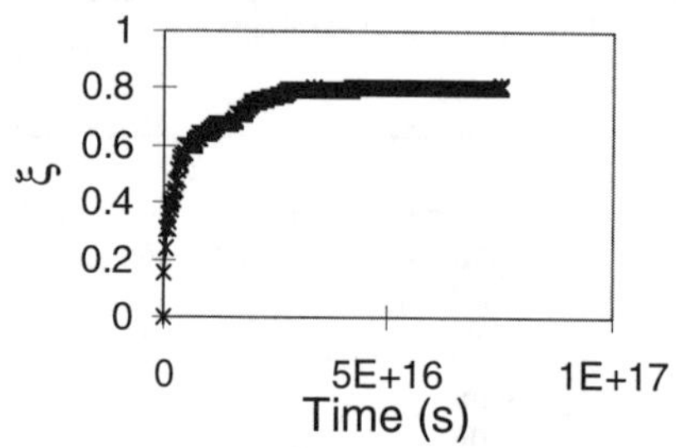

T = 380K, $\tau_{KMC.}$=8.1 10^{15}s, (extrap. 1.5 10^{16}s)

Figure 1 : KMC precipitation advancement curve in Fe-1.34%Cu.

The evolution of ξ obtained by simulation can be seen Fig. 1. Also, the corresponding τ (τ_{KMC}) are compared to the extrapolated values, and are found to be of the same order of magnitude. The value for the factor n (see eq. 4) is found to be below 1, in agreement with the experiments and with the simulations of Soisson et al. [4] who used pair interaction potentials. At 300K, vacancy trapping prevented us from obtaining the kinetics precipitation curve in a reasonable computational time.

EVOLUTION OF THE PRIMARY DAMAGE

A typical image of the primary damage after MD simulations of 20 keV cascades in Fe-0.2%Cu is presented in Fig. 2. The " usual " vacancy clusters as well as some isolated defects can be observed (the interstitials were removed). After a few millions of KMC steps at 300K, one can observe vacancy clusters, V-Cu complexes and V-Cu clusters (with up to 13 coppers atoms).

For high energy cascades (20 keV) with 0.1%Cu and 0.2%Cu, 300K KMC simulations on debris cascades produce, in a small span of time, small vacancy clusters and very mobile V-Cu complexes (one or two vacancy bound to one Cu atom) which scavenge other copper atoms to form mixed V-Cu clusters. The copper distributions in mixed clusters are very similar to these observed by Pareige [16] by tomographic atom probe (TAP) as can be seen Fig. 3 (vacancies cannot be observed by TAP). The complexes move and scavenge other Cu atoms. Cu is mostly transported by divacancies: V-Cu are more mobile than VV-Cu, however in V-Cu the bond is weaker and they dissociate too easily to be able to transport Cu far enough. The mixed clusters are mobile as well and their mobility decreases with size while the pure vacancy clusters are much less mobile. Most of the " objects " described above remain in the cascade area. A small part (on the order of 10%) of the vacancies escape far from the cascade (more than 15nm).

In addition, we wished to compare our results with experimental data of Pareige et al. [16] of 300 keV Fe$^+$ irradiation at room temperature on a 0.1% Cu alloy. We computed the observation frequencies in our Fe-0.1%Cu alloy simulations for a given number of copper atoms within 1000 blocks of 432 atoms each situated in the debris field (Table II). For comparison, the results from Pareige et al. [16] are shown in Table III.

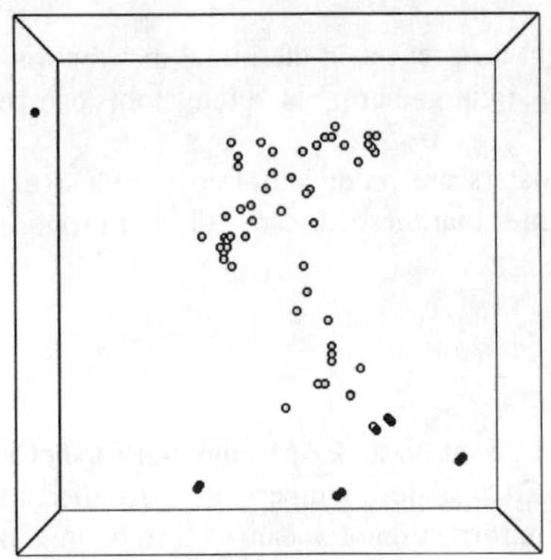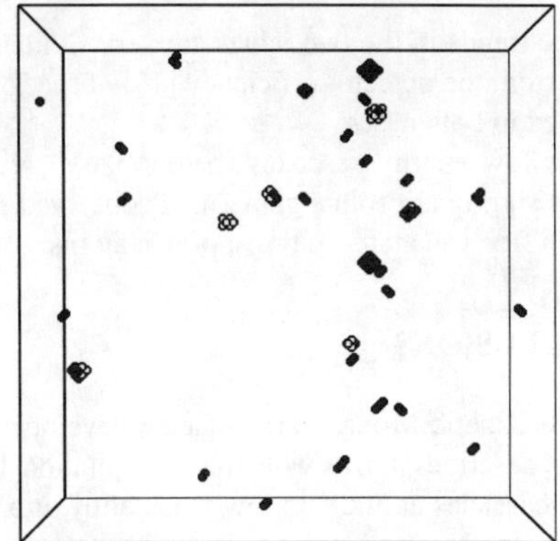

Figure 2 (a) Typical debris from a 20 keV cascade in Fe-0.2%Cu at the end of the recombination phase. Only the vacancies (white) and copper atoms (black) with a vacancy or at least a copper as first neighbor are represented. (b) After 10^8 KMC steps at 300K (t_{KMC} ~50 days). The box size is 18.6 nm.

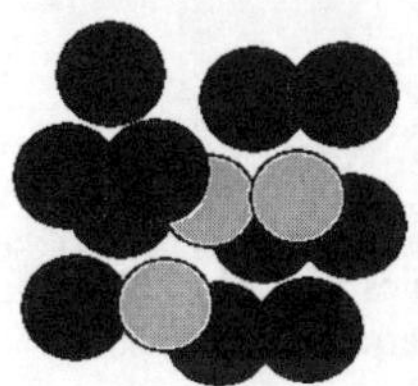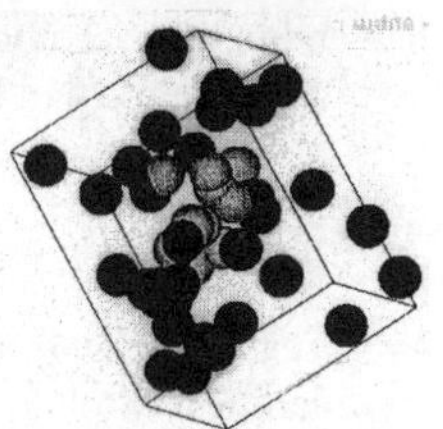

Figure 3 (a) A typical mixed cluster (Cu = black atoms, vacancies = grey atoms) obtained in 300K KMC simulation from 20 keV cascade debris. (b) Magnified view of the copper cluster (grey atoms) experimentally observed in [16] after a 300 keV Fe$^+$ irradiation at room temperature on a 0.1% Cu alloy (the black atoms are Fe^{54} isotopes, the grey atoms are Cu)

Table II Calculated observation frequencies (in %) of a given number of copper atoms per block of 432 atoms averaged over 1000 blocks for a 20 keV cascade debris (0.1%Cu).

Number of copper atoms per block	0	1	2	3	4	5	6	7	8	9
Box % before KMC	61.8	31.3	6.2	0.7	-	-	-	-	-	-
Box % after 10^7 KMC steps	63.0	30.1	5.6	0.7	-	0.1	0.1	0.1	-	-

Table III Experimental observation frequencies (in %) of a given number of copper atoms per block of 400 ions averaged over 1000 blocks, after [16].

Number of copper atoms per block	0	1	2	3	4	5	6	7	8	9
Unirradiated box %	43.6	35.8	14.3	4.6	1.6	0.1	-	-	-	-
Irradiated box %	52.0	28.0	12.6	5.2	1.6	0.3	0.1	-	-	0.2

The trends in the two tables are very similar (the slight differences in the initial distributions come from the detector efficiency [16] of the TAP) and a slight gathering of solute atoms can be deduced in both cases.

For low energy cascades (below 5keV), vacancy clusters are produced almost exclusively and no significant solute gathering is observed. We can infer that these cascades did not produce enough free vacancies to transport Cu atoms.

CONCLUSIONS

The Kinetic Monte Carlo scheme developed here and based on an EAM potential for FeCu alloys, describes fairly well the precipitation kinetics of Cu at low temperatures. Applied to debris cascades at 300K in low FeCu alloy, it produces the formation of vacancy clusters, mixed clusters and V-Cu complexes in a manner very similar to the defects inferred from experimental data. It has been observed that the formation of mixed clusters is linked to the stability and mobility of vacancy copper complexes which are able to gather. A similar mechanism may be responsible for copper atmospheres observed in irradiated pressure vessel steels.

This paper presents the first results of an ongoing research effort and is very far from fully describing the evolution of the primary damage. While interstitials were not taken into account, resistivity measurements in FeCu alloys have shown that there is a weak but positive binding between solute and interstitial atoms [18]. Also, strains and relaxation effects were not included. The influence of temperature will be examined in the future.

REFERENCES

1   R.S. Averback and T. Diaz de la Rubia, Solid State Physics **51**, 281 (1998)

2   W.J. Phythian, R.E. Stoller, A.J.E. Foreman, A.F. Calder and D.J. Bacon, J. Nucl. Mater. **223**, 245 (1995)

3   W.M. Young and E.W. Elcock, Proc. phys. Soc. **89**, 75 (1966)

4   F. Soisson, A. Barbu and G. Martin, Acta mater. **44**, 3789 (1996)

5   H.E.Schaefer, K.Maier, M.Weller, D.Herlach, A.Seeger, J.Diehl, Scripta Met.**11**,803 (1977)

6   A. Vehanen, P. Hautojärvi, J. Johansson, J. Yli-Kauppila and P. Moser, Phys. Rev. B **25**, 762 (1982)

7   M.Ludwig, D.Farkas,D.Pedraza,S.Schmauder, Model.Simul. Mater. Sci. Eng. **6**, 19 (1998)

8   J.M. Raulot, Master of Science Thesis, Université de Marne la vallée (1998)

9   G.J. Ackland, D.J. Bacon, A.F. Calder and T. Harry Phil. Mag. A **75** 713 (1997)

10   M.H. Mathon, PhD. Thesis, CEA-5701, ISSN 0429-3460 (1995)

L.D. Schepper, D. Segers, L. Dorikens-Vanpraet, M. Dorikens, G. Knuyt, L.M. Stals and P. Moser, Phys. Rev. B **27**, 5257 (1983)

12   G. Brauer and K. Popp, Phys. Status Solidi **102**, 79 (1987)

13   Yu.N. Osetsky, A.G. Mikhin and A. Serra, J. Nucl. Mater. **212-215,** 236 (1994)

14   W.J. Phythian, A.J.E. Foreman, C.A. English, J.T. Buswell, M. Hetherington, K. Roberts and S. Pizzini, 15th International Symposium on Effects of Radiation on Materials, ASTM STP 1125, R.E Stoller,A.S. Kumar and D.S. Gelles, Eds., ASTM, Philadelphia, 131 (1992)

15   T.N. Lê, A. Barbu, D. Liu and F. Maury, Scripta metall. mater. **26**, 771 (1992)

16   P. Pareige, S. Welzel and P. Auger, Jour. de Phys. IV, C **5**, 229 (1996)

17   G.R. Odette, B.D. Wirth, J. Nucl. Mater. **251**, 157 (1997)

18   F.Maury, A. Lucasson, P. Lucasson, P. Moser and F. Faudot, J. Phys. C **2**, 9291 (1990)

ATOMISTIC SIMULATION OF MOBILE DEFECT CLUSTERS IN METALS

Yu.N. OSETSKY[1] , D.J. BACON[1] and A. SERRA[2]

[1] Materials Science and Engineering, Department of Engineering, The University of Liverpool, Liverpool L69 3BX, UK, (osetsky@liverpool.ac.uk).
[2] Dept. Mat. Aplicada III, Universitat Politecnica de Catalunya, Jordi Girona 1-3, E-08034 Barcelona, Spain.

ABSTRACT

The structure, stability and thermally-activated motion of interstitial and vacancy clusters in Fe and Cu have been studied using atomic scale computer simulation. All studied interstitial clusters and perfect interstitial loops (PILs) in Fe are mobile whereas their mobility in Cu can be suppressed at large sizes (bigger than 49-61 self-interstitials depending on the temperature) due to dissociation. A comparative study of relaxed configurations has shown that the structure of small perfect dislocation loops of vacancy and self-interstitial nature is very similar. Molecular dynamics simulation has demonstrated that small perfect vacancy loops (PVLs) in Fe consisting of more than 37 vacancies are stable over a wide temperature range and produce atomic displacements by a thermally-activated movement in the direction of the Burgers vector. The mechanism is qualitatively similar to that of SIA clusters studied earlier. Motion of vacancy loops in Cu does not occur because they transform into sessile configurations similar to stacking fault tetrahedra. These results point to the possibly important contribution of vacancy loop mobility to the difference in radiation damage between bcc and fcc metals, and between fcc metals with different stacking fault energy.

INTRODUCTION.

Within the last decade numerous molecular dynamics (MD) simulations have been used to study displacement cascades generated by energetic atom recoils in a variety of metals (see [1] for recent reviews). Clear evidence has emerged that clusters of self-interstitial atoms (SIAs) form directly within the cascade volume in all metals, and simulation has further established the fact that a large fraction of these SIA clusters is glissile and highly mobile. It is interesting to note that indirect experimental evidence for intracascade clustering and for one-dimensional glide of SIA clusters has existed in the literature for some time [2]. These features and the fact that the SIA clusters are thermally stable clearly imply that the problem of defect accumulation under damage production by displacement cascades cannot be treated within the framework of homogeneous reaction kinetics using the mean-field approach, and so the concept of the production bias model (PBM) was proposed [3]. The PBM in its present form [4] provides a theoretical framework within which the accumulation of defects can be calculated and it has been applied with some accuracy for copper. Nevertheless, uncertainty as to the cause of the difference in damage production between bcc and fcc metals in general, and between copper (Cu) and alpha-iron (Fe) in particular, remains [5]. The fact that SIA clusters are mobile is not sufficient to explain details of the difference.

Mat. Res. Soc. Symp. Proc. Vol. 538 © 1999 Materials Research Society

Possible reasons can be found in atomic-scale computer simulation of the behaviour of defect clusters in Cu and Fe. These are the dynamical interactions between SIA clusters in Cu and Fe [6], different size dependence of cluster mobility in Cu and Fe [7] and/or thermally-activated mobility of vacancy clusters [8].

Simulations to investigate the properties of individual SIA clusters of up to 17 SIAs in Fe and Cu have shown that their one-dimensional motion is the result of individual jumps of their constituent interstitials [9-11]. Sufficiently large clusters (more than about 30 SIAs) can be described as dislocation loops [6]. The mobility of such loops has not been described as yet, although it is expected that their motion is similar to that observed for the smaller clusters. Comparison of the atomic configuration of vacancy and interstitial dislocation loops simulated in Fe [12,6] shows that their structure is very similar, and so it is reasonable to expect that small perfect vacancy dislocation loops can be mobile as well. In this paper, we present the results of an MD study of the thermally-activated mobility of vacancy and interstitial loops containing more than 20 defects. The principal purposes are to study mobility of perfect dislocation loops, to compare vacancy and interstitial loops of the same size and shape, and to demonstrate that vacancy loops can be mobile and that their mobility is different in the bcc and fcc lattices.

CALCULATION MODEL

The mobility of vacancy and interstitial dislocation loops was studied by MD in rectangular crystallites with periodic boundary conditions. For simplicity of analysis, the Burgers vector was oriented along the Z-axis of a coordinate frame in which the X, Y, Z axes were along $[1\bar{1}0]$, $[11\bar{2}]$, $[111]$ and $[100]$, $[01\bar{1}]$, $[011]$ for the bcc and fcc crystallites respectively. The total number of atoms in a mobile crystal varied between 15000 and 36000. The evolution of clusters was studied over the temperature range 260-1100K for times of between 0.9 and 2.0ns. We have studied planar compact clusters of size N_D=28-91 in Fe and N_D=25 - 100 in Cu, where N_D is the number of vacancy or SIA defects. Previous studies have shown that the most stable SIA clusters consist of defects distributed over three {111} and two {110} adjacent planes in bcc and fcc lattices, respectively [8]. PILs consist of crowdions oriented along the close-packed direction perpendicular to the habit plane. The structure of PVLs is similar to that of perfect interstitial loops, in that after relaxation, both types exhibit a perfect lattice separated from the matrix by the loop of the corresponding dislocation.

To study the atomic transport associated with thermally-activated motion of the loops, the square of the X, Y and Z displacements of all atoms was calculated as function of time :

$$R_\alpha^2(t) = \sum_{i=1}^{N_a} \left[r_{i,\alpha}(t) - r_{i,\alpha}^0 \right]^2$$

where α indicates the direction (X, Y or Z), N_a is total number of atoms in the crystallite, $r_{i,\alpha}^0$ and $r_{i,\alpha}(t)$ are the components of the initial and instantaneous radius-vector of the i-th atom. To formulate and parameterise the movement of dislocation loops, we have applied the diffusional approximation to their centre of mass, as in [6,10], and thereby calculated the jump frequency for loops at different temperatures. The displacement of the centre of mass of a cluster was calculated and when the displacement reached the magnitude of the Burgers vector, the number of jumps of the cluster was increased by one. During the simulation time we observed typically 600-800 jumps of any one cluster, which was sufficient to estimate the jump frequency with a statistical error $\approx$2-4%. The exception to this was the PVL with $N_D = 37$ in Fe at temperature 1010K. After about 0.95ns it transformed into a weakly mobile cluster which was identified as being similar in form to a perfect <100> loop. Before transformation about 300 jumps were observed and the jump frequency was estimated with a statistical error $\approx$10%.

Long-range, pair interatomic potentials [13] for Cu and Fe were used for most simulations, but some were duplicated with many-body potentials (MBP) of Finnis-Sinclair type [14] in order to test the sensitivity of the results to the potential form.

RESULTS

It was found that both types of loop produce significant displacements of atoms along the Burgers vectors direction during annealing. This is illustrated in fig.1, where $R_z^2(t)$ is presented for vacancy and interstitial loops containing 37 defects (PIL-37 and PVL-37) in Fe. It can be seen that both loops produce significant displacements along the Z-axis. The displacements along the X and Y-axes were negligible for both loops. A strong anisotropy was observed in the distribution of displacements for different atoms in the simulated crystallites. Analysis of the distribution of the displaced atoms shows that in both cases only the atoms inside the glide cylinder were displaced. This is clearly seen in fig.2, where we superimpose sets of the instantaneous positions of atoms projected on to the $(1\bar{1}0)$ plane through the centre of crystallites containing the PIL-37 and PVL-37. The positions of the atoms were recorded during the MD simulation every 10 ps for about 200 ps. It can be seen that the picture is qualitatively the same for vacancy and interstitial loops.

The results of estimation of jump frequency of some loops are presented in fig.3 as functions of the reciprocal temperature. Treatment of the jump frequency using the Arrhenius-like approach shows a very low activation energy for all loops (table 1), similar to those of individual <111> crowdion [10].

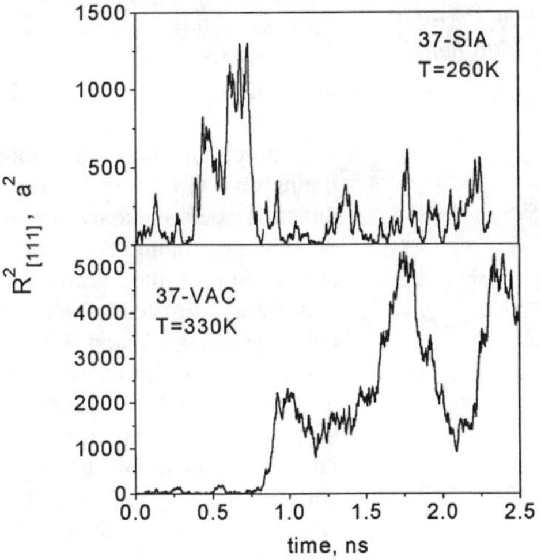

The results above were obtained with the long-range pair potential for Fe. For the many-body potential for this metal, perfect interstitial loops show the same qualitative behaviour. Perfect vacancy loops, however, are found to be unstable up to a size equivalent to $N_D \approx$ 60. Clusters of this size range form uncollapsed vacancy platelets in static simulation, and transform into disordered three-dimensional sets of vacancies in MD. However, MD simulation of the collapsed and stable PVL-61 at T = 650K shows that its gliding features are qualitatively the same as for the long-range pair potential. For comparison, we present the jump frequency of its centre of mass in fig.3.

Fig.1 *Square of atomic displacements along the Burgers vector during evolution of vacancy and interstitial loop of 37 defects.*

Turning now to Cu, it was observed earlier [6] and confirmed in the present study that interstitial loops in Cu with $N_D \leq 25$ execute thermally-activated one-dimensional glide along <110> directions, in a similar fashion to the motion of SIA clusters along <111> in α-Fe. Data for the jump frequency of PIL-25 are presented in table 1. However, it is impossible to parameterise the motion of larger PILs ($49 \leq N_D \leq 81$) due to their dissociation on the glide prism into partial dislocations (see [6]). Such loops produce atomic displacements but their motion cannot be decomposed into separate jumps of the centre of gravity. Interstitial loops with $N_D > 81$ are found to be immobile, i.e. during annealing they do not produce displacements of atoms in the crystallite.

The behaviour of vacancy clusters in Cu is found to be qualitatively different from than that in Fe. Collapsed vacancy loops in Cu are structurally unstable and dissociate and transform during annealing into sessile configurations similar to stacking fault tetrahedra.

One-dimensional motion of dislocation loops cannot be described as random walk. It was observed that each loop jumps several time in the same direction before changing its direction of motion. For the above loops, the correlation factor was estimated to be in the range between 1.7 and 3.7. In general it increases at low temperature and decreases for bigger clusters. The correlation factor is larger in Fe than in Cu and for vacancy clusters it is slightly lower than for interstitial ones. Separate simulations show that the value of the correlation factor does not depend on the size of the crystallite along the Burgers vector direction.

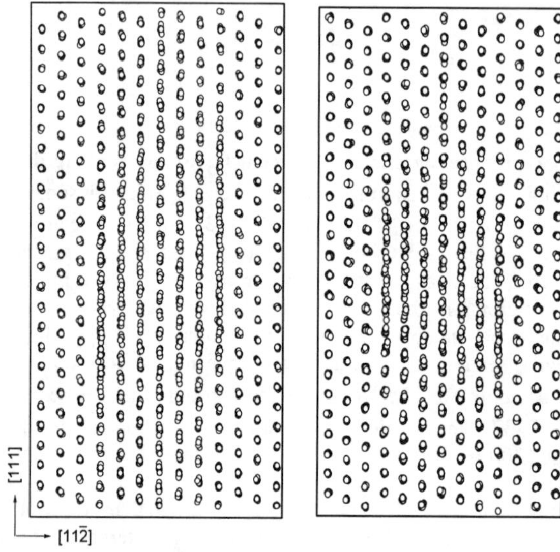

Fig.2 Instantaneous positions of atoms projected on to $(1\bar{1}0)$ plane during evolution of PIL-37 at 260K (left) and PVL-37 at 330K (right).

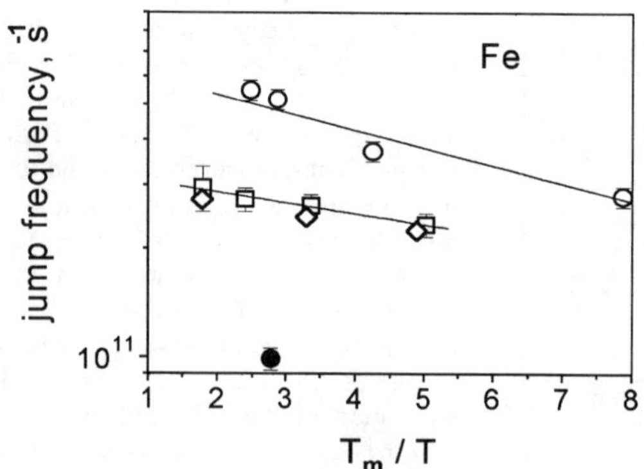

Fig.3 Jump frequency versus reciprocal temperature for vacancy and interstitial loops in Fe : open circle – PIL-37, square and diamond – PVL of 37 and 43 vacancies, respectively, and solid circle – PVL–61 simulated with the MBP.

<table>
<tr><td>

Metal	*Type, N_D*	v_o $[10^{12}$ $s^{-1}]$	E^m $[eV]$
Bcc-Fe	SIA, 28	0.87±0.01	0.021±0.003
	SIA, 37	0.67±0.01	0.018±0.003
	SIA, 91	0.43±0.01	0.021±0.004
	VAC, 37	0.33±0.01	0.011±0.004
	VAC, 43	0.31±0.01	0.010±0.004
Fcc-Cu	SIA, 25	0.85±0.01	0.025±0.004

</td><td>

Table 1. *Results of the treatment of the jump frequency for stable clusters in Fe and Cu using an Arrhenius-like dependence. E^m is the effective migration energy and v_o is the jump frequency pre-exponential factor.*

</td></tr>
</table>

DISCUSSION

Analysis by MD of thermally-activated, one-dimensional motion of SIA clusters in bcc Fe has shown that the movement of the centre of mass of clusters with $N_D \leq 17$ can be described as near-independent jumps of individual crowdions along a <111> direction [11]. Although the treatment of PILs applied here allows all crowdions within the simulated loops to be recognised, it is unlikely that the model based on the stochastic jumps of crowdions can be applied for the glide of large loops with N_D =37-91. Furthermore, the results obtained for the glide of vacancy loops cannot be represented as movement of crowdions. On the other hand, both types of loop exhibit similar behaviour, indicating that the mechanism of thermally-activated glide of small dislocation loops does not depend on their vacancy or interstitial nature.

Another problem is that of establishing a formal description and parameterisation of loop motion. As in [6,10], we have applied the diffusional approximation and, while recognising that this does not formally describe thermally-activated glide, obtained parameters for comparing the behaviour of loops of different size and nature. Thus, in general, bigger loops are less mobile than smaller ones and vacancy loops are less mobile than interstitial ones. The characteristic described here as "jump frequency" of the centre of mass is more appropriately related to the speed of glide. The physical meaning of the effective "migration energy" obtained in the Arrhenius-like treatment of the temperature dependence for the jump frequency in not clear. It is as small as the barrier to motion of a crowdion along a close packed direction, but is lower than the thermal energy of atoms at almost all temperatures considered.

Regardless of difficulties in the description and parameterisation of the glide process, it may be concluded that vacancy and interstitial loops in bcc metals can produce atomic displacements and, therefore, atomic transport. This one-dimensional glide occurs due to the thermal vibration of atoms and does not require the application of stress. It is related to the lattice dynamics and cannot be studied within the continuum approximation

Finally, a different situation has been observed for dislocation loops in Cu. Thus, although PILs are found to be one-dimensionally mobile, at least for $N_D \leq 49$, the equivalent vacancy clusters always dissociate into sessile configurations similar to stacking fault tetrahedra. This effect probably occurs when stacking fault energy is low, as in Cu, Ag and Au. For metals with high stacking fault energy, one would expect PVLs to be more stable and thus behave like PILs i.e. to execute one-dimensional, thermally-activated glide. This effect could possibly be one of the reasons for different radiation damage characteristics of fcc metals with different stacking fault energy. In addition, the difference in the glide properties of vacancy clusters could contribute to the observed difference in radiation damage of Cu and Fe. These possibilities are now receiving further attention.

ACKNOWLEDGEMENTS

This research has been carried out using the facilities of CESCA and CEPBA (Barcelona, Spain) under the coordination of C4. The authors also acknowledge financial support from the University of Liverpool and the U.K. Engineering and Physical Sciences Research Council.

REFERENCES

1. T.Diaz de la Rubia T., *Ann. Rev. Mater. Sci.*, **26**, p. 213 (1996); D.J.Bacon, A.F.Calder and F.Gao,1997, *J.Nucl.Mater.*, **251**, p. 1 (1997); D.J.Bacon, F.Gao and Yu.N.Osetsky, *Proc. Int.Conf.:* COSIRES-98, to be published in *Nucl. Instr. And Methods.*

2. H.Trinkaus, B.N.Singh and A.J.E.Foreman, *J.Nuc.Mater*, **199**, 5 (1992); *Ibid*, **206**, 200 (1993).

3. C.H.Woo and B.N.Singh, *Phys. Stat. Sol. (b),***159**, 609 (1990); *Philos.Mag.A,***65**, 889 (1992); B.N.Singh and A.J.E.Foreman, *Philos.Mag.*, A, **66**, 975 (1992); H.Trinkaus, B.N.Singh and A.J.E.Foreman, *J.Nuc.Mater*, **199**, 5 (1992); *Ibid*, **206**, 200 (1993).

4. B.N.Singh, S.I.Golubov, H.Trinkaus, A.Serra, Yu.N.Osetsky and A.V.Barashev, *J. Nucl. Mater.*, **251**, 107 (1997).

5. S.I.Golubov, B.N.Singh and H.Trinkaus, (1998), private communication.

6. Yu.N.Osetsky, V.Priego, A.Serra, B.N.Singh and S.I.Golubov, *Philos. Mag.*, A, (1998) in press.

7. Yu.N.Osetsky, A.Serra and V.Priego, *Proc. Int. Workshop on: Basic Aspects of Differences in Irradiation Effects Between fcc, bcc and hcp Metals and Alloys,* Cangas de Onis (Spain) October 15-20, 1998.

8. Yu.N.Osetsky, D.J.Bacon and A.Serra, *Philos. Mag.Lett.,* (1998) submitted.

9. B.D.Wirth, G.R.Odette, D.Maroudas and G.E.Lucas, *J.Nucl.Mater.,***244**, 185 (1997).

10. Yu.N.Osetsky, A.Serra and V.Priego, *Mat.Res.Soc.Symp.Proc.* **527**, 59 (1998).

11. A.V.Barashev, Yu.N.Osetsky and D.J.Bacon, *these Proc. Symp. N.*

12. V.G.Kapinos, Yu.N.Osetsky and P.A.Platonov, *Phys. Stat. Sol.*(b), **155**, 373 (1989); *J.Nucl.Mater,* **170**, 66 (1990); *Ibid* **173**, 229 (1990).

13. Yu.N.Osetsky, A.G.Mikhin and A.Serra, *Philos. Mag.*,A, **72**, 361 (1995).

14. G.J.Ackland, G.Tichy, V.Vitek and M.V.Finnis, *Philos. Mag.*, A, **56**, 735 (1987); G.J.Ackland, D.J.Bacon, A.F.Calder and T.Harry, *Philos. Mag.*, A, **75**, 713 (1997).

EFFECTIVE INTERACTIONS APPROACH TO PHASE STABILITY IN ALLOYS UNDER IRRADIATION

RAÚL A. ENRIQUE, PASCAL BELLON

Univ. of Illinois, Dept. of Materials Science and Engineering, Urbana, IL, 61801 USA.

ABSTRACT

Phase stability in alloys under irradiation is studied considering effective thermodynamic potentials. A simple kinetic model of a binary alloy with phase separation is investigated. Time evolution in the alloy results from two competing dynamics: thermal diffusion, and irradiation induced ballistic exchanges. The *dynamical* (steady state) phase diagram is evaluated exactly performing Kinetic Monte Carlo simulations. The solution is then compared to two theoretical frameworks: the effective quasi-interactions model as proposed by Vaks and Kamishenko, and the effective free energy model as proposed by Martin. New developments of these models are proposed to allow for quantitative comparisons. Both theoretical frameworks yield fairly good approximations to the dynamical phase diagram.

INTRODUCTION

An alloy subjected to a continuous external driving force may reach a steady state; the properties of this steady state will, in general, be different from the equilibrium properties [1]. Labeled under the term *driven alloys*, examples include materials under continuous plastic deformation, such as cyclic fatigue or ball milling, and materials under irradiation, such as nuclear reactor core components or compounds produced by ion implantation.

In particular, irradiation can affect phase stability in alloys in many different -and seemingly contradictory- ways. For example, while for certain experimental conditions irradiation-induced solute segregation can take place, in some other conditions enhanced solubility has been observed [2].

For equilibrium systems, thermodynamics and statistical mechanics provide a framework to describe the phase diagram. From an applications point of view, it would be very useful to count with an equivalent theoretical framework to construct the *dynamical* (steady state) phase diagram for driven alloys. Conceptually, this dynamical phase diagram would have an extra axis, describing the intensity of the driving force.

Driven alloys are, however, systems far from equilibrium, for which thermodynamical concepts cannot be directly applied. The temporal evolution of the driven alloy is described by a non Hamiltonian (dissipative) dynamical system, so the steady state probability distribution cannot be written using the equilibrium Gibbs distribution. We then turn into the construction of *effective* thermodynamical potentials in order to describe the steady state properties of the driven alloy. Contrary to equilibrium potentials, these effective potentials depend explicitly on the forcing parameters.

Several levels of description are possible, ranging from the atomistic to the continuum. Each description level provides with a different strategy to define effective thermodynamical potentials. In the macroscopic description, the natural path is to define an effective free energy (which is actually a Lyapunov function of the system) by interpreting terms in the diffusion equations [3]. In the microscopic description, on the other hand, the idea is to keep the form of the Gibbs distribution for the probabilities of each microscopic state, but with the definition of an effective Hamiltonian, written in terms of pair, triplets, etc, effective quasi-interactions [4, 5].

229

Mat. Res. Soc. Symp. Proc. Vol. 538 © 1999 Materials Research Society

The construction of effective thermodynamical potentials is not guaranteed to be successful. In fact, numerical examples have shown that binary alloys under continuous shearing [6], or under arbitrary length random exchanges, can exhibit mesoscopic structures. These mesoscopic structures can not be explained by appealing to a *local* effective free energy. These examples, however, involve some kind of long range mixing or disorder, which we will not consider in this paper.

This paper is focused on the study of the effect of low energy irradiation in a binary mixture that exhibits phase separation. We consider the case of dominant *ballistic exchanges*, where irradiation produces mainly local disorder in the matrix [3]. At a given irradiation flux, this disorder enhances solubility, and at low enough temperatures, it causes solubility through the whole concentration range. For simplicity, the concentration of point defects is assumed to be have reached a steady value, the contribution of interstitials is neglected, as well as defect sinks and clusters. Under these assumptions, the proposed model excludes phenomena related to macroscopic solute migration, such as irradiation induced heterogeneous precipitation [2].

The work presented in this paper is structured in three parts. In the first part we perform a direct evaluation of the dynamical phase diagram using Kinetic Monte Carlo simulations. In the second part, we use the same code to directly evaluate the effective pair quasi-interactions, in the theoretical framework of Vaks and Kamishenko [4], and compute an effective equilibrium phase diagram from these. In the third part, we compute a mean field phase diagram using the macroscopic model introduced by Martin [3].

BASIC MODEL

The model under investigation is a binary (AB) alloy, with a low vacancy concentration and a positive heat of mixing, on a rigid FCC lattice. Atomic interactions are of an Ising-type: first nearest neighbors bonds with a positive ordering energy $w_{ab} = 2e_{ab} - e_{aa} - e_{bb}$. Thermal diffusion is modeled by considering a frequency for vacancy exchange, proportional to a Boltzmann factor of the energy required to break the atomic bonds. Low energy irradiation is modeled by considering a frequency for random exchange of nearest neighbors atoms, Γ_b. A detailed description of the model can be found in reference [8].

The time evolution for the probability $P(\alpha)$ to find the system in a given microscopic state α is governed by the master equation:

$$\frac{dP(\alpha)}{dt} = \sum_{\beta} W(\alpha, \beta)P(\beta) - W(\beta, \alpha)P(\alpha) \equiv \hat{S}P \tag{1}$$

where the W's are the transition probabilities between states, and are the sum of thermal and ballistic transition rates. The steady state probability distribution is given by the homogeneous solution of Eq. (1).

For systems in thermodynamical equilibrium, the steady state probability distribution is known, and one writes the probability of each microscopic state as a Boltzmann factor. For the driven alloy under study, the steady state probability distribution can not be found easily, and no formal general method exists. At present, only solutions in the mean-field approximation [4, 7], and for systems of low dimensionality [5] have been found.

A residence time algorithm is used for the Monte Carlo simulations (a review of this method can be found in [1]). In this way, the system evolves according to Eq. (1) following a temporal trajectory. This property allows for a meaningful comparison between the frequencies of random and vacancy exchanges at each simulation step.

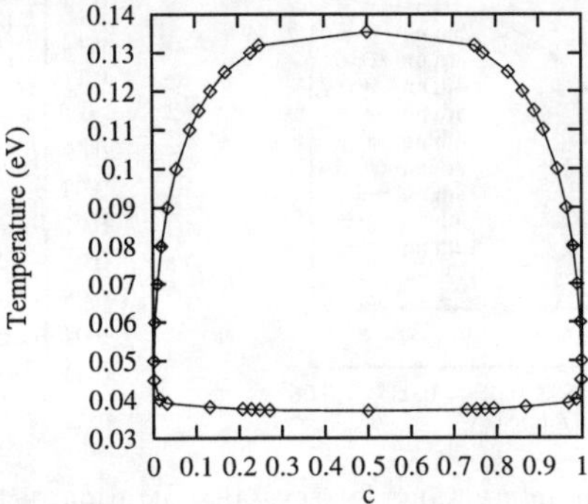

Figure 1: Dynamical phase diagram. $C_v = 1/64^3$, $\Gamma_b = 100s^{-1}$.

The simulation domain is a 64×64×64 rhombohedric crystal with periodic boundary conditions. The faces of the rhombohedron are {111} planes in the FCC crystal. A single vacancy is placed in the simulation domain. The competition between the frequencies for vacancy and ballistic exchanges determines which transition takes place at each time step.

The materials parameters used in the simulation are the following: $e_{aa} = e_{bb} = -723.3meV$, $w_{ab} = e_{ab} - e_{aa} - e_{bb} = 55.3meV$, $e_{av} = e_{bv} = -255meV$. The ordering energy was chosen so as to reproduce an estimated critical temperature of $T_c = 1573K = 135.5meV$ in the Cu-Co system. The vacancy migrates with an attempt frequency $\Gamma_v = 10^{14}s^{-1}$, and an average migration energy of $800meV$. All the results presented in this paper have been obtained at a ballistic frequency $\Gamma_b = 100s^{-1}$.

DYNAMICAL PHASE DIAGRAM

The dynamics of the model we have described is conservative in the concentration. The miscibility gap is built by letting the system evolve into steady state separate phases and then measuring the local concentration in each phase. The lower and upper limits of the miscibility gap are determined using the fourth order cumulant method [9], yielding these temperatures with an excellent accuracy ($\approx 0.3\%$). The finite size analysis of the fourth order cumulant is made with statistical analysis of small boxes of sides L_b=4,6,8,10,12 contained in the simulation domain. The statistical analysis for these small boxes showed to be unaffected by the constraint of conserved global concentration. The computed dynamic phase diagram is showed in Fig. 1.

EFFECTIVE PAIR INTERACTION ENERGIES

Vaks and Kamyshenko [4] have studied binary driven alloys using effective quasi-interactions. Starting from the master equation that describes the time evolution of the system, they point out that the most general expression for the steady state probability distribution can be written in the form:

$$Pn_i = exp(A + \sum_i \lambda_i n_i - \hat{Q})$$

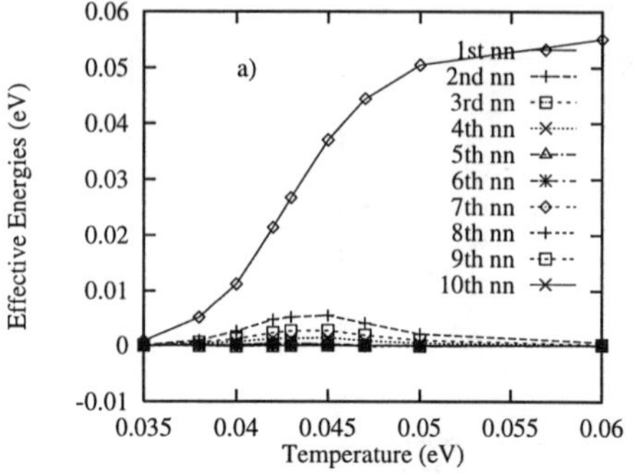
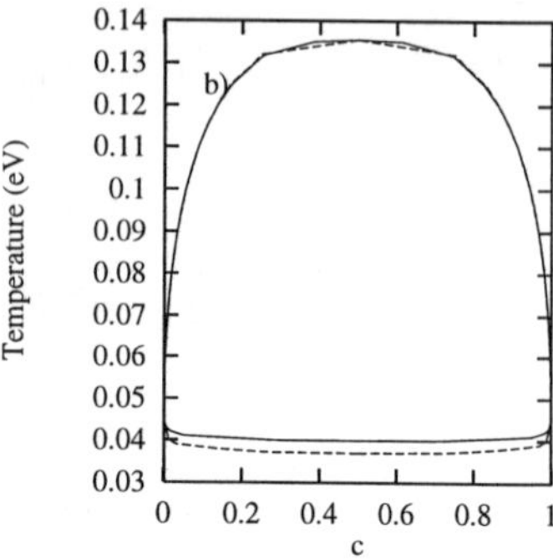

Figure 2: a) Effective pair interactions for several separation distances (times k_BT). b) Dynamical phase diagram using effective pair interactions (solid line). The exact dynamical phase diagram (dash line) is also shown for comparison.

$$\hat{Q} \;=\; \sum_{i<j} a_{ij}n_i n_j + \sum_{i<j<k} a_{ijk}n_i n_j n_k + ... \tag{2}$$

Where, using the notation n_i for the occupancy of a lattice site by a B atom, $\hat{Q}$ plays the role of an *effective* Hamiltonian, and the previous equations can be interpreted as a generalized Gibbs distribution (notice that the a's are the ordering energies over k_BT).

If the steady state probability distribution for an arbitrary configuration is known, evaluation of the quasi-interactions can be performed by inverting Eqs. 2. Given two states i and j, the individuals probabilities are written $p_i \propto N_i exp(-q_i)$ and $p_j \propto N_j exp(-q_j)$, where q_i represents the value of $\hat{Q}$ for the corresponding configuration and N_i the multiplicity. Keeping in mind that the proportionality constants for the previous equations are the same, we can write the difference in *effective energy* as:

$$q_i = q_j - \log(\frac{p_i N_j}{N_i p_j}) \tag{3}$$

In particular, the pair quasi-interactions (first term in the expression for $\hat{Q}$) can be evaluated, using Eq.3, from the probabilities of the separation distance of two B atoms in a pure A matrix. Proceeding in this way we obtain the effective energies showed in Fig. 2a. It is interesting to notice that the measured effective energies extend beyond the first shell, and nonzero effective ordering energies are observed up to the fourth nearest neighbors shell. Effective interactions are temperature dependent, decreasing to zero as the ballistic exchanges become more dominant at lower temperatures.

From these effective energies, we can evaluate the corresponding equilibrium phase diagram, using standard grand canonical equilibrium Monte Carlo simulations. This phase diagram compares surprisingly well with the original phase diagram, as we can see in Fig. 2b, even though we restricted ourselves to pair effective quasi-interactions.

MACROSCOPIC MODEL

An heuristic, macroscopic description of phase separation under continuous random exchanges due to irradiation was introduced by Martin [3]. The description is based on

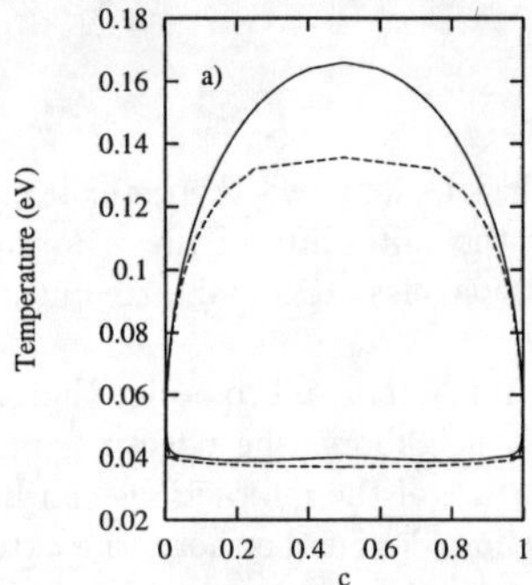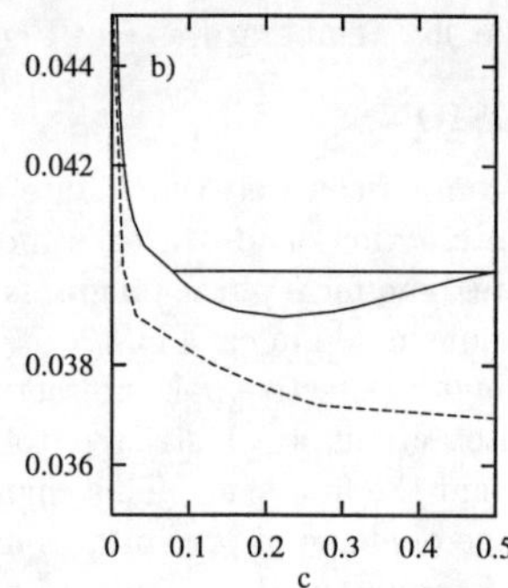

Figure 3: a) Phase diagram computed using Martin's model (solid line). The exact dynamical phase diagram is also shown (dash line) for comparison. b) Close up at low temperatures showing a three phase coexistence line.

Cahn's model for diffusion in a binary mixture, where a irradiation related diffusion term is added. The basic idea is to absorb the extra term so as to rewrite the equation in the same form it has in the absence of irradiation. A new *effective* free energy ϕ (actually a Lyapunov function), is introduced, defined on its second derivative through the expression:

$$\phi''(c) = f''(c) + \frac{kT}{c(1-c)} \frac{\mathcal{D}_b}{\tilde{D}} \tag{4}$$

Here, f is the equilibrium free energy, $\mathcal{D}_b$ is the ballistic diffusion coefficient, c the molar fraction of B and $\tilde{D}$ the chemical diffusion coefficient.

Evaluation of the effective free energy requires, in the first place, knowledge of the equilibrium free energy, and in the second place, knowledge of the concentration dependence of the chemical diffusion coefficient. In this work, as in Martin's original paper [3], we use for $f(c)$ the expression obtained in the Braggs-William approximation.

The chemical diffusion coefficient is written $\tilde{D} = cD_1^* + (1 - c)D_2^*$, where D_i^* are the individual tracer diffusion coefficients. These can be expressed as $D_i^* = \frac{1}{12}a_o^2\Gamma_i f_i$, where Γ_i is the exchange frequency and f_i the correlation factor, for and atom of type i.

The correlation factor is an important contribution to the diffusion coefficient when trapping occurs. At low temperature and low concentrations, a vacancy exchanging with a solute atom will exchange with it again with a great probability. For f_i, we use the expression recently derived by Nastar et al [10]. The exchange frequencies for a vacancy with an atom is evaluated in a mean field approximation, assuming that each particle is surrounded by an average environment determined by the composition.

Using the model as described, we proceed to evaluate the effective free energy, and from there the dynamical phase diagram, making use of the common tangent rule. Since ϕ is only a Lyapunov function, the common tangent rule may not apply strictly [3], but it should be a reasonable approximation. The computed phase diagram is showed in Fig. 3.

At high temperatures, we observe well known discrepancies due to the limitations of the Bragg-Williams approximation. The low temperature closure of the phase diagram is, on the other hand, well modeled by the macroscopic model. The macroscopic model, however, presents a feature not observed so far in the simulations: the existence of a three phase line at low temperatures (see Fig. 4b).

CONCLUSIONS

Direct comparison between Monte Carlo simulations and two theoretical approaches (effective interactions and Martin's model), shows the suitability of these approaches to approximate dynamical phase diagrams. Small discrepancies are found, attributable to the approximations made in each case.

Regarding the effective pair interaction energies, it is important to notice that, although physical interactions are restricted to first nearest neighbors, the effective interactions extend beyond the first shell. This change in the nature of the interactions can be related to the change of shape of precipitates under irradiation [8]. Furthermore, the extension in the range of the interactions can lead us to speculate that for the cases where mesoscopic structures are observed, effective interactions must become long range (or decreasing slower than $1/r^3$).

Regarding Martin's model, the use of a concentration dependent diffusion coefficient results in the fact that irradiation effects are not simply equivalent to a higher effective temperature [3]. Furthermore, this new evaluation of Martin's model suggests that three phases may coexist in dynamical equilibrium at low temperatures. This prediction, although not observed so far in our Monte Carlo simulations, deserves some further investigation, especially in view of experimental results for ball milling in the Ag-Cu system [11], where the dynamical coexistence of a 50% composition phase with the two solid solutions has been proposed to rationalize X-ray diffraction results.

ACKNOWLEDGMENTS

Stimulating discussions with G. Martin, M. Nastar and B. Legrand are gratefully acknowledged. This work was partly supported by the US Department of Energy grant DEFG02-96ER45439 through the University of Illinois Materials Research Laboratory. The present simulation work has greatly benefited from the MRL Center for Computation.

REFERENCES

1. G. Martin and P. Bellon, *Solid State Physics* **50**, p. 189 (1997)

2. K.C. Russel, *Prog. Mat. Sci.*, **28**, p. 229-434 (1984).

3. G. Martin, *Phys. Rev. B*, **30**, p. 1424 (1984).

4. V.G. Vaks and V.V. Kamyshenko, *Phys. Let. A*, **177**, p. 269-274 (1993)

5. P.L. Garrido and J. Marro, *Phys. Rev. Let.*, **62**, p. 1929-1932 (1989).

6. P. Bellon and R.S. Averback, *Phys. Rev. Let.*, **74**, p. 1819-1822 (1995).

7. F. Soisson, P. Bellon, and G. Martin, **46**, p. 11332-45 (1992).

8. P. Bellon, *Phys. Rev. Let.*, **81**, p. 4176 (1998).

9. K. Binder and D.W. Heermann, *Monte Carlo Simulations in Statistical Physics*, Springer, Heidelberg, 1997, p.44-56.

10. M.Nastar, V. Yu. Dobretsov, G. Martin, Phil. Mag. A, submitted.

11. T. Klassen, U. Herr and R. S. Averback, *Acta Mater.*, **7** p. 2921-2930 (1997).

Oxygen Vacancy Migration in Disordered $Gd_2(Zr_xTi_{1-x})_2O_7$

R.E. Williford*, W.J. Weber*, R. Devanathan*, and J.D. Gale**
*Materials Department, Pacific Northwest National Laboratory, Richland, WA 99352
**Department of Chemistry, Imperial College, South Kensington, London, UK SW7 2AY

Abstract

Oxygen vacancy formation and migration energies in the pyrochlores $Gd_2(Zr_xTi_{1-x})_2O_7$ were computed using atomistic energy minimization methods, with particular attention to the role of cation antisite disorder. The interatomic potentials recognize oxygens in distinct crystallographic sites (polarizations) as distinct species, and permit assessment of the directional anisotropy in vacancy migration energies. Oxygen vacancy migration is determined by O_{48f} - O_{48f} transitions along the shortest edges of the TiO_6 octahedra, in agreement with previous results. Oxygen vacancies migrate towards disorder to reduce local stresses. In contrast to previous work, the V_{O8a} and $V_{O8a} + I_{O8b}$ Frenkel defects are favored over the V_{O48f} and $V_{O48f} + I_{O8b}$ Frenkel. The transition $V_{O8a} \rightarrow V_{O48f}$ is energetically favored for the ordered structure, and activates the three dimensional O_{48f} - O_{48f} vacancy migration network by filling empty 8a sites with 48f oxygens. Local regions of order and moderate disorder must both be present for optimum oxygen conductivity. High disorder retards vacancy migration.

Introduction

In addition to their adjustable lattice parameters and flexible cation chemistries, the pyrochlores $Gd_2(Zr_xTi_{1-x})_2O_7$ (GZT) are effective oxygen conductors at temperatures lower than many other solid oxide electrolytes. GZT thus offers advantages for fuel cells, oxygen sensors, and ion pumps [1]. $Gd_2Ti_2O_7$ (herein denoted GTO) and related pyrochlores are also being considered for the immobilization of nuclear waste actinides and weapons plutonium [2].

Extensive experimental work [1, 3, 4] has shown that the substitution of Zr for Ti is associated with a large increase in oxygen vacancy conductivity due to an increase in the carrier concentration. As x increases, GZT reverts from the pyrochlore towards the fluorite structure via cation antisite disordering, along with partial emptying/filling of some oxygen sites that are normally filled/empty in the pyrochlore. The onset of disorder occurs at about x=0.25. Saturation is approached at about x=0.6, where defect clustering causes the migration activation energy to increase. Cation antisite disorder is <5%, ~10%, and ~35% at x=0.25, 0.6 and 1.0, respectively [3]. Such disorder can also be caused by ion bombardment during ion-beam experiments or from incorporation of decayed radionuclides in the case of nuclear waste immobilization [2].

The disorder and defect clustering that control the ionic conductivity of GZT occur on a localized basis. Because there are no means to directly measure, on the localized scale of several atoms, how such specific *local* disorder affects the vacancy migration energetics, it is therefore necessary to employ atomistic computer simulations. Two previous simulations of these materials [5, 6] used energy minimization with interatomic pair potentials, but neither addressed cation disorder. In recent work [7], the x=0 endpoint of the GZT composition range was addressed, and a new set of interatomic potentials was developed. With further refinements, those potentials are used here to simulate oxygen vacancy migrations for the x=0.6 composition and the x=1 endpoint.

Pyrochlore Crystal Structure

The crystal structure of pyrochlores ($A_2B_2O_7$, or $A_2B_2O_6O'$) has been previously described in detail [e.g., 8]. Each B atom is coordinated with six oxygens, forming slightly distorted BO_6 octahedra with edges of two lengths. The so-called 'basic pyrochlore structural units' (four corner-sharing BO_6 octahedra) are arranged about rows of A atoms (Figure 1). The 8a oxygens are aligned along the A rows, and define the AO_8 dodecahedra. The 48f oxygens occupy the corners of the BO_6 octahedra. Of eight oxygen vacancy migration paths considered in [7], two were found to be important: a 0.265 nm, O48f - O48f, path that forms shared edges between the TiO_6 octahedra and the the GdO$_8$ dodecahedra; and a 0.307 nm, O8a - O48f, path that forms shared edges between dodecahedra.

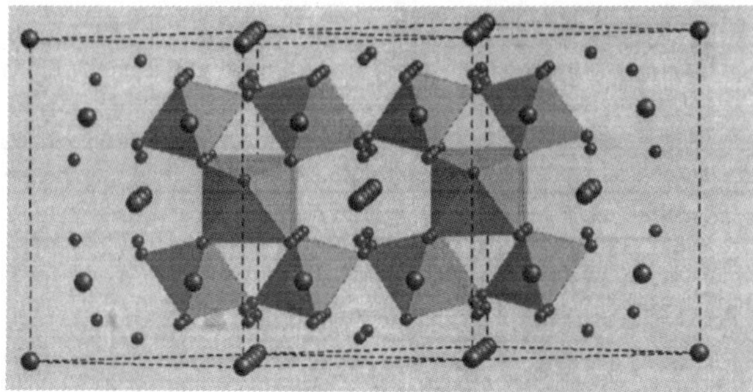

Figure 1. Polyhedral representation of the pyrochlore structure, showing BO$_6$ octahedra in a 2x1x1 supercell along the <110> axis. B (Ti or Zr) atoms are not shown for clarity. The small spheres at each octahedron corner are the 48f oxygen atoms. The larger isolated spheres are A (Gd) atoms. The small isolated spheres above and below the A rows are oxygen atoms in the 8a positions. The 8b positions are above and below the 8a positions, in the interstices between the BO$_6$ octahedra.

Computational Methods and Interatomic Potentials

The General Utility Lattice Program (GULP) [9, 10] was used to simulate the energetics and structures of perfect and defective lattices. GULP uses an energy minimization method based on the Born model, and is unique in its use of crystal symmetry to speed up calculations. Isolated defects in extended solids were addressed with the Mott-Littleton approximation [11], with Region 1 sizes of 100-450 atom cores. Ionic polarization was treated using the Dick-Overhauser shell model [12], including the important repulsion - polarization coupling which prevents excessive polarization. Defect energies were calculated by standard methods [e.g., 13], and transition state searches [9] were used to find energetic 'saddle points' for calculating the activation energies of vacancy migrations.

In previous work [7], a set of potentials was developed based on the Catlow O-O [14] and the Lewis and Catlow [15] Gd-O potentials. In contrast to other work [5, 6], the Ti shell model was replaced with a rigid-ion Ti-O potential developed by fitting to rutile TiO$_2$. Since anion polarizability depends strongly on the local crystal environment [15, 16], the O48f and O8a atoms were treated as distinct species in a second stage of fitting to the three Gd$_2$(Zr$_x$Ti$_{1-x}$)$_2$O$_7$ structures (denoted GTO, GZT, and GZO for x=0.0, 0.6, and 1.0, respectively). This is reasonable since the O48f has two Gd and two Ti atoms as nearest neighbors (total cationic charge of +14), while the O8a has four Gd nearest neighbors (total cationic charge of +12). Consequently, the O48f should have a higher polarization than the O8a, as reflected in the ratio of polarizabilities: α(O48f)/α(O8a)=1.57. Although this 'distinct oxygen species' idea has been used for structural models of other materials [17], it has not been employed for pyrochlores, and apparently has not been used with energy minimization to determine the directional bias of vacancy migration energies. The new potentials predict unit-cell dimensions, elastic constants, and static dielectric constants for GTO, GZT, and GZO that are in excellent agreement with experimental data.

Defect Formation Energies

Predictions of defect formation energies for GTO are shown in Table 1, without defect charges for ease of notation. The notation 1X (2X) denotes one pair (two pairs) of cation antisite defects that are nearest neighbors to the oxygen defects. Note that the 2X case was for disorder tightly clustered around the migration path, in contrast to other results presented below. The Region 1 radius in these calculations included 315-450 ion cores for improved accuracy [7]. Results have been corrected for the disorder effect by subtracting the appropriate disorder energy.

Table 1. Defect formation energies (eV) for GTO, corrected for disorder energies. Frenkel and Split Vacancy energies are on a per defect basis.

Defect	0X	1X	2X
V_{O48f}	19.37	17.37	20.12
V_{O8a}	18.81	13.45	12.20
I_{O8b}	-8.38	-9.03	-7.22
Frenkel $V_{O48f} + I_{O8b}$	5.49	4.17	6.45
Frenkel $V_{O8a} + I_{O8b}$	5.21	2.21	2.49
Split Vacancy	7.56	-----	-----

Although other work [5, 6] predicted that the vacancy formation energy for V_{O48f} was less than that for V_{O8a}, the reverse is true for the potentials developed here. The latter is the more reasonable result when the cationic charge around each site is considered: a larger vacancy formation energy is expected for the anion surrounded by the larger cationic charge. Similar results and conclusions were obtained for GZT and GZO.

The oxygen interstitial in the 8b site is normally vacant in pyrochlores, but partially occupied in fluorites. The O8a shell properties were used for the 8b interstitials for two reasons: (1) Although using the 48f (instead of 8a) properties in the 8b interstitial gave Frenkel defect energies lower by about a factor of two, the second Frenkel in Table 1 was still energetically preferred over the first, so that conclusions were not changed; (2) The $V_{O48f} + I_{O8b}$ Frenkel was stable only if the two point defects were widely separated, whereas the $V_{O8a} + I_{O8b}$ Frenkel was stable as near neighbors. Consequently, it appears that the second Frenkel is more likely to occur. The O48f split vacancy [5] (along a TiO_6 <110> octahedron edge, near an empty 8b site) is also not energetically favored (see also Reference 7).

Oxygen Vacancy Migration Energies

In previous work [7], several important trends in defect energetics were found as cation antisite disorder increased. These trends included the separation of the oxygen vacancy migration energies into two groups representing the shorter (shared) polyhedral edges with lower energies, and the longer (unshared) edges with higher energies. Each group approached a unique migration energy as disorder increased, as expected because each anion site is surrounded by a +14 cationic cloud at 2X disorder.

A total of eight O-O vacancy migration paths were investigated in [7], and showed that the most likely migration paths were 48f-48f transitions along the shortest edges of the TiO_6 octahedra, in agreement with [1, 3, 5, 6]. These edges are shared with the GdO_8 dodecahedra. The longer (unshared) 0.285 nm octahedra edges may participate in oxygen vacancy migration, but do not form a complete network and are thus of lesser importance. Each O48f atom is also neighbored by one O8a atom at 0.307 nm, a migration path forming a shared edge between dodecahedra. Migration energies for these two paths are shown in Table 2. Region 1 sizes were ~100 ion cores here, for computational expediency in this screening exercise. Care was taken to subtract the proper vacancy energy from the saddle point energy [7]. Notations such as V_{O48f} -> V_{O8a} in Table 2 indicate the vacancy migration direction, rather than ion migration.

The disorder in Table 2 was more 'open' than in Table 1, in that the cation antisite defects were oriented *away* from the migration path, rather than tightly clustered around the path. The open configuration was investigated when it was found that the tightly clustered configuration gave much higher vacancy migration energies than experimentally observed. These calculations revealed the differences between vacancy migration away from or towards the cation antisites in 1X disorder cases. It appears that the V_{O48f} -> V_{O48f} path has the lowest energy for 1X disorder when the vacancy moves towards the disorder, probably to relieve compressive stresses there. The V_{Oa} -> V_{O48f} path has lowest energy for the ordered structure (0X), and can activate the 0.265 nm V_{O48f} -> V_{O48f} path by inserting vacancies in that 3D network (Figure 2). The reverse path (V_{O48f} -> V_{O8a}) has lowest energy at 2X disorder, and removes vacancies from the 48f network. These results support experimental observations that moderate disorder enhances oxygen vacancy migration, but defect clustering at high disorder retards it.

More accurate results (300-315 ion cores in Region 1) are shown in Table 3 for the above preferred migration paths. The GZT calculations here used a mean field model for the combined $Zr_{0.6}$ and $Ti_{0.4}$ potentials. For consistency with experimental observations, only Zr was substituted for Gd. The conclusions are the same for all three pyrochlores. Figure 3 shows that the results are in good agreement with experimental data from Figure 8 of Reference 18. The 1.5X disorder notation for GZO is an average of results for 1X (0.59 eV) and 2X (1.40 eV).

Table 2. Oxygen vacancy migration barrier energies (eV) for GTO.

Disorder	Description	V_{O48f} -> V_{O48f}	V_{O8a} -> V_{O48f}	V_{O48f} -> V_{O8a}
0x	ordered	1.17	**0.23**	4.19
1x	towards disorder	**0.63**	3.20	3.88
1x	away from disorder	1.76	3.26	1.80
2x	high disorder	3.00	2.68	**0.35**

Table 3. Oxygen vacancy migration energies (eV) for three pyrochlores, for the three preferred cases in bold print in Table 2.

Migration Path		GTO	GZT	GZO
V_{O48f} -> V_{O48f}	(1X)	0.60	0.79	0.59
V_{O8a} -> V_{O48f}	(0X)	0.23	0.25	0.30
V_{O48f} -> V_{O8a}	(2X)	0.14	------	~0.01

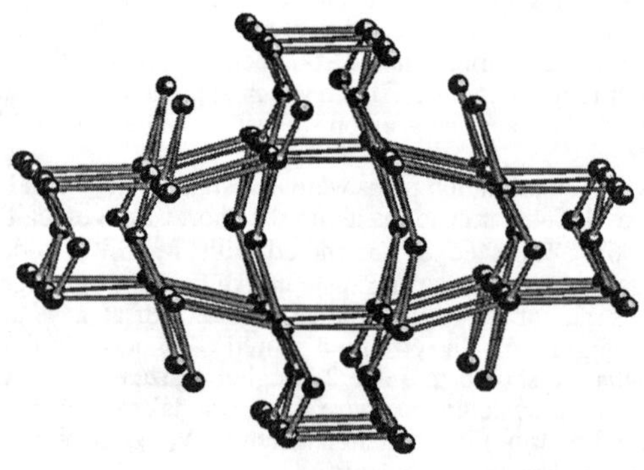

Figure 2. Three dimensional network of the oxygen migration paths along the shared, 0.265 nm TiO_6 octahedra edges of a 2x2x2 supercell. Spheres are O48f atoms. Viewed along <110>.

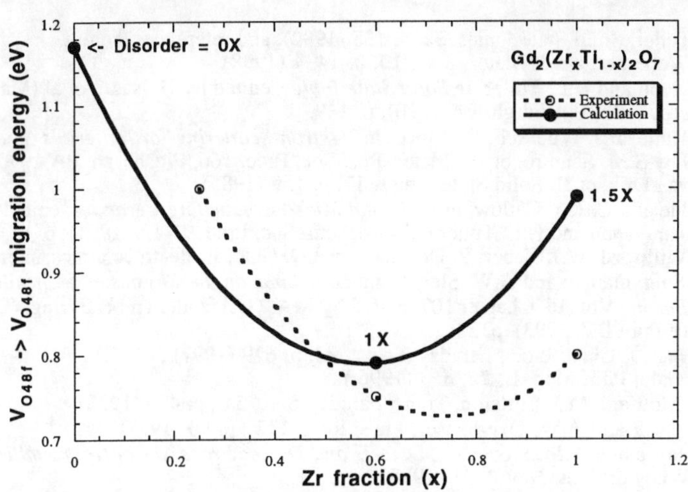

Figure 3. Comparison to data [18]. The disorder notations 0X, 1X, and 1.5X refer to 0, 1, and 1.5 pairs of cation antisite defects, respectively. The lower case x refers to Zr fraction.

Conclusions

1. Treating anions in crystallographically distinct sites (48f vs 8a) as distinct species, with appropriate polarizabilities, results in defect formation energies that are consistent with the local cationic charge environment. This 'distinct oxygen' concept also permits assessment of anisotropies in vacancy migration energetics, a unique capability for energy minimization methods.
2. For these pyrochlores, the O8a vacancy and the $V_{O8a} + I_{O8b}$ Frenkel defects are energetically preferred over the O48f vacancy and the $V_{O48f} + I_{O8b}$ Frenkel.
3. Oxygen vacancy migration is controlled by 48f-48f transitions along the shortest edges of the TiO$_6$ octahedron, in agreement with previous results.
4. The preferred direction for V_{O48f} -> V_{O48f} transitions is for vacancies to migrate towards the cation antisite disorder, probably to relieve internal compressive stresses.
5. The V_{O8a} -> V_{O48f} transition is energetically favored at moderate disorder, and activates the 48f-48f migration network by inserting vacancies there.
6. The V_{O48f} -> V_{O8a} transition is favored at high disorder, and assists in retarding oxygen vacancy migration by filling vacancies in the 48f-48f network.
7. Moderate disorder enhances oxygen vacancy migration, but high disorder retards it. Closely clustered antisite disorder increases migration barrier energies, while dispersed or open disorder reduces it.
8. Regions of local disorder and regions of local order must both exist simultaneously for enhanced oxygen vacancy migration, in agreement with experimental observations.

Acknowledgements

This work was supported by the Division of Materials Sciences, Office of Basic Energy Sciences, U.S. Department of Energy under Contract DE-AC06-76RLO 1830. The Pacific Northwest National Laboratory is operated by Battelle Memorial Institute for the U.S. Department of Energy.

References

[1] H.L. Tuller, Solid State Ionics, **52**, p. 135 (1992), and references therein.
[2] W.J. Weber, et. al., J. Mater. Res., **13**, p. 1434 (1998).
[3] P.K. Moon and H.L. Tuller, in *Solid State Ionics*, edited by G. Nazri, et al (Mater. Res. Soc. Proc. 135, Pittsburgh, PA 1990), p. 149.
[4] S.M. Haile, B.J. Wuensch, E. Prince, in *Neutron Scattering for Materials Science*, edited by S.M. Shapiro, et al (Mater. Res. Soc. Proc. 166, Pittsburgh, PA 1990) p. 81.
[5] M.P. van Dijk, et al, Solid State Ionics, **17**, p. 159 (1985).
[6] P. Wilde and C.R.A. Catlow, in *Ionic and Mixed Conducting Ceramics*, edited by T.A. Ramanarayanan and H.L. Tuller (Electrochem. Soc. Proc. 91-12, 1991), p. 18.
[7] R.E. Williford, W.J.Weber, R. Devanathan, J.D. Gale, J. Electroceramics, submitted.
[8] M.A. Subramanian and A.W. Sleight, in *Handbook on the Physics and Chemistry of Rare Earths*, Vol. 16, Chapter 107, edited by K.A. Gschneider and L Eyring, (Elsevier Science Publ BV, 1993) p. 225.
[9] J.D. Gale, J. Chem. Soc., Faraday Trans., **93**, p. 629 (1997) .
[10] J.D. Gale, Phil. Mag. B, **73**, p. 3 (1996) .
[11] N.F. Mott and M.J. Littleton, Trans. Faraday Soc., **34** , p. 485 (1938) .
[12] B.G. Dick and A.W. Overhauser, Phys. Rev., **112** , p. 90 (1958) .
[13] W. Hayes and A.M. Stoneham, *Defects and Defect Processes in Nonmetallic Solids*, John Wiley & Sons, New York, 1985.
[14] C.R.A. Catlow, in *Defects in Solids: Modern Techniques*, Plenum, NY, 1986, p. 264.
[15] G.V. Lewis and C.R.A. Catlow, J. Phys. C: Sol. State Phys., **18**, p. 1149 (1985).
[16] M.J.L. Sangster and A.M. Stoneham, Phil. Mag., **43**, p. 597 (1980).
[17] A. Dietrich, Molec. Simul., **11(5)**, p. 251 (1992).
[18] M. Spears, S. Kramer, H.L. Tuller, P.K. Moon, in *Proc. 1st Int'l. Symp. on Ionic and Mixed Conducting Ceramics*, edited by T.A. Ramanarayan and H.L. Tuller (Mater. Res. Soc., Pittsburgh, PA 1991), p. 32.

THE ROLE OF CLUSTER SIZE AND TOPOLOGY ON THE RIPENING OF DEFECT AGGREGATES IN CRYSTALLINE Si

ANTONINO LA MAGNA*, SALVATORE COFFA* AND LUCIANO COLOMBO**
*CNR-IMETEM, Stradale Primosole 50 , I-95121 Catania Italy
**Dipartimento di Scienza dei Materiali, INFM and Università degli Studi di Milano,via Emanueli 15, I-20126 Milano, Italy

ABSTRACT

·We present an 'on lattice' Monte Carlo study of defect clustering in a Si matrix. The simulations are based on local interaction models, which allows us to reliably include the energetics of defects obtained by tight binding molecular dynamic calculations. In particular the dependence of cluster stability and reactivity on cluster size and geometrical configuration is considered. Proper accelerated algorithms have been implemented in order to extend simulations to macroscopic time scales. Our simulations show that the stability of specific cluster configurations strongly affects the kinetic evolution of the defect ensemble.

INTRODUCTION

The outstanding development of computational techniques, occurred in the last few years, allows us to gain access to atomic scale phenomena by quantum mechanical calculations. Molecular Dynamics (MD) simulations provide precious information on several key parameters (e.g. diffusivities, binding energies, migration paths, etc.) ruling the evolution of a system of interacting impurities or defects[1-3]. This approach can only be used to analyze fast phenomena (lasting few ten of ps) as the early evolution of the damage caused by one implanted ion [4].

A different method is required when long lasting phenomena need to be modeled. One important example is the nucleation and growth of vacancy cluster which is observed in crystalline silicon after processes such as implantation, irradiation, large crystal growth and thermal annealing [5]. Therefore it is appealing to develop an approach which, while reproducing reliably the defects energetics obtained by MD calculation, allows to extend the simulations over macroscopic time scales.

On lattice Kinetic Monte Carlo (KMC) simulations based on local Ising type interaction models [6] are promising candidate to fulfill this task. We have shown [7-8] that a modified version of Ising model, including second neighbour vacancy interaction, can recover the main details of vacancy cluster energetics in Si emerging from Tight Binding Molecular Dynamics calculation [3]. In particular this model is able to reproduce remarkably well the non monotonic behaviour of the binding energy for the last added particle and its dependence on the assumed pattern of growth.

In this paper we will introduce two Ising type diffusion models which are useful to study the dynamics of a system of substitutional impurities or vacancies in Si. We will demonstrate that the use of accelerated KMC algorithms allows to follow structural evolution of defects in Si up to long times. Finally, by comparing the result obtained using the two different model, we will show that a reliable implementation of the MD calculations in a KMC code is mandatory in order to correctly elucidate the mechanism of defects aggregation in Si.

BINDING MODELS AND ACCELERATED MONTE CARLO ALGORITHMS

Two Ising-type models

The Ising binding model (IBM) provides a simplified atomic description of the interaction for a system of substitutional defect or impurities in a host Si lattice; its formal expression is:

$$E = -\sum_{<i,j>} \frac{E_b^1}{2} S_i S_j \qquad\qquad (1)$$

where $<i,j>$ denotes that i and j are nearest-neighbor sites. $S_i = 1$ if the site labeled by i is occupied, $S_i = 0$ if it is empty. Therefore a bond forms with binding energy E_b^1 for each pair of nearest neighbour sites occupied by two particles (see Fig.1 for the specific case of vacancy in Si). We have shown [7,8] that the energetics of a vacancy cluster in Si, as calculated by TBMD approach, is better described by an extension of IBM which includes a fully screened second neighbour (sn) interaction. This interaction should derive from a rearrangement of the dangling bonds connecting two or more Vs through Si occupied site (fig 1). Then we consider the following Extended Ising effective Model (EIM) for V coalescence:

$$E = -\sum_{<i,j>} \frac{E_b^1}{2} S_i S_j - \sum_k E_b^2(k)(1-S_k) \qquad\qquad (2)$$

where if we indicate with $n_{db}(k)$ the number of dangling bond coming from the Si ion sittings at the k site, we have:

$$E_b^2(k) = \begin{cases} 0 & \text{if } n_{db}(k) = 0,1 \\ n_{db}(k)E_b^2 & \text{if } n_{db}(k) \geq 2 \end{cases} \qquad\qquad (3)$$

Note that the model (2) leads to a screened sn effective interaction between vacancies. We choose the following values for the parameters of the models (1) and (2): $E_b^2 = 1.4$eV and $E_b^2 = 0.4$. Using these values the model (2) gives an energetics of a vacancy clusters which is in a good agreement with that calculated by means of TBMD codes [3,7,8].

Kinetic Monte Carlo algorithms

In order to explore the effects of different energetics on vacancy cluster evolution we have implemented KMC codes for the description of vacancy agglomeration.
KMC simulation for a diffusive system ruled by the models (1) and (2) is possible using a Metropolis algorithm[9]. The probability for one atom to perform a transition (i.e. a jump) during a MC step (MCS) from an initial site i to a final one f is:

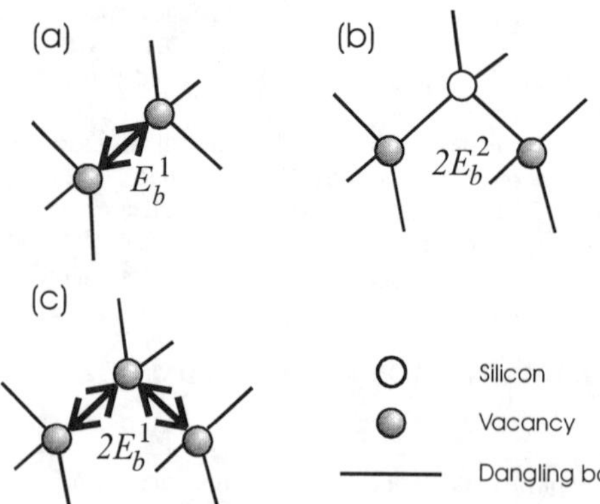

Fig.1 *Schematic description of the interaction between vacancies sitting (a) at nearest neighbour position; (b) at second neighbour position and (c) for a system containing three bound vacancies. E_b^1, $2E_b^2$ and $2E_b^1$ are the total binding energies of the three systems.*

$$p_{if} = \begin{cases} 0 & \text{if f is occupied} \\ 1/4N & \text{if f is empty and } E_f \leq E_i \\ 1/4N \times \exp[-(E_f - E_i)/kT] & \text{if f is empty and } E_f > E_i \end{cases} \qquad (4)$$

here E_i and E_f are the total energies for a system of free particles in the two configuration which differ for a jump of a particle from site i to site f, $4xN$ is the product of the four equivalent jumping direction and the total number of particles. A MCS per particle corresponds to a real time derived from the estimated particle diffusivity D: 1 MCS ~ $(a^2/6D)$, where a is the nearest neighbour distance. We have performed MC simulations of a system of vacancy ruled by IBM and EIM using Metropolis algorithm. We have considered a cubic simulation box with periodic boundary conditions. A random distribution of N MC particles has been chosen as initial condition and, when a comparison between different approaches or models will be shown, the same initial distribution has been considered. We have observed that Metropolis dynamics is able to readily follow the evolution of a system ruled by IBM or EIM till to the nucleation stage. At larger times the rejections (i.e .transitions not allowed) dominate with respect to accepted transitions and the dynamics became very slow. The huge amount of unsuccessful attempts makes Metropolis algorithm inefficient when E_b ~ 1 eV. Since this is the typical energy scale for the binding of defects and impurities, a more efficient approach is mandatory.

Using Metropolis algorithm the calculation of the trapping time $\tau(\sigma)$ spent in a given configuration σ is performed by counting repeatedly rejected moves. However, in order to compute $\tau(\sigma)$ avoiding rejections it is possible to sample it from its *a priori* known probability distribution $P[\tau(\sigma)]$. For each configuration obtained during the dynamical evolution the escape probability $p_e(\sigma)$ (i.e the probability for leaving σ in one MC step) is the sum of the probabilities p_t for all the possible transitions. Thus the probability distribution for $\tau(\sigma)$ will be

$$P[\tau(\sigma)] = p_e(\sigma)[(1 - p_e(\sigma)]^{\tau-1} \qquad (5)$$

and $\tau(\sigma)$ can be statistically evaluated. Then one of the final states is picked from its conditional probability $p_t/p_e(\sigma)$. Not all p_t (t=1,...4N) need to be calculated for each step. In fact due the locality of the models (1) and (2) this calculation could be avoided since p_t fall in a finite number of classes (i.e. they can assume a finite number of values), and a transition will produce only a 'topological' change of the p_t ensemble (n-fold way method).

Rejections are not the only source of slow dynamics. It is noteworthy to observe that a large amount of transitions performed during a dynamical evolution lead to repeated visit of a restrict number of quasi-degenerate configurations. A method which allows to take in consideration a new configuration per iteration will circumvent the problem. This task can be achieved computing the escaping time from a set $\Sigma \equiv \{\sigma\}$ of configurations (cluster algorithm). Assuming that the probability for the system, during its permanence in the cluster, to visit a given configuration σ is Bolzmann like:

$$p_{eq}(\sigma) = \frac{\exp[-E(\sigma)/kT]}{\sum_{\sigma} \exp[E(\sigma)/kT]}$$

Then the probability to move from any σ to outside the cluster Σ is

$$p_e(\Sigma) = \sum_{\sigma} p_{eq}(\sigma) p_{out}(\sigma)$$

where $p_{out}(\sigma)$ is the sum of all the probability related to transition from the configuration given to outside the cluster. Now it is obvious that the probability distribution for the escaping time from the cluster is given by the analogous of the Eq.(5), changing $p_e(\sigma)$ with $p_e(\Sigma)$. Then it is possible to sample the escaping time from the cluster and pick a new configuration in a straightforward manner.

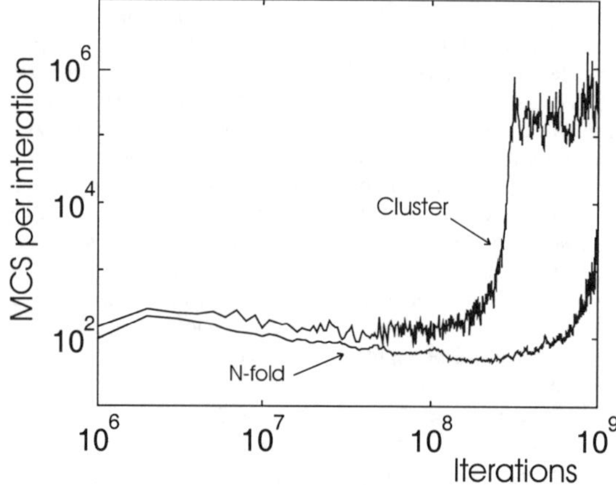

Fig.2 *Number of Monte Carlo steps per iterations corresponding to one iterations for n-fold and cluster algorithms.*

The number of MCS corresponding to one iteration of n-fold or cluster algorithm depends on the evolution stage of the system (see fig.2). Of course one iteration of the accelerate code requires a bigger CPU time with respect the Metropolis Algorithm. However the optimized cluster algorithm is, in the late evolution stage, $\sim 10^5$ faster than Metropolis in the case of IBM and $\sim 10^3$ faster in the case of the EIM.

RESULTS

The accelerated KMC algorithms described in the previous section have been used in order to follow vacancy agglomeration. In order to obtain a reliable representation of the differences between the kinetic evolution in the IBM and EIM cases, in Fig. 3 we report a comparison between long time simulation of two systems (T=900°C $\rho=10^{19}$ cm^{-3}) described by the two models. The system ruled by the IBM is frozen in a state dominated by hexagonal closed ring clusters: note that the only change in the configurations showed in Fig. 3(a,c) is the exchange of few vacancies from cluster to cluster. The system ruled by the EIM shows instead a clear tendency toward a self organization of vacancies in bigger clusters (Fig. 3 b.d). This feature is strictly related to the mobility of the aggregates [7] which is also well visible in the snapshots. Second neighbour vacancy interaction is responsible of this effective mobility of the aggregates as a result of successive detachment and recapture of single vacancies lying at the cluster surface. Cluster size distributions (see fig.4 for a system with $\rho = 10^{20}$ cm^{-3} and T =900°C) show significant difference between the two models as result of the different ripening mechanism. The emerging features of distribution for the IBM are the clear peaks at sizes (6,10,14...) corresponding to the close ring aggregates. This behavior, which produces a severe bottleneck for ripening of voids, appears during the time evolution soon after the dissolution of smaller

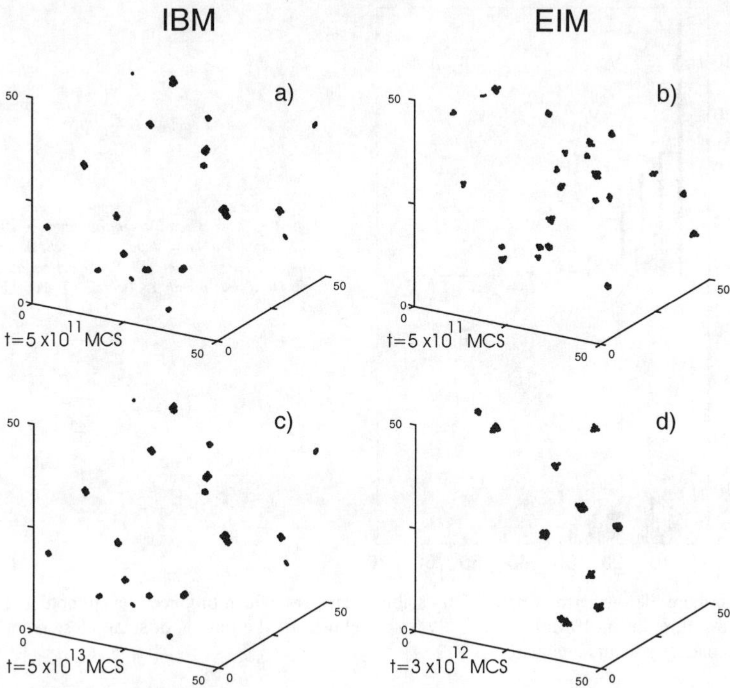

Fig.3 Snapshots of vacancy cluster obtained by Ising Binding Model (a,c) and by Extended Ising Model (b-d) for a system with $\rho=10^{19}cm^{-3}$ and $T=900°C$.

cluster (up to V_5) when the clusters with size n= 7,8,9,11,12... become the main source of free particles which drive the ripening process.

The kinetics appears strongly modified in the EIM case. The distribution shows an increased spreading in cluster size and reveals the presence of very large clusters (containing as much as 63 vacancies). We stress again that the cause of this features is the mobility of small vacancy-aggregates which have an essential role in the cluster self-organization and it boosts considerably the ripening process. Note that the main features of vacancy agglomeration reported above remain unchanged when changing the vacancy density and the annealing temperature [7].

CONCLUSIONS

We have developed accelerated KMC algorithms that can be reliably used to simulate vacancy agglomeration in c-Si up to long times. Using these novel codes we have analyzed the ripening mechanism achieved when using two different binding models: a simple Ising type model (IBM) and a refined Ising model (EIM) taking into account second neighbour interaction. It has been found that void evolution does not occur through a standard Ostwald ripening but rather through a *self organization* of micro voids mediated by the effective mobility of small clusters.

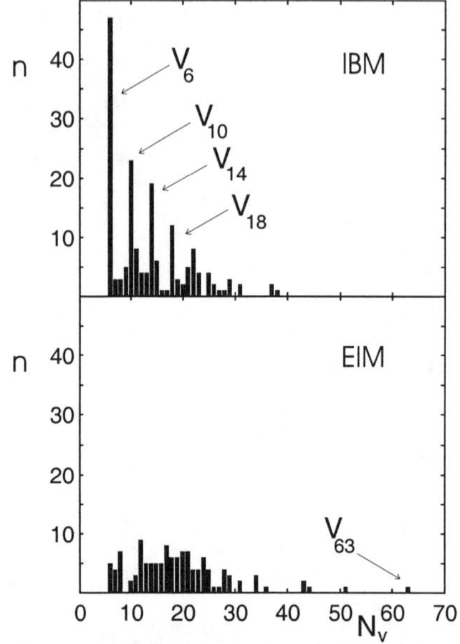

Fig. 4 *Cluster size distribution after* $\sim 10^{12}$ *MCS using an Ising type model (IBM) and extended Ising model (EIM). The simulation refers to a system with* $\rho = 10^{20}$ cm^{-3} *T=900°C*

Furthermore the presence of particularly stable V clusters, which produces severe bottleneck for void evolution using IBM, is completely overwhelmed by the motion of small clusters induced by second neighbours interaction.

ACKNOWLEDGEMENTS

We would like to acknowledge E. Rimini for several useful discussion and Nicolò Parasole for expert technical assistance. This work has been partially supported by Progetto 5% Microelettronica.

REFERENCES

[1] M. Tang, L. Colombo, J. Zhu and T. D. de la Rubia, Phys. Rev. B **55**, 14279 (1997).

[2] N. Arai, S. Takeda, M. Kohyama, Phys. Rev. Lett. **78**, 4265 (1997).

[3] A. Bongiorno, L. Colombo and T. Diaz de la Rubia, Europhysics Letters **43,** 695 (1998).

[4] G.H. Gilmer, T. Diaz de la Rubia, D.M. Stock and M Jaraiz, Nucl.Instr.Meth. B **102**, 247 (1995)

[5] S. M. Myers, G. A. Petersen, and C. H. Seager, J. Appl. Phys. **80**, 3717 (1996); V. Raineri, M. Saggio, Appl. Phys. Lett. **71**, 1673 (1997)

[6] M. Strobel, K. H. Heinig, W. Moller, Computational Materials Science **494**, 138(1997)

[7] A. La Magna, S. Coffa and L. Colombo, submitted to PRL.

[8] A. La Magna, S. Coffa and L. Colombo Nucl.Instr.Meth. B in press.

[9] N. Metropolis, A. W. Rosenbluth, M. N. Rosenbluth, A. H. Teller, E. Teller, J. Chem. Phys. **21**, 1087 (1953).

Molecular dynamics simulations of crystal growth from melted silicon: Defect formation processes

Manabu Ishimaru and Teruaki Motooka
Department of Materials Science and Engineering, Kyushu University,
Hakozaki, Fukuoka 812-8581, Japan

Abstract

Molecular dynamics calculations have been performed to simulate crystal growth from melted silicon (Si) and defect formation processes based on the ordinary Langevin equation employing the Tersoff interatomic potential. The findings of this investigation are as follows: (i) The [110] bonds at the solid-liquid interface induce the eclipsed configurations or hexagonal Si structures which stabilize microfacets composed of the {111} planes. (ii) Defect formation during crystal growth processes is due to misorientations at the {111} interfaces which result in an "elementary" grown-in defect structure including five- and seven-member rings. (iii) The "elementary" grown-in defect migrates in c-Si by bond-switching motions during further crystal pulling or annealing.

Introduction

As the size of Si integrated circuits (ICs) has shrunk down, the single-crystal Si (c-Si) wafers on which ICs are fabricated have grown in size to reduce the cost of the production. The most common sizes for the Si wafers now are 6 and 8 inches in diameter. However, in the future, the usage of 12 and 16 inch wafers are expected to be the norm. These c-Si wafers are mostly produced from the liquid phase by using the pulling or floating-zone method, and various factors such as crystal growth rates, temperature gradients at the c-Si and liquid Si (l-Si) interface and impurity concentrations are known to affect the quality of the Si ingots produced from the melted Si.

The technique of wafer production has relied on the trial and error method. As the sizes of the wafers increase, the cost of a single pulling trial as well as the investment in the facilities for the pulling become non negligible. Understanding the crystal growth mechanism is therefore of technological importance to establish the optimum parameters for obtaining good quality c-Si. However, the details of the crystal growth from the melted Si remain an open question, because high melting temperature and intensive reactivity of l-Si have inhibited extensive experimental studies. To overcome this problem, the application of computer simulations is very useful to obtain atomistic information on the crystal growth from l-Si. Several investigators[1-7] have actually performed molecular dynamics (MD) simulations of crystal growth from melted Si. However, no investigations have been conducted for the defect formation processes during crystal growth in the [001] direction which is the most popular pulling direction in production of commercial Si wafers.

In the present study, we carried out MD simulations based on Langevin equations in order to clarify the defect structures as well as their formation processes during the crystal growth from melted Si.

Simulation procedures

Since the enthalpy is essentially the same as the internal energy at atmospheric pressure for condensed systems as l-Si and c-Si, we have performed MD simulations under a constant volume. The MD cell with the constant volume of $62.0 \times 62.0 \times 42.0$ Å^3 (tetragonal prism) includes 9216 Si atoms, which gives the density of l-Si, 2.58 g/cm^3. For the calculation of the atomic trajectories, our MD simulations utilizes interatomic potentials developed by Tersoff[8] as well as the Langevin equations which consist of interatomic force, random force, and viscosity terms.[9,10] The Tersoff potential has been known to reproduce the structural and dynamical properties of amorphous Si and l-Si very well.[11-13] We employed the scheme developed by van Gusteren and Berendsen[14] for numerical integration of Langevin equations using a time step of 2×10^{-3} ps and a friction constant of 5.0 ps^{-1}.

For MD simulations of crystal growth on (001) faces, the MD cells were immersed in a thermal bath with a temperature gradient in the [001] or Z direction and these cells were pulled in the Z direction with a constant speed. Periodic boundary conditions were employed in the X and Y directions. In the Z direction, the atoms in the top two layers were fixed and the bottom was

Mat. Res. Soc. Symp. Proc. Vol. 538 © 1999 Materials Research Society

assumed to be a free surface. A reflection wall was set at 2 Å below the bottom plane to prevent atoms from escaping from the MD cells. The details of the computational procedure are described elsewhere.[9,10]

Results and discussion

Figure 1 shows the atomic arrangements in (a) the initial stage of the crystallization process, (b) after 267 ps, and (c) after 533 ps. These figures were obtained by projecting all atomic positions in selected regions of the MD cells on the ($\bar{1}$10) plane. The pulling rate 12 m/s and temperature gradient 15 K/Å of the present simulations correspond to the cooling rate of 1.5×10^{12} K/s. Although melted Si transforms into an amorphous phase at this quenching rate,[13] crystallization occurs due to the existence of the c-Si part.

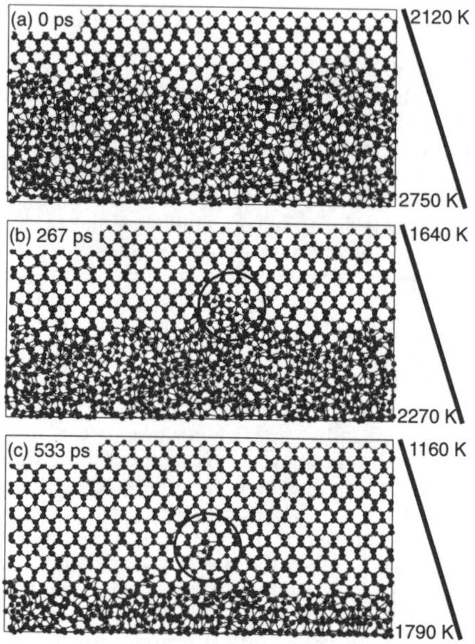

The crystal-melt interface is initially located around 2300-2400 K [Fig. 1(a)] and during the crystal pulling the interface is stabilized by forming the {111} faces as illustrated in Fig. 1(b). In Fig. 1(b), the solid-liquid interface exists around ~2000 K and l-Si is supercooled. It should be noted that the crystallized region enclosed in Fig. 1(b) includes the atomic configuration which is different from that of (001)Si. This region possesses a different structure from that of the surrounding c-Si region and causes microfacets composed of primarily {111} planes. Most of the misoriented configurations are annihilated during the further growth due to back melting or *dynamical self-annealing* as pointed out by Landman *et al.*[3] In Fig. 1(c), l-Si transforms into glass for rapid quenching rates and the crystallization can no longer occur. In previous

Figure 1. Three snapshots taken during crystal growth in the [001] direction: (a) initial stage of the crystallization process, (b) after 267 ps, and (c) after 533 ps. All atomic positions are projected on the ($\bar{1}$10) plane. Solid lines of the right-hand sides denote temperature distributions at each stage of the crystallization.

MD simulations using ~1000 Si atoms,[1-3] no isolated defects were created in c-Si. In the present study, on the other hand, we succeeded in generating an "elementary" grown-in defect for the first time by using large-scale MD calculations (~10000 Si) [Fig. 1(c)]. This defect is due to the misoriented configurations described above.

To obtain more detailed structures of the crystal-liquid interfaces, magnified pictures of the regions encircled in Fig. 1(b) are given in Fig. 2(a). It should be noted that the bonds along the [110] direction are formed at the solid-melt interface. The cubic Si possesses the staggered configuration [Fig. 2(b)], while there exists the eclipsed configuration [Fig. 2(c)] near the solid-liquid interface. These eclipsed configurations induce the hexagonal Si structure indicated by the black circles. This hexagonal structure is locally stable, because the nearest neighbor structure of hexagonal Si is the same as that of cubic Si. Therefore, these regions disturb further (001) crystallization and induce prominent {111} microfacets as seen in Fig. 1(b).

A magnified view of the "elementary" grown-in defect in Fig. 1(c) is illustrated in Fig. 3(a). It should be noted that the obtained defect is not the single-vacancy which is considered as an elementary defect in c-Si. It can be seen that several atoms shift from the single c-Si sites and five- and seven-member rings exist in Fig. 3(a). The stacking sequence around the defect possesses the eclipse configuration, suggesting that the defect formation during crystal growth processes is due to misorientations at the {111} interfaces as described above. The bond lengths and angles between the first nearest neighbors around the defects were confirmed to be almost equal to those of c-Si,

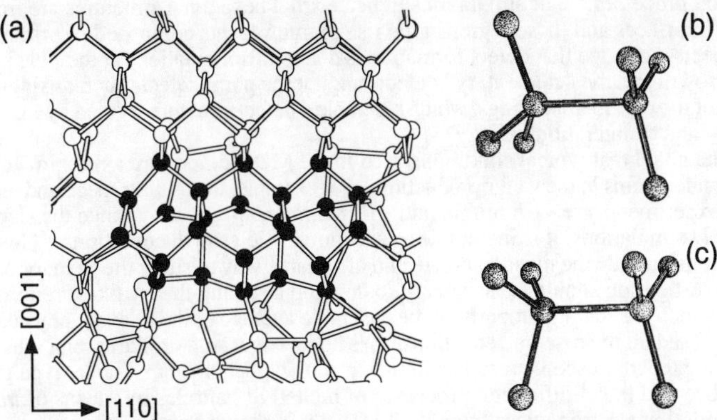

Figure 2. (a) Atomic arrangements near the crystal-melt interface encircled in Fig. 1(b). A bond is drawn if its distance is within a cutoff of 2.90 Å. (b) Staggered and (c) eclipsed configurations. It can be seen that the epitaxially grown hexagonal Si structures (black circles) appear near the solid-liquid interfaces.

thus the tetrahedral structure is still maintained around each atom. It was confirmed that the number of atoms is smaller than that of the lattice site,[10] thus this defect is not recovered during further crystallization or annealing. Figure 3(b) displays the atomic arrangement obtained by the further pulling. It can be seen that the bond between A and B in Fig. 3(a) is broken and atom A occupies the diamond lattice site, suggesting that the defect migrates in c-Si by bond-switching motions.

It is known that the vacancy-type defects, such as Crystal Originated Particle (COP), Flow Pattern Defect (FPD), and Laser Scattering Tomography Defect (LSTD), exist in c-Si grown from melted Si. It is generally considered that these defects are formed by the aggregation of single-vacancies. However, electron-spin-resonance signals of single-vacancy have never been observed in Czochralski-grown c-Si wafers, therefore the formation processes of vacancy-type defects are still unclear. The "elementary" grown-in defect obtained by the present MD simulation can explain the formation mechanisms of vacancy-type defects in c-Si grown from the melt.

Summary

Molecular-dynamics simulations were performed to investigate the defect structures as well

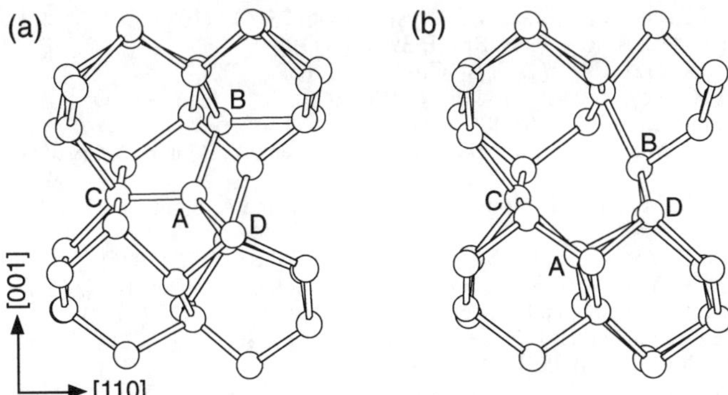

Figure 3. Magnified pictures around the "elementary" grown-in defect.. After the crystallization for (a) 533 and (b) 600 ps. It shoul be noted that the "elementary" grown-in defect is not annhilated during the further crystallization, and migrates in c-Si by bond switching motions.

as their formation processes. Our simulations indicated that hexagonal structures are formed near the solid-liquid interfaces and these regions give rise to microfacets composed of primarily {111} planes. It has been suggested that defect formation is due to misorientations at the {111} interfaces during crystal growth and the "elementary" defects are not the point defects such as single vacancy but they consist of the odd-member rings, which resemble a defect structure formed by a combination of the divacancy and di-interstitial pair.[15,16]

It should be noted that typical conditions used in the MD methods are approximately 10^4-10^7 off those of the real systems in the wafer production. For example, the pulling speed and temperature gradient in the experiments are ~1 mm/min and ~10 K/mm, respectively. Since the size and time are limited in MD simulations, it is inevitable to use unrealistic growth conditions. Nevertheless, the MD simulations provide the most complete and affordable way to study the Si ingot production. Thus, further investigation should be necessary to develop a scaling theory like a renormalization group theory in which the scaling property of the ensemble average of the relevant physical quantity can be obtained based on microscopic equations. We believe that self-similarity plays an important role for connecting the microscopic results with macroscopic phenomena. Fractal type growth has been actually observed in solidification processes of melted Si particles by means of *in-situ* high resolution transmission electron microscopy.[17] It is also important to investigate oxygen effects on crystal growth and defect formation, since ~10^{18} /cm^3 oxygen is included in Czochralski grown Si. More extensive study is currently underway.

Acknowledgments

This work was conducted as JSPS Research for the Future Program in the Area of Atomic-Scale Surface and Interface Dynamics under the project of "Dynamic Behavior of Silicon Atoms, Lattice Defects and Impurities near Silicon Melt-crystal Interface". We would like to thank T. Kumamoto, K. Yoshida, S. Munetoh, and K. Nishihira for their assistance on molecular-dynamics calculations. Part of this work is the results of "Technology for Production of High Quality Crystal" which is supported by the New Energy and Industrial Technology Development Organization through the Japan Space Utilization Promotion Center in the program of the Ministry of International Trade and Industry. One of us (MI) also acknowledges the partial support from the Sumitomo Foundation and Iketani Science and Technology Foundation.

References
[1] U. Landman, W. D. Luedtke, R. N. Barnett, C. L. Cleveland, M. W. Ribarsky, E. Arnold, S. Ramesh, H. Baumgart, A. Martinez, and B. Khan, Phys. Rev. Lett. **56**, 155 (1986).
[2] U. Landman, W. D. Luedtke, M. W. Ribarsky, R. N. Barnett, and C. L. Cleveland, Phys. Rev. B **37**, 4637 (1988).
[3] W. D. Luedtke, U. Landman, M. W. Ribarsky, R. N. Barnett, and C. L. Cleveland, Phys. Rev. B **37**, 4647 (1988).
[4] J. Q. Broughton and F. F. Abraham, J. Cryst. Growth **75**, 613 (1986).
[5] M. D. Kluge and J. R. Ray, Phys. Rev. B **39**, 1738 (1989).
[6] F. X. Kelly and L. H. Unger, J. Cryst. Growth **102**, 658 (1990).
[7] Q. Yu and P. Clancy, J. Cryst. Growth **149**, 45 (1995).
[8] J. Tersoff, Phys. Rev. B **49**, 16349 (1994).
[9] M. Ishimaru, S. Munetoh, T. Motooka, K. Moriguchi, and A. Shintani, J. Cryst. Growth **194**, 178 (1998).
[10] M. Ishimaru, S. Munetoh, T. Motooka, K. Moriguchi, and A. Shintani, Phys. Rev. B **58**, 12583 (1998).
[11] M. Ishimaru, K. Yoshida, and T. Motooka, Phys. Rev. B **53**, 7176 (1996).
[12] M. Ishimaru, K. Yoshida, T. Kumamoto, and T. Motooka, Phys. Rev. B **54**, 4638 (1996).
[13] M. Ishimaru, S. Munetoh, and T. Motooka, Phys. Rev. B **56**, 15133 (1997).
[14] W. F. van Gunsteren and H. J. C. Berendsen, Mol. Phys. **45**, 637 (1982).
[15] T. Motooka, Phys. Rev. B **49**, 16367 (1994).
[16] T. Motooka, S. Harada, M. Ishimaru, Phys. Rev. Lett. **78**, 2980 (1997).
[17] R. Oshima, F. Hori, T. Ueno, N. Yaguchi, M. Komatsu, and H. Mori, Proc. of the Fifty Third Annual Meeting of the Japanese Society of Electron Microscopy, p. 263 (in Japanese).

A STUDY OF GRAIN GROWTH AND MICROSTRUCTURE CONTROL
IN SILICON NITRIDE BY COMPUTER SIMULATION

Y. OKAMOTO, N. HIROSAKI AND H. MATSUBARA

Synergy Ceramics Research Laboratory, Fine Ceramics Research Association

2-4-1, Mutsuno, Atsuta-ku, Nagoya-shi, Aichi-ken 456-8587, Japan

ABSTRACT

A grain growth simulation technique is developed for microstructure design of ceramic materials by extending the Potts model. The simulation model has successfully reproduced the self-reinforced microstructure which is peculiar to silicon nitride, consists of large elongated grains and small equiaxed grains.

To control the microstructure of silicon nitride further, a seeding technique is effective and often employed. Grain growth behavior of seed particles in the simulation is studied. It is found that the seed grains grow with a relatively high growth rate during the early stage of grain growth. In contrast in the later stage, the difference between microstructures with and without seed particles decreases. This behavior agrees well with that of actual silicon nitride. These results demonstrate the consistency of the simulation.

INTRODUCTION

During sintering of silicon nitride, several grains grow preferentially to form coarse rod-like grains, and these elongated grains act like a toughening second phase. The diameters of elongated coarse grains (typically 5 to 20 μm) are extremely large compared to that of the matrix grains ($\cong 1$ μm). This phenomenon improves the mechanical properties of silicon nitride and is often referred to as "self-reinforcement" or "*in-situ* toughening" [1-4]. To promote self-reinforcement, seeding technique is often employed [3,5-7]. Artificially added large grains grow preferentially over small grains, and the addition results in self-reinforced microstructures within a shorter sintering time. To optimize the effect of seeding, precise microstructure design is required. However, attempts at microstructure design in the past have depended empirical, experimental methods.

Grain growth simulation is a promising technique for analyzing grain growth behavior and for designing microstructure design tool. The Potts model [8,9] is one of the most versatile grain growth models for Monte Carlo simulation. Furthermore, the present authors have extended the Potts model to include a liquid phase between grains and anisotropy of crystal growth [10-12], which are thought to be the principal characteristics of silicon nitride. Calculations with this model have successfully reproduced the self-reinforced microstructure.

In this paper, the grain growth behavior of artificially added large grains (seed particles) in the simulation was investigated. The consistency of the grain growth simulation was also discussed.

SIMULATION METHODOLOGY

Simulation model and algorithm

The simulation model is based on the Potts model [8,9]. Additional features are presence of liquid phase and anisotropy of interface energy (fig. 1). Details of the simulation algorithm are given elsewhere [11,12]. The algorithm essentially consists of four steps: (1) Selection of a cell at random from the simulation lattice (2) Change in the orientation number of the selected cell. If the cell is adjacent to liquid phase, also change the cell's position via a diffusion process (random walk) until it hits another solid cell. (3) Calculation of the change of total interface energy. (4) If the total energy decreases or does not change, the orientation / position change of the cell is successful.

Mat. Res. Soc. Symp. Proc. Vol. 538 © 1999 Materials Research Society

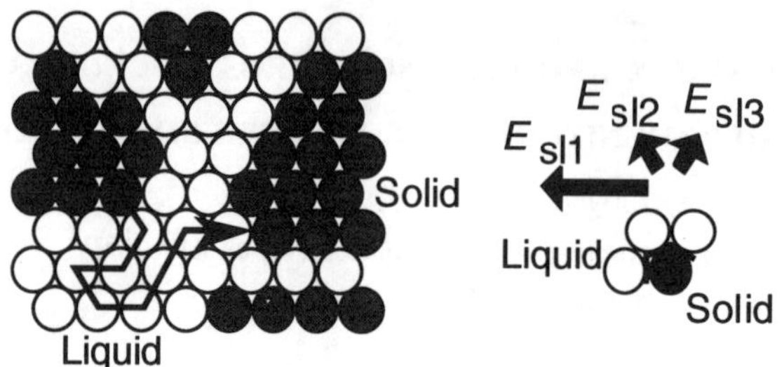

Fig. 1. Simulation model used in this study. Diffusion through the liquid phase is modeled by a random-walk process through liquid phase cells. Anisotropy is introduced by assigning three different interface energies E_{sl1}, E_{sl2}, E_{sl3} for the three possible directions [12].

In step (2) of the above procedure, when a cell hits another solid cell after the random walk, the cell position immediately before the collision is changed to a solid cell and given the same orientation number as the one that was hit. The cell from which the random walk originated is replaced by liquid phase.

Actual silicon nitride contains two main types of interface; one is a triple junction and the other is liquid thin film. In this simulation, thin liquid film was assumed to be present at interfaces between solid cells. Thus the mass transport through a solid to solid interface was assumed to be a solution - reprecipitation process, similar to mass transport through triple junctions. When a solid cell surrounded by solid cells was selected, its new orientation number was chosen from one of the surrounding cells. These two operations are the major differences from the original Potts model.

Calculation

Simulations were carried out using a 2-dimensional triangular lattice. Periodic boundary conditions were imposed. Orientation numbers Q, ranging from 1 to 64, were distributed to each cell at random. These cells corresponded to solid phase. Cells of orientation number 0 were also introduced into the lattice. These cells corresponded to liquid phase.

First, "matrix textures" were prepared from the random structure by carrying out a short simulation for 10 MCS (Monte Carlo Steps). For this calculation, the interface energies assigned were $E_{ss} = 1$ and E_{sl1}, E_{sl2}, E_{sl3} equal to 0.5, 0.5, 0.5, respectively, which gave isotropic grain growth. E_{ss} is the interface energy between solid and solid. Average grain diameters after 10 MCS were 5.1 cells (liquid phase 5%), 4.9 cells (10%) and 4.8 cells (20%). After that, we embedded large grains (seed grains) in the middle of each matrix.

The simulations of grain growth behavior in the seeded material started from these microstructures. Interface energies were assigned as follows, $E_{ss} = 1$ and E_{sl1}, E_{sl2}, $E_{sl3} = 0.1$, 0.1, 0.5, respectively. The direction of the higher energy for solid - liquid interface 0.5 was assigned according to the orientation number, Q. The size of the simulation lattice was 500 cells by 500 cells. It should be noted that no special grain, $e.g.$ lower interface energies or higher boundary mobility, was assumed in this simulation.

RESULTS AND DISCUSSION

First, simulations with only one seed grain were carried out. An oval shaped seed grain was embedded in matrices of various liquid phase fractions. The diameter of the seed was 20 cells, which was nearly four times larger than the average grain size of the matrix. This situation is not so different from the actual process [5,13]. Length and width of the growing seed were measured every 20 MCS.

A typical example of microstructure development is shown in fig. 2. The embedded seed grain grew preferentially. Figure 3 shows the result of quantitative analysis (net length increase). For small amount of liquid phase (5%), the growth was hindered after 50 MCS. For higher liquid phase fraction (10%), the seed grains grew smoothly and most rapidly. When the amount of liquid phase was increased further (20%), the growth rate was reduced, indicating some saturation. However, average diameters over all grains at 500 MCS were 7.4, 7.4, 7.5 cells for 5% liquid phase fraction (for aspect ratio = 2, 5, 10, respectively), 8.1, 8.1, 8.2 cells for 10% and 11.9, 12.1, 11.9 cells for 20%. The ratio of net length increase to average diameter was the largest for 10% liquid phase fraction. This means that the effect of seeding is optimized for a certain amount of liquid phase (around 10%). This result is consistent with our previous study, in which self-reinforcement was most remarkable around 10% liquid phase fraction [11,12].

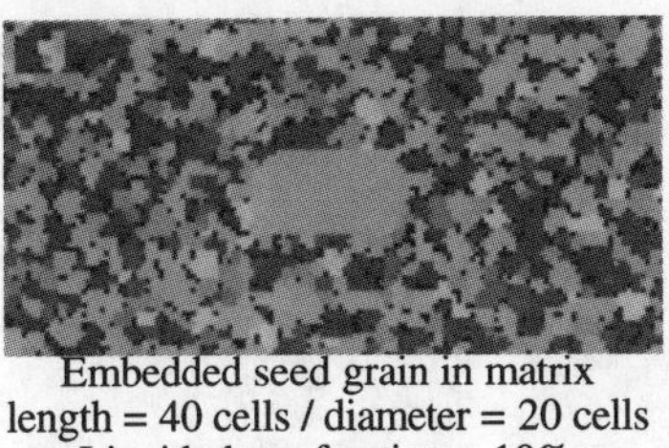

Embedded seed grain in matrix
length = 40 cells / diameter = 20 cells
Liquid phase fraction = 10%

after 500 MCS
length = 84 cells / diameter = 44 cells

Fig. 2. Growth of a seed grain in 10% liquid phase fraction matrix

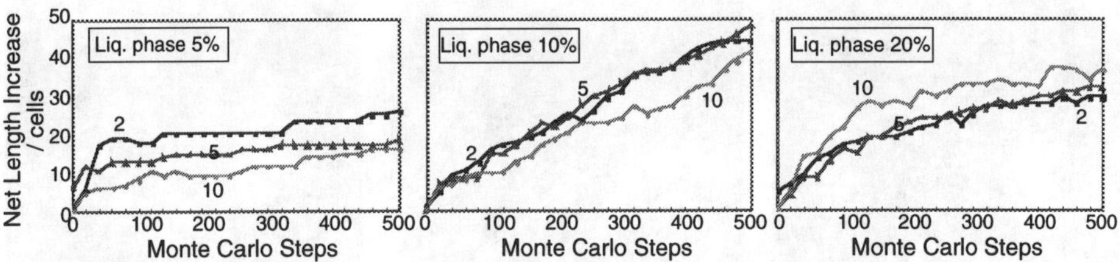

Fig. 3. Growth behavior of seed grains. Parameters indicate the aspect ratio of the seeds.

The next problem considered was that of microstructures containing multiple seeds, as a study of microstructure control. Seed grains of diameter 20 cells and length 100 cells were placed in the matrix. Positions and directions of each seed were assigned at random. The liquid phase fraction chosen was 10 %. A large number of seed grains grew preferentially and promoted the self-reinforcement (fig. 4). As stated in the introduction, a mixture of elongated large grains and equiaxed small grains improves the mechanical properties of silicon nitride. In other words, inhomogeneity in grain size distribution is a significant factor of microstructure. In the present study, we calculated the second order moment for evaluation of inhomogeneity. The second order moment, μ_2, is defined as;

$$\mu_2 = \int_0^\infty (\frac{r}{r_0})^2 \cdot f(r)dr \tag{1}$$

where r is grain diameter (or width), r_0 is average diameter and $f(r)$ is distribution function. The larger value of μ_2 indicates a more inhomogeneous microstructure, and the minimum value is unity (completely homogeneous microstructure). It is clearly revealed in fig. 5 that the addition of certain amount of seed grains promoted self-reinforcement in the simulation, in particular, during the early stage of grain growth.

It has also been found experimentally that addition of a small amount of coarse particles (seeds) also changes the microstructure drastically. For example, addition of 5 mass% of coarse particles in silicon nitride raw powder resulted in a bimodal grain size distribution, and the fraction of large grains reached 40% or more (after sintering at 2173 K, 4 h) [13]. Since the grain size distribution of the reference sample (no seed) was nearly log-normal and the fraction of large grains was less than 5%, it is obvious that the added coarse particles grew preferentially and promoted self-reinforcement in the real material. The results of our simulation are consistent with these experimental results. The area fraction of large grains in 500 MCS were 4%, 15%, 26% and 46% for 0%, 4%, 8% and 16% seed addition respectively, where the "large" grain was defined in the same manner as in ref.[13]. Finally, it should be noted that the self-reinforcement occurred after a long time, even if no seed was added. This behavior is also consistent with real silicon nitride.

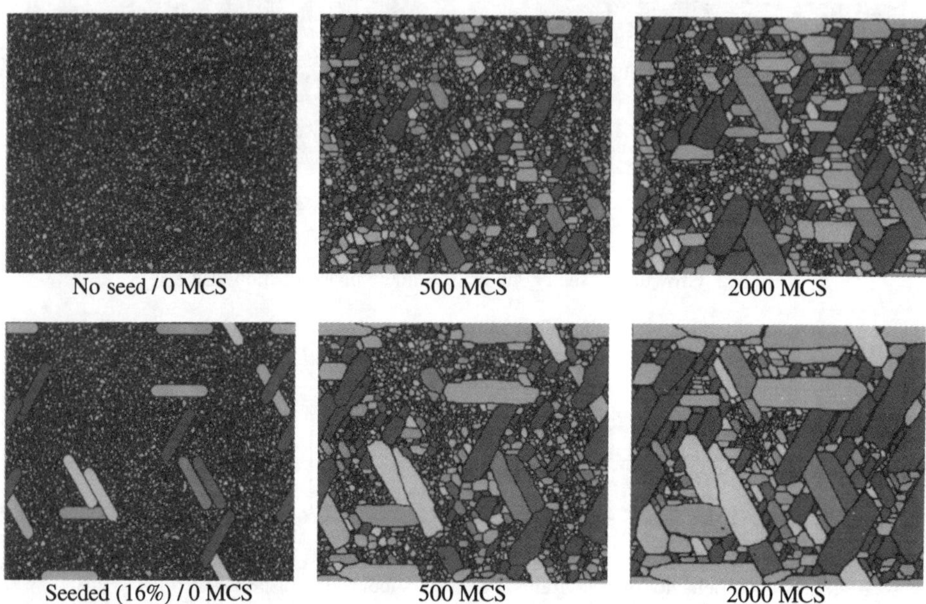

Fig. 4. Evolution of microstructures with and without seed particles. Seed grains promoted self-reinforcement during the early stage (middle), however the difference between cases with and without seed decreased during the later stage (right).

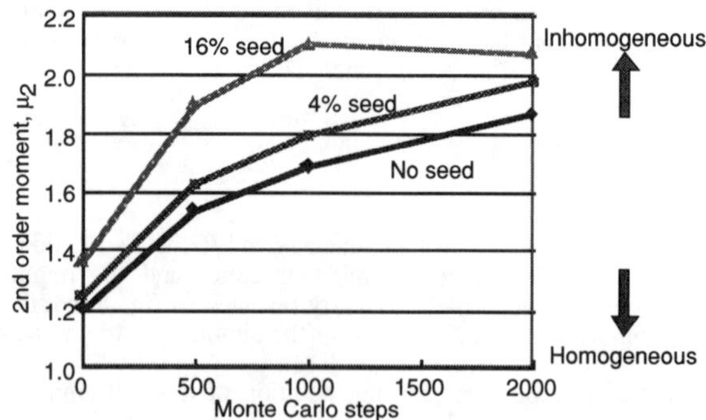

Fig. 5. Inhomogeneity of microstructures for various seed contents. Retardation of self-reinforcement was observed for higher seed contents during the later stage.

CONCLUSION

A grain growth simulation technique has been proposed for microstructure design of silicon nitride, and grain growth behavior of artificially added large grains was studied. It was found that the seed grains grow with a relatively high growth rate during the early stage of grain growth. The growth was retarded during the later stage, and the difference between microstructures with and without seed particles became smaller. This behavior agrees well with that of actual silicon nitride. These results demonstrate the consistency of the simulation, and furthermore, the possibility of using simulation methods for designing complex microstructures in ceramics.

ACKNOWLEDGMENT

This work has been entrusted by NEDO as part of the Synergy Ceramics Project under the Industrial Science and Technology Frontier (ISTF) Program promoted by AIST, MITI, Japan.

REFERENCES

1. E. Tani, S. Umebayashi, K. Kishi, K. Kobayashi and M. Nishijima, *Am. Ceram. Soc. Bull.*, **65**, 1311-15 (1986).
2. M. Mitomo and S. Uenosono, *J. Am. Ceram. Soc.*, **75**, 103-08 (1992).
3. D. E. Wittmer, D. Doshi and T. E. Paulson, in *Proceedings of 4th International Symposium on Ceramic Material and Components for Engines*, Elsevier Science Publishers, (1992) pp. 594-602.
4. A. J. Pyzik and D. R. Beaman, *J. Am. Ceram. Soc.*, **76**, 2737-44 (1993).
5. H. Emoto and M. Mitomo, *J. Euro. Ceram. Soc.*, **17**, 797-804 (1997).
6. K. Hirao, T. Nagaoka, M. E. Brito and S. Kanzaki, *J. Am. Ceram. Soc.*, **77**, 1857-62 (1994).
7. H. Feuer, G. Wötting and E. Gugel, in *Key Engineering Materials*, **89-91**, (Trans Tech Publications, 1994) pp. 123-28.
8. M. P. Anderson, D. J. Srolovitz, G. S. Grest and P. S. Sahni, *Acta metall.*, **32**, 783-91 (1984).
9. D. J. Srolovitz, M. P. Anderson, P. S. Sahni and G. S. Grest, *Acta metall.*, **32**, 793-802 (1984).
10. H. Matsubara and R. J. Brook, in *Mass and Charge Transport in Ceramics*, edited by K. Koumoto, L. M. Sheppard and H. Matsubara, (Ceramic Transactions **71**, Westerville, OH, 1996) pp. 403-17.
11. Y. Okamoto, N. Hirosaki and H. Matsubara, *J. Ceram. Soc. Japan*, **107**, 109-14 (1999).
12. Y. Okamoto, N. Hirosaki and H. Matsubara, in *Computational and Mathematical Models of Microstructural Evolution*, edited by J. W. Bullard, R. Kalia, M. Stoneham and L-Q. Chen, (Mater. Res. Soc. Proc. **529**, Warrendale, PA, 1998) pp. 85-88.
13. N. Hirosaki, Y. Akimune and M. Mitomo, *J. Am. Ceram. Soc.*, **77**, 1093-97 (1994).

A UNIFIED COMPUTATIONAL APPROACH TO OXIDE AGING PROCESSES

Harold P. Hjalmarson, Peter A. Schultz, Duane J. Bowman, and Daniel M. Fleetwood
Sandia National Laboratories, Albuquerque, NM 87111

ABSTRACT

In this paper we describe a unified, hierarchical computational approach to aging and reliability problems caused by materials changes in the oxide layers of Si-based microelectronic
devices. We apply this method to a particular low-dose-rate radiation effects problem.

INTRODUCTION

The oxide SiO_2 is both an electrical insulator and a protective coating for Si-based microelectronics devices. This oxide has been an ongoing source of technical problems involving
defects, radiation effects, and degradation. [1] Defects are intrinsic to the oxide because the
thermal growth process produces a non-crystalline, amorphous oxide. The main defects are
oxygen vacancies and P_b centers which are Si dangling bonds at the Si/SiO_2 interface. [2]
Both types of defects have been extensively studied.

The main radiation problems are due to holes which can become trapped to form charged
defects, and the generation of interface traps via the interaction of hydrogen related species
at the Si/SiO_2 interface. The resultant trapped charge can alter the device properties. For
example, it can change the turn-on voltage of a transistor.

Recently, a low-dose-rate radiation problem has been identified in certain types of devices
such as bipolar transistors in linear integrated circuits used in spacecraft, satellite, and
weapons applications. [3–7] This is a problem which may take years to develop and thus
we classify it as an aging problem. Studies have shown that for these devices a given total
ionizing radiation dose is more harmful if the radiation rate is reduced to a level which is
much lower than that used for baseline testing. [3]

The low-dose-rate problem seems to occur in the passivation or field oxides covering the
emitter-base junction in lateral or substrate bipolar transistors. [8–10] Device modeling has
shown that increased net positive oxide-trap charge may lead to increased surface recombination in the base of a transistor. [8] In turn, this affects the transistor gain, an effect
consistent with the measured modification of the device characteristics. However, the physical mechanism responsible for the buildup of excess trapped positive charge at low dose
rates is not clear.

The absence of a clearcut mechanism leads to time-consuming and expensive testing.
Because of the slowness of the radiation exposure, a comprehensive test of a component can
take up to six months or more to complete.

APPROACH TO THE PROBLEM

The essence of the computational problem is to compute the radiation generated trapped
charge, compute the effect of this charge on the performance of microelectronic devices such
as transistors, and compute the effect of device performance on the circuit performance.
Our approach to these problems is an integrated, multiple level approach illustrated in Fig.

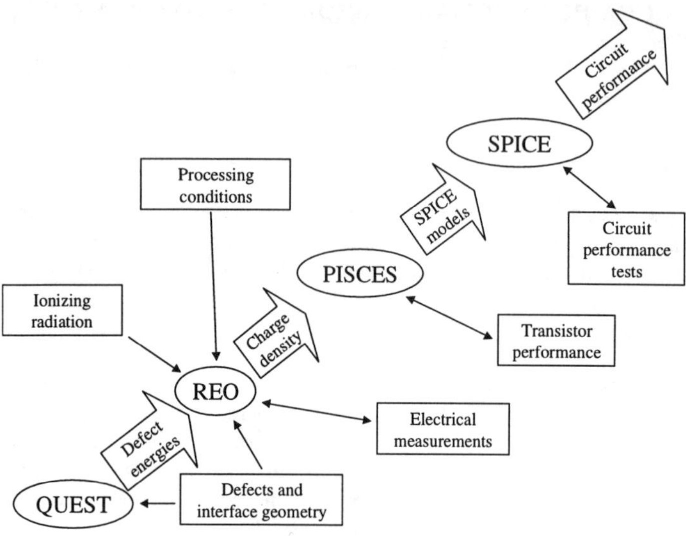

Figure 1: Shows the computational approach.

1. This approach takes into account the fact that the aging problem has multiple length and time scales; these will be described in more detail. Furthermore, the problem has incomplete information which must be taken into account in a pragmatic way. The approach we have crafted couples or links a variety of computational tools. As can be seen in the figure, our approach involves several levels ranging from the microscopic to the circuit level and a different computational tool is used for each level. The essence of the physics problem and the need for levels of approach will now be discussed.

The computation of the trapped charge requires solving a set of equations governing electrons and holes in semiconductors, the semiconductor transport equations. [11,12] The governing equations for electron density $n(\mathbf{r},t)$ and hole density $p(\mathbf{r},t)$ as a function of position r and time t are the continuity equations:

$$
\begin{aligned}
\partial n(\mathbf{r},t)/\partial t &= R_n(\mathbf{r},t) + \nabla \mathbf{J}_n(\mathbf{r},t)/q \\
\partial p(\mathbf{r},t)/\partial t &= R_p(\mathbf{r},t) - \nabla \mathbf{J}_p(\mathbf{r},t)/q.
\end{aligned}
\tag{1}
$$

In these equations, $\mathbf{J}_n(\mathbf{r},t)$ and $\mathbf{J}_p(\mathbf{r},t)$ are, respectively, the electron and hole current densities and $R_n(\mathbf{r},t)$ and $R_p(\mathbf{r},t)$ are generation-recombination terms for electron and holes, respectively.

A prototypical contribution to the generation-recombination term arises from hole trapping. This process can be written as a reaction

$$
T + p \Longleftrightarrow Tp
\tag{2}
$$

involving traps T and trapped holes Tp, and it leads to a generation-recombination term

$$R_{gr} = -k_f[T][p] + k_r[Tp] \tag{3}$$

in terms of rate constants k_f and k_r. A series of equations with terms of this form need to be solved to compute response to ionizing radiation. [12]

A computational approach to an aging problem is more challenging. One aspect of the challenge can be seen by considering the trapping reaction, Eq. (2). In an aging problem, the total concentration of traps can also become time-varying on a slow time scale. This leads to new equations which are similar to those which would govern the preparation and processing of the oxide. More challenging is the fact that the time scales of these new equations is vastly different than that of the transport equations. For example, the electrons and holes can be captured in 10^{-11}s but the aging process of interest may take place in 10^8 s. This leads to equations with multiple time scales; such equations are called 'stiff' and they require special solution techniques. [13]

Another aspect of the problem is incomplete information. The best and most studied phenomena are those which take place rapidly during transient radiation studies. These processes certainly need to be considered in the reactions. However, the more important reactions for this problem are those which take place slowly and these have not been well-studied. Thus it is likely that some of the information characterizing the trap reactions during device aging is incomplete.

The problem of incomplete information led us to work on this problem on both a continuum and a microscopic level as shown in the lower left of Fig. 1. The continuum level involves the solution of the semiconductor transport and kinetic equations as a function of time and radiation dose for a given set of processing conditions. These equations are solved using software REO (Radiation Effects in Oxides). For the initial calculations, the parameters are either obtained from the literature [12] or estimated. The initial series of continuum calculations were done to explore which mechanisms have the potential to explain the low-dose-rate effect. These calculations led to insight about candidate mechanisms such as the hydrogen de-passivation mechanism described in this paper. However, many of the quantities which govern the reaction rates are not well known and this fact stimulated a series of more microscopic calculations, electronic structure calculations of defect energies.

The electronic structure calculations are performed using QUEST, software which uses density functional theory to compute the total energy of defects in the SiO_2. [14] Two types of calculations are important for the reaction rates. First, the equilibrium energy of the various defects such the E' center is needed to define the steady state or equilibrium defect densities using REO. Second, the transition state or barrier energy is needed to compute the rate of the defect reactions to be used in REO. At this time, the second set of calculations has not been undertaken. Instead, the rates are estimated by approximating the relevant partition functions. [15]

Given the trapped charge as a function of time, another set of calculations is needed to define the device response. These device level calculations use the information from the continuum level calculations to compute the device response. We use PISCES for these calculations. [16] In this level the semiconductor transport equations are solved for particular devices such as bipolar transistors and they take into account the trapped charge in the oxide and at the interface.

Finally, to draw conclusions about circuit performance, SPICE is used to compute the electrical performance of a circuit consisting of transistors, resistors, capacitors and other components. [17] For these calculations, the information from the PISCES calculations is conveyed in a physics-based transistor model. [18]

APPLICATION TO THE LOW-DOSE-RATE PROBLEM

Several candidate mechanisms can be postulated to explain the low dose rate effect. The nonlinear nature of many of the defect reactions leads to this conclusion. To understand this point, consider the reaction shown in Eq. (2). If the trap density were constant, then this reaction would be linear. If all such reactions were linear, there would be no true dose rate effects. However, if the trap density $[T]$ also varies due to atomic migration, then the reaction becomes non-linear and it has the potential to contribute to a dose rate effect. Perhaps the most simple reaction which yields a dose rate effect is the reaction of an electron and a hole to form an exciton, a neutral hydrogen-like entity. The formation of excitons is more prevalent at high dose rates because the reaction is bi-molecular. Thus, if these excitons were to recombine without generating trapped charge, then this reaction would lead to reduced charge trapping at high dose rate, consistent with the experimental situation.

Several reactions are included in the simulations. These were all chosen for their role in controlling the density of P_b and E' centers, the main defects which play a role in radiation-induced trapped charge. In addition, reactions involving hydrogen were also included because it has been implicated both in passivating and creating defects in SiO_2. [2, 19–21]

The importance of hydrogen in passivating P_b-centers led us to formulate a mechanism in which de-passivation of traps leads to a low-dose-rate effect. The essence of this mechanism is contained in the reactions

$$T^0 + H^0 \Leftrightarrow TH^0 \tag{4}$$
$$TH^0 + H^0 \Leftrightarrow T^0 + H_2^0 \tag{5}$$
$$T^0 + p \Leftrightarrow Tp^+ \tag{6}$$

which govern the densities of holes (p), neutral hydrogen (H^0), neutral, un-passivated traps (T^0), trapped holes (Tp^+), passivated traps (TH^0), and molecular hydrogen (H_2^0). The first

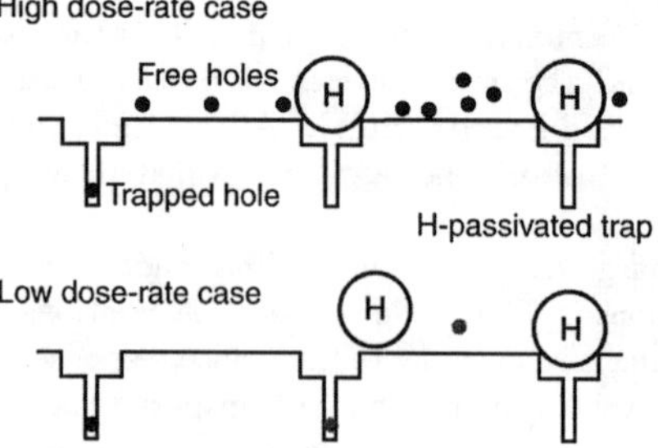

Figure 2: Schematic showing the de-passivation mechanism.

two reactions are stimulated by the well-known role of H in passivating P_b-centers. [2, 19] At dose rates comparable with the hydrogen release rate (reverse reaction of 4), additional un-passivated traps are created at a rate comparable with the hole capture rate (forward reaction of 6). The de-passivation reaction is slow, which leads to a dose rate effect at low dose rates. However, at high dose rates, holes are captured only at the initially un-passivated traps. This mechanism is illustrated in Fig. 2.

The predicted net charge of the de-passivation mechanism is shown in Fig. 3. For this calculation, $[T^0] \cong 3.3\times10^{10}$ cm^{-3}, $[TH^0] \cong 10^{13}$, and the oxide thickness is 1 μm. [15] By inspection, this mechanism predicts a low dose rate effect which is in qualitative agreement with the experimental observations. [3–7, 22] However, the observed cross-over occurs at somewhat higher dose rates; for this reason, we are revising the kinetic parameters of this mechanism as well as considering other contributions to the effect. [15]

A series of calculations concerning this defect reaction and similar reactions are underway. The kinetics are sensitive to defect formation energies which are not very well known. Thus QUEST calculations are being undertaken to compute the key defect formation energies to further examine whether this mechanism may contribute to the low dose rate problem.

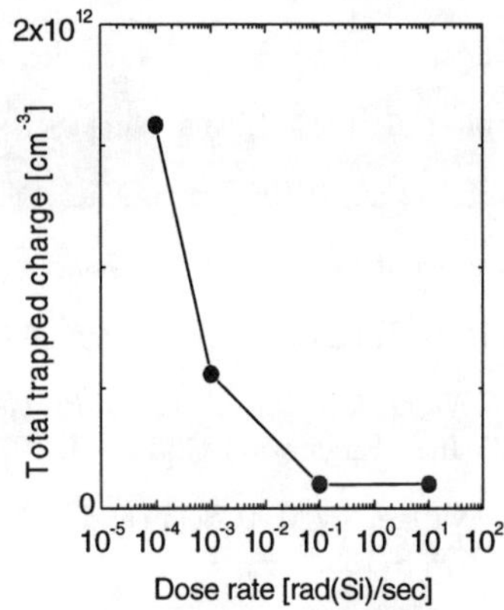

Figure 3: Shows the trapped charge dependence on dose rate for the hydrogen de-passivation mechanism.

CONCLUSIONS

We have described a computational approach to oxide aging problems in which multiple time and distance scales are incorporated by using several linked computational methods. The approach is illustrated by discussing a mechanism which may contribute to the problem of low-dose-rate radiation effects in bipolar devices.

ACKNOWLEDGMENTS

Sandia is a multiprogram laboratory operated by Sandia Corporation, a Lockheed Martin Company, for the United States Department of Energy under contract DE-AC04-94AL85000.

REFERENCES

[1] E. H. Nicollian and J. R. Brews, *MOS Physics and Technology* (Wiley, New York, 1982).

[2] P. M. Lenahan and J. J. F. Conley, J. Vac. Sci. Tech. B **16**, 2134 (1998).

[3] E. W. Enlow *et al.*, IEEE Trans. Nucl. Sci. **NS-38**, 1342 (1991).

[4] R. N. Nowlin *et al.*, in *Proceedings 1991 IEEE Bipolar Circuits and Technology Meeting* (IEEE, NY, 1991), pp. 174–177.

[5] A. H. Johnston, B. G. Rax, and C. I. Lee, IEEE Trans. Nucl. Sci. **42**, 1650 (1995).

[6] A. H. Johnston, C. I. Lee, and B. G. Rax, IEEE Trans. Nucl. Sci. **43**, 3049 (1996).

[7] R. L. Pease *et al.*, IEEE Trans. Nucl. Sci. **44**, 1981 (1997).

[8] D. M. Schmidt *et al.*, IEEE Trans. Nucl. Sci. **42**, 1541 (1995).

[9] R. D. Schrimpf *et al.*, IEEE Trans. Nucl. Sci. **42**, 1641 (1995).

[10] D. M. Fleetwood *et al.*, IEEE Trans. Nucl. Sci. **43**, 2537 (1996).

[11] S. M. Sze, *Physics of Semiconductor Devices* (J. Wiley and Sons, New York, 1981).

[12] R. Sokel and R. C. Hughes, J. Appl. Phys. **53**, 7414 (1982).

[13] C. W. Gear, *Numerical Initial Value Problems in Ordinary Differential Equations* (Prentice-Hall, Inc., Englewood Cliffs, NJ, 1971).

[14] Developed and written by M. P. Sears and P. A. Schultz at Sandia National Laboratories, Albuquerque, NM 87185.

[15] H. P. Hjalmarson (unpublished).

[16] M. R. Pinto, C. S.Rafferty, and R. W. Dutton, Technical report, Stanford Electronics Laboratories, Stanford University (unpublished).

[17] L. W. Nagel, Technical Report No. Memorandum No. ERL-M520, Electronic Research Laboratory, College of Engineering, University of California, Berkeley (unpublished).

[18] T. A. Fjeldly, T. Ytterdal, and M. Shur, *Introduction to Device Modeling and Circuit Simulation* (John Wiley and Sons, Inc., New York, NY, 1998).

[19] R. E. Stahlbush, A. H. Edwards, D. L. Griscom, and B. J. Mrstik, J. Appl. Phys. **73**, 658 (1993).

[20] K. L. Brower, Phys. Rev. B **38**, 9657 (1988).

[21] K. L. Brower, Phys. Rev. B **42**, 3444 (1990).

[22] S. C. Witczak *et al.*, IEEE Trans. Nucl. Sci. **44**, 1989 (1997).

MODELING OF FAILURE IN METALLIC THIN FILMS INDUCED BY STRESS AND ELECTROMIGRATION: A MULTISCALE COMPUTATIONAL ANALYSIS

M. R. GUNGOR[1], L. J. GRAY[2], S. J. ZHOU[3] and D. MAROUDAS[1,a)]
[1]Department of Chemical Engineering, University of California, Santa Barbara, CA 93106
[2]Computer Sciences & Mathematics Division, Oak Ridge National Laboratory, Oak Ridge, TN 37831
[3]Applied Theoretical and Computational Physics Division, Los Alamos National Laboratory, Los Alamos, NM 87545

ABSTRACT

A common failure mechanism in metallic thin–film interconnects is void propagation driven by electric fields and thermomechanical stresses. In this paper, a multiscale computational analysis is presented for predictive modeling of transgranular void dynamics. The modeling approach is hierarchical and involves atomistic simulations for property database development, molecular–dynamics simulations for understanding of void–tip mechanisms, and self–consistent mesoscopic simulations based on boundary–element methods and techniques for moving boundary propagation. An extremely rich void dynamical behavior is predicted, which includes faceting, facet selection, propagation of slits from the void surface, as well as formation of fine–scale crack–like features on the void surface, in agreement with recent experimental data.

INTRODUCTION

Failure of polycrystalline aluminum and copper thin films, which are used for device interconnections in integrated circuits, is one of the most serious materials reliability problems in microelectronics [1]. The interconnect microstructure is complex due to the presence of several interfaces, such as grain boundaries (GBs), material interfaces, and surfaces of microvoids, all of which are fast paths for mass transport. Failure mechanisms in these systems are driven mainly by electromigration [1]. Ultra–large–scale integration (ULSI) and miniaturization of devices demands interconnect cross–sectional dimensions of submicron scale. As a result, these films have to tolerate increasingly higher current densities and levels of thermomechanical stress. In passivated metallic films that are mechanically confined due to their encapsulation by a dielectric material, thermomechanical stresses are induced during film cooling after passivation. Voids usually nucleate at the film edges as a mechanism of thermal stress relaxation. After cooling and aging, the metallic films are in a state of hydrostatic tension [1,2]. This residual stress is another important driving force of void morphological evolution that may cause film failure. It has been established experimentally that transgranular voids are common sources of failure in bamboo films, where grain boundaries are oriented almost perpendicularly to the length–direction of the film [3,4]. Such voids are not intersected by GBs and may form after detachment from GBs and further migration into the grain under the action of an applied electric field. In spite of recent theoretical studies of electromigration–induced transgranular void dynamics [5–10] and of our understanding of surface instabilities in mechanically stressed solids [11], the complex nonlinear phenomena associated with transgranular void evolution under the combined action of stresses and electric fields are not well understood.

In this paper, a multiscale computational analysis is presented for modeling transgranular void dynamics in metallic thin films under mechanical loading and surface electromigration conditions. Our modeling approach involves atomistic simulations for property database development, molecular–dynamics simulations for understanding of void–tip mechanisms, and self–consistent mesoscopic simulations of void surface morphological evolution. An extremely rich void dynamical behavior is predicted under the combined action of the electric field and mechanical stress, which is consistent with recent experimental data. Special emphasis is placed on understanding atomic–scale void–tip phenomena and upgrading the present state of the mesoscopic modeling accordingly.

MULTISCALE MODELING METHODOLOGY

Failure of metallic interconnects is a very complex problem that is ideally suited for multiscale modeling linking atomistic and mesoscopic length scales, from Angstroms to

Mat. Res. Soc. Symp. Proc. Vol. 538 © 1999 Materials Research Society

microns, and a wide range of time scales, from picoseconds to hours. The modeling strategy that we have developed to address the problem is highlighted in Fig. 1, which focuses on failure mechanisms mediated by the dynamics of void surfaces. The backbone of our modeling approach consists of self–consistent numerical simulations of surface morphological evolution based on boundary–element methods (BEM) and techniques for moving boundary propagation [10,12]. A thermophysical property database for the heterogeneous thin–film materials, that is needed as input for the mesoscopic modeling, is provided by atomistic simulations based on the embedded–atom method (EAM) [13]. Atomic–scale void–tip phenomena are explored through EAM–based molecular–dynamics (MD) simulations. Void dynamics is explored in a multidimensional parameter space as determined by dimensional analysis of the mesoscopic field equations. Systematic comparisons with experimental measurements are used to test and improve the predictive capabilities of the modeling. Additional modeling predictions motivate new experiments aiming at improving interconnect reliability.

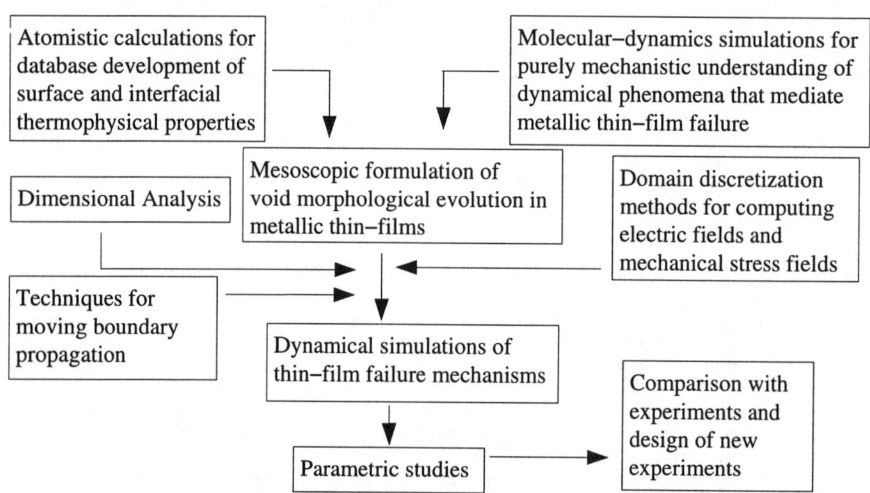

FIG. 1. Schematic representation of the multiscale modeling strategy employed for predictive modeling of failure mechanisms in metallic thin films mediated by void dynamics.

Our mesoscopic analysis is based on a continuum formalism of surface mass transport under the action of external fields [7,10]. The total mass flux, J_s, on the void surface is given by a Nernst–Einstein equation that includes contributions from curvature–driven surface diffusion, stress–induced surface diffusion, and drift induced by electromigration. The evolution of the local displacement normal to the void surface, u_n, is given by the continuity equation, according to which the local rate of change of u_n is proportional to the surface divergence of J_s. The electromigration and stress contributions to J_s provide a strong coupling between the void morphological evolution and the stress and electric field distributions in the metallic film [7,10]. The electric field, E, in the metallic conductor can be written as $E \equiv -\nabla\Phi$, where the electrostatic potential, Φ, obeys Laplace's equation, i.e., $\nabla^2\Phi = 0$. Cauchy's mechanical equilibrium equation, $\nabla \cdot \sigma = 0$, also is satisfied, where σ is the local stress tensor. Here the mechanical deformation of the solid is addressed within the framework of isotropic linear elasticity in the limit of infinitesimal displacements: $\varepsilon = (1/2)[\nabla u + (\nabla u)^T]$, where ε and u are the local strain and displacement fields, respectively, and T denotes the transpose of a tensor. Our analysis is two–dimensional in x (length) and y (width), for voids that extend throughout the film thickness (in z) [10]. In this 2–D model, plane strain conditions are assumed. The void surface is considered to be traction free, while the domain boundaries are subjected to a constant applied stress, σ_o, imposed as hydrostatic tension. This loading resembles the strain state of interconnect films after thermal processing and aging. The void surface and the film's edges are modeled as electrically insulating boundaries, while a constant electric field, $E_\infty = E_\infty \hat{x}$, is imposed away from the void. The diffusivity of surface atoms is expressed by $D_s = D_{s.min} f(\theta)$, where $D_{s,min}$ is the minimum surface diffusivity corresponding to a specific surface orientation, and $f(\theta)\geq 1$ is an anisotropy function of the angle θ formed by the local tangent to the surface and E_∞. $f(\theta)=1+A\cos^2[m(\theta+\phi)]$ has been adopted as the anisotropy function in the present treatment, where A, m, and ϕ are dimensionless parameters that determine the strength of the anisotropy, the grain symmetry, and the misorientation of a symmetry direction of fast surface diffusion with respect to E_∞, respectively. On the other hand, the surface tension, γ is assumed to be isotropic; this is justified by recent atomistic simulations, according to which the dependence of γ on θ is weaker by orders of magnitude than that of D_s [14]. Specifically, for the materials and temperatures of interest A can reach values as high as 10^4; A increases with decreasing temperature [14]. Finally, the effective surface charge, q_s, that determines the electromigration force [1] also is assumed to be independent of surface orientation. Detailed EAM–based analyses of the $D_s(\theta)$ and $\gamma(\theta)$ dependencies are currently underway.

Dimensional analysis of Eqs. (1) – (3) yields three important dimensionless parameters: $\Gamma \equiv E_\infty q_s^* w^2/(\gamma\Omega)$, $\Sigma \equiv \sigma_0^2 w/(E\gamma)$, and $\Lambda \equiv w_t/w$. Γ scales electric forces with capillary forces and Σ scales elastic strain energy with surface energy; Ω is the atomic volume, E is the Young's modulus of the material, w is the width of the film, and w_t is the initial extent of the void across the film. The resulting time scale is $\tau \equiv k_B T w^4/(D_{s,min}\delta_s\gamma\Omega)$. In an Al film of $w\approx 1$ μm, a value of $\Gamma=50$ corresponds to a current density of about 2 MA/cm^2, which is typical of accelerated electromigration experiments [3,4]. For the same film, a value of $\Sigma=1$ corresponds to $\sigma_0\approx 140$ MPa; this is typical of residual stresses in interconnect lines after cooling and aging [1,2]. Finally, for the same film τ is estimated on the order of 10^4 hours. The parameters Γ, Σ, Λ, A, m, and ϕ define a 6–D hyperspace that is indicative of the complexity of the void dynamics problem.

The electrostatic potential and the displacement vector on the domain boundary are computed using a Galerkin BEM formulation. The corresponding electric and stress fields on the boundary are computed using the formulation of Ref. 12. The surface flux divergence is calculated based on a centered finite–difference scheme and the continuity equation is integrated using an Adams–Bashforth algorithm to compute the evolution of the normal surface displacement, $u_n(t)$. Our BEM discretization is adaptive and employs typically several hundred nodes along the void surface. In the simulations, the initial void shape is taken to be semi–circular, i.e., a configuration that includes all possible surface orientations with respect to E_∞. Additional details of our mesoscopic simulation methods can be found in Refs. 10.

RESULTS AND DISCUSSION

First, we examine the void dynamical response under the simultaneous action of mechanical stresses and electric fields; electromigration–driven void dynamics in unpassivated films has been examined in detail in Refs. 10. We focus on <111>–oriented grains with 6–fold symmetry, $m=3$ [6,10]. An interesting case of electromigration–induced failure at this grain symmetry is demonstrated in Fig. 2(a) for parameters $\Gamma=150$, $\Lambda=0.5$, $A=10$, $\phi=-15^o$, and $\Sigma=0$. Initially, the void shape becomes faceted and then a facet selection process is initiated, where the facets grow at the expense of the right tilted facet leading to a wedge–like void shape. Subsequent elongation of the left tilted facet and complete destabilization of the right vertical facet result in the formation of a faceted slit that propagates fast causing the failure of the film. Figures 2(b)–(f) demonstrate how the void dynamics changes as the applied hydrostatic tension in the film increases, i.e., along the Σ–axis of the 6–D parameter space. In Fig. 2(b), $\Sigma=0.3$, and the dynamical sequence resembles closely that of Fig. (2a): faceting is followed by facet selection, wedge formation, and faceted slit formation and propagation. In this case, however, after the slit propagates a certain distance, a finer–scale crack–like feature forms and its tip propagates at a speed greater by about two orders of magnitude than the average void migration speed. Thus, open–circuit failure is caused by the fast propagation of the crack–like feature's tip; this feature is narrower than the original slit by at least an order of magnitude. Fig. 2(c) shows the corresponding dynamics at $\Sigma=0.7$. The void shape evolves again into a wedge but, interestingly, subsequent faceted slit formation is not observed; instead, a fine–scale crack–like feature forms at the tip of the wedge and propagates fast to cause failure. As Σ keeps increasing, not even the facet selection process can be completed toward wedge formation. Instead, fatal crack–like features appear at corners of the faceted void, where the strain energy density increases abruptly. Such dynamics is shown in Fig. 2(d) at $\Sigma=1.05$.

The dynamical behavior of the transgranular void changes drastically if the applied stress is increased slightly to Σ=1.075, as demonstrated in Fig. 2(e). Specifically, there is a narrow range of stress, $1.05 \leq \Sigma \leq 1.1$, where neither electromigration–induced facet selection nor stress–induced crack formation are initiated. Instead, the faceted void is stable and it migrates along the film at constant speed maintaining its semi–hexagonal shape. This response implies that film failure due to electromigration–induced slit–like morphological instabilities of the void surface can be inhibited completely by the simultaneous action of a mechanical stress. The applied stress level must be just enough to balance the destabilizing effect of the electric field without causing crack–like morphological instabilities that lead to a faster mode of failure. Increasing the applied stress above the narrow range of stability leads to formation of crack–like features that emanate from the faceted void surface. This is shown in Fig. 2(f) for Σ=1.1 following facet selection. At Σ=1.5, crack–like features form at increasingly earlier stages before even the initiation of a facet selection process. At such high stresses, the crack propagation directions are almost symmetric with respect to both principal stress directions, resembling the purely stress–induced instability behavior shown below. Generally, the location of crack formation is correlated strongly with the distribution of elastic strain energy on the faceted void surface. The predicted general mixed mode of failure, where the void surface exhibits electromigration–induced features, such as faceted slits, coexisting with stress–induced crack–like features is consistent with recent experimental observations [4].

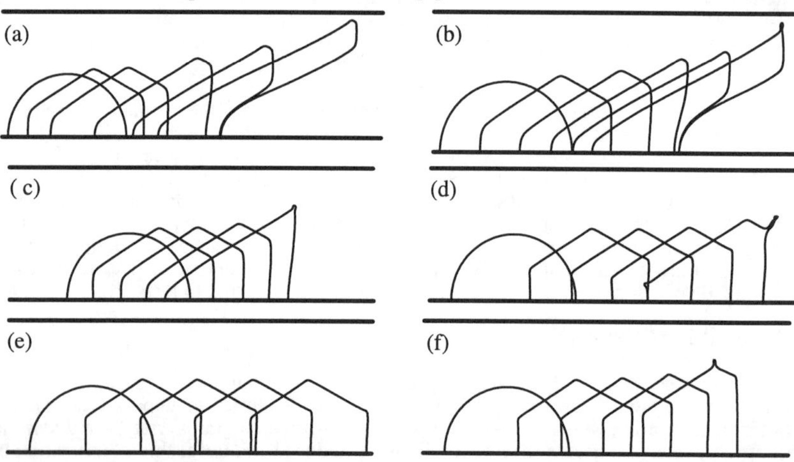

FIG. 2. Evolution of transgranular voids in a <111>–oriented grain of a metallic film under the action of both hydrostatic tension and an electric field directed from left to right. The corresponding parameters are Γ=150, Λ=0.5, A=10, m=3, ϕ=-15° and (a) Σ=0, (b) Σ=0.3, (c) Σ=0.7, (d) Σ=1.05, (e) Σ=1.075, and (f) Σ=1.1. The times corresponding to the different morphologies from left to right are: (a) t=0, 0.574, 1.272, 2.486, 3.395, 3.929 $\times 10^{-4}$ τ, (b) t=0, 1.047, 2.061, 2.776, 3.228, 3.608 $\times 10^{-4}$ τ, (c) t=0, 0.672, 1.453, 2.176, 2.773 $\times 10^{-4}$ τ, (d) t=0, 1.901, 2.974, 4.012, 4.959 $\times 10^{-4}$ τ, (e) t=0, 1.314, 2.702, 4.089, 5.477 $\times 10^{-4}$ τ, and (f) t=0, 1.778, 2.882, 3.976, 5.125 $\times 10^{-4}$ τ.

Next, the effects of mechanical stress on transgranular void dynamics are examined without the simultaneous action of an electric field. It is demonstrated in Fig. 3 that failure can occur under the action of hydrostatic tension alone even for high–symmetry grains, m=3, if the applied stress level is higher than a critical one; this is the outcome of a stress–induced surface morphological instability [11]. For a stress level Σ=1.2, Fig. 3 shows that the initially semi–circular void morphology becomes faceted and, later, fine–scale crack–like features start emanating from corners of the faceted shape; propagation of these crack–like features provides the only mode of failure in this case. The location of crack formation is determined by the distribution of the elastic strain energy on the faceted void surface. At a misorientation angle

$\phi=0^\circ$, the directions of crack propagation are symmetric with respect to both principal stress directions.

The characteristic length scale of the crack–like features on the void surfaces shown in Figs. 2 and 3 is very fine, on the order of nanometers. This has motivated detailed MD simulation studies for the mechanistic investigation of purely stress–induced dynamical phenomena at void tips. Some preliminary results are shown in Fig. 4 for loading that resembles the conditions that led to the dynamics of Fig. 3. In our MD simulations, an EAM parametrization for Cu is used to describe the interatomic interactions [13]. The initial configuration includes a perfectly cylindrical void in the middle of the simulation supercell and the simulations are performed at constant temperature, T, and applied hydrostatic strain, ε_0, with periodic boundary conditions in all three directions. The number of atoms in the supercell is on the order of 10^5. In the case of Fig. 4, the grain is [111]–oriented to model 6–fold symmetry, $m=3$, and the conditions are $T=400$ K and $\varepsilon_0=5\%$; the high applied strain is required to accelerate the dynamics. In Fig. 4, the atoms shown are marked according to their squared displacement, $\Delta r^2 \equiv [\mathbf{r}(t)-\mathbf{r}(0)]^2$, where $\mathbf{r}$ is the atomic position vector; Δr^2 is a measure of atomic mobility and, thus, it is useful for the visualization of the diffusional mechanisms that drive void evolution. Although the times that correspond to the snapshots of Fig. 4 are on the order of picoseconds, qualitative comparisons with the mesoscopic results of Fig. 3 are possible, since the diffusional time scale, τ, is proportional to the fourth power of the characteristic void size.

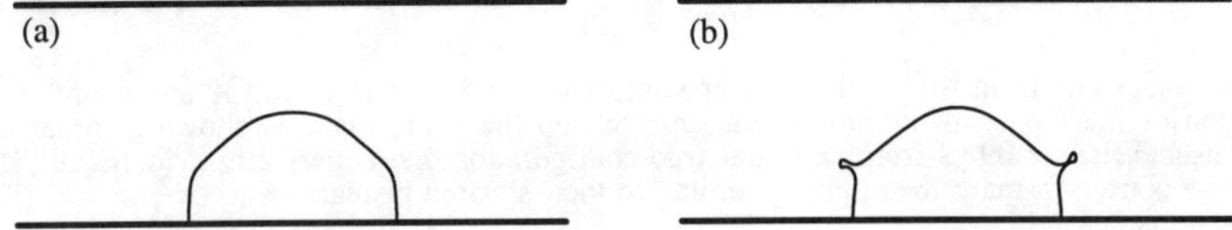

FIG. 3. Evolution of an initially semi–circular void in a metallic film under the action of hydrostatic tension. The values of the parameters are $\Gamma=0$, $\Sigma=1.2$, $\Lambda=0.5$, $A=1000$, $m=3$, and $\phi=0^\circ$ and the corresponding times are (a) $t=1.411\times10^{-6}\,\tau$, and (b) $t=2.983\times10^{-6}\,\tau$.

The MD results of Fig. 4 demonstrate that within a few picoseconds, the void shape has become faceted, Fig. 4(a), resembling the morphology of Fig. 3(a). The atomic mobility is highest at the void surface and especially at the corners of the hexagonal faceted shape, where the surface gradient of the strain energy density increases most abruptly. After a few more picoseconds the void has evolved to the morphology of Fig. 4(b) that is characterized by the presence of cusp–like features, which have been blunted by the emission of dislocations from the void surface. Again, the atomic mobility is highest at the blunted corners and cusp–like features of the void surface; this increased mobility around these regions is partly due to pipe diffusion along dislocation cores. The configuration of Fig. 4(b) corresponds to a time that is almost twice as long as that of Fig. 4(a), i.e., the dynamical sequence is analogous to that of Fig. 3.

SUMMARY

In summary, a multiscale modeling approach has been presented for the analysis of failure mechanisms in metallic thin films mediated by void morphological evolution that is induced by mechanical stresses and electric fields. The approach combines atomistic simulations for surface and interface property calculations and mechanistic understanding of void–tip phenomena with mesoscopic simulations of surface morphological evolution over a multidimensional parameter space. The general mode of failure due to void propagation is characterized by the coupling of two modes of instability: one is electromigration driven and leads to slit formation, while the other is stress driven and leads to formation of fine–scale crack–like features. Our mesoscopic simulations have demonstrated that in textured films of high grain symmetry, failure prevention is possible through thermal processing that can tailor the resulting state and level of stress in films that will be subjected to strong applied electric fields. MD simulations have highlighted the possible important role of atomistic void–tip phenomena in the failure process; a detailed account of these phenomena will be presented in a forthcoming publication. Such dynamical phenomena are currently being incorporated into the mesoscopic dynamical formalism through dislocation field equations. In addition, the effects of the film grain structure on void dynamics are currently under investigation and will be presented in subsequent publications.

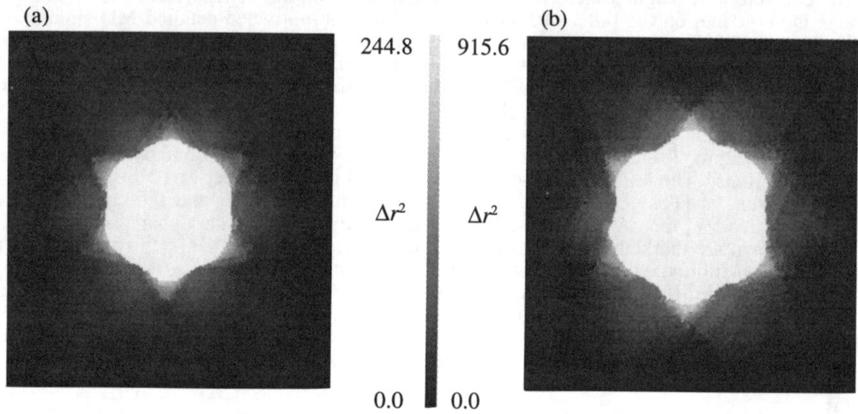

FIG. 4. Snapshots from MD simulation of void dynamics in Cu at T=400 K and ε_0=5%. A projection of the atoms in the simulation supercell on the (111) plane is shown at times (a) t=2.84 ps and (b) t=6.09 ps from an initial void configuration that is perfectly cylindrical. The gray scale is used to mark the atoms according to their squared displacement, Δr^2, in Å^2. The area shown is 298×312 Å^2.

ACKNOWLEDGMENTS

The authors acknowledge fruitful discussions with C. V. Thompson, B. C. Larson, and D. R. Clarke. This work was supported by the Frontiers of Materials Science Program of the UCSB Materials Research Laboratory and Los Alamos National Laboratory (Award No. STB–UC:97–63), by the National Science Foundation through a CAREER Award to DM (ECS–95–01111) and grant PHY94–07194 with the Institute for Theoretical Physics, UCSB, by ORISE through a Faculty Research Fellowship to DM, and by the US Department of Energy under contract DE–AC05–96OR22464 with Lockheed Martin Energy Research Corporation.

REFERENCES

a) To whom correspondence should be addressed; E–mail address: dimitris@calypso.ucsb.edu
1. P. S. Ho and T. Kwok, Rep. Progr. Phys. **52**, 301 (1989); C. V. Thompson and J. R. Lloyd, MRS Bulletin **18**, No. 12, 19 (1993), and references therein.
2. see, e.g., A. I. Sauter and W. D. Nix, J. Mater. Res. **7**, 1133 (1992).
3. J. E. Sanchez, Jr., L. T. McKnelly, and J. W. Morris, Jr., J. Electron. Mater. **19**, 1213 (1990); J. Appl. Phys. **72**, 3201 (1992); J. H. Rose, Appl. Phys. Lett. **61**, 2170 (1992); O. Kraft, S. Bader, J. E. Sanchez, Jr., and E. Arzt, Mater. Res. Soc. Symp. Proc. **309**, 199 (1993); E. Arzt, O. Kraft, W. D. Nix, and J. E. Sanchez, Jr., J. Appl. Phys. **76**, 1563 (1994).
4. Y.–C. Joo and C. V. Thompson, J. Appl. Phys. **81**, 6062 (1997).
5. Z. Suo, W. Wang, and M. Yang, Appl. Phys. Lett. **64**, 1944 (1994); W.Q. Wang, Z. Suo, and T.–H. Hao, J. Appl. Phys. **79**, 2394 (1996).
6. O. Kraft and E. Arzt, Appl. Phys. Lett. **66**, 2063 (1995); Acta Mater. **45**, 1599 (1997).
7. D. Maroudas, Appl. Phys. Lett. **67**, 798 (1995); D. Maroudas, M. N. Enmark, C. M. Leibig, and S. T. Pantelides, J. Comp.–Aided Mater. Des. **2**, 231 (1995).
8. L. Xia, A. F. Bower, Z. Suo, and C. F. Shih, J. Mech. Phys. Solids **45**, 1473 (1997).
9. M. Schimschak and J. Krug, Phys. Rev. Lett. **80**, 1674 (1998).
10. M. R. Gungor and D. Maroudas, Appl. Phys. Lett. **72**, 3452 (1998); Surf. Sci. **418**, L1055 (1998); J. Appl. Phys., in press.
11. see, e.g., W. H. Yang and D. J. Srolovitz, Phys. Rev. Lett. **71**, 1593 (1993).
12. L. J. Gray, D. Maroudas, and M. N. Enmark, Comp. Mech. **22**, 187 (1998).
13. S. M. Foiles, M. I. Baskes, and M. S. Daw, Phys. Rev. B **33**, 7983 (1986).
14. C.–L. Liu, J. M. Cohen, J. B. Adams, and A. F. Voter, Surf. Sci. **253**, 334 (1991).

COUPLED SIMULATIONS OF MECHANICAL DEFORMATION AND MICROSTRUCTURAL EVOLUTION USING POLYCRYSTAL PLASTICITY AND MONTE CARLO POTTS MODELS

C.C. BATTAILE, T.E. BUCHHEIT, E.A. HOLM, G.W. WELLMAN, AND M.K. NEILSEN
Sandia National Laboratories, Albuquerque, NM 87185-1411, ccbatta@sandia.gov

ABSTRACT

The microstructural evolution of heavily deformed polycrystalline Cu is simulated by coupling a constitutive model for polycrystal plasticity with the Monte Carlo Potts model for grain growth. The effects of deformation on boundary topology and grain growth kinetics are presented. Heavy deformation leads to dramatic strain-induced boundary migration and subsequent grain fragmentation. Grain growth is accelerated in heavily deformed microstructures. The implications of these results for the thermomechanical fatigue failure of eutectic solder joints are discussed.

INTRODUCTION

Thermal fluctuations during the normal course of service or storage can accelerate the failure of solder joints in electronic components [1]. This thermomechanical fatigue (TMF) of solder joints involves a number of different processes acting in concert. Thermal expansions and contractions of the surrounding pieces generate a complex state of stress in the joint. The local details of that stress state, and the properties of the solder microstructure, create regions of high localized strain in the joint. These localized deformations induce coarsening in the microstructure, which weakens the material. These weakened regions fatigue as the thermal fluctuations continue, eventually leading to component failure. This phenomenon has been observed, for example, as satellites pass into and out of the sun's rays, or when stored components are subjected to daily and seasonal temperature changes.

Understanding TMF requires an examination of how each process operates individually and in conjunction with the other processes. In this paper, we examine the effects of the localized deformation on the evolution of a polycrystal microstructure. Solder joints are typically composed of eutectic PbSn, but we will use pure polycrystalline Cu as a proof-of-concept system for the present study. The deformation of the material is simulated using the finite element method (FEM) with a constitutive law for polycrystal plasticity fit to experimental data for Cu. This polycrystal plasticity model predicts the local mechanical state of the system, and this information is combined with the Monte Carlo Potts model to simulate grain growth (i.e., microstructural coarsening) in the deformed material. The polycrystal plasticity and Potts models will be outlined briefly, and the extension of the Potts model to account for stored strain energies will be discussed. The effects of deformation on the evolution of the microstructure will be presented, and the implications of these results for the TMF of solder joints will be addressed.

SIMULATION METHODS

Polycrystal Plasticity

The polycrystal plasticity model will be described here only in brief, and the reader is referred elsewhere [2,3] for a more detailed discussion. The local states of stress and strain in the Cu polycrystal are simulated using FEM with a constitutive law for polycrystal plasticity fit to experimental data for Cu. The finite element mesh is generated by discretizing a polycrystal microstructure onto a cubic lattice. Each lattice site is associated with a spatial orientation, and a cluster of lattice sites with the same orientation represents a grain. Each lattice site represents an element in the finite element mesh, and the nodes of the mesh form a cubic grid occupying the corners of the elements. Ideally, the polycrystal microstructure would correspond directly to the real Cu sample to which the plasticity model was fit. However, for the purpose of the present study, the microstructure was taken from a conventional 3D Potts model simulation, and the

Mat. Res. Soc. Symp. Proc. Vol. 538 © 1999 Materials Research Society

orientations of the grains are assigned at random.

The constitutive model for the mechanical response of a polycrystal tracks the spatial orientations of the material at each element, and the material deformations on each of the twelve FCC slip systems. The elastic response of the material is anisotropic, and the plastic response is strain rate dependent such that $\tau \propto \dot{\gamma}^m$, where τ is the shear stress, $\dot{\gamma}$ is the shear strain rate, and m is the strain rate sensitivity exponent. In this study, an exponent of $m = 10$ was used. Each of the twelve slip systems hardens isotropically according to a power law,

$$d\tau = \tau_o + Ad\gamma^n,$$

(1)

where τ_o is the initial critical resolved shear stress, γ is the effective plastic shear strain, and A and n are parameters fit to the experimental stress-strain response of a Cu polycrystal. All slip systems are initialized to the same critical resolved shear stress and harden equally. However, the polycrystal evolves an anisotropic mechanical response during deformation due to the variations in geometric constraints between and within grains.

The plasticity simulation is performed on a 54-grain polycrystal microstructure mapped onto a 30x30x30-element mesh. The average grain size is about 500 elements. Each element is cubic before deformation. The initial element orientations in each grain are the same (random) value, and the stress and strain values at each element set to zero. Periodic boundary conditions are imposed in all three directions. Deformation is simulated by a prescribed elongation of the simulation cell in one direction, with zero net force in the other two directions. The elongation of the simulation cell is increased in increments of 0.001%. At each strain increment, the mechanical state of the system is equilibrated. The stress and strain increments on each slip system in each element are used to calculate the stored plastic energy, which is input to a Monte Carlo Potts model of microstructure evolution, as described below. (Stored elastic energy is relatively small at the deformations of interest, and is therefore neglected in the present study, though it may be important under certain circumstances.)

Microstructural Evolution

The evolution of the deformed microstructures is simulated using a Monte Carlo Potts model [4,5] modified to account for the plastic energy stored in the material during deformation. The finite element mesh is input directly to the Potts simulation. Each grain on the mesh is assigned a unique integer identity, or "spin," and all the elements in a grain assume the spin of that grain. Periodic boundary conditions are imposed in all directions. Each element contains twelve slip systems for which the twelve shear stress and strain increments are taken from the polycrystal plasticity model.

The system energy is described by a modified Potts Hamiltonian,

$$H = \sum_{i=1}^{N} \left\{ \frac{E_o}{2} \sum_{j=1}^{26} \left[1 - \delta\left(s_i, s_j\right) \right] + \frac{1}{V_i \tau_o} \sum_{\Delta l=0}^{\Delta l_f} \sum_{\alpha=1}^{12} \tau_i(\alpha, \Delta l) \Delta \gamma_i(\alpha, \Delta l) \right\}.$$

(2)

The sum over i includes all of the N elements in the mesh, and the sum over j includes all of the 26 elements that share at least one node with element i. (Recall that the mesh is initially cubic, and thus the element connectivity of the deformed mesh is that of a cubic grid.) The sum over Δl includes all elongation increments up to the elongation of interest, Δl_f, and the sum over α includes all twelve slip systems in element i. E_o is a measure of the grain boundary energy, and is assumed to be unity for all boundaries in this study. s_i is the spin of element i, and $\delta(s_i, s_j)$ is the Kronecker delta, i.e., $\delta(s_i, s_j) = 1$ if $s_i = s_j$ and 0 otherwise. V_i is the volume of element i, τ_o is the initial critical resolved shear stress, $\tau_i(\alpha, \Delta l)$ is the average plastic shear stress on slip system α in element i at elongation Δl, and $\Delta \gamma_i(\alpha, \Delta l)$ is the plastic shear strain increment on slip system α in element i at elongation Δl. The first term in Equ. 2 represents the grain boundary energy, and the second describes the stored plastic energy.

The Monte Carlo Potts simulation proceeds by choosing an element at random and attempting to change its spin to that of one of its neighbors. When an element changes its spin to that of a different grain, the stored plastic energy in that element is changed to the average of the values at the neighboring elements that belong to that different grain. This implies that, when a grain boundary moves, nearly all of the dislocations adjacent to the boundary are extended into the material through which the boundary migrates. However, this is not always the case in reality. On the other hand, if no dislocations were propagated with the boundary, then a new grain would be introduced and the material would recrystallize. Though recrystallize undoubtedly occurs, and is certainly important in the materials of interest and at these levels of deformation, it is not the only option. A grain boundary can simply migrate in a heavily deformed polycrystal [6]. The migration of a grain boundary in a deformed polycrystal involves the propagation of some, but not always all, of the dislocations that are adjacent to it. However, a meaningful quantitative description of how the dislocation structure propagates with a migrating grain boundary is not currently available, and thus we assume that nearly all the dislocations are carried with a boundary when it moves.

The change to the system energy that results from an attempted spin "flip" is computed using Equ. 2, and the spin flip is accepted if the energy change is less than or equal to zero, and rejected otherwise. This is repeated until no more energy-decreasing spin flips are available. This pseudo-kinetic approach corresponds to the zero-temperature Metropolis Monte Carlo algorithm [7], though in practice we use a more efficient version [8,9]. The evolution of the microstructure (without deformation) toward its equilibrium reproduces the kinetics of normal grain growth remarkably well [10].

RESULTS AND DISCUSSION

The temporal evolution of the microstructure after 31% elongation is shown pictorially in Fig. 1. The black lines denote the grain boundaries. The shading corresponds to the stored plastic energy in each element, with dark regions representing low stored energy values and light regions indicating high values. As evident in Fig. 1, the grain boundaries roughen as they migrate and the grains fail to remain compact. This phenomenon is also observed experimentally [6] in deformed polycrystals, and is commonly termed strain-induced boundary migration (SIBM). SIBM occurs when two adjacent grains have very different dislocation densities (i.e., stored plastic energies). If the difference is much larger than the grain boundary energy, then the only significant driving force for grain boundary migration is the reduction in stored energy provided by the consumption of high-strain material by adjacent low-strain grains. Under these circumstances, grains are not motivated to remain compact and grain boundaries roughen, as seen in Fig. 1 and observed experimentally [6].

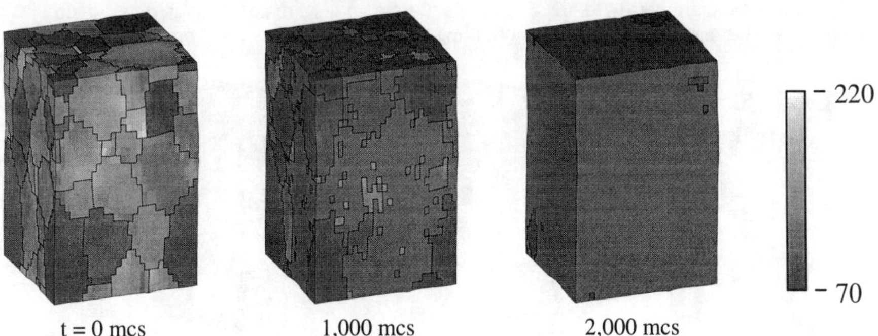

Figure 1. Evolution of the stored plastic energy during grain growth after 31% elongation. The stored energy distribution is shown at three simulation times in Monte Carlo steps (mcs). Dark areas have low stored energy and light areas have high energy. The scale bar at the right indicates normalized strain energy densities as per the second term in Equ. 2.

At elongations larger than 31%, SIBM becomes even more pronounced. At lower elongations, grains remain compact and the evolution of the microstructure more closely resembles normal grain growth, but boundary motion is still influenced by the stored energy in the material. This strain-enhanced boundary migration (SEBM) occurs when the driving forces due to grain boundary energy and stored plastic energy are comparable. This situation applies at 11% elongation, as depicted in Fig. 2, where grain growth appears normal but is dominated by material with relatively little stored plastic energy.

The grains that dominate the growth process at higher deformations aren't always the ones with the least stored plastic energy, since the plastic driving force for grain growth relies on the differences in stored energy between adjacent grains, and not on the absolute value of the stored energy. Therefore, the large grain in the last frame of Fig. 1 is lighter in shading (i.e., higher in stored plastic energy) than the grains that it surrounds. The same can be true for lower deformations, except that the significance of the curvature driving force (represented by the first term in Equ. 2) complicates the issue further.

By examining the time values in Figs. 1 and 2, one can readily see that the 31% elongated polycrystal coarsens much more quickly (by an order of magnitude) than does the 11% elongated structure. This is shown graphically in Fig. 3 which contains plots of the grain size evolution and the time required to reach an average grain radius of 10 elements, for various elongations. At large elongations in Fig. 3a, the average grain size decreases at early times as SIBM causes grains become non-compact and to fragment. At later times, the rate of grain growth increases with increasing elongation, so that the time required to achieve an average grain radius of ten elements decreases with increasing elongation, as in Fig. 3b.

Post mortem analyses of failed solder joints suggests that accelerated coarsening in regions of high local deformation leads to mechanical weakening and eventually failure of the joint [1]. The results presented in Fig. 3 support a part of this evidence by indicating that coarsening in single-phase polycrystals can be dramatically accelerated by deformation. As mentioned above, the plastic response of the simulated polycrystal is modeled after pure Cu. However, the material used in many real-world components is eutectic PbSn solder. Coarsening in eutectics generally involves interphase diffusion, which cannot be captured by a model of single-phase grain growth. Therefore, these results should be taken primarily as a demonstration of the modeling approach.

SUMMARY

Microstructural evolution in a 54-grain region of a deformed Cu polycrystal was simulated by coupling a constitutive finite element model of polycrystal plasticity with the Monte Carlo Potts model of grain growth. Heavy deformation introduces high stored plastic energies into the material which dominate the grain growth process. In this regime, grains are not motivated to remain compact, and strain-induced boundary motion leads to grain boundary roughening and grain

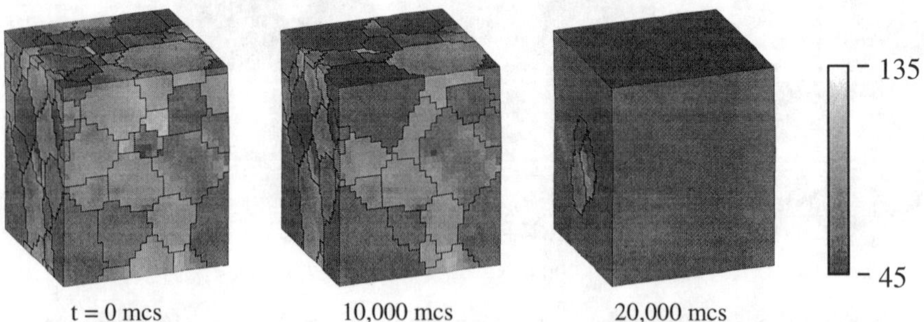

Figure 2. Evolution of the stored plastic energy during grain growth after 11% elongation. The stored energy distribution is shown at three simulation times in Monte Carlo steps (mcs). Dark areas have low stored energy and light areas have high energy. The scale bar at the right indicates normalized strain energy densities as per the second term in Equ. 2.

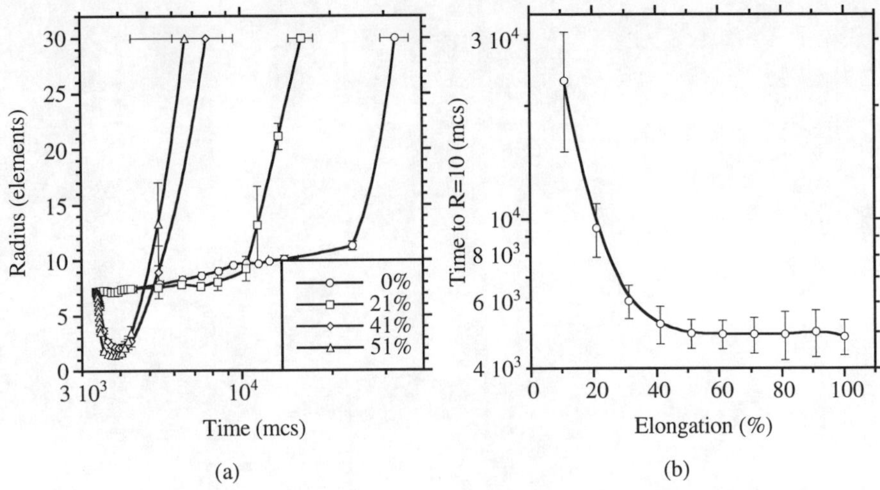

Figure 3. Grain growth kinetics: a) the average grain radius as a function of Monte Carlo time for four elongation values, and b) the Monte Carlo time required to achieve an average grain radius of ten elements as a function of the elongation. Error bars correspond to standard deviations of forty grain growth simulations at each elongation.

fragmentation. The additional driving force of the stored plastic energy serves to increase the coarsening rate with increasing deformation. This provides some insight into the thermomechanical fatigue failure of solder joints, which is precipitated by accelerated coarsening in locally deformed regions of the microstructure.

ACKNOWLEDGMENTS

This work was performed at Sandia National Laboratories under US DOE contract DE-AC04-94AL85000.

REFERENCES

[1] D.R. Frear, in *Solder Mechanics: A State of the Art Assessment,* edited by D.R. Frear, W.B. Jones, and K.R. Kinsman (TMS, Warrendale, PA, 1991), pp. 191-237.
[2] T.E. Buchheit, R.J. Bourcier, G.W. Wellman, and M.K. Neilsen, *Mater. Sci. Eng.* 5, pp. 421-437 (1997).
[3] U.F. Kocks, C.N. Tom, and H.-R. Wenk, *Texture and Anisotropy: Preferred Orientations in Polycrystals and Their Effect on Materials Properties.* 1998, Cambridge University Press: New York, NY. p. 676.
[4] M.P. Anderson, D.J. Srolovitz, G.S. Grest, and P.S. Sahni, *Acta Metall.* 32, pp. 783-91 (1984).
[5] D.J. Srolovitz, M.P. Anderson, P.S. Sahni, and G.S. Grest, *Acta Metall.* 32, pp. 793-802 (1984).
[6] P.A. Beck, in *Metal Interfaces,* edited by R.M. Brick (ASM, Cleveland, OH, 1952), pp. 208-47.
[7] N. Metropolis, A.W. Rosenbluth, M.N. Rosenbluth, A.H. Teller, and E. Teller, *J. Chem. Phys.* 21, pp. 1087-92 (1953).
[8] A.B. Bortz, M.H. Kalos, and J.L. Lebowitz, *J. Comp. Phys.* 17, pp. 10-8 (1975).
[9] G.N. Hassold and E.A. Holm, *Comp. in Phys.* 7, pp. 97-107 (1993).
[10] M.P. Anderson, G.S. Grest, and D.J. Srolovitz, *Phil. Mag. B* 59, pp. 293-329 (1989).

SURFACE CHEMISTRY OF CVD DIAMOND: LINKING THE NANOSCALE AND MESOSCALE MODELLING HIERARCHIES

I.I. Oleinik, D.G. Pettifor, A.P. Sutton*, C.C. Battaile**, D.J. Srolovitz***
and J.E. Butler****
*Department of Materials, University of Oxford, Parks Rd, Oxford OX1 3PH, UK
**Sandia National Laboratories, Albuquerque, NM
***Department of Materials Science and Engineering, University of Michigan, Ann Arbor, MI
****Gas/Surface Dynamics Section, Naval Research Laboratory, Washington, DC

ABSTRACT

The β-scission growth mechanism at the diamond (100)(2x1) surface is studied by a combination of nanoscale ab-initio LDA/GGA and semiempirical tight-binding techniques to provide the necessary input into the mesoscale variable time step Kinetic Monte-Carlo (KMC) simulations of CVD diamond growth. The reaction path of the beta-scission reaction is critically examined and the activation barrier of the reverse etching of the methylene adsorbate is deduced. Our quantum mechanical calculations support a previous semiempirical PM3 study confirming that the molecular mechanics values for the entalphy of the reaction are a factor of 2 wrong. This conclusion provides strong support for the preferential etching mechanism introduced into KMC to predict experimentally measured growth rates.

INTRODUCTION

Diamond chemical vapor deposition science and technology has developed dramatically over the last decade due to the highly promising and unusual properties of CVD diamond films. The applications are diverse and include hard coatings, tools, optical and electronic components, thermal management, corrosion protection, radiation detection [1,2]. Experimental studies of the growth processes have revealed complex chemistry and physics inside the CVD reactor. In the CVD environment diverse physico-chemical processes take place including activation of reactive species in the gas phase, transport of the reactive species to the growth surface, adsorption, nucleation and subsequent incoporation of the carbon species into the growing diamond film, see Fig. 1. Theoretical and experimental investigations have identified the important growth species, studied the nature and structures of the diamond surfaces, and measured and predicted the growth rates as a function of temperature and reactant fluxes [3,4,5].

However, there is still a gap between experiment and theory in the sense that the large experimental database on morphology, crystalline texture, twinning, extended and point defects, and impurity incorporation can not be rationalized consistently within a unified and self-contained theoretical framework. This is primarily due to a lack of understanding of the basic growth mechanisms on diamond surfaces, where, on one hand, the severe chemical environment precludes the monitoring of elementary reaction steps by surface science techniques, and on the other hand, the complexity and size of the system limit the scope and

Mat. Res. Soc. Symp. Proc. Vol. 538 © 1999 Materials Research Society

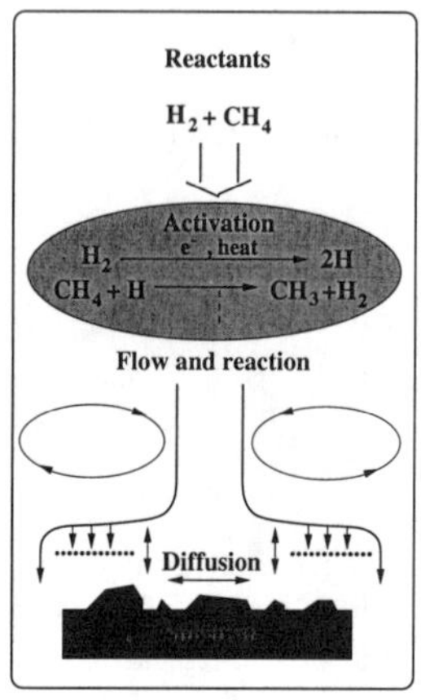

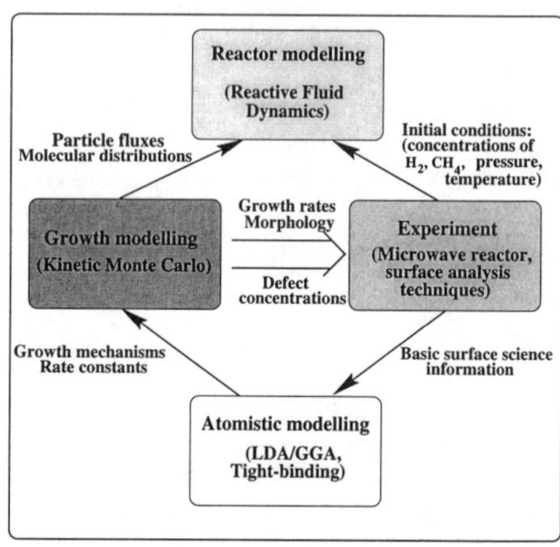

Figure1: Schematic of CVD process and multiscale modelling of CVD diamond growth

capability for modelling the growth. In principle this gap can be bridged by separating the system into subsystems with different time and length scales and applying appropriate modelling techniques to describe properly the phenomena at the specific level of detail. In the case of CVD diamond growth these hierarchies include the gas phase simulations of the CVD reactor, the mesoscale modelling of the crystal growth and the nanoscale modelling of the surface chemistry during deposition as illustrated in Fig. 1. During the last few years progress has been made in the mesoscale modelling of CVD processes by the kinetic Monte-Carlo (KMC) method [6,7,8,9]. Such a lattice-based model, which is fully three-dimensional, enables a detailed study at the atomic level of the structure and morphology of the surface during the growth. An important requirement for the KMC model is that it must simulate crystal growth over macroscopic time scales (seconds to minutes) so that the data on growth rate or film composition may be compared directly to existing experimental results. This requirement of long time scales can be accomplished by the use of the variable time step kinetic Monte Carlo method known as the N-fold Way [10]. This method introduces the temporal evolution into the growth sequence and discriminates between fast and slow reactions by proper increments of the time step.

The variable-time step KMC method has been successfully applied to the modelling of the chemical vapor deposition of diamond thin films [6,7,8,9]. Simulations corresponding to *hours* of growth for surfaces with *thousands* of atoms per layer were performed on desktop workstations. The calculated growth rates, on both (111)- and (110)- oriented surfaces, were in good agreement with experiment for temperatures ranging from 800 to 1500K. Furthermore, by simulating the incorporation of point defects, optimum growth temperatures for the production of high-quality films were predicted.

However, the ability of KMC to model realistically three-dimensional deposition processes depends crucially on a detailed understanding of the appropriate growth mechanisms and a reliable database of the rates of the surface chemical reactions. This was clearly demonstrated during the KMC growth modelling of the (100)(2x1) diamond surface. Simulations based on the surface kinetics of Harris and Goodwin [11] predict growth on the (100)(2x1) surface that is much faster than on the (110) and (111) surfaces in contradiction with experiment. In addition, atomic scale KMC produces atomically rough morphologies in disagreement with scanning probe and other experimental observations [12,13,14]. Apparently, some steps in the reaction sequence are missing or proceed with different rates. Thus, the nanoscale modeling of the surface chemistry on (100) diamond surface is crucial for reconciling the KMC modelling with experiment.

In this paper we show how quantum-mechanical modelling can aid in building the link between the individual chemical events and the mesoscopic growth model. We particularly focus on the issue of (100) surface growth and consider in detail the β-scission reaction mechanism which feeds into the KMC modelling chain. This will allow the proper kinetics for predicting the experimentally observed growth rates. In section II we outline the general methodology of quantum-mechanical modelling. In section III we consider the kinetics of the β-scission mechanism and its role in the preferential etching of CH_2 groups. This etching mechanism slows down the growth rate on (100) surface by allowing the removal of the undercoordinated carbon atoms and promoting the growth of the smooth regions of the (100)(2x1) surface.

METHODOLOGY OF QUANTUM-MECHANICAL MODELLING OF THE SURFACE CHEMICAL REACTIONS

To model successfully the chemical reactions that take place on a surface several requirements have to be fulfilled. Firstly, the process of bond making and bond breaking at the surface is accompanied by substantial rearrangements of the atoms around the reactive region. Even for covalently bonded systems, where the bonds are quite localized and the nature of the chemical interaction between the atoms can be modelled via a cluster, the surface relaxation and change in the strain during the course of the reaction have a strong influence on the reaction dynamics. Therefore, in order to take into account the geometrical constraints due to the presence of the surface and the underlying crystal lattice we have to employ quantum-mechanical methods that decribe properly the surface structures. Secondly, the chemical interactions at the transition state have to be modelled at the appropriate level of accuracy. The prediction of the transition state is a very complex and demanding matter due to the delicate balance of numerous factors such as a proper treatment of correlation, spin, size of basis set etc. Obviously, these two requirements of high accuracy yet large simulation cells are difficult to meet due to the high demand on the computer resources required for high quality calculations.

In order to find a compromise between these requirements, we employ a 'multiscale' methodology and divide the system under study into two regions with different length scales. The small area around the reaction zone is treated with high level ab-initio calculations but the large area surrounding this small cluster is treated with low level semi-empirical quantum mechanics but with the proper treament of the solid state environment. Within the small reactive region we employ the local density approximation with the generalized gradient corrections implemented in the code DMol, the MSI software package. Density functional theory has its own limitations and drawbacks especially for transition state calculations. However, the accuracy of 5-10 kcal/mol in barrier heights for hydrocarbon reactions achieved by LDA with gradient corrections [15] is acceptable for our purposes, when the appropriate data base of the gas-phase chemical reactions is available for comparison and guidance. Influence of the size of the basis set, the integration grid and other parameters were carefully studied to ensure the proper quality of the results. In addition, spin was taken into account for the correct description of the process of bond-making and bond-breaking. The size of the cluster is limited by the computational resources available and in our case it is 20-30 atoms.

The large area surrounding the reactive cluster consists of a slab of 5 atomic layers with 16 atoms in each layer. The strength of the carbon-carbon bonds as well as the local nature of the hydrocarbon chemical interactions ensures that the finite size of the slab has negligible influence on the reactive region. The quantum-mechanical forces are calculated within the two-centre, orthogonal tight-binding (TB) model. We use the Xu et al's parametrization for carbon-carbon interactions [16] and Davidson and Pickett [17] and Horsfield *et al* [18] paramentrization for C-H interactions. The H-H parameters were fitted to give the proper description of the dissociation of the hydrogen molecule. The tight-binding Hamiltonian is diagonalized at every geometry optimization step and forces are calculated using the Helman-Feynman theorem. An important ingredient of the tight-binding model is the local charge neutrality (LCN) constraint [19]. It is well known from ab-initio calculations that charge redistribution in hydrocarbon systems is quite small with the total charges of atoms close to zero. Therefore, the LCN constraint reflects the proper chemical behaviour of hydrocarbon bonding and also allows us to maintain the homolytic nature of the bond rupture. The tight-binding parametrization of Horsfield *et al* has been highly succesful in describing the energetics and geometries of the hydrocarbon molecules and diamond surfaces, so that we would expect a good overall performance of TB in predicting the energetics and bond geometries of *stable* adsorbate complexes at the diamond surface.

The interface between the tight-binding and LDA/GGA regions is accomplished according to the following scheme. The initial geometric configuration of atoms is determined by the TB structural relaxation of the slab with the constraints imposed on the bottom atoms saturated with hydrogens as well as constraints on the reacting atoms to follow the chosen reaction path. Then the LDA/GGA cluster of fixed size is cut from the TB slab and the C-C bonds that are broken are saturated with hydrogen atoms according to the standard procedure which takes into account the proper change of hybridization of the carbon atom during the structural relaxation. Thus, the shape of the cluster changes during the course of the reaction in order to match the movements of the atoms in the slab. The next step is the structural optimization of the LDA/GGA cluster by DMOL. The consistency with the TB geometry is enforced by keeping all the carbon atoms at the boundaries as well as the 'ghost' hydrogen atoms fixed in addition to the constraints imposed on the reacting atoms to follow the chosen reaction path. The constrained geometry optimization is accomplished with the use of the Perdew & Wang (1992) LDA functional [20]. In addition, the energy of the final geometry is calculated using a Becke-Perdew generalized gradient-corrected functional [20,21]. We have tested the difference in geometries optimised by LDA and GGA and found minor differences in the bond lengths and bond angles as was expected. However, the energy differences reflect the well known tendency of LDA to overbinding. The post-GGA corrects this problem of overbinding and gives and error of about 5 kcal for the binding energies of the hydrocarbon molecules and 5-10 kcal/mol for the activation barriers. To make reliable predictions of the transition state we refined the transition state geometry with pure GGA forces.

A very important question has to be answered: how consistent is the TB geometry with the geometry obtained by LDA? The answer is that the structures obtained by TB and LDA are almost the same except in the region in the vicinity of the transition state. It takes one to two geometry optimization steps to refine the minimum energy structure starting from the TB geometry. The substantial difference in the vicinity of the transition state can be explained by the deficiency of the TB parametrization to treat the long range C-C interactions that are not screened by the presence of other atoms. At the transition state the distance between the adsorbed methylene carbon and the carbon atom at the other end of the dimer bond is around 2.2 Å. This is very close to the TB cut-off so that the TB predicts zero interaction between these atoms. This is in contrast to the small but nonzero C-C interaction which can be seen in the DMol simulations from the change of the sp^2 hybridization of the methylene carbon to an intermediate hybridization which distorts the planar geometry of the methylene. Therefore, it is very important for predicting transition state structure to refine the cluster geometry by an LDA rather than a TB calculation.

QUANTUM MECHANICAL MODELLING OF THE β-SCISSION GROWTH MECHANSIM AT THE DIAMOND (100) SURFACE

The chemical vapor deposition of diamond proceeds through the adsorption and subsequent incorporation of various hydrocarbon fragments at the diamond surface. The characteristic feature of the CVD environment is the presence of a high flux of atomic hydrogen ($10^{19} - 10^{21}$ $1/cm^2sec$) which is created by the plasma, hot fialment or combustion flame. As a result of the rich abundance of hydrogen the diamond surface is fully hydrogenated although dangling bond radical vacancies appear from time to time as a result of atomic hydrogen abstraction. Several reaction mechanisms have been proposed to account for growth on the (100) 2x1 reconstructed surface. The most plausible one was proposed by Brenner, Garrison and collaborators based on the addition of a methyl radical and the subsequent insertion of a CH_2 group into the dimer bond following the hydrogen abstraction [22], see Fig. 2. (The reaction pathway for growth from methyl radicals at (100) surface). The most important reaction step in this sequence is the so called beta-scission reaction when the carbon in the methylene group forms a double bond with the dimer carbon atom and the dimer bond is broken as a result of the formation of this double bond. The insertion of the CH_2 group into the opened dimer is the key step for adsorbate incorporation into the diamond lattice.

The variable step kinetic Monte-Carlo simulations of diamond (100) growth based on the Harris-Goodwin reaction kinetics faced several difficulties in reproducing the well-known experimental observations. Firstly, the growth rate at (100) is too fast compared to the growth at the (110) and (111) faces. The diamond growth on (100), (110) and (111) faces requires the incorporation of one, two and three carbon atoms respectively. The KMC growth rate would reflect this trivial mechanistic rule giving the (100) face as the fastest and the (111) as the slowest growing faces. Obviously, this discrepancy between experiment and KMC kinetics questions the rates of the reactions in the Harris-Goodwin table or indicates

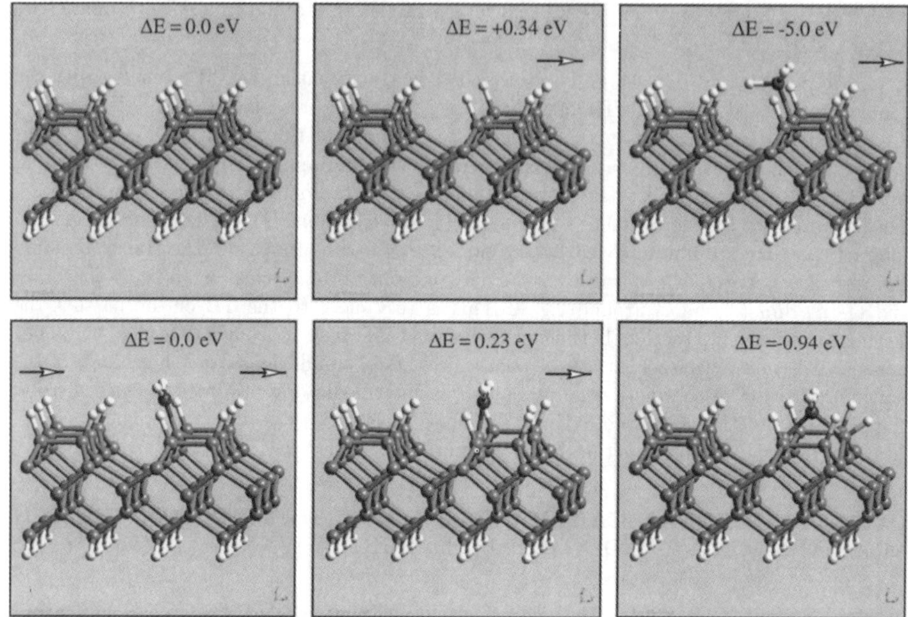

Figure 2: Growth mechanism on diamond (100)(2x1) surface.

the presence of other reactions which were not considered previously. The preferential etching mechanism was introduced in order to slow down the growth rate on the (100) surface. It was assumed in the KMC modelling that the Brenner- Garrison reaction of methylene insertion into the dimer bond is totally irreversible. However, the reverse sequence can make possible the de-insertion of the methylene and the subsequent abstraction of this group by atomic hydrogen. A two step etching rate can be deduced from the Harris-Goodwin reaction set by assuming that the rate limiting step of etching is the hydrogen abstraction of the methylene group. Therefore, the fast reactions of insertion and de-insertion are in equilibrium. This gives the rate of etching per carbon site $R_{etch} = A_e exp(-28.3/RT)$. The activation barrier of 28.3 kcal/mol comes from the equilibrium constant of reactions (a) and (b) which is determined by the difference in enthalpies of the products and reactants of the direct insertion reaction. The prefactor A_e is equal to $A_e = 10^{14}[H]$, where $[H]$ is the atomic hydrogen concentration. At typical CVD conditions $[H] = 10^{-9} mol/cm^3$ and the absolute rate of etching at 1200 K is 0.76 sec^{-1}. The hydrogen abstraction rate at the diamond surface occurs at the rate of 10^7 sec^{-1}, which provides the natural time scale in the system. Compared with this time, the etching rate according to Harris and Goodwin kinetic model is much too slow.

It was thought likely that the molecular mechanics data of Harris and Goodwin are in error. Therefore, the KMC modelling introduced an etching rate to match the experimental data. The KMC "experimetal etching rate" at standard CVD conditions is $2 \cdot 10^4$ sec^{-1} which is several orders of magnitude larger than the Harris/Goodwin number. Obviously, a thorough reinvestigation of the whole reaction sequence from the nanoscale modelling was required by the KMC in order to provide a firm quantum-mechanical basis for the proposed etching mechanism.

There are two critical parameters in the Harris-Goodwin data which determine the kinetics of the Brenner-Garrison dimer insertion reaction. The first one is the activation barrier of the β-scission reaction, i.e the breaking of the dimer bond and the subsequent insertion of methylene into the dimer bond. The second one is the heat of the beta-scission reaction which determines the rate of the reverse process for deinsertion of the methylene group from the dimer. Therefore, we have reexamined critically the reaction path of the beta-scission reaction by using the multiscale quantum-mechanical approach outlined in the preceding section. This particular surface reaction has a single natural reaction coordinate, that is the coordinate of the methylene carbon along the direction of the dimer bond. Therefore, we implemented a simple scheme for the one-dimensional search for the transition state. We constrained the carbon atom to be in the plane perpendicular to the reaction coordinate. All the other atoms in the reactive region were allowed to move and the system was optimized to get the energy minimum at every point along the reaction coordinate.

The potential energy profile along the reaction coordinate is shown in Fig. 3 with the appropriate structures of the reactants, products and transition state. The activation barrier E_a and the heat of the beta-scission reaction H are 13.6 kcal/mol and 12.3 kcal/mol. These numbers should be compared with the Harris-Goodwin data $E_a = 8.8$ kcal/mol and $H = 28.3$ kcal/mol. The Harris-Goodwin activation barrier is quite close to our barrier and the reason is quite simple. This number has been obtained in the work of [23] by employing high quality CI tehniques. However, the heat of the reaction has been obtained with the use of molecular mechanics. Therefore, the 100 % error is not surprising. The deficiency of the MM3 force field in this particular situation is explained by the application of the method beyond the region of its validity. Interestingly, the activation barrier and the heat of the reaction are in good agreement with PM3 calculations by S. Skokov *et al* [24], that is $E_a = 15.3$ kcal/mol and $H = 14.6$ kcal/mol.

What are the implications of the results for KMC modelling? Firstly, we predict the activation barrier for the etching rate to be two times smaller than the Harris-Goodwin barrier. Our absolute value of the etching rate is in a good agreement with the KMC "experimental" value ($6\ 10^4 \text{sec}^{-1}$ compared to $2\ 10^4 \text{sec}^{-1}$ respectively). Thus, our calculations support the preferential etching mechanism which was suggested by KMC. Secondly, the difference in the activation energies will result in a different temperature dependence for the net etching rate. A deviation of the KMC growth rate from experimental data has been observed during the KMC runs, so that this issue has to be reexamined more thoroughly with the use of the newly derived etching rates.

Another interesting aspect of preferential etching is that it can explain the growth of the atomically smooth 2x1 dimer reconstructed areas. The ordering would be present of the atomic hydrogen can etch the isolated methylene groups inserted into dimer, thus disallowing the island growth mode. The formation of ordered domains would then be possible if the etching of already assembled etching groups were highly prohibited from the thermodynamical point of view. For this purpose we calculated the heat of removal of the CH_2 group from the middle of the three bridge methylene complex and from the side of the three bridge complex. We have obtained the following energies for the etching reactions : from

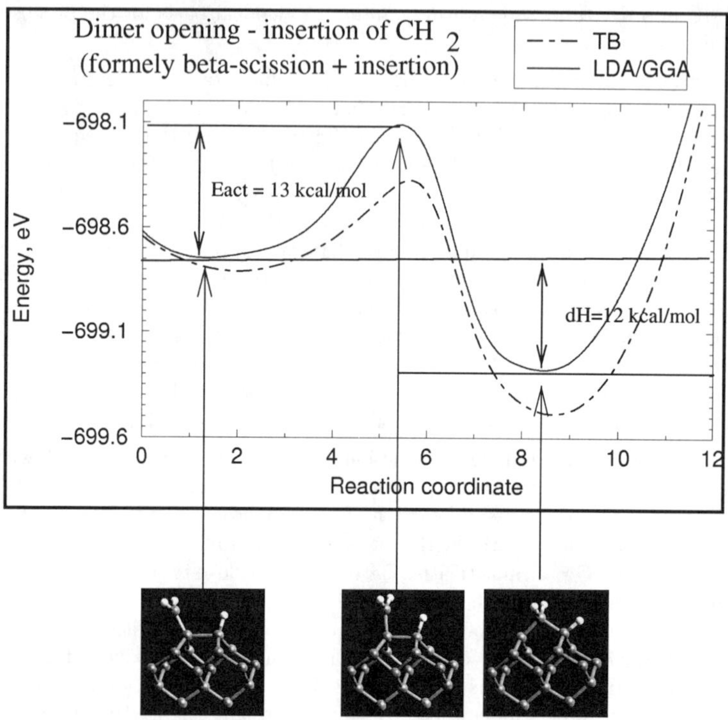

Figure 3: Reaction path of the β-scission reaction.

the middle - 100.8 kcal/mol, from the side - 39.4 kcal/mol. These numbers can be compared with the heat of etching of the isolated CH_2 unit - 15.68 kcal/mol. Two conclusions can be drawn from these results: (i) the etching of the isolated CH_2 unit is much more feasible as compared to etching from an already assembled domain (ii) etching from the side end of the domain is thermodynamically more preferable as compared to etching from the middle of the island, thus disallowing the formation of voids.

CONCLUSIONS

We have considered the β-scission mechanism of the diamond (100) growth and have demonstrated that the quantum-mechanical modelling has been successful in predicting the key parameters governing the kinetics at the atomic scale. Feeding the information into the KMC variable time step kinetic Monte Carlo model we have achieved agreement with the experimentally measured growth rates. This demonstrates the importance of linking together different modelling hierarchies for realistic modelling of the CVD growth of diamond films. The overall success of this approach encourages us to apply this methodology to other areas of vapour phase processing.

ACKNOWLEDGEMENTS

We gratefully acknowledge the support of the Defence Advanced Research Projects Agency and the Naval Research Laboratory under contracts N00014-96-1-G000 and N00014-96-1-G001. The computations were performed with the use of the computer facilities in the Materials Modelling Laboratory at the Department of Materials, University of Oxford.

References

[1] *Diamond Films: Recent Developments*, edited by D.M Gruen and I. Buckley-Golder, MRS Bulletin **23**, (9) (1998).

[2] *Handbook of Industrial Diamonds and Diamond Films,* edited by M.A. Prelas, G. Popovici and L.K. Bigelow (Mracel Dekker Inc., New York, 1998) p. 527.

[3] D.G. Goodwin and J.E. Butler, in *Handbook of Industrial Diamonds and Diamond Films,* edited by M.A. Prelas, G. Popovici and L.K. Bigelow (Mracel Dekker Inc., New York, 1998).

[4] M.P. D'Evelyn, in *Handbook of Industrial Diamonds and Diamond Films,* edited by M.A. Prelas, G. Popovici and L.K. Bigelow (Mracel Dekker Inc., New York, 1998) p. 89.

[5] M. Frenklach and S. Skokov, J. Phys. Chem. B **101**, 3025 (1997).

[6] C.C. Battaile, *Atomic-scale Kinetic Monte Carlo Simulations of Diamond Chemical Vapor Deposition,* PhD thesis, University of Michigan, 1998.

[7] C.C. Battaile, D.J. Srolovitz and J.E. Butler, J. Appl. Phys. **82**, 6293 (1997).

[8] D.J. Srolovitz, D.S. Dandy, J.E. Butler, C.C. Battaile and Paritosh, JOM **49**, 42 (1997).

[9] M. Frenklach, J. Chem. Phys. **97**, 5794 (1992).

[10] A.B. Bortz, M.H. Kalos and J.L. Ledowitz, J. Comp. Phys **17**, 10 (1975).

[11] S.J. Harris and D.G. Goodwin, J. Phys. Chem. **97**, 23 (1993).

[12] Y. Kuang, Y. Wang, N. Lee, A. Badzian, T. Badzian and T.T. Tsong, Appl. Phys. Lett. **67**, 3721 (1995).

[13] B.D. Thoms and J.E. Butler, Surf. Sci. **328**, 291 (1995).

[14] H. Sasaki, M. Aoki and H. Kawarada, Diam. Rel. Mater. **2**, 1271 (1993).

[15] A.C. Scheiner, J. Baker and J.W. Andzelm, J. Comput. Chem. **18**, 775 (1997).

[16] C.H. Xu, C.Z. Wang, C.T. Chan and K.M. Ho, J. Phys. Condens. Matter **4**, 6047 (1992).

[17] B.N. Davidson and W.E. Pickett, Phys. Rev. B **49**, 11253 (1994).

[18] A.P. Horsfield, P.D. Goodwin, D.G. Pettifor and A.P. Sutton, Phys. Rev. B **54**, 15773 (1996).

[19] A.P. Sutton, M.W. Finnis, D.G. Pettifor and Y. Ohta, J. Phys C **21**, 35 (1988).

[20] J.P. Perdew and Y. Wang, Phys. Rev. B **45**, 13244 (1992).

[21] A.D. Becke, J. Chem. Phys. **88**, 2547 (1988).

[22] B.J Garrison, E.J. Dawnkaski, D. Strivastava and D.W. Brenner, Science **225**, 835 (1992).

[23] C.B. Musgrave, S.J. Harris, W.A. Goddard, Chem. Phys. Lett. **247**, 359 (1995).

[24] S. Skokov, B. Weiner and M. Frenklach, J. Phys. Chem. **98**, 8 (1994).

SELF-CONSISTENT KINETIC LATTICE MONTE CARLO

ANDREW HORSFIELD * SCOTT DUNHAM ** HIDEAKI FUJITANI [†]
*Fujitsu European Centre for Information Technology, 2 Longwalk Road, Stockley Park, Uxbridge, UK
**Boston University, Department of Electrical and Computer Engineering, 8 St. Mary's Street, Boston, MA 02215
[†]Fujitsu Laboratories Ltd., 10-1 Morinosato-Wakamiya, Atsugi 243-01, Japan

ABSTRACT

We present a brief description of a formalism for modeling point defect diffusion in crystalline systems using a Monte Carlo technique. The main approximations required to construct a practical scheme are briefly discussed, with special emphasis on the proper treatment of charged dopants and defects. This is followed by tight binding calculations of the diffusion barrier heights for charged vacancies. Finally, an application of the kinetic lattice Monte Carlo method to vacancy diffusion is presented.

INTRODUCTION

Diffusion of dopants and point defects in semiconductors is characterized by rapid single events interspersed by long periods of no important activity. In this regime stochastic methods are applicable. The central argument is that during the long inactive periods, atomic vibrations randomize the positions and momenta of the atoms (within a highly restricted phase space), after which a single important event then occurs. Thus we can treat separate events as independent, and occurring at random. The numerical method for studying such processes is Monte Carlo.

To implement these methods, it is necessary to know the diffusion barrier heights and interaction energies between particles. Some values can be found from experiment, but many others can only be obtained by means of computation.

In the following sections we present: an outline of the theory of kinetic lattice Monte Carlo (KLMC), including the underlying statistical mechanical equations and approximations, the additional features required to make the method kinetic, and a discussion of how to evaluate energies and charge distributions. This is followed by tight binding calculations of the diffusion barrier height of silicon vacancies. We conclude with some results from a KLMC simulation involving vacancies.

KLMC THEORY

Statistical Mechanics

The following arguments will all be developed within the framework of classical statistical mechanics. The same results can obtained from quantum mechanical statistical mechanics. The general picture is of atoms vibrating with small amplitudes (compared with the inter-atomic spacing) about fixed sites, and then occasionally (that is, with a time scale much greater than the atomic vibrational period) hopping to new sites. We are thus free to treat the atoms within the harmonic approximation while they vibrate about their lattice sites. Hopping between sites is clearly not accurately described by the harmonic approximation. But since these events are so rare we can neglect them initially.

The classical Hamiltonian (H) is a function of the momenta (represented by $\vec{p}$) and positions (represented by $\vec{r}$) of the particles. The probability density for finding the system in the neighborhood of the phase space point $(\vec{p}, \vec{r})$ is $P(\vec{p}, \vec{r}) = \exp\left(-\beta H(\vec{p}, \vec{r})\right)/Z$, where $\beta = 1/k_B T$, k_B is Boltzmann's constant and T is the temperature. The partition function Z is defined by $Z = \int d\vec{p} \int d\vec{r} \exp\left(-\beta H(\vec{p}, \vec{r})\right)$.

For the purposes of a Monte Carlo simulation, the momentum is considered to be randomized between hopping events. Since we are not interested in the momentum, we can integrate it out to obtain the probability of the system being in the neighborhood of some set of spatial coordinates. If we assume that the Hamiltonian separates as $H(\vec{p}, \vec{r}) = K(\vec{p}) + V(\vec{r})$, we get the simple result $P(\vec{r}) = \exp\left(-\beta V(\vec{r})\right)/Q$, where the classical configurational integral Q is defined by $Q = \int d\vec{r} \exp\left(-\beta V(\vec{r})\right)$.

We now introduce the harmonic approximation for the potential energy $V(\vec{r})$. This allows us to further simplify the equations, and to transform the problem from one involving continuous variables ($\vec{r}$) to one

Mat. Res. Soc. Symp. Proc. Vol. 538 © 1999 Materials Research Society

containing discrete variables. First we define the set of sites about which the atoms vibrate to be $\vec{R}_i$ for some configuration i. These sites correspond to a local energy minimum so when the atoms occupy these sites they experience no force. Let displacements from these sites be given by $\vec{u}$, so that $\vec{r} = \vec{R}_i + \vec{u}$. Within the harmonic approximation the potential energy can be written as $V(\vec{r}) \approx V(\vec{R}_i) + \frac{1}{2}\vec{u}\mathbf{D}_i\vec{u}$, where $\mathbf{D}_i$ is the dynamical matrix for configuration i.

The next approximation is to replace the integral over $\vec{r}$ with the combination of a sum over i and an integral over $\vec{u}$. The justification for this is as follows. Values of $\vec{u}$ that take the system far from the energy minimum lead to large values of energy for the system. These will contribute little to configurational averages provided the temperature is not too high (in which case the harmonic approximation no longer holds). Thus the original integral and the combination of sum and integral will give approximately the same weight to each configuration, and thus produce similar configurational averages quantities.

For Monte Carlo simulations consisting of transitions between discrete configurations, we need the total probability that the system lies in the neighborhood of a specific energy minimum. To obtain this, we carry out the integral over $\vec{u}$. If we make the final approximation that the dynamical matrix is independent of configuration (which is not guaranteed to be a good approximation in general), we get the probability that the system is in configuration i to be

$$P(\vec{R}_i) = \frac{\exp\left(-\beta V(\vec{R}_i)\right)}{\displaystyle\sum_j \exp\left(-\beta V(\vec{R}_j)\right)}. \tag{1}$$

<u>Kinetics</u>

The result obtained above allows us to obtain configurational averages (static quantities), but does not tell us anything about the time evolution of configurations, which is important for calculating dynamical quantities, as well as providing us with the history of a system.

If we were equipped with a table giving the transition rates between all configurations we could generate a history in the following manner. We start from some particular configuration (call it i). It can make transitions to its neighboring configurations (indexed by j) with rates $W(i \rightarrow j)$. The probability that any one transition will occur within any given short time (much smaller than the time required for the fastest process) is proportional to the rate for that transition. The average time required for a transition to occur is $\tau_i = \left[\sum_j W(i \rightarrow j)\right]^{-1}$. Thus a transition is chosen at random from the list of allowed transitions with a probability proportional to the rate. This takes us to configuration j. The process is then repeated starting with configuration j, and so on until a complete history has been created.

To reduce the number of rates that have to be explicitly tabulated, we can use detailed balance, which states $P(\vec{R}_i)W(i \rightarrow j) = P(\vec{R}_j)W(j \rightarrow i)$. Combining this with Eq. (1) we obtain $W(i \rightarrow j)/W(j \rightarrow i) = \exp\left(-\beta\left[V(\vec{R}_j) - V(\vec{R}_i)\right]\right)$. To obtain absolute rates, some form must be assumed for the rates. If we assume that the energy barrier to diffusion varies linearly with the energy difference between the initial and final states we get

$$W(i \rightarrow j) = W_0 \exp\left(-\frac{\beta}{2}\left[V(\vec{R}_j) - V(\vec{R}_i)\right]\right), \tag{2}$$

where W_0 is a rate which corresponds to a transition with no energy change, but which is also closely related to the transition whose rate is required. It should be noted that (for example) a vacancy/silicon exchange will have a different value of W_0 from a vacancy/dopant exchange. Thus large contributions to the energy barrier height are treated carefully.

<u>Energy Expansion and Charge Distribution</u>

The energies can be calculated according to a simple many-center perturbation expansion about the perfect lattice. For example, adding one vacancy to the crystal raises the energy by an amount $\epsilon^{(1)}(v)$. Adding N well separated vacancies will thus raise the energy by an amount equal to $N\epsilon^{(1)}(v)$. As the vacancies get closer together they begin to interact. This interaction will be pairwise initially, but higher order terms could become important as the density increases. In general we have

$$V = \sum_i \epsilon^{(1)}(d_i, q_i, \vec{x}_i) + \frac{1}{2}\sum_{i \neq j} \epsilon^{(2)}(d_i, q_i, \vec{x}_i; d_j, q_j, \vec{x}_j) + \dots \tag{3}$$

where i and j are site indices, d_i is the type of defect on site i, $\vec{x}_i$ is the position of site i and q_i is the charge on the defect on site i. Determination of the charges is discussed below.

There are a number of contributions to the energy that can be isolated. These include: the energy associated with covalent bond formation; elastic energy from strain fields; the energy associated with the Coulomb interaction between charged defects; and the energy associated with the population of gap states. These separate contributions behave in different ways. The first (covalent bond formation) is very short ranged and is associated with energies of the order of 1eV. The next two (elastic and Coulomb energy) are longer ranged but weaker. The population of gap states is of unlimited range, and has a value of order the size of the electron energy gap.

The energy associated with covalent bonding is complicated and can best be handled by being tabulated as a function of cluster type (such as isolated defect, defect pairs with atoms on different sites and so on) in precisely the way encapsulated by the many-site expansion. The elastic energy can be represented by a pair potential, and thus is also easily contained within the many-site expansion.

The Coulomb energy and the energy associated with the population of the gap states must be treated together in a self-consistent manner, as the charge on a defect is a result of the population of the corresponding gap state. The distribution of charges needed to evaluate this expression is found by minimizing the free energy (found by adding a configurational entropy to the total energy given in Eq. (3)) with respect to the charge distribution. This gives:

$$\langle n_{dq}(\vec{x})\rangle = \langle n_d(\vec{x})\rangle \, \frac{\exp\left(-\beta[\varepsilon_{dq}(\vec{x}) - \mu q]\right)}{\sum_{q'} \exp\left(-\beta[\varepsilon_{dq'}(\vec{x}) - \mu q']\right)}, \tag{4}$$

where $\langle n_{dq}(\vec{x})\rangle$ is the density of defects of type d and charge q located at position $\vec{x}$, $\langle n_d(\vec{x})\rangle$ is the density of defects of type d at position $\vec{x}$ summed over all charges, $\varepsilon_{dq}(\vec{x}) = \delta V/\delta n_{dq}(\vec{x})$, and μ is the chemical potential for the electrons, and is determined by imposing charge neutrality on the whole system.

The problem of including excitations from the valence band and into the conduction band is difficult, so some simple scheme must be used instead. To motivate the scheme we note the phenomena that need to be described: the most important excitations are from the valence band into acceptor states and from donor states into the conduction band; band bending produced by a positively (negatively) charged dopant will tend to cause electrons in the conduction band (holes in the valence band) to accumulate near the dopant. These observations suggest that we can treat the valence and conduction band as a set of localized states uniformly distributed throughout the system. In the absence of dopants they are degenerate, but the degeneracy is broken by local electrostatic fields produced by charged dopants. Empty conduction states and occupied valence states are treated as neutral. The conduction states are allowed to acquire a charge of -1 when they are populated. The valence states are allowed to acquire a charge of $+1$ when they are depopulated. These localized states can now be treated in the same way as the dopant and defect states. The density of these states is fixed by requiring that the number of holes in the valence band and electrons in the conduction band be correct.

TIGHT BINDING CALCULATIONS

To obtain the most accurate numbers, a very well converged *ab initio* computation is required[1, 2]. However, these are very expensive calculations. Tight binding has proven to be a very successful model for a number of covalent materials, including silicon. Here we use the tight binding model of Bowler *et al.*[3], with self-consistency applied by means of a modified version of the scheme of Elstner *et al.*[4].

Self-consistency is introduced by Elstner *et al.* by means of the addition of the following term to the usual tight binding expression for the total energy: $\Delta E = \frac{1}{2}\sum_{i,j} q_i q_j \gamma_{i,j}$, where $\gamma_{i,j} = \int d\vec{r} d\vec{r}' F_i(|\vec{r} - \vec{X}_i|) F_j(|\vec{r}' - \vec{X}_j|)/|\vec{r} - \vec{r}'|$. This expression ignores any contribution from exchange and correlation, which is small. The function $F_i(r)$ is a spherical charge density, normalized to 1. In the original paper this was taken to have an exponential form. Here a Gaussian is used as this leads to simple closed form expressions for both clusters and periodic systems. Using Gaussians $(F_i(r) = (a_i/\pi)^{3/2}\exp(-a_i r^2))$ we get (with energies in rydbergs, and lengths in Bohr radii) $\gamma_{i,j} = (2.0/x_{ij})\mathrm{erf}\left(\sqrt{a_i a_j/(a_i + a_j)}x_{ij}\right)$. The values of the the the decay constants a_i are found from isolated atoms using the expression $a_i = \pi U_i^2/8$, where U_i is the second derivative of the total atomic energy with respect to charge. U_i is found from the experimentally determined first ionization energy (I_i) and electron affinity (A_i): $U_i = I_i - A_i$.

The net charges on each site are evaluated using Mulliken population analysis: $q_i = \sum_{\alpha, j\beta, n} C^{(n)}_{i\alpha} f_n C^{(n)}_{j\beta} \times S_{j\beta, i\alpha} - Z_i$, where α and β are orbital indices, and n is an index running over eigenstates. The occupancy of state n is f_n, and the eigenvector is $C^{(n)}_{i\alpha}$. The overlap integral between two orbitals on sites i and j is $S_{i\alpha, j\beta}$. The core charge on site i is Z_i.

We now define the energy shift w_i on site i due to the charges on all sites to be $w_i = \sum_j q_j \gamma_{i,j}$. The self-consistent tight binding Hamiltonian now becomes: $H_{i\alpha, j\beta} = H^{(0)}_{i\alpha, j\beta} + \frac{1}{2}(w_i + w_j) S_{i\alpha, j\beta}$, where $H^{(0)}_{i\alpha, j\beta}$ is the non-self-consistent tight binding Hamiltonian.

The force on atom k is given by $\vec{F}_k = \vec{F}^{(0)}_k - \sum_{i\alpha, j\beta, n} C^{(n)}_{i\alpha} f_n C^{(n)}_{j\beta} \frac{1}{2}(w_i + w_j) \partial S_{i\alpha, j\beta} / \partial \vec{r}_k - \frac{1}{2} \sum_{i,j} q_i q_j \partial \gamma_{i,j} / \partial \vec{r}_k$, where $\vec{F}^{(0)}_k$ is the force evaluated using the expression for non-self-consistent tight binding. Thus we see that this form of self-consistency can be bolted on to any tight binding model.

For solids w_i must be evaluated in reciprocal space. If we define $\tilde{\gamma}_{i,j}(q) = (8\pi/q^2 V_{cell}) \exp(-(a_i + a_j)q^2/4a_i a_j)$, then we get $w_i = \sum'_{j,\vec{G}} q_j \tilde{\gamma}_{i,j}(G) \exp(i\vec{G} \cdot (\vec{d}_i - \vec{d}_j))$, where i and j span only one unit cell, V_{cell} is the volume of a unit cell, $\vec{G}$ is a reciprocal lattice vector, and the prime on the sum means that the $G = 0$ term is dropped.

<u>Barrier Height Calculations</u>

For charged defects there is a problem with using periodic boundary conditions: the interaction of a charged cell with its periodic images leads to an infinite electrostatic energy. This is conventionally overcome by adding in a uniform background charge that makes the unit cell charge neutral and is the approach taken here. The total energy then varies with respect to cell size as the inverse of the linear dimension of the cell[5].

The diffusion barrier heights are found using the following simple procedure. Each hopping process consists of the movement of one atom (the hopping atom) from one known site to another. The hopping atom is placed at several points along the line joining these two points (the reaction coordinate), and the unit cell is relaxed. During a relaxation the hopping atom is constrained to lie in a plane perpendicular to the reaction coordinate. One atom far from the hopping atom is held fixed to prevent a translation of the whole unit cell. The imposition of periodic boundary conditions prevents rotation of the cell. For each point the total energy is recorded, and the set of points forms a curve with an energy minimum near each end, and an energy maximum near the middle. The difference between the maximum and minimum energies is the energy barrier height for the transition. Since the barrier heights are found as small differences between large numbers (the energy of a simulation cell), it is important that the total energy minimization be carried out very carefully. A relaxation is stopped when the largest atomic force is below 0.0001 Ry/a_0.

As a test of convergence with respect to cell size the barrier height for the neutral vacancy is evaluated for two cell sizes. A cubic cell of silicon (with one vacant site) in the diamond structure is used containing either 63 atoms or 215 atoms. Only the Γ point is used to sample the Brillouin zone, and self-consistency is not used. The barrier height for the 63 atom cell is 0.30eV and for the 215 atom cell is 0.36eV. The experimental value is 0.45eV [6]. The larger unit cell gives better accuracy, and so is used for the barrier calculations.

N	$\epsilon^f_v(0)$(eV)	$\epsilon^f_v(-1)$(eV)
64	3.81 (3.67)	4.40 (4.07)
216	3.60 (3.31)	4.28 (3.88)

Table I: Convergence of energy with cell size. The numbers in parenthesis are obtained from the well converged planewave calculations described in Ref. [2]. For the negative vacancy the chemical potential of the electrons is taken to coincide with the top of the valence band.

Given that we are studying charged defects, for which the convergence with respect to cell size is expected to be slow, a test of the rate of convergence with respect to cell charge was carried out. For this test, only the vacancy formation energy was used. The sampling of the Brillouin zone was carried out with a $2 \times 2 \times 2$ and a $3 \times 3 \times 3$ mesh of k points for the 216 and 64 atom unit cells respectively. This gives the same k point density for each calculation. From Table I we can see that charged cells appear to converge faster than the neutral cells, which must be due to a cancellation of errors. We can conclude from this that charged cells do not converge any more slowly than neutral ones.

q	Tight binding	Experiment
-2	-	0.18
-1	-	-
0	0.39	0.45
+1	0.30	-
+2	0.21	0.32

Table II: The vacancy diffusion barrier height (in eV) as a function of charge state. The experimental values are taken from Ref. [6]

The values of the calculated and experimental barrier heights are given in the Table II. Calculated barrier heights are not shown for the negative vacancies because of problems with the calculations. For the double negative vacancy a *negative* barrier was found. This follows from the fact that within the model used here the lowest energy state occurs when the diffusing atom lies midway between its starting site, and the vacant site towards which it is diffusing. This state is stable because the diffusing atom can form six bonds at this position. For the single negative vacancy the diffusion barrier appears to have more than one peak, a result we have little confidence in given that the symmetry of the relaxed negative vacancy as found from these tight binding calculations disagrees with experiment and more accurate calculations.

The neutral and positive vacancies appear to be well described, however. The trend in variation of the barrier heights with charge state agrees with experiment, and the values are very respectable. It appears therefore that this orthogonal tight binding model can give a reliable description of the behavior of neutral and positive vacancies, but not of negative vacancies.

KLMC SIMULATION

As an illustration of the method described above, a simulation has been carried out of the annealing of a high concentration of vacancies at 600K. A simulation cell containing 8×10^6 lattice sites, and 100 vacancies distributed randomly throughout the cell was set up. This is equivalent to a concentration $6.2 \times 10^{17} \mathrm{cm}^{-3}$. The equilibrium concentration at 600K is $9 \times 10^{-7} \mathrm{cm}^{-3}$. However, much higher concentrations can be created by implantation or nitridation of the surface[11]. Charges were assigned at random to the vacancies so as to give the equilibrium distribution described by Eq. (4). There were three single negative vacancies, with the remainder being neutral. The three electrons excited into the vacancies came from the valence band. There were no excitations directly from the valence band into the conduction band. The experimental band gap for 600K of 1.04eV was used. The simulation was carried out for 2×10^6 Monte Carlo steps, equivalent to a time of $25\mu s$.

The values of the formation energy for the neutral vacancy was taken from a density functional calculation[7]. The values for the charged vacancies ($E_v^f(q)$) were found by adding the appropriate experimental ionization level values[8, 9, 10]: $E_v^f(-2) = 5.29\mathrm{eV}$, $E_v^f(-1) = 4.23\mathrm{eV}$, $E_v^f(0) = 3.63\mathrm{eV}$, $E_v^f(1) = 3.58\mathrm{eV}$ and $E_v^f(2) = 3.45\mathrm{eV}$. The binding energy for a neutral vacancy pair was taken from a density functional calculation[1], and has the value 1.6eV. The same value was used for all vacancy-vacancy interactions regardless of charge state. This is clearly an approximation, but should not change the final result. The binding is sufficiently large that neutral pairs will not break up during the course of the simulation. For singly charged systems the binding energy increases, whereas for doubly charged systems the decrease is small (about 0.1eV).

The hopping barriers for the double positive, double negative and neutral vacancy were taken from experiment[6], and the singly charged vacancy barrier heights were obtained by taking the average of the neutral and the corresponding doubly charged vacancy barrier heights. The barrier heights as a function of charge state ($E_v^b(q)$) are thus: $E_v^b(-2) = 0.18\mathrm{eV}$, $E_v^b(-1) = 0.315\mathrm{eV}$, $E_v^b(0) = 0.45\mathrm{eV}$, $E_v^b(1) = 0.385\mathrm{eV}$ and $E_v^b(2) = 0.32\mathrm{eV}$.

After $0.24\mu s$ of the anneal, two of the charged vacancies had paired up with neutral vacancies, and four neutral vacancies had paired up to form two neutral pairs. After $0.6\mu s$ all three charged vacancies had paired up with neutral vacancies, and six neutral vacancies had paired up to form three neutral pairs. By the end of the anneal many of the neutral vacancies had formed clusters, and several of the small clusters had grown to form larger clusters containing as many as four vacancies.

These results are easily understood in terms of the variation in barrier height with charge state. The negative vacancies have a lower diffusion barrier height than the neutral vacancies, so diffuse faster. Thus, in a given time they explore more of the simulation cell and so become trapped into clusters sooner.

CONCLUSION

The study of the diffusion of dopants and defects in semiconductors can be treated at the atomic level by means of a Monte Carlo based method as the important motion of mobile particles can be thought of as random hops between lattice sites. A method for evaluating the hopping rate has been presented which takes into account the influence of neighboring particles and the charge state of the mobile particles.

A simple tight binding model has been described and used to evaluate the diffusion barrier height for silicon in a range of charge states. The model is found to give respectable agreement with experiment for positive and neutral vacancies, but breaks down for negative vacancies.

The diffusion of a high concentration of vacancies, starting from a random distribution, has been studied. It is found that the charged vacancies combine with neutral vacancies faster than the neutral vacancies do. This is easily understood from the facts that negative vacancies have a lower diffusion barrier height than the neutral vacancy, and vacancies can lower their energies by clustering.

ACKNOWLEDGEMENTS

Useful discussions with Prof. T. Frauenheim on modeling charged defects with periodic boundaries are gratefully acknowledged, as are helpful conversations with Dr. M. Fearn about the properties of point defects.

REFERENCES

1. M. Pesola, J. von Boehm, S. Pöykkö and R. M. Nieminen, Phys. Rev. B **58**, 1106 (1998).

2. M. J. Puska, S. Pöykkö, M. Pesola and R. M. Nieminen, Phys. Rev. B **58**, 1318 (1998).

3. D. R. Bowler, M. Fearn, C. Goringe, A. P. Horsfield and D. G. Pettifor, J. Phys.: Condens. Matter **10**, 3719 (1998).

4. M. Elstner, D. Porezag, G. Jungnickel, J. Elsner M. Haugk, Th. Frauenheim, S. Suhai and G. Seifert, Phys. Rev. B **58**, 7260 (1998).

5. G. Makov and M. Payne, Phys. Rev. B **51**, 4014 (1995).

6. G. D. Watkins, in *Deep Centers in Semiconductors*, edited by S. T. Pantelides (Gordon and Breach, New York, 1986).

7. J. L. Mercer, J. S. Nelson, A. F. Wright and E. B. Stechel, Modelling Simul. Mater. Sci. Eng. **6**, 1 (1998).

8. G. D. Watkins, in *Lattice Defects in Semiconductors 1974*, edited by F. A. Huntley (Institute of Physics, London, 1975).

9. G. D. Watkins, J. R. Troxell and A. P. Chatterjee, in *Defects and Radiation Effects in Semiconductors 1978*, edited by J. H. Albany (Institute of Physics, London, 1979).

10. J. L. Newton, A. P. Chatterjee, R. D. Harris and G. D. Watkins, Physica **116B**, 219 (1983).

11. P. M. Fahey, P. B. Griffin and J. D. Plummer, Rev. Mod. Phys. **61**, 289 (1989).

LINKING *ab initio* ENERGETICS TO EXPERIMENT: KINETIC MONTE CARLO SIMULATION OF TRANSIENT ENHANCED DIFFUSION OF B IN Si

Silva K. Theiss*, M.-J. Caturla*, T. Diaz de la Rubia*, M.C. Johnson**, Ant Ural[†], and P.B. Griffin[†]
*Lawrence Livermore National Laboratory, Livermore CA 94550
**Avant! Corp., Silicon Business Unit, Fremont CA 94538
[†]Electrical Engineering Dept., Stanford University, Stanford CA 94305

ABSTRACT

We have developed a kinetic Monte Carlo (kMC) simulator that links atomic migration and binding energies determined primarily from first principles calculations to macroscopic phenomena and laboratory time scales. Input for the kMC simulation is obtained from a combination of *ab initio* planewave pseudopotential calculations, molecular dynamics simulations, and experimental data. The simulator is validated against an extensive series of experimental studies of the diffusion of B spikes in self-implanted Si. The implant energy, dose, and dose rate, as well as the detailed thermal history of the sample, are included. Good agreement is obtained with the experimental data for temperatures between 750 and 950° C and times from 15 to 255 s. At 1050° C we predict too little diffusion after 105 s compared to experiment: apparently, some mechanism which is not adequately represented by our model becomes important at this temperature. Below 1050° C, the kMC simulation produces a complete description over macroscopic time scales of the atomic level diffusion and defect reaction phenomena that operate during the anneals. This simulator provides a practical method for predicting technologically interesting phenomena, such as transient enhanced diffusion of B, over a wide range of conditions, using energetics determined from first-principles approaches.

INTRODUCTION

The continuing miniaturization of integrated circuits places ever more strict constraints on the size of the source and drain regions of field effect transistors (FETs). These highly doped regions of a silicon wafer are usually created by ion implantation of the dopant. After ion implantation, wafers must be annealed to remove damage caused by the implant and to electrically activate the dopant. During this annealing, the excess population of defects in the silicon can cause the dopant to diffuse orders of magnitude faster than it does under equilibrium conditions. As gate lengths decrease far below 1 μm, such transient enhanced diffusion (TED) can become a very serious problem, as the dopant moves both deeper into the substrate and laterally, changing the gate length. The energy and dose of the implant, the highest temperature achieved, and the detailed thermal history (e.g., temperature ramp rates) all influence the final dopant profile. Thus it is important to know what combination of these variables leads to the most favorable ultimate dopant profile.

Experiments to separate the influence of each variable for each new generation of technology are expensive and time-consuming. The ultimate goal of our work is to develop a truly predictive model of dopant TED. Thus, given the implant species, energy, dose, and dose rate, the detailed thermal history, and the impurity concentration, the model would produce the correct three dimensional final dopant profile, as well as the degree of activation of the dopant and the damage remaining in the silicon. Previous approaches to this problem have typically relied on a set of parameters that had to be fitted in order to reproduce particular experimental observations. In order to be more fully predictive, we base our model on an extensive set of fundamental physical parameters for diffusion and clustering. If we accurately capture the essential physical processes, the model should be predictive over a very wide range of conditions.

The development of such a model is complicated by the huge range of time and length scales involved. Phenomena such as the diffusion of dopants require long time scales, on the order of minutes to hours. However, during ion implantation a dopant atom deposits its energy into the lattice in just a few picoseconds. Device dimensions are on the order of a micron, yet the fundamental diffusive jumps are a fraction of a nanometer. Therefore, it is necessary to use an atomic-scale model with laboratory-scale predictive capabilities.

Our approach has been to develop a kinetic Monte Carlo simulator, which takes as input energies and prefactors from *ab initio* planewave pseudopotential calculations (for the dopant, impurity, and defect energetics), molecular dynamics simulations (for Si interstitial and vacancy diffusion and cluster binding energies), and fundamental experiments (for diffusion prefactors and very large Si interstitial cluster binding energies). It produces a three dimensional, atomic-scale description of dopant, defect, and impurity diffusion and clustering, on laboratory time and length scales. In this study, we apply the simulator to TED of B in Si, an important system in the production of FETs. The model is validated against an extensive series of experimental studies of the diffusion of B spikes in self-implanted Si. Details of the model and experiments are given in the next section. The results are discussed, and the areas in which the model currently falls short are presented along with those in which it succeeds remarkably well. Finally we conclude and indicate future directions for this work.

METHOD

Simulation

The basic input data necessary for the simulation of dopant diffusion are migration energies and binding energies of vacancies (Vs), self-interstitials (Is) and dopants. These values can be obtained from different sources, both theoretical and, in some cases, experimental. Recent computer simulations have provided a better understanding of defect production, diffusion, and clustering in Si. V and I formation and migration energies have been obtained using *ab initio* simulations[1-3], tight binding molecular dynamics (MD),[4] and empirical MD.[5] Binding energies for I clusters and for V clusters have been obtained using Stillinger-Weber MD and tight binding MD models.[4, 5] The interactions between dopants, impurities (e.g., C), and defects have also been studied using *ab initio* calculations. These can provide accurate information about the migration path of B,[1, 3] about its binding energy in clusters with Is, and about the trapping of Is by C.[6, 7]

B diffuses via an interstitial kick-out mechanism. After an I binds with a substitutional B atom to form a BI cluster, the Si atom may replace the B atom on the substitutional site, "kicking out" a mobile B interstitial, Bi. Our clustering model assumes that I and Bi spontaneously join any clusters they encounter, without an energy barrier. Clusters with two B atoms and three Si atoms (B2I3) and with three B atoms and two Si atoms (B3I2) are unstable with respect to emitting a Si interstitial to form B2I2 and B3I, respectively. The B3I cluster is particularly stable, and in this simulation essentially all of the inactive B fraction is found in B3I clusters. Clusters containing more than four B atoms were not considered, as they do not appear to be important for the B concentrations we are studying here.

Diffusion prefactors for Is and Vs were obtained from MD simulations.[5] The diffusion prefactor for Bi was found by running a kMC simulation in a box with one I and one (initially substitutional) B, and adjusting the prefactor until the B diffusivity matched the experimental value. The prefactor for I diffusion was used for clusters that breakup by emitting an I; likewise that for Bi was used when clusters emit a Bi. These are reasonable estimates. However, these prefactors for cluster dissolution are not independently calculated or experimentally determined, and it is not unreasonable to treat them to a limited extent as adjustable parameters.

The dopant and damage profiles after implantation can be accurately modeled by binary collision codes, such as UT-Marlowe.[8] UT-Marlowe allows simulation of different implanted species, energies, angles and doses in Si. In the binary collision approximation, two atoms interact according to a simple repulsive interatomic potential. The implanted ion undergoes a series of binary collisions with the lattice atoms. The recoiling lattice atoms collide with other lattice atoms, and so on. Thus the simulation also provides information about the cascade of defects produced by each implanted ion; that is, the location of all the Vs and Is produced during irradiation. The defect distribution obtained from this model is valid for irradiation with light ions, when the damage is primarily in the form of V-I pairs (Frenkel pairs) and no direct amorphization is produced by the implanted ions. Only the defects produced by ion implantation are considered in our kMC simulation. At the temperatures studied here, the concentration of thermally excited defects in Si is expected to be negligible compared to that produced by cluster dissolution after the implant.

Our computationally efficient kMC model is based on that of Heinisch.[9] As a function of time, it tracks the locations of defects, dopants, impurities, and extended defects (clusters). The various species are all treated as point particles with basic attributes that include size and diffusivity. During the simulation the particles participate in certain events. The possible events are: the dissociation of a particle from a cluster, with rate determined by the binding energy of the cluster; the diffusive jump of a particle, where the rate depends on the migration energy; and the introduction of a new cascade, that is, a new implanted ion and all its associated Vs and Is. The rate of cascade introduction is the dose rate of the simulated implantation. At each time step, we randomly choose among all possible events, ensuring that events occur at the proper rate by assigning each event a probability proportional to its rate. Following each chosen event, we perform all events that occur spontaneously as a result of that event. For example, an I that jumps within the capture radius of an I cluster then spontaneously joins the cluster. The simulation time is incremented by the inverse of the sum of the rates for all possible events in the simulation box. (Thus the model takes longer time steps when fewer events can occur.) The process then repeats. Because kMC focuses only on important particles and events, time scales of hours can be reached with these simulations. Clearly, great care must be taken to completely enumerate the relevant particles and events.

Experiment

The simulator was tested against a series of experimental studies of B TED. The test structures differed from FET structures in two important ways. First, rather than doping the sample by implanting B ions, the test structures were grown by chemical vapor deposition (CVD). They consisted of several B concentration spikes of amplitude 1×10^{19} cm^{-3} created during the deposition of a 1 μm thick epitaxial Si layer. Implant damage was created by subsequently bombarding the structure with 5×10^{13} Si ions/cm^2 at 50 keV, up to a dose of 5×10^{13} ions/cm^2. In such a structure the B layers can serve as markers for I diffusion. The second difference from FET structures is that the test samples are uniform in the plane of the wafer. Thus only depth information is relevant. A rapid thermal anneal (RTA) was performed at temperatures between 750 and 1050° C for times between 15 and 255 s. Temperature ramp rates for the RTAs were 150° C/s for the 750° C anneals and 100° C/s for the others. Additional furnace anneals were done at 750° C for times up to 10 hours. The total B concentration as a function of depth was determined by secondary ion mass spectrometry (SIMS).

RESULTS

We assume a uniform background C concentration of 1×10^{16} cm^{-3}, which is reasonable for CVD Si. Additionally, we assume that the front surface of the sample is a perfect sink for V, I, and Bi. As discussed above, it is reasonable (within limits) to treat the prefactors for cluster dissolution as fitting parameters. We adjusted the prefactor for the dissolution of the BI pair down by a factor of one-third from the prefactor for I diffusion, and that for the B3I dissolution up by a factor of two from the prefactor for Bi diffusion, in order to improve the match to the experimental data. The same prefactors were used in all the simulations presented here.

Results of the simulations are shown in Fig. 1. The as-grown B profile, taken as input to the simulations, is shown by the dashed line. The experimentally determined total B profile after annealing is shown as the solid line in each plot. The simulated total B profile is marked by filled circles, and the concentration of B3I clusters is marked by xs. Fig. 1a) and b) show good agreement between the simulations and experiments for both 15 and 255 s at 750° C. Note that the concentration of B3I clusters is the same at 15 s and at 255 s. In fact, all B3I clusters in the simulation form during the initial temperature ramp. Fig. 1c) shows excellent agreement between experiment and simulations after 195 s at 850° C. The predicted profile of B3I clusters is the same as in the 750° C simulations. Reasonable agreement is also obtained for 150 s at 950° C (not shown), although the simulation predicts too little diffusion in the deepest B peak. Some of the

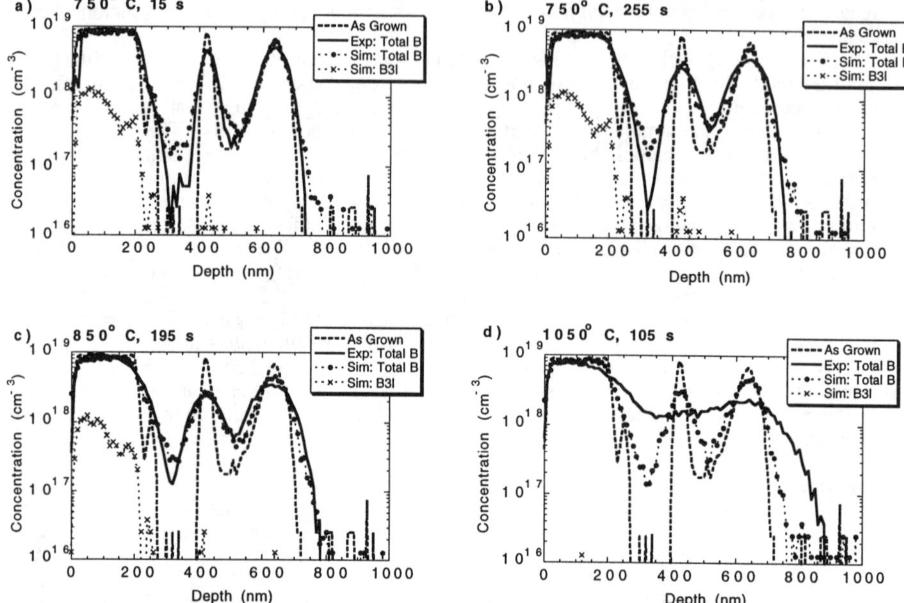

Fig. 1. Comparison of simulations and experimental results. As grown total B profiles shown by dashed line. Experimental results for total B concentration after RTAs shown by solid lines. Simulation results for total B shown by filled circles, for B3I clusters shown by Xs. a) 750° C anneal, 15 s. b) 750° C anneal, 255 s. c) 850° C anneal, 195 s. d) 1050° C anneal, 105 s.

B3I clusters that form during the initial temperature ramp have dissolved after 150 s at 950° C. Unfortunately, the 1050° C simulation grossly underestimates the experimental diffusion, as shown in Fig. 1d). Experimentally, the deeper two B peaks have completely merged and become a shoulder on the first peak, whereas the model still predicts three distinct peaks. Further, all B and I clusters have dissolved and all defects have been eliminated from the simulation box by 60 s simulation time. It appears that some additional mechanism becomes important at this temperature which is not accounted for in the model.

CONCLUSIONS AND FUTURE DIRECTIONS

We are approaching a predictive model for TED of B in Si. Good agreement has been obtained with RTA data for temperatures between 750 and 950° C and times between 15 and 255 s. The only parameters adjusted to improve this fit are the prefactors for the breakup of the BI and the B3I clusters.

The B3I clusters are predicted to form during the initial ramp to high temperature, and not to dissolve at temperatures up to 850° C. The final fraction of B which is electrically active is influenced by the temperature ramp rate, even though in these samples all the B is initially substitutional.

Further work is required to improve the behavior of the model at 1050° C. It is possible that at this temperature the equilibrium concentration of defects -- although there would be an average of less than one thermally excited defect in a volume the size of the simulation box at any time -- may be so mobile as to have a significant effect. We are in the process of investigating the influence of the low concentration of thermally excited defects. Dislocation loops can be formed after ion implantation, although generally not at the doses studied here. However, dislocation loops change the kinetics of interstitial clustering and dissolution, extending the time over which interstitials are available. Thus we plan to add dislocation loop formation to our kMC code in the future. Finally, the concentration of Si vacancies and interstitials predicted by UT-Marlowe has not been as extensively tested as its predictions of dopant profiles, and may be causing problems in our simulation. We plan to check the defect concentrations produced by UT-Marlowe against MD simulations of Si implants at the energies of interest.

Work is currently underway to improve our B clustering model using improved *ab initio* total energy calculations. Additionally, we are incorporating prefactors for Si V and I diffusivities determined from experimental results.[10] In the future, as we simulate lower energy implants and higher local dopant concentrations, we will need to incorporated the effects of shifting the Fermi level within the sample. This will have two effects. First, the diffusion rate and direction of charged species will be influenced by electric field gradients caused by a spatially varying Fermi level. Second, the relative populations of differently-charged species will be changed.

In future experimental work, we plan to validate the B clustering model using a combination of spreading resistance profiling and Hall measurements. Finally, additional time evolution information at higher temperatures will provide a check on our activation energies.

Although many refinements remain to be made, the results of this study are highly encouraging. We are approaching a model with enough of the necessary physics in it that a good approximation of reality is produced, over a range of conditions and with a minimum of fitting. Additional work should enable us to accurately link microscopic time and length scales to a fuller range of the conditions of interest in the laboratory or fabrication facility.

ACKNOWLEDGMENTS

Martin Giles of Intel and Majeed Foad of Applied Materials provided us with invaluable advice and encouragement during the development of the kinetic Monte Carlo simulator. We thank Aaron Lilak and Mark Law of the University of Florida for numerous helpful discussions. This work was performed under the auspices of the U.S. Department of Energy by Lawrence Livermore National Laboratory under contract W-7405-Eng-48.

REFERENCES
1. C. S. Nichols, C. G. Van de Walle, S. T. Pantelides, Phys. Rev. **B40**, 5458 (1989).
2. W.A. Harrison Mat. Res. Soc. Symp. Proc. in *Defects and Diffusion in Silicon Processing* , Vol. **469**, p. 211, Ed. T. Diaz de la Rubia, S. Coffa, P. A. Stolk, C. S. Rafferty, (1997).
3. J. Zhu, T. Diaz de la Rubia, L. H. Yang, C. Mailhiot, G. H. Gilmer, Phys. Rev. **B54**, 4741 (1996).
4. M. Tang, L. Colombo, J. Zhu, T. Diaz de la Rubia, Phys. Rev. **B55**, (1997).
5. G. H. Gilmer, T. Diaz de la Rubia, D. M. Stock and M. Jaraiz, Nucl. Instrum. and Methods **B102**, 247 (1995).
6. J. Zhu, Mat. Res. Soc. Symp. Proc. in *Defects and Diffusion in Silicon Processing* , Vol. **469**, p. 151, Ed. T. Diaz de la Rubia, S. Coffa, P. A. Stolk, C. S. Rafferty, (1997).
7. O. Pankratov, H. Huang, T. Diaz de la Rubia, C. Mailhiot, Phys. Rev. B (1997).
8. S. Tian, S. J. Morris, B. Obradovic, M. F. Morris, G. Wang, G. Balamurugan, A. F. Tasch, C. Snell, UT-Marlowe, Version 4.0 (1996).
9. H. L. Heinisch, J. Nucl. Mater. **117**, 46 (1983).
10. Salvatore Coffa and Sebania Libertino, Appl. Phys. Lett. **73**, 3369 (1998).

Part III

Crystal Defects and Interfaces

ATOMIC-SCALE MODELING OF THE DEFORMATION OF NANOCRYSTALLINE METALS

J. SCHIØTZ[1,2,*], T. VEGGE[1,2], K. W. JACOBSEN[1]
[1]Center for Atomic-scale Materials Physics and Department of Physics, Technical University of Denmark, DK-2800 Lyngby, Denmark
[2]Materials Research Department, Risø National Laboratory, DK-4000 Roskilde, Denmark.
*Email: schiotz@fysik.dtu.dk

ABSTRACT

Nanocrystalline metals, i.e. metals with grain sizes from 5 to 50 nm, display technologically interesting properties, such as dramatically increased hardness, increasing with decreasing grain size. Due to the small grain size, direct atomic-scale simulations of plastic deformation of these materials are possible, as such a polycrystalline system can be modeled with the computational resources available today.

We present molecular dynamics simulations of nanocrystalline copper with grain sizes up to 13 nm. Two different deformation mechanisms are active, one is deformation through the motion of dislocations, the other is sliding in the grain boundaries. At the grain sizes studied here the latter dominates, leading to a softening as the grain size is reduced. This implies that there is an "optimal" grain size, where the hardness is maximal.

Since the grain boundaries participate actively in the deformation, it is interesting to study the effects of introducing impurity atoms in the grain boundaries. We study how silver atoms in the grain boundaries influence the mechanical properties of nanocrystalline copper.

INTRODUCTION

In recent years, advanced production techniques have made it possible to create metals, alloys and ceramics with grain sizes down to 5 nm. For metals, this represents a reduction of the grain size by approximately four orders of magnitude compared to most conventionally produced metals. As can be expected, such a dramatic change in the microstructure leads to significant changes in the mechanical properties of the metals. For example, the hardness of typical nanocrystalline metals is far higher than what is seen in their coarse-grained counterparts [1, 2, and references therein].

Nanocrystalline metals are an attractive group of metals to model, as the small grain size provides a "cut-off" of the typical length scales, where structures appear during deformation. In coarse-grained materials structures appear on vastly different length scales, making it very difficult to model the properties of these materials. The models must include processes that occur on length scales from the sub-nanometer scale of the atomic processes in dislocation cores, to the micro- or even millimeter scale of grain and subgrain structures [3, 4]. Atomic-scale simulations of systems of these sizes are beyond the reach of even the most powerful of todays supercomputers. One is thus forced to split the system into sub-problems at different length-scales, and only treat sub-problems at the atomic scale with atomistic models. On coarser length scales other modeling paradigms must be used, such as Dislocation Dynamics and continuum plasticity calculations.

Dividing the problem into sub-problems at different length scales often results in a better understanding of the problem, as it draws attention to the structures which are relevant at a given length scale (for example, when studying the formation of dislocation structures it is clearly more relevant to focus on dislocations as the fundamental concept rather than on individual atoms).

Mat. Res. Soc. Symp. Proc. Vol. 538 ©1999 Materials Research Society

On the other hand, one becomes dependent on this understanding, when creating the coarse-grained models, as many assumptions about the relevant phenomena at different length-scales will by necessity be built into the multi-scale model. The low grain size of nanocrystalline metals "compresses" this range of length scales to a range, where the whole deformation problem can be modeled at the atomic scale, as many grains in the polycrystalline material can be handled in an atomic-scale simulation. This makes it possible to perform unbiased simulations of the deformation process, where no *a priori* assumptions are made about the deformation mechanisms.

In recent papers, we have presented simulations of the plastic deformation of nanocrystalline Cu and Pd [5–7]. Other authors have presented simulations of the structure and elastic properties of nanocrystalline metals and semiconductors [8–12], and of the plastic deformation of Ni under constant stress loading [13, 14]. In this paper we review our simulations of the deformation mechanisms in nanocrystalline metals, and present simulations of the effects of impurities in the grain boundaries.

SIMULATIONS OF PURE METALS

Simulation setup

We have attempted to generate three-dimensional systems with a microstructure similar to that observed experimentally in samples generated by inert gas condensation (IGC). The grains appear to be randomly oriented, approximately equiaxed, and dislocation free. The grain size distribution is close to log-normal [1, 15, and references therein]. We try to create systems that match this description.

The grains are produced using a three-dimensional Voronoi tesselation: random grain centers are chosen, and space is divided into regions in such a way that each region consists of the points in space closer to a given grain center than to any other grain center. Each region is then filled with a randomly oriented fcc lattice. An example of a two-dimensional Voronoi tessellation is shown in Figure 1.

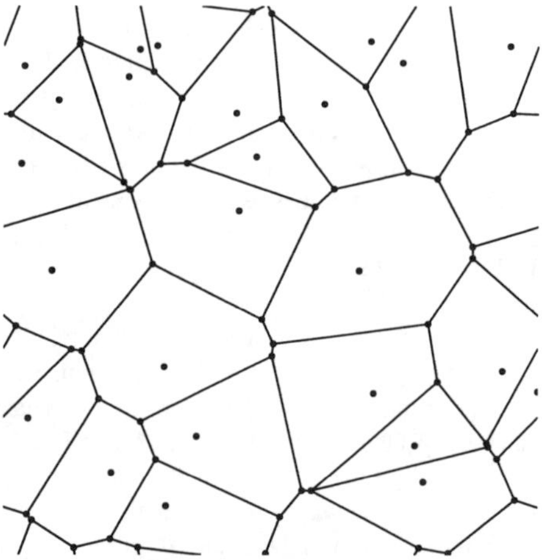

The generated samples are annealed for 50 ps at 300 K to relax the grain boundary structure. We found that the annealing time and temperature are uncritical, but that the properties of the system are different if *no* annealing is done.

The interactions between the atoms are modeled using the Effective Medium Theory

Figure 1: An example of a two-dimensional "grain structure" generated by a Voronoi construction. Periodic boundary conditions are used. A two-dimensional nanocrystalline material could be obtained by filling each cell with atoms in a regular lattice. A three-dimensional version of this algorithm is used when generating the systems studied in this paper.

(EMT) [16, 17]. EMT is a many-body potential providing a realistic description of the metallic bonding, in particular in face-centered cubic (fcc) metals and alloys of fcc metals. Computationally, EMT is not much more demanding than pair potentials, but provide a significantly more realistic description of the metallic bonding.

The systems were deformed using a molecular dynamics (MD) procedure. A conventional MD simulation was performed, but at each timestep the atomic coordinates in the pulling direction (the z coordinates) were rescaled in order to deform the system gradually. At the same time, the box dimensions in the transverse directions are allowed to shrink to keep the transversal components of the stress (σ_{xx} and σ_{yy}) close to zero [7]. The change in system dimensions is slow at the timescale of the simulation, the relative elongation is 2.5×10^{-6} each timestep. With a timestep of 5 fs, this nevertheless results in a very high strain rate ($\dot{\varepsilon} = 5 \times 10^{8} s^{-1}$). The results presented here are not very sensitive to the strain rate, although some dependence is seen (typical stresses increase by 20% when the strain rate is changed from $2.5 \times 10^{7} s^{-1}$ to $1 \times 10^{9} s^{-1}$) [7].

Results

Figure 2 shows the deformation of a typical sample with an average grain size of 5.21 nm. The atoms have been color coded according to the local crystal structure [18, 19]. White atoms are in local fcc order, and thus situated inside the grains. Light grey atoms are in local hexagonal close-packed (hcp) order, these atoms are at stacking faults. Atoms in all other local environments are colored dark grey. These are typically atoms at grain boundaries and in dislocation cores.

Some dislocation activity is seen in the system, as witnessed by the generation of stacking faults. The dislocation activity is not sufficient to account for the observed plastic deformation. A detailed analysis of the deformation shows that the main deformation mode is sliding in the grain boundaries [5, 7].

As the volume fraction of atoms in the grain boundary increases with decreasing grain size, one would expect that increasing the grain size increases the strength of the material as long as the grain boundaries remain the carriers of the deformation. This is indeed what we see in the simulations. Figure 3 shows the stress-strain curves obtained from simulations of systems with various average grain sizes. A "reverse Hall-Petch effect", i.e. a softening of the material with decreasing grain size, is observed.

Discussion

The simulations of the deformation of nanocrystalline metals show a reverse Hall-Petch effect in Cu and Pd for the grain sizes studied. This softening of the material, when the grain size is reduced is caused by plastic deformation in the grain boundaries. There appears to be two different deformation mechanisms active at different grain sizes. In metals with the very small grain sizes studied here, the dominating deformation mechanism is sliding in the grain boundaries through a large number of essentially uncorrelated events, where a few atoms move with respect to each other at each event [5].

At much coarser grain sizes, dislocations are known to be the dominating carriers of deformation. In that regime, a hardening of the material is seen, when the grain size is reduced, as the grain boundaries act as barriers to the dislocation motion. Experimentally, this behavior is seen to continue far down into the nanocrystalline range.

As the grain size is reduced, dislocation-mediated deformation becomes more and more difficult. On the other hand, the volume fraction of the grain boundaries increases, favoring a deformation mechanism where the grain boundaries carry the deformation. Furthermore, as the grain size is approaching the grain boundary thickness, it becomes geometrically easier for slip to occur on more than one grain boundary, without large stress concentrations where the grain

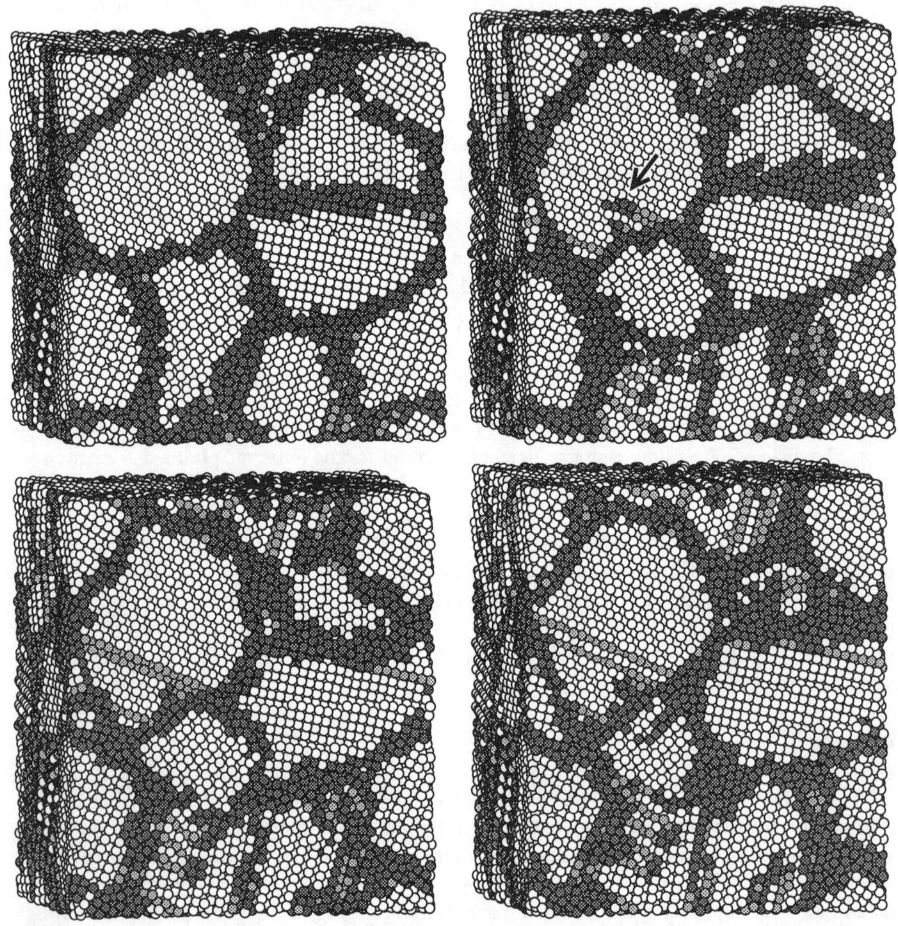

Figure 2: Simulation of the deformation of nanocrystalline copper at $T = 300K$ and $\dot{\varepsilon} = 5 \times 10^8 s^{-1}$. The four panels show the system after 0, 5, 7.5 and 10% deformation. Little dislocation activity is seen. In the second panel a Shockley partial dislocation (indicated by the arrow) is seen moving through the grain, leaving an intrinsic stacking fault behind. Between 7.5 and 10% deformation another Shockley partial has moved through the grain on an adjacent $\{111\}$ plane, transforming the intrinsic stacking fault to an extrinsic one.

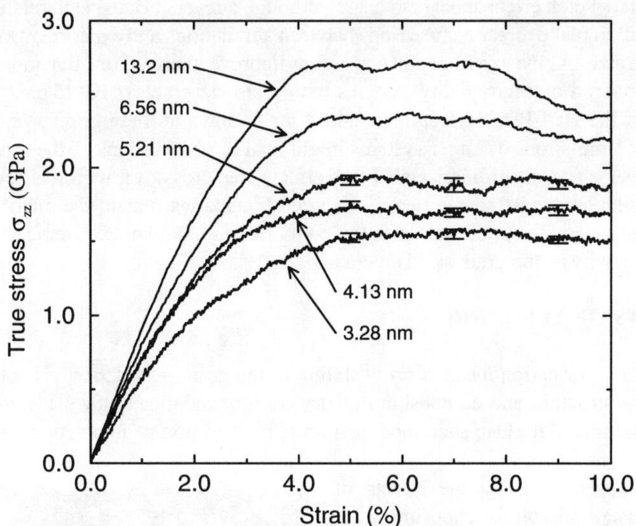

Figure 3: Stress-strain curves for nanocrystalline copper at 300K. Four simulations were performed at each grain size below 6 nm, the curves shown are average stress-strain curves for the simulations. The error bars are 1σ uncertainties on the mean value. The larger grain sizes were simulated using a larger system (approximately 10^6 atoms instead of 10^5), only one simulation was made with $d = 6.56$ nm, and two with $d = 13.2$ nm. Adapted from Ref. [6].

boundaries meet [20].

The emerging picture is one where two deformation mechanisms compete. One is active at very small grain sizes, another at larger grain sizes. This leads to a maximum in the yield stress and hardness of nanocrystalline metals at intermediate grain sizes, see Figure 4.

There have been many reports in the literature of a "reverse Hall-Petch effect" at sufficiently small grain sizes. However, the hardness measurements are very sensitive to sample defects, and in particular to sample porosity [21–23]. In high-quality Cu samples, the Hall-Petch effect is seen to continue at least down to grain sizes around 15 nm [21].

There does not appear to be any *unequivocal* experimental evidence for a reverse Hall-Petch relationship in porosity free nanocrystalline metals. There are, however, indications of a break-down of the ordinary Hall-Petch relation at grain sizes below 15 nm in high quality copper samples produced by inert gas condensation [21].

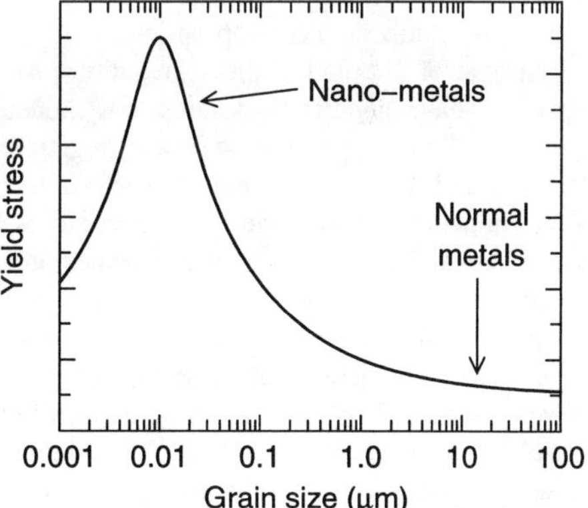

Figure 4: A *qualitative* view of the relation between the yield stress and the grain size in metals.

The ordinary Hall-Petch effect appears to cease, although a reverse Hall-Petch relation is not seen. It is still difficult to make direct comparisons between simulations and experiment, as there is little overlap in grain size. As the experimental techniques improve, there is hope that more experimental data will be gathered on high-quality samples with grain sizes below 10–15 nm. It should then become clear if the Hall-Petch effect does indeed break down at these grain sizes. At the same time, computer simulations of larger systems might lead to observations of the cross-over region between the reverse and the ordinary Hall-Petch effect. That cross-over region is beyond the reach of the largest simulations presented here. The grain boundaries remain the main carriers of the deformation even when the grain size is increased to 13.2 nm, but be *do* observe a slight increase in dislocation activity as the grain size is increased.

THE EFFECTS OF ALLOYING

As the main deformation mechanism is sliding in the grain boundaries, it could be expected that altering the structure and composition of the grain boundaries might have an effect on the properties of the material. One such modification is the addition of impurity atoms in the grain boundaries.

We have chosen to study the effects of silver impurities. A reason for choosing silver was that silver and copper are immiscible, low concentrations of silver impurities in nanocrystalline copper can therefore be expected to segregate to the grain boundaries. We have not studied the segregation process itself, as it is beyond the scope of this study. Segregation is a slow, diffusional process that cannot be studied directly with MD simulations due to the timescales involved. Other approaches (such as Monte Carlo or Kinetic Monte Carlo simulations) may be more appropriate.

Instead of simulating the segregation process, we have generated systems that model nanocrystalline copper after such a segregation has occurred. It is done by replacing 25% of the atoms in the grain boundaries with silver atoms, see Figure 5. The system is then annealed and deformed in the same way as the pure systems.

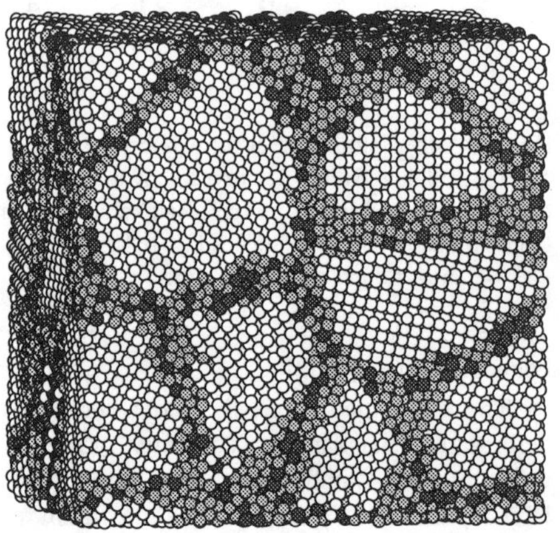

Figure 5: Typical initial configuration for a silver-containing simulation. White and light gray atoms are Cu atoms in the bulk and in grain boundaries, respectively. Dark atoms are Ag atoms.

Figure 6 shows how the stress-strain curves have changed, when silver is introduced in the grain boundaries. The general trend seems to be a slight softening of the material, although the effect is very weak and not seen in all systems. Figure 7 shows the flow stress levels for the different simulations, i.e. the stress level at the horizontal part of the stress-strain curve (for simplicity the flow stress was defined as the average stress for $6\% \leq \varepsilon \leq 10\%$). We again see the tendency for the silver-containing systems to be softer.

In the simulations of impurities in the grain boundaries, the mechanical behavior is close to what is seen in the pure systems.

As the major part of the deformation happens in the grain boundaries, one could expect that adding impurities in the grain boundaries could have a relatively large effect on the mechanical

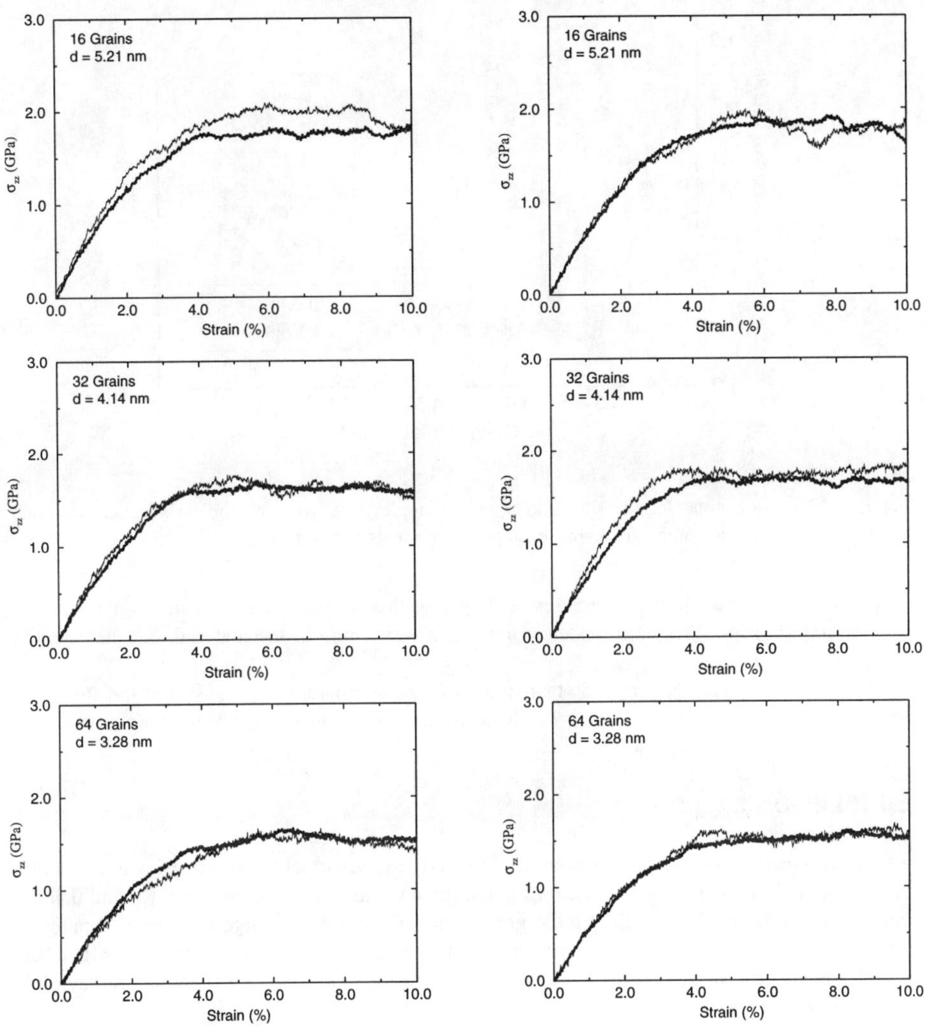

Figure 6: Comparison between simulations with and without silver atoms in the grain boundaries. Each graph shows the stress-strain curve of a simulation with 25 at-% Ag in the grain boundaries (thick lines), and the same system (i.e. the same grain structure) without the silver (thin lines). For each grain size, two different sets of simulations were performed with different random grain structures.

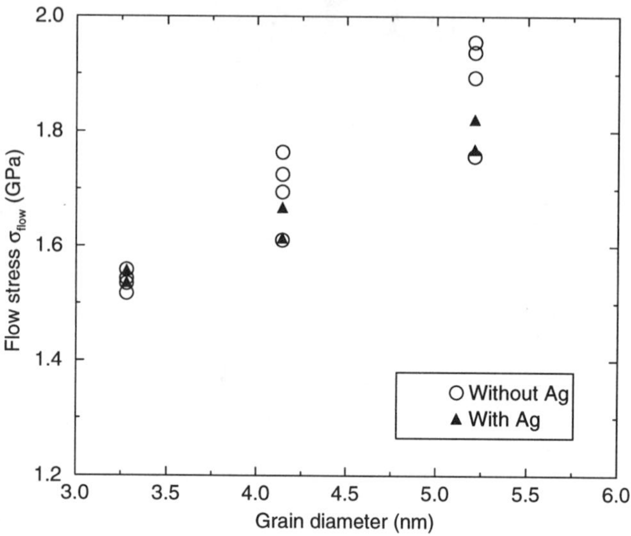

Figure 7: The flow stress for the simulations with (triangles) and without (circles) silver in the grain boundaries, as a function of the grain size. Each point represents one simulation.

properties. The simulations presented here show that this is not the case, when silver is used as impurities in copper. We see a tendency towards a weakening of the material, but the effect is barely detectable.

The atomic bonding in copper and silver are of a similar nature, and the size of the atoms are not very different. This may account for the absence of a stronger effect of alloying in the grain boundaries.

CONCLUSION

Atomic-scale simulations have been used to study the deformation mechanisms in nanocrystalline copper with and without impurities in the grain boundaries. In both cases, we find that the main deformation mode is sliding in the grain boundaries through a large number of apparently uncorrelated events, each involving only a few atoms. Some dislocation activity is seen in the grains, the dislocations are probably necessary to allow the grains to deform a little, as they glide past each other.

We observe a *reverse Hall-Petch effect*, i.e. a hardening of the material as the grain size is increased: as the amount of grain boundary atoms is decreased, the deformation becomes harder. At the same time the dislocation activity is seen to increase a little with grain size. This effect is seen in the entire range of grain sizes that we have studied (3 to 13 nm), but at some point we expect that dislocation motion will begin to dominate the behavior of the system. When that happens, the yield strength should begin to *decrease* with increasing grain size.

Adding silver to the grain boundaries has only a weak effect on the properties of the material. The strength of the material is seen to decrease marginally, when the silver is introduced. We expect that other elements, which chemically behave in a way that is more different from copper will have a larger effect, but such simulations remain to be made.

ACKNOWLEDGMENTS

This work was financed by The Danish Technical Research Council through Grant No. 9601119. Parallel computer time was financed by the Danish Research Councils through Grant No. 9501775. Center for Atomic-scale Materials Physics is sponsored by the Danish National Research Council.

REFERENCES

[1] R. W. Siegel and G. E. Fougere, in *Nanophase Materials: Synthesis — Properties — Applications*, edited by G. C. Hadjipanayis and R. W. Siegel, vol. 260 of *NATO-ASI Series E: Applied Sciences*, p. 233 (Kluwer, Dordrecht, 1994).

[2] D. G. Morris and M. A. Morris, Mater. Sci. Forum, **235-238**, p. 861 (1997).

[3] J. V. Carstensen, T. Leffers, T. Lorentzen, O. B. Pedersen, B. F. Sørensen, and G. Winther, eds., *Modelling of Structure and Mechanics of Materials from Microscale to Product*, Proceedings of the 19th Risø International Symposium on Materials Science, Risø National Laboratory, Roskilde (1998).

[4] A. E. Carlsson and R. Thomson, Solid State Physics, **51**, p. 233 (1998).

[5] J. Schiøtz, F. D. Di Tolla, and K. W. Jacobsen, Nature, **391**, p. 561 (1998).

[6] J. Schiøtz, T. Vegge, F. D. Di Tolla, and K. W. Jacobsen, in Carstensen *et al.* [3], p. 133.

[7] J. Schiøtz, T. Vegge, F. D. Di Tolla, and K. W. Jacobsen, *Atomic-scale simulations of nanocrystalline metals*, (to be published).

[8] D. Chen, Comput. Mater. Sci., **3**, p. 327 (1995).

[9] S. R. Phillpot, D. Wolf, and H. Gleiter, J. Appl. Phys., **78**, p. 847 (1995).

[10] S. R. Phillpot, D. Wolf, and H. Gleiter, Scripta Met. Mater., **33**, p. 1245 (1995).

[11] H. Zhu and R. S. Averback, Materials and Manufacturing Processes, **11**, p. 905 (1996).

[12] P. Keblinski, R. Phillpot, D. Wolf, and H. Gleiter, Acta Mater., **45**, p. 987 (1997).

[13] H. Van Swygenhoven and A. Caro, Appl. Phys. Lett., **71**, p. 1652 (1997).

[14] H. Van Swygenhoven and A. Caro, NanoStructured Materials, **9**, p. 669 (1997).

[15] R. W. Siegel, J. Phys. Chem. Solids, **55**, p. 1097 (1994).

[16] K. W. Jacobsen, J. K. Nørskov, and M. J. Puska, Phys. Rev. B, **35**, p. 7423 (1987).

[17] K. W. Jacobsen, P. Stoltze, and J. K. Nørskov, Surf. Sci., **366**, p. 394 (1996).

[18] H. Jónsson and H. C. Andersen, Phys. Rev. Lett., **60**, p. 2295 (1988).

[19] A. S. Clarke and H. Jónsson, Phys. Rev. E, **47**, p. 3975 (1993).

[20] H. Hahn and K. A. Padmanabhan, Phil. Mag. B, **76**, p. 559 (1997).

[21] P. G. Sanders, C. J. Youngdahl, and J. R. Weertman, Mater. Sci. Eng. A, **234-236**, p. 77 (1997).

[22] P. G. Sanders, J. A. Eastman, and J. R. Weertman, Acta mater., **46**, p. 4195 (1998).

[23] S. R. Agnew, B. R. Elliott, C. J. Youngdahl, K. J. Hemker, and J. R. Weertman, in Carstensen *et al.* [3], p. 1.

AB-INITIO BASED CALCULATIONS OF VACANCY FORMATION AND CLUSTERING ENERGIES INCLUDING LATTICE RELAXATION IN Fe₃Al

Leonid S. Muratov, Bernard R. Cooper, West Virginia University, Dept. of Physics, Morgantown, WV; John M. Wills, Los Alamos National Laboratory, Los Alamos, NM

ABSTRACT

Vacancy formation and clustering significantly affect structural properties of transition-metal aluminides. *Ab-initio* quantum mechanical total-energy calculations using a full-potential linear combination of muffin-tin orbitals (LMTO) technique provide a convenient method of studying relevant characteristics such as changes in density of states, and charge redistribution around defects. Augmented with Hellmann-Feymann forces, LMTO allows calculations of relaxation geometries and relaxation energies. We have performed such calculations for vacancies and antisite substitutional point defects in $Fe_3A\ell$ with DO_3 crystallographic structure. There are two limiting factors complicating calculations of defect formation energies directly from *ab-inito* calculations. The first is that a single defect, due to the lattice periodicity necessitated by the use of *ab-inito* total energy techniques, cannot be considered as an isolated defect, even in the maximum computable simulation cell. Unlike previous calculations [1], which did not find a dependency on the size of the simulation cell, our calculations have shown a significant difference in results for 32- and 16- atom cells. This difference provides information about vacancy clustering since it can be explained by a relatively small attractive interaction energy ~0.2 eV between two vacancies located in adjacent simulation cells and separated by the lattice constant distance (5.52Å) [2]. By comparing the internal energies for two configurations of 30 atom cells (32 atom – 2 vacancies) we were able to estimate that the attractive interaction between two vacancies could reach 1.2 eV. The second complication is the fact that chemical potentials of elements cannot be directly extracted from the total energy calculations for the compound. To deal with this problem, we considered two possible approximations and compared results, which were found to be quite similar for iron vacancies.

INTRODUCTION

Transition-metal aluminides and in particular iron-aluminides have attracted considerable attention due to their excellent oxidation and corrosion resistance at elevated temperature and in hostile environment. Recently, much research has been directed to enhance room- and high-temperature mechanical properties of iron-aluminides such as ductility and creep resistance. Improvement of material mechanical properties often requires a basic physical understanding of processes involved in formation and interaction of various point defects, since vacancies, antisite atoms, impurities and various clusters of defects determine the thermodynamic and kinetic behavior of these metals.

$Fe_3A\ell$ is considered as a promising structural material by itself and as a base for several commercial alloys [3,4]. Here we present a consideration of $Fe_3A\ell$ in DO_3 crystal structure. Although this is a low temperature phase, the temperature of the transition can be increased by use of various additives. Above 550 °C pure $Fe_3A\ell$ transforms to imperfectly ordered B2 structure and then to a disordered solid solution. A DO_3 (space group *Fm3m,* FCC with 4 atoms per unit cell, sometimes called ordered BCC) crystal structure is shown on Figure 1. The three iron atoms of a formula unit of $Fe_3A\ell$ are located on two different types of lattice sites. Two atoms on γ sites, which are shown as open circles on Fig .1, have four iron and four aluminum nearest neighbors at a distance of $\sqrt{3}a/4$, where a is a lattice constant. The remaining one iron atom of the formula unit occupies the β site, which is shown as a gray filled circle on

Mat. Res. Soc. Symp. Proc. Vol. 538 © 1999 Materials Research Society

Fig.1. This β type of iron atoms has all eight irons as nearest neighbors. All aluminum atoms, shown as crosshatched circles α have eight neighboring irons of type γ.

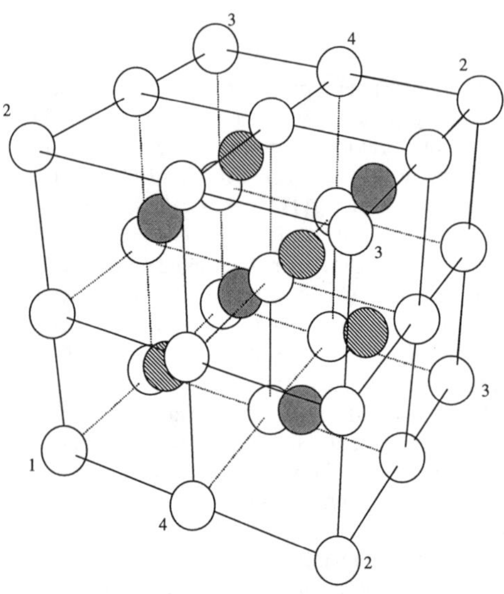

Fig 1. Structure of DO$_3$ Fe$_3$Aℓ. It is an FCC lattice with 4 atoms per unit (formula) cell. One atom of aluminum (α sites) at (1/4,1/4,1/4), is shown as a crosshatched circle; one iron atom (type β) at (3/4,1/4, 1/4) is shown as a gray-filled circle, and two crystallographically identical atoms of iron (type γ) at the origin and at (1/2, 0, 0) are shown as open circles. Atoms of the unit cell are then repeated in accordance with the FCC structure. Iron of type β has 8 iron atoms as nearest neighbors at the distance $d=0.25\sqrt{3}a$, while iron of type γ has 4 iron neighboring atoms and 4 aluminum atoms. Some of atoms located on faces of the large cube belong to adjacent supercells. 32-atom supercell also includes atoms from the cube just above the one shown here.

For the 16-atom cell, a vacancy created at the origin "1" is repeated due to periodic boundary conditions at sites marked by "2" and "3". For 32 atom cells vacancies are located at sites marked as "1", "3" while sites "2" are occupied by irons. For divacancy consideration two 32 atom cells have been compared, one with an additional vacancy at sites "2" i.e. repetition of two 16-atom cells, and one with additional vacancy at sites marked by "4"

THERMODYNAMICS OF VACANCY FORMATION

Following [5,6] we have employed the grandcancnical formalism for a description of ordered compound binary systems. Recently V.Schott and M.Fähnle [7], have provided thorough derivations for this approach, while showing the equivalence of the canonical and the grand-canonical formalism.

In the ground state all (fixed number) atomic sites are occupied by atoms of corresponding types. Energy of an elementary excitation is equal to the total energy change in the system when the point defect is created, i.e. it is a difference in the total energy of the simulation cell with corresponding defect and the energy of the initial, perfect cell. If we neglect the interaction between point defects, for the system under consideration there are six possible elementary excitations: ε_V^α – aluminum vacancy, ε_V^β –vacancy on iron site β, ε_V^γ vacancy on iron site γ, ε_{Fe}^α antisite iron atom at aluminum site, ε_{Al}^β antisite aluminum atom at iron site β, ε_{Al}^γ antisite aluminum atom at iron site γ.

The grandcanonical potential Φ for a system of M formula cells is equal to:

$$\Phi = E_N - TS - \mu_{Fe}N_{Fe} - \mu_{Al}N_{Al}, \tag{3}$$

where temperature T is measured in energy units, N_{Fe} and N_{Al} are number of iron and aluminum atoms in simulation cell and entropy S can be taken in Bragg-Williams approximation. For the perfect lattice, in the simulation cell of M formula units of Fe$_3$Aℓ there are $N_{Al} = M$ and $N_{Fe}=3M$ atoms. Minimizing (3) with respect to numbers of of defects one can derive for concentrations of corresponding point defects:

$$C_V^\alpha = 0.25e^{-E_V^\alpha/k_BT}/(1-e^{-E_V^\alpha/k_BT}-e^{-E_{Fe}^\alpha/k_BT}), \quad C_V^\beta = 0.25e^{-E_V^\beta/k_BT}/(1-e^{-E_V^\beta/k_BT}-e^{-E_{Al}^\beta/k_BT}),$$

$$C_V^\gamma = 0.5e^{-E_V^\gamma/k_BT}/(1-e^{-E_V^\gamma/k_BT}-e^{-E_{Al}^\gamma/k_BT}), \quad C_{Fe}^\alpha = 0.25e^{-E_{Fe}^\alpha/T}/(1-e^{-E_V^\alpha/T}-e^{-E_{Fe}^\alpha/T}), \qquad (4)$$

$$C_{Al}^\beta = 0.25e^{-E_{Al}^\beta/k_BT}/(1-e^{-E_V^\beta/k_BT}-e^{-E_{Al}^\beta/k_BT}), \quad C_{Al}^\gamma = 0.5e^{-E_{Al}^\gamma/k_BT}/(1-e^{-E_V^\gamma/k_BT}-e^{-E_{Al}^\gamma/k_BT}).$$

In Eq (4) effective formation energies are defined as the following: $E_V^\alpha = \varepsilon_V^\alpha + \mu_{Al}$;

$E_V^\beta = \varepsilon_V^\beta + \mu_{Fe}$; $E_V^\gamma = \varepsilon_V^\gamma + \mu_{Fe}$; $E_{Fe}^\alpha = \varepsilon_{Fe}^\alpha + \mu_{Al} - \mu_{Fe}$; $E_{Al}^\beta = \varepsilon_{Al}^\beta + \mu_{Fe} - \mu_{Al}$;

$E_{Al}^\gamma = \varepsilon_{Al}^\gamma + \mu_{Fe} - \mu_{Al}$. It can be seen that at the low concentration limit, effective formation en

ergies represent Boltzmann exponential factors for concentrations. Chemical potentials are still unknown, therefore in six equations (4) there are eight unknowns (six concentrations plus two chemical potentials). For the seventh equation a zero pressure thermodynamic relation $\Phi=0$ can be used. For Fe$_3$Aℓ it gives us:

$$\frac{E_N}{M} = \mu_{Al} + 3\mu_B - T\ln[64(0.25 - C_V^\alpha - C_{Fe}^\alpha)(0.25 - C_V^\beta - C_{Al}^\beta)(0.5 - C_V^\gamma - C_{Al}^\gamma)^2] \qquad (5)$$

At zero temperature, Equation (5) represents the fact that energy of formula unit is equal to the sum of chemical potentials of its components. In an element, with one atom per unit cell, Equation (5) would uniquely determine the chemical potential. For ordered composite materials Equation (5) does not provide a way to decouple μ_{Fe} and μ_{Al}, therefore an additional approximation is needed; or, in other words, we need an eighth equation in addition to (4) and (5) to find equilibrium concentrations and chemical potentials. There is a physical reason for this uncertainty in the value of chemical potential, because its values depend on conditions during the crystal growth such as excess of one or another element, pressure etc. Postulating the exact chemical composition throughout the whole lattice (i.e. neglecting grain boundaries, precipitates of components, etc) is sufficient to solve system (3) and (4), since the stoichiometry equation is added:

$$\frac{N_{Fe}}{N_{Al}} = \frac{C_{Fe}^\alpha + C_{Fe}^\beta + C_{Fe}^\gamma}{C_{Al}^\alpha + C_{Al}^\beta + C_{Al}^\gamma} = 3 \qquad (6)$$

Resulting solution would provide good estimates for chemical potentials in almost perfect singe-crystal environment, and has been used in the present publication.

In "real" systems with multiple defects we only can put bounds on the values of chemical potentials using the following consideration. The lower bound can be found using the heat of alloy formation. However, caution should be advised, since *ab-initio* calculations often do not provide accurate values of alloy formation heat [8] since without the gradient correction, LDA calculations do not produce the correct value for the iron ground state [9]. In addition, evaluation of the heat of formation would require comparison between calculations for several different structures, and therefore subtraction of two large absolute values of energies calculated for different structures. Such subtractions would produce significant errors, and using the experimental value is questionable for the same reason. Another estimation for the chemical potential is available from atomic energy in element precipitates with lower interface energy. In such precipitates all atoms of the host material lattice are substituted with one element (this would result in BCC for DO_3). For comparison, these chemical potentials are also used in Table 1.

CALCULATIONS.

Fully quantum-mechanical, self-consistent electronic structure calculations presented here are based on density functional theory and used local density approximation (LDA). We have utilized the full-potential linear-combination-of-muffin-tin-orbitals (LMTO) technique. This is an *ab initio* all-electron technique with relativistic corrections included. To assure high accuracy we have implemented an augmented basis through use of multiple κ's (interstitial kinetic energies of basis states) and energy windows. To treat relaxation effects around point defects for this transition metal system, we have implemented a force methodology based on Hellmann-Feynman theorem.

Great attention has been given to stability and consistency of calculations. Due to limitations of computers it is impossible to performed unrestricted calculations. Compromises on number of k-points in Brillouin zone, number of basis states, dimension of Fourie transformation, radii of muffin-tin spheres, etc. are always required. In order to minimize impact of these compromises energy values used in equations (3)-(5) have been obtained from the calculations of systems with the same symmetry (i.e. results for 16 and 32 atom calculations were not mixed together, and when the origin was shifted, the system energy of the perfect system was recalculated too). This approach significantly reduces the number of k-points needed for accurate calculations, since only relative changes are included. However, for the purpose of verification, calculations with larger number of k-points are underway (32-atoms cells).

Our initial calculations have involved a 16-atom supercell, which has a simple cubic superlattice structure. This supercell is shown on Figure 1. Superlattice vectors are aligned with edges of the large cube shown on figure. For visual convenience, atoms located on faces of the large cube and belonging to adjacent simulation cells are also shown. In order to take maximum advantage of the supercell symmetry, and therefore reduce computational cost, a single vacancy has been created by removing atom located at the origin (this atoms is marked as "1" on Fig.1). When other vacancies (Al and β Fe) have been considered the whole supercell has been shifted, in order to align the corresponding atom with the origin. To avoid an additional error, total energies of perfect simulation cells with different origins have been recalculated (absolute differences between their values have not exceeded 0.03eV, the same can be said for 32-atom simulations) and used in corresponding calculations. Due to periodic boundary conditions, identical vacancies have been created on sites marked by "2" and "3" on Fig.1, at the ends of superlattice vectors. It can be easily seen that there are six neighboring vacancies, which are separated from the original vacancy by only one atom.

Results of our calculations are presented in Table 1. Relaxation energies obtained for this system are significantly higher (0.3 eV for 16 atom cell and 0.65eV for 32-atom cell) than typical relaxation energies calculated for elemental metals which are usually in order or below of 0.1 eV (e.g. aluminum and magnesium [10]). The vacancy on γ sites of iron were found to have the lowest formation energy, while antisite aluminum atoms at iron β sites and antisite iron atoms at aluminum sites could be placed without any loss of energy. Although these results are in a good agreement with other calculations [1], visual assessment of separations of vacancies on Fig.1 and zero effective substitutional energy for antisite atoms suggests that 16 atom cell is not sufficient for consideration of an isolated vacancy. In fact zero substitutional energy indicates that the DO_3 structure with ordered vacancies does not represent the ground state of this compound - $x(Fe_{15}A\ell_4)$.

Therefore 32-atom simulation cells have been considered. For the 32-atom cell, the superlattice has an FCC structure. The vacancy at origin (site "1" on Fig.1) is repeated at ends of superlattice vectors on sites marked by "3" on Fig.1. In addition to 16 atoms previously included in the 16-atom cell, another 16 atoms from the cube just above the one shown on Fig.1 have been included in the 32-atom simulation cells. Results of 32-atom calculations are also presented in Table 1. Unlike results reported in [1] where differences between 16- and 32-atom cells calculations were found to be insignificant, effective formation energies calculated the 32 atom simulation cell are much higher than obtained for the 16-atom cell. In addition, it has been found that for the 32-atom cell energies of antisite substitutions are of the same magnitude as vacancy formation energies, this guarantees a correct ordering of atoms. Additional degrees of freedom for atoms to move are responsible for the higher lattice relaxation in the 32-atom cell where the energy of relaxation reaches 0.65 eV and is comparable to relaxation energies observed only in semiconductor systems.

It can be seen that chemical potentials of iron obtained from Eq.6 and bulk calculations are similar (energies for iron vacancies). Unlike results for aluminum, therefore iron vacancy formation energies would not depend on local environment.

The differences in the vacancy formation energies for 16 and 32 atom cells can be explained by interaction between vacancies. Judging from our results for two supercells one can estimate that attractive interaction between two vacancies is about -0.2 eV {(1.25-2.58)/6, where 6 is number neighboring vacancies}. In order to further investigate interaction between vacancies two 32-atom cells (with 30 atoms and 2 vacancies) have been considered. One cell in addition to the vacancy at the origin has another vacancy at the site "2", therefore it represents a repetition of two 16-atoms cells with one vacancy in each. In another cell second vacancy have been located at site marked as "4". The energy of the second cell where distance between vacancies

Table 1 Effective formation energies (in eV) of various point defects.
These energies can be used in Eq. 4 for equilibrium defect concentrations.

System	Relaxation energy	Changes to the first neighbor distance	Formation energy, using element energy	Formation energy using Eq.5
16 atom simulation cell				
γFe vacancy	0.29	-10.2% Al; +0.9% Fe	1.26	1.25
βFe vacancy	0.0	0.0	2.36	2.35
antisite Al at γ site	0.46	+3.3% Al; -0.2% Fe	0.25	1.63
antisite Al at β site	0.14	+1.3%	-1.5	~0.0
Al vacancy	0.60	-6.25%	3.23	1.78
Fe at Al site	0.0	0.0	1.35	~0.0
32 atom simulation cell				
γFe vacancy	0.65	-7.8% Al; -2.7%Fe	2.35	2.58
βFe vacancy	0.04	-1.2%	4.03	4.21
antisite Al at γ sites	0.19	+3.0% Al; -0.2% Fe	1.91	4.35
antisite Al at β sites	0.07	+1.2%	0.18	2.58
Al vacancy	0.60	-2.4%	5.8	2.45
Fe at Al site	0.0	0.0	4.9	3.53

is equal to one half of the lattice constant (2.76 Å) found to be slightly higher (by 0.13 eV) than for the first cell with 5.52 Å separation. However, in the first cell there are 6 vacancies on the same distance of 5.52 Å, while for the second cell distance to the next pair of vacancies is equal to 7.8 Å. For a divacancy in the second cell (this is the smallest separation between two iron vacancies on the same sublattice γ) we can estimate divacancy formation energy by neglecting interaction between divacancies from different simulation cells and by comparing results to 32-atom cell simulation with one vacancy. Such comparison give divacancy formation energy (from two monovacancies) of about -1.2 eV. However, judging from difference between 16- and 32- atom cell calculations interaction between divacancies from different simulation cells cannot be neglected, and this value of divacancy formation energy might be significantly overestimated. Due to reduced symmetry calculations of divacancies are very computer extensive, there are 16 different types of atoms in the simulation cell, in comparison to 6 for a single vacancy. As a consequence, the lattice relaxation for this system has not been considered.

Summarizing, we can stipulate that the calculations presented indicate strong attractive interaction between vacancies with clustering formation energy on the order of fraction of an electron volt per vacancy. To obtain more quantitative results larger simulation cells are required. As the next step calculations for 64 atom have been initiated.

CONCLUSION

Point defect simulations have shown that lattice relaxation plays an extremely significant role in defect formation processes in DO_3 phase of $Fe_3A\ell$. Energy gained by the lattice relaxation is comparable to the effective defect formation energies by themselves. It has been found that preferred point defects are iron vacancies on sites γ (with four aluminum and four iron neighbors) and aluminum antisite atoms at sites of both iron sublattices. Results obtained for a 16 atom unit cell are in very good agreement with previously reported calculations [1]. However unlike [1], we have found significant differences between 16- and 32-atom simulation cells. For 32-atom simulation cell nonzero antisite substitutional energies have been found. Therefore the correct atomic ordering has been predicted.

Significantly for the possibility of vacancy clustering, our calculations in the 32-atom supercell in comparison with those for a 16-atom supercell and divacancy calculations indicate an interaction between nearest-neighbor vacancies on type γ sites is between 0.2 and 1.2 eV. Based on our systematic study of the individual vacancy formation energies, we do not view these results as a completely reliable measure of the isolated divacancy binding energy (since 16 atom cell is not sufficient for single vacancy consideration, 32-atom cell is not sufficient for divacancy). However, the calculated strong vacancy clustering tendency can probably explain the discrepancy between calculated 2.6 eV iron vacancy formation energy and the experimental value of 1.2 eV for the iron vacancy [11]. A formation energy very close to the experimental value has been found for 16-atom cell, where the lowering of this energy apparently is due to the interaction between vacancies from the neighboring cells. The same vacancy formation energy (about 1.2eV) has been found for each of two vacancies in divacancy complex. This serves as a strong indication of vacancy clustering, i.e. either for a divacancy or for a substantially larger cluster.

The next step is to extract and use atomistic potentials to predict the energetics of divacancy formation and vacancy clustering. Therefore we have initiated that calculation.

ACKNOWLEDGMENTS

We want to thank Prof. D.Price for useful discussion. This research has been supported by the U.S.D.O.E. through subcontract 19x-ST547C from Lockheed Martin Energy System Inc.

REFERENCES

[1] J.Mayer, B.Meyer, J.S.Oehrens, G.Bester, N.Börnsen, and M.Fähnle, Intermetalics **5** 597 (1997).
[2] For calculations presented here the calculated lattice constant has been used. Although calculations at the experimental lattice constant better reproduce bulk modulus of materials, they

also result in significantly larger lattice relaxation around vacancy. L.S.Muratov and
B.R.Cooper, presented at MRS Fall Meeting, Boston MA, 1997 (unpublished).

[3] S.C.Deevi, V.K.Sikka Intermetalics **3** 357 (1995).

[4] C.T.Liu and K.S.Kumar, J. Minerals, **45** 38 (1993).

[5] S.M.Foiles and M.S.Daw J.Mater Res. 2, 5 (1987).

[6] J.Mayer, C.Elsässer, and M.Fähnle Phys. Status Solidi B **191** 283 (1995).

[7] V.Schott and M.Fähnle Phys. Status Solidi B **204** 617 (1997).

[8] J.Zou and C.L.Fu Phys. Rev. B. **51** 2115 (1995) [9] T.C.Leung, C.T.Chan, and
B.N.Harmon Phys.Rev.B **44** 2923 (1991)

[10] N.Chetty, M.Weinert, T.S.Rahman, and J.W.Davenport, Phys.Rev.B. **52** 6313 (1995).
G.Bester, B.Meyer, and M.Fähnle Phy.Rev.B **57** R11019 (1998);

[11] H.E.Schaefer R.Würschum, M.Šob, T.Zák, W.Z.Yu, W.Eckert, and F.Banhart,
Phys.Rev.B. **41** 11869 (1990)

VACANCY MIGRATION BARRIER ENERGETICS AND PATHWAYS IN SILICA

L. R. CORRALES[*], R.M. VAN GINHOVEN[*,**], J. SONG[*], H. JÓNSSON[**]
[*]Pacific Northwest National Laboratory, Richland, WA 99352
[**]Department of Chemistry, The University of Washington, Seattle, WA 98195

ABSTRACT

A study of vacancy defect migration pathways and energetics in α-quartz is carried out using an empirical simulation methodology that is coupled with the nudged elastic band method. Results from this study indicate that the migration pathway for migration is between adjacent sites. We anticipate the results will guide modifications to empirical potentials for use in the study of point defect formation of more complex systems.

INTRODUCTION

With the advent of powerful computers, materials scientists have gained the ability to study the formation of defects, and their diffusion and aggregation properties at atomistic and molecular scales. Multi-component systems with large unit cells are difficult to study with anything other than empirical potentials. This is especially true when lattice relaxations and long-range interactions play a role in stabilizing the formation and migration of defects. Unfortunately, empirical potentials generally display shortcomings when used to model surfaces and defects, owing to the lack of electronic structure detail.[1,2] Therefore, a clear understanding of how well empirical potentials perform in evaluating defect structures and energetics is needed.

In this work, a study of the oxygen vacancy defect migration pathways and energetics in α-quartz is carried out using an empirical potential, namely the BKS potential.[3] We have coupled the nudged elastic band (NEB) method with a molecular dynamics (MD) simulation methodology, to study the migration pathways and barrier energetics of vacancy defects.[4,5] The cost of performing the same calculations using a MD density functional theory (DFT) approach limits the system size and requires considerably more computational resources. In the current study, we calculate the predicted structure of the vacancy defect using the BKS potential. Results of more extensive details comparing results obtained with the BKS potential with those obtained from DFT-MD will be reported elsewhere.

The BKS potential was chosen for several reasons over other empirical potentials that are also used to study point defects in crystalline silicates. This potential does an excellent job of reproducing the structural and physical properties of the crystalline polymorphs of silica. It is a reasonably straightforward potential to implement in a MD simulation because the model is a continuous function without cutoffs. Finally, upon removing an oxygen from the lattice network, the net charge of +1.2 is left in the lattice that is effectively the charge of an E_1' center. Hence, the vacancy structure predicted by the BKS potential is similar to the puckered structure predicted by ab initio, semi-empirical and DFT calculations.[6]

SIMULATION MODEL AND PROCEDURE

The Empirical Potential

The BKS potentials is a Coulombic potential that includes a modified Buckingham potential for the repulsion and dispersion interactions, respectively, that is given by

Mat. Res. Soc. Symp. Proc. Vol. 538 © 1999 Materials Research Society

$$E = \sum_{i>j}\left[A_{ij}\exp\left(-b_{ij}r_{ij}\right) - \frac{C_{ij}}{r_{ij}}\right] + \sum_{i>j}\frac{q_i q_j}{r_{ij}^6} \tag{1}$$

It has been parameterized using partial charges obtained from ab initio calculations of a tetrahedral silica cluster. The parameters have been further modified using bulk empirical data and are listed in Table 1. The Coulombic interaction is between all atom pairs, whereas the exp-6 interaction is only between Si-O and O-O atom pairs. Because of the low energy barrier at small r, a harmonic function is spliced in to maintain a repulsive interaction. This is generally required for high temperature simulations and is also necessary when determining the diffusion pathways of the atoms using the method described below.

Table 1

Atom pair	A_{ij} (eV)	$b_{ij}(\text{Å}^{-1})$	C_{ij} (eV Å^6)	Charges
O-O	1388.773	2.760	175.000	$q_O = -1.2$
Si-O	18003.757	4.873	33.538	$q_{Si} = \sigma\,2.4$

In this work, the potential is implemented in a MD computer simulation code using Ewald sums for the long range Coulombic interaction. Minimization of a structure is carried out using a velocity Verlet algorithm in which each simulation is started at zero velocity. Atoms that are not at the potential minimum slide down the well and gain velocity. As they turn around at the minimum the velocity reaches a maximum. At this point, the velocities are reset to zero and the simulation continued until convergence is reached.

The α-quartz lattice is set up using orthogonal periodic boundary conditions. We have created the vacancy and relaxed the structure on system sizes consisting of 144, 576, 768 and 1536 atom systems. The results presented are for the optimized 576 atom system consisting of 384 oxygen atoms and 192 silicon atoms.

The Nudged Elastic Band Method

The nudged elastic band (NEB) method is a modified chain of states method described as follows. A chain of images, or replicas of the full system, is created which maps out a path from an initial to a final state. Each image is linked to its nearest neighbors in the chain by a harmonic potential. The system energy of the simple chain consisting of P images along the elastic band (PEB), without nudging, is given by

$$S^{PEB}\left(\vec{R}_1,...,\vec{R}_{P-1}\right) = \sum_{i=0}^{P}V\left(\vec{R}_i\right) + \sum_{i=1}^{P}\frac{Pk}{2}\left(\vec{R}_i - \vec{R}_{i-1}\right)^2 \tag{2}$$

where k is the spring constant between images. During 'nudging', the component of the true force parallel to the path, and the component of the spring force perpendicular to the path, are set to zero. Consequently, the force on each image is given by

$$\vec{F}_i^0 = -\vec{\nabla}V\left(\vec{R}_i\right)\Big|_\perp + \vec{F}_i^s\cdot\hat{\tau}_{||}\hat{\tau}_{||} \tag{3}$$

with the forces parallel to the path are $\vec{\nabla}V\left(\vec{R}_i\right)\cdot\hat{\tau}_{||}\hat{\tau}_{||} = 0$, and the spring forces perpendicular to the path are $\vec{F}_i^s\Big|_\perp = 0$. The τ's are the three component directions of which one is perpendicular and the other two are parallel components relative to a specified vector direction. The spring force between images is given by

$$\vec{F}_i^s = k_{i+1}\left(\vec{R}_{i+1} - \vec{R}_i\right) - k_i\left(\vec{R}_i - \vec{R}_{i-1}\right) \tag{4}$$

In the NEB method, the spring force keeps the images spaced evenly along the path without interfering with the relaxation to the minimum energy path (MEP).

RESULTS AND DISCUSSION

Starting from a perfect a-quartz structure, as modeled by the BKS potential, a single oxygen is removed from the lattice and the structure allowed to relax to its minimum energy state. The resulting structure is a "puckered" configuration in which one silicon, that had formed a long bond with the removed oxygen, moves a considerable distance to form a new "back" bond with an oxygen that is now three-fold coordinated. The other silicon, that formed a short bond with the removed oxygen, goes into a nearly planar configuration with respect to its remaining three bonded oxygens. This is one of the stable structures predicted by electronic structure calculations.

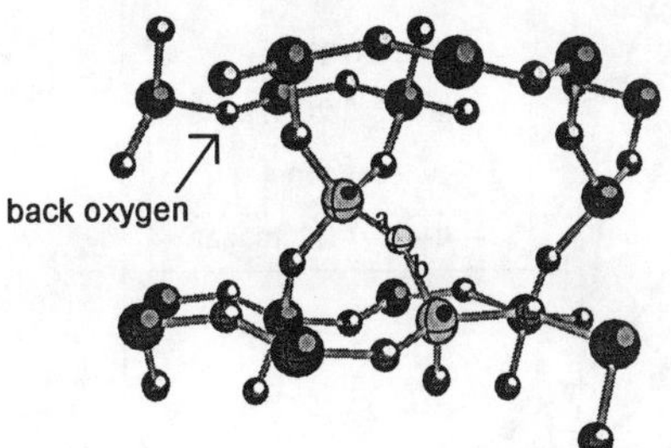

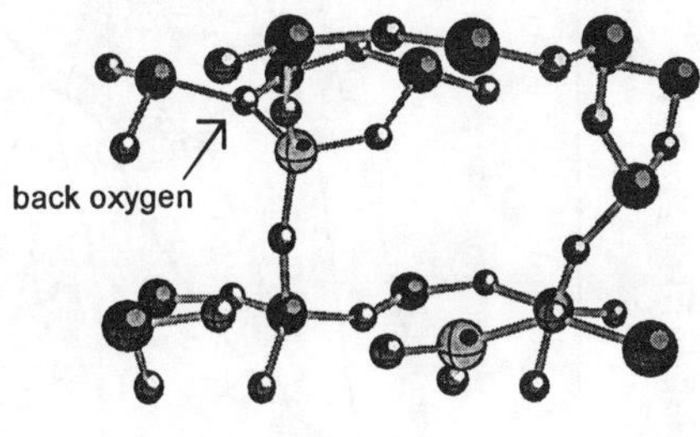

Figure 1. A cut away view of the crystal configuration. The figure on the left is the perfect crystal in which the light color atom in the center is the oxygen atom that is removed. The bond labeled with an *a* is the long bond, and that one labeled *b* is the short bond. The figure on the right shows the structure with the oxygen vacancy after relaxations. The arrow in both figures points to the oxygen atom that forms the back bond (three fold coordinated) after relaxation.

Starting from the puckered configuration, there are six oxygen atoms that are bonded to each of the two silicon atoms that are moved into the vacancy position. In addition, a next nearest neighbor and a next-next nearest neighbor are also moved into the vacancy position. The

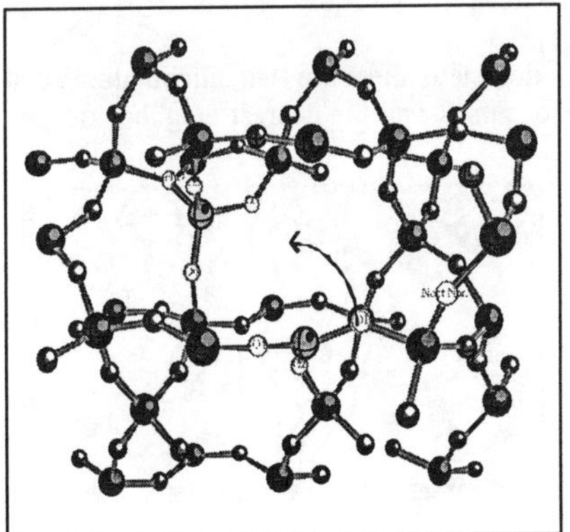

Figure 2. The figure shows which neighboring atoms are separately moved into the vacancy position.

Table 2

Barrier energies for 10 (20) image chains

Oxygen	Barrier Energy (eV)
1	2.999 (3.031)
2	3.274
3	3.048 (2.819)
4	2.292 (2.435)
5	3.008
6	2.346

chains consist of ten or twenty images where each image is allowed to relax to its minimum energy state. The lowest barrier energies are for the nearest neighbor oxygen atoms. The energies are slightly different owing to the asymmetry of the oxygen atoms and to the relaxation response of the lattice. The estimated barrier energies for each of the pathways are shown in Table 2 for both the ten and twenty image chains.

The migration pathway energetics are plotted in Figure 3 for three of the low energy paths. The chains with ten images are compared with the chains of twenty images. There is an increase in the resolution in going from ten to twenty images that also results in a longer pathway being sampled.

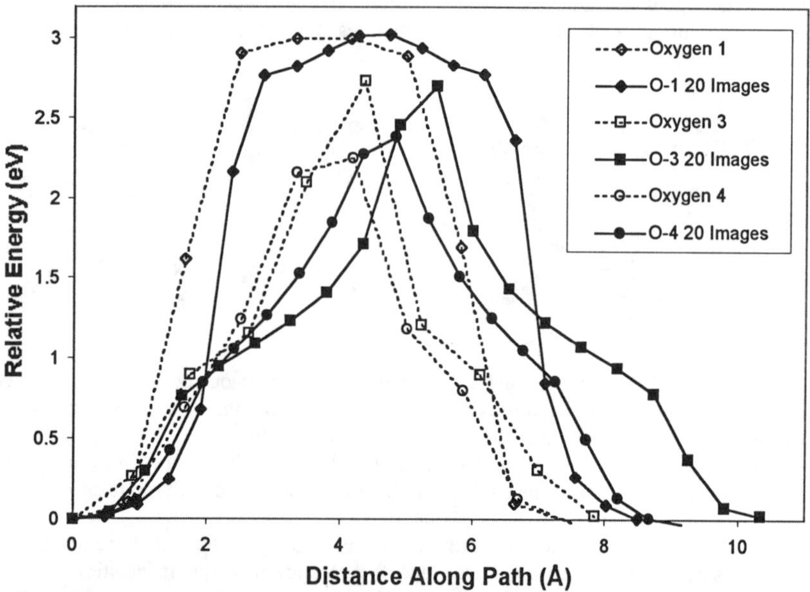

Figure 3. Migration pathway energetics of the low barrier energy chains. The paths with 20 images show greater resolution than those with only 10 images.

As an oxygen atom migrates through the transition state, there are two silicon atoms that are in the puckered state, as shown in Figure 4. This occurs for all the nearest neighbor oxygen atoms as they move to the vacancy.

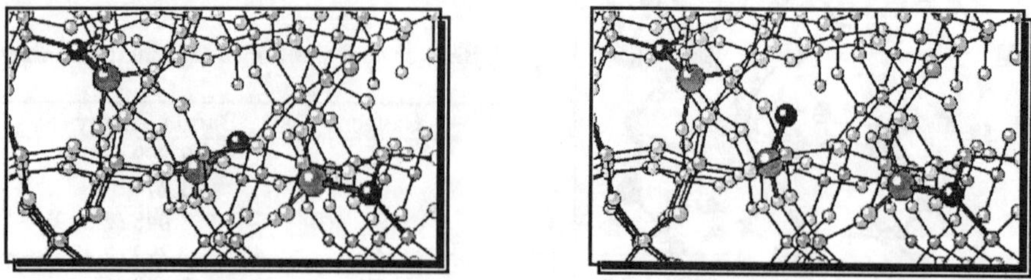

Figure 4. As an oxygen goes through the transition state there are two puckered silicon atoms.

CONCLUSIONS

We have calculated the predicted relaxed structure of the α-quartz network about an oxygen vacancy using the BKS empirical potential. Using the nudged elastic band method, we have also calculated certain pathways for the migration of the oxygen vacancy and determined the migration barrier energies for those specific paths. Several authors are currently using empirical potentials to calculate migration energy barriers of atoms in ceramic materials. The objective of this work is to put into perspective the quality of the migration barrier energetics and to compare the migration pathways with ab initio molecular dynamics simulations. This work presents the empirical potential calculations that will be used to compare with the density functional theory molecular dynamics (DFT-MD) calculations that are currently underway in our laboratory.

The use of MD as a method to find the relaxed structure may appear to be an inefficient approach. However, coupling the images via the NEB method requires a technique that is computationally efficient. Thus, a method that can relax a single image more efficiently is not generally efficient when the images are coupled. Moreover, the MD method is much easier to implement in a parallel computational environment and with a variety of simulation potentials. The former is valuable as the system size is increased, the number of images is increased, and when implementing the method in parallel decreases the computational time of the potential.

The advantage of using empirical potentials is that they are easy to use in larger more complex systems and are computationally inexpensive. Comparing the empirical and ab initio results can lead to improving the empirical potentials for use in calculating defect migration energetic and pathways.

ACKNOWLEDGEMENT

The authors would like to thank W. J. Weber who has provided helpful discussions and insight into this problem. This work was supported by the Environmental Management Science Program, Office of Environmental Management, U.S. Department of Energy under Contract DE-AC06-76RLO 1830.

REFERENCES

1 N. Capron, A. Lagraa, S. Carniato and G. Boureau, J. Non-Cryst. Solids, 216 (1997) 10.

2 N. C. Pyper and P. Popelier, J. Phys. Condens. Matter, 7 (1995) 5103; J. H. Harding and N. C. Pyper, Philop. Mag. Lett., 71 (1995) 113.

3 B. W. H.van Beest, G. J. Kramer and R. A. van Santen, Phys. Rev. Lett. 64 (1990) 1955; G.J. Kramer, N.P. Farragher, B.W.H. van Beest and R.A. van Santen, Phys. Rev. B 43 (1991) 5068.

4 H. Jönsson, G. Mills and K.W. Jacobsen, Classical and Quantum Dynamics in Condensed Phase Simulations, B.J. Berne, G. Cicotti and D.F. Coker, editors (World Scientific, 1998).

5 G. Mills, H. Jönsson and G.K. Schenter, Surface Science, 325 (1995) 305.

6 W.B. Fowler and A.H. Edwards, J. Non-Cryst. Solids, 222 (1997) 33.

MULTISCALE SIMULATIONS OF THE RF DIODE SPUTTERING OF COPPER

H. N. G. WADLEY, W. ZOU, X. W. ZHOU, J. F. GROVES
Intelligent Processing of Materials Laboratory
Department of Materials science and Engineering
University of Virginia
Charlottesville, VA 22903
S. DESA, R. KOSUT, E. ABRAHAMSON, S. GHOSAL, A. KOZAK
SC Solutions
Santa Clara, CA 95054
D. X. WANG
Nonvolatile Electronics, Inc.
Eden Prairie, MN 55344-3617

ABSTRACT

The morphology and microstructure of RF diode sputter deposited materials is a complicated function of many parameters of the reactor operating conditions. Using a combination of computational fluid dynamics (CFD), RF plasma, molecular dynamics (MD) sputter, and direct simulation Monte Carlo (DSMC) transport models, a multiscale approach has been used to analyze the RF diode sputtering of copper. The CFD model predicts the velocity and pressure distribution of the working gas flows in the deposition chamber. The plasma model uses these CFD results to compute ion energies and fluxes at the target and substrate. The MD model of sputtering is used to determine the initial energy distribution of sputtered atoms and reflected neutral working gas atoms and both of their angular distributions. A DSMC transport model then deduces the target atom deposition efficiency, the spatial distribution of the film thickness, the target and reflected neutral atoms energy and impact angle distributions given reactor operating input conditions such as background pressure, temperature, gas type, together with the reactor geometry. These results can then be used in atomistic growth models to begin a systematic evaluation of surface morphology, nanoscale structure, and defects dependences upon the reactor design and its operating conditions.

INTRODUCTION

Physical vapor deposition processes are being used to produce increasingly complex devices whose performance is critically dependent upon atomic scale features of their structure. A good example is giant magnetoresistive (GMR) metal multilayers (e.g., NiFe/Cu/NiFe) which exhibit large drops in their electrical resistance when a magnetic field is applied [1-3]. These materials can be used for making new magnetic field sensors [4-6], read heads for disk drives, and magnetic random access memories (MRAM) [5,6]. GMR-based MRAM has many potentially attractive features, such as non-volatility, radiation hardness, low power consumption, high memory densities (comparable to those of dynamic random access memory), and high access speed. The key issue for this technology is to design an economical deposition process that can produce thermally stable GMR multilayers with high GMR ratios (defined as the maximum resistance change divided by the resistance at magnetic saturation) at low magnetic field. Both theoretical and

323

experimental work indicated that the best GMR properties are achieved when the atomic scale interfacial roughness and interlayer chemical mixing are both minimized. Molecular dynamics simulation of GMR multilayer deposition identified that adatom incident energies in the range between 0.1 to 5.0 eV are needed to minimize both interfacial roughness and intermixing [7]. Sputter deposition methods can result in incident adatom energies in the range of 0.1 to 20 eV, and have been widely explored for GMR multilayer deposition [8-23]. However, because the experimental atomic scale characterization is difficult, and the dimensions of the processing parameter space is large, the experimental search for an optimized sputter deposition process for synthesis of GMR multilayers has been prolonged. A multiscale reactor simulation tool relating the morphology and microstructure of multilayers to the processing conditions of a diode sputter deposition system has been developed to help the optimization of the process.

DIODE SPUTTER DEPOSITION

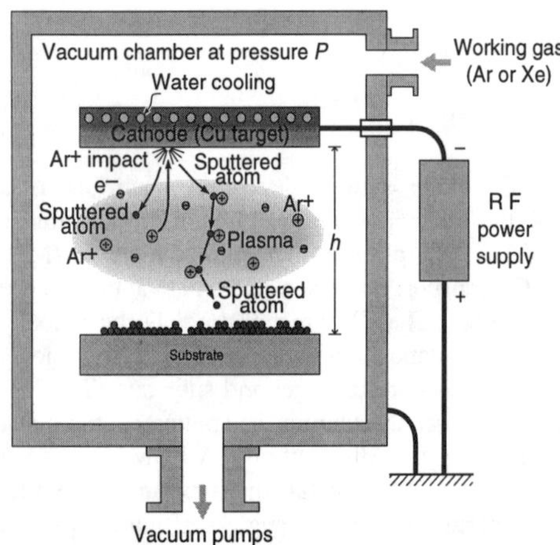

Figure 1. Schematic of A Diode Sputter System

Atomistic simulations of vapor deposited films have shown that film structure is a function of the key deposition conditions including substrate temperature, deposition rate, adatom incident energy, and incident angle [7,24]. These key deposition conditions are in turn controlled by many process parameters. To illustrate, a diode sputter deposition system is schematically shown in Figure 1. An inert gas plasma is initiated and maintained between target and substrate by an RF power. The inert gas ions created in the plasma are accelerated to the metal target under the bias voltage. The high energy bombardment of inert gas ions on the target surface results in the sputtering of metal atoms from the target. These sputtered atoms are then transported to the substrate and deposited on the substrate surface. Generally, the system geometry and the RF power determine the densities, the energies, and the angles of the atoms emitted from the target. Scattering with the lower energy working gas modifies these quantities during the transportation of the sputtered metal atoms to the substrate. The final distributions of these quantities depend on the target-substrate distance, the working gas temperature, and pressure. Some inert gas ions are neutralized after their bombardment of the target and can be reflected toward the substrate. These reflected neutral particles can also modify the morphology and the structure of a growth film. To develop a better understanding of the process and to be able to predict the uniformity, morphology, and structure of thin films, it is essential to model the deposition efficiency (defined by the ratio of the deposited material at the substrate to the sputtered materials at the target), the distributions of density, energy, and angle of depositing atoms, as well as inert ions at the substrate, all as a function of reactor scale pressure, temperature, system geometry, power, etc.

EXPERIMENTS

Vapor deposited copper films were grown on silicon wafer using a Randex Model 2400-6J diode sputter deposition system at fixed Ar pressure and target-substrate distance but with different plasma power. The surface morphology was characterized using a Pico SPM MS 300 atomic force microscope. The morphology of two typical samples are shown in Figure 2.

Pressure = 20 mTorr, target-substrate distance = 1.5 in, Thickness = 2000 Å

Power = 50 W   Power = 350 W

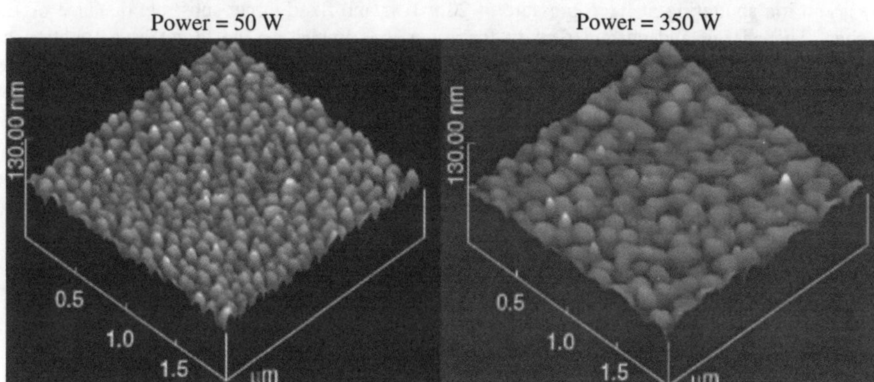

Figure 2. Atomic Force Microscope of Sputter Deposited Copper Films

Figure 2 indicates that an increase of the power from 50 to 350 W dramatically increases the grain size and reduces the surface roughness.

SIMULATION METHODOLOGY

The multiscale reactor model is illustrated in Figure 3. A CFD finite element model was used to calculate the velocity and pressure distribution of inert gas flow in the chamber; a steady-state plasma model was used to simulate the density and energy of Ar ions striking the target and substrate; a molecular dynamics (MD) model was used to determine the sputtering yield of target by the inert ions, the energy and angular distribution of the sputtered atoms at the target; and a Direct Simulation Monte Carlo (DSMC) model was used to trace the change of density, energy, and angle of the sputtered atoms as they transported through the low-pressure inert gas to the deposition substrate. The individual models for gas flow, plasma discharge, sputtering, and atom transport were then

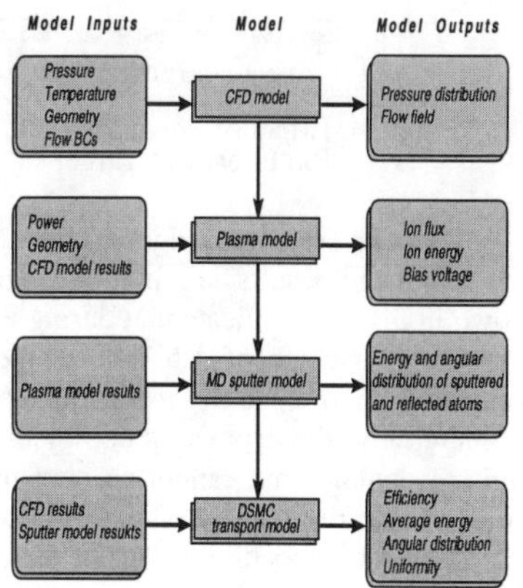

Figure 3. Reactor Scale Integrated Model

integrated to create a detailed, steady-state, input-output model capable of predicting incident energy, incident angle, deposition-rate, and uniformity as a function of the process input variables: power, pressure, gas temperature, and electrode spacing. These results can in turn account for the morphology and microstructure of vapor deposited films [7,24].

RESULTS

The multiscale model was used to explore the effects of power during deposition of copper by argon ion sputtering at fixed pressure of 20 mTorr and fixed target-substrate distance of 1.5 inches. The energy and current density for argon ions impacting the target as a function of power are given in Figures 4 and 5. The deposition rate as a function of power is shown in Figure 6. The energy and current density of depositing fluxes at the substrate as a function of power are drawn in Figures 7 and 8. It can be seen that increasing the power from 50 to 350 W increases the average ion energy from 0 to 600 eV, and the ion current density from 0 to 7×10^{19} atoms/m^2. Higher ion energy at the target leads to higher energy sputtered atoms and a higher sputtering yield [25]. As a result, increase of power not only increases the deposition rate, Figure 6, but also increases the incident energy of the atoms deposited at the substrate [26], Figure 7.

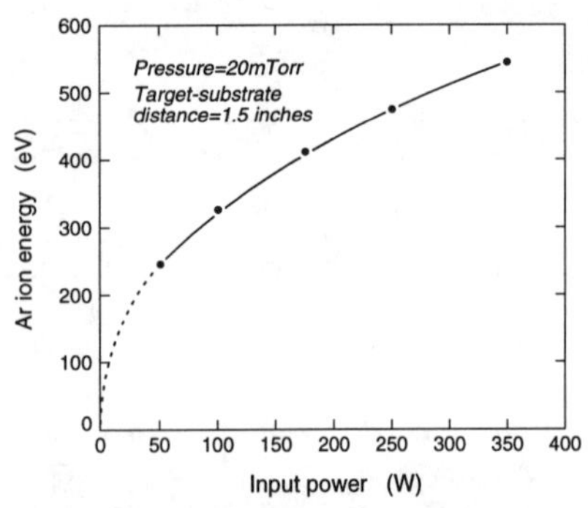

Figure 4. Effect of Power on Inert Ion Energy at Target

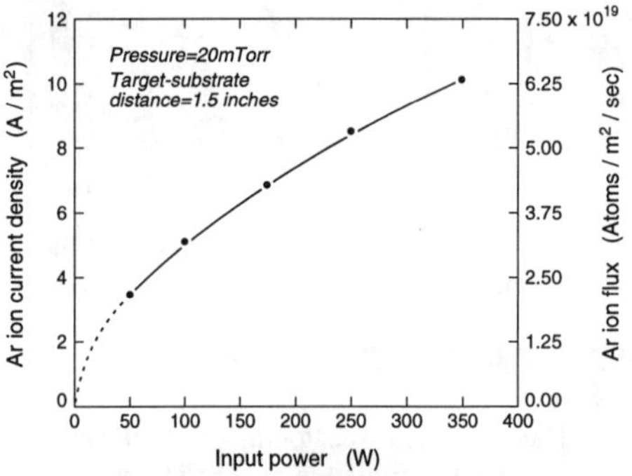

Figure 5. Effect of Power on Inert Ion Current Density at Target

It can be seen from Figure 6 that the deposition rate is almost linearly related to the power. The simulated results are in good agreement with experimental measurements. The results shown in Figure 7 indicate that during RF diode sputter deposition at a pressure of 20 mTorr and a working distance of 1.5 inches, the scattering with the background working gas almost completely thermalize the metal flux. A net increase of power of 350 W results in only a 0.04 eV increase in the depositing energy of copper. This is unlikely to significantly change the thin film morphology and cannot account for the results shown in Figure 2. Interestingly, Figure 8 indicates that the density of the reflected inert gas atoms at the substrate is comparable to the density of the depositing flux. On the other hand, Figure 7 reveals that increasing input power significantly increases the energy of the reflected inert flux to 100 eV range. Molecular dynamics simulations indicated that 100 eV inert atom bombardment on the substrate can cause a significant transient local heating that results in surface atom athermal diffusion and a flattening of a growth surface. It appears that the neutral flux is a powerful means for modifying the surface morphology. However, it may be undesirable for growing thin metal multilayers used for GMR devices because it induces interlayer mixing. The results obtained by the integrated model allow us to use atomistic simulations [7,24] to quantify the microstructure and morphology of deposited films as a function of reactor scale processing parameters.

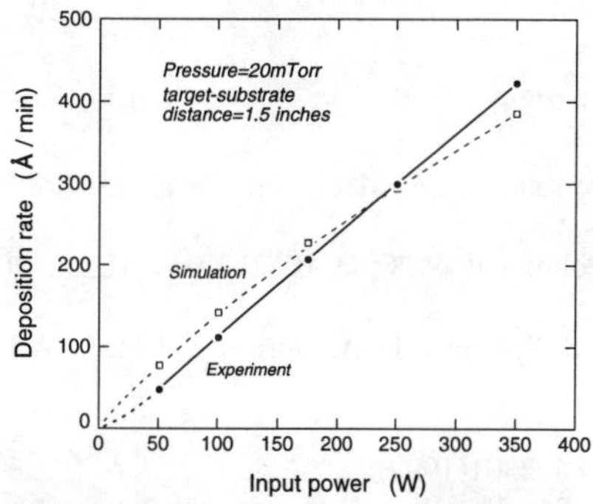

Figure 6. Effect of Power on Deposition Rate

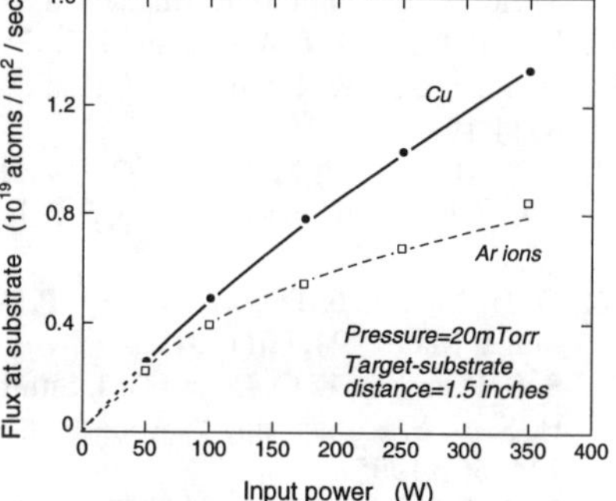

Figure 7. Effect of Power on
Flux Energies at Substrate

CONCLUSIONS

A multiscale reactor model has been developed to simulate the effects of processing conditions (power, pressure, temperature, target-substrate distance, etc.) on the deposition rate, the density and the energy of depositing atoms, and reflected inert atoms. These results can be in turn used in atomistic growth models to simulate surface morphology and microstructure of deposited films. Our results indicate that:

1. Increasing the RF power increases the inert gas ion density and energy at the target.

2. Increasing power linearly increases the metal flux density at the substrate and therefore the deposition rate, but causes little change in metal incident energy at pressure of 20 mTorr.

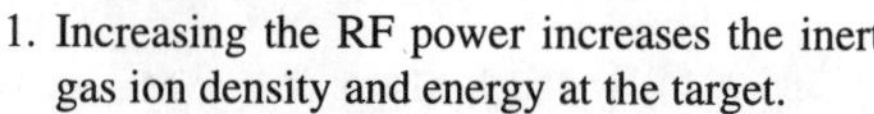

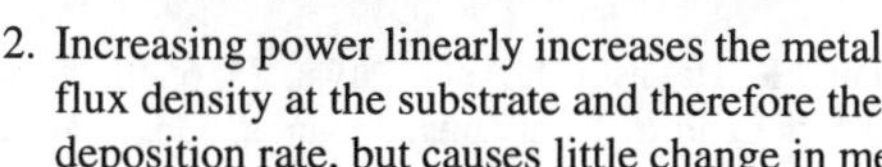

Figure 8. Effect of Power on Flux
Current Densities at Substrate

3. The flux of reflected inert gas neutral flux is comparable to that of the depositing flux. The neutral energy increases rapidly with power.

4. The energetic bombardment of the substrate by reflected neutral Ar is the apparent origin of the smoother surfaces observed after higher power deposition. However, this may lead to multilayer intermixing of metal like those of interest for GMR devices.

ACKNOWLEDGEMENTS

We are grateful to the Defence Advanced Research Projects Agency (A. Tsao, Program Manager) and the National Aeronautics and Space Administration for support of this work through NASA grants NAGW1692 and NAG-1-1964.

REFERENCES

1. P. M. Levy, J. Magn. Magn. Mater., 140-144, 485(1995).
2. M. D. Stiles, Phys. Rev., B48, 7238(1993).
3. W. H. Butler, X. G. Zhang, D. M. C. Nicholson and J. M. MacLaren, Phys. Rev., 52B, 13399(1995).
4. J. Daughton, J. Brown, E. Chen, R. Beech, A. Pohm and W. Kude, IEEE Trans. Magn., 30, 4608(1994).
5. C. Tsang, R. E. Fontana, T. Lin, D. E. Heim, V. S. Speriosu, B. A. Gurney and M. L. Williams, IEEE Trans. Magn., 30, 3801(1994).
6. J. L. Simonds, Phys. Today, April, 26(1995).
7. X. W. Zhou and H. N. G. Wadley, J. Appl. Phys. 84, 2301(1998).
8. E. E. Fullerton, D. M. Kelly, J. Guimpel and I. K. Schuller, Phys. Rev. Lett., 68, 859(1992).
9. M. L. Yan, W. Y. Lai, Y. Z. Wang, S. X. Li and C. T. Yu, J. Appl. Phys., 77, 1816(1995).
10. T. L. Hylton, K. R. Coffey, M. A. Parker and J. K. Howard, J. Appl. Phys., 75, 7058(1994).
11. S. Honda, S. Ohmoto, R. Imada and M. Nawate, J. Magn. Magn. Mater., 126, 419(1993).
12. R. J. Pollard, M. J. Wilson and P. J. Grundy, J. Magn. Magn. Mater., 146, L1(1995).
13. H. Zhang, R. W. Cochrane, Y. Huai, M. Mao, X. Bian and W. B. Muir, J. Appl. Phys., 75, 6534(1994).
14. T. R. McGuire, J. M. Harper, C. Cabral Jr. and T. S. Plaskett, J. Appl. Phys., 76, 6601(1994).
15. J. D. Kim, A. K. Petford-Long, J. P. Jakubovics, J. E. Evetts and R. Somekh, J. Appl. Phys., 76, 6513(1994).
16. D. H. Mosca, F. Petroff, A. Fert, P. A. Schroeder, W. P. Pratt Jr. and R. Laloee, J. Magn. Magn. Mater., 94, L1(1991).
17. S. S. P. Parkin, Z. G. Li and D. J. Smith, Appl. Phys. Lett., 58, 2710(1991).
18. H. Sato, Y. Kobayashi, Y. Aoki, R. Loloee and W. P. Pratt Jr., J. Magn. Magn. Mater., 140-144, 567(1995).
19. R. J. Highmore, W. C. Shih, R. E. Somekh and J. E. Evetts, J. Magn. Magn. Mater., 116, 249(1992).
20. K. Kagawa, H. Kano, A. Okabe, A. Suzuki and K. Hayashi, J. Appl. Phys., 75, 6540(1994).
21. K. Meguro, S. Hirano, M. Jimbo, S. Tsunashima and S. Uchiyama, J. Magn. Magn. Mater., 140-144, 601(1995).
22. J. C. S. Kools, J. Appl. Phys., 77, 2993(1995).
23. T. C. Anthony, J. A. Brug and S. Zhang, IEEE Trans. Magn., 30, 3819(1994).
24. Y. Yang, R. A. Johnson and H. N. G. Wadley, Acta Mater., 45, 1455(1997).
25. H. H. Andersen, H. L. Bay and H. E. Roosendaal, in Topics in Applied Physics, Sputtering by Particle Bombardment I, edited by R. Behrisch (Springer-Verlag, Berlin Heidelberg, 1981), Vol. 47, p. 145; 219.
26. W. O. Hofer, in Topics in Applied Physics, Sputtering by Particle Bombardment III, edited by R. Behrisch and K. Wittmaack (Springer-Verlag, Berlin Heidelberg, 1991), Vol. 64, p. 15.

MEASUREMENT OF PLANAR FAULT ENERGIES IN
Ni_3Ge-Fe_3Ge INTERMETALLIC ALLOYS

Mukul Kumar,* T. J. Balk, and K. J. Hemker

Department of Mechanical Engineering, Johns Hopkins University, Baltimore, MD 21218-2686
*Now at Lawrence Livermore National Laboratory, University of California, L-370, Livermore, CA 94550

ABSTRACT

A combination of transmission electron microscopy (TEM) and image simulations has facilitated a highly quantitative measure of superdislocation dissociations. The experimental observations have been corrected for image shifts within the framework of anisotropic elasticity by comparison with simulated images. This allows experimental quantification of planar fault energies, thus providing a benchmark for first principles and atomistic simulations and fundamental insight towards alloy modeling and design. Such measurements of superdislocation dissociations on the order of 1–15 nm have been recorded for the pseudobinary Ni_3Ge-Fe_3Ge alloy system. These detailed measurements of fault widths obtained by weak-beam TEM observations of deformation structures will be presented and discussed as a function of alloy composition. The transition from anomalous to normal temperature dependence of yielding behavior in these alloys will also be discussed in terms of observed dislocation structures and planar fault energies calculated using the above measurements.

INTRODUCTION

The anomalous increase of flow strength with increasing temperature exhibited by many $L1_2$ intermetallic compounds has attracted considerable attention in the literature [1]. Ni_3Al in particular has been the focus of numerous studies that have attempted to understand and model the flow strength anomaly. In general, the anomaly is associated with the thermally activated formation of Kear-Wilsdorf (KW) locks [2], during which the <110>{111} superdislocations cross-slip from a {111} octahedral plane, where they are mobile, to a {010} cube cross-slip plane, where they are sessile. The dissociation of a superdislocation in this ordered crystal structure results in the formation of an antiphase boundary (APB) bounded by two similar $^1/_2$<110> superpartials, and further sub-dissociation of the superpartial results in a complex stacking fault (CSF) bounded by dissimilar Shockley partials. The geometry of the dissociated core, and in particular, the width of the APB and the CSF that is governed by the respective fault energies, has a profound effect on the dislocation mobility and, consequently, the gross deformation behavior [1]. For this reason, studies of the dislocation core geometry allow one to bridge the length scale between microscopic deformation mechanisms and macroscopic mechanical behavior.

While the anomalous macroscopic mechanical properties of Ni_3Al have been shown to be governed by the mobility of superdislocations the fact that several $L1_2$ alloys do not exhibit anomalous flow behavior is of special interest. In order to study the mechanical behavior of $L1_2$ intermetallics in general, the model system $(Ni_xFe_{1-x})_3Ge$ was chosen. This system is pseudobinary and exhibits complete solid solubility if the Ge content is held constant at 25at% and Fe is substituted for Ni as the composition varies from Ni_3Ge to Fe_3Ge [3]. The $L1_2$ compounds Ni_3Ge and Fe_3Ge show dramatically different mechanical behavior. Ni_3Ge has been reported to have a strong anomaly [3,4], whereas Fe_3Ge shows a more normal decrease of yield strength with increasing temperature [3,5]. Suzuki *et al.* [3] have reported that the temperature dependence of flow strength changes from anomalous to normal as the Fe content is increased, and this has been corroborated by the current authors [4].

This study of the dislocation configurations that are characteristic of the Ni_3Ge-Fe_3Ge model system has been undertaken to explicitly measure and calculate the effect of Fe content on the fault energies and consequently on the yielding behavior of these alloys. Thus, it focuses on the microscopic dislocations in the context of the attendant fault energies and the influence they exert on macroscopic deformation behavior. Specifically, the faults that govern the dissociation of dislocation cores will be investigated through experimental measurement of the dissociation widths and calculation of the fault energies within the framework of anisotropic elasticity.

DEFORMATION MICROSTRUCTURES

Detailed TEM investigations were conducted on samples compressed to 1% strain to better understand the change in yielding behavior as a function of Fe content and temperature. A dramatic difference could be seen in the deformation microstructures of binary Ni_3Ge deformed at room

Mat. Res. Soc. Symp. Proc. Vol. 538 © 1999 Materials Research Society

temperature (fig. 1a) and below, and an alloy containing 32at%Fe that was compressed at room temperature (fig. 1b). Straight, screw character <110> superdislocations, as seen in fig. 1a, dominated the microstructure in binary Ni_3Ge. Tilting experiments determined that they were in the KW locked configuration with the APB lying on the cube cross-slip plane. Deformation at 77 K also produced a significant minority of superdislocations of mixed orientation that were gliding on the octahedral plane.

In contrast, ambient temperature compressive deformation of a Fe-rich (32at%Fe) alloy produced a microstructure of curved <110> superdislocations with random line orientations (fig. 1b). It was determined that the superdislocations were gliding on the cube plane. The curved nature of the dislocations suggested that they were not in locked configurations, unlike the case in $Ni_{75}Ge_{25}$ (Ni_3Ge).

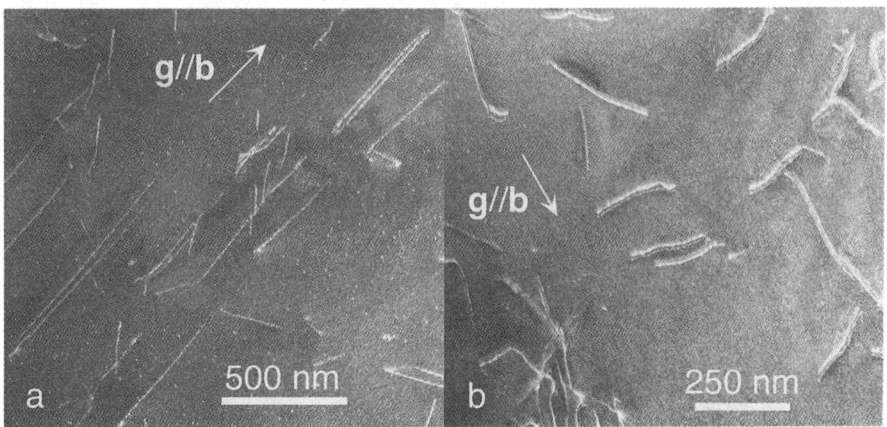

Fig. 1: Weak-beam TEM images of the microstructure after room temperature deformation in, a) $Ni_{75}Ge_{25}$ and b) $Ni_{43}Fe_{32}Ge_{25}$.

With the addition of 15at%Fe, the dislocations in $Ni_{60}Fe_{15}Ge_{25}$ although still generally screw-oriented with a high degree of KW locking, showed considerable localized bowing on the cube plane. The observed increase in flow stress [4] suggests that cross-slip locking occurs more frequently in this alloy.

Figure 2 is a micrograph from $Ni_{50}Fe_{25}Ge_{25}$ deformed at room temperature. In some of the grains it was observed that the average orientation of dislocations was in the screw direction, but the KW locked segments can clearly be seen bowing out on the cube cross-slip plane. The presence of superkinks and switch-over kinks clarifies the operation of K-W lock by-passing mechanisms [6], and suggests that the cross-slip locking process is dynamic in nature at this test temperature. The increased bowing on the cross-slip plane points to a decrease in lattice friction on this plane. These observations are also representative of the microstructure that develops in $Ni_{43}Fe_{32}Ge_{25}$ on deformation at 77K.

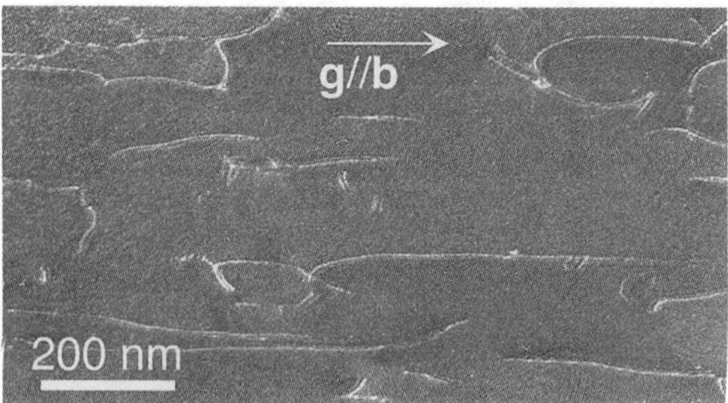

Fig. 2: Weak-beam TEM image of microstructure developed in some grains in $Ni_{50}Fe_{25}Ge_{25}$ after room temperature deformation.

An even stronger indication of this increase in cube plane mobility is given in fig. 1b, which is representative of the room temperature deformation microstructure in the majority of grains in $Ni_{50}Fe_{25}Ge_{25}$ samples and for all grains in $Ni_{43}Fe_{32}Ge_{25}$ samples. Here, mixed dislocations dominate the microstructure and the mixed character of these dislocations is much closer to edge than to screw. Tilting experiments have shown that these dislocations are APB dissociated on the cube plane and that cube glide has replaced octahedral glide as the dominant deformation mechanism. These microstructural observations are very similar to those made for $Fe_{75}Ge_{25}$ (Fe_3Ge) [5], where deformation was seen to occur through cube glide of mixed dislocations. It appears that these two compositions are similar to Fe_3Ge both in the normal decrease of flow strength with temperature and in the dominance of cube glide during plastic deformation.

IMAGE SHIFT CORRECTIONS AND CALCULATION OF FAULT ENERGIES

The precise determination of fault energies is of intrinsic value towards modeling the anomalous flow stress behavior as it allows determination of the thermodynamic driving forces for cross-slip and also indicates the stability of a planar fault on a particular plane with respect to other close-packed planes. It should be noted, however, that the experimental observations do not directly bring forward the equilibrium width of the planar fault and it is necessary to correct for image shifts to obtain these values as input for the force balance equations to compute the fault energy. Image shifts, or the difference in the distance between intensity peaks in a TEM micrograph and the actual separation of the partial dislocations are inherent during imaging of dislocations. The shift arises because the intensity peak does not correspond to the exact location of the dislocation but instead is related to the local curvature of the lattice planes near the defect where Bragg's law is satisfied for diffraction. Moreover, close proximity of the partials bounding a stacking fault leads to overlapping of their strain fields and hence the shift of the intensity peaks from the actual positions are non-symmetric. These effects are particularly acute during weak-beam imaging of narrowly dissociated superdislocations in ordered lattices with magnitudes of Burgers' vectors larger than those encountered in normal fcc solid solutions.

An isotropic correction to this image shift was developed by Cockayne *et al.* [7]. It has, however, been found recently [8-11] that the approximations used to develop this correction cause significant departures (or overestimations) in the case of highly elastically anisotropic materials like intermetallic alloys. Hence, image simulations within the framework of anisotropic elasticity were carried out to determine the true dissociation distances in the manner described by Hemker and Mills [10]. In the particular case of Ni_3Ge-Fe_3Ge alloys, the elastic constants for the binary Ni_3Ge (experimental measurements of Yasuda *et al.* [12]) and Fe_3Ge (first principles calculations of Mryasov and Freeman, private communication) alloys were used to approximate for the intermediate compositions using the rule of mixtures approach.

Weak-beam image simulations were carried out by using CUFOUR [13], a many-beam simulation program within the framework of anisotropic elasticity that is based on the earlier two-beam program of Head *et al.* [14]. Intensity profiles of all simulated images were averaged over the length of the dislocations, thus incorporating numerous oscillations. Five fundamental diffracted beams in the systematic row of the operating reflection and the transmitted beam were considered in all cases. Extinction distances and absorption coefficients for the relevant beams for the different alloy compositions were calculated using the program EMS developed by Stadelmann [15] and the scattering factors reported by Weickenmeier and Kohl [16].

The results obtained from image simulations and matching are given in Table 1, where the experimentally measured and corrected values for the APB dissociation widths have been compiled for the 4 alloys studied. In a number of cases several line orientations were analyzed to obtain a consistent measure of the fault energies since dissociation widths are a function of the character of the dislocation, whereas the fault energy is a constant. The typical scatter in the data is on the order of 5–10%, with somewhat higher uncertainty in measurement for the fault widths on the octahedral plane. The errors arise due to experimental measurement of the dissociation widths from the TEM micrographs and, in particular, in those instances where the dislocation segments have a strongly curved nature. This uncertainty is also reflected in making the corrections by matching with computer simulated images.

The noteworthy feature of the weak-beam observations is the dramatic increase in the extent of the cube plane width and, hence, the strong indication that the cube plane APB energy decreases as the concentration of Fe (substituting for Ni) is increased. It is observed that corrected fault widths for screw APB-dissociated superdislocations on the cube plane increase from about 2.7 nm to a value of about 5.4 nm, with the width being even higher in the edge orientation. In contrast, the increase in the width of the APB on the octahedral plane is much smaller; the dissociation width in edge orientation only increases from 3.6 nm to 4.2 nm.

Table 1: APB energies in the Ni_3Ge–Fe_3Ge system as a function of alloy composition and temperature calculated on the basis of corrected and uncorrected dissociation widths.

Alloy Composition (Temperature)	Fault Plane	Line Orientation	$d_{measured}$ (nm)	$d_{corrected}$ (nm)	Antiphase Boundary Energy (mJ/m^2)	
					$\gamma_{measured}$	$\gamma_{corrected}$
$Ni_{75}Ge_{25}$	{001}	0° (screw)	3.8±0.4	2.7±0.1	212±22	296±11
(77K)	{111}	90° (edge)	5.5±0.2	3.6±0.3	220±8	339±29
	{111}	30°	3.5±0.2	2.7±0.2	259±15	337±25
$Ni_{75}Ge_{25}$	{001}	45°	4.2±0.2	3.7±0.2	261±13	292±12
(300K)	{001}	0° (screw)	3.7±0.5	2.7±0.2	219±29	297±22
$Ni_{60}Fe_{15}Ge_{25}$	{001}	0° (screw)	5.0±0.2	3.8±0.1	154±6	202±6
(77K)	{111}	90° (edge)	5.3±0.3	3.8±0.5	223±13	315±41
	{111}	60°	4.1±0.3	3.4±0.3	264±20	320±28
$Ni_{60}Fe_{15}Ge_{25}$	{001}	90° (edge)	8.9±0.3	8.2±0.2	141±5	153±4
(600K)	{001}	0° (screw)	6.2±0.2	5.5±0.3	124±4	140±7
$Ni_{50}Fe_{25}Ge_{25}$	{001}	90° (edge)	8.2±0.2	7.1±0.1	151±7	172±5
(300K)	{001}	0° (screw)	5.7±0.3	4.2±0.4	132±7	180±17
$Ni_{43}Fe_{32}Ge_{25}$	{001}	0° (screw)	5.8±0.3	5.4±0.3	127±5	132±8
(77K)	{111}	90° (edge)	6.8±0.3	4.2±0.2	168±7	272±13
$Ni_{43}Fe_{32}Ge_{25}$ (300K)	{001}	90° (edge)	12.9±0.4	9.0±0.3	95±3	137±4

Based on the corrected dissociation widths, that have been compiled in Table 1, the APB energies were calculated by treating the superpartials as individual dislocations and by balancing the repulsive force of the superpartials with the surface tension associated with the planar fault within the framework of Stroh's [17] formalism. The value of image simulations is quite apparent from Table 1 when a comparison is made of the fault energies calculated from the measured spacings and corrected spacings, as is also shown in fig. 3. The variation of elastic force with dislocation character and experimental diffraction conditions is taken into account through this analysis, and a convergence in the values of fault energies and a more consistent assessment is achieved.

The cube plane energy decreases from a value of about 295 mJ/m^2 to about 135 mJ/m^2 as the Fe concentration increases to 32at%, with a dramatic drop of about 95 mJ/m^2 seen for the alloy with 15at% of Fe from the value for Ni_3Ge. As a point of contrast, the octahedral plane APB energy decreases from an average value of 335 mJ/m^2 to 272 mJ/m^2, with only a marginal decrease of about 15 mJ/m^2 seen for the 15at%Fe containing alloy.

The above-mentioned observations obviously suggest that the driving force for the octahedral to cube plane cross-slip process, represented by the APB energy anisotropy [18], increases as a function of Fe content in the alloys. This can be inferred from Table 2 where the elastic anisotropy modification to this driving force represented by the equal and opposite elastic interaction force on the two superpartials, or the so-called Yoo torque [19], has also been compiled. Based on these calculations it is possible to get a measure of the driving forces represented by the changes in the APB energies and elastic anisotropy. Since the elastic anistropy changes only marginally from 0at%Fe to 32at%Fe, it is seen that the value of parameter 'a', which is defined in Table 2 and reflects the elastic anisotropy [20], changes gradually too as the Fe content is increased. However, the parameter, 'z', the ratio of APB energies on the octahedral to cube planes and a measure of the APB energy anisotropy [20], almost doubles in value. As a consequence, we can observe the considerable influence that the APB anisotropy exerts on the cross-slip driving force (F_c), which increases dramatically with increasing Fe content.

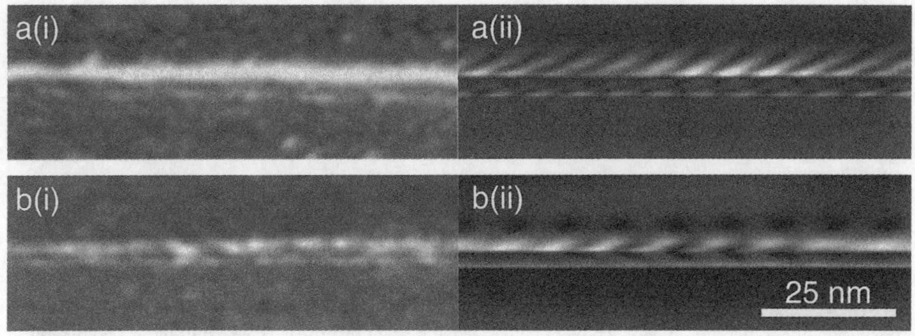

Beam Direction	Line Orientation	Experimental Observations		Image Simulations	
		d (nm)	γ^{APB} (mJ/m^2)	d (nm)	γ^{APB} (mJ/m^2)
~[001]	45°	4.2 (fig. 3a)	260	3.8	288
~[111]	45°	2.8 (fig. 3b)	390	3.8	288
~[001]	0° (screw)	3.6	221	2.7	291
~[111]	0° (screw)	2.3	343	2.7	291

Fig. 3: Comparison of experimental (a,b(i)) and simulated (a,b(ii)) weak-beam images for different diffraction conditions: a) $\mathbf{B}$~[001] and b) $\mathbf{B}$~[111] to show the efficacy of computer simulations for correcting image shifts and accounting for changes in elastic forces with a change in dislocation character. The example shows that the experimentally observed width of the APB lying on the cube plane, and associated with the 45°-oriented dissociated superdislocation, decreases from 4.2 nm (fig. 3a) to 2.8 nm (fig. 3b) as the fault plane is tilted about 50° away from the cube plane normal. However, image simulations with an input width of 3.8 nm predict the experimentally measured separations for both cases by accounting for the changing foil geometry. This has been shown in the inset table.

Table 2: Effect of elastic and fault energy anisotropy on the driving force for cross-slip of screw superdislocations from octahedral to cube planes under zero-stress conditions. Values for APB energies are taken from Table 1.

Alloy Composition	Elastic Constants c_{11}, c_{12}, c_{44} (GPa)	Elastic Anisotropy $A=2c_{44}/(c_{11}-c_{12})$, $a=(A+2)/A\sqrt{3}$	APB Anisotropy $z=\gamma^{111}/\gamma^{001}$	Cross-slip Force $F_c=\gamma^{001}\{(z-a)/a\}$ (mJ/m^2)
Ni$_{75}$Ge$_{25}$	263, 143, 103	1.717, 1.25	1.154±0.16	-60.7 — +15.1
Ni$_{60}$Fe$_{15}$Ge$_{25}$	248, 139, 103	1.89, 1.188	1.59±0.186	+36.8 — +99.9
Ni$_{50}$Fe$_{25}$Ge$_{25}$	238, 136, 102	2.0, 1.155	—	—
Ni$_{43}$Fe$_{32}$Ge$_{25}$	231, 134, 102	2.103, 1.126	2.009±0.17	+86.6 — +128.1
Fe$_{75}$Ge$_{25}$	189, 123, 102	3.091, 0.951	—	—

SUMMARIZING COMMENTS

The striking feature in this study is the athermal evolution of deformation microstructure with increasing concentrations of Fe in the alloys. The TEM observations described above indicate that, as Fe content changes from 0 to 32at%, the microstructure changes from screw dislocations locked in the KW configuration to edge and mixed dislocations that glide on the cube plane. There is a general shift of the transition to cube glide towards lower temperatures as Fe content increases; cube glide is observed at 600K in Ni$_{60}$Fe$_{15}$Ge$_{25}$, but appears at room temperature in Ni$_{50}$Fe$_{25}$Ge$_{25}$. This transition in dislocation microstructure is concurrent with the change in macroscopic yield behavior; the disappearance of anomalous behavior in $Ni_{50}Fe_{25}Ge_{25}$ coincides with low temperature strengthening by enhanced cross-slip and with the observed onset of cube glide. The microstructural evolution described here is analogous to what is observed as a function of temperature in $L1_2$ alloys that exhibit the anomalous rise in yield stress with temperature [1]. The concomitant contribution of solid solution strengthening has been considered to be negligible in view of the significant changes that are observed in the geometry of the core of the superdislocations.

The thermodynamic driving force cannot be considered to be the key factor that governs the cross-slip process even though the APB and elastic anisotropies are observed to increase with increasing Fe in the alloys. It has been argued [10,21-23] that the key parameter is the value of the CSF energy as the kinetics of the cross-slip process are governed by the activation barrier to the recombination of the Shockley partials of the leading superpartial. This, however, could not be determined even for Ni_3Ge due to experimental limitations on the resolving power of the weak-beam dark field technique that permit a resolution of about 1 nm. It can be only hypothesized at present that the balance between the CSF energy and the cross-slip driving force may be such that overcoming the activation barrier is not rate-controlling for the cross-slip process. Determination of the CSF energy by TEM and first principles atomistic calculations is the subject of on-going investigations.

ACKNOWLEDGMENTS
This work was supported by the US Air Force Office of Scientific Research (#F49620-95-1-0280) and the NSF-NYI program (#DMR-9457964). The authors also wish to thank Dr. D. M. Dimiduk of Wright-Patterson Air Force Base for help in alloy preparation for this study.

REFERENCES
1.	Nabarro, F. R. N., and Duesbery, M. S. (eds.), 1996, *Dislocations in Solids*, **10**, North-Holland, Amsterdam.
2.	Kear, B. H., and Wilsdorf, H. G. F., 1962, *Trans AIME*, **224**, 382.
3.	Suzuki, T., Oya, Y., and Wee, D.-M., 1980, *Acta Metall.*, **28**, 301.
4.	Balk, T. J., Kumar, M., and Hemker, K. J., 1998, *Scripta Mater.*, **39**, 577.
5.	Ngan, A. H. W., Jones, I. P., and Smallman, R. E., 1992, *Phil. Mag.* A, **65**, 1003.
6.	Hirsch, P. B., 1992a, *Phil. Mag.* A, **65**, 569; 1992b, *Prog. Mater. Sci.*, **36**, 63.
7.	Cockayne, D. J. H., Ray, I. L. F., and Whelan, M. J., 1969, *Phil. Mag.*, **20**, 1265.
8.	Baluc, N., Karnthaler, H. P., and Mills, M. J., 1991, *Phil. Mag.* A, **64**, 137.
9.	Baluc, N., Schäublin, R., Hemker, K. J., 1991, *Phil. Mag. Lett.*, **64**, 327.
10.	Hemker, K. J., and Mills, M. J., 1993, *Phil. Mag.* A, **68**, 305.
11.	Hemker, K. J., 1997, *Phil. Mag.* A, **76**, 241.
12.	Yasuda, H., Takasugi, T., and Koiwa, M., 1992, *Acta Metall. Mater.*, **40**, 381.
13.	Schäublin, R., and Stadelmann, P. A., 1993 *Mater, Sci. Engg.*, **A164**, 373.
14.	Head, A. K., Humble, P., Clarebrough, L. M., Morton, A. J., and Forwood, G. T., 1973, *Computed Electron Micrographs and Defect Identification*, North-Holland, Amsterdam.
15.	Stadelmann, P., 1987, *Ultramicroscopy*, **21**, 131.
16.	Weickenmeier, A., and Kohl, H., *Acta Crystallog.* A, **47**, p. 590 (1991).
17.	Stroh, A. N., 1958, *Phil. Mag.*, **3**, 625.
18.	Flinn, P. A., 1960, *Trans. AIME*, **218**, 125.
19.	Yoo, M. H., 1986, *Scripta Metall.*, **20**, 915; 1987, *Acta Metall.*, **35**, 1559.
20.	Veyssière, P., and Saada, G., 1996, *Dislocations in Solids* (eds. F. R. Nabarro and M. S. Duesbery), **10**, North-Holland, Amsterdam.
21.	Dimiduk, D. M., Thompson, A. W., and Williams, J. C., *Phil. Mag.* A, **67**, 675.
22.	Baluc, N., and Schäublin, R., 1996, *Phil. Mag.* A, **74**, 113.
23.	Karnthaler, H. P., Mühlbacher, E. T., and Rentenberger, C., 1996, *Acta Mater.*, **44**, 547.

MULTISCALE ANALYSIS OF INTERFACIAL STABILITY AND MISFIT DISLOCATION FORMATION IN LAYER-BY-LAYER SEMICONDUCTOR HETEROEPITAXY

L. A. ZEPEDA-RUIZ, D. MAROUDAS[a)], and W. H. WEINBERG
Dept. of Chemical Engineering, University of California, Santa Barbara, CA 93106-5080

ABSTRACT

A theoretical analysis based on continuum elasticity theory and atomistic simulations is presented of the interfacial stability with respect to misfit dislocation formation, the strain fields, and the film surface morphology during layer-by-layer semiconductor heteroepitaxy. The energetics of the transition from a coherent to a semicoherent interface consisting of a misfit dislocation network, the structure of this semicoherent interface, the resulting strain fields and the morphological characteristics of the epitaxial film surfaces are calculated for InAs/GaAs(111)A. Continuum elasticity is found to describe the atomistic simulation results very well. Our theoretical results are discussed in the context of recent experimental data.

INTRODUCTION

The growth of heterostructures consisting of lattice-mismatched epitaxial layers of semiconductor materials is generating tremendous interest due to a wide range of potential applications in electronic and optoelectronic device fabrication technologies. A critical issue in semiconductor heteroepitaxy is the effect of strain due to lattice mismatch between the film and the substrate materials. An excellent prototypical system for the study of semiconductor heteroepitaxy is InAs/GaAs, since it is characterized by a high compressive lattice mismatch strain, $\epsilon_m = -6.64\ \%$; $\epsilon_m \equiv (a_s^o - a_f^o)/a_f^o$, where a_f^o and a_s^o are the equilibrium lattice constants of the film and substrate materials, respectively. In addition, the growth mode in this system is known to depend strongly on the surface orientation, giving rise to different strain relaxation mechanisms, such as formation of interfacial misfit dislocations or formation of three-dimensional (3-D) islands [1]. In particular, epitaxy on the (001) surface is characterized by a transition in the growth mode from a layer-by-layer or two-dimensional (2-D) to 3-D for coverages just higher than one monolayer (ML) [1]. On the other hand, it has been established that the formation of 3-D islands is completely suppressed during InAs epitaxy on GaAs(110) and (111) surfaces [2,3]. In both cases, the InAs film remains 2-D even after strain relaxation.

This paper aims at a comprehensive theoretical analysis of the energetics, interfacial stability, strain fields, and film surface morphology in layer-by-layer heteroepitaxy by combining continuum elasticity theory and atomistic simulations of structural relaxation within an empirical description of interatomic interactions. The more general case of epitaxy on a compliant substrate of finite thickness is considered and the common case of epitaxy on an infinitely thick substrate is derived as an asymptotic limit.

THEORETICAL AND COMPUTATIONAL APPROACH

Our theoretical analysis addresses separately each possible equilibrium state of strain in the heteroepitaxial system, where the film/substrate interface can be either coherent or semicoherent with the biaxial strain due to lattice mismatch relaxed either partly or fully through interfacial misfit dislocation (MD) formation. Our theoretical approach follows the linear elastic analysis developed in Refs. 4 and 5 and it is summarized as follows: the elastic strain energy, U, stored in the coherently strained film/substrate system is given by

$$U = \int \frac{1}{2} tr \left(\underline{\underline{\sigma}} \cdot \underline{\underline{\epsilon}} \right) dV, \tag{1}$$

where $\underline{\underline{\sigma}}$ and $\underline{\underline{\epsilon}}$ are the corresponding stress and strain fields and the integral is taken over the entire volume, V, of the system. Assuming homogeneous deformation and solving the compatibility condition for perfect interfacial coherence and the condition for zero net force

Mat. Res. Soc. Symp. Proc. Vol. 538 ©1999 Materials Research Society

on any atomic plane perpendicular to the interface, Eq. (1) can be expressed as

$$\tilde{U}_1 = \frac{M_f \epsilon_m^2 h_f}{1 + \Lambda(h_f/h_s)}, \tag{2}$$

where M_f and M_s are the corresponding biaxial moduli defined as $M \equiv 2\mu(1+\nu)/(1-\nu)$: μ and ν are the shear modulus and Poisson's ratio, respectively, $\Lambda \equiv M_f/M_s$ is the ratio of the biaxial moduli of the constituents, and $\tilde{U}_1$ is used to denote energy per unit area in the coherently strained system.

Here, we limit the analysis of 2-D heteroepitaxial systems with semicoherent interfaces to those characterized by a fully relaxed state of strain, which is the case for InAs/GaAs(111)A heteroepitaxy [5]. A fully developed network of interfacial MDs characterizes the corresponding semicoherent interface. In such case, the contributions to the total energy of the system should include all the possible MD types present. Thus, the total energy per unit interfacial area, $\tilde{U}_2$, can be written as

$$\tilde{U}_2 = \sum_j \left\{ N_j \frac{L_i}{A} \left(\hat{U}_{c,j} + \hat{U}_{d,j} \, ln \left[\frac{1}{b_j} \frac{h_s h_f}{(h_f/h_s)} \right] \right) + \right.$$

$$\left. \sum_k N_k \frac{L_k}{A} \left(\hat{U}_{c,k} + \hat{U}_{d,k} \, ln \left[\frac{1}{b_j} \frac{h_s h_f}{(h_f/h_s)} \right] \right) + \frac{\gamma_{sf} A_{sf}}{A} \right\}, \tag{3}$$

where the tildes and the hats are used to denote energy per unit area and unit length, respectively, and the index j runs over all the principal crystallographic directions of the interface plane. For the (111) plane these are the $[1\bar{1}0], [10\bar{1}]$, and $[01\bar{1}]$ directions. The index k within the sum over j is used to take into account the possible dissociation of perfect interfacial dislocations into partials that can occur in order to reduce further the total energy. L_j and L_k are the corresponding perfect and partial dislocation lengths, N_j/A and N_k/A are the number densities per unit interfacial area of the j-type and k-type of the MD lines, respectively, $\hat{U}_c$ is the corresponding dislocation core energy, $\hat{U}_d$ is an energetic parameter characteristic of each MD type [4,5], and γ_{sf} and A_{sf} are the energy per unit area and the area, respectively, of the stacking fault that the partial dislocation may bound at the semicoherent interface.

Our calculations of the energetics, strain fields, and interfacial structures of InAs epitaxial films on GaAs(111)A substrates are based on a valence force field (VFF) description of the interatomic interactions in covalent solids. In our calculations, the energy of the system is minimized based on a conjugate gradient scheme with respect to the atomic coordinates for a given state of strain: biaxial, uniaxial, or fully relaxed. In our analysis, the film thickness is the dynamical variable, while the substrate thickness is the corresponding parameter. The details of our simulation methodology have been presented in Refs. 4 and 5. Our interfacial stability analysis is based on an energy criterion: we compare all the possible equilibrium states of strain for given film and substrate thickness and identify the one with lowest energy as the stable state.

RESULTS AND DISCUSSION

First, we focus on coherently strained epitaxial systems and demonstrate the beneficial effects of using finite-thickness compliant substrates in layer-by-layer heteroepitaxy. Specifically, the dependence of the elastic strain energy, $\hat{U}_1$, on film thickness, h_f, is calculated over a wide range of substrate thicknesses, h_s. In Fig. 1, results are presented for $h_s \to \infty$ and $h_s = 10$ ML for film thicknesses up to 100 ML. The insert to the figure highlights the energetics over the range 1 ML $\leq h_f \leq$ 10 ML. It is evident from Fig. 1 that using a compliant substrate of finite thickness serves to reduce substantially the elastic strain energy stored in the system compared to the case where $h_s \to \infty$. The continuous curves are fits to the simulation results according to the continuum elasticity theory, Eq. (2). The excellent fits shown in Fig. 1 provide evidence for the validity of the continuum theory all the way down to the monolayer thickness limit.

Next, we examine the equilibrium state where the strain in the epitaxial film/substrate system relaxes fully with respect to the coherently strained system through formation of a fully developed network of interfacial MDs. The computed dependence of the elastic strain energy on film thickness is shown in Fig. 2 for $h_f \leq 10$ ML, $h_s = 10$ ML and $h_s \to \infty$. For comparison, the energetics is included of both states of strain where the InAs/GaAs(111)A interface is either coherent or semicoherent. Open circles and diamonds correspond to the coherently strained and fully relaxed states of strain, respectively, for $h_s = 10$ ML. Filled circles and open squares correspond to the coherently strained and fully relaxed states of strain for $h_s \to \infty$. For $h_s = 10$ ML, the coherently strained system is stable for $h_f < 5$ ML, while the fully relaxed system gains stability for $h_f > 5$ ML. For $h_s \to \infty$, the transition from a coherent to a semicoherent interface occurs earlier, at $h_f = 4$ ML. This result is in very good agreement with the experimental data of Yamaguchi, *et al.* [6], according to which an interfacial dislocation network consisting of intersecting MDs has almost developed fully for $h_f > 3$ ML. The continuous curves in Fig. 2 represent fits to the simulation results according to the continuum theory, Eqs. (2) and (3). The computed dislocation parameters in the fitting procedure are $\hat{U}_c = 10.5$ eV/nm and 5.4 eV/nm for the core energies of the perfect and partial dislocations (see below), respectively, and $\gamma_{sf} = 275$ mJ/m^2 for the interfacial stacking fault energy.

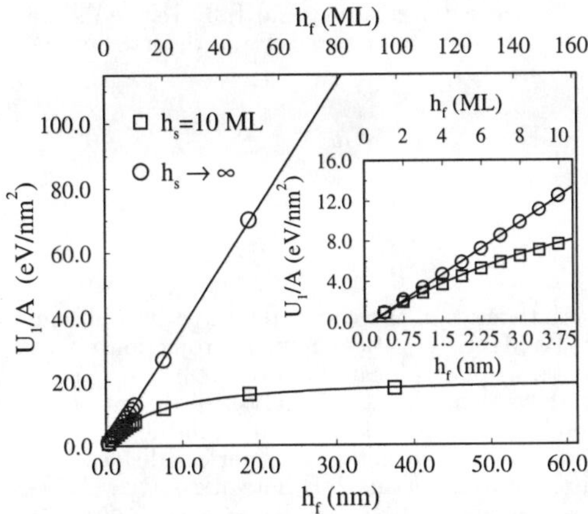

FIG. 1. Computed dependence on film thickness, h_f, of the elastic strain energy per unit interfacial area, U_1/A, in InAs/GaAs(111)A that is coherently strained for $h_s \to \infty$ (o) and $h_s = 10$ ML ($\square$). In both cases, the continuous curves are fits to the simulation results according to Eq. (2) and its limit when $h_s \to \infty$. The insert to the figure emphasizes the U_1/A dependence on h_f for $h_f \leq 10$ ML.

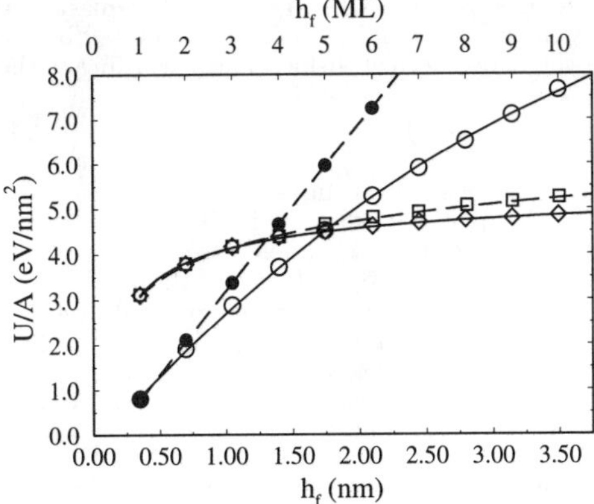

FIG. 2. Computed dependence on film thickness, h_f, of the elastic strain energy per unit interfacial area, U/A, of InAs/GaAs(111)A under different states of strain: coherently strained (o) and fully relaxed ($\Diamond$) at $h_s = 10$ ML and for coherently strained ($\bullet$) and fully relaxed ($\square$) at $h_s \to \infty$. The solid and dashed continuous lines are fits to the simulation results according to Eqs. (2) and (3) and their limits as $h_s \to \infty$, respectively. The transition from a coherent to a semicoherent interface is completed for $h_f > h_{f,c}$, where $h_{f,c} = 4$ and 3 ML for $h_s = 10$ ML and $h_s \to \infty$, respectively.

A parametric study has been carried out to investigate the effect of the substrate thickness, h_s, on the coherent-to-semicoherent interface transition as predicted from the interfacial stability analysis of Fig. 2. The calculated dependence of the critical film thickness for completion of the transition, $h_{f,c}$, on compliant substrate thickness, h_s, is shown in Fig. 3. Specifically, the computed values of $h_{f,c}$ are shown for three different substrate thicknesses: $h_s = 6, 10$, and 20 ML. It is evident from Fig. 3 that $h_{f,c}$ decreases monotonically as the substrate thickness increases, until it reaches the corresponding value predicted for $h_s \to \infty$. The continuous curve through the simulation results of Fig. 3 represents the solution to the nonlinear algebraic equation that expresses thermodynamic coexistence of the coherently strained and fully relaxed states in InAs/GaAs(111)A. Thus, the continuous curve represent the loci of the state coexistence points as calculated by the continuum theory, Eqs. (2) and (3), quantified by the computed parameters from the fitting of the atomistic simulation results of Fig. 2.

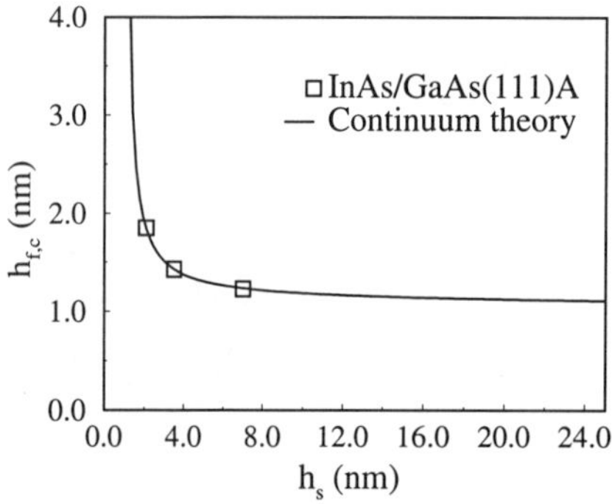

FIG. 3. Dependence of critical film thickness, $h_{f,c}$, on compliant substrate thickness, h_s, for InAs/GaAs(111)A epitaxy. The continuous curve represents coexistence of the coherently strained and fully relaxed states according to the continuum theory, Eqs. (2) and (3).

Figure 4 shows some of the structural characteristics of the computed fully relaxed semicoherent interface for $h_f = 5$ ML at $h_s = 10$ ML. Fig. 4(a) shows a projection on the geometrical plane of the interface of the two atomic planes located above the interface plane and two atomic planes located below the interface plane. Detailed analysis of the interfacial structure reveals the presence of triangular faulted regions at the interface, in the neighborhood of every other dislocation trijunction of a hexagonal network of perfect dislocations. For clarity, Fig. 4(b) shows a schematic representation of this MD network on the (111)A plane. Strain relaxation along $\langle 110 \rangle$ directions results in the formation of perfect 90^0 MDs extending along $\langle 112 \rangle$ directions. These perfect MDs intersect to form trijunctions that provide the nodes of the hexagonal MD network. Every other MD trijunction is replaced by triangular faulted regions that are bounded by Shockley partials, which are generated by the energetically favorable dissociation of each of the above perfect dislocations according to the dissociation reaction

$$\frac{a}{2}\langle 1\bar{1}0 \rangle \to \frac{a}{6}\langle 2\bar{1}\bar{1} \rangle + \frac{a}{6}\langle 1\bar{2}1 \rangle. \tag{4}$$

Further characterization of the fully relaxed state of strain in InAs/GaAs(111)A includes the analysis of the strain field and film surface morphology associated with the interfacial structure shown in Fig. 4. A detailed characterization of the strain field associated with the fully developed network of interfacial MDs is shown in Fig. 5 for $h_f = 6$ ML and $h_s = 10$ ML. This strain field is representative of the film thickness range where the fully relaxed state is stable. The computed distribution of $\epsilon_{xx}(x, y)$ is shown in Fig. 5(a) for the first monolayer above the interfacial plane. Clearly, the strain field is determined completely by the structure of the dislocation network at the semicoherent interface and can be used as means of visualization of this interfacial structure. Specifically, the regions of highest compression are located directly above the dislocation trijunctions and the regions of highest tension directly above the dislocation lines. Fig. 5 (b) shows a measure, ϵ^*, of the local strength of the interfacial MD field: ϵ^* corresponds to the value of ϵ_{xx} right above the dislocation trijunction and is plotted as a function of the distance, z, from the interface for different film thicknesses, h_f. It is clear from Fig. 5 (b) that the strength of the dislocation field decays abruptly with distance from the interface. Furthermore, it is evident that the absolute value of the compressive ϵ^* increases with h_f.

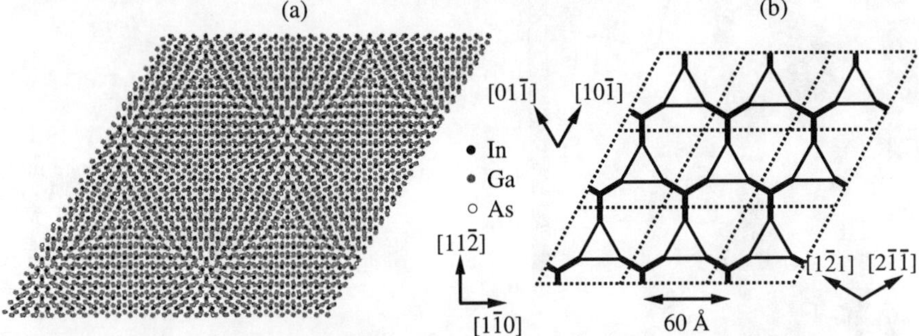

FIG. 4. (a) Fully relaxed semicoherent interface structure of InAs/GaAs(111)A for $h_f = 5$ ML at $h_s = 10$ ML for an interfacial area of four (2×2) interfacial unit cells. (b) Schematic representation of the dislocation network in the interfacial structure shown in (a) for an area of nine (3×3) interfacial unit cells. Thick lines denote perfect 90^0 MDs, while thinner lines denote Shockley partials. The triangular regions bounded by the partial dislocations are faulted, while the rest of the interface is unfaulted.

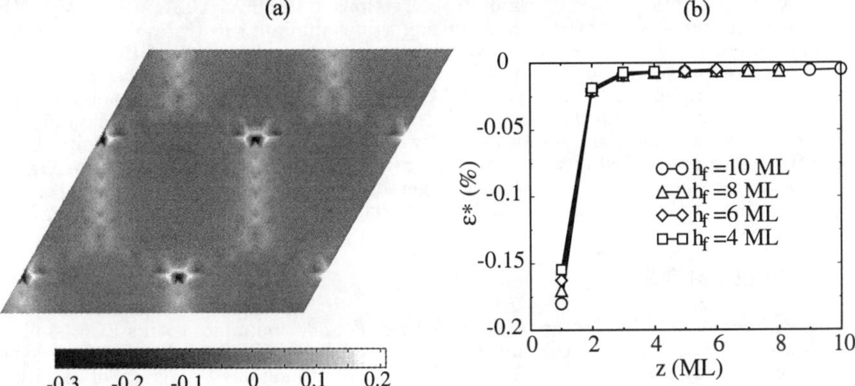

FIG. 5. (a) Computed strain field $\epsilon_{xx}(x,y)$ for the fully relaxed state of InAs/GaAs(111)A at $h_f = 6$ ML and $h_s = 10$ ML, where x and y correspond to the [1$\bar{1}$0] and [11$\bar{2}$] crystallographic axes, respectively. An area consisting of four (2×2) unit cells is shown for the first monolayer above the interface. (b) Variation of the local strength, ϵ^*, of the interfacial MDs with distance from the interface, z, for different film thicknesses.

Finally, we discuss the morphological characteristics of the surface of the deposited InAs film on GaAs(111)A. A typical surface morphology for $h_s = 10$ ML is shown in Fig. 6 for $h_f = 6$ ML, for which the fully relaxed state of strain is thermodynamically stable. The surface shown in Fig. 6 corresponds to an area of eight (2×4) interfacial unit cells and the shading of the surface indicates depth with respect to a flat surface. The surface morphological characteristics also are determined fully by the semicoherent interface structure, through the strain field of the MDs. Specifically, the minima in the surface profile occur directly above the location of perfect MD trijunctions at the interface, while the maxima occur directly above the centers of the interfacial stacking fault regions.

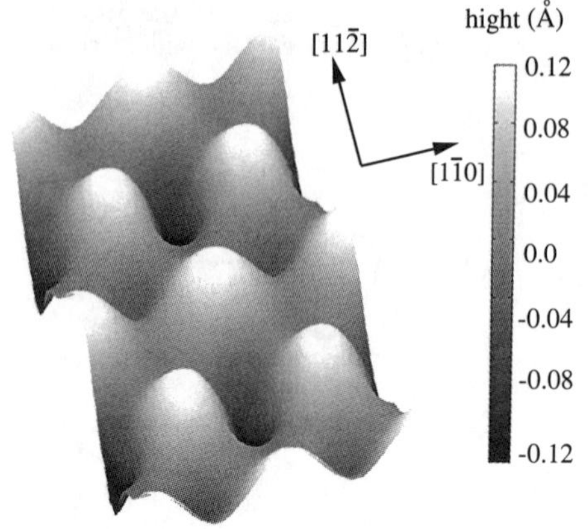

FIG. 6. Computed film surface morphology of InAs/GaAs(111)A for h_f = 6 ML and h_s = 10 ML. The shading of the surface indicates hight with respect to a flat surface.

CONCLUSIONS

In summary, a comprehensive analysis has been carried out of the energetics, structural stability, strain fields, and surface morphological characteristics in InAs/GaAs(111)A. The analysis is based on a combination of continuum elasticity theory and atomistic simulations within a VFF description of the interatomic interactions in the InAs/GaAs system [4,5]. Our analysis demonstrates the beneficial role of using thin compliant substrates in heteroepitaxy as a mechanism to relieve the elastic energy stored in the film. In addition, it predicts successfully the completion of the transition from a coherent to a semicoherent interface consisting of a complete network of perfect MDs and triangular faulted regions bounded by partial MDs at a critical film thickness in agreement with recent experimental data [6]. The predictions of our atomistic simulations are fully consistent with continuum elasticity theory; specifically, continuum elasticity theory can provide an excellent guide to the energetics of heteroepitaxy when quantified through targeted atomistic calculations. Our theoretical analysis also has applicability in other heteroepitaxial systems characterized by 2-D or layer-by-layer growth.

ACKNOWLEDGMENTS

This work was supported by the National Science Foundation under grants ECS-95-01111 through a CAREER Award to DM and DMR-95-04400 to WHW. LAZR also acknowledges financial support by the IIM-DGAPA of the National Autonomous University of Mexico.

REFERENCES

a) To whom correspondence should be addressed. E-mail address: dimitris@calypso.ucsb.edu
[1] V. Bressler-Hill, A. Lorke, S. Varma, P. M. Petroff, K. Pond, and W. H. Weinberg, Phys. Rev. B **50**, 8479 (1994); V. Bressler-Hill, S. Varma, A. Lorke, B. Z. Nosho, and W. H. Weinberg, Phys. Rev. Lett. **74**, 3209 (1995).
[2] X. M. Zhang, D. W. Pashley, L. Hart, J. H. Neave, P. N. Fawcett, and B. A. Joyce, J. Crystal Growth **131**, 300 (1993).
[3] J. G. Belk, J. L. Sudijono, X. M. Zhang, J. H. Neave, T. S. Jones, and B. A. Joyce, Phys. Rev. Lett. **78**, 475 (1997).
[4] D. Maroudas, L. A. Zepeda-Ruiz, and W. H. Weinberg, Surf. Sci. **411**, L865 (1998).
[5] L. A. Zepeda-Ruiz, D. Maroudas, and W. H. Weinberg, J. Appl. Phys. **85**, in press.
[6] H. Yamaguchi, J. G. Belk, X. M. Zhang, J. L. Sudijono, M. R. Fahy, T. S. Jones, D. W. Pashley, and B. A. Joyce, Phys. Rev. B **55**, 1337 (1997).

FIRST-PRINCIPLES STUDY OF SI(111) HOMOEPITAXY

KYEONGJAE CHO*, EFTHIMIOS KAXIRAS**

*Mechanics and Computation Division, Department of Mechanical Engineering, Stanford University, Stanford, CA 94305, kjcho@stanford.edu

**Department of Physics and Division of Engineering and Applied Sciences, Harvard University, Cambridge, MA 02138, kaxiras@cmtek.harvard.edu

ABSTRACT

Epitaxial growth on the Si(111) surface is studied using first-principles total-energy pseudopotential calculations. The energetics of added Si atoms essentially determines epitaxial growth modes under different growth conditions (surface temperature, Si flux rate, and surface step density). We have determined the surface adatom diffusion barriers and cluster formation energies; we use these microscopic energy parameters to address the possibilities for macroscopic morphological evolution of the surface under different conditions.

INTRODUCTION

Epitaxial growth is a widely used experimental method to grow single crystals on a substrate. From numerous experimental works, it has been established that the nature of epitaxial growth is a very sensitive function of many growth parameters for a chosen pair of substrate and growing materials. These parameters include growth temperature, incoming atom flux, presence of auxiliary atoms (e.g., surfactants), and surface defect density. Typically, an amorphous film is obtained for low-temperature epitaxial growth, while either island growth or step-flow growth is observed with high-temperature conditions. [1]

The formation of different macroscopic phases of the epitaxially grown thin films is controlled by atomic scale dynamics and energetics of the added atoms on the substrate during epitaxial growth. Specifically, the energy barrier for diffusion of the added atoms and the critical island size (below which islands spontaneously disintegrate by detachment of atoms) control the kinetics and the relative importance of island formation vs. step-flow.

In this paper, we report a detailed analysis of the energetics of the critical island size for Si homoepitaxial growth on the reconstructed Si(111) surfaces. We have chosen Si homoepitaxy as a representative system where the diffusion barrier and island formation energies can be isolated from other possible complications such as misfit strains in heteroepitaxy and kinetically induced defects in epitaxial growth of compound materials.

SIMULATION METHODS

In order to determine the energy parameters of homoepitaxial growth of Si on Si(111), we have used the first-principles total energy pseudopotential method based on density functional theory (DFT) and the local density approximation (LDA). [2] The Kohn-Sham single electron wave functions are expanded on a plane wave basis with 8 Ry cutoff energy. We use a supercell consisting of (4×4) surface unitcells and a slab of five atomic layers with 12Å vacuum. The Brillouin-zone integration is approximated by Γ-point sampling. [3] The non-local pseudopotential is implemented in the Kleinmann-Bylander separable form [4] with s and p non-local components. The accuracy of the simulation has been tested as a function of cutoff energy and k-point sampling, and the error bar of the calculated energy differences is ±0.05 eV. The atomic geometries are fully relaxed until the forces on the atoms are smaller than 0.1 eV/Å. [3]

Mat. Res. Soc. Symp. Proc. Vol. 538 ©1999 Materials Research Society

The simulations are performed in three steps. First, the surface energies of different reconstructions of the (111) surface are calculated to determine the surface energy as a function of surface adatom density (Fig. 1). Second, the energy of a single extra-adatom on the reconstructed surface is calculated. Third, the formation energies of small clusters on the reconstructed surfaces are calculated. We have considered clusters of sizes from one to six in order to investigate critical island size issues.

RESULTS

Figure 1 shows the surface energies of the reconstructed (111) surfaces as a function of the surface adatom density relative to the ideal bulk-terminated surface. At zero coverage, the surface undergoes a sequence of reconstructions leading to the (5×5) Tagayanagi Dimer-Adatom-Stacking fault (DAS) reconstruction. The global minimum energy structure is the (7×7) DAS reconstruction. The addition of Si atoms increases the surface energy linearly as indicated in Fig. 1. If the deposition rate is not too fast, the sytem can reach its equilibrium state at each surface adatom coverage, and the surface will form domains of mixed reconstructions (i.e., 7×7 and 2×2 or 2×2 and √3× √3). Therefore, the stability of small clusters on a large terrace can be determined by comparing the relative stability of cluster formation energy against the energy lines connecting the 7×7, 2×2 and √3× √3 reconstructions.

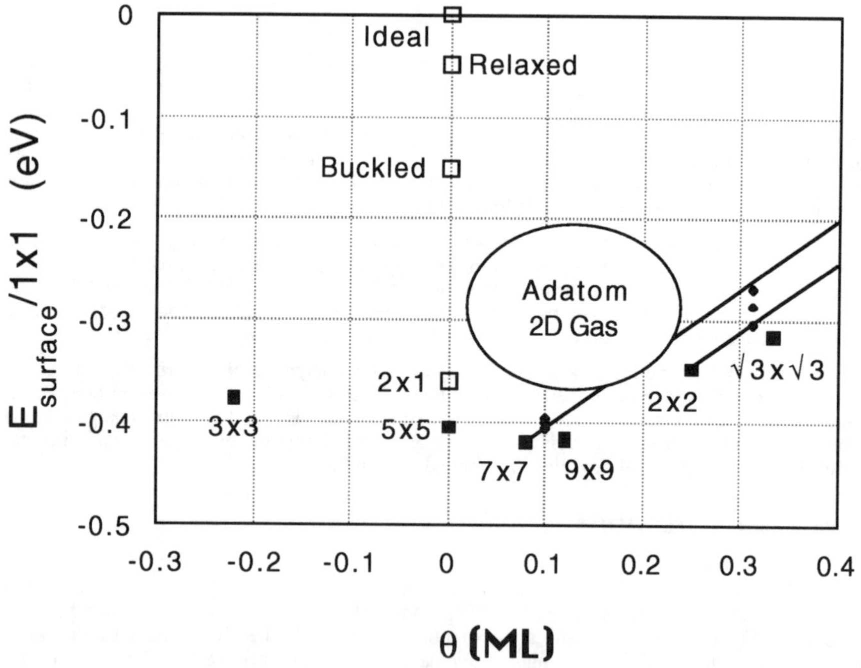

FIGURE 1   Surface energy of the (111) surfaces a function of adatom density. θ = 0 corresponds to an ideal bulk-terminated surface which undergoes a sequence of reconstructions dwon to the 5x5 DAS reconstuction. The filled squares represent the local minimum energy structures at the corresponding surface adatom densities.

Figure 2 shows the local geometry of six clusters of sizes 1-6, whose formation energies are listed in Table I. Figure 3 shows the cluster formation energies as a function of the cluster size. The formation energy is relative to the reference energy of the (2x2) reconstructed surface energy plus N-1 bulk Si atom energy where N is the cluster size. In this plot E = 0 corresponds to the absolute stability energy of the clusters, and all the clusters have positive formation energies indicating that none of them are absolutely stable. Two linear energy lines are also shown in Fig. 3, and they are the energy of a single extra-adatom (0.73 eV/adatom) and the energy line connecting the 2x2 and $\sqrt{3}\times\sqrt{3}$ reconstructions. It is interesting that the dimer and trimer formation energies are above both lines while tetramer, pentamer, and hexamer formation energies are below both lines. This comparison shows that the minimum stable cluster is the tetramer. This is independent of the surface relaxation rate: if the surface dynamics is slow, the extra-adatom energy line is the reference energy; if, on the other hand, the surface dynamics is fast, the 2x2 to $\sqrt{3}\times\sqrt{3}$ transition energy line is the reference energy.

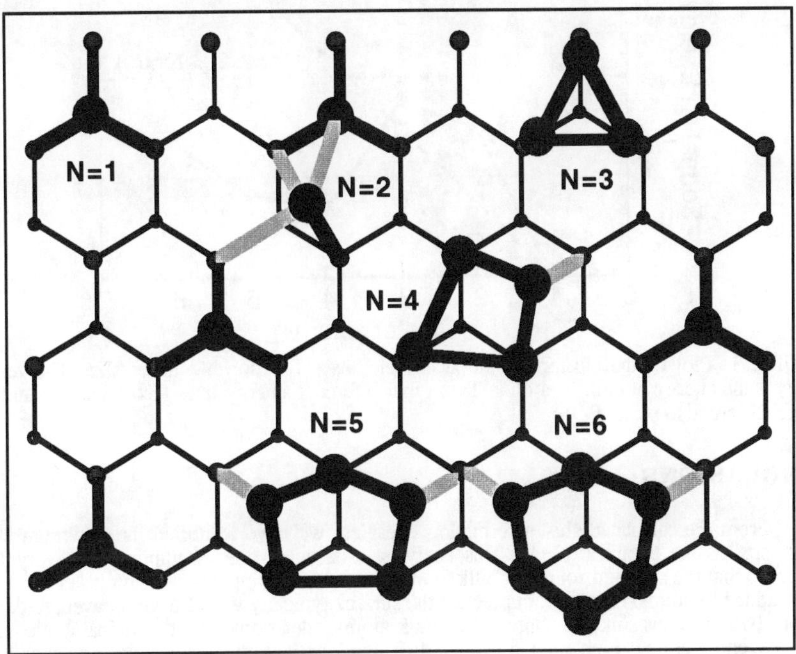

FIGURE 2   Ball-Stick model representation of the clusters size, N = 1, ..., 6.

N	1	2	3	4	5	6
E (eV)	0.0	0.734	1.746	0.688	0.822	1.330
E' (eV)		1.473		2.136		

TABLE I   Formation energies of the clusters. E and E' correspond to the energies of alternative metastable structures of the dimer and the tetramer, rescpetively.

A detailed analysis of the local bonding structure and geometry relaxation of the clusters has been performed. This analysis indicates that the large decrease of the formation energy between the trimer and the tetramer is due to the capacity of the fourth atom in the tetramer to form an extra bond with the neighboring surface rest-atom (indicated as light gray line in Fig. 2). We have also calculated the formation energy of another tetramer structure with a Si atom on top of the trimer structure, and its formation energy is 1.5 eV higher than the stable tetramer formation energy. Figure 2 also shows the local bonding of the pentamer and hexamer with gray lines representing bonding between the added Si atoms and the neighboring rest-atoms.

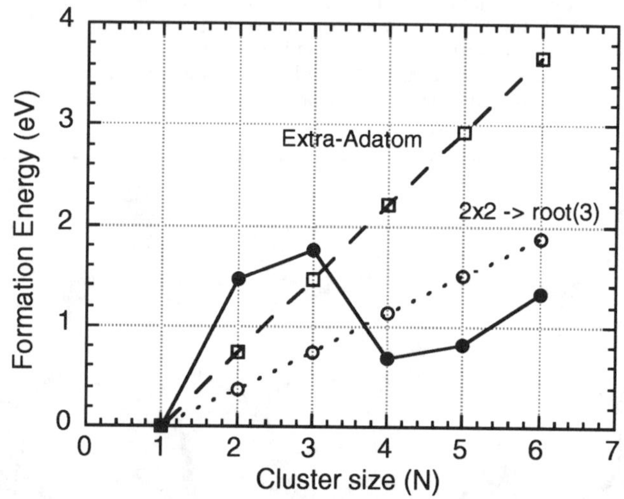

FIGURE 3 Plot of the cluster formation energies as a function of cluster size. The zero of energy is the energy of bulk Si atoms. Two other reference energy lines (Extra-adatom and 2x2 → √3x√3) are also plotted.

CONCLUSIONS

From the calculated cluster formation energies, we have identified the minimum stable cluster size as the tetramer (N=4). This analysis is based on the assumption of a very large terrace, so that the sink and source of bulk Si atoms are not playing any role in cluster formation by the added Si atoms. However, in practice the surface typically would have an average density of steps (which act as sink and source of Si bulk atoms), determined by the vicinal angle. If the substrate temperature is high enough that the diffusion length of adatoms becomes comparable to the terrace size, the reference energy of the cluster formation is that of bulk Si atoms. As noted in Fig. 3, all the clusters have positive formation energy relative to bulk Si atoms, and they are unstable against the dissociation followed by diffusion and incorporation into surface steps.

From this analysis and the results of our previous work, [5] we can construct the following model of different growth modes as a function of temperature. At low temperature, the extra-adatom diffusion is very slow so that the adatoms form amorphous clusters around their surface arrival sites; these clusters are not necessarily coherent with the substrate, and the ensuing film is amorphous. As the surface temperature is increased, the adatoms become more mobile and begin to form clusters which have a size of at least N=4, and are coherent with the substrate; these clusters play the role of nucleation sites for larger islands. As the temperature is further increased, the adatoms become more mobile and establish an equilibrium with surface steps. In this temperature range, cluster formation is not likely and the added atoms dominantly diffuse to nearby surface steps to be incorporated as bulk atoms, leading to step-flow growth.

The calcualted energies in this work can be combined with a growth model using the kinetic Monte Carlo simulation (e.g., Ref. 6) to perform a direct simulation of epitaxial growth processes at different growth conditions (i.e., surface step density, adatom flux and temperature). To obtain a complete set of energy parameters for the simulation of Si(111) hompepitaxial growth, we will also need to determine the adatom diffusion energies over the (7x7) reconstructed surface as well as the energy barriers of cluster formations and dissociations. We are currently performing a simulation to determine the extra-adatom diffusion energetics over the (7x7) reconstructed surface unit cell boundaries.This additional information will allow us to determine the relative importance of the extra-adatom diffusion over a wide range of possible surface structures during the growth process.

ACKNOWLEDGMENTS

KC acknowledges a partial support from the Frederich E. Terman Award of Stanford University from The Packard Foundation and the NSF-MRSEC (Stanford Center for Materials Research).

REFERENCES

[1] Köhler et al., J. Vac. Sci. Technol. A **7**, 2860 (1989).

[2] M.C. Payne et al., Rev. Mod. Phys. **64**, 1045-97 (1992).

[3] The convergence of the energy difference as a function of **k**-point sampling has been performed, and Γ-point sampling provides energy convergence within 0.05 eV. Typical spring constant of Si-Si bond is about 20 eV/ Å^2, and the corresponding residual energy due to 0.1 eV/Å would be smaller than 1 meV per atom (less than 0.05 eV for the supercell).

[4] L. Kleinman and D.M. Bylander, Phys. Rev. Lett. **48**, 1425-8, (1982).

[5] K. Cho and E. Kaxiras, Europhys. Lett. **39**, 287 (1997).

[6] M. Itoh, Phys. Rev. B **57**, 14623 (1998).

AN EFFECT OF HYDROSTATIC COMPRESSION ON DEFECTS IN ENERGETIC MATERIALS: *AB INITIO* MODELING

MAIJA M. KUKLJA AND A. BARRY KUNZ

Electrical Engineering Department, Michigan Technological University, 1400 Townsend Drive, Houghton, MI, 49931-1295, USA

ABSTRACT

First-principle theoretical investigation of the basic defects such as a molecular vacancy, a vacancy dimer, an edge dislocation, and a micro-crack in organic explosive molecular crystals is presented. As an example we considered solid RDX ($C_3H_6N_6O_6$) which is well studied unstable solid. It was established that external hydrostatic pressure changes optical properties of defect-free RDX as well as of the crystal with defects narrowing the band gap. The lattice defects (especially dislocations) are identified with the so-called "hot spots." The nature of local electronic states introduced in the band gap by the edge dislocation and formed mainly by molecular orbitals of $N-NO_2$ group is analyzed. Favorable conditions for molecular dissociation due to electronic excitation are shown.

INTRODUCTION

Most of explosive materials are molecular solids having complicated crystal structures and composed of polyatomic organic molecules. Macroscopic properties of energetic substances are fairly well understood, especially in gas or liquid phases. Microscopic processes and behavior of the explosive solids are poorly known due to complexity of their structures and ultrafast velocity of the detonation initiation process. A great deal of information is currently available about the influence of impact impulse on the detonation front parameters for various explosive materials. However, the spectral and time resolution of modern techniques is still insufficient for complete understanding of the initial stages of chemical decomposition. Hence, computer multi-scale simulations of structure and properties of the materials are highly desirable.

It is generally believed that lattice defects provide sites for so-called "hot spots [1]," localizing energy of impact wave and developing detonation. The reaction can start as a result of internal stress introduced by inhomogeneities such as pores, cracks, inclusions, grain boundaries, and dislocations along with other structural defects in the materials. Although there are several different realistic mechanisms of "hot spots" formation and detonation initiation [2], the question of what defects (or their associations) are responsible for sensitivity of the explosive solids to detonation still remains open. The existed models can not explain all experimental observations of shock front and some contradictions in the theory. After Dremin [3], the main problem of the detonation theory development lies in the lack of detailed information about micro-structure of the materials and various defects, their interactions and influence on crystal properties.

In this paper, we have studied electronic structure of RDX crystal containing point and extended defects using Hartree-Fock (HF) and Many-Body Perturbation-Theory (MBPT) techniques. Properties of the ideal and defective material excited by external pressure due to impact wave propagation were also modeled. Experimental facts supporting theoretically obtained results are discussed in some details.

Mat. Res. Soc. Symp. Proc. Vol. 538 © 1999 Materials Research Society

METHODOLOGY

The method of computation used in this paper is based on standard Hartree-Fock technique as implemented in the CRYSTAL95 program package [4]. Electron correlation corrections based on the second-order MBPT are included in the calculations. For this purpose we used the LOPAS computer code [5] based on the local orbital approach [6].

The basis set was optimized to provide a balance between computational feasibility and a good representation of the properties of perfect RDX and TATB ($C_6H_6N_6O_6$) crystals [7] and defects in RDX. In this work we used the 6-21g split valence set with the modified scaling factor of 1.10 for outer (most diffuse) Gaussian basis vector on all atoms. In all calculations internal geometry parameters of the RDX molecule such as bond lengths, angles, and torsion angles were taken from experimental data [8], while the lattice constants were theoretically optimized. All results reported here correspond to this theoretically determined geometry of perfect RDX.

To simulate a molecular vacancy in RDX we remove one of the eight molecules from the orthorhombic RDX unit cell (see Fig.1). Applying periodic boundary conditions (supercell model) results in a crystal with a concentration of defects as high as 12.5%. One can consider this as a model of a porous RDX material investigated experimentally [13-14]. By doubling the size of the supercell and changing positions of vacancies we were able to study the effect of the defect distribution, in particular formation of vacancy dimers, on the properties of the solid. In a similar manner we modeled two-dimensional defects such as fine micro-cracks placed close to each other.

The modeling of dislocations is more challenging. The crystal is assumed to be composed of stacked parallel layers. Each layer is one unit cell thick. Then, the edge dislocation most likely has the Burgers vector of [001] [9, 10]. This is because the shortest distance between molecules is in c direction that corresponds to the smallest magnitude (|b|=1.0709 nm) among all possible Burgers. A (010) primary slip plane was first proposed by May and Connick [9] and confirmed later by Halfpenny et. all [11]. Hence, the (010)[001] is most likely configuration for edge dislocation in RDX crystals. To produce such a dislocation one can slide one crystal plane (say the plane containing the molecules numbered 2-8-1-7 in Fig.1) with respect to another part of the solid along the z direction. This slip produces a plastic deformation in the material and the boundary between the slipped and unslipped regions is the edge dislocation. Its position is marked by the termination of an extra vertical half-plane of the RDX molecules crowded into the upper half of the crystal. In our calculations we modeled the dislocation core by a one-dimensional (polymer) model. The periodic translation was applied along the dislocation line d [010]. The crystal unit cell consisted of 10 RDX molecules, is schematically shown in Fig.2.

The effect of the hydrostatic compression on ideal and defective RDX crystals was modeled by using the rigid molecule approximation. This is based on the assumption that the energy required for a distortion of the molecular geometry is much higher than that required for mutual displacements of the molecules in solid. Thus, isotropic compressibility of the crystal in terms of Volume-Energy dependence was investigated by decreasing all lattice constants in proportion and keeping the structure of the RDX molecules in crystal unaltered. The pressure was obtained using the low temperature formula P=-dU/dV which is the negative volume derivative of the ground-state energy. The pressure-volume plot for the defect-free solid RDX [12] is in good qualitative agreement with the experimental impact adiabatic curves [13-14].

Lattice relaxation induced by defects was neglected in this study. The band gap calculations have been made at a HF level for defective RDX were then corrected for electron-polarization effects in defect-free RDX using MBPT. For defective crystal under compression we use corrections analytically obtained for ideal compressed material. Thus, obtained in this study band gaps can be directly compared to the relevant experimental data.

RESULTS AND DISCUSSION

The main objective of this work was to study the influence of defects on the electronic band gap. Most calculations have been performed in equilibrium crystal structure as well as under external hydrostatic compression. The band gap width as a function of the degree of the crystal compression V/Vo is shown in Fig.3, which allows us to make two important observations. First, most studied defects induce reduction of the band-gap even at zero pressure. The band-gap is most sensitive to the presence of dislocations due to a strong internal stress induced. Second, usually volumetric contraction leads to the additional narrowing of the gap. The amount of narrowing strongly depends on the nature and spatial orientation of defects in the solid. For instance, in crystals containing zigzag-like micro-cracks (micro-crack (2NN) line in Fig.3) the gap weakly depends on pressure, while in crystals with edge dislocations the gap drops to 1.5eV at relatively small compression about 92% of normal crystal volume Vo.

It is well known that strong hydrostatic compression of the solid leads to the electronic insulator-metal phase transition, which is characterized by critic pressure (Herzfeld-Mott metallization). According to the detonation theory of Zeldovich [15], an impact wave propagating through the crystal, is characterized by high temperature and pressure. Following Gilman et.all [16], the origin of the large pressure spike at the detonation front is associated with the valence electrons being suddenly able to expand in the anti-bonding states. The authors argue that in this moment the explosive substance is undergoing the electronic phase transition. At constant volume this produces pressures in the 10^{11} Pa range [17]. Our theoretical estimates fall into this range. The pressure corresponding to zero band gap for defect-less RDX comes out to be about 180GPa. Simple vacancies presented in RDX decrease this pressure about 1.5 times corresponding to the pressure of 120GPa. One may attempt to relate this tend to the changes in detonation pressure of RDX with different densities. Then, our conclusion is supported by a deduction based on experimental data on solid and porous RDX samples (see Ref. 13, pp.117-121; Ref.14, pp.313-322 and Fig. 8.30). In particular, it is shown, that the density decrease from $\rho_o=1.72$g/cm^3 (which is close to the theoretical perfect crystal density) to 1.46 g/cm^3 (which is $0.85\rho_o$ and perfectly well agree with $0.87\rho_o$ in our model of the crystal with vacancies) leads to the detonation pressure decrease about 1.5 times. This rate is in remarkable agreement with the obtained in present study predictions with respect to metallization pressures. The fact, that under the same pressures of the initiation wave a length of the zone before the reaction start decreases with the decrease of the explosive density (see Ref. 14, p.286-292), also lends some confirmation to our results. This theoretical simulation along with further modeling of large vacancy aggregates can provide some insight for the accumulation mechanism of detonation initiation [3] and for the model based on the collapse of pores [2].

Our calculations demonstrate that edge dislocation make strong influence on optical properties of RDX (Fig.3 and 4). A significant distortion in the electronic structure of RDX crystal caused by local deformations in the lattice, (which represent an element of a dislocation) has been described [18]. The Hartree-Fock embedded cluster calculations show increased molecular polarization (up to 20%) along different directions resulting in a scenario where some of the molecules tend to attract each other while others tend to repel. The lattice deformation has been approximately represented by dipol moments of the RDX molecules displaced from their regular crystalline positions.

It is explicitly shown here that interactions among molecules in the dislocation core produce local electronic states in the band gap even at zero pressure. Applying external pressure to this already stressed crystal results in the further reduction of the band gap. From our results one may also conclude that edge dislocations decrease the metallization pressure more essentially than other defects because of dramatic narrowing of the gap, which is related to the splitting of local levels from both: the top of the valence band and bottom of the conduction band. This can be perfectly seen in Fig.4 where the total electronic density of states (DOS) for the compressed crystal with the edge dislocation is presented in comparison with the ideal material. An eigenvectors analysis proves that these local states are related to the weakest N-NO$_2$ bond in a RDX molecule located near the dislocation core and have bonding and antibonding character, respectively.

From this observation very important conclusions follow. First, the band gap reduction increases the probability of the optical electronic transition. And the second, the lowest energy electronic excitation, i.e. promotion of electron from HOMO to LUMO, would lead to the breaking the N-NO$_2$ bond. If this excited state lives longer than characteristic time of nuclear vibrations, then dissociation can occur. The nature of the HOMO to LUMO excitation and its relation to the molecular decomposition has been established in earlier theoretical [19] and experimental [20] studies. In particular, it was found that the strong intramolecular absorption at wavelengths below 340 nm leads to the formation of the NO$_2$ radicals.

As has been shown above, in the presence of defects, such as dislocations, the pressure reduces the energy of electronic transition. Thus, favorable conditions are created for molecular dissociation by a thermally induced HOMO-LUMO transition.

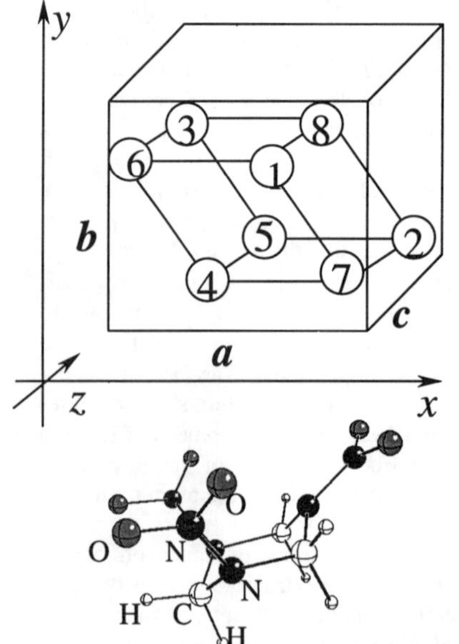

FIG.1. The crystal structure of the orthorhombic (space group Pbca) solid RDX. Open circles indicate positions of the RDX molecules in the crystal lattice. Atomic structure of individual molecules is shown below.

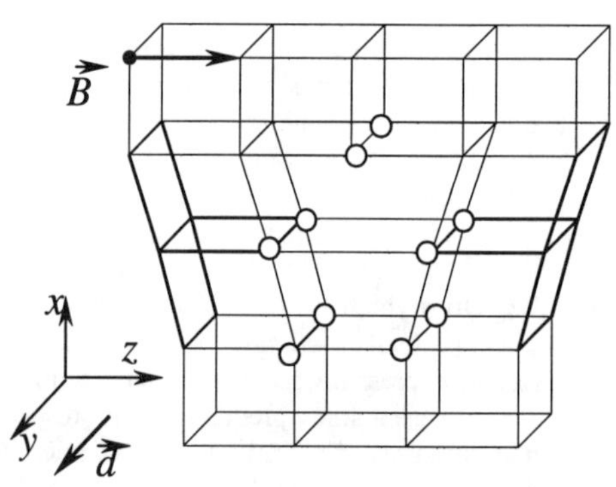

FIG.2. The edge dislocation with the Burgers vector [001] and the slip plane (010) in the 1D polymer model. The slip plane is indicated by vector **d** perpendicular to the plane of the figure. Each parallelogram corresponds to one unit cell of the RDX crystal lattice. Open circles represent RDX molecules that were explicitly included in calculations. The crystal is build of molecular layers (only one layer is shown) repeated in the **d** direction.

RDX

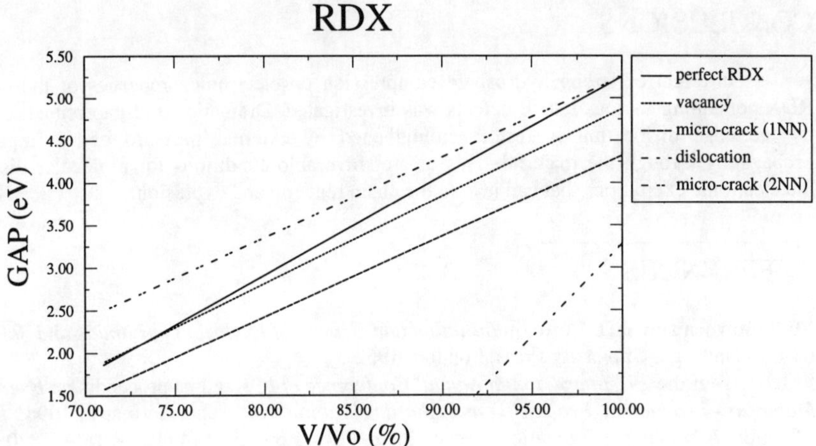

FIG.3 The band gap of RDX as a function of the degree of crystal compression V/Vo

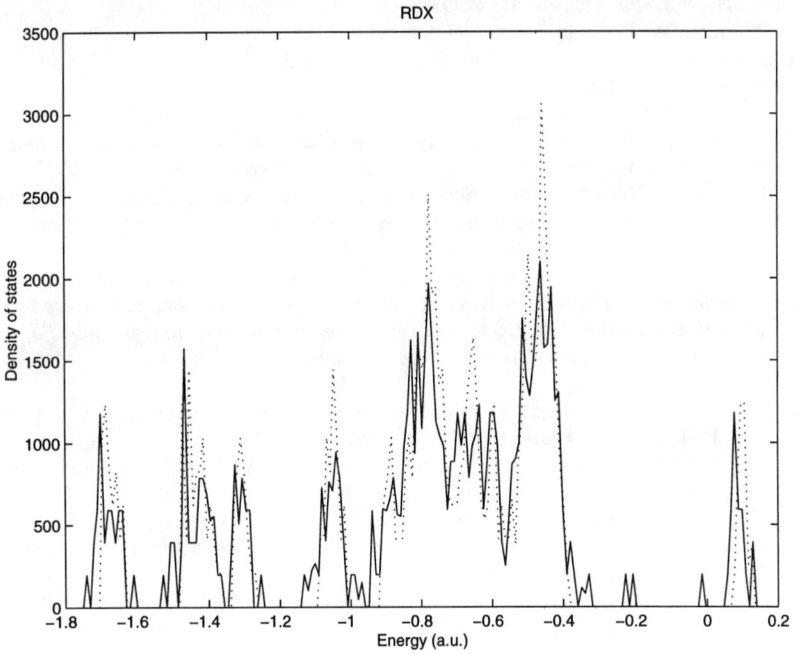

FIG.4 The total electronic density of states (DOS) for the perfect RDX crystal (dotted line) in comparison with the crystal contained the [001](010) edge dislocation and compressed to 71% of its perfect volume Vo (solid line).

CONCLUSIONS

An effect of strong hydrostatic compression on electronic properties of the crystalline RDX containing various lattice defects was investigated. Dramatic band gap reduction induced by the edge dislocation is further strengthened by external pressure of an impact wave propagating through the material. This creates favorable conditions for molecular dissociation via electronic excitations that can lead to the chain reaction and explosion.

REFERENCES

[1] F.P. Bowden and Y.D. Yoffe, in *Initiation and Growth of Explosion in Liquids and Solids,* pp. 64-65. Cambrige, University Press, London (1952).

[2] Materials of the *29th International Annual Conference of ICT* and its proceedings: *Energetic Materials - Production, Processing and Characterization*, Karlsruhe, Germany, 1998.

[3] Dremin, A.N., On the detonation theory, Chem. Phys. Reports, v.14 (12), p.1851-1870, 1995.

[4] Dovesi R.,. Saunders V.R, Roetti C., Causà M., Harrison N.M, Orlando R., and Aprà E., in *CRYSTAL95 User's Manual*, University of Torino, Torino (1996).

[5] Kunz A.B., *J.Phys. Condens. Matter* **6**, 1233 (1994).

[6] Kunz A.B., *Theor. Chim. Acta* **84**, 353 (1993).

[7] Kunz A.B., *Phys. Rev. B*, **53**, N15, 9733-9738 (1996).

[8] Choi C.S. and Prince E., *Acta Crystallogr. B*, **28**, 2857 (1972).

[9] W.Connick and F.G.J. May, J. of Crystal Growth **5**, 65-69 (1969).

[10] W.L.Elban, J.C. Hoffsommer, and R.W.Armstrong, J. of Materials Science **19**, 552-566 (1984); W.L. Elban and R.W.Armstrong in *Proceedings of the 7th Symposium (International) on Detonation*, Annapolis, MD, 1981 (Silver Spring, MD, 1982, NSWC MP 82-334), p.976.

[11] P.J. Halfpenny, K.J. Roberts, and J.N. Sherwood, J. of Mater. Science **19**, 1629-1637 (1984).

[12] M.M.Kuklja and A.B.Kunz, Hydrostatic compression of the perfect RDX and the crystal with the vacancy, J.of Appl.Phys. (submitted), 1998.

[13] A.N. Dremin, S.D. Savrov, V.S. Trofimov, and K.K. Shvedov, in *Detonacionnye volny v kondensirovannyh sredah (Detonation waves in condensed matter)*, Izd. Nauka, Moscow (1970).

[14] G.I. Kanel, S.V. Razorenov, A.V. Utkin and V.E. Fortov, in *Udarno-volnovye yavlenija v kondensirovannyh sredah (Impact-wave phenomena in condensed matter)*, Yanus-K, Moscow (1996).

[15] Zeldovich, Ia.B., A.S.Kompaneets, *Theory of Detonation*, Academic Press, New York (1960).

[16] J.J. Gilman, Philosophical Magazine B, **71**, N6, 1057-1068 (1995).

[17] J.J. Gilman, in *Shock Compression of Condensed Matter - 1989*, ed. by S.C. Schmidt, J.N. Johnson, and L.W. Davidson (New York, Elsevier), p. 267.

[18] G. Gao, R. Pandey, and A.B. Kunz, in *Structure and Properties of Energetic Materials, Materials Research Society Symposium Proceedings*, ed. by D.H. Liedenberg, R.W. Armstrong, and J.J. Gilman, (Pittsburgh, Pennsylvania, 1993) v. **296**, 149-155.

[19] M.K.Orloff, P.A.Mullen, and F.C.Rauch, J.Phys. Chem, v.74 (10), p.2189-2192 (1970).

[20] P.L.Marinkas, J. of Luminescence **15**, p.57-67 (1977).

UNDERSTANDING STRUCTURE AND ELECTRONIC PROPERTIES OF EXTENDED SELF-INTERSTITIAL DEFECTS IN SILICON

P. ALIPPI and L. COLOMBO
INFM and Dept. Materials Science,
via Cozzi 53, I-20126 Milano, Italy
paola.alippi@mater.unimi.it
luciano.colombo@mater.unimi.it

Abstract

The results of an atomistic investigation on the coalescence mechanisms of self-interstitial {311} defects are presented. Formation energies and equilibrium configurations of defect structures are determined by tight-binding molecular dynamics simulation. We focus on the characterization of the lattice strain field around the defect complex: By means of the determination of the atomic stress distribution, we discuss how it may influence the formation mechanisms of the planar {311} structures. We also attempt a correlation between structural features and electronic properties through the analysis of defect-related orbitals occupations and inverse participation ratios.

Introduction

In this work, we have addressed the study of extended {311} defects in silicon, observed in ion implanted (as well as electron irradiated) samples in a temperature range 300° C < T < 800° C. Such *rod-like* defects show up in transmission electron microscopy (TEM) observations as interstitial structures elongated in the <110> direction, having the {311} plane as their habit plane. The {311} defects are unstable at T > 800° C, when they transform into dislocation loops. Understanding the coalescence mechanism of the {311} and their stability is a key issue in modelling the microstructural evolution of Si: in fact, their dissolution during thermal annealing has been proved to affect the diffusion mechanism of the dopant species in Si samples. Experimental investigations [1] have in fact correlated the {311} evaporation during annealing with the detection of an enhanced transient diffusion (TED) of boron in low-energy ($\simeq$ 10 keV) implanted samples and have strongly indicated the {311} defects as the source of interstitials.

Atomic models of the {311} defects have been built up since their first discovery [2], mostly through the analysis of the images obtained in TEM experiments [3, 4, 5]. In the attempt to characterize the basic building block of the defect, a given interstitial structure is inserted into the perfect crystal matrix and the nearby atomic bonds are locally rearranged through educated guesses. The atomic model of the extended {311} defect is then created repeating this basic unit on the {311} plane. Tan [2] was the first to point out that the introduction of a <110> chain of interstitial atoms in the perfect diamond structure requires a minimum number of dangling bonds. The repetition of <110> chains (hereafter referred to as I-chains) along the < 233 > direction was then proposed as the model of the {311} stacking fault defect, with five- and seven-membered rings formed on the {110} plane around the interstitial chain. The presence of eight-membered rings (in the following denoted as O-rings) inserted between groups of two or three <110> chains with no periodicity along the <233> direction was detected by Takeda [3] from the analysis of TEM images. This

Mat. Res. Soc. Symp. Proc. Vol. 538 © 1999 Materials Research Society

structure can be also seen as a tiny precipitate of the hexagonal phase of Si on the {311} plane [4].

Theoretical studies [6, 7, 8] have calculated the formation energies of several defect structures, modelled as periodic repetitions of unit blocks containing different numbers of I-chains and O-rings. The introduction of a O-ring is found to be energetically favoured both in classical molecular dynamics simulations [3] and within tight–binding frameworks [7, 8]. Discussion of the structure stability and arguments on the formation mechanisms of the {311} have been given in those works.

Our goal is to investigate the stability of some of the defect models presented in literature, through the analysis of the stress field around the defect complex. In particular, we choose the structures that the work of Kim *et al.* [7] has shown to be the most stable ones among those containing the same number of interstitial chains, and having no periodicity in the <233> direction.

Computational details

We perform tight–binding molecular dynamics [9] (TBMD) simulations of different structural models of the {311} defect, with the TB representation developed by Kwon *et al.* [10]. We use an orthorhombic unit cell with the $\hat{x}$, $\hat{y}$ and $\hat{z}$ axes corresponding to the $<0\bar{1}1>$, $<2\bar{3}\bar{3}>$ and $<311>$ directions of the diamond lattice, taking n_x, n_y, n_z replica along the three directions. The TBMD simulations presented here require the use of large unit cells, ranging between $\simeq 500$ and $\simeq 800$ atoms, depending on the number of interstitial chains inserted. In order to handle such large cells, we use the linear scaling O(N)–TBMD method introduced by Goedecker and Colombo [9, 11] and efficiently implemented on the Cray T3E. We also mention that a first qualitative insight on the stability of the atomic configurations used as starting models of the {311} defect structures is first gained via molecular dynamics simulations employing the Stillinger–Weber interatomic potential [12]. TBMD simulations within the O(N)-scheme are performed on these structures, applying a finite–temperature annealing procedure (up to 600 K). Accurate defect formation energies are obtained from total energy calculations performed with the $O(N^3)$-TB code on the structures annealed and relaxed at zero temperature.

Results and discussion

In Fig. 1, the relaxed structures of two defect linear configurations are presented: the defect contains a single interstitial $<0\bar{1}1>$ chain forming two five-membered (left panel) and six-membered (right panel) rings with the adjacent atoms in the $<2\bar{3}\bar{3}>$ direction. The formation energy E_f of a defect structure is calculated from the calculated total energies of supercells containing N_B bulk atoms and N_I interstitial atoms as:

$$E_f = \left(E_{BI} - \frac{(N_B + N_I)}{N_B} E_B \right) \quad , \tag{1}$$

where E_{BI} is the total energy of a cell containing $N_B + N_I$ atoms and E_B the energy for a bulk cell with N_B atoms. The formation energy per interstitial, E_f^I, is given by $E_f^I = E_f/N_I$. The formation energies per interstitial of the two structures showed in Fig. 1 are 2.04 eV and 1.6 eV, in agreement with the calculations of Kim *et al.* [7]. The structure on the right is the one proposed by Tan [2] as basic building block of the {311} defect, and has in fact the lower formation energy between configurations containing a single interstitial chain.

The stress field around the defect has been calculated following the approach of Marks *et al.* [13], where the total stress σ_{ij} in the system is expressed as the average value of the quantity σ_{ij}^n, that measures the contribution of the atom n to σ_{ij}. The atomic stress σ_{ij}^n is given by:

$$\sigma_{ij}^n = \frac{1}{V} \sum_{m=1}^{N} \frac{\mathbf{F}^{(nm)} \cdot \mathbf{x}^{(nm)}}{|\mathbf{x}^{(nm)}|^2} x_i^{(nm)} x_j^{(nm)} \tag{2}$$

where $\mathbf{F}^{(nm)}$ is the force between atoms n and m, projected on the direction $\mathbf{x}^{(nm)}$ of the distance vector between the two atoms, which has components $x_i^{(nm)}$. A color code is used in Fig. 1 to distinguish between atoms under tensile and compressive stress $< \sigma^n > = \sum_{i,j} < \sigma_{ij}^n >$ (we took the temporal average value of σ_{ij}^n). The striking difference between the two pictures resides in the different orientation of the crystalline planes containing atoms under tensile stress (light grey atoms): while these are {110} planes in the left picture, the defect structure on the right shows that tensile stress contributions come from the atoms lying on {2$\bar{3}\bar{3}$} planes, thus perpendicular to the defect habit plane. The stress in both case is localized few atomic planes around the I-chain. The formation mechanism of the {311} defects has been proposed [7] as composed of two steps: first the formation of I-structures elongated in the <110> direction, followed then by the clustering of I-chains along the <233> direction. The stress field analysis presented here suggests that the I-chains may come close along particular crystalline directions, relieving elastic energy.

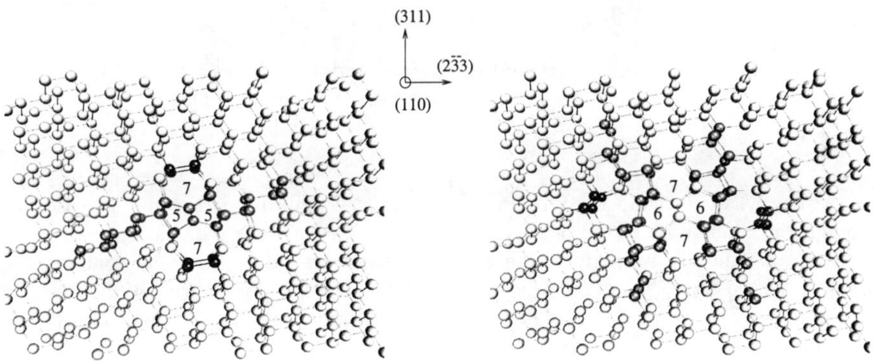

Figure 1: Relaxed defect structures for a single <110> interstitial chain forming five-membered (left) and six-membered (right) rings on the {110} plane. The atomic stress tensor is represented by atoms colored in dark grey (compressive stress) and light grey (tensile stress).

The same stress-field analysis has been carried out for two configurations that contain two I-chains: those were obtained repeating the most stable configuration for the single

I-chain (i.e., right hand side of Fig. 1) after $a\sqrt{11}/2\sqrt{2}$ and $a\sqrt{11}/\sqrt{2}$ along the $\hat{y}$ direction (the latter is L_y, the unit cell dimension along $\hat{y}$). The corresponding relaxed structures are shown, respectively, in the left and right hand side pictures of Fig. 2, and have formation energies per interstitial of 1.23 eV and 1.5 eV. Here also, the color code shows clearly that in both cases atoms on the $\{2\bar{3}\bar{3}\}$ planes perpendicular to the defect habit plane are under tensile stress: however, while all adjacent $\{2\bar{3}\bar{3}\}$ planes are under tensile stress in the left hand side configuration, planes of almost zero stress can be seen in the picture on the right, nearby the position of the eight-membered ring. This is in agreement with the idea that the insertion of an eight-membered ring in the defect is preferred in order to relief the strain induced by adjacent interstitial chains [4, 6].

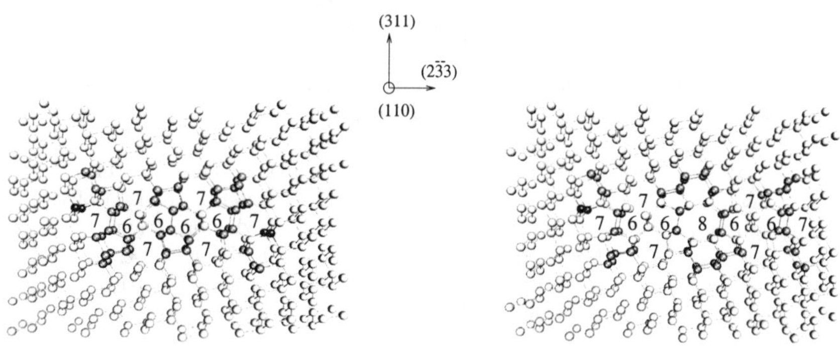

Figure 2: Relaxed defect structures for two <110> interstitial chains separated by $L_y/2$ and L_y on the $<2\bar{3}\bar{3}>$ direction. The atomic stress tensor is represented by the same color code as in Fig. 1.

We have also correlate the structural and the electronic properties of these I-chains, through the calculation of the density of states and the inverse participation ratio (IPR). For a given structure we perform the electronic structure calculation, and compute then $I(\psi_n) = N \sum_{i=1}^{N} a_{ni}^4 / (\sum_{i=1}^{N} a_{ni}^2)^2$, which measures the degree of localization of the electronic wave function [14] $\psi_n = \sum_i a_{ni} \phi_i$: the two limiting cases are in fact $I \simeq 1$ for an extended state and $I \simeq N$ for a localized state.

The density of states of the two structure considered in Fig. 1 is shown in Fig. 3, together with the IPR values, drawn as vertical bars (different scales are being used on the $\hat{y}$ axis, in order to draw both quantities). We choose to discuss only the structures of Fig. 1, since the results for defects containing two I-chains (Fig. 2) do not differ appreciably. It should be noted that there is no electronic state that shows a localized character. All the calculated $I(\psi_n)$ values are near the limiting case of $I_n \simeq 1$. This however is to be expected for the defect structures considered, since we consider as building block of the defect an infinitely long I-chain, where all atoms are 4-fold coordinated. We may however expect to find considerable differences when turning to the case of I-chain of finite dimension along the <110> direction, which are presently under investigation.

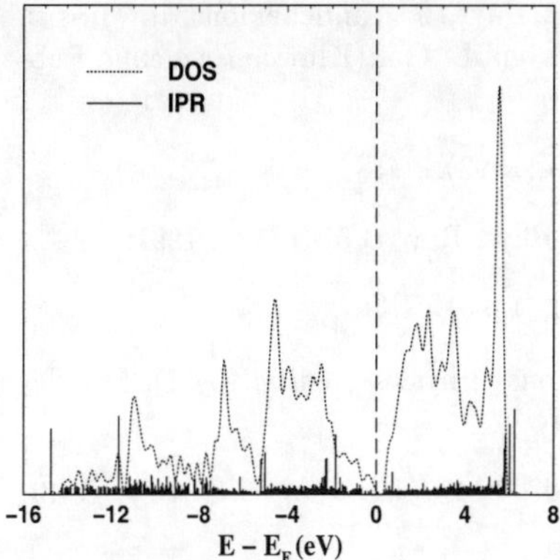
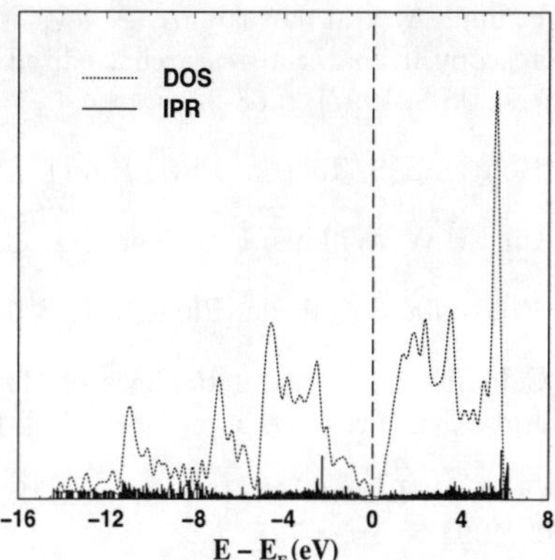

Figure 3: Density of states (solid lines) and IPR values (vertical bars) for the two structures in left and right panels of Fig. 1.

Conclusions

We have characterized the atomic stress field around chosen atomic models of the *rod-like* {311} defect, relaxed at zero temperatures by means of TBMD calculations. We have shown that a local rearrangement of the atomic bonds around the defect complex may have decisive influence on the stress redistribution, suggesting a way to discuss the formation mechanism of the {311} defect. The analysis of the electronic structures does not reveal "anomalous" features, in accordance with the 4-fold coordination characters of all the structures analyzed.

Acknowledgements

We acknowledge computational support on the CRAY–T3E computer at CINECA (Casalecchio di Reno, Italy) by INFM under initiative "Progetti di Supercalcolo" (project "From point to extended self-interstitials in silicon: a TBMD investigation").

References

[1] D. J. Eaglesham, P. A. Stolk, H.-J. Gossmann, and J. M. Poate, Appl. Phys. Lett. **65**, 2305 (1994); A. Agarwal, T. E. Haynes, D. J. Eaglesham, H.-J. Gossmann, D. C: Jacobson, J. M. Poate, and Y. E. Erokhin, ibid. **70**, 3332 (1997).

[2] I. G. Salisbury and M. H. Loretto, Philos. Mag. A **39**, 317 (1979); T. Y. Tan, Philos. Mag. A **44**, 101 (1981).

[3] Seiji Takeda, Jpn. J. Appl. Phys. **30**, L639 (1991).

[4] A. Parisini and A. Bourret, Philos. Mag. A **67**, 605 (1993).

[5] L. Fedina, A. Gutakovskii, A. Aseev, J. Van Landuyt, J. Vanhellemont, in "In-situ microscopy in material research", edited by Pratibha L. Gai (Kluwer Academic Publishers, USA, 1997), p.63-92.

[6] M. Kohyama, S. Takeda, Phys. Rev. B **46**, 12305 (1992).

[7] J. Kim, J. W. Wilkins, F. S. Khan, A. Canning, Phys. Rev. B **55** 16186, 1997.

[8] M. Kohyama, S. Takeda, Phys. Rev. B **51**, 13111 (1995).

[9] L. Colombo, in "Annual Reviews of Computational Physics", edited by D. Stauffer (World Scientific, Singapore, 1996), Vol. IV, p.147.

[10] I. Kwon, R. Biswas, C. Z. Wang, K. M. Ho, and C. M. Soukoulis, Phys. Rev. B **49**, 7242 (1994).

[11] S. Goedecker and L. Colombo, Phys. Rev. Lett. **73**, 122 (1994).

[12] F.H. Stillinger and T.A. Weber, Phys. Rev. B **31**, 5262 (1985)

[13] N. A. Marks, D. R. McKenzie, and B. A. Pailthorpe Phys. Rev. B **53**, 4117 (1996).

[14] Jianjun Dong and D. A. Drabold, Phys. Rev. Lett. **80**, 1928 (1998).

A STUDY OF VACANCIES IN PURE ALUMINIUM AND THEIR ROLE IN THE DIFFUSION OF LITHIUM IN A DILUTE AL-LI ALLOY USING THE EMBEDDED ATOM MODEL

P. M. Derlet *, R. Høier *, R. Holmestad *, K. Marthinsen **, N. Ryum **

* Department of Physics, Norwegian University of Science and Technology (NTNU), N-7034 Trondheim, Norway.

** Department of Metallurgy, Norwegian University of Science and Technology (NTNU), N-7034 Trondheim, Norway.

ABSTRACT

Constant temperature and pressure molecular dynamics are employed to study bulk Al and the substitutional Al-Li alloy using the embedded atom model. The appropriate embedding energy functionals and pair potentials have been determined using known experimental data of bulk Al and Li, together with the results of density functional theory studies of small Al_n-Li_m clusters and the ordered Al_3Li crystallographic phase. The primary goal is to study the role vacancies play in the early stages of the nucleation and growth of the Al_3Li crystallographic phase. To this end we present the preliminary results of a study of vacancy diffusion in Al, and Li diffusion through the Al matrix via a vacancy mechanism. In the present work the results of vacancy diffusion are analysed using a modified inverse Laplace transform, producing a continuous distribution of vacancy hopping times which reproduces the characteristic times for both the single and double vacancy hopping events.

INTRODUCTION

Alloys constitute a complex and dynamical environment where the growth and precipitation of various stable and meta-stable phases can lead to drastic improvements in macroscopic properties. Such phases necessarily begin at the atomic level through homogeneous and heterogeneous nucleation dynamics. It is at this early stage that the diffusion properties of single atoms and their complexes play a vital role in determining the eventual large-scale phase morphology. For example, it is known that in the dilute Al-Li system, lithium diffuses via a vacancy mechanism towards edge dislocations. Here the ordered meta-stable phase Al_3Li (δ'-phase) is known to nucleate without destroying the dislocation [1]. In addition, lithium diffusing along the dislocation line (pipe diffusion) is believed to play an important role in the precise growth kinetics [2].

The microscopic modeling of such material processes via the techniques of Molecular Dynamics (MD) and Monte-Carlo must, at its most fundamental level, be able to realistically describe the nature of the interactions between the constituent atoms. The Embedded Atom Model [3,4] (EAM) provides one such description, providing a physically appealing picture of the interactions between atoms that is fully transferable to inhomogeneous systems and especially suited to the inter-metalics [5]. In the present work we develop an EAM description for the Al-Li system and, using MD, we investigate vacancy diffusion in bulk aluminium and the diffusion of the lithium-vacancy complex through a homogeneous aluminium matrix. Our preliminary investigation of lithium diffusion is a precursor to a more detailed MD study of lithium diffusion in the presence of an edge dislocation.

Mat. Res. Soc. Symp. Proc. Vol. 538 © 1999 Materials Research Society

THEORY

The Embedded Atom Method For Alloys

Within EAM [3,4], the total energy of a system containing atoms of type t_i is given by the sum:

$$E = \sum_i F^{t_i}[\rho_i] + \frac{1}{2} \sum_{ij, i \neq j} \phi^{t_i t_j}(r_{ij})$$ (1)

where $\rho_i = \sum_{j, j \neq i} f^{t_j}(r_{ij})$. Here $f^{t_j}(r_{ij})$ can be viewed as the electronic density that atom i sees due to atom j, and $F^{t_i}[\rho_i]$, the corresponding embedding energy of atom i when placed in the electronic density arising from the surrounding atoms. $\phi^{t_i t_j}(r_{ij})$ can be interpreted as the effective inter-atomic potential. The precise form of the embedding energy functional, the inter-atomic electronic density function and the inter-atomic potential (between atoms of the same kind) can be determined using the universal scaling properties of solids under tension and compression [6]. These functions can be chosen to exactly reproduce the bulk equilibrium lattice constant and cohesive energy [7,8] with the remaining free parameters usually fitted to the bulk elastic constants and the unrelaxed vacancy formation energy

Johnson [7] found that by assuming Eq. 1 to have a particular symmetry [9], the inter-atomic potential between atoms of a different kind can then be uniquely given as

$$\phi^{12}(r) = \frac{1}{2}\left[\frac{f^2(r)}{f^1(r)}\phi^1(r) + \frac{f^1(r)}{f^2(r)}\phi^2(r)\right].$$ (2)

At first sight it may seem that this potential is precisely defined through the EAM description of the pure substances. However the embedding energy functionals turn out to be independent of the equilibrium charge densities of the individual materials. Eq. 2 on the other hand contains ratios of inter-atomic electron density functions and therefore the *relative* equilibrium charge densities can be adjusted (in addition to the other free parameters) to reproduce the properties of a bulk and composite systems.

In fitting the pure substances, the unrelaxed vacancy formation energy is required. Since this is not experimentally attainable, we must rely on *ab initio* Density Functional Theory (DFT) calculations for bulk aluminium [10-14] and bulk lithium [10,11,15]. In previous work [16] we find that for aluminium, a value of $E_{vf}^u \approx 0.7\text{eV}$ is needed to reproduce the experimental [17] relaxed vacancy formation energy: $0.67 \pm 0.05\text{eV}$. For lithium, the corresponding experimental value for the relaxed vacancy formation energy is less well known ranging between 0.34eV [18] to 0.48eV [19]. To reproduce this range using EAM, the parameter E_{vf}^u must range respectively between $\approx 0.5\text{eV}$ and $\approx 0.85\text{eV}$. Following the results of *ab initio* DFT studies we use both $E_{vf}^u = 0.52\text{eV}$ [15] and $E_{vf}^u = 0.81\text{eV}$ [10] in the present work.

The relative equilibrium charge densities of the materials are obtained via Eq. 2 by numerical fitting to the energetics of small Al_nLi_m clusters [20,21] and/or the cohesive energy of the Al_3Li ordered alloy phase [22]. These energies are again obtained by DFT methods. In general it was difficult to obtain fits that accurately reproduced both the cluster and bulk alloy phases. Despite the transferability of the EAM, it is thought that the cluster physics may well be at variance with that of the bulk; the former showing significant sp-hypridization. In the present work the chosen potentials are those which most accurately reproduce the Al₃Li phase cohesive energy. Table 1 summarises the parameters used in obtaining the final inter-atomic potentials and embedding energy functionals for the aluminium-lithium system.

	C_{11}	C_{12}	C_{44}	E_{vf}^{u} (eV)	$E_{cohesive}$ (eV)	a_0 (Å)
Al	0.920	0.693	0.395	0.70	3.39	4.05
Li	0.148	0.125	0.108	0.52 & 0.81	1.63	3.491
Al₃Li	-	-	-	-	0.277	3.96

Table 1 Experimental and *ab initio* parameters used to obtain the inter-atomic potentials and embedding energy functionals for the aluminium-lithium system. The experimental elastic constants where obtained from Simmons and Wang [23] and other bulk data from Harrison [24]. The *ab initio* data for the Al₃Li phase was obtained from Mehl [22].

Fig. 1a displays the resulting pure inter-atomic potentials for the system. The Al-Al inter-atomic potential extends out to between the third and fifth nearest neighbour, whilst the Li-Li potential extends out to beyond the ninth nearest neighbour shell (within the bulk BCC lithium lattice). It is evident that the Li-Li potential is a strong function of E_{vf}^{u}, however this dependence has been found to have little effect on vacancy relaxation in bulk lithium [16]. Fig. 1b shows the Al-Li potentials. For comparison, we also show the Al-Li potential obtained using lattice-inversion techniques [25] which assumes that the cohesive energy is constructed entirely from a sum of pair potentials. The shallow and extended nature of the Li-Li potential reveals it to be the "weakest" of the bonds whilst the much deeper Al-Li potential indicates it to be the strongest of the three inter-atomic interactions; in agreement with previous observations [21].

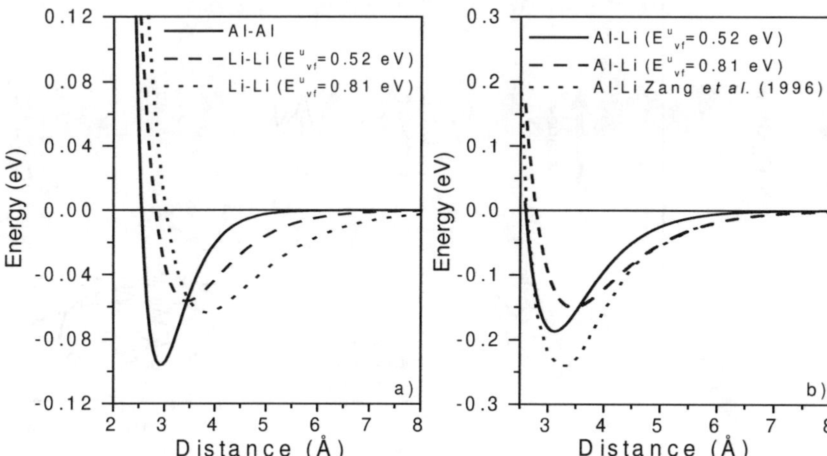

Figure 1 a) The Al-Al and Li-Li EAM inter-atomic potential for $E_{vf}^{u}(Al)$ and the two fitted values of $E_{vf}^{u}(Li)$. **b)** The corresponding Al-Li EAM inter-atomic potentials. Also shown is the Al-Li potential obtained using lattice-inversion techniques [25].

RESULTS

Vacancy Diffusion In Aluminiun

Early MD studies of vacancy diffusion in aluminium established that so-called double vacancy hops play an important role in deviations away from linearity in the Arrhenius plots at temperatures approaching the melting point [26]. Fig. 2 shows the particle trajectories in both space and time, of such an event, and we see that the double vacancy hop is a result of the correlated movement of two neighbouring atoms. This correlation is thought to be induced by anharmoticity and lattice distortion due to the presence of the vacancy. Upon closer inspection of fig. 2 we see that, depending on how we define the act of an atom departing and arriving, one atom may lag closely behind that of another. Thus the double hop can be interpreted as two closely spaced single hops. From the point of view of a distribution of single vacancy hops there then exists two dominant time scales: that of the short time hop which leads to double hopping and the long time hop which contributes to the linearity of the Arrhenius plot and the bulk of the vacancy's diffusion properties.

To investigate such a distribution, a large sample of vacancy hops is required and a correspondingly large number of MD time steps must be performed. Thus it becomes important to automatically track the position of the vacancy without too much computational effort. Since no (computer) variable exists [27] to define the position of the vacancy, we characterise the actions of many of the atoms in a way that will include all possible vacancy hopping events.

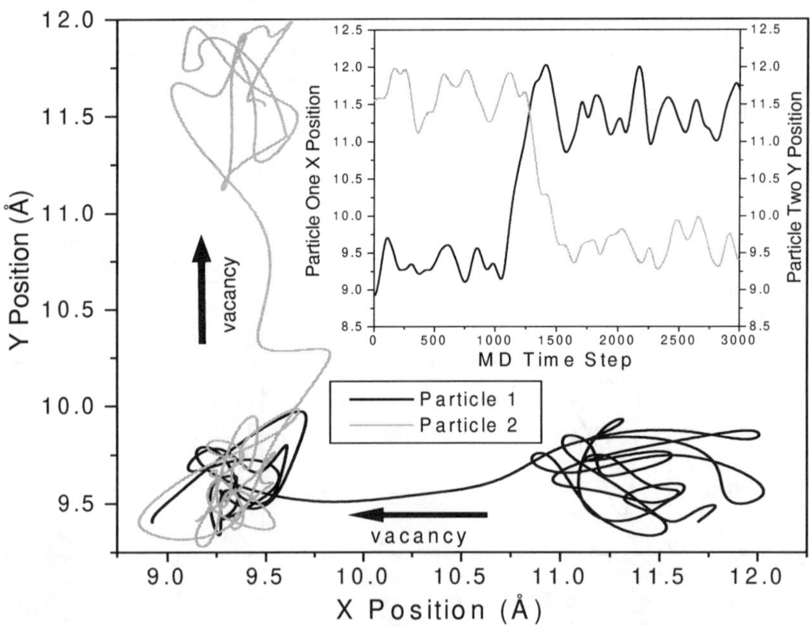

Figure 2 A two-dimensional projection of a double (vacancy) hopping event within aluminium. The insert shows the positions of the two atoms as a function of time.

We found that the most rigorous algorithm involved first checking if any atom had moved away from its mean location defined through an imaginary lattice with a volume shadowing that of the super-cell. If such an event is found, a time of flight variable is created and incremented until either it returns to its original mean position or to within a new mean lattice position. The current position of the vacancy is then defined as the imaginary lattice site corresponding to the position of the last atom that moved to a new site. Thus the time between two successive vacancy hops is defined as the time between two atoms arriving successively at new equilibrium positions. This definition is similar to the cavity analysis method [28] and also allows one to probe the time of flight statistics; necessary to extract the short characteristic hopping times. In a typical MD simulation we find that a unique vacancy path constructed from single hops always arises.

We study vacancy diffusion by undertaking a zero pressure and constant temperature ($T = 750\text{K}$) MD simulation on a 256-site super-cell using a time step equal to 1.06×10^{-15} seconds. Such a size is required to completely eliminate any interactions between the vacancy and its images. Fig. 3a shows the measured distribution of single hops taken from an MD simulation consisting of thirty million MD time steps. Viewing this distribution through a vertical natural logarithmic scale we obtain approximately a straight line, verifying that a roughly exponential distribution is seen. For such an isotropic system, the diffusion process can be represented by a random walk through the FCC lattice and the fundamental measurable quantity is then the hopping rate, Γ, between nearest neighbours. For three-dimensional diffusion, the diffusion coefficient D is then given by

$$D = \Gamma \frac{r_0^2}{6}.$$ (3)

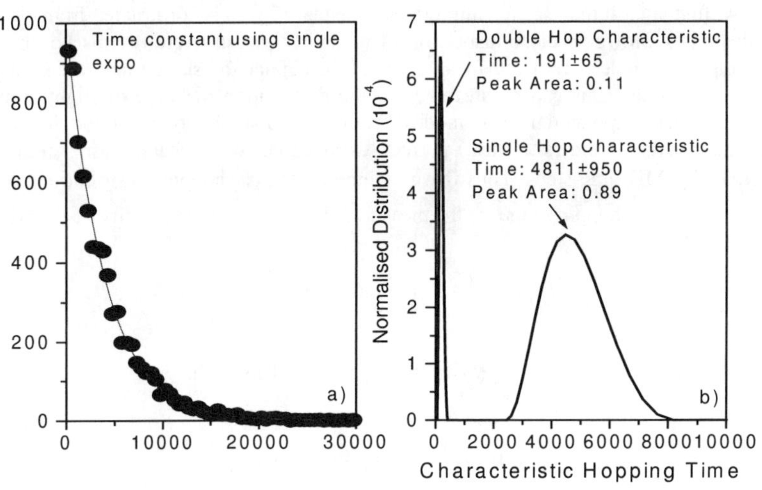

Figure 3 Distribution obtained by CONTIN using the kernel described in Eq. 6. A long time single-hop component and a short time double hop component are evident.

To obtain D from Fig. 3a we must extract the characteristic time of the exponential distribution. Performing a simple least squares fit we obtain a mean hopping time (ie $1/\Gamma$) equal to 3870 ± 58 MD time steps corresponding to $D = 2.0 \pm 0.03 \times 10^{-4}$ cm^2/s. This is in agreement with [26] and slightly higher than that derived from the experimental self-diffusion properties of aluminium ($\approx 10^{-4}$ cm^2/s) due to the MD simulation being performed at zero-pressure.

The hopping rate can be directly related to the Gibbs migration energy, G_m, via

$$\Gamma = z\nu \exp\left(-\frac{G_m}{k_b T}\right) = z\nu \exp\left(\frac{S_m}{k_b}\right)\exp\left(-\frac{H_m}{k_b T}\right), \tag{4}$$

where z is the nearest neighbour coordination number, ν is the hopping attempt rate and H_m and S_m are the corresponding migration enthalpy and entropy. For aluminium ($z=12$, $\nu \approx 5 \times 10^{12}$ and $S_m/k_b \approx 1$ [26]) we obtain, using our value of Γ, a migration energy (zero-pressure enthalpy) of 0.43 ± 0.01eV which is in agreement with that of [26] and also close to the experimental values 0.44eV [29] and 0.52eV [30].

To investigate the structure of the distribution for shorter time scales we perform a numerical inverse Laplace transform using the CONTIN program developed by S. Provencher [31]. From this we can obtain a continuous rather than discrete distribution of characteristic hopping times. This algorithm inverts noisy data with respect to a general kernel, which for the inverse Laplace transform is an exponential function. Such problems are typically ill-conditioned and CONTIN uses a constrained regularised inversion technique to find the optimal solutions subject to the constraint of positivity and the principle of parsimony: that the solution contains the minimum amount of information required to describe the input data.

At first sight it may seem natural to assume that Fig. 3a is constructed from a distribution of exponentials. However close inspection of its short time structure reveals no obvious short time component, indeed rather than an increase in counts for short times the data reveals a reduction in counts. This can be attributed to the distribution of "time of flight" times which convolute the true exponential decay modes. From the MD simulation we found that the "time of flight" distribution follows a near perfect exponential with characteristic time equal to $\tau_{of} = 403 \pm 12$ MD time steps. Thus if we assume a vacancy hopping distribution containing a single characteristic hopping time τ, the measured distribution will in reality be the convolution:

$$P_{hop}(t) = \int_0^t dt' P(t') P_{tof}(t-t') \tag{5}$$

where $P(t) = \exp(-t/\tau)/\tau$ and $P_{tof}(t) = \exp(-t/\tau_{of})/\tau_{of}$. This evaluates to

$$P_{hop}(t) = \frac{1}{\tau - \tau_{of}}\left[\exp\left(-\frac{t}{\tau}\right) - \exp\left(-\frac{t}{\tau_{of}}\right)\right]. \tag{6}$$

Thus the appropriate kernel to use in the inversion of the MD hoping time data of Fig. 3a is Eq. 6 rather than a simple exponential. Clearly for the single hop time scale such a modification is negligible, however for the shorter double-hop time scale, the "time of flight" effect will significantly effect any such component (note that for $t=0$, Eq. 6 is equal to zero).

Fig. 3b displays the corresponding normalised solution and we see two components, that of a long time single hop component at 4811 ± 950 and also a shorter double hop component at 191 ± 65. These values have been determined via the first and second moments of each peak. The somewhat large width of the long and short time components does not indicate an intrinsic distribution of hopping times, rather it is due to statistical noise inherent in the raw data. Indeed, distributions of much smaller widths are obtainable by CONTIN, however the algorithm will always choose that solution containing the least amount of information or equivalently the least defined peak(s). The area under each peak reveals that at $T = 750$K approximately 10% of the vacancy hops arise from the double hop component.

For fig. 3b the bin size is 500 MD time steps. If a similar CONTIN analysis is performed for the smaller bin sizes of 200 and 100 MD time steps we find that the double hop component largely maintains the shape and location we see in fig. 3b. However the single hop component significantly broadens due to the increasingly worse statistics. It may seem surprising that for a bin size of 500 MD time steps one can resolve significantly smaller time scales. However, the nature of the kernel used (Eq. 6), is such that any characteristic time component present which is less than $\tau_{of} = 403$ will extend well beyond the 500 MD time step limit as evidenced by the short time anomaly identified in fig. 3a; a regime in which the counting statistics are optimal.

Lithium Diffusion In Aluminium

In the ideal Al-Li substitutional alloy, lithium can only diffuse through the aluminium matrix when in the presence of vacancies. If a vacancy happens to be in the first nearest neighbour shell of the lithium, it is only then that the lithium atom can hop to it and begin to move through the lattice. At some point the vacancy will move away and it or another vacancy will return to a different location within the first nearest neighbour shell and again the possibility of the lithium atom hopping will arise. Through this process the lithium atom diffuses throughout the lattice via some type of random walk. If the motion of vacancies and lithium atoms are un-correlated, then the diffusion coefficient of lithium will be a simple function of vacancy density. In reality the situation is far more complex: the presence of both the lithium atom and a vacancy distorts the lattice causing a strain field which induces an effective interaction between the two "particles".

Fig. 4 displays the relaxed zero-temperature interaction between lithium and a single vacancy as a function of nearest neighbour shell for both values of $E_{vf}^{u}(Li)$. We see for both cases that the interaction energy is at a minimum if the vacancy is in the first nearest neighbour shell. In addition there are maximums at the third and fifth nearest neighbour shells. Monte-Carlo simulations reveal that this structure survives at $T = 750$ K. In terms of temperature, the depths of the wells are ≈ 950 K for $E_{vf}^{u}(Li) = 0.52$eV and ≈ 3000 K for $E_{vf}^{u}(Li) = 0.81$. If we ignore the dynamical effects of temperature on the lattice, this indicates that for a system at a temperature of 750K, a vacancy can become strongly bound to the $E_{vf}^{u}(Li) = 0.81$eV lithium defect. Whilst for $E_{vf}^{u}(Li) = 0.52$eV, the vacancy will become only loosely attached to the lithium, since the energy difference between the first and second nearest neighbour shell is approximately 750K and that between the second and third, is 200 K.

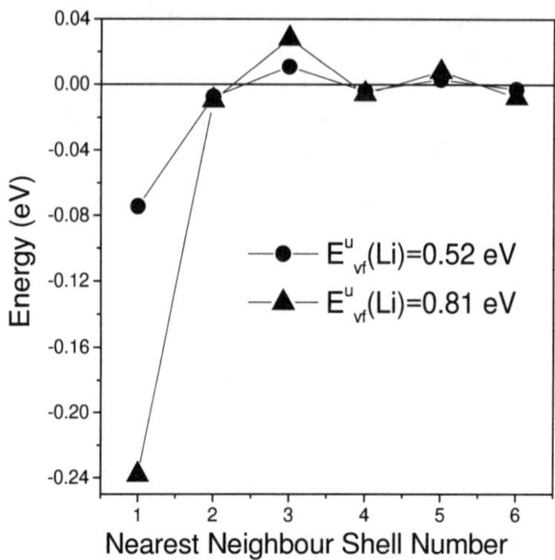

Figure 4 Zero temperature lithium-vacancy interaction energy for both values of $E_{vf}^{u}(Li)$ using a 256-site super-cell.

These two scenarios may severely effect the diffusion characteristics of the lithium complex, for how frequently a vacancy visits a lithium atom will determine how often the lithium atom can undergo a useful lattice hop. Crudely speaking, an extremely deep well will form a well defined lithium-vacancy complex, however if the vacancy cannot easily move around the lithium (via the third neighbour shell) the lithium will simply hop back and forth on the same two sites contributing little to its overall diffusion coefficient. To investigate these two scenarios we perform zero-pressure MD simulations on a super-cell containing a single lithium atom and a vacancy. As in the case of vacancy diffusion in bulk aluminium, we perform the simulation at 750 K to obtain good statistics. Using a 256-site super-cell we have a vacancy concentration of one part in 256 which is significantly higher than the equilibrium case. Furthermore the super-cell approximation excludes the possibility for more than one vacancy being near the lithium atom (an important consideration if the vacancy-vacancy interaction is prevalent). Like wise any thermodynamic quantities will suffer from finite size effects. Nevertheless for the present purpose of studying the high-temperature dynamics of the lithium-vacancy interaction, such a simulation is an adequate starting point.

Fig. 5 displays the normalised pair correlation function obtained from the MD simulations between the lithium and vacancy defects for both values of $E_{vf}^{u}(Li)$. We see that in both cases there is a significant probability that the vacancy is located in the first nearest neighbour shell. For $E_{vf}^{u}(Li) = 0.81\text{eV}$ the probability is the largest and there is very little structure beyond this shell indicating that the lithium vacancy complex is extremely well defined. However for $E_{vf}^{u}(Li) = 0.52\text{eV}$, we see that peaks exist at distances slightly less than the fourth and sixth nearest neighbour shells, corresponding to the small troughs in Fig. 5. For both $E_{vf}^{u}(Li)$ a significant distribution of lithium-vacancy distances exists at distances less than the first nearest neighbour shell. With the vacancy being so close to the lithium atom and so often, a local isotropy exists and the random thermal motion of the lithium atom will tend to be weighted towards the direction of the vacancy.

Each MD simulation ran for a total of 10^7 time steps and for $E_{vf}^u(Li) = 0.52$eV the lithium defect recorded 525 hops with a mean time between hops of 18502 ± 2400 time steps. If we assume a simple isotropic random walk model for lithium's three dimensional diffusion (Eq. 3), then the diffusion constant D is equal to $7.3 \pm 0.3 \times 10^{-5}$ cm^2/s. Such a model is of course a gross simplification due to the presence of the vacancy. In fact up to seventy five percent of the hops observed simply involved the lithium hopping back and forth between itself and the stationary vacancy; these contributing little to its overall diffusion through the aluminium matrix. If however we use the more general definition for the diffusion coefficient:

$$D = \frac{1}{6}\frac{d}{dt}\left\langle\left(\tilde{r}(0) - \tilde{r}(t)\right)^2\right\rangle \tag{7}$$

(Einstein's equation) we obtain $D = 1.3 \pm 0.2 \times 10^{-5}$ cm^2/s. For $E_{vf}^u(Li) = 0.81$eV the lithium defect recorded 3101 hops with a mean time between hops of 3224 ± 105 time steps corresponding to $D = 4.2 \pm 0.1 \times 10^{-4}$ cm^2/s when using Eq. 3. In this case over ninety percent of the hops involved hoping back and forth. On the other hand, using Eq. 7 we obtain the value of $D = 9.5 \pm 1.3 \times 10^{-6}$ cm^2/s, which is in close agreement with the value obtained for $E_{vf}^u(Li) = 0.52$eV.

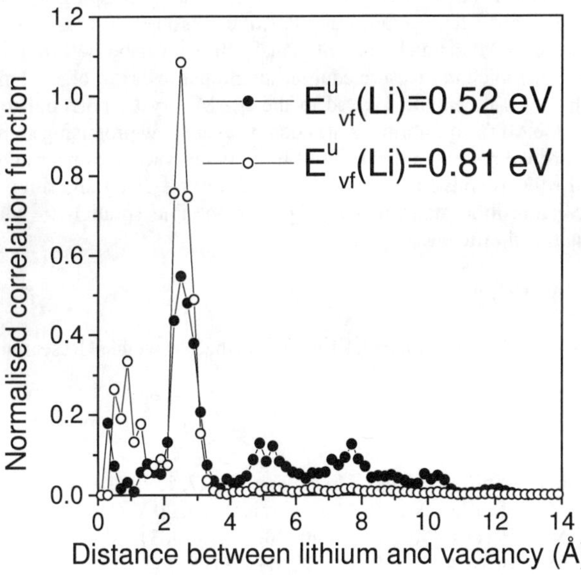

Figure 5 The normalised pair correlation function between the lithium and vacancy defects in an aluminium matrix for both values of $E_{vf}^u(Li)$. The zero pressure MD simulation was performed at 750K using a 256-site super-cell, containing one lithium atom and a vacancy.

CONCLUSIONS

In the present work we have applied the EAM to vacancy diffusion in aluminium and lithium diffusion in aluminium via a (single) vacancy mechanism. For vacancy diffusion we have employed the CONTIN algorithm to analyse the measured distribution of vacancy hopping times. By taking into account the convoluting effect of the "time of flight" effect in the process of vacancy hopping, we have demonstrated that such a method of analysis is an extremely sensitive tool to investigate the short hopping time structure which manifests itself as the double hop. This will allow for a detailed study of the effect of temperature on this component, where it is known to greatly increase as the melting temperature of bulk aluminium is approached. The determination of the double hop migration enthalpy will also then be possible. For lithium diffusion we have considered the rather contrived case of an aluminium super-cell containing a lithium atom and a single vacancy. From this we have found that despite the differences in the effective vacancy-lithium interaction potential for the two chosen values of $E_{vf}^{u}(Li)$ (fig. 4), the corresponding diffusion coefficient for lithium turns out not to be such a sensitive function of the unrelaxed vacancy formation energy. For this system only 256 sites were needed (ie. a $4 \times 4 \times 4$ super-cell), however for the more realistic case of lithium diffusion at equilibrium vacancy concentrations much larger super-cells would be needed to even begin to include more than one vacancy and thus, the effects of their subsequent interaction. The study of lithium diffusion at a dislocation edge would require similarly large super-cells.

The primary disadvantage is that an inordinate number of MD time steps must be performed to obtain a sufficient number of hops. In the case of vacancy diffusion thirty million time steps ($\approx 3ns$) were needed to record only 7765 vacancy hopping events. The situation becomes far worse for lower temperatures and positive pressures. This is of course the age-old problem of time scale in MD; much computational effort must be performed to re-create the thermal environment in which to measure equilibrium properties. New algorithms such as hyper-MD [32] in which a bias potential is added to the (EAM) crystal potential to "speed" up the transitions between the MD's equilibrium states do however show promising signs.

In future work we study the diffusion of lithium and vacancies near an edge dislocation using conventional MD. We also investigate the feasibility of using algorithms such as hyper-MD and dynamical transition state theory [33] as a possible solution to what is, largely, a problem of computational efficiency.

ACKNOWLEDGMENTS

P.M.D. acknowledges the financial support of the Norwegian Research Council (Grant no. 116559/431)

REFERENCES

1. Z. M. Wang and G. J. Shiflet, Metall. Mater. Trans A **27**, 1599 (1996)
2. Z. M. Wang and G. J. Shiflet, Metall. Mater. Trans A **29**, 2073 (1998)
3. M. S. Daw and M. I. Baskes, Phys. Rev. Lett. **50**, 1285 (1983)
4. M. S. Daw and M. I. Baskes, Phys. Rev. B **29**, 6443 (1984)
5. A. F. Voter, *Intermetallic Compounds: Vol. 1, Principles*. Edited by J. H. Westbrook and R. L. Fleischer, 1994 John Wiley and Sons Ltd, 77
6. J. H. Rose, J. R. Smith, Francisco Guinea, and John Ferrante, Phys. Rev. B **29**, 2963 (1984)
7. R. A. Johnson, Phys. Rev. B **37**, 3924 (1988)
8. J. Mei, J. W. Davenport and G. W. Fernando, Phys. Rev. B **43**, 4653 (1991)

9. For atoms of the same type Eq. 1 is invariant with respect to the composite transformation $F[\rho] \to F[\rho] + \kappa\rho$ and $\phi(r) \to \phi(r) - 2\kappa f(r)$.

10. W. Frank, U Breier, C. Elsässer, and M. Fähnle, Phys. Rev. B **48**, 7676 (1993)

11. R. Benedek, L. H. Yang, C. Woodward, and B. I. Min, Phys. Rev. B **45**, 2607 (1992)

12. M. J. Gillan, J. Phys: Condens. Matter **1**, 689 (1989)

13. M. J. Mehl and B. M. Klein, Physics B **172,** 211 (1991)

14. N. Chetty, W. Weinert, T. S. Rahman, and J. W. Davenport, Phys. Rev. B **52**, 6313 (1995)

15. G. Jacucci and R. Taylor, J. Phys. F: Metal Phys. **9**, 1489 (1979)

16. P. M. Derlet, R. Høier, R. Holmestad, K. Marthinsen, N. Ryum, submitted to Phys. Rev. B.

17. P. Ehrhart, P. Jung, H. Schulta, and H. Ullmaier in *Atomic Defects in Metals*, edited by H. Ullmaier, Landolt-Börnstein, New Series, Group III, Vol. 25 (Springer-Verlag, Berlin, 1990)

18. R. Feder, Phys. Rev. B **4**, 828 (1970)

19. H. Schultz, Mater. Sci. Eng **4**, 1 (1991)

20. C. Majumder, G. P. Das, S. K. Kulshrestha, Vaishali Shah and D. G. Kanhere , Chem. Phys. Lett. **261**, 515 (1996)

21. V. Shah, D. G. Kanhere, C. Majumder and G. P. Das, J. Phys.: Condens. Matter **9**, 2165 (1997)

22. M. J. Mehl, Phys. Rev. B **47**, 2493 (1993)

23. G. Simmons and H. Wang, *Single crystal elastic constants and calculated aggregate properties: a handbook* (MIT Press, Cambridge, Mass., 1971)

24. W. A. Harrison, *Electron Structure and the Properties of Solids*, (W. H. Freeman and Company, 1980) Section~17-b

25. W. Zhang, Q. Xie and X. Ge, J. Appl. Phys. **82**, 578 (1996)

26. A. Da Fano and G. Jacucci, Phys. Rev. Lett. **39**, 950 (1977)

27. One may follow the vacancy if an imaginary particle is inserted into the void and a force is devised with which it interacts, but does not influence, the surrounding lattice. See M. Manninen and Häkkinen, J. of Comp. Phys. **100**, 197 (1992)

28. J. Laakkonen and R. M. Nieminen, J. Phys. C **21**, 3663 (1988)

29. F. Bradshaw and S. Pearson, Phil. Mag. **2**, 570 (1957)

30. W. DeSorbo and D. Turnbull, Acta Met. **7**, 83 (1957)

31. S. W. Provencher, Comput. Phys. Commun. **27**, 213 (1982), Comput. Phys. Commun. **27**, 229 (1982), *CONTIN user manual*, EMBL technical report DA05m (European Molecular Biology Laboratory, 1982)

32. A. F. Voter, J. Chem. Phys. **106**, 4665 (1996), Phys. Rev. Lett. **78**, 3908 (1997), Phys. Rev. B **57**, R13985 (1998).

33. J. B. Anderson, Adv. Chem. Phys. **91**, 381 (1995).

FIRST-PRINCIPLES STUDY OF π-BONDED (100) PLANAR DEFECTS IN DIAMOND

Peter Zapol, Larry A. Curtiss and Dieter M. Gruen

Materials Science and Chemistry Divisions, Argonne National Laboratory, Argonne, IL 60559

ABSTRACT

A periodic density functional study of the high-energy π-bonded (100) stacking fault in diamond that can serve as a prototype of a twist grain boundary has been carried out. Information on formation energies, geometries and the electronic structure has been obtained. A single point electronic structure calculation of a $\Sigma 5$ twist grain boundary based on the geometry taken from a molecular dynamics simulation has also been performed.

INTRODUCTION

Recent successes in growth of nanocrystalline diamond from hydrogen poor plasmas[1] has attracted attention due to a number of highly desirable mechanical and electronic properties of the new materials. It also raised an interest in effect of the grain boundaries on diamond properties. Indeed, for a typical grain size in the synthesized materials of 3-10 nm, a significant portion of diamond atoms (several percent) is located in the grain boundaries. Therefore, grain boundaries effectively control many properties. Understanding grain boundary formation and structure in the nanocrystalline diamond can help in tailoring new diamond materials in a systematic fashion. A number of molecular dynamics studieshave investigated the energetics and structures of grain boundaries.[2] It was found that twist (100) grain boundaries are usually high-energy structures having a large portion of sp^2 –bonded atoms in the interface region which is corroborated by TEM studies.[3] It should be noted that the (100) surface can be considered to represent a general diamond surface because atoms on this surface have two broken bonds which is a general situation with only exceptions for two special surfaces, namely (111) and (110) surfaces. Therefore, we concentrate our attention on (100) interfaces.
First-principles calculations give reliable results for atomic structure and chemical bonding. However, their use in grain boundary studies in diamond is limited due to the low symmetry and a large size of such systems. We performed a periodic density functional study of a stacking fault in diamond. This system is characterized by π-bonded dicarbon moieties in the interface region and a high formation energy , similar to the twist grain boundaries. In addition, we considered more complicated structure that could occur in the nanocrystalline diamond such as graphite-like planar structures that interconnect two diamond grains. In principle it is desirable to calculate a more realistic grain boundary structure at *ab initio* level but the size of the problem, disordered geometry and a large number of possible metastable states makes it computationally intractable. Instead we have used the equilibrium geometry of a $\Sigma 5$ twist grain boundary geometry obtained in tight-binding molecular dynamics calculation for a single-point density functional calculations.

THEORETICAL METHOD

The periodic density functional method implemented in CRYSTAL95[4] was used to calculate the properties of a model for the grain boundary in nanocrystalline diamond. Crystalline orbitals in this approach are constructed from the localized atomic orbitals, which are linear combinations of Gaussian orbitals. Becke exchange[5] and Perdew-Wang correlation[6] functionals were employed in this study. The standard 6-31G* basis set for carbon was modified by using a

Mat. Res. Soc. Symp. Proc. Vol. 538 © 1999 Materials Research Society

scaling factor of 1.1 for outer sp and d functions to account for less diffuse atomic orbitals in the crystalline environment. A large number of k-points in the irreducible part of Brillouin Zone (approximately 600) was used for reciprocal space integrations. Optimized lattice constants for diamond (3.58 Å) and graphite ($a = 2.47$ Å and $c = 6.71$ Å) are in a very good agreement with experimental data (3.57 Å for diamond and $a = 2.46$ Å and $c = 6.7$ Å for graphite). We have calculated the difference in total energies between diamond and graphite to be 0.9 kcal/mol per one carbon atom. As far as the electronic structure is concerned, we have obtained a value of 4.1 eV for the indirect forbidden energy gap width of diamond as compared to the experimental result of 5.45 eV. The position of the conduction band minimum is in the Γ-X direction about ¾ of the way to the X point. For graphite, our calculations produce a zero width band gap at the K point in agreement with other studies.

STACKING FAULT RESULTS

The model for the diamond stacking fault comprises a 6-layer slab periodically repeated in the z-direction. The two innermost layers of the slab represent a (100) stacking fault, which is separated by 4 layers of diamond from the next equivalent defect. The geometry of the system is shown in Fig. 1. There are 12 atoms in the unit cell, only 3 of which are symmetry unique. The positions of atoms in the xy-plane were restricted to be those in bulk diamond, leaving three interplanar spacing parameters to be optimized: the distances between the two interface planes (d_1 in Fig. 1), between the surface plane and the second diamond plane (d_2 in Fig. 1), and between the second and third diamond planes (d_3 in Fig. 1). The remaining interplanar spacing, d_4, was fixed at the bulk diamond value of 0.895 Å. The optimized values of these parameters are 1.34 Å, 0.828 Å and 0.918 Å. The distance between interface atoms is smaller than that in graphite rings (1.42 Å) and the interlayer distances between diamond layers are slightly different from the defect-free diamond interplanar spacing of 0.895 Å. The volume increase per unit surface area is $0.1a$. As far as the system energetics is concerned, the formation energy per one interface atom is 0.84 eV. This translates into a defect energy per surface area of 4.2 J/m^2.

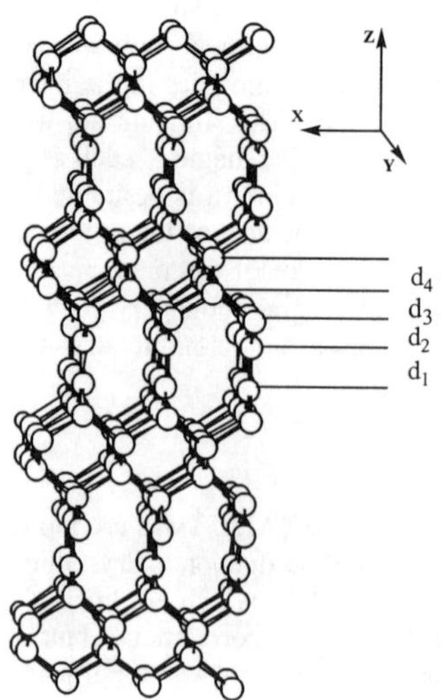

Figure 1. Geometry of a periodic planar defect in the diamond structure. Inequivalent distances between interface layers (d_1), interface layer and second layer (d_2), as well as between the next two diamond layers (d_3 and d_4) are shown.

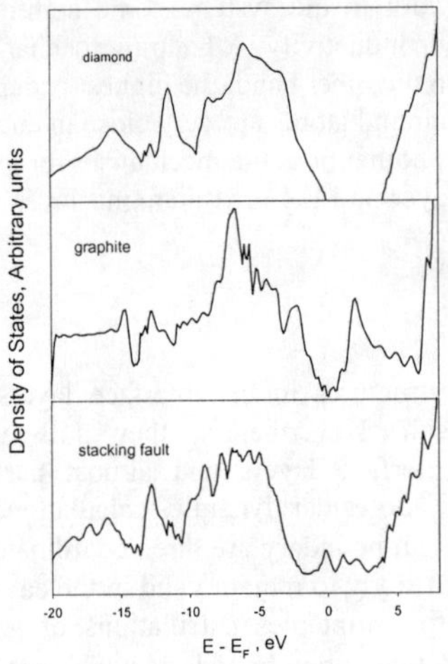

Figure 2. Calculated total density of states plots for diamond, graphite and diamond with stacking faults. Each plot is normalized by the number of carbon atoms in the respective unit cell. No smoothing or thermal broadening have been applied.

The density of states (DOS) plot for diamond with stacking faults is compared to diamond and graphite DOS plots in Fig.2. Peaks that are similar to graphite π-states appear in the band gap of the diamond as a result of the presence of sp^2-bonded atoms. The height of the peaks is strongly dependent on the concentration of planar defects. Since our model has too high a defect concentration (one interface atom per two bulk atoms), it is expected that the density of states in the energy gap of real nanocrystalline material should be several times lower. The three most prominent peaks are at -0.1 eV, 1.1 eV and 2.1 eV with respect to the Fermi level. The appearance of defect states in the fundamental gap of diamond is attributed to the electronic states of atoms in the interfacial layer. The DOS projected on the orbitals of atoms in the interface (first) layer and other layers are shown in Fig.3. It is evident that electronic states in the gap are localized in the interface layer of the stacking fault. Further analysis reveals that the states in the gap are π-states formed by p-orbitals lying in the y-direction in the defect plane. These states might play a role in grain boundary electric conductivity.

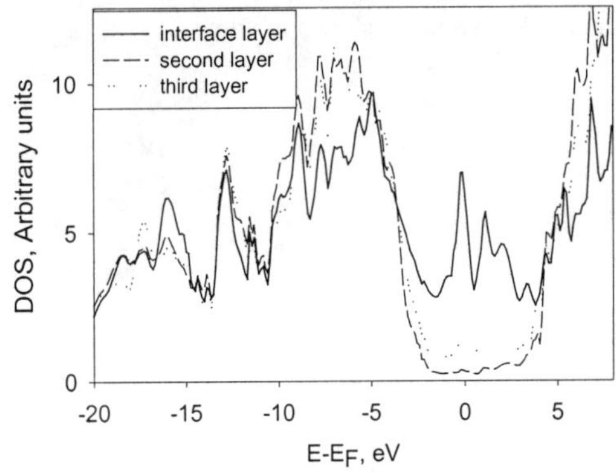

Figure 3. Density of states projected to electronic orbitals of atoms in different layers of the stacking fault in diamond.

Since sp^2-bonded carbons do not form a continuous network in the system, some additional explanations are required for mechanism of grain boundary conductivity. A hopping mechanism for grain boundary conductivity was recently suggested[7]. On the other hand, the highest occupied states in the valence band belonging to four-coordinated diamond atoms are very close in energy to the unoccupied states of sp^2-bonded atoms. This leads to another possible mechanism for grain boundary conductivity involving atoms in the interfacial layer and the next diamond layer that together form a continuous path along the interface.

Σ5 TWIST GRAIN BOUNDARY

 Real grain boundaries are expected to have disordered structures in the interface layers in contrast to our model calculation of the stacking fault. Nevertheless, they still have predominantly three-coordinated carbon atoms in the interface layers and almost perfect diamond structure throughout the rest of the crystal. Indeed, molecular dynamics calculations of Σ29 twist grain boundary found that 80% of atoms in the grain boundary are three coordinated[2]. Therefore, we expect our conclusions from the stacking fault study to remain valid in the case of more complicated grain boundaries. As a next step in first-principles calculations of grain boundaries, we have studied the electronic structure of a Σ5 twist grain boundary, which has the smallest possible planar unit cell with five carbon atoms in each layer. We have performed a single-point calculation taking the equilibrium geometry from the tight-binding molecular dynamics calculations of Cleri[8]. We have reduced the 24-layer 120-atom unit cell that has two Σ5's in it to a smaller unit cell of 40 atoms as illustrated in Fig. 4. The latter structure comprises two diamond-like layers and two interface layers periodically repeating in z-direction. Coordinates of the atoms in the unit cell were taken from molecular dynamics without changes. The average distance between two interface layers is 1.26 Å A smaller STO-3G basis set was used in this calculation and 36 k-points in the first Brillouin Zone were used in reciprocal space integration. The formation energy of the grain boundary was calculated to be 1.4 eV/atom, which translates into 6.5 J/m^2. It is a higher value compared to our stacking fault results but geometry optimization using density functional method would bring it down somewhat.

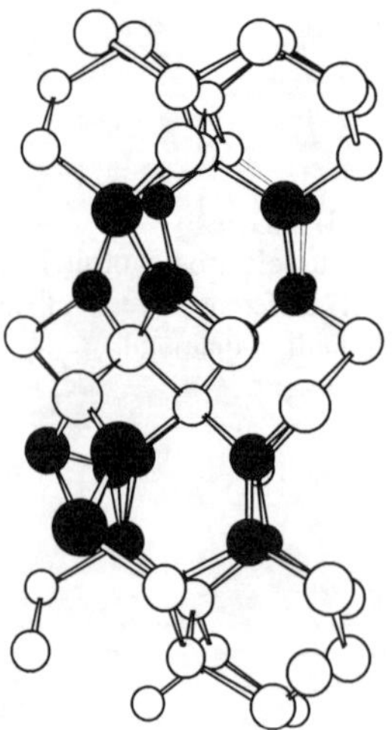

Figure 4. Geometry of Σ5 (100) twist grain boundary used in density functional calculations. Atoms in the interface layers are shown in black, atoms in the diamond-like layers are shown in white.

The calculated density of states plot for the Σ5 grain boundary is shown in Fig. 5. The main peaks in the diamond band gap region are attributed to π-like states on sp^2 atoms and have energies 0.1 eV, 1.1 eV and 3 eV with respect to the Fermi level. A large number of smaller peaks is explained by structural disorder in the grain boundary region. The incomplete convergence of DOS calculation also introduces some spurious features. Thus, the presence of electronic states in the diamond band gap caused by three-coordinated atoms in the interface is a common feature for both high-energy twist grain boundaries as well as stacking faults.

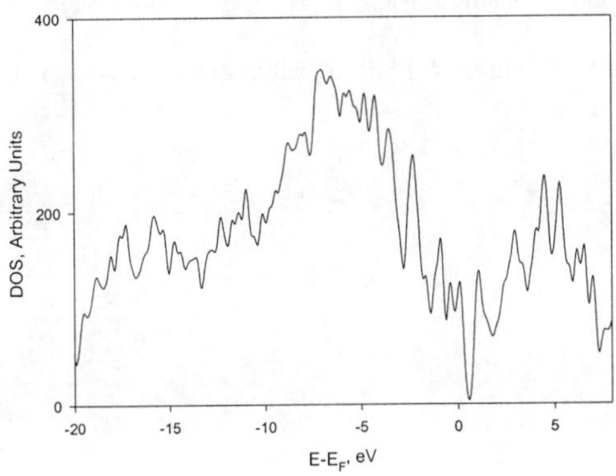

Figure 5. Total density of states for Σ5 twist grain boundary.

CONCLUSIONS

Periodic density functional calculations of high-energy planar defects in diamond adequately reproduce properties of (100) twist grain boundaries. The formation energy of a π-bonded stacking fault is 4.2 J/m^2. The defect region is very thin, having a thickness of one-two atomic layers.The electronic structure of both the stacking fault and the Σ5 grain boundary is characterized by the presence of π-states in the forbidden gap localized on the interfacial atoms of diamond with defects. These states produce several broad peaks in the gap and might play a key role in both grain boundary conductivity and mechanical properties. These results have also shown that single point density functional calculations using molecular dynamics geometry can be useful in studying the electronic structure of the grain boundaries.

ACKNOWLEDGMENT

We gratefully acknowledge use of the advanced computing resources at Argonne's Center for Computational Science and Technology. This work is supported by the U.S. Department of Energy, BES-Materials Sciences, under Contract W-31-109-ENG-38.

REFERENCES

1. C. Zuiker, A. R. Krauss, D. M. Gruen, X. Pan, J. C Li, R. Csencsits, A. Erdemir, C. Bindal, and G. Fenske, Thin Solid Films, **270**, p. 154 (1995).

2. P. Keblinski, D. Wolf, S. R. Phillpot, and H. Gleiter, J. Mater.Res. **13**, p. 2077 (1998).

3. L.C.Qin, D. Zhou, A. R. Krauss and D. M. Gruen. Nanostructured Materials, **10**, p. 649 (1998).

4. R. Dovesi, V.R. Saunders, C. Roetti, M. Causa, N.M.Harrison, R. Orlando, E. Apra, CRYSTAL95 User's Manual, University of Torino, Torino, 1996.

5. A. D. Becke, Phys. Rev. A **38**, p. 3098 (1988).

6. J.P. Perdew and Y. Wang, *Phys. Rev. B* **33**, 8800, 1986; J.P. Perdew and Y. Wang, Phys. Rev. B **40**, 3399, (1989); J.P. Perdew and Y. Wang, Phys. Rev. B **45**, p. 13244, (1989).

7. F. Cleri, P. Keblinski, L. Colombo, D. Wolf and S. R. Phillpot, submitted to Phys. Rev. Let.

8. F. Cleri, private communication.

SUBSTITUTION BEHAVIOR IN NiAl—A FIRST PRINCIPLE PREDICTION CONSIDERING LATTICE RELAXATION

D. S. XU, D. LI, Z. Q. HU
Institute of Metal Research, Chinese Academy of Sciences, Shenyang 110015, P. R. China

ABSTRACT

The substitution behavior of alloying elements in NiAl is investigated on the basis of electronic structure calculation using the discrete variational $X\alpha$ cluster method. The method proposed by the authors to predict the substitution behavior is extended to considering lattice relaxation around an alloying atom. A diagram is drawn using the binding energy of clusters when each alloying element occupies either Ni or Al sites as parameters. Two straight lines on the diagram with slope of unity and passing the point of Ni or Al separate the alloying elements into three groups. The elements above the upper line will take Al sites and those below the lower line will take Ni sites regardless of the composition. The substitution behavior of elements in between the two lines will be affected by the alloy composition. The driving force for an element to order on one sublattice is related to its position on the diagram. The effects of temperature, composition and multi-element alloying are discussed. The concepts of site competition and stepped ordering are put forward.

INTRODUCTION

The ordering of component atoms in intermetallic compounds makes this large group of materials possess many special properties. For example, some have special magnetic, optic or superconductive properties, others have attractive mechanical properties. Addition of alloying elements in these compounds plays an important role in improving their properties. The alloying elements may have different effect when it substitutes for atoms on different sublattice. Knowing the law of the substitution behavior, we may have more degree of freedom of controlling the site occupation of alloying elements, and improving the performance of the materials. The knowledge of the substitution behavior is also essential to understand the alloying effect on the electronic structure level.

The NiAl compound with B2 structure has very attractive high temperature properties, but the room temperature brittleness hinders its application [1]. Some alloying elements, Fe or Ga for example, was found to enhance the room temperature ductility greatly [2], but the mechanism is still not well understood. The enhancement may come from the electronic structural effect when these elements occupy specific sublattice sites. The substitution behavior of a few alloying elements in NiAl compounds has been studied by many researchers using different techniques [3,4,5], but no agreement between different results exists. For example, Duncan *et al.*, using APFIM (Atom Probe-Field Ion Microscopy), show that Fe substitute preferentially for Al, independent on Al to Ni ratio [3]. Anderson *et al.* studied the site occupation of Fe in NiAl using ALCHEMI (Atom Location by CHanneling Enhanced MIcroanalysis). They found that Fe has no strong site preference in NiAl, and tends to reside on the site of the stoichiometrically deficient element [4]. Therefore, a systematic study of the substitution behavior in NiAl is still necessary.

In this paper, the substitution behavior of various alloying element (M) in NiAl compounds is studied based on the discrete variational (DV) $X\alpha$ cluster electronic structure calculation [6]. The site occupancy prediction method proposed by the present authors [7, 8] is extended to

Mat. Res. Soc. Symp. Proc. Vol. 538 © 1999 Materials Research Society

include the lattice relaxation around an alloying atom. A diagram is drawn using two lines to classify the alloying elements into three groups with different substitution behavior. The effects of composition, temperature and multi-element alloying are discussed. The concepts of site competition and stepped ordering are put forward.

ELECTRONIC STRUCTURE AND BINDING ENERGY CALCULATION

The DV-Xα cluster method is used to calculate the electronic structure. Two 15-atom clusters, as shown in Fig. 1, containing two shells of neighbors are employed to calculate the electronic structure when M substitute for Ni or Al sites. Slater's Xα exchange potential is employed with α fixed at 0.70. The single site orbital of atoms with all occupied orbital is chosen as basis functions. The self-consistent charge approximation is introduced and the inner shells of the atomic orbital are frozen to save computer time. The integrations are calculated by a weighted sum over 30,000 sampling points.

To study the tendency of M substituting Ni or Al, the binding energies of clusters with different substitution are calculated. The binding energy of the cluster is defined as the difference between the total energy of the cluster and the sum of the energy of the constituent atoms:

$$Eb(M) = \sum_i E^i_{atom} - E_{tot} .$$

The total energy of the cluster is defined as usual [9]. If the binding energy with M taking some site is larger, the total energy is lower, therefore, such a substitution is more favored.

For alloying elements with different atomic radii, the local atomic relaxations around it may give different contributions to the binding energy. Since the relaxations are mainly on the first shell of neighbors, that of the second nearest neighbor is very small. We will fix the second nearest neighbors and let the first neighbors relax in the radial direction. The configuration with the maximum binding energy is considered to be the relaxed one.

RESULTS AND DISCUSSIONS

Low Temperature Substitution Behavior Classification

In general, the equilibrium configuration is determined by the free energy. At temperature low enough, the enthalpy term dominate the free energy. So, we can predict the site occupation at low temperature by comparison of binding energy of clusters with different substitution.

The binding energy of relaxed clusters with M taking both Ni and Al sites, $Eb_{Ni}(M)$ and $Eb_{Al}(M)$, are calculated. The binding energies are used as parameters to draw a diagram, as shown in Fig. 2. The two straight lines on the diagram are the lines with slopes of unity passing through the points of Ni or Al respectively. These two lines separate the elements into three groups. For the elements located on the upper line, the binding energy change due to substitution of Al with M is the same as the energy difference between M and Al on Ni site:

$$Eb_{Al}(M)-Eb_{Al}(Al)=Eb_{Ni}(M)-Eb_{Ni}(Al).$$

This means that M on the upper line has the same tendency to take an Al site compared with Al. For the M above the upper line, the

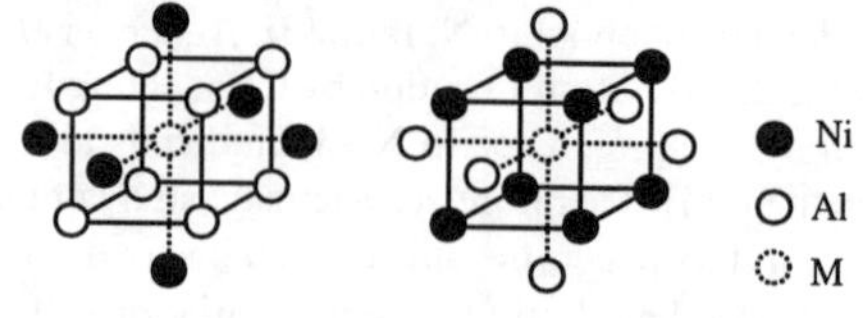

Fig. 1 Clusters employed in the calculation

tendency to take Al site is even larger than Al itself. This means that for elements above the upper line, the Al sites are so favored that even in Al-rich compounds, M will still occupy Al sites and push some Al atoms into Ni sites. This is the disordering effect of alloying elements. Therefore, the elements above the upper line, such as Nb, Ti, etc., will occupy Al sites, no matter what the Ni to Al ratio is. Similarly, the elements located below the lower line, such as, Pt, Pd, Au, etc., will occupy Ni sites, regardless of the composition.

Between the two lines, the situation is more complicated. The substitution depends on the Ni to Al ratio and the amount of M itself. In a stoichiometric compound, the elements located near the upper line will prefer Al sites, whereas the elements close to the lower line will prefer Ni sites. For the elements midway between the two lines, there is no driving force for them to take one set of sublattice site, they will distribute randomly. In a nonstoichiometric compound, the elements between the two lines will take the sites of the insufficient element first, and then distribute according to which line is nearer.

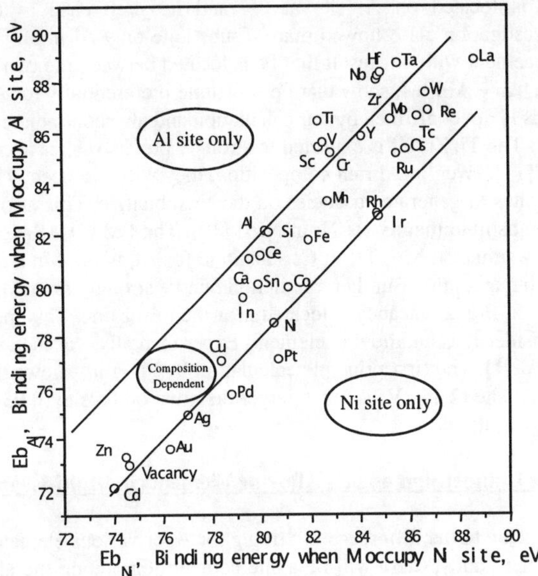

Fig.2 Substitution behavior classification in NiAl

Temperature Effects on the Substitution and Ordering Speed

In Fig. 2, the element most distant from the middle of the two lines has the largest ordering tendency to one sublattice. The driving force is large enough for the ordering process to complete in a short time. Due to the large energy difference of alloying elements on either Ni or Al sites, the ordering of alloying elements is very stable up to very high temperature. For the elements between the two lines, the driving forces are not enough for them to be ordered immediately, their ordering might need longer annealing time. The ordering of these elements will vanish as temperature raises, due to the entropy effect. For the element almost mid-way between the two lines, the driving force is so small that even after annealing, it is still far away from fully ordered. The behavior above can be seen in the ordering of alloying elements in TiNi compounds [7]. This understanding may help when we design a heat treatment process.

Comparison with experiments and other theoretical results

A number of results are found for Fe in NiAl. Anderson *et al.* [4] studied the site occupation of Fe in NiAl using ALCHEMI. Their results show that Fe has no strong site preference and tends to reside on the site of the stoichiometrically deficient host element. The first principle calculation by Fu *et al.* show that in Al rich NiAl, Fe occupy Ni sites exclusively, but in the Ni rich case, the Fe is located on Al site at 0K and no preference as temperature goes up. The EXAFS investigation also showed that Fe substitute only Al in Ni rich NiAl [10]. These findings are in agreement with our result that Fe is located between the two lines on Fig. 2. However, Duncan *et al.*, using APFIM, show that Fe substitute preferentially for Al, independent on Al to Ni ratio [3]. This is not confirmed by our calculation and also contradicted with most other results.

The Ti, Zr, Hf is expected to strongly prefer Al sites according to the analysis by Kitabjian *et al.* [11]. Even in Al rich composition, they will still occupy Al sites and put some Al atoms into Ni sites or generate vacancies on the Ni sublattice. This also agrees with our results. Co is found to substitute mainly for Ni in NiAl [10]. The LMTO-ASA calculation by Medvedeva *et al.* [12] shows that Zr, Mo, Ti, V, Cr substitute for Al, while Mn slightly tends to Al sites. In agreement with our results. But their prediction that Fe strongly favored Ni site is not agree with ours.

In Fig. 2, vacancy is located near the lower line. This means that it will occupy Ni site if we consider it as an alloying element. Experimentally, vacancy is found mainly on Ni site in Al rich NiAl [1]. The first principle calculation by Fu *et al.* shows that the formation energy of vacancy on Al site (2.14 eV) is much larger than that on Ni site (0.93 eV) [13]. These findings agree with our results.

Site Competition among Alloying Elements in Multi-Element Alloy

Due to the difference in driving force for various elements to order on the lattice, in a multi-element alloy, there will be a site competition among the alloying elements. The element most distant above the upper line will substitute Al first. The element most distant below the lower line will substitute Ni first. The substitution behavior of elements nearer or between the two lines will be affected by the distant elements. For example, in a Ni rich compound, Mn mainly occupies Al sites. If some Ti is added, since Ti has a larger tendency to Al sites than Mn, the Ti atoms will take Al sites first. Some Mn atom will be pushed to the Ni sites. If enough Ti is added that the sum of atomic fraction of Al and Ti exceeds 0.5, then all the Mn atoms will be pushed to Ni sites. By this competition, we can control the site occupancy of elements in between the two lines using the elements outside the two lines as we wish. The experimental verification of the site competition is in progress.

The Stepped Ordering

The elements in between the two lines in Fig.2 have smaller ordering tendency compared with the matrix elements. They will disorders at a lower temperature (T_C') than perfect NiAl. In this way, the disordering process will be divided into two steps, corresponding to two critical temperatures, one related to the matrix NiAl, the other to the alloying element. There is some evidence of stepped ordering transformation in Fe_3Al based alloys, but with a higher T_C' [14]. The stepped ordering process might be useful in heat treatment. For instance, we can control the ordering state and antiphase domain size in order to achieve better properties.

CONCLUSION

The substitution behavior of alloying elements in NiAl compounds are predicted based on electronic structure calculations using the DV-Xα method. The lattice relaxation is taken into account when calculating the binding energy. The agreement with experiments is good. The concepts of site competition and stepped ordering are proposed.

ACKNOWLEDGMENTS

The financial support by the Ministry of Science and Technology of China via the Pandeng Project is gratefully acknowledged.

REFERENCES

1. R.D. Noebe, P.R. Bowman and M.V. Nathal, Intern. Mater. Rev. **38**, 193 (1993).
2. R. Darolia, D. Lahrman and R. Field, Scripta Metall. Mater. **26**, 1007 (1992).
3. A.J. Duncan, M.J. Kaufman, C.T. Liu, M.K. Miller, App. Surf. Sci. **76**, 155 (1994).
4. I.M. Anderson, A.J. Duncan and J. Bentley, in *High-Temperature ordered Intermetallic Alloys VI*, edited by J.A. Horton *et al.* (MRS Proc. **364**, Pittsburgh, PA, 1995), p.443.
5. C.L. Fu and J. Zou, Acta Metall. **44**, 1471 (1996).
6. D.E. Ellis, And G.S. Painter, Phys. Rev. B2, 2887 (1970).
7. D.S. Xu, Y. Song, D. Li and Z.Q. Hu, Phil. Mag. A**75**, 1185 (1997).
8. D.S. Xu, Y. Song, D. Li and Z.Q. Hu, Mater. Sci. Eng. A**234**, 230 (1997).
9. B. Delley, D.E. Ellis, A.J. Freeman *et al.*, Phys. Rev. B**27**, 2132 (1983).
10. M. Balasubramanian, D.M. Pease, J.I. Budnick T. Manzur and D.L. Brewe, Phys. Rev. B**51**, 8102 (1995).
11. P.H. Kitabjian and W.D. Nix, Acta Mater. **46**, 701 (1998).
12. N.I. Medvedeva, Y.N. Gornostyrev, D.L. Novikov, O.N Mryasov and A.J. Freeman, Acta Mater. **46**, 3433 (1998).
13. C.L. Fu, Y.Y. Ye and M.H. Yoo, Phys. Rev. B**48**, 6712 (1993).
14. D.G. Morris, Acta Metall. Sinica (Eng. Lett.), **8**, 393 (1995).

Atomistic Studies of Generic Tilt
Grain Boundary Structures

D. N. Pawaskar[1], R. Miller[2], R. Bai[1], A. Schwartzman[1], R. Phillips[1], and
C. L. Briant[1]

[1]Division of Engineering, Brown University, Providence, RI 02912
[2]Department of Mechanical Engineering, University of Saskatchewan, Canada, S7N 5A9

Abstract

We investigate the atomic-scale structures of high-sigma (Σ) tilt grain boundaries in aluminum using lattice statics calculations. In particular, we examine the efficacy of the structural unit model (SUM) in the context of long-period boundaries. Our investigation of both the equilibrium and metastable structures for certain high Σ boundaries may necessitate a revision of the SUM formalism. We also consider further departures from high symmetry boundaries by considering the structural rearrangements induced by steps on boundaries.

Introduction

Interfacial structure is known to strongly influence the thermomechanical properties of polycrystalline materials. The atomic structure of grain boundaries (GBs) has been shown to influence phenomena such as segregation, diffusion, migration, sliding at elevated temperatures and hardening mechanisms like the Hall-Petch effect [1, 2], which in turn significantly alter the macroscopic properties of materials.

While many atomistic studies of GBs have been performed, they have often been confined to a number of special boundaries such as simple symmetric tilt or twist boundaries with a high degree of symmetry and with a low sigma (Σ) number. This has been primarily because of the fact that such boundaries exhibit special properties such as low GB energy [3]. However it has been found that GBs which are not of this special type are also present in real microstructures [4]. Such boundaries include those with mixed tilt and twist character, aperiodic boundaries, those with steps and microfacets, and those in which the grain boundary plane wanders [3]. In general, such boundaries have low periodicity and symmetry. In this paper, steps are made to depart from the paradigm of high symmetry boundaries and to look for regularities in the structures of boundaries, either of very long period, or containing steps.

Conceptual Background

The structural unit model (SUM) [5] is a powerful tool in understanding the structure of either tilt or twist boundaries with arbitrary misorientation angles. The SUM allows for the prediction of the structure of a boundary of any misorientation angle for a given tilt axis, providing that one knows the structure of special low energy boundaries with the same tilt axis. It is found, normally by atomistic simulation, that the structure of these special low energy boundaries is simply a string of short, identical units (labeled as A, B, C, etc.). These special boundaries are often referred to as *delimiting boundaries* since they delimit a range of misorientation angles. Any boundary whose misorientation angle lies in this range will have a structure that is a predictable pattern (linear combination) of the structural units found in the two delimiting boundaries. Tilt boundaries which are periodic along the direction normal to the tilt axis and in the plane of the boundary will be made up of a periodically repeating pattern of these structural units. For the so-called *irrational* boundaries which exhibit periodicity only along the tilt axis, the pattern will be quasiperiodic [6].

Experimental observations [3, 7] reveal that the GB is sometimes nonplanar. A frequent instance is that of a step in the GB. Fig. 1 shows such a $\Sigma 3\langle 110\rangle\{111\}$ boundary [7].

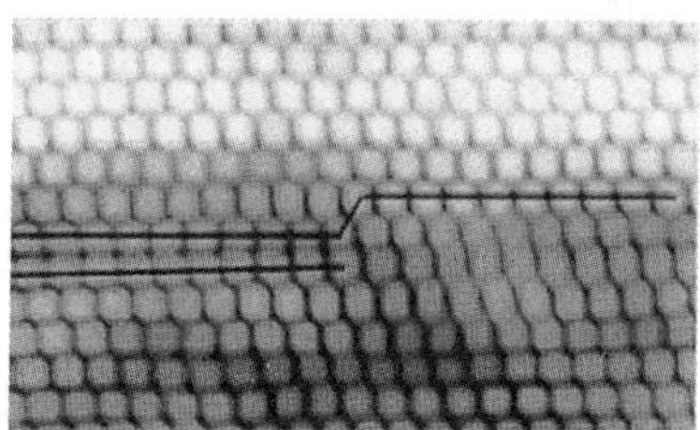

Figure 1: $\Sigma 3\langle 110\rangle$ stepped boundary

It has been conjectured that this step is not an equilibrium structure and is characterized geometrically by being two atomic layers high. Thus it is expected to transform into another structure on relaxation. Our objective is to examine these claims by simulating the stepped structure. An equilibrium step [8], has been simulated in the past along with studies involving its interaction with a dislocation. At large, our strategy is to to take a step towards generic boundaries by considering tilt boundaries either with very long periods, or disturbed by the presence of steps.

Simulation Methodology

The computational cell consists of a bicrystal. As shown in fig. 2, the boundary plane XZ is at the center of the cell. The atomic positions within the cell are generated using the coincident site lattice (CSL) model [9]. The energy of a bicrystal is calculated using the embedded atom method (EAM) formalism [10] with potentials for FCC aluminum developed by Ercolessi and Adams [11]. The energy minimization is carried out using a conjugate gradient algorithm.

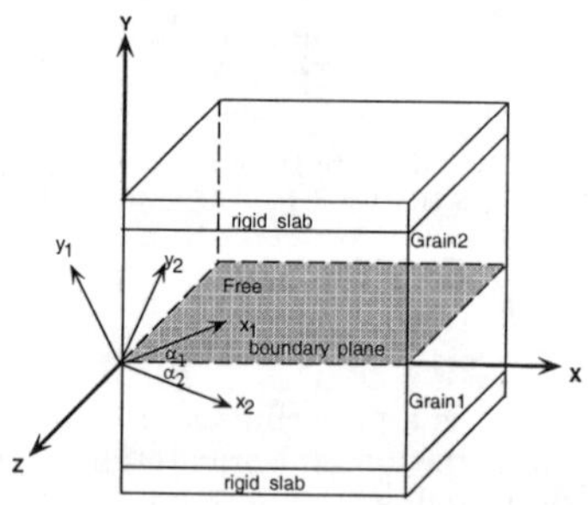

Figure 2: Computational cell

The computational cell is periodic along the X and Z directions to eliminate surface effects. In the Y direction, the cell is divided into a *free* region and into two *rigid slab* regions. The atoms in the free region are allowed to individually move as a part of the relaxation process while those within each fixed region move as one unit. Hence only three translational degrees of freedom are associated with each rigid slab. Because the rigid slabs are at least twice as thick as the range of the atomic potentials used, the atoms in the free region essentially behave as though they were surrounded by two semi-infinite perfect crystals. The usual technique is to use periodicity along all three directions. Our method allows us to model asymmetric boundaries and considerably reduces the number of degrees of freedom involved in the computation. It also ensures volumetric expansion in the direction perpendicular to the boundary plane.

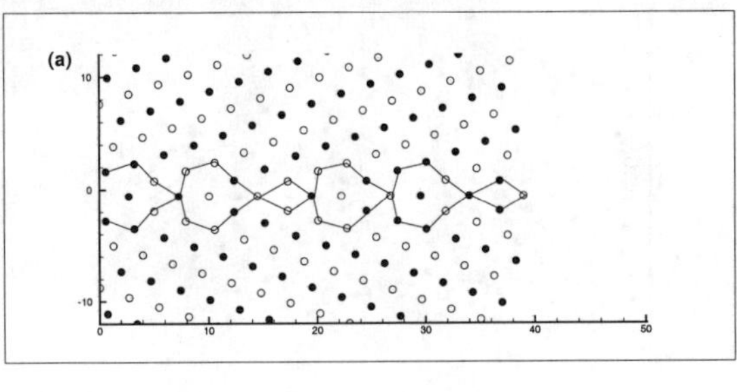

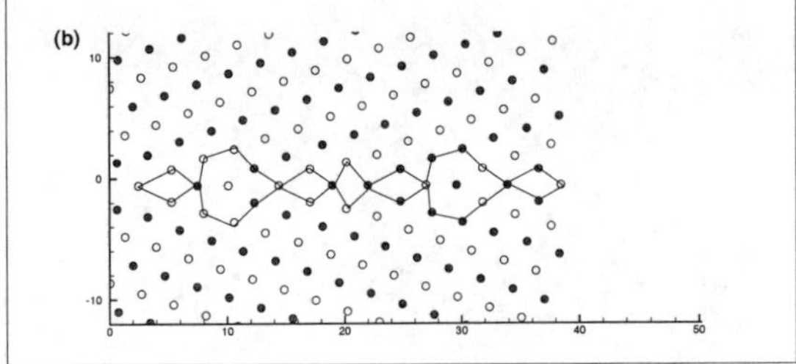

Figure 3: Structure of the $\Sigma 187\langle 110\rangle$ boundary (a) SUM prediction (b) relaxed structure

It is well-known [13] that lattice statics energy minimization of grain boundaries can lead to many different boundary structures depending on the initial configuration of the two grains. This is due to the multiwelled nature of the energy function in configuration space. To find the global minimum, the procedure is to introduce a relative displacement between the two grains along the boundary plane prior to relaxation. The global energy minimum and hence the associated structure is then chosen from the set of local minima. It has been shown [12] that it is possible to represent all possible initial configurations by the infinite set of points in the so-called cell of nonidentical displacements (CNID) which consists of the rectangle bounded by the periodic vectors defining the boundary plane. By taking a reasonably fine grid of initial configurations from within this CNID one can be fairly confident of achieving the global minimum energy structure. We have found that a grid spacing of about 1 Å is sufficient. However, such a fine spacing considerably increases the computational time for high Σ boundaries since the period $p \sim \sqrt{\Sigma}$. One way around this problem is to approximate the CNID of the long period boundary by that of an appropriate short period one and then use the corresponding displacements of the short period boundary as rigid translations for the long period one. Another method [13], is to use the displacements of the long period boundary but by using a computational cell only a few cutoff radii thick and then use those displacements which yield low energies in this tiny unit cell as displacements for the original unit cell.

Results

One way to analyze the structure of an irrational boundary is to approximate that misorientation by a rational approximant with a very high Σ number. As examples of this strategy, we have chosen two such high Σ symmetric tilt boundaries, $\Sigma187(36.15°)\langle110\rangle$ $\{10,10,43\}$ and $\Sigma697(37.5°)\langle110\rangle\{25,25,104\}$. Note that because of the large Σ numbers associated with these boundaries, they have long periods along the boundary plane.

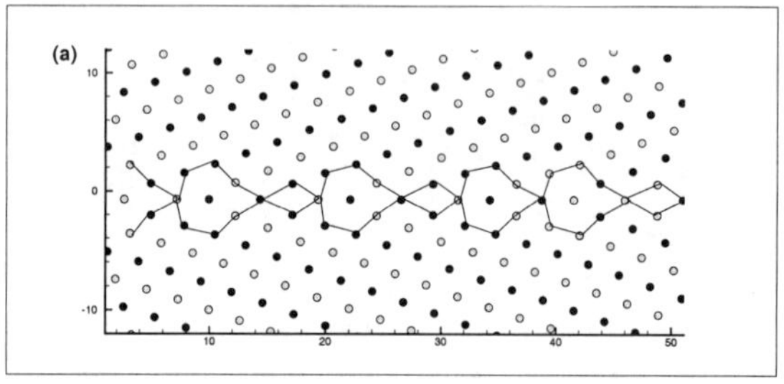

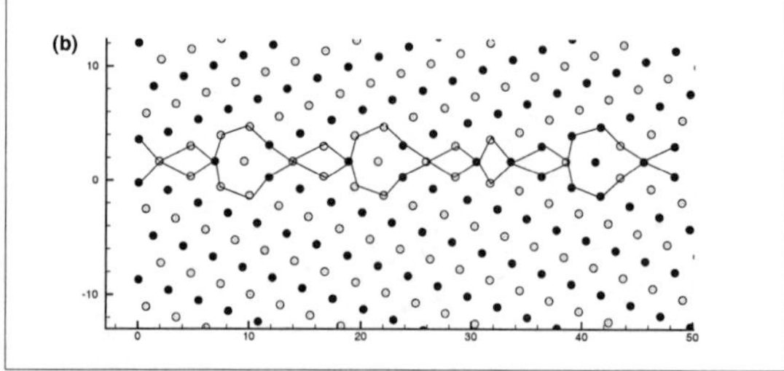

Figure 4: Structure of the $\Sigma697\langle110\rangle$ boundary (a) SUM prediction (b) relaxed structure

According to the SUM [13], these boundaries lie between the the two delimiting misorientations of $26.53°$ and $50.48°$, consisting of B and C units respectively, where $B \equiv \frac{1}{4}\langle552\rangle$ and $C \equiv \frac{1}{4}\langle332\rangle$. So their structures should consist of only B and C units. The sequence of the B and C units can be worked out from the algorithm in [5] and it yields a structure CBB for the $\Sigma187$ boundary and CBCBBCBCBBCB for the $\Sigma697$ boundary. However, we have obtained structures with lower energies for these boundaries-CBCA for the $\Sigma187$ and ACBCBCACBCBCBC for the $\Sigma697$ boundaries, where $A \equiv \frac{1}{4}\langle220\rangle$. Figs. 3 and 4 show the structures of these boundaries. It is important to note that for each of the two boundaries the two conflicting structures are crystallographically equivalent but energetically different. It should be mentioned that the energy difference between the observed structures and those predicted by the SUM is quite small (less than 5%). The potentials used for the simulations may have some bearing on this difference. This may call for a need to revise some of the details, energetics in particular, of the SUM.

In order to study the stepped boundary, we build a bicrystal containing a stepped GB and then relax it. Fig. 5 shows the initial and final configuration of this boundary. It is seen that the plateau of the step relaxes to a structure corresponding to a regular $\Sigma3$ boundary. The dimensions and hence the geometry of the step remains unchanged as a part of this process. It can be seen that the step moves parallel to the boundary plane as a part of the relaxation process. This may suggest a shearing mechanism at work. It is significant to note that even such small structural changes can cause a large change in the energies and can thus significantly affect the microstructural properties of a material.

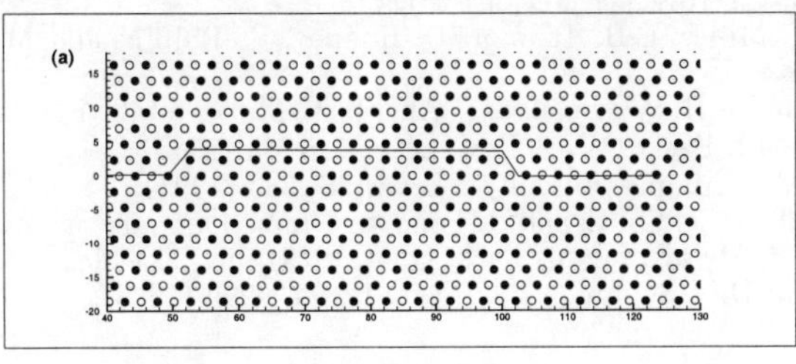

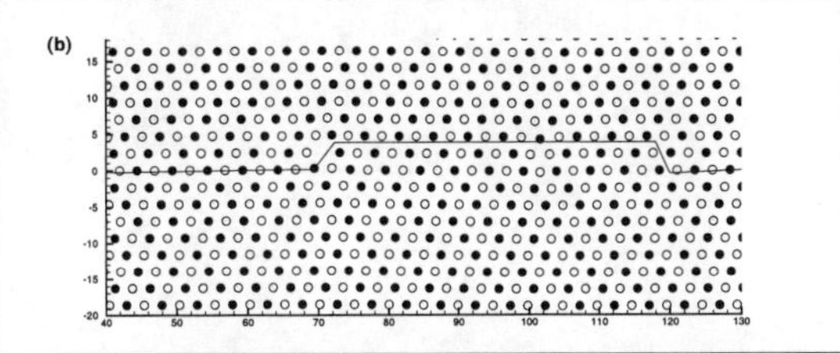

Figure 5: Structure of the $\Sigma 3\langle 110\rangle$ stepped boundary (a) unrelaxed (b) relaxed

Conclusion

In this paper, we have reported on preliminary results aimed at exploring those grain boundary structures which fall outside of the purview of the traditional low Σ boundary paradigm. We have considered a wide range of long period boundaries as well as simpler boundaries disturbed by the presence of steps. It appears that for the potentials used here, long period boundaries in Al fail in some cases to have those structures expected on the basis of the SUM. The simulation on the stepped boundary reveals that the structure at the step decomposes into a lower energy structure coupled with a translation of the step along the boundary plane. Future work will have to exploit finite temperature algorithms such as Monte Carlo to search for the presence of even lower energy structures for such boundaries. This would lay a groundwork to further study the deformation mechanisms of generic boundaries and their interactions with other microstructural defects.

Acknowledgments

We thank M. Daw and S. Foiles for use of their DYNAMO code. This work was supported by the NSF through grants CMS-9414648 and DMR-9632524 and by the DOE through grant DE-FG02-96ER45578. This support is gratefully acknowledged.

References

1. A. P. Sutton and R. W. Balluffi, *Interfaces in Crystalline Materials*, Oxford University Press, Oxford, 1995.
2. D. Wolf and S. Yip, *Materials Interfaces- Atomic-level structure and properties*, 1st ed. (Chapman & Hallm London, 1992).
3. K. Merkle, J. Phys. Chem. Solids, **55**, 991 (1994).
4. A. Garbacz, B. Ralph and K.J. Kurzydlowski, Acta. Metall. Mater., **43**, 1541 (1995).
5. A.P. Sutton and V. Vitek, Phil. Trans. R. Soc. Lond. A, **309**, 1 (1983).

6. A.P. Sutton, Prog. Mat. Sci., **36**, 167 (1992)
7. R. Bai, Ph.D. thesis, Brown University, 1998.
8. V.B. Shenoy, R. Miller, E.B. Tadmor, D. Rodney, R. Phillips and M. Ortiz, J. Mech. Phy. Solids, in press, (1998).
9. M.L. Kronberg and F.H. Wilson, Trans. Am. Inst. Min. Metall. Engrs., **185**, 50 (1949).
10. M.S. Daw and M.I. Baskes, Phy. Rev. Lett., **1**, 1285 (1983).
11. F. Ercolessi and J. Adams, Europhys. Lett., **26**, 583 (1993).
12. V. Vitek, A. Sutton, D. Smith and R. Pond, in *Grain Boundary Structure and Kinetics*, (American Society of Metals, Metals Park, Ohio, 1979), pp. 115-148
13. J.D. Rittner and D.N. Seidman, Phy. Rev. B, **54**, 6999 (1996).

AB INITIO CALCULATIONS OF POINT DEFECTS IN SILICON

JIANJUN XIE, S.P. CHEN
Los Alamos National Laboratory, T-11, MS-B262, Los Alamos, NM 87545

ABSTRACT

We use the *ab initio* plane wave pseudopotential method and the density functional theory (DFT) to study the arsenic(As)-vacancy interactions in silicon. The detailed lattice distortions surrounding the As-vacancy defect and the energetics of As-vacancy reaction around the six-fold ring are investigated. We find that the As displaces its neighboring silicon atoms outward while the vacancy attracts its neighboring atoms inward. The binding energy and the formation energy of an As-vacancy pair are 1.21 eV and 2.37 eV, respectively. Once the vacancy and As binds together, the highest migration barrier for the whole complex is 1.19 eV, which is in good agreement with the experimental measurement of 1.07 eV. The calculated activation energy for the vacancy mediated diffusion of the neutral As in silicon is 3.56 eV. The nature of the binding between As and vacancy is explained from the lattice distortions introduced by the As-vacancy complex.

INTRODUCTION

Arsenic has been regarded as an important n-type dopant in silicon-based integrated circuits. The implanted As atoms dissolve in the silicon lattice almost exclusively on substitutional lattice sites [1]. The diffusion of these dopant atoms is determined by the interactions with the host atoms and the native defects (such as vacancies and interstitials) in silicon.

It is widely accepted [1] that the diffusion of the substitutional dopant in silicon is mainly mediated by vacancies and interstitials. The vacancy mechanism [2] requires that a dopant atom forms a pair with an adjacent vacancy. The vacancy moves away from the dopant atom around a six-fold ring to at least a third-neighbor position. It can then return by a different path, placing itself next to the dopant. The vacancy and the dopant exchange positions and the process can repeat itself in different rings for long-range diffusion of dopant. The interstitial-mediated diffusion [3] mainly occurs through the "kick out" or "coordinated push" mechanism via interaction of dopant atom with an interstitial silicon atom.

In our study, we focus on the As-vacancy interaction in silicon [4]. An early potential-energy diagram of the dopant-vacancy interaction in silicon was proposed by Hu [5] in 1973, which was very useful for analyzing the vacancy mechanism for dopant diffusion. Due to the lack of *ab initio* calculations, this was a schematic picture without detailed data on the potential. Later *ab initio* studies [3,6,7] provided more information such as the formation energy of an As-vacancy pair, the energy barrier for As-vacancy exchange, as well as the binding energy between As and vacancy. However, to our knowledge, there still lacks a detailed potential-energy diagram of As-vacancy interaction in silicon. Particularly, there is little information about the lattice distortion around the As-vacancy complex. Furthermore, the nature of the binding between As and vacancy is not yet very clear, i.e., Coulombic interaction or non-Coulombic interaction. Here we report a thorough investigation of the structure and energetics of the As-vacancy complex in silicon, using the first-principles total energy calculations. Both As and vacancy are in their neutral states. The nature of the at-

Mat. Res. Soc. Symp. Proc. Vol. 538 © 1999 Materials Research Society

tractive interaction between the neutral As and vacancy can be understood from the lattice distortions around the As-vacancy complex.

COMPUTATIONAL DETAILS

The calculations were performed using the $FHI96MD$[8] plane wave package. The ground state total energy was obtained within density functional theory (DFT) in the local-density approximation (LDA) to the exchange-correlation energy functional [9,10]. Nonlocal and norm-conserving pseudopotentials constructed by the scheme of Hamann [11] were used to describe the valence electron interactions with the atomic core. Periodic supercell with 64 atomic sites was employed to model the considered system as in other *ab initio* calculations [6,7,12,13]. The integration of **k** points over Brillouin zone (BZ) was performed by summing a $2 \times 2 \times 2$ mesh in the first BZ. The kinetic energy cut-off of the plane wave basis is taken as 12 Ry. The iterative damped Joannopoulos algorithm [14] was used to obtain the self-consistent one-electron Kohn-Sham equation. The Hellmann-Feynman theorem was used to evaluate the forces on all atoms which were allowed to move. Atomic relaxations were considered to the point where the magnitude of the calculated Hellmann-Feynman forces on each atom was smaller than 0.4 mRy/Bohr. Some results are checked by the all electron linear augmented plane wave (LAPW) method. The energy difference of the checked cases is less than 0.1 eV.

RESULTS AND DISCUSSIONS

We first calculated the static equilibrium lattice constant of bulk silicon. The obtained lattice constant was $a_0 = 5.40$ Å, which is 0.5% smaller than the experimental value of 5.43 Å due to the LDA approximation. This theoretical lattice constant is used in the following calculations. We consider a vacancy moving near the As dopant. One Si atom is replaced by an As impurity and another atomic site is left empty representing a vacancy. The potential energy of the As-vacancy pair is calculated as a function of the As-vacancy separation. The vacancy is placed on different sites in the six-fold ring as shown in Figure 1(a). For each As-vacancy configuration, the first-nearest-neighbor and the second-nearest-neighbor atoms around the defects are fully relaxed. We also calculated the diffusion of the vacancy along the six-fold ring. Note that the diffusion of vacancy is equivalent to the diffusion of silicon atoms in the opposite direction. The diffusion barriers are obtained by moving the hopping atom along the diffusion path and relaxing its position in a plane perpendicular to the diffusion direction. Other atoms up to the second-nearest-neighbors of the defects as well as the hopping atom are relaxed simultaneously. The obtained potential-energy diagram of As-vacancy interaction is shown in Figure 1(b). The vacancy coordinates labeled as 1, 2, 3 etc. refer to the atomic sites shown in Fig. 1(a). However, after the lattice relaxation, they are not at the ideal lattice site. Details of the relaxation will be discussed below. It can be seen in Fig. 1 (b) that the energy barrier for the vacancy moving from site 3 (third-nearest-neighbor of As) to site 2 (second-nearest-neighbor of As) is 0.23 eV. Once the vacancy is located at site 2, it will move quickly to site 1 (first-nearest-neighbor of As) which experiences a very small energy barrier of only 0.08 eV. The position of this small energy barrier is very close to site 2. The barrier between sites 1 and As is 0.55 eV high, which describes the As $\leftrightarrow$ vacancy exchange. After exchanging position with As, the vacancy may move away along a different ring which connected with As (see Fig. 1(a)). The vacancy first overcomes an energy barrier of 0.92 eV arriving at a second-nearest-neighbor site of As. Then it moves to the third-nearest-neighbor site by passing over another barrier with height 0.35 eV. Once the vacancy overcomes the third-nearest-neighbor site, it can return to the As via a different path. By repeating such a process, diffusion of the As-vacancy complex occurs. The migration barrier $E_m(\text{AsV})$ for the whole As-vacancy complex is 1.19 eV as shown in Fig. 1(b), which is close to the experimental result of 1.07 eV [5]. We also calculated the first-neighbor binding energy of an As-vacancy pair (the vacancy is located next to As) relative to the As-vacancy separation at 9.35 Å (vacancy is located at the 12th-nearest-neighbor of As), where the As-vacancy interaction is negligible. The obtained result is $E_b(\text{AsV}) = 1.21$ eV, which is in a good agreement with other *ab initio* calculations of 1.20 eV [3] and the experimental measurements of 1.23 eV [15]. It is found that the attractive potential of As-vacancy is beyond the third-nearest-neighbors since the binding energy of the As-vacancy pair is still 0.25 eV (1.21 minus 0.96) when the vacancy is located at the third-nearest-neighbor site of As. For comparison, we calculated the formation energy and the migration barrier of a vacancy in pure silicon. The obtained formation energy is 3.57 eV which is in good agreement with the experimental measurement of 3.6 eV and other *ab initio* calculations [6,16]. The calculated migration barrier is 0.31 eV. Comparing with the migration barriers of the vacancy moving near an As, it is found that the As lowers (increases) the migration barrier of the vacancy hopping toward (away from) it.

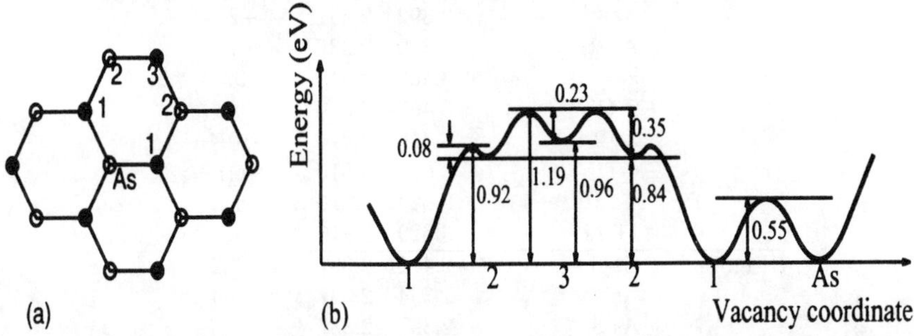

Figure 1: (a) Different As-vacancy separations along the six-fold ring. The atomic sites are projected onto the (111) surface. The filled and the empty circles represent the atomic sites from different layers. 1, 2 and 3 denote the first-nearest-neighbor, second-nearest-neighbor and third-nearest-neighbor sites of As, respectively. (b) Potential-energy for As-vacancy complex as a function of vacancy coordinate.

We calculated the activation energy Q for As diffusion via the vacancy mechanism by summing the formation and migration energies for the As-vacancy complex. The formation energy of an As-vacancy pair is calculated as [3]

$$E_f(\text{AsV}) = E(\text{AsV}) - E_s(\text{As}) + \frac{1}{64}E(\text{Si}_{64}), \tag{1}$$

where $E(\text{AsV})$ is the calculated total energy of the 64-atomic sites supercell containing an As-vacancy pair, $E_s(\text{As})$ is the total energy of the supercell containing a substitutional As impurity, and $E(\text{Si}_{64})$ is the total energy of the 64-atomic sites supercell of pure silicon.

The calculated result for $E_f(\text{AsV})$ is 2.37 eV. The corresponding activation energy $Q = E_f(\text{AsV}) + E_m(\text{AsV}) = 3.56$ eV. This value is in good agreement with the experimental result of 3.44 eV [17] for the diffusion of neutral As in silicon and is smaller than that of 4.05 eV [17] for the charged As. Masters and Fairfield [18] observed an activation energy of 4.2 eV for As diffusion in silicon, but the authors pointed out at the same time that the effective activation barrier would be 3.4 eV in the high doping level and the lower temperature range.

Table 1: Relaxation properties of the atoms neighboring the As-vacancy complex. Atoms are labeled as illustrated in Fig. 2. All distances are in Å. $\Delta d/d_0 = (d - d_0)/d$, where d_0 denotes the distance in the perfect silicon crystal. The symbols + and - denote expansion and shrinkage, respectively.

system	pair	d	d_0	$\Delta d/d_0$ (%)
Fig. 2 (a)	As-i (i=6,7,8)	2.415	2.337	+3.4
	As-i (i=3,9,10)	3.498	3.816	-8.3
	6-8, 7-8, 6-7	3.749	3.816	-1.8
	3-9,9-10,10-3	3.631	3.816	-4.9
Fig. 2 (b)	As-8	2.392	2.337	2.4
	As-7	2.406	2.337	+2.9
	As-6	2.392	2.337	+2.4
	As-2	2.440	2.337	+4.4
	2-4	3.463	3.816	-9.3
	2-11	3.463	3.816	-9.3
	2-12	3.418	3.816	-10.4
	4-11	3.620	3.816	-5.1
	4-12	3.623	3.816	-5.1
	11-12	3.620	3.816	-5.1
Fig. 2(c)	As-6	2.411	2.337	+3.2
	As-7	2.394	2.337	+2.4
	As-8	2.397	2.337	+2.6
	As-2	2.411	2.337	+3.2
	13-14	3.615	3.816	-5.3
	13-3	3.571	3.816	-6.4
	13-5	3.571	3.816	-6.4
	14-3	3.512	3.816	-8.0
	14-5	3.512	3.816	-8.0
	3-5	3.514	3.816	-7.9

Table 1 lists the detailed relaxation properties around the As-vacancy complex in the process of a vacancy moving along the six-fold ring. The corresponding atom sites around the As-vacancy complex are shown in Figure 2. The arrows indicate the relaxation direction and the length of the arrow is proportional to the displacement of the corresponding atom. It can be seen from Table 1 and Fig. 2 that when a vacancy is located next to an As (Fig. 2(a)), As atom moves significantly towards the vacancy site. The distance between As and its nearest-neighbor silicon atoms (6,7,8 in Fig. 2 (a)) is 3.4% larger than the bond length of pure silicon reflecting the repulsion between As and silicon. On the other hand, the vacancy attracts the surrounding atoms (including As) inward. This geometry possess a C_{3v} symmetry. When the vacancy moves away from As (Fig. 2(b)), atoms neighboring the vacancy relax inward and those neighboring the As relax outward. There is then no symmetry maintained. The displacement of atom 2 is most significant due to the repulsion of As and the attraction of vacancy. When the vacancy is located at the third-nearest-neighbor site of As (Fig. 2(c)), the lattice relaxation around vacancy is still affected by the As. Two silicon atoms, 3 and 5, which are the second-nearest-neighbors of As, move more significantly than atoms 13 and 14.

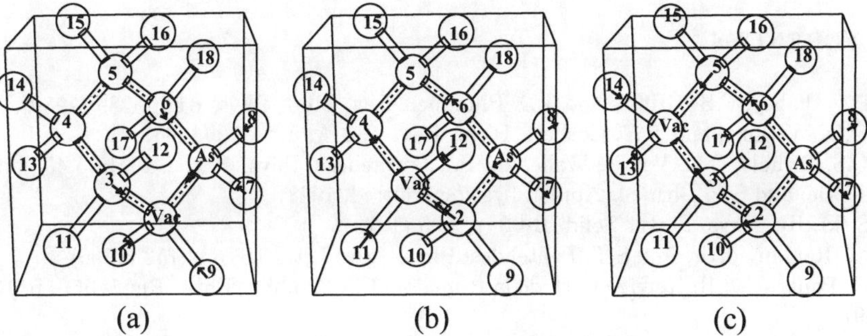

(a)   (b)   (c)

Figure 2: Lattice relaxation of an As-vacancy complex for different As-vacancy separations. The circles denote the atomic sites corresponding to the perfect silicon. The arrows point in the relaxation direction and the length of the arrow is proportional to the displacement of the atom in 3-dimensions. The dashed line shows the six-fold ring: (a) vacancy is next to As, (b) vacancy is the second-nearest-neighbor of As, (c) vacancy is the third-nearest-neighbor of As.

The outward relaxation surrounding As and the inward relaxation surrounding the vacancy is very helpful for understanding the binding properties of the As-vacancy complex. When an As impurity is implanted into silicon, it pushes the surrounding atoms outward and increases the elastic energy of the lattice. The existence of a neighboring vacancy provides the space for As and releases the elastic energy by inward relaxation. Since the As dopant and the vacancy considered here are in their neutral states, the Coulomb interaction between them should be negligible. Therefore, the binding energy of As-vacancy arises mainly from the release of the elastic energy.

SUMMARY

In summary, we have provided a detailed potential-energy diagram for As-vacancy interactions in silicon based on the *ab initio* total energy calculations. Within the theoretical error bar of about 0.1 eV due to the finite **k** points and the kinetic energy cut-off, we find that the formation and binding energies of a As-vacancy pair are 2.37 eV and 1.21 eV, respectively. The energy barriers for a vacancy moving away from As along the six-fold ring

393

are 0.92 eV and 0.35 eV. While for a vacancy moving toward As, the barriers are 0.23 eV and 0.08 eV. The exchange barrier of As-vacancy pair is 0.55 eV. The migration barrier for the whole As-vacancy complex is 1.19 eV. The calculated activation energy for the vacancy mediated diffusion of the neutral As impurities in silicon is 3.56 eV. The origin of the binding for the As-vacancy pair can be understood as the release of the elastic energy accumulated around As and vacancy.

ACKNOWLEDGMENTS

We would like to thank Martin Giles, Scott Dunham, Sheldon Aronwitz, Richard Fair and Wolfgang Windl for helpful discussions. This work was supported by the Department of Energy (DOE) CRADA between LANL and the SRC under contract W-7405-ENG-36.

REFERENCES

1. P.M. Fahey, P.B. Griffin, and J.D. Plummer, Rev. Mod. Phys. **61**, p. 289 (1989).
2. G.D. Watkins, and J.W. Corbett, Phys. Rev. **134** A, p. 1,359 (1964).
3. C.S. Nichols, C.G. Van de Walle, ans S.T. Pantelides, Phys. Rev. B**40**, p. 5,485 (1989).
4. J. Xie, and S.P. Chen, J. Appl. Phys. (submitted, 1998).
5. S. M. Hu, Phys. Status Solidi B **60**, p. 595 (1973).
6. M. Ramamoorthy and S.T. Pantelides, Phys. Rev. Lett. **76**, p. 4,753 (1996).
7. O. Pankratov, H. Huang, T. D. de la Rubia, and C. Mailhiot, Phys. Rev. B **56**, p. 13,172 (1997).
8. M. Bockstedte, A. Kley, J. Neugebauer and M. Scheffler, Comput. Phys. Commun. **107**, p. 187 (1997).
9. D.M. Ceperley and B.J. Alder, Phys. Rev. Lett. **45**, p. 567 (1980).
10. J.P. Perdew and A. Zunger, Phys. Rev. B **23**, p. 5,048 (1981).
11. D.R. Hamann, Phys. Rev. B**40**, p. 2,980 (1989).
12. O. Sugino and A. Oshiyama, Phys. Rev. B **46**, p. 12,335 (1992)
13. A. Antonelli, E. Kaxiras, and D.J. Chadi, Phys. Rev. Lett. **81**, p. 2,088 (1998).
14. M.C. Payne *et al.*, Phys. Rev. Lett. **56**, p. 2,656 (1986).
15. M. Hirata, M. Hirata, and H. Saito, J. Phys. Soc. Jpn. **27**, p. 405 (1969).
16. J.L. Mercer, J.S. Nelson, A.F. Wright and E.B. Stechel, Modelling Simul. Mater. Eng. **6**, p. 1 (1998).
17. R.B. Fair, in *Impurity Doping Process in Silicon*, edited by F.F.Y. Wang (North-Holland Publ. Comp., 1981), p. 315-442.
18. B.J. Masters and J.M. Fairfield, J. Appl. Phys. **40**, p. 2,390 (1969).

THE INFLUENCE OF HEAT EXCITATIONS, VACANCIES AND IMPURITIES ON THE ENERGY ELECTRONIC BAND-STRUCTURE OF METALLIC LITHIUM

V. A. POPOV
Altai State Technical University, 656099 Barnaul, Russia

ABSTRACT

The Korringa-Kohn-Rostoker method with Green's function averaged over the atomic configurations in a complex Ising lattice and a muffin-tin potential was used to calculate the electronic-band structure in lithium containing vacancies and s, p, and d impurities. It is shown that substantial changes in the profile of the Fermi surface do not lead to necking, as was postulated previously, but cause splitting of the electronic states at the face of the Brillouin zone. This is attributed to the reduced symmetry of the crystal lattice with impurity excitation of the electronic-subsystem.

INTRODUCTION

Lithium is probably the most popular material for research into the physics of metals. The abundance of experimental and theoretical data on lithium offers extensive scope for using this material to check out newly developed research techniques. At the same time, intensive testing of new methods using lithium has occasionally resulted in the discovery of unexpected properties of this material [1-5]. Some of these, especially the properties of the Fermi surface [3-7], have formed the subject of active discussion for decades.

In the author's opinion, the problem of the shape of the Fermi surface of lithium, formulated back in 1970 in Ref. 5, has still not been settled despite the apparently unambiguous conclusion reached in the fundamental study reported in Ref. 7, that this surface is close to spherical with a maximum deviation of 4.6±1.0%. This is confirmed by all known theoretical studies aimed at calculating the shape of the Fermi surface. Thus, the present paper attempts to identify, at least in principle, the factors which could cause substantial deviations from sphericity and then assess them.

RESULTS AND DISCUSSION

The first assumption made was that die shape of the Fermi surface in lithium may undergo considerable changes as a result of thermal excitation. This assumption was based on the fact that the electron-phonon interaction in this material is fairly strong and the sample becomes intensely hot during an experiment [4]. However, a thorough analysis of the temperature dependence of the band structure of lithium [8] revealed that thermal action cannot be responsible for such substantial changes in the shape of the Fermi surface which lead to necking or at least contact with the face of the Brillouin zone. Some differences do exist, and we discuss them below.

We discovered that the outer-shell electrons are redistributed according to symmetry types: $N(s) = 0.32$ and $N(p) = 0.68$ at the standard deviation of the atoms $\langle u^2 \rangle = 0$; and $N(s) = 0.15$ and $N(p) = 0.85$ at $T = 0$ and $\langle u^2 \rangle = 0.16(a.u.)^2$. This transition of electrons from s-symmetry states to p-symmetry states can be interpreted as the strengthening of directional bonds. This tendency is observed with the increases in temperature, but the strengthening is insignificant. Fig. 1 depicts the density of outer-shell states, with the partial densities of states designated by the

Mat. Res. Soc. Symp. Proc. Vol. 538 © 1999 Materials Research Society

letters s, p, and d. Thermal excitation has practically no effect on the shape of these curves. In particular, at the Fermi energy we have $D(s)=0.69$, $D(p)=5.67$, and $D(d)=0.00$ at $\langle u^2 \rangle = 0$; $D(s)=0.58$, $D(p)=5.69$, $D(d)=0.02$ at $T=0$ and $\langle u^2 \rangle = 0.16(a.u.)^2$; and $D(s)=0.54$, $D(p)=5.40$, and $D(d)=0.02$ at $T=450$K and $\langle u^2 \rangle = 1.00(a.u.)^2$. These values show that the variations in the densities of states are small and are different for each symmetry type. The width of the band occupied by electrons increases from 0.267 Ry to 0.274 Ry with temperature. On the whole, though, our results of calculations of the energy band structure of the electrons prove to be satisfactory when compared with previous calculations cited in the literature and with the experimental data [9].

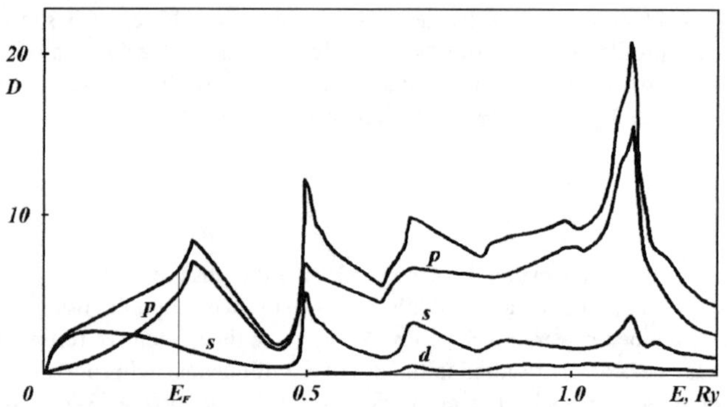

FIG. 1. The total density D of the outer-shell states and its s-, p-, and d-components for lithium electrons measured in number of states per atom per Rydberg unit.

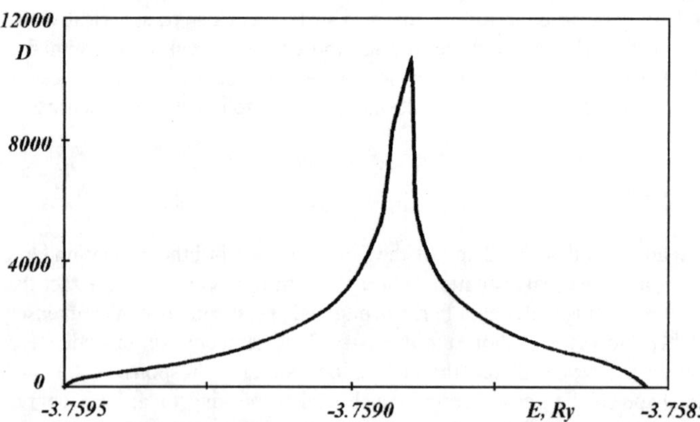

FIG. 2. The total density D of the core states for lithium electrons measured in number of states per atom per Rydherg unit.

The temperature dependence of the core states is quite different, however. The density of the core states at $\langle u^2 \rangle = 0$ are depicted in Fig.2. All these states have the $1s$-symmetry, which does not change under thermal excitation. But the width of the band increases from 0.0010 Ry to 0.0013 Ry (at $T=0$) and then to 0.0063 Ry (at $T=450$ K). This phenomenon (broadening of the band of the core states caused by an increase in temperature) corresponds to the current ideas and leads to an experimentally observed broadening of spectral lines and weakening of the intensity of the peaks in x-ray spectra.

The calculated position of the Fermi level, 4.036 Ry, in relation to the peak in the density of core states fully agrees with the experimentally discovered value 4.03 Ry of the position of the edge of x-ray spectra [10]. The temperature dependence of the K-edge of the x-ray emission spectrum in lithium is depicted in Fig. 3, where the solid curve represents the results of calculations. These results coincide (within experimental errors) with the experimental data [11], depicted by dots. Note that in Fig. 3 the position of the K-edge of the x-ray spectrum is measured from the value of ΔE with $\langle u^2 \rangle = 0$.

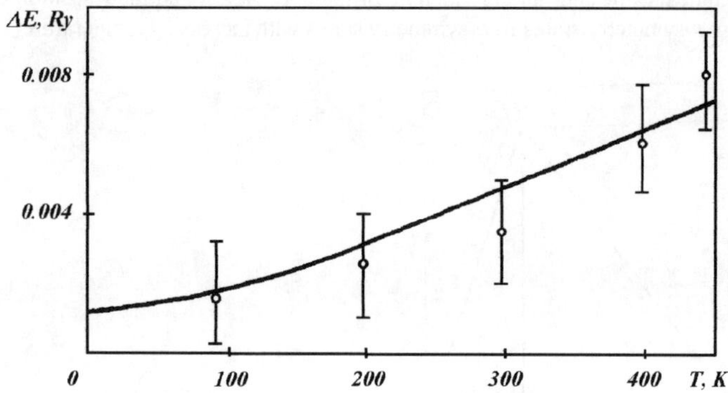

FIG. 3. The temperature dependence of the K-edge of x-ray emission spectrum in lithium. The solid curve represents the results of calculations and the dots correspond to the experimental values.

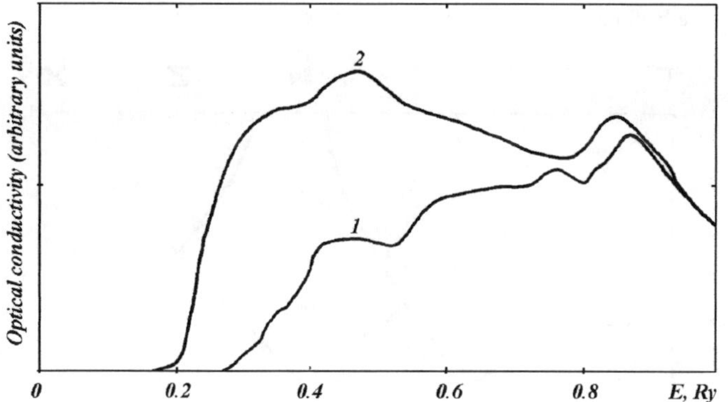

FIG. 4. The optical conductivity σ of lithium electrons obtained in the models of direct (curve 1) and indirect (curve 2) interband transitions.

The calculations of optical conductivity were done with the model of direct interband transitions. The results are depicted in Fig. 4 (curve 1). The resulting value of 0.27 Ry for the lower edge of the direct transitions is somewhat lower on the energy scale than the values calculated by other researchers [9]. The results in the literature suggest that the beginning of interband transitions is located even lower on the energy scale than the present result. The discrepancy between the calculated results and the experimental data is believed to be caused by indirect transitions. The results of calculations of optical conductivity done with the model of indirect interband transitions is also depicted in Fig. 4 (curve 2). The resulting value of the gap in indirect interband transitions is in good agreement with the experimental data: ~0.18 Ry at 125 K. Thus, inelastic electron-phonon scattering has proved to play an important role in the formation of the optical conductivity spectrum of lithium under thermal excitation.

On the whole, atomic vibrations in the lattice (averaged over the displacement ensemble) lead to small changes in the energy band structure of the electron in lithium. The following changes are the most significant: the shift and broadening of the core bands, and the transition of outer-shell electrons from s-symmetry states to p-symmetry states with increasing temperature.

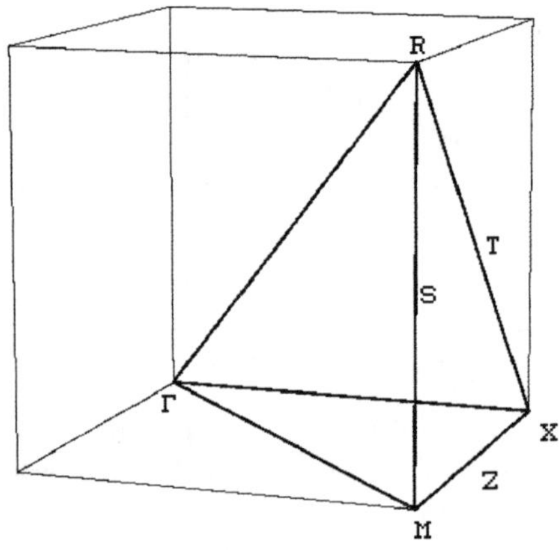

FIG.5. Irreducible part of the Brillouin zone of a simple cubic lattice.

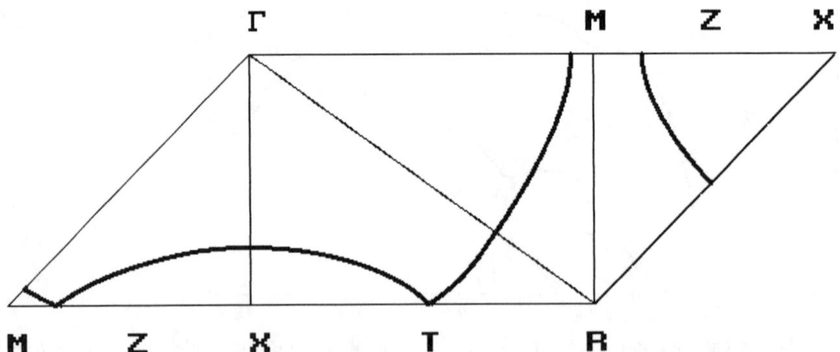

FIG. 6. Cross section of the Fermi surface with faces of the irreducible part of the Brillouin zone in high-purity lithium.

The next step was to study the shape of the Fermi surface as a function of the lithium vacancies and impurities. For this purpose the Korringa-Kohn-Rostoker method [12-13] was used with the Green's function [14] averaged over the atomic configurations in a complex Ising lattice. The muffintin potential was constructed as in Ref.15. The constant a = 6.597 a.u. of the body-centered cubic lattice [3] was the same in all the calculations. The cross sections of the Fermi surface on the faces of the irreducible part of the Brillouin zone (Fig. 5) for high-purity metallic lithium are shown in Fig. 6. In this case, the Fermi surface consists of near-spherical segments. The deviations from spherical in our calculations are almost 1.5 times bigger than those in other theoretical studies, although they do not exceed 6%. The deviations remain almost unchanged after small quantities of vacancies and impurities are introduced into the lithium. The shape of the Fermi surface is only severely distorted above certain critical values of the concentrations (Fig. 7) caused by splitting of states reaching the XMR face of the Brillouin zone.

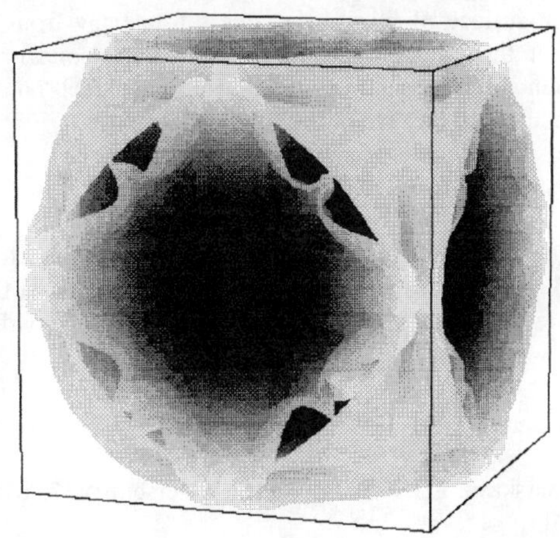

FIG. 7. Fermi surface of lithium containing 0.1 at. % sodium impurities

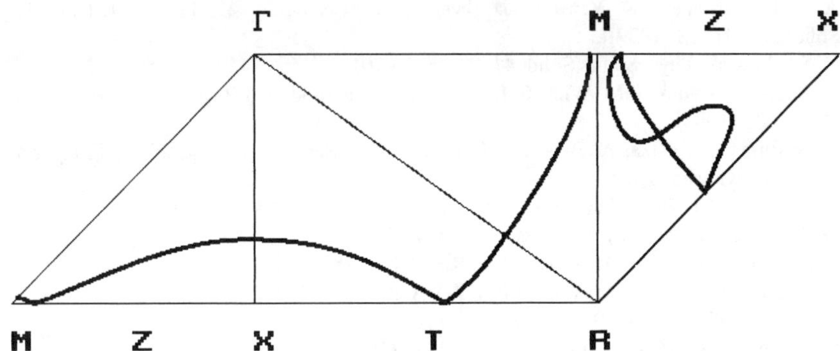

FIG. 8. Cross sections of the Fermi surface of lithium containing 0.1 at. % sodium impurities.

The cross sections of the Fermi surface,in Fig. 7 at the faces of the irreducible part of the Brillouin zone are shown in Fig. 8. In this case, the shape of the Fermi surface is almost independent of whether vacancies or s, p, and d impurities are introduced into the material. It changes sharply with increasing concentration, above certain critical values. For vacancies this critical value is 0.1%. However, under normal conditions lithium contains significantly fewer vacancies. Thus, there is no reason to assume that vacancies, and vacancies alone, may influence the shape of the Fermi surface in lithium.

When 0.1% sodium or potassium impurities are introduced into lithium, the states undergo splitting at the surface, as shown in Fig. 7. This splitting may be achieved by introducing only 0.04 at. % iron. However, these concentrations of such a commonly used metal as iron are encountered extremely rarely in lithium samples intended for research purposes. Sodium and potassium are the most commonly occurring impurities with concentrations frequently exceeding 0.1%.

Unfortunately, the quantitative impurity composition is not given in those experimental studies which indicate that the Fermi surface of lithium deviates substantially from spherical. Attention is merely drawn to the fact that samples (including single-crystal ones) containing below-critical concentrations of sodium and potassium impurities, not only exhibit no necking at the Fermi surface but also reveal no contact of the face of the Brillouin zone.

CONCLUSION

To conclude, a substantial change in the shape of the Fermi surface does not lead to necking, as has been postulated in various studies, but causes splitting of the electronic states at the face of the Brillouin zone. This is attributed to the reduced symmetry of the crystal lattice with impurity excitation of the electron subsystem.

REFERENCES

1. O. V. Stal'gorova and E. L. Gromitskaya, Fiz. Tverd. Tela (St. Petersburg) **37**, 1671 (1995) [Phys. Solid State **37**, 908 (1995)].
2. E. L. Gromitskaya and O. V. Stal'gorova, Zh. Eksp. Teor. Fiz. **106**, 1453 (1994) [JETP **79**, 785 (1994)].
3. J. Gallaway, X. Zou, and D. Bagayoko, Phys. Rev. B **27**, 631 (1983).
4. A. A. Lysykh and I. K. Yanson, Fiz. Tverd. Tela (Leningrad) **21**, 117(1979) [Sov. Phys. Solid State **21**, 68 (1979)].
5. H. Stachowiak, Phys. Status Solidi **41**, 2, 599 (1970).
6. R. Ahuja. S. Auluck, J. M. Wills, 0. Eriksson, P. Sodderliod, and B. Johansson, Phys. Rev. B **50**, 18 003 (1994).
7. V. Sakurai, Y. Tanaka, A. Bansil, S. Kaprzyk, A. T. Slewart, Y. Nagashima, T. Hyodo, S. Nanao, H. Kawata, and N. Shiotani, Phys. Rev. Lett. **74**, 2252 (1995).
8. V. A. Popov, Zh. Eksp. Teor. Fiz. **110**, p. 1474 (1996) [JETP **83**, 815(1996)].
9. W. Ching and J. Callaway. Phys. Rev. **B9**. 5115 (1974).
10. T. A. Callcott and E. T. Arakawa. Phys. Rev. **B16**. 5185 (1977).
11. T. A. Callcott and E. T. Arakawa. Phys. Rev. Lett. **38**, 442 (1977).
12. J. Korringa, Physica **13**, 392 (1947).
13. W. Kohn and N. Rostoker, Phys. Rev. **94**, 1111 (1954).
14. A. I. Nazhalov, V. F. Nyavro, N. I. Fedyainova, V. E. Egorushkin, and V. P. Fadin, Izv. Vyssh. Uchebn. Zaved. Fiz. **6**, 114 (1978).
15. I. Nazhalov, V, F. Nyavro, N. I. Fedyainova, V. E. Egorushkin, and V. P. Fadin, Izv. Vyssh. Uchebn. Zaved. Fiz. **7**. 12 (1978).

MICROSCOPIC DESCRIPTION OF PLASTICITY IN COMPUTER GENERATED METALLIC NANOPHASE SAMPLES

M. SPACZÉR*, H. VAN SWYGENHOVEN*, A. CARO**
* Paul Scherrer Institute, CH-5232, Villigen PSI, Switzerland, helena.vs@psi.ch
** Centro Atómico, 8400 Bariloche, Argentina, caro@cab.cnea.edu.ar

ABSTRACT

We report simulations on the plastic behaviour of nanocrystalline Ni and Cu with grain sizes in the range of 3-12 nm. We observe a change in deformation mechanism in both materials: at the smallest grain sizes all deformation is accommodated in the grain boundaries, while at higher grain sizes we observe intragrain deformation. Analysis of the atomic configurations shows that stacking faults are produced by the passage of partial dislocations generated and absorbed in opposite grain boundaries. In Cu, we observe the stacking faults at smaller grain sizes than in Ni (8 and 12nm, respectively), which is attributed to the lower stacking fault energy of copper. Dislocations appear on slip systems that are not necessarily those favoured by the Schmid factor. Atomic displacement analysis shows deformation starts at triple points, with grain boundary sliding followed by the creation of intragrain partial dislocations.

INTRODUCTION

The relation between yield stress and grain size has been the subject of intensive research in recent years due to the complex behaviour observed in nanophase materials. Overviews of experimental data are given in [1]. Most of the results confirm the validity of the classical Hall-Petch relation down to mean grain sizes of a few tens of nanometers [2]. At grain sizes below 20nm the results are controversial: whereas some results indicate a yield stress independent on the grain size, or even a reverse Hall-Petch relation, others confirm an increasing yield stress with decreasing grain size [3]. The controversy can, in some cases, be ascribed to (i) sample imperfections which have particular influence on the measured methods, and (ii) uncertainty in the determination of the grain size distribution using different methods.

The reasons to expect a departure of the Hall-Petch relation can be explained from dislocation theory. However, when the grain size becomes comparable to grain boundary width, like in nanophases, all approaches become questionable and the possibility of a new regime of deformation, associated to the highly disordered interface regions, is open. For a review of the different approaches see [4]. Other deformation mechanisms that has been invoked is sliding at the grain boundaries. This process has several microscopic contributions and modelling, and is at the basis of the explanation of superplastic behaviour.

Computer study of plastic deformation is a powerful tool to provide evidences of possible mechanisms of plastic behaviour. Simulations are usually performed in two different frames: tensile tests (constant strain rate) and creep (constant stress). Unfortunately they use so large stresses and strain rates that departures of linear behaviour are to be expected. Recent computer simulation results [5-10] point toward the conclusion that around 10 nm there is a limit for dislocations existence inside nanograins and therefore a change in regime is to be expected around these sizes.

In previous works [5,6] we reported simulation results of creep tests under a variety of conditions of stress, temperature, grain size, and texture, and provided a quantitative explanation of the results in terms of a non-linear viscosity originated by sliding at the interfaces. However,

Mat. Res. Soc. Symp. Proc. Vol. 538 © 1999 Materials Research Society

as the grain size increases, a departure from this regime was observed and a new mechanisms came into play, namely intragrain dislocation activity.

In the present work we present simulations designed to address some of the following questions: Is there dislocation activity? Where are dislocations created? Which are the sources? How are the stresses involved? By an exhaustive analysis of Cu and Ni samples, we determine the conditions under which partial dislocations are created. They appear on slip systems that are not necessarily those favoured by the Schmid factor. Atomic displacement analysis shows that deformation starts at triple points, with grain boundary sliding followed by the creation of intragrain partial dislocations.

THE SAMPLES

Nanocrystalline Cu and Ni samples with at least 15 or more grains and with mean grain sizes of 3.4, 5.2, 8.0, 10.0 and 12.0 nm were created using a stochastic procedure: The simulation cell was filled with grains starting from seeds with random location and random crystallographic orientations, and filling the space according to the Voronoi construction [11], resulting in samples with mainly high angle grain boundaries. The number of atoms in the samples ranges from 10^5 to $1.2 \cdot 10^6$. The samples were relaxed to a free enthalpy minimum at 300 K, using a parallel molecular dynamics code with a tight binding potential [12] in the Parrinello-Rahman approach, with periodic boundary conditions and fixed orthorhombic angles. Inadequacy of the Parrinello-Rahman approach to treat inhomogeneous systems is not critical in these simulations, as checked by comparing results obtained using the algorithm suggested by Zhu et al. [13]. Although deformation rates are high ($\sim 10^7$/sec), the small sample size ($\sim 10^2$ nm) implies that relative velocity between atoms or planes induced by the strain are still several orders of magnitude smaller than the sound velocity; creep is therefore stationary.

We have chosen Cu and Ni because both are fcc metals but with quite a substantial difference in stacking fault energy. The tight binding potential predicts stacking fault energies for both materials (Ni: 320 mJ/m^2, Cu: 20 mJ/m^2) within the experimental values reported [14]. After relaxation, the final density of the samples ranges from 96.4 to 97.5% of the perfect crystal value for Cu and from 95.2 to 97% for Ni, the highest values are obtained for the largest grain size. It is important to point out that the Cu and Ni samples at a given mean grain size have the same microstructure, as a result of the same set of parameters in the sample creation program.

Strain-time curves were recorded at 300 K and snapshot configurations were regularly stored for detailed analysis at several values of applied constant uniaxial stress of 1 and 2.6 GPa. The Young's modulus of all samples was calculated applying small loads (<60 MPa): the samples with mean grain size of 12 and 10 nm have a Young's modulus of 200 and 90 GPa for Ni and Cu respectively, values which agree well with those reported for polycrystalline material. The 8 nm sample shows a reduction in Young's modulus of about 15%, while the reduction in the 3.4 nm sample is about 25% compared to the polycrystalline phase.

Characterisation of the microstructure is made in several ways: by determining atomic energy and coordination number distributions and by the local crystalline order in terms of a bond analysis technique based on [15] and described in details in [8]. Here we just note that this technique allows distinguishing the fcc and hcp structures even taking into account only nearest neighbour. On this basis, we define five categories of atoms : (i) *perfect fcc*, atoms having a local fcc order till fourth neighbours, (ii) *good fcc*, atoms having a local fcc order till first neighbours, (iii) *hcp atoms* having a local hcp order till first neighbours, (iv) atoms having *other 12 coordinated* combinations, and (v) *non-12 coordinated* atoms. The presence of hcp atoms is relevant because they help to visualise the existence or activity of dislocations, since two nearest neighbours good hcp planes indicate a stacking fault in the fcc lattice.

Although the initial structure of the samples is the same, after relaxation to a minimum enthalpy, the Cu samples have higher grain boundary density and consequently higher sample density, and a greater % of atoms in the grain boundary than the corresponding values for Ni Analysis of the coordination number distributions indicates that at the larger grain sizes the fraction of over-coordinated atoms decreases, which could explain the reduced grain boundary density. The mean atomic energy of the fcc atoms is close to the perfect crystal value in both materials. A small excess energy due to elastic strain is present only at the smallest grain sizes (<5.2 nm). The *good fcc* atoms and the atoms in the grain boundary have an energy higher than the perfect crystal value, the excess being greater in Ni than in Cu. The mean energy of the grain boundary atoms seems to increase slightly with increasing mean grain size. The excess enthalpy is higher in the Ni samples than in the corresponding Cu samples and increases in Ni from 20% to 54% and in Cu from 16% to 39% when grain size is decreased from 12 nm to 3.4 nm. The differences in density and energy distribution between Ni and Cu can be explained on the basis of the relatively higher cohesive energy, melting temperature and heat of melting, and the relatively lower density of the melt in Ni compared to Cu.

RESULTS AND DISCUSSION

Influence of the grain size on the strain rate

It has been shown in previous papers [5,6], that in the early stages of deformation of Ni samples the strain could be approximated by a linear behaviour in time, reflecting the absence of damage accumulation. We also showed that when Ni is deformed with a 1.5 GPa uniaxial tensile load, two regimes could be observed in the strain rate versus inverse of the grain size plot. Similar strain rate dependence is obtained for Cu, when it was deformed with 1 GPa. (Fig. 1).

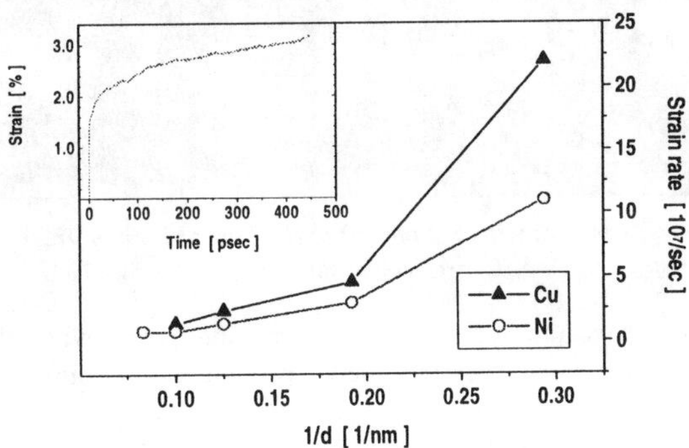

Fig. 1. *Strain rate as a function of mean grain size*
(Cu : 1.0 GPa, Ni : 1.5 GPa uniaxial tensile stress. The inset shows a typical strain-time curve)

The strain-time curve for Cu_5.2 nm in the inset of the figure shows the validity of the linear approach in this time scale. For grain sizes larger than 10 nm, the strain rate is independent of the grain size; while for smaller grain sizes, the strain rate increases considerably. A transition for d < 8 nm between an inter-grain channelled slip behaviour and an intermediate mechanism controlled by a decreasing contribution from GB sliding and an increasing contribution of intra-grain dislocation slip is suggested in [6].

Deformation mechanism on atomic level

We have analysed the atomic structure of the Ni and Cu samples deformed at about the same strain rate and up to the same plastic deformation. In that way we take into account the different elastic contribution due to the reduction of the Young's modulus in very fine grained samples. The elastic part is obtained using the calculated values of the Young's modulus. The elastic strain values are in good agreement with the values obtained from the elastic recovery calculated from unloading strain-time curves.

Fig. 2 shows slices of Ni_5.2 and Ni_12, both deformed with a strain rate of about $3 \cdot 10^7$/sec using a tensile load of 1.5 GPa for the 5.2 nm sample and of 2.6 GPa for the 12 nm sample. The total deformation of the sample in this plot is 2.7% for Ni_12 and 2.3% for Ni_5.2. At that strain, both samples have a similar plastic deformation of about 1.4%. We observe:

• In Ni_5.2 small hcp clusters grow in the grain boundary during deformation. No intra-grain defects can be observed (Fig. 2a).

• In Ni_12 (Fig.2b) stacking faults inside the grains can be observed since two hcp planes cross the grain from a grain boundary to the opposite one. This indicates a transition into another deformation mechanism based on partial dislocation, as is well known for coarse grain polycrystalline material.

(a)   (b)

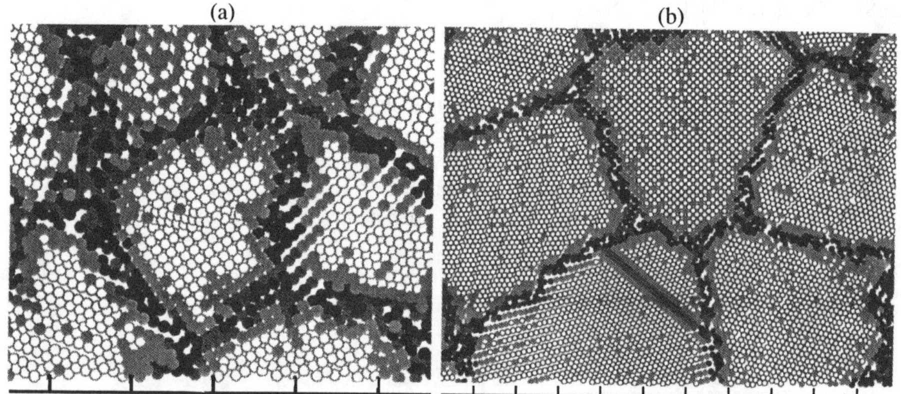

Fig. 2. Slices Ni samples with 5.2 nm (a) and 12 nm (b) mean grain sizes.
(The strain rate is 3.10^7/sec, level of plastic deformation is 1.4%)

In polycrystalline materials intra-grain plastic deformation occurs when the resolved shear stress reaches a critical value, which represents the yield strength of the single crystal.

$$\tau_R = \tau_A \cos \theta \cos \phi \qquad (1)$$

In the above equation, the term $\cos\theta \cos\phi$ is the Schmid factor, with θ the angle between the load axis and the normal to the slip plane, and ϕ is the angle between the load axis and the slip direction, τ_A and τ_R are the applied and resolved stresses, respectively. The applied stress is the sum of the external plus the internal stresses. Yielding will then first occur on the slip system with the greatest Schmid factor. The slip system in fcc structures is the $\{111\}\langle110\rangle$ system composed of twelve possible combinations. We have calculated the Schmid factor for the grain in which we observe the first stacking fault and found that slip occurs in the (111) plane, which is indeed for this grain the most favourable slip plane. Due to the orientation of the particular grain relative to the strain direction, the two slip directions $\langle0,1,1\rangle$ and $\langle1,-1,0\rangle$ have both a

Schmid factor of 0.37, which is the largest possible value in this grain. This results in a curved partial dislocation travelling trough the grain. The partial is emitted from the grain boundary and travels across the grain; when it has reached the opposite grain boundary, two hcp planes are left behind. When deformation is continued, no second partial is observed but stacking faults appear in other grains.

Partial dislocations travelling across a grain were already observed in our previous work [14] in a textured Ni sample with mean grain size of 5.2nm. We suggested that the larger degree of coherence in the interfaces in a textured sample helps transferring the shear component of the applied stress into the grain interior, which results in the creation of the partial dislocations.

A similar change in deformation mechanism with increasing grain size is observed in Cu, but starting already at a lower grain size.

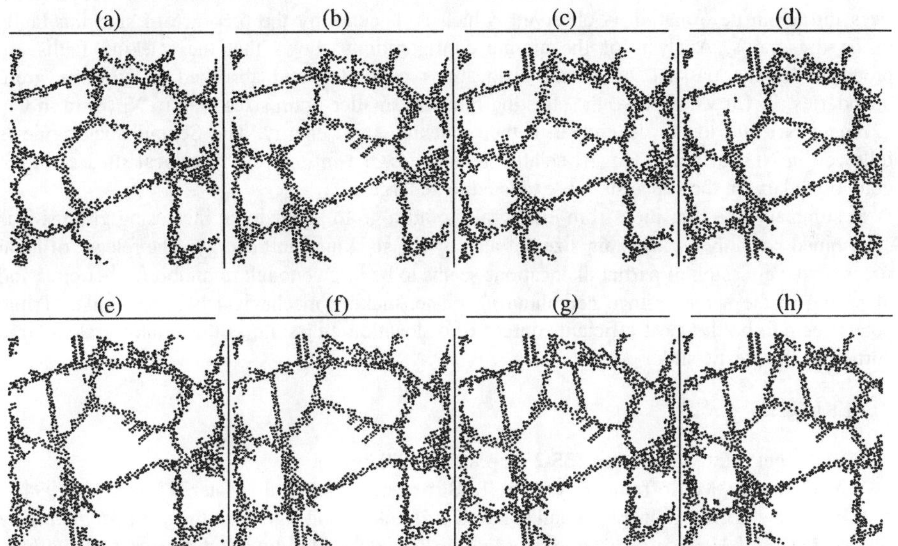

Fig. 3. *Slices of nanophase Cu with mean grain size of 8 nm at different strain levels.*

- Cu_5.2 was deformed at a strain rate of $4\cdot10^{7}$/sec (1 GPa) until a total deformation level of 3.2%, which represents a plastic deformation of 1.7%. Analysis of the configurations of the samples only shows small hcp planes in the boundaries and that plastic deformation is mainly accommodated in the grain boundaries.

- Deformation of Cu_8.0 up to the same plastic deformation level, (a total strain of 3.7% in this case), shows evidences of intra-grain deformation. Fig. 3 presents the time evolution of a slice of the sample. The first picture is the configuration without deformation. The following pictures are at increasing deformation levels between 3.2%, and 4.1%. At a strain level of 3.2%, a small cluster of hcp atoms can be observed at a triple point. As the deformation proceeds, sliding occurs along a grain boundary, then slip in a grain, and later on slip is observed in two other grains. In all cases, slip is produced by the motion of a partial dislocation emitted from a grain boundary to the grain interior, leaving behind two hcp planes. Analysis of the sample revealed that plastic deformation in Fig. 3 is initiated at the grain boundary in a triple point.

All the observed slip belongs to the $\{111\}\langle110\rangle$ systems. However, the Schmid law is not followed. The first slip plane is a $\{1,1,-1\}$ plane and the slip direction is $\langle1,-1,0\rangle$. The Schmid factor is 0.12, the highest for this slip plane, but not the highest possible for that grain orientation. In other grains, slip does not go along the directions that are the most favoured by the Schmid factor either. This reflects the existence of local stresses in the surrounding grain boundaries. This influence is expected to become more important at reduced grain size.

CONCLUSIONS

We simulated the plastic behaviour of two model fcc structures Ni and Cu, with high and low stacking fault energies. It was found that the grain boundaries in Ni are less dense and more energetic. In both materials we observed a change in deformation mechanism on atomic level: at the smallest grain sizes all deformation is accommodated in the grain boundaries. At larger grain sizes intragrain deformation is observed, which is detected by the presence of stacking faults inside the grains. Analysis of the atomic configurations shows that the stacking faults are produced by the passage of partial dislocations generated and absorbed in opposite grain boundaries. In Cu we observe the stacking faults at smaller grain sizes than in Ni (8 nm in Cu, 12 nm in Ni) due to the difference in their stacking fault energy. The Schmid factor rule is followed in Ni, but not in Cu. At smaller grain sizes, the influence of the local stresses at the grain boundary on the slip system is expected to be larger.

In summary, the transition from inter-grain to intra-grain plasticity at increasing grain size is determined not only by the grain size but also by the stacking fault energy. The release of local stress by the emission of partial dislocations seems to be large enough to inhibit further emission of partial in the same or in a neighbouring plane, making mechanical twining unlike. Triple points seem to be the most efficient sources of dislocation emission, as they concentrate a large portion of the plastic slide.

REFERENCES

1. R.W. Siegel, Mat. Sci. Forum **235-238,** p.851 (1997).
2. V.Y. Gertsman, M. Hoffman, H. Gleiter, R. Birringer, Acta Metall. Mater. **42**, p.3539 (1994).
3. S.R. Agnew, B.R. Elliot, C.J. Youngdahl, K.J. Hemker, and J.R. Weertman, in *Modelling of Structure and Mechanics from Microscale to Product*, edited by J.V. Carstensen, T. Leffers, T. Lorentzen, O.B. Pedersen, B.F. Sorensen, and G. Winther (Risø National Laboratory, Roskilde, Denmark, 1998, p. 1, and A.M. El-Sheiric, U. Erb, G. Palumbo, K.T. Aust, Scripta Metall. Mater. **27**, p.1185 (1997).
4. E. Arzt, Acta Metall Mater. **221**, p.5611 (1998).
5. H. Van Swygenhoven and A. Caro, Appl. Phys. Lett. **71**, p.12 (1997).
6. H. Van Swygenhoven and A. Caro, Phys. Rev. B **58,** p.11246 (1998).
7. H. Van Swygenhoven, M. Spaczér, A. Caro, Nanostruct. Mater. **10**, p.819 (1998).
8. H. Van Swygenhoven, M. Spaczér, A. Caro, D. Farkas, Proc. of the Fourth International Conference on Nanostructured Materials, NANO'98, Stockholm, Sweden, June 1998
9. J. Schiøtz, F.D. DiTolla, and K.W. Jacobsen. Nature **391**, p.561 (1998).
10.Z. Chen and J. Ding, Nanostruc. Mater. **10**, p.205 (1998).
11.G.Z. Voronoi, Reine Angew. Math. **134**, p.199 (1908).
12.F. Cleri and V. Rosato, Phys. Rev. **B 48**, p.48 (1993).
13.H. Zhu and R.S. Averback, Phys. Rev. **B 51**, p.15559 (1995).
14.J.P. Hirth and J. Lothe in *Theory of Dislocations*. Ed. by J. Wiley & Sons, NY 1982.
15.D.J. Honeycutt and H.C. Andersen, J. Phys. Chem **91**, p.4950 (1987).

PRIMARY AND SECONDARY GRAIN BOUNDARY DISLOCATIONS IN SYMMETRIC TILT GRAIN BOUNDARIES OF FINITE LENGTH

ANNAMALAI LAKSHMANAN and ALEXANDER H. KING

Department of Materials Science and Engineering, State University of New York at Stony Brook, Stony Brook, NY 11794-2275, U. S. A.

ABSTRACT

Secondary grain boundary dislocations in large-angle grain boundaries (also called dsc-dislocations) can be described as local variations in the density of the primary (or lattice) dislocations that make up the boundary. We present a simple meso-scale simulation in which the interactions of primary dislocations with each other and with the crystal lattice produce secondary dislocations with smaller Burgers vectors and larger spacing. We use the model to explore the interactions of the primary and secondary defects with a free surface terminating the grain boundary, and demonstrate cases in which the primary dislocations dominate the interactions, forcing the secondary dislocations to increase their energy. Other cases are found, for which all of the dislocations can respond in such a way that their energy decreases. The creation of interfacial disclinations is also demonstrated.

INTRODUCTION

Dislocation models for grain boundaries were first suggested by Burgers [1], and subsequently used by Read and Shockley to determine the interfacial energy of small-angle boundaries [2]. These models have been highly successful over a long period of time, successfully predicting, *inter alia*, mechanical behaviors of grain boundaries [3] and the variation of grain boundary diffusivity with misorientation, in the small-angle regime [4]. Dislocations are also observed in large-angle grain boundaries [5] where they are usually associated with special misorientations that produce coincidence-site lattices (CSLs) [6]. In these cases, the dislocations are often found to have small Burgers vectors, relative to those of lattice dislocations, and they are variously called "grain boundary dislocations," "dsc-dislocations" or "secondary dislocations." As demonstrated by Kvam and Balluffi [7], secondary dislocations actually co-exist with primary (or lattice) dislocations of much finer spacing, and the secondary dislocations can be interpreted as perturbations in the spacing of the primary dislocation array.

A curious interplay of length-scales exists in grain boundaries that contain both primary and secondary dislocations: the secondary dislocations have smaller Burgers vectors, but larger spacing and longer-ranged strain fields than the primary dislocations. In many cases, adequate models of the grain boundary behavior can be constructed using the coarser length-scale of the secondary dislocation network, but this occasionally produces misleading results. In this paper, we develop a model that allows us to track both the primary and secondary dislocations in a simple grain boundary, and apply it to the problem of the interactions of grain boundary dislocations with a free surface.

Most models of grain boundaries assume that the interface has infinite extent, and these "infinite boundary models" produce results that are applicable in many practical situations. On the other hand, a grain boundary may be categorized functionally as having finite length when the end effects caused due to the termination of the interface (at a surface or a junction with another boundary) become significant. Such boundaries are found in nanocrystalline materials, thin films and fine wires, and the results of the present study demonstrate extensive end effects in the case of thin films and fine wires.

Mat. Res. Soc. Symp. Proc. Vol. 538 © 1999 Materials Research Society

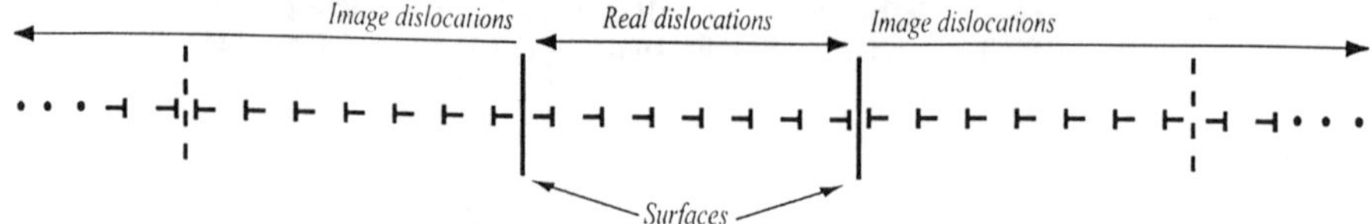

Figure 1. A finite grain boundary traversing a thin film, represented by an array of edge dislocations. These dislocations produce image dislocations about a surface, which further produce more image dislocations about the other surface, *etc.*

SIMULATION METHOD

We have considered a symmetric tilt boundary of finite extent which is terminated by two free surfaces. The array of edge dislocations which constitute the grain boundary produce image dislocations about a surface; which further produce more image dislocations about the other surface (Fig. 1). The force exerted by the surface on any grain boundary dislocation can be imagined to be the force exerted by all these image dislocations on individual dislocations constituting the grain boundary. The image forces tend to move the dislocations toward the surface. Forces are also applied to each dislocation by its interactions with all of the other dislocations in the array, and the net of these forces may be toward or away from the surface. In an infinite grain boundary, they produce a zero net force on every grain boundary dislocation, if the array is equispaced. Finally, there is a lattice interaction force which is analogous to Peierls force and is assumed to be sinusoidal in nature, and resists the motion of the dislocations. It should be noted that Peierls force is strictly applicable only to dislocation glide and not dislocation climb; so the lattice resistance force is not exactly Peierls force. The sinusoidal assumption of the lattice resistance force is helpful in explaining coincident site lattice (CSL) boundaries. The boundary is a CSL boundary when the dislocation spacing is commensurable with the lattice resistance periods.

The code for the simulations was written in C programming language. The array under investigation initially contains 100, equally-spaced primary dislocations. It was assumed that the dislocations constituting the grain boundary experience forces which have the following two components: (i) forces due to interactions with all other dislocations (real and image) (ii) lattice interactions. The simulations proceed by computing the total force acting on each primary dislocation in the grain boundary, and then moving the dislocation by an amount proportional to the force, and repeating the process on the new configuration. The magnitudes of the displacements were chosen to ensure that the most active dislocation in the array would move by no more than one tenth of the mean dislocation spacing in any time step.

EXPERIMENTS AND RESULTS

A coincidence-related grain boundary is one in which the primary dislocation structural period is equal to a lattice vector. We model this by creating a boundary in which the primary dislocation spacing is equal to three lattice-interaction periods. Near-coincidence boundaries are modeled by starting with a spacing of 2.9 lattice periods (*i.e.* a smaller primary dislocation spacing, and thus a greater misorientation angle) or 3.1 lattice periods (a larger primary dislocation spacing, and a smaller misorientation angle, θ). All of our simulations start with the primary dislocations uniformly spaced. In this brief report, we can address only one case in detail, so we present the results for the case of an initial primary dislocation spacing of 3.1 lattice periods.

The number of lattice resistance periods between two successive dislocations was assumed to be 3.1; which would imply that most of the dislocations would be in a non-equilibrium position

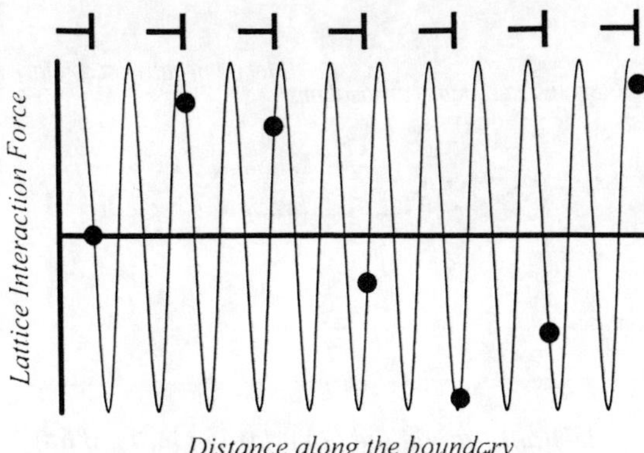

Figure 2. Schematic illustration of how the lattice resistance force acts on a uniformly spaced array of primary dislocations, when the array spacing is incommensurate with the lattice spacing. Relaxation in response to this force creates non-uniform spacings, in multiples of the lattice spacing. Perturbations in such relaxed arrays correspond to secondary dislocations.

with respect to the lattice resistance force (Fig 2). The forces on the dislocations move them such that they attain a stable configuration. A stable configuration would be where the lattice resistance force on the dislocations is zero.

It was seen that after about 20 time-steps, there were groups of dislocations with a spacing of 3 lattice resistance periods and pairs of dislocations with a spacing of 3.5 lattice resistance periods. (Fig 3) These pairs of dislocations are perturbations in the array of dislocations; and are actually secondary dislocations or dsc-dislocations. These perturbations maintain their stability by increasing the spacing to 3.75 and then to 4 (Fig 4).

Dislocations in the central part assume a stable configuration, but the rest of the dislocations are in a dynamic state. This is because the dislocations near the surface are dominated by the image forces and forces due to rest of the dislocations in the array; but the dislocations in the central re-

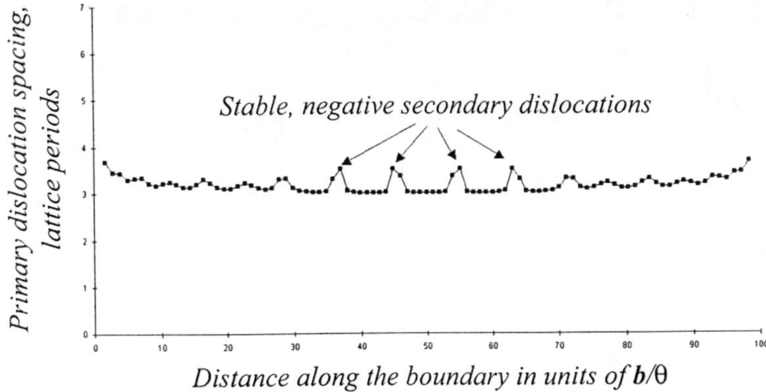

Figure 3. Configuration after 20 time steps with four secondary dislocations with a perturbation of about 3.5 lattice resistance periods. The unit of length for the abscissa is the initial primary dislocation spacing, $S = b/\theta$, where b is the Burgers vector and θ is the misorientation angle.

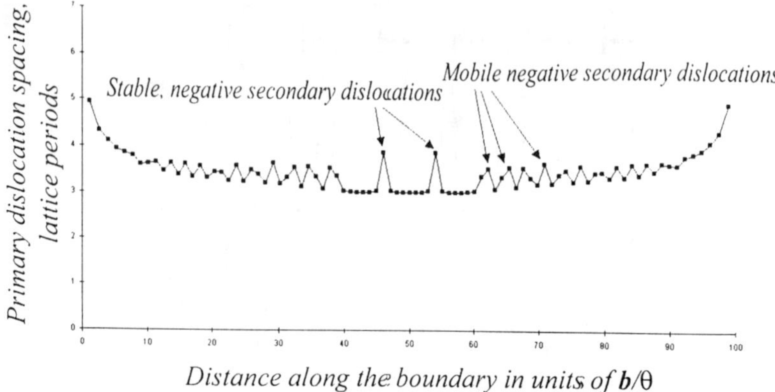

Figure 4. Negative secondary dislocations moving toward the central region after 90 computational time steps.

gion are screened from the surface by rest of the dislocations in the array and thus dominated by the lattice resistance force. Dislocations near the surface are attracted more strongly toward the surface than other dislocations which leads to their higher dislocation spacing (Fig. 4). Primary dislocations are constantly absorbed by the surface which leads to a constant increase in the average dislocation spacing of the array, and a corresponding reduction in the misorientation of the grain boundary.

Negative secondary dislocations (*i.e.* perturbations with an above-average primary dislocation spacing) form at the surface and move towards the center as shown in Fig 4. They start crowding near the central stable secondary dislocations. This leads to an increase in the energy of the secondary dislocations which are moving in. This suggests that the primary dislocations dominate the interactions and the secondary dislocations are forced to increase their energy.

Some dislocations arrange themselves into a stable configuration with a spacing of 3.5 lattice resistance periods. These configurations are found between the crowded region of secondary dislocations and the dislocations near the surface. (Fig 5). These configurations are unstable; be-

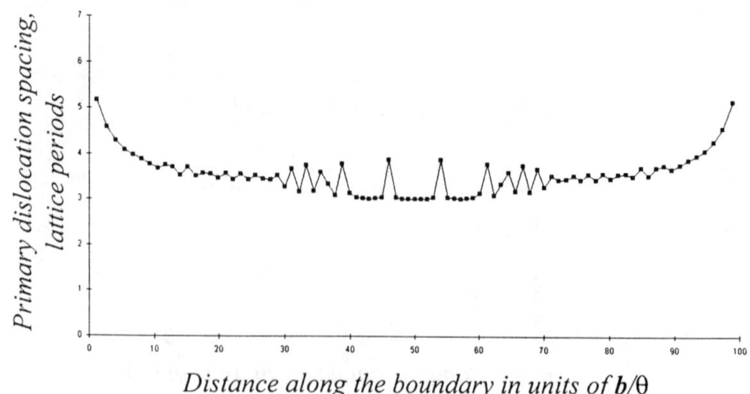

Figure 5. After 115 time steps some dislocations have arranged themselves in a configuration with a dislocation spacing of 3.5

410

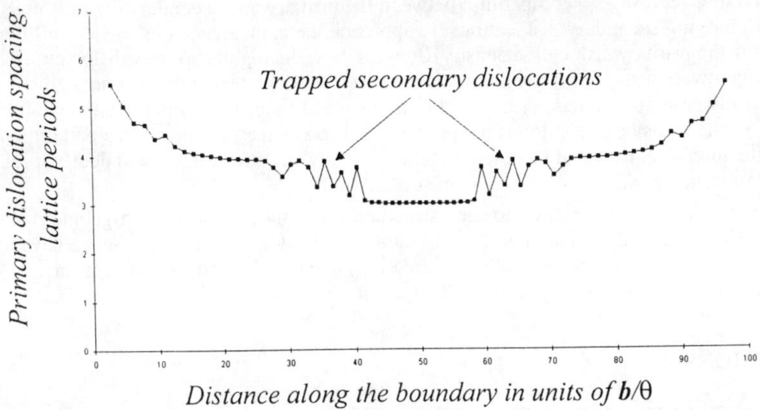

Figure 6. After 320 time steps dislocations arrange themselves into lattice resistance spacing of three and four; with the secondary dislocations trapped in between.

cause a dislocation spacing of 3.5 lattice resistance periods means that every alternate dislocation is at an unstable equilibrium, on a lattice interaction energy maximum and a slight fluctuation can disturb the configuration.

After about 300 time steps some dislocations arrange themselves into a stable configuration with a spacing of four lattice resistance periods. There are secondary dislocations trapped between the configuration with a dislocation spacing of four and the configuration with a dislocation spacing of three (Fig. 6). The trapped secondary dislocations act as positive secondary dislocations for the part of the boundary, where the dislocation spacing is 4 lattice resistance periods; and start moving out thereby canceling with the negative secondary dislocations which are moving toward the center; thus reducing the overall energy.

Ultimately, the whole grain boundary is divided into parts; with each part having its own, locally uniform, dislocation spacing. Since a uniform dislocation spacing corresponds to a fixed misorientation, there is a change of misorientation where the spacing changes, and these changes thus correspond to disclinations. Dislocations near the surface are widely spaced, non-uniform, and unstable. The dislocation spacings of each part in the specimen center are whole numbered multiples of the lattice resistance periods. The dislocation spacing of the part nearest to the surface is the maximum and that of the one in the center is the minimum.

DISCUSSION

Our simulations are effectively models of the crystallite rotation experiments pioneered by Gleiter and his co-workers [8], and carried out with *in-situ* observation of the interfacial dislocation structure by Chan and Balluffi [9]. Chan and Balluffi, in particular, observed cases in which crystallites rotated through coincidence misorientations, apparently ignoring the energy minima that they were presumed to represent. The rotation of the crystallites has been rationalized as a response to the gradient of the interfacial energy as a function of misorientation [10] and also as the removal of secondary interfacial dislocations, due to their interactions with the free surface [11].

Our investigations also include cases in which the changing misorientation of the bicrystal is effected by the removal of both primary and secondary dislocations, analogous to the model of Pond and Smith. In the case described in this paper, however, the rotation of one crystal with respect to the other represents a competition between the primary and secondary dislocations in the interface. Removing secondary dislocations (to approach the coincidence misorientation) requires an increase in the primary dislocation density. Conversely, reducing the primary dislocation density (thus rotating toward a single-crystal orientation) necessitates increasing the secondary dislocation density. For the case considered here, in addition to several other interesting details, we find that the loss of primary dislocations drives the process, and forces the injection of secondary dislocations into the interface. It should be noted that this is purely an "edge effect" and does not relate to the energy of large boundaries of the same misorientation.

This work is closely related to the "structural unit" models developed geometrically by Bishop and Chalmers [12] and atomistically by Sutton and Vitek [13]. In our case we have identified dislocations with different types of structural units, and thereby deduce a simple form of interaction potential between them, and also with a free surface.

CONCLUSIONS

1.   Dislocations close to the surface are dominated by the image forces and dislocations in the central region are dominated by the lattice resistance force, as would be expected. This further leads to the fact that dislocations near the surface are dynamic and the configurations there keep changing; whereas dislocations in the central region assume a stable configuration.

2.   Negative secondary dislocations which form at the surface keep crowding toward the central region thereby increasing the overall energy. Negative secondary dislocations form at the surface irrespective of whether the initial spacing is 2.9 or 3.1 lattice interaction periods. These negative secondary dislocations spend most of their time in the dynamic region of the bicrystal which is affected more by the surface than by the lattice resistance.

3.   The final structure breaks down into equally spaced primary dislocations of various dislocation spacings and interfacial disclinations.

Acknowledgment: This work was supported by National Science Foundation grant number DMR 9530314.

REFERENCES

1.   J.M. Burgers, Proc. Kon. Ned. Akad. V. Wet. Amsterdam **42** , 293 (1939).
2.   W.T. Read and W. Shockley, Phys. Rev. **78** , 275 (1950).
3.   J. Washburn and E.R. Parker, Trans. A.I.M.E. **194** , 1076 (1952).
4.   D. Turnbull and R.E. Hoffman, Acta Metall. **2**, 419 (1954).
5.   For a thorough review, see A.P. Sutton and R.W. Balluffi, *Interfaces in Crystalline Materials*, Oxford University Press, Oxford (1995).
6.   W. Bollmann, *Crystal Defects and Crystalline Interfaces*, Springer, Berlin (1972).
7.   E.P. Kvam and R.W. Balluffi, Phil. Mag. A **56** , 137 (1987).
8.   G. Hermann, H. Gleiter and G. Baero, Acta Metall. **24** , 353 (1976).
9.   S.W. Chan and R.W. Balluffi, Acta Metall. **33** , 1113 (1985); *ibid.* **34** , 2191 (1986).
10.   R.M. Allen and P.J. Goodhew, Scripta Metall. **11** , 37 (1977).
11.   R.C. Pond and D.A. Smith, Scripta Metall., **11** , 77 (1977).
12.   G.H. Bishop and B. Chalmers, Scripta Metall., **2**, 133 (1968).
13.   A.P. Sutton and V. Vitek, Phil. Trans. R. Soc. Lond A, **309** , 1 (1983); *ibid.* **309** , 37 (1983); *ibid.* **309** , 55 (1983).

SILICON SELF–INTERSTITIAL CLUSTERS

L. COLOMBO[1,†], A. BONGIORNO[1] and M. ROSATI[2]
[1]INFM and Dept. Materials Science, via Emanueli 15, I-20126 Milano, Italy
[2]CASPUR, Univ. Roma "La Sapienza", P.le Moro 5, 00185 Roma, Italy
[†]E-mail: luciano.colombo@mater.unimi.it

Abstract

A tight–binding molecular dynamics investigation on the structure and energetics of self–interstitial clusters in silicon is presented. We discuss how a small number of self–interstitial atoms give rise to the formation of tedrahedally–shaped clusters, while a larger number of defects exhibit a self–organization mechanism driving the system to form rod–like defects.

Introduction

Interactions among self–interstitial (SI) atoms affect the microstructural evolution of bulk silicon under ion–beam processing. In this framework, SI interaction dynamics and SI–cluster formation represent the two most relevant atomic–scale mechanisms[1, 2, 3]. The aim of the present work is to study at the relevant atomic scale such a coalescence mechanism, driving several nearby interacting SI defects to form a cluster, and to predict the energetics of the resulting complexes.

Our study is based on tight–binding molecular dynamics (TBMD) simulations. TBMD was used to anneal defect structures at finite temperature and to characterize the most stable complexes, as well as to compute their equilibrium energetics. We made use of the TB representation by Kwon *et al.*[4] and the O(N) formulation of TBMD by Goedecker *et al.*[5]. Present simulation cells contain as many as 512 atoms (in addition to the atoms forming the SI cluster).

Results and discussion

In Tab. 1 we report the formation energy E_f and binding energy $E_b = \frac{1}{n}\left[nE_f(1) - E_f(n)\right]$ for the smallest SI clusters I1, I2, and I3 containing $n = 1, 2, 3$ self–interstitial atoms, respectively.

Table 1: Formation E_f and binding E_b energy for SI–clusters in units of eV.

	I1	I2	I3
E_f	3.87	4.91	6.69
E_b	-	1.41	1.63

Fig.1 shows the atomic structure of the above SI clusters. The stick–and–ball picture was drawn by using a bond cutoff distance of 2.8Å. In the following we term as Si defective

Mat. Res. Soc. Symp. Proc. Vol. 538 ©1999 Materials Research Society

atoms (DAs) only those atoms that are filled in light–gray in the figure, while bulk–like atoms are represented in dark–gray. In I1 (the well known < 110 > dumbbell) the DAs are 4–fold coordinated as is the Si atom in the bulk, while by increasing the cluster size the coordination raises from 5– in I2 to 6–fold in I3. Up to and including I3, all DAs in a given SI cluster exhibit the same connectivity. The structural evolution emerging from Fig.1 is towards the formation of a tetrahedral cage symmetrically embedded into the crystalline environment, the "nucleation catalyst" being a dumbbell defect.

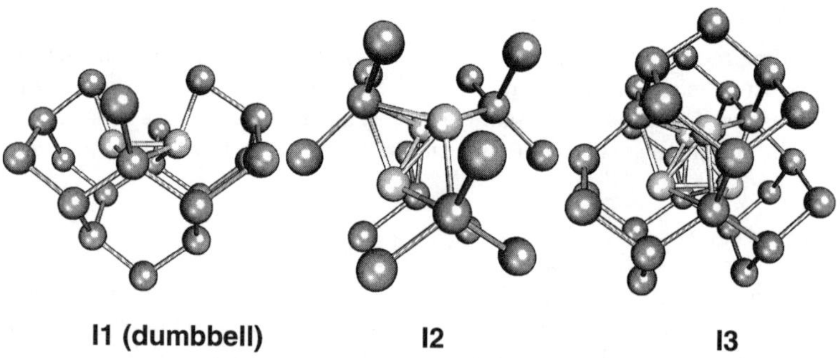

Figure 1: Equilibrium atomic structure of I1 (left), I2 (center), and I3 (right) after TBMD annealing. Atomic connectivity is computed by using a bond cutoff distance of 2.8Å. Dark–gray: bulk–like atoms; light–gray: Si defect atoms (DAs).

This noticeable symmetry is broken in I5, as shown in Fig.2, following the transition from three–dimensional to rod–like defect structures discussed below.

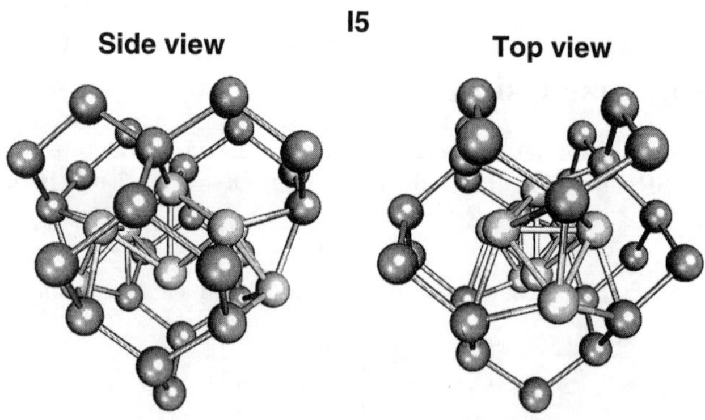

Figure 2: Equilibrium atomic structure of I5. Left panel: side view; right panel: top view. Dark–gray: bulk–like atoms; light–gray: Si defect atoms (DAs).

Indeed in I5 the two central DAs (a bulk–like atom and a tetrahedral interstitial, respectively) have a different coordination (5–fold and 7–fold) while the four DAs linked to them (forming two dumbbells at both sides) are 6–fold coordinated. Finally the two outermost DAs (sitting at tetrahedral interstitial positions) have a 5–fold coordination. The average coordination in I5 is 5.75, a value somewhat smaller than that found in I3, which has the highest average connectivity in the investigated SI cluster series.

Self-organization of the I8 cluster

Initial configuration

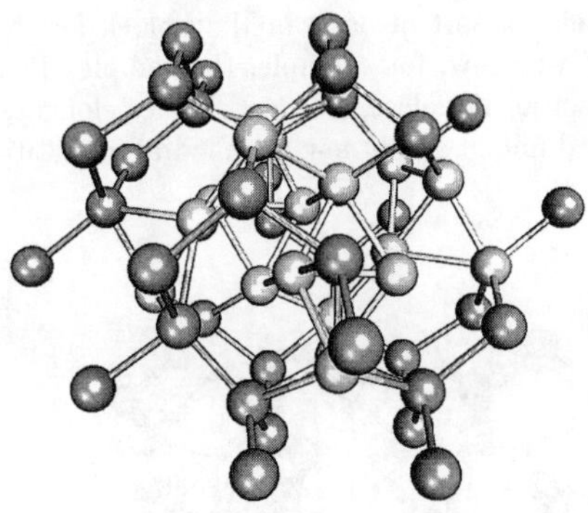

Final configuration (side view)

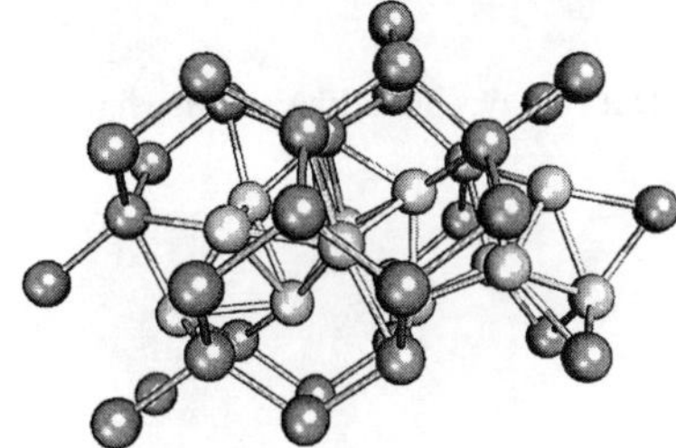

Final configuration (top view)

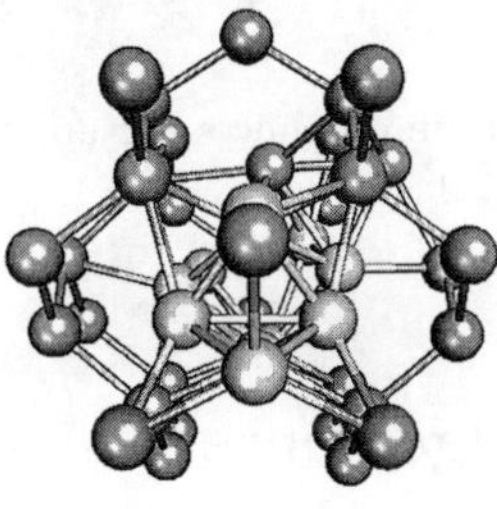

Figure 3: Initial (top) and equilibrium final (bottom) atomic structure of I8. The initial random distribution of eight self–interstitial defects was annealed by TBMD at 600K. Dark–gray: bulk–like atoms; light–gray: Si defect atoms (DAs).

In I2 and I3 clusters, the DAs first neighbors show the same coordination as the DAs, while in I1 and in I5 their average connectivity is respectively larger (4.5) and smaller (5) than found for the associated DAs.

As for the energetics of larger complexes (i.e. containing more than 3 SI defects), while formation energy increases linearly with cluster size, the binding energy data indicate that the average energy content per self–interstitial is about 1.55 eV, in rather good agreement with the 1.8 eV experimental data provided by Chason *et al.*[1]. Moreover, larger complexes behave differently as far as cluster shape is concerned. We observed that during the TBMD finite–temperature annealing process SI cluster self–organize so to form rod–like defects with an intriguing structure. The I5 SI defect is in fact the smallest rod–like structure of this kind. The driving mechanism is summarized in Fig.3: here we show the initial random distribution of 8 SI atoms (top panel) and the final equilibrium configuration after annealing of I8 (bottom panels). The annealing procedure was performed at 600K.

A careful analysis on atomic coordination and bonding[6, 7] provides a rationale for the evolution of SI cluster shape towards rod–like structures. The atomic coordination increases linearly with cluster size up to 6 for I3. Above this value Si atoms cannot form new bonds (the maximum average coordination is 6.5, as observed in liquid silicon[8]) and therefore a new growth pattern for larger complexes must be followed. Our conclusion is that I5 cluster plays a special role, being a sort of basic bulding block for the formation of more extended complexes. In Fig.4 we show, for example, the complex I9 where it is apparent that the core structure is in fact an I5 defect, which is further elongated in both directions by alternatively adding a new dumbell and a new tetrahedral interstitial.

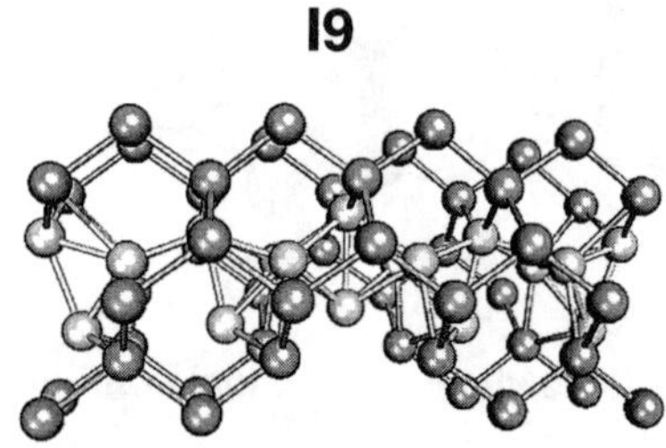

Figure 4: Equilibrium atomic structure of I9. Dark–gray: bulk–like atoms; light–gray: Si defect atoms (DAs).

Acknowledgements

We acknowledge computational support on the CRAY–T3E computer at CINECA (Casalecchio di Reno, Italy) by INFM under initiative "Progetti di Supercalcolo" (project "From point to extended self-interstitials in silicon: a TBMD investigation") and by CASPUR (Rome). We also acknowledge financial support by CNR under project "Progetto 5% Microelettronica"

References

[1] E. Chason *et al.*, J. Appl. Phys. **81**, 6513 (1997).

[2] J.L. Benton *et al.*, J. Appl. Phys. **82**, 120 (1997).

[3] S. Libertino *et al.*, Appl. Phys. Lett. **71**, 389 (1997).

[4] I. Kwon *et al.*, Phys. Rev. B**49**, 7242 (1994).

[5] S. Goedecker and L. Colombo, Phys. Rev. Lett. **73**, 122 (1994).

[6] A. Bongiorno *et al.*, Proceedings of the 24$^\text{th}$ International Conference on the Physics of Semiconductors (Jerusalem, August 1–6, 1998).

[7] A. Bongiorno *et al.*, in preparation.

[8] I. Stich *et al.*, Phys. Rev. B **44**, 4262 (1991).

POINT DEFECT INTERACTIONS
WITH EXTENDED DEFECTS IN SILICON

J. F. JUSTO*, A. ANTONELLI**, and A. FAZZIO*

*Instituto de Física - USP, CP 66318, CEP 05315-970, São Paulo - SP, Brazil

**Instituto de Física Gleb Wataghin - UNICAMP, CEP 13083-970, Campinas - SP, Brazil

ABSTRACT

We investigated the interaction of point defects (vacancy and self-interstitials) with an intrinsic stacking fault in silicon. The calculations were carried out using *ab initio* total energy methods. The results show that point defects at an intrinsic stacking fault display a different behavior as compared to the same defect in the crystalline environment. This is evidenced by differences in formation energies and electronic structures. These results suggest that there is migration of point defects from the bulk to a stacking fault. This could affect the dislocation mobility in the crystal.

INTRODUCTION

Dislocations in semiconducting materials have been a source of major concern over the last decade. This is because they affect both the mechanical and the electronic properties of the material [1]. In a zincblende semiconductor, full dislocations belonging to the {111} glide planes dissociate into partial dislocations having a stacking fault (SF) connecting the partials [2]. This dissociation is energetically favorable and facilitates the dislocations to glide conservatively. Theoretical work on dislocations has focused on the core properties of dislocations in silicon using empirical [3, 4] as well as *ab initio* methods [5, 6]. Stacking faults, on the other hand, have received attention only more recently [7, 8, 9].

The plasticity of the material is controlled by dislocation motion inside the crystal. Dislocations glide conservatively by thermal kink mechanisms [3]. However, point defects, such as vacancies, may play an important role in dislocation mobility, mainly in pinning the dislocations [2]. Theoretical [6] and experimental [10] work have studied the interactions of vacancies with dislocations. However, the role of stacking faults in the mechanisms of dislocation motion, as result of the interaction of point defects with the SF, has not been explored yet.

Here we investigated the electronic and structural properties of intrinsic point defects (vacancies and self-interstitials) in silicon at a stacking fault and compared them to the respective defects in the crystalline environment. We showed that at the stacking fault the formation energy of these defects is smaller than in the crystalline environment. Therefore, there should be migration of point defects to the stacking fault, which could affect the mobility of the dislocations. There has been a growing interest in modeling dislocation motion at mesoscopic level using information from the processes at atomistic level [11]. Therefore, it is important to identify all the microscopic mechanisms involved in those processes.

Mat. Res. Soc. Symp. Proc. Vol. 538 ©1999 Materials Research Society

COMPUTATIONAL METHOD

We computed the structural and electronic properties of the defects using the density functional theory and local density approximation framework [12]. The one-electron Kohn-Sham equations were solved using the Car-Parrinello scheme [13] with pseudopotentials [14] in the Kleinmann-Bylander form [15]. The basis-set were expanded in plane-waves, with kinetic energy up to 10 Ry. The sampling in the Brillouin zone was performed using the Γ-point.

A reference orthorhombic supercell, shown in figure 1 comprises 120 atoms. The supercell consists of five double planes stacked along the [111] direction, and simulates one infinite stacking fault in the [111] glide plane. It is important to point out that the cell consists of 120 atoms, and this may not be large enough to release all the long range strain fields resulting from the point defects. However, using of the same reference cell to compute all the defects allowed us to compare formation energies among the defects. The optimization of the atomic structure was performed by allowing atoms to move until the Hellmann-Feynman forces were smaller than 10^{-4} eV/Å.

INTERACTION OF POINT DEFECTS WITH STACKING FAULTS

A stacking fault is an irregularity in the stacking sequence of the material. In zincblende materials, the normal stacking sequence is in the {111} (close-packed) direction. An intrinsic SF (ISF) is equivalent to removing a {111} double layer and gluing together the remaining material. An extrinsic SF (ESF), on the other hand, is is equivalent to adding a {111} double layer. Here, we only focused on the ISF which has been identified as the prevailing fault in silicon [16].

In Fig. 1, the ISF is perpendicular to the [111] direction. Atoms at the fourth double-layer in Fig. 1 are at a distance of 7.8Å from the ISF (or of its image resulting from the boundary conditions) and their surrounding environment is that of an atom in a perfect crystal up to the fourth nearest-neighbor. These atoms can be considered as bulk-like atoms. All the calculations we performed using the 120-atom supercell with an ISF as reference and placing the point defects at the ISF and in the bulk-like site. In the crystalline environment, an atom (site B in Fig. 1) has four first neighbors in a tetrahedral configuration at 2.35Å apart, and twelve second nearest neighbors at 3.84Å apart. An atom at an ISF (site A in Fig. 1) has the same first nearest neighbor geometry. However, it has, besides the twelve second nearest neighbors, an additional neighbor at 3.91Å. Only based on those geometrical considerations, we conclude that the properties of any defect at the SF may differ from an equivalent defect in the crystal.

We first studied the vacancy in its neutral charge state. Fig. 1 shows the four nearest neighbors (gray atoms) of an atom (in black) at an ISF site and in a crystal-like site. We computed the vacancy at the SF or in the crystalline environment by removing atoms A or B respectively from the 120-atom supercell. The formation energy of a vacancy in the bulk-like site is 3.06 eV, which is in reasonably good agreement with other theoretical calculations (3.60 eV)[17]. However, a vacancy at the ISF has formation energy lower by 0.23 eV. In both cases, there is an inward relaxation of the first neighboring atoms of the vacancy and a pairing of atoms. As mentioned earlier, the first neighborhood of a vacancy is equivalent at the SF site and in the crystal-like site. The differences appear for second

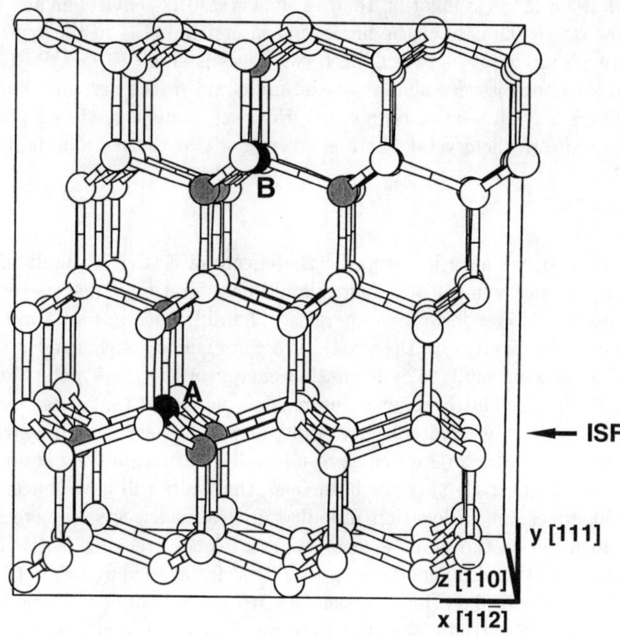

Figure 1: Structure consisting of an orthorhombic large unit cell with 120 atoms used as a reference cell in our calculations. x, y, and z axes are respectively parallel to the $[11\bar{2}]$, $[111]$, and $[\bar{1}10]$ directions of the diamond cubic lattice of the Si single crystal. The cell has one ISF in the xz plane. Black atoms represent the atoms at the SF (A) and in the crystalline environment (B), while the grey atoms represent the four nearest neighbors of atoms A and B. Vacancy and interstitial configurations at the ISF are studied by respectively removing or adding an atom in the plane which comprises the fault (site A) in the simulation cell. They are compared to respective defects in a crystal-like configuration (site B).

(and higher order) nearest neighbors. Therefore, the difference in formation energy comes from long range interactions.

We now present the results of the study on self-interstitials. The [110]-split dumbbell in Si is regarded to be the interstitial with the lowest formation energy [18]. The geometry of a [110]-split dumbbell consists of an empty lattice site with two atoms aligned along the [110] direction forming a dumbbell. For the [110]-split interstitial, the environment outside and at the fault are considerably different even at the first neighborhood of the dumbbell. The formation energy in the crystalline environment is 3.10 eV, which is consistent with theoretical calculations of 3.20 eV using a different cell geometry [19]. At the ISF, the formation energy decreases by 0.61 eV. This large difference in the formation energy can be explained by the local structure close to the fault. Although the planar {111} packing is identical in the crystal and at the stacking fault, the local structure around the defect at the ISF allows a larger relaxation than in the crystal-like environment.

We now study the case of the hexagonal interstitial (I_H) in the crystal and at the SF. The difference in energy between these two cases is small (0.15 eV). This is because the hexagonal structure is very similar at the fault and in the crystal. The six-atom ring of first neighbors is equivalent in both cases. However, while there is a T_d interstitial (I_T) site next to a I_H site in the crystal, there is no such site at the stacking fault.

CONCLUSIONS

In conclusion, we have investigated the effects of a stacking fault in the formation of intrinsic point defects in silicon. There should be migration of defects towards the stacking faults as result of lower formation energies. The difference in formation energy between a point defect in the fault or in the crystal is a consequence of changes in the local geometry close to the stacking fault. The formation energy of the three point defects studied here (vacancy, [110]-split and hexagonal interstitials) suggests that these defects will migrate to the fault. In highly irradiated materials, in which there is high concentration of point defects, the stacking faults, formed as result of dislocation splitting and motion, should not be defect-free as generally suggested. Instead, the faults will have concentration of defects. This would affect the dislocation mobility in several aspects. There might be pinning of the dislocation as the point defects migrate to the stacking fault. Recent theoretical calculations found binding energies up to 2.0 eV for a vacancy inside the core of a partial dislocation in silicon [6]. This is consistent with our finding of lower formation energies at the fault. Since partial dislocations are on the edge of a stacking fault, defects, which eventually migrate towards the dislocation cores, may first migrate to the stacking faults. Therefore, the stacking fault would play an important role in the pipe diffusion of point defects through the dislocations [20].

Results of several experiments on the electrical activity of dislocation related defects in semiconductors have been reviewed recently [1, 10]. Deformation induces several active centers, which have been associated to specific structures related to vacancies in the dislocation cores [10]. Our results suggest that this picture is considerably more complex, and vacancy centers in the stacking fault may also be electrically active and provide additional peaks in the spectra related to active centers.

ACKNOWLEDGMENTS

The authors acknowledge partial support from Brazilian agencies FAPESP and CNPq.

References

[1] H. Alexander, in *Dislocation in Solids*, ed. by F. R. N. Nabarro, Vol. 7, (North Holland, Amsterdam, 1986), p. 115.

[2] J. P. Hirth and J. Lothe, *Theory of Dislocations* (Wiley, New York,1982).

[3] V. V. Bulatov, S. Yip, and A. S. Argon, Phil. Mag. A **72**, 453 (1995).

[4] V. V. Bulatov, J. F. Justo, W. Cai, and S. Yip, Phys. Rev. Lett. **79**, 5042 (1997).

[5] G. Csányi, S. Ismail-Beigi, and T. A. Arias, Phys. Rev. Lett. **80**, 3984 (1998).

[6] N. Lehto and S. Öberg, Phys. Rev. B **56**, 12 706 (1997).

[7] M. Y. Chou, M. L. Cohen, and S. G. Louie, Phys. Rev. B **32**, 7979 (1985).

[8] N. Lehto, Phys. Rev. B **55**, 15 601 (1997).

[9] C. Stampfl and C. G. Van de Walle, Phys. Rev. B **57**, 15 052 (1998).

[10] J. Weber, Solid State Phenomena **37-38**, pp. 13-24 (1994).

[11] V. Bulatov, F. Abraham, L. Kubin, B. Devincre, and S. Yip, Nature **391**, 669 (1998).

[12] W. Kohn and L. J. Sham, Phys. Rev. **140**, 1133A (1965).

[13] R. Car and M. Parrinello, Phys. Rev. Lett. **55**, 2471 (1985).

[14] G. B. Bachelet, D. R. Hamann, and M. Schluter, Phys. Rev. B **26**, 4199 (1982).

[15] L. Kleinmann and D. M. Bylander, Phys. Rev. Lett. **48**, 1425 (1982).

[16] K. Wessel and H. Alexander, Phil. Mag. **35**, 1523 (1977).

[17] A. Antonelli, E. Kaxiras, and D. J. Chadi, Phys. Rev. Lett. **81**, 2088 (1998).

[18] P. E. Blochl, E. Smargiassi, R. Car, D. B. Laks, W. Andreoni, and S. T. Pantelides, Phys. Rev. Lett. **70**, 2435 (1993).

[19] J. Zhu, T. Diaz de la Rubia, L. H. Yang, and C. Mailhiot, Phys. Rev. B **54**, 4741 (1996).

[20] J. Huang, M. Meyer, and V. Pontikis, Phys. Rev. Lett. **63**, 628 (1989).

Part IV

Novel Methods for Material Modelling

ACCELERATING ATOMISTIC SIMULATIONS OF DEFECT DYNAMICS: HYPERDYNAMICS, PARALLEL REPLICA DYNAMICS, AND TEMPERATURE-ACCELERATED DYNAMICS

Arthur F. Voter and Mads R. Sørensen

Theoretical Division, Los Alamos National Laboratory, Los Alamos, NM 87545

ABSTRACT

Obtaining a good atomistic description of diffusion dynamics in materials remains a daunting task due to the time-scale limitations of the molecular dynamics method. We discuss new methods, derived from transition state theory, for accelerating molecular dynamics simulations of these infrequent-event processes. Two of these methods (hyperdynamics and parallel replica dynamics) have been presented previously, and are briefly reviewed here. The third, temperature-accelerated dynamics (TAD), is presented in detail. In TAD, the system temperature is raised to stimulate more rapid escape out of each potential basin, but attempted transitions are filtered to allow only those that would have occurred at the normal temperature. The characteristics of the methods are compared.

I. INTRODUCTION

Many of the important dynamical processes in materials are diffusive in nature. Examples include dislocation climb, pipe diffusion along a dislocation core, segregation of minority species to an interface, void growth, grain growth, and surface shape evolution. The individual, microscopic events that combine sequentially to evolve the system have an infrequent-event nature, meaning that the system vibrates in a potential basin many times before each new transition occurs. This characterization is valid even when the process is driven by an applied stress, unless the stress is extremely high. These processes could be well described with modern atomistic simulation methods were it not for the time-scale problem. Molecular dynamics (MD) simulations are limited to nanoseconds, while the time between successive diffusive events can be microseconds, milliseconds, or longer. A commonly used alternative to direct MD, kinetic Monte Carlo, can reach these longer times, but it requires pre-cataloging all the diffusive events that may take place. For realistic systems, the dynamical events are often complicated, so obtaining this advanced information about the individual diffusive events is difficult or impossible. Discovering all the possible events could require running the unfeasibly long MD simulation that the kinetic Monte Carlo was supposed to replace.

Recently, some new approaches to this atomistic time scale problem have been proposed [1–3]. The basic idea is to exploit the infrequent-event nature of the dynamics in a way that shortens the required simulation time between diffusive events. For example, in the hyperdynamics method [1,2], the potential energy surface is modified with a bias potential that raises the energy within each basin, so that the trajectory spends less time near the minimum and finds an escape path sooner. The accelerated simulation time, which becomes a statistical property of the system, is estimated as the trajectory proceeds. This hyperdynamics method is briefly reviewed in Section II below. Another recent development in accelerating infrequent events is the parallel replica method [3], in which the power of parallel processing is applied to extend the MD simulation time. This is in contrast to the usual parallel MD algorithms, which extend the length scale. This parallel replica

Mat. Res. Soc. Symp. Proc. Vol. 538 © 1999 Materials Research Society

method, which can be used in combination [4] with the hyperdynamics method, is briefly described in Section III.

The hyperdynamics method is a powerful approach; simulation times in the microseconds have been demonstrated. However, designing an effective bias potential requires some experimentation. Moreover, the iterative procedure for computing the bias potential and its derivatives must be tuned somewhat for each new type of system. Consequently, there is a motivation to develop simpler methods.

One way to achieve this simplicity, as suggested by Steiner *et al.* [5], is to employ a simpler form for the hyperdynamics bias potential. They studied an extremely simple form, in which all potential energies below a fixed value (V_b) are set to V_b. This approach, which requires no computational overhead and very little tuning, appears promising for low-dimensional systems, e.g., tens of atoms, and may be effective for larger systems when the active region can be reduced to this size [6].

In this paper, we present an alternative to the hyperdynamics method that is in some ways simpler and should be advantageous in many situations. This new approach is based on the well-known concept that events occur more frequently when the temperature is raised. The key is to filter the events (i.e., rejecting many events) in a way that gives the proper evolution of the system. We have developed an extrapolation procedure that extracts the correct transition, and the correct transition time, from the high-temperature run. The method, which we call temperature-accelerated dynamics (TAD), thus gives the correct state-to-state dynamical evolution. Assuming a few requirements are met, TAD is easy to implement and achieves boost factors roughly comparable to hyperdynamics. The method should be easily generalized to various types of condensed-phase systems and different interatomic potentials. It should also be possible to combine TAD with parallel replica dynamics, although this is not demonstrated here.

This paper is organized as follows. The next two sections give brief reviews of transition state theory, hyperdynamics and parallel replica dynamics. These descriptions provide a foundation for the derivation of the new temperature-accelerated dynamics, which is given in Section IV. A few example calculations using TAD are presented in Section V, and the characteristics of the three acceleration methods are then compared in Section VI.

II. BRIEF REVIEW OF TRANSITION STATE THEORY AND HYPERDYNAMICS

We seek to characterize the dynamics of a system that makes infrequent transitions from one potential basin to another. In transition state theory (TST), the rate constant for escape from one state (A) to another (B) is given by the flux through the dividing surface separating the two states. This is an equilibrium property of the system, in the following sense. We could, in principle, run a dynamical simulation for a very long time, so that the system makes many transitions back and forth between states A and B, giving an average time spent in each state that is proportional to the partition function for that state. Analyzing this trajectory, we can count the number of forward (A to B) crossings of the dividing surface. This count, divided by the total time the trajectory resided in state A, gives the TST rate constant $k_{A \to B}^{TST}$. If the successive crossings of the dividing surface are not dynamically correlated, i.e., if many vibrational periods occur between crossings, then the TST rate constant is equivalent to the exact rate constant. However, the real beauty of TST is that $k_{A \to B}^{TST}$ can be computed without ever running a trajectory. In the canonical ensemble, $k_{A \to B}^{TST}$ is proportional to the ratio of the partition function of the dividing surface "state" to the partition function of state A. For solid state systems, TST tends to be a very good approximation to the exact rate. Even the harmonic approximation to TST [7] is reasonably accurate; this is employed below for the temperature-accelerated dynamics.

In the traditional application of TST, one must first define the dividing surface, or find the saddle point (in the case of harmonic TST), after which the rate constant is computed. However, transition state theory, and its implications for the dynamical properties of the system, are far more general. The methods described here exploit these general properties of TST to give accelerated escape from each basin without advanced knowledge of the dividing surfaces or the events that might occur.

In hyperdynamics, the dynamical simulation is run on a potential energy surface that has been modified in a specific way. A nonnegative "bias" potential $\Delta V_b(\mathbf{R})$, which is zero at all dividing surfaces and also meets a few other requirements [1], is added to the original potential. A schematic example of this type of modification is shown in Fig. 1.

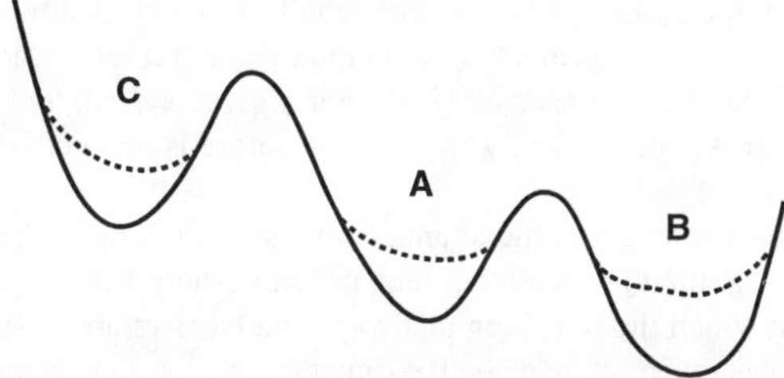

Figure 1. Illustration of the hyperdynamics method. A trajectory is evolved on the biased potential (dashed line) rather than the original potential (solid line). Relative escape probabilities are preserved because the bias potential is zero at the dividing surfaces. The accelerated time scale is calculated as the simulation proceeds.

As discussed in detail elsewhere [1], appealing to TST, one can show that the dynamical evolution on this modified potential surface has some desirable properties. First, and most obvious, is the fact that escape from each state will occur, on average, more quickly than for the regular dynamics (dynamics on the unbiased potential surface). Second, the relative escape rates to different possible adjacent states (e.g., A→B vs. A→C in Fig. 1) are exactly the same as for the original surface. This means that the system evolves from state to state in a sequence that is indistinguishable from the regular dynamics. Third, an estimate of the (accelerated) time that the system has evolved after n_{step} MD steps can be computed from

$$t_{hyper} = \sum_{i=1}^{n_{step}} \Delta t_{MD} \exp[\Delta V_b(\mathbf{R}_i)/k_B T] \, , \qquad (2-1)$$

where Δt_{MD} is the MD time step, $\mathbf{R}_i$ is the configuration-space position of the trajectory at the ith MD step, k_B is the Boltzmann constant, and T is the temperature. While this definition for the accelerated time (or "hypertime") is meaningless on times comparable to a vibrational period, it converges on the exact value (with vanishing relative error) at long times. It thus gives a realistic estimate, with error bars, for the time at which each transition occurs. Note that Eq. (2-1) reduces to the expression for normal MD if ΔV_b is zero everywhere.

Defining an effective bias potential is a key requirement for application of the hyperdynamics method. $\Delta V_b(\mathbf{R})$ must be zero (or nearly so, for the method to be useful) at all the dividing surfaces, even though we do not know in advance where the dividing surfaces are. As shown elsewhere [2], the distance from a position in **R**-space to a dividing surface (or a possible dividing surface) can be determined by examining the local slope and curvature properties. A suitable bias potential is created by designing a function of these properties that goes smoothly to zero as the dividing surface is approached and gives a controllable maximum value within the basin. The necessary slope and curvature quantities (based on the gradient and eigenvalues of the Hessian) are evaluated using an iterative procedure that requires only first derivatives of the potential. In this approach, the bias potential, and its derivatives, are computed using a few tens of calls to the MD force routine at each position of the hyperdynamics trajectory.

This implementation of hyperdynamics requires a roughly 30 times more work per integration step than direct MD. If the average boost factor (the ratio of hypertime to MD time) is greater than this extra work factor, hyperdynamics is computationally faster than direct MD. In tests so far, mostly on metal systems using embedded atom method (EAM) interatomic potentials [8], boost factors from 10^0 to 10^6 have been observed [2,4]. For a given system and a fixed bias potential definition, the boost increases dramatically as the temperature is lowered, due to the exponential dependence in Eq. (2-1).

For a system at fixed temperature, the attainable boost is limited by the lowest barrier in the system. Loosely speaking, the bias potential must not raise the potential higher than this lowest barrier. For a system in which the transition pathways and barriers are unknown, one can choose a safe strength for the bias potential by using the time the simulation has progressed since the last transition to place a lower bound on the lowest barrier in the system. As the simulation proceeds, accumulating more hypertime, this estimate of the lowest barrier height can be successively increased. This, in turn, causes the hypertime to accumulate more rapidly [4]. In this approach, the bias strength must be reset to zero, or a very low value, after each transition.

III. BRIEF REVIEW OF PARALLEL REPLICA DYNAMICS

In parallel replica dynamics, we appeal to the first-order nature of infrequent-event dynamics to develop a parallel method [3] for extending the time scale. In this approach, a replica of the entire system is placed on each processor, and each trajectory is evolved independently until a transition occurs. This is in contrast to the usual parallel implementation in large-scale MD programs, in which the system is divided up over a grid of processors.

For an M-processor computer, the method works as follows: On each processor, an independent trajectory for the entire system is propagated, while monitoring for a transition. Transition monitoring typically consists of simply interrupting the trajectory every few ps to follow a steepest descent path to the minimum of the current potential basin. The geometry thus obtained will abruptly change if a transition has occurred since the last check. When a transition to a new basin is detected (on processor i, say), the other processors are stopped, and the master simulation clock is advanced by the sum of the trajectory times accumulated on all the processors up to the time of the transition. The trajectory on processor i is then continued for a short time (a picosecond or two) to allow completion of correlated dynamical events. This additional correlation time is added to the simulation clock, and the procedure is repeated by replicating the new geometry of processor i onto all M processors.

As shown elsewhere [3], this method gives the correct probability distribution for the time between escapes, and preserves the relative probabilities of the different possible escape paths. The state-to-state dynamical evolution is thus exact, and proceeds nearly M times faster than a single-processor trajectory, provided the average time between transitions is much longer than the time required to evolve the correlated events on one processor. Communication time is typically insignificant, although it could become important in implementations on hundreds of processors.

Combining this approach with the hyperdynamics method gives a total boost (the increase in simulation time per wall-clock time) that is the product of the hyperdynamics boost with the parallel boost. Simulations on hundreds of processors have been performed in this fashion, acheiving millisecond time scales [4].

IV. TEMPERATURE-ACCELERATED DYNAMICS

As in hyperdynamics and parallel replica dynamics, the goal in temperature-accelerated dynamics (TAD) is to stimulate rapid escape from each potential basin. In TAD, this is accomplished by raising the temperature of the system from the desired temperature (T_{low}) to some higher temperature (T_{high}). Although raising the temperature is a well-known and often-used approach for accelerating diffusive events, it does not give the correct state-to-state evolution for temperature T_{low} because the relative escape probabilities change with temperature. However, if, from among the transitions occurring rapidly at T_{high}, the correct T_{low} transitions can be selected and the others prevented, then the system will evolve from state to state appropriately for temperature T_{low}. Here we show how to accomplish this and how to recover the exact time at which each transition occurs.

Consider a system trapped in a potential basin with n escape paths, and a rate constant, k_i, associated with each escape path. The total escape rate (k_{tot}) is given by the sum of these rates,

$$k_{tot} = \sum_{i=1}^{n} k_i . \qquad (4-1)$$

Our first approximation is that this system obeys transition state theory. Because this is also an infrequent-event system, escape from this basin is a first-order process. The probability distribution for the time until the first escape is thus given by the exponential distribution

$$p(t) = k_{tot} \exp(-k_{tot}t) . \qquad (4-2)$$

Similarly, if we consider each of the escape paths individually, the probability distribution for the time to first escape for each escape path i is given by

$$p_i(t) = k_i \exp(-k_i t) . \qquad (4-3)$$

In essence, the goal of any accelerated dynamics method is to answer the question: which path will the system choose to escape from the current basin, and when will that escape occur? If we have the set of rates $\{k_i\}$, one way to find the answer is to draw one random number from each of the n distributions in Eq. (4-3). Each of these numbers corresponds to the time at which that transition would first occur. We then select the lowest of these times and the pathway to which it corresponds. The remaining times can be discarded, as the system moves to a new state at the instant of the first transition. Because we are considering an infrequent event system that obeys transition state theory, in which all memory of the previous transition is lost before the next one occurs, this is an *exact* way to determine the state-to-state dynamics. (In fact, although inefficient, this would be one way to construct a kinetic Monte Carlo algorithm.)

Now consider evolving this system using molecular dynamics. After a number of vibrational excursions (perhaps a huge number), the trajectory passes through a dividing surface, taking the system into a new state. In essence, the simulation "picks" one of the escape paths. The time of this transition can be precisely determined by dropping steepest-descent paths from successive positions along the trajectory to identify the time at which these steepest descents begin to fall into the new state instead of the old state. The key point is this: the escape path and escape time determined in this way are indistinguishable from the random-number method described in the previous paragraph. In either method the probability of choosing a particular escape path is k_i/k_{tot}. Applied repeatedly, either method yields a distribution of first passage times given by Eq. (4-2), and an average escape time of $1/k_{tot}$.

We can extend this concept by considering a "basin constrained" trajectory, in which all transitions out of the basin are blocked, e.g., by reflecting the trajectory back into the basin at the instant of each attempted escape. In a basin-constrained trajectory, the set of all escape attempts and associated times (the first escape time for each pathway) will be indistinguishable from a set of times chosen randomly from exponential distributions as described above. In this sense, a molecular dynamics simulation can be viewed as a very expensive random number generator.

We now make our second approximation. Assume that each escape rate k_i is well described using the harmonic approximation to TST [7]. The rate constant then takes on an Arrhenius form,

$$k_i = \nu_i \exp[-\Delta E_i/k_B T] \,, \qquad\qquad (4-4)$$

where ΔE_i is the energy difference between the basin minimum and the saddle point, k_B is the Boltzmann constant, and ν_i is a *temperature-independent* pre-exponential factor that depends on the vibrational spectrum at the minimum and the saddle point [7]. This is an excellent approximation for many diffusive processes in materials, although if the temperature is raised high enough, the anharmonicities cause a deviation from this ideal behavior. A consequence of Eq. (4-4) is that the ratio of rate constants at two temperatures (T_{high} and T_{low}) is given by

$$k_i(T_{high})/k_i(T_{low}) = \exp[-\Delta E_i(1/k_B T_{high} - 1/k_B T_{low})] \,, \qquad\qquad (4-5)$$

The above discussion leads us to the following key concept: the set of first-escape times $\{t_i^{high}, i = 1, 2, ..., n\}$ resulting from a basin-constrained simulation at temperature T_{high}, maps onto a valid set of low-temperature first-escape times $\{t_i^{low}, i = 1, 2, ..., n\}$, via the following transformation:

$$t_i^{low} = t_i^{high} \exp[\Delta E_i(1/k_B T_{low} - 1/k_B T_{high})] \,. \qquad\qquad (4-6)$$

In other words, by running a simulation at high temperature, we can generate a list of first-escape times for the low temperature case from Eq. (4-6) that is indistinguishable (in a probabilistic sense) from a list generated by a simulation at low temperature. No statistical test could distinguish whether the list was generated by an actual low-temperature simulation or by the transformation of a high-temperature list.

The transformation (4-6) is easily performed if the barrier height for the escape path is known. Thus, in this method, as each attempted transition is detected during the high temperature simulation, the trajectory is momentarily interrupted to perform a search for the corresponding saddle point. Finding this saddle, a requirement for the success of this method, is not a trivial task. We discuss this in more detail elsewhere [9], but we note that finding the saddle point to a given precision in the energy typically requires an amount of computational work that scales as N^x, where N is the number of atoms and x is between one and two. In contrast, computing the pre-exponential factor, which is not necessary in this algorithm, would require work scaling as N^3, in addition to construction of the Hessian matrix.

As the high-temperature, basin-constrained trajectory proceeds, a list of first-escape times $\{t_i^{high}\}$, accumulates, each one mapping onto a low-temperature first-escape time, t_i^{low}. The low-temperature values are always larger, and in general the order will not be the same, because the barriers and pre-exponentials vary. (This is why simply running a simulation at elevated temperature gives the wrong state-to-state sequence). Our goal is to determine which event would occur first at the low temperature. In principle, we could run the constrained simulation long enough to detect all possible events at least once (some will occur many times before we reach that point). However, this would be problematic, as in general we don't know how many possible events there are. Luckily, finding all the events is unnecessary, as we will now show.

Appealing to a visual construct, an Arrhenius plot, we note that the ordinate can be thought of as either the logarithm of the rate or as the logarithm of inverse time. In the inverse-time view, the time progression of a simulation (at temperature T) corresponds to logarithmic motion *downward* along a vertical line. On such a plot, the transformation of Eq. (4-6) maps a point on the T_{high} time line onto a point on the T_{low} time line by a simple extrapolation along a line with slope $-\Delta E_i/k_B$. Figure 2 shows an example after two attempted transitions have occurred at T_{high}. We return to this picture momentarily.

Considering a particular event i, with rate constant k_i, the average time until the first occurrence of that event is

$$\tau_i = 1/k_i \ . \tag{4-7}$$

Integrating the probability distribution in Eq. (4-3), it is easy to show that for a confidence value f (e.g., $f=0.99$ for 99% confidence), the time we have to wait to be "sure " to observe event i is

$$\tau_i^f = \tau_i \ln(1/(1-f)) \ . \tag{4-8}$$

Using Eqs. (4-4) and (4-7), we can rewrite Eq. (4-8) as

$$\frac{1}{\tau^f} = \nu_i^f \ \exp[-\Delta E_i/k_B T] \ . \tag{4-9}$$

where ν_i^f is the confidence-adjusted pre-exponential factor

$$\nu_i^f = \frac{\nu_i}{\ln(1/(1-f))} \ . \tag{4-10}$$

On the inverse-time Arrhenius plot, Eq. (4-9) is a line with slope $-\Delta E_i/k_B$ and intercept $\ln(\nu_i^f)$. The intersection of this line with the time line for some temperature gives the simulation time at which we can be f-confident that the first occurrence of event i will have already occurred. I.e., if we plot points, $\{(1/T, \ln(1/t_i))\}$, at the time of the first occurence of event i for each of a large number of independent simulations, a fraction f of these point will be above the intersection of Eq. (4-9) with the time line.

We now introduce the third and last approximation. We assume that the pre-exponential factor for every rate in the system is greater than ν_{min}. Because pre-exponential factors in real systems are typically grouped in the vicinity of $\sim 10^{12} - 10^{13} s^{-1}$, we can easily pick a value for ν_{min} (e.g., $10^{11} s^{-1}$) that is almost certainly below all the pre-exponentials in the system. With this assumption, a simple calculation tells us the time at which we can stop running the constrained trajectory and select the event corresponding to the lowest value of t_i^{low} observed so far.

Following Eq. (4-10), we define the f-adjusted minimum pre-exponential factor ν_{min}^f by

$$\nu_{min}^f = \frac{\nu_{min}}{\ln(1/(1-f))} \ . \tag{4-11}$$

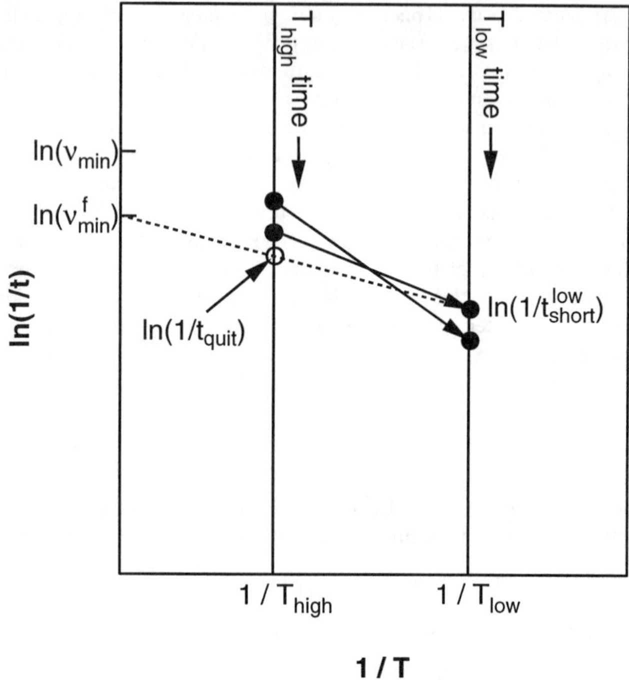

Figure 2. Schematic illustration of the temperature-accelerated dynamics method. Progress of the high-temperature trajectory can be thought of as moving down the vertical time line on the left. For each transition detected during the run, the trajectory is reflected back into the basin, the saddle point is found, and the time of the transition (solid dot on left time line) is transformed (arrow) into a time on the low-temperature time line on the right. Plotted in this Arrhenius form, this transformation is a simple extrapolation along a line whose slope is the negative of the barrier height for the event. The dashed termination line connects the shortest-time transition recorded so far on the low temperature time line (solid dot) with the confidence-modified minimum pre-exponential ($\nu^f_{min} = \nu_{min}/\ln(1/(1 - f))$) on the y axis. The intersection of this line with the high-T time line gives the time (t_{quit}) at which the trajectory can be terminated. With confidence f, we can say that any transition observed after t_{quit} could only extrapolate to a shorter time on the low-T time line if it had a pre-exponential lower than ν_{min}.

We also define t^{low}_{short} as the shortest time (the highest point) on the low-temperature time line yet observed for this basin. Consider the dashed line in Fig. 2, which passes through t^{low}_{short} on the T_{low} time line and has intercept $\ln(\nu^f_{min})$. This line, which has a slope of $-\Delta E_{term}/k_B$, will be referred to as the "termination line." The intersection of the termination line with the T_{high} time line defines the time (t_{quit}) at which the high-T trajectory can be terminated, as we now prove.

Assume that the high-temperature basin-constrained trajectory has been run to time t_{quit}, so that any future transition would give a point on the T_{high} time line below the intersection with the termination line. Drawing on the argument given above, we know with confidence f that any escape pathway that would first reveal itself after t_{quit} must have a rate lower than $\nu_{min}\exp[-\Delta E_{term}/k_B T_{high}]$. Because no pathway can have a pre-exponential lower than ν_{min}, the rate can only be lower by having a higher activation barrier. But having a barrier higher than ΔE_{term} guarantees that the extrapolation (4-6) will create a point on the low-T time line that lies below t_{short}^{low} (i.e., at longer time), and hence cannot change the chosen transition.

The intersection of the termination with the high-temperature time line defines the quitting time as

$$t_{quit} = \frac{1}{\nu_{min}^f}\left(t_{short}^{low}\nu_{min}^f\right)^{T_{low}/T_{high}}. \qquad (4-12)$$

Note that in the limit as T_{high} approaches T_{low}, Eq. (4-12) reduces to $t_{quit} = t_{short}^{low}$, which is identical to direct molecular dynamics – every transition is immediately accepted. Thus, the TAD approach is completely compatible with direct MD, in the sense that the high temperature, basin-constrained sampling can be turned on and off at any time without corrupting the state-to-state dynamics.

The computational boost obtained during a temperature-accelerated dynamics simulation in one basin will consist of two factors: the ideal boost, which is the ratio of t_{short}^{low} to t_{quit}, and the inefficiency arising from the need to detect transitions and search for saddle points.

Inspection of Eq. (4-6) shows that low barriers limit the available boost in this method, a problem also seen in hyperdynamics. In TAD, it results from the fact that t_{low}/t_{high} depends exponentially on the barrier height. We show elsewhere [9] that this problem can be substantially reduced in TAD by applying a special treatment to transitions that have occurred many times. This modification still gives the correct dynamics (to the extent that the basic approximations in TAD are valid), but reduces the time wasted on these fast transitions.

V. APPLICATIONS OF TEMPERATURE-ACCELERATED DYNAMICS

Here we present results of two demonstration calculations using temperature-accelerated dynamics. First, we discuss a relatively simple case, the diffusion of a surface vacancy on Ni(001) under anisotropic strain. The system consists of one moving layer of atoms (with one atom removed) resting on two fixed layers. Constraining the system to one moving layer eliminates the possibility of the vacancy moving by an exchange event, keeping the example simpler. To create two simple processes with differing barrier heights, the lattice is strained anisotropically – it is compressed 2% in the [110] (x) direction and expanded by 2% in the [1$\bar{1}$0] (y) direction. We employ an EAM interatomic potential [10,11] which was fit to properties of bulk Ni and the diatomic molecule.

The barrier for the basic hop event is 0.6745 eV in the x direction and 0.6202 eV in the y direction. The Vineyard pre-exponentials (computed from the normal modes of the 35 moving atoms at the minimum and the saddle) are $1.49\times10^{13}\mathrm{s}^{-1}$ (x) and $1.33\times10^{13}\mathrm{s}^{-1}$ (y).

For this demonstration, we chose $T_{low} = 400K$, $T_{high} = 1000K$, and $\nu_{min}=4.0\times10^{12}$ s^{-1}. (An aggressive value for ν_{min} was chosen based on the known pre-exponential factors.) During the simulation, which lasted 5 days on an SGI R10000 workstation, a total of 2000 transitions were observed during a total (low-temperature) time of 4.28×10^{-3} s. Of these 2000 transitions, 424 were hops in the x direction, and 1566 were hops in the y direction. (Ten events were more complicated, as discussed below.) At the high temperature, there were a total of 10564 attempted x-hop events and 14454 attempted y-hop events. The ratio of x hops to y hops at the high temperature (0.731) was noticeably higher than at the low temperature (0.271), as expected from the higher barrier for the x-hop event.

The rates at the low temperature are in reasonable agreement with the Vineyard predictions. The average time between x-hop events was $1.01(\pm0.05)\text{x}10^{-5}$ s, which is 4% faster than the Vineyard prediction ($1.05\text{x}10^{-5}$ s), while the average time between y-hop events was $2.73(\pm0.07)\text{x}10^{-6}$ s, 10% slower than the Vineyard prediction ($2.46\text{x}10^{-6}$ s). We attribute the deviations from the Vineyard prediction largely to anharmonicity, as they are similar to the deviations observed in the average escape times at the high temperature (7% fast for the x hop and 11% slow for the y hop). This error is projected onto the low temperature when the time extrapolation is performed.

It is also interesting to examine what event occurred first during each basin-constrained trajectory – i.e., what transitions would have occurred had the simulation simply proceeded at T_{high}. Of the 2000 transitions, the first events consisted of 729 x-hops, 1072 y-hops, 44 double x-hops (in which two atoms move in concert, in the same direction, to fill in the vacancy), 142 double y-hops, 1 triple x-hop, 11 triple y-hops, and 1 more complicated event. Of these exotic events, only the double y-hop (whose barrier is 0.861 eV) was ever selected (10 times) for the low-temperature transition.

The raw boost for this run was $4.9\text{x}10^{3}$, while the computational boost (the total low-temperature simulation time divided by the simulation time that could have been achieved with the same computer time in a direct MD simulation), was $4.3\text{x}10^{3}$. A more conservative choice for ν_{min} would have lowered these values somewhat. The computational overhead due to the transition detection and saddle-point searches was about 15%.

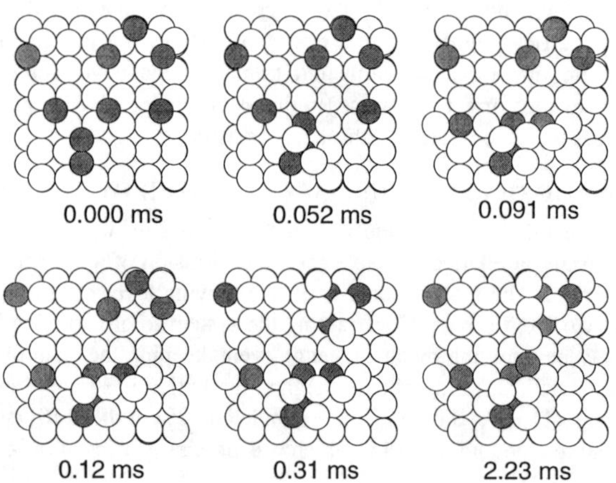

Figure 3. Selected snapshots from temperature-accelerated dynamics simulation of the T=150K evolution of 1/4 monolayer of Cu deposited on Ag(100). The time at which each configuration was first visited is shown.

The second example, which demonstrates the power of the TAD method for more complex problems, is a simulation of 0.25 monolayer of Cu atoms on a perfect Ag(100) surface at T=150K. This mimics an experiment in which 0.25 monolayer is deposited at a very low temperature (e.g., 77K), after which the system is quickly warmed to 150K and held there.

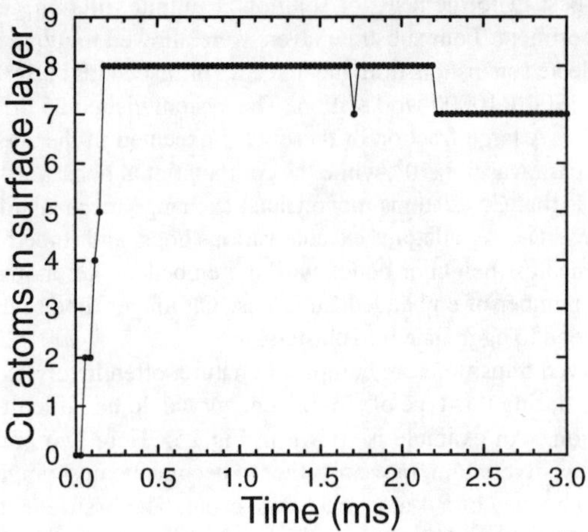

Figure 4. Time evolution of the number of copper atoms that become embedded in the top substrate layer of the Ag(100). The intial configuration had nine copper adatoms, none embedded.

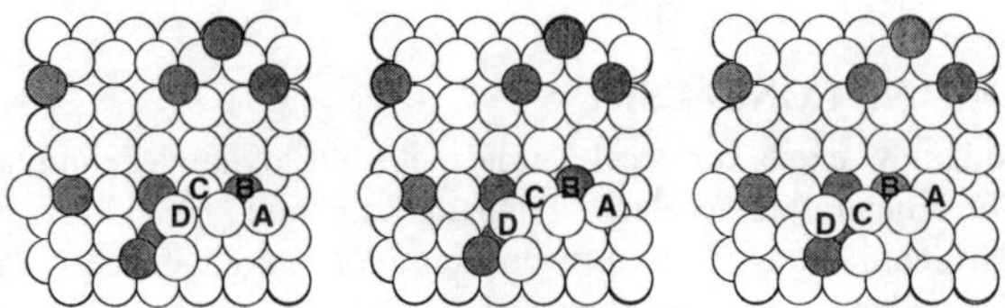

Figure 5. Selected event from the temperature-accelerated dynamics simulation of the T=150K evolution of 1/4 monolayer of Cu deposited on Ag(100). The left frame shows the initial state, the center frame shows the saddle point (0.216 eV), and the right frame shows the final state (-0.710 eV). The process, which brings two embedded copper atoms together, involves a total of four atoms, two in the substrate and two on the surface.

In this simulation, we set f=0.99 and ν_{min}=1.0x10^{12} s^{-1}. This slightly aggressive value for ν_{min} was chosen to produce results more quickly for this example. Reducing ν_{min} by a factor of 10 (a very safe value) would increase the computational overhead by about a factor of three (see Eq. (4-12)). T_{high} was adjusted automatically for each basin, to achieve a value of around 6 for the ratio of the lowest barrier height to the temperature. Also, the speed of the simulation was increased by applying the special treatment of low-barrier transitions [9] mentioned above. The (EAM) interatomic potentials employed for Cu and Ag were very similar to those presented previously [3,12], except that a smoother cutoff was employed. The cross potential was fit using the method proposed by Johnson [13], and a scale factor of 1.285 was applied to the density function for Ag to obtain the best fit to the heats of solution at infinite dilution, resulting in values each within 0.02 eV of experiment. Four substrate layers were allowed to move in the simulation.

Figure 3 shows selected snapshots from the first 2.23 milliseconds of the simulation, which took about five days on the SGI R10000 workstation. The system visited 25 different states, making a total of 474 transitions. A large fraction of these were executed in the special low-barrier mode. The raw boost in this case was 9.8×10^4, while the computational boost was 2.4×10^4.

The overall trend is that Cu adatoms move via an exchange mechanism into the surface layer, generating Ag adatoms. The Ag adatoms execute various hops, and sometimes further exchanges, but tend to maintain nearest-neighbor bonds with the embedded Cu atoms. Figure 4 shows the time evolution of the number of embedded Cu atoms. On longer time scales, we believe that the embedded Cu atoms tend to aggregate into clusters.

Many of the observed transitions are complex in nature, often involving exchange [14]. These events occur naturally during this type of simulation, but would be difficult to include in a kinetic Monte Carlo simulation. An example is shown in Fig. 5. Here two embedded copper atoms, seperated by a substrate silver atom, move together to become nearest neighbors (an energetically favored configuration) by way of a 4-atom exchange event. The basic event is similar to the long-distance exchange event described by Cohen [15], in which an adatom (A) pushes a substrate atom (B), which pushes another substrate atom (C), causing it to emerge as an adatom. In the variation oberved here, A is silver, B is copper, C is silver, and the emerging silver atom pushes another silver adatom (D) over one site, so that four atoms are directly involved. In addition, there is a silver-adatom spectator, which maintains nearest-neighbor bonds to A, B, and C throughout the process. This is a key event, moving two embedded copper atoms together and lowering the energy by 0.71 eV.

VI. COMPARISON AND CONCLUSIONS

A set of methods now exists for accelerating molecular dynamics simulations of infrequent events in solids. The temperature-accelerated dynamics method (TAD) looks particularly promising for systems in which the anharmonic effects are minor or not a concern. (At the high temperatures used in TAD, we have seen rates increase by a factor of two or more due to anharmonicity, especially for exchange events.) Application of TAD to new system types should be relatively easy (so far we have almost exclusively examined systems modeled with EAM potentials). In contrast, applying hyperdynamics to new systems can require more serious tuning of both the bias potential and the iterative procedure for computing it. On the other hand, hyperdynamics may yet prove to be more powerful as new, computationally efficient forms for the bias potential are developed. Defining the bias potential for a subspace of the system can also help [6]. This local approach is not an option for TAD.

Both hyperdynamics and TAD give intrinsic boost factors that increase exponentially as the temperature is lowered or as the lowest barrier is raised. (For a few systems on which we have performed a comparison, hyperdynamics and TAD give boost factors that are of the same order of magnitude, but this may be very case specific.) This intrinsic boost factor is reduced by an efficiency factor. In hyperdynamics, the inefficiency arises from the need to compute the bias potential and its derivatives. In TAD, the main inefficiency is due to the extra time, after detecting the transition that will ultimately be chosen, that the basin-constrained trajectory must be integrated to reach t_{quit}, In TAD, some additional inefficiency arises from detecting transitions and finding saddle points.

In parallel replica dynamics, the boost factor is the number of processors. The computational overhead is minimal unless events occur so rapidly that the communication time or correlated-event trajectory time interferes. This limit is approached as the temperature is raised or the number of processors is increased.

In principle, all three methods methods suffer from the low-barrier problem. In the parallel replica approach the problem is typically not severe unless events are very rapid. For the TAD method, the low barrier problem can be reduced substantially using an extension presented elsewhere [9].

The parallel replica method can be combined with either of the other two methods to obtain multiplicative boosts in simulation time. While hyperdynamics and TAD assume the system obeys TST, the parallel replica method (though derived using TST) does not make this requirement. Thus, it may be useful on its own to accelerate the dynamics in system where correlated events are significant, or are important to the physics under study.

ACKNOWLEDGEMENTS

This work was supported by the United States Department of Energy, Office of Basic Energy Sciences.

REFERENCES

1. A.F. Voter, J. Chem. Phys. **106**, 4665 (1997).
2. A.F. Voter, Phys. Rev. Lett. **78**, 3908 (1997).
3. A.F. Voter, Phys. Rev. B **57**, 13985 (1998).
4. A.F. Voter and T.C. Germann, Mat. Res. Soc. Symp. Proc. **528**, 221 (1998).
5. M. M. Steiner, P.-A. Genilloud, and J. W. Wilkins, Phys. Rev. B **57**, 10236 (1998).
6. W.G. Rudd and A.F. Voter, elsewhere in this MRS proceedings volume.
7. G.H. Vineyard, J. Phys. Chem. Solids **3**, 121 (1957).
8. M.S. Daw, S.M. Foiles, and M.I. Baskes, Mater. Sci. Reports **9**, 251 (1993).
9. M.R. Sørensen and A.F. Voter, to be published.
10. A.F. Voter and S.P. Chen, Mat. Res. Soc. Symp. Proc. **82**, 175 (1987).
11. A.F. Voter, in *Intermetallic Compounds: Principles and Practice*, edited by J.H. Westbrook and R.L. Fleischer, John Wiley and Sons, Ltd, 1995), Vol. 1, p. 77.
12. A.F. Voter, in *Modeling of Optical Thin Films*, M.R. Jacobson, Ed., Proc. SPIE **821**, 214 (1987).
13. Johnson, R.A., Phys. Rev. B **39**, 12554 (1989).
14. P.J. Feibelman, Phys. Rev. Lett. **65**, 729 (1990).
15. J.M. Cohen, Surf. Sci. Lett. **306**, L545 (1994).

ATOMISTIC ASPECTS OF FRACTURE MODELLING IN THE FRAMEWORK OF CONTINUUM MECHANICS

F. CLERI
Divisione Materiali Avanzati, ENEA, Centro Ricerche Casaccia, C.P. 2400, 00100 Roma (Italy)

ABSTRACT

The validity and predictive capability of continuum models of fracture rests on basic informations whose origin lies at the atomic scale. Examples of such crucial informations are, e.g., the explicit form of the cohesive law in the Barenblatt model and the shear-displacement relation in the Rice-Peierls-Nabarro model. Modern approaches to incorporate atomic-level information into fracture modelling require to increase the size of atomic-scale models up to millions of atoms and more; or to connect directly atomistic and macroscopic, e.g. finite-elements, models; or to pass information from atomistic to continuum models in the form of constitutive relations. A main drawback of the atomistic methods is the complexity of the simulation results, which can be rather difficult to rationalize in the framework of classical, continuum fracture mechanics. We critically discuss the main issues in the atomistic simulation of fracture problems (and dislocations, to some extent); our objective is to indicate how to set up atomistic simulations which represent well-posed problems also from the point of view of continuum mechanics, so as to ease the connection between atomistic information and macroscopic models of fracture.

INTRODUCTION

The traditional approach to the study of the mechanical properties of materials is based on the classification of the various types of macroscopic response of a solid body to an external stress. The corresponding theoretical descriptions of this complex phenomenology, such as linear isotropic and anisotropic elasticity, linear-elastic and elastic-plastic fracture mechanics, dislocation theory, elastodynamics and so on, are based on the continuum mechanics of solid bodies. For example, macroscopic plastic deformation near the tip of a moving crack may involve very high dislocation densities up to 10^8 dislocations/mm^2, thus making irrelevant the description of single dislocations while a continuum treatment is the most natural approach. Many useful results for the average mechanical response, failure analysis and lifetime prediction for a particular material can routinely be obtained through numerical techniques based on macroscopic continuum-mechanics theories [1]. The literature in this field is immense and dates back at least to the first theoretical study of crack extension in an infinite plate, carried out by A. A. Griffith in 1920 [2].

The mechanical response of a material to a state of stress is a typical example of a multiscale phenomenon in physics, being the result of competing processes (*e.g.*, interplanar cleavage, dislocation generation and motion, self-diffusion, segregation, grain-boundary sliding and migration) that although originating at the atomic scale have macroscopic consequences. The ability to develop a science of "materials engineering", therefore, requires a deeper experimental and theoretical understanding of such processes than is currently available. Indeed, the theoretical description of many macroscopic mechanical properties (*e.g.*, fracture toughness) lies well beyond the current possibilities of either analytic methods or computer simulations, owing

441

to the extreme complexity of real materials whose final properties depend on the interplay of (at least) chemistry, microstructural evolution and processing history. This complexity has encouraged the development of more fundamental approaches, aimed at a systematic and progressive clarification of the issues involved in the mechanical response of materials at the various length scales separately. Therefore, it is not surprising that there is general agreement that detailed theoretical understanding of the atomic-level mechanisms of deformation and failure, for which atomistic simulation techniques should be the most appropriate tool, can represent the basic level of description, from which information can then be integrated by progressively "coarsening" the space and time meshes [3].

Atomistic simulations and continuum mechanics must make contact with each other in order to deal with the multiple length- and time-scales over which the elementary phenomena combine into the macroscopic response. For example, recent nanoindentation experiments [4] show that a critical irreversible deformation under compressive load is carried by less than 100 dislocations within a volume of about 3×10^5 nm^3. Such deformation phenomena are too complex to be approached by present-day atomistic simulation techniques, but still require a considerably more detailed description than available to continuum mechanics. Solutions to this and many other problems in mechanical behavior can then be sought in either (a) a fully atomistic approach based on large-scale simulations, trying to attain the continuum limit by progressively inceasing the system size, (b) hybrid techniques, in which atomic-scale and continuum-mechanics-based computational methods are "tightly-and-smoothly" coupled, or (c) a hierarchic approach based on the use of *constitutive relations*, in which atomic-scale simulations provide the input to mesoscopic or macroscopic techniques in the form of a specific constitutive law (*e.g.*, a stress-strain relation). Clearly, in all of the above approaches atomistic simulations will play a central role, despite their intrinsic time- and length-scale limits which seem to exclude the possibility of an entirely atomistic approach, irrespective of the computer power available.

In this paper we firstly review the different possibilities of linking the information obtained at the atomic scale to the framework of macroscopic continuum mechanics, in the perspective of describing phenomena bridging different length (and possibily time) scales; secondly, we wish to provide a basis for the critical assessment of different methods for atomistic simulations of fracture problems.

LINKING ATOMISTIC SIMULATIONS AND CONTINUUM MECHANICS

The current capabilities of atomistic simulations are typically limited to lengths of some 100 nm and times of some 10^{-9} seconds. Unfortunately, the study of extended defects in solids, and thus of mechanical properties, requires the simulation of very large systems, necessitating a compromise between physical precision and computational feasibility.

In the area of fracture simulations, a fully atomistic treatment was attempted as early as about twenty years ago [5] and pursued in a number of successive studies. Only recently, however, has the sizeable increase in computing capacity provided by massively parallel computers allowed such a treatment to be systematically exploited: various groups [6-10] have reported results for system sizes in the range 10^6-10^8 atoms, corresponding to linear dimensions approaching the μm scale.

Large-scale atomistic simulations undoubtly represent the most rich and detailed tool for the study of complex phenomena originating at the atomic scale. The approach is conceptually simple, requiring only the increase of the system size to lengths much larger than the intrinsic length scale of the problem (*e.g.*, the length of a microcrack or the distance between two dislocations). This implies that no artificial border constraints are present, and the system is free to evolve following the most natural path in the phase space. However, the connection with continuum-mechanics models and with experimental results may be ambiguous, due to the fact that the simulation conditions are difficult to control.

One possible way around the need for indefinitely increasing the atomistic-system size is to couple the atomistic simulation to a continuum-mechanics model in a far away region, meaning far enough that the small-displacements assumption of linear elasticity holds. Starting in the early 1970's, "hybrid" techniques were therefore developed in which the region enclosing the extended defect (dislocation, microcrack), fully described at the atomic level, is embedded in a continuum medium. The most successful techniques in this group are based on the direct coupling of an inner fully atomistic region to an outer finite-element (FE) region [11,12]. Implementation of the coarsening of the spatial mesh going from the atomistic to the continuum region deserves particular care because information is integrated out in the process. The connection between the atomistic and FE regions was provided in earlier models [11] by a "transition" region described by one or more rows of volume elements containing several atoms each. More recent versions of the coupled atomistic-FE technique [12] make use of a transition region described by a non-local continuum-elasticity theory in order to improve the smoothness of the connection in presence of strain gradients.

A related, promising, atomistic-FE approach is the so-called "quasicontinuum" technique [13]. In this approach the formulation of the problem rests completely on a FE methodology, with a spatial mesh adaptively refined around highly energetic regions, becoming atomic-scale around defects. Interatomic interactions are explicitly accounted for in those elements describing individual atoms, while elements comprising several atoms are deformed according to the response of "representative" atoms usually chosen at the center or at the edges of the element.

A third approach to the problem of linking the atomic and mesoscopic or macroscopic length scales is the use of constitutive relations (CR), which has long been widespread in continuum mechanics. The basic idea of the CR approach is to assume a standard constitutive model (such as linear elasticity) for the bulk region away from the defect, while prescribing a specific material behavior (such as a particular stress-strain relation, *e.g.*, in the Peierls model of a mobile dislocation, or a cohesive equation, *e.g.*, in the Barenblatt cohesive-zone model of a crack tip) in the vicinity of the defect. Such a procedure allows the introduction of atomic-level information into a continuum description in a natural way.

Phenomenological CRs in the context of continuum-mechanics models are usually obtained on the basis of some simple and reasonable assumption relative to the underlying physics of the process. Recent work demonstrates that CRs obtained from atomistic simulations can effectively incorporate atomic-scale information into a continuum model [14-16]. Such atomistic simulations must be carefully connected with continuum mechanics in such a way that a common set of variables (*e.g.*, the stress and displacement fields) can be meaningfully established. The microscopic theory of elasticity [17] gives the exact relationships between lattice dynamics and the elastic properties of a crystal providing, for example, the definition of an atomic-level stress tensor that can be directly compared to a continuum-mechanical stress field. In order to make such a comparison meaningful, one must set up atomistic simulations that are well-defined problems from the point of view of continuum-mechanics analogs. When such a capability is firmly established, one can use atomistic simulations to derive CRs for more complex problems.

CONSISTENCY BETWEEN ATOMISTIC SIMULATIONS AND CONTINUUM MECHANICS

Although several atomistic studies of fracture have been already published little attention has been paid to such crucial issues as: what kind of border conditions can meaningfully represent an external load, what is the meaning of plane strain and plane stress in the atomistic context, how the different methods for initiating a microcrack or a notch in the atomistic system affect the results, how can the quality of the results of atomistic simulations be assessed. In the following we will discuss in some detail all these issues [18].

Border conditions

A central problem in any atomistic treatment of a system under external loading is to ensure that the simulation-cell border conditions properly represent the type and magnitude of the external load. This problem is closely connected with the suppression of system-size effects, such that the far-away stress and strain fields surrounding the crack converge to the proper linear-elastic fracture mechanics (LEFM) limit for a particular type of loading. The most frequently-used border conditions in atomistic simulations of static cracks can be grouped in two broad classes:

(a) *constant-displacement* border condition, in which a few rows of border atoms are kept fixed in the strained configuration or, equivalently, periodicity is preserved across the border for the whole duration of the simulation;

(b) *constant-traction* border condition, in which periodicity is removed all around the borders in the plane containing the crack and the forces (surface tractions) necessary to preserve the state of stress are computed and applied to the border atoms during the simulation.

The border conditions (a) and (b) correspond, respectively, to the displacement-boundary value (or Dirichlet) and to the stressed-boundary value (or Neumann) formulations of continuum-mechanics problems, in which either displacements or tractions are prescribed along the system border. In continuum mechanics it is also possible to use a combination of the two, by specifying tractions along some portion of the system border and displacements along the rest. For atomistic simulations, this can be shown to correspond to the removal of periodicity only in some directions, while keeping the others periodic.

Our aim is to set up a scheme to apply a general external load to a finite-size atomistic system reproducing the loading conditions of an infinite continuum. To this purpose, we consider the case in which a static external load, ideally applied at infinite distance from the region of the system where a microcrack will be subsequently placed, is represented by a homogeneous deformation tensor ε_{ij} applied to the atomistic system as a whole. Then, the analogy between constant-displacement (or constant-traction) borders and the presence (or absence) of periodicity across a border can be understood with the help of Fig. 1.

Figure 1a represents the constant-displacement border condition: the horizontal straight line (indicating a periodic border) defines the limit between the portion of the system explicitly simulated and its periodic replica. If some initial deformation ε_{ij} is imposed homogeneously to the system, equal and opposite forces will be induced on each border atom i (black) from its equally displaced neighbors j, ending up with a zero net force and thus with no additional displacement with respect to the imposed deformation; *i.e.*, although the atoms experience a state of non-zero stress they keep the deformed configuration because the net force on each atom is zero.

Figure 1b represents the constant-traction border condition: after imposing the same homogeneous deformation ε_{ij} to the system as above, periodicity in the direction normal to the

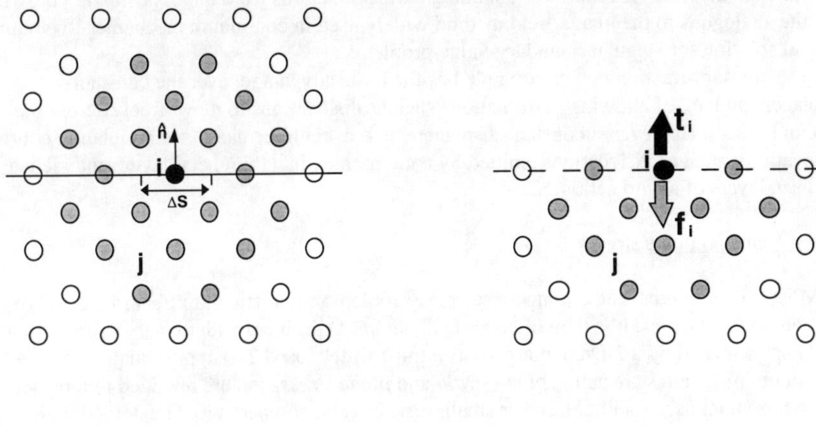

(a)   (b)

Fig 1 - Schematic for the definition of the surface traction in terms of the missing force on the border atom (black) when periodicity (indicated by the continuous line) is removed (indicated by the dashed line.

border is suppressed; *i.e.*, atoms in the region above the dashed line (indicating a free border) have been "removed". As we will show, the effect of these missing atoms can be mimicked by a suitable force (the equivalent of a surface traction in continuum mechanics) applied to the border atoms for the whole duration of the simulation. Because one half of their neighbors are missing, atoms at the border experience a non-zero resulting force f_i^α

$$f_i^\alpha = \sum_j \left(-\frac{1}{r_{ij}} \frac{\partial V}{\partial r_{ij}} \right) r_{ij}^\alpha \tag{1}$$

Using the virial definition of the stress [19], and noting that at temperature T=0 the kinetic contribution to the stress is zero,

$$\sigma_i^{\alpha\beta} = \frac{1}{\omega_i} \sum_j \left[\frac{1}{r_{ij}} \left(\frac{\partial V}{\partial r_{ij}} \right) (r_{ij}^\alpha \cdot r_{ij}^\beta) \right] \tag{2}$$

(where ω_i is the atomic volume) it can be shown that the *opposite* of this force, $-f_i^\alpha$, is equivalent to a surface force parallel to the surface normal n_β which tends to preserve the state of deformation, *i.e.*, a surface traction t_i^α

$$-f_i^\alpha = \int_{\Delta S} ds\, (\sigma_i^{\alpha\beta} \cdot n_\beta) = t_i^\alpha \tag{3}$$

445

We note that this procedure for obtaining surface tractions from the opposite of a bulk force is quite analogous to the image-field method widely used in continuum mechanics to obtain a special solution for singular boundary-value problems.

The constant-traction b.c. in principle has the basic advantage, over the constant-displacement b.c., of allowing perturbations such as dislocations to move freely across the system borders. However, a constant-displacement b.c. could be useful for simulating confined dislocations, such as dislocations emitted by a microcrack in a thin metal layer sandwiched between layers of a hard ceramic.

<u>Plane strain and plane stress</u>

Most 3-D problems, and among these crack problems, are difficult to treat in the context of continuum mechanics. It is then often argued that a 3-D problem with planar symmetry can be well approximated by a 2-D counterpart. The most widely used 2-D approximations in continuum mechanics are called *plane strain* and *plane stress*, and are intended to represent a slab whose thickness is either large or small, respectively, compared to the lateral dimensions. Although atomistic simulations can be carried out in fully 3-D geometries, the only limitation being the size of the system with respect to the available computer hardware, it is nevertheless useful to establish the conditions under which such commonly used continuum-mechanics approximations can be reproduced with atomistic simulations. This equivalence is especially necessary when comparing atomistic results to 2-D continuum solutions, or when studying an atomic-scale problem with the aim of extracting a CR to be used in a 2-D continuum model.

Both the plane-strain and plane-stress 2-D approximations can be represented with corresponding *quasi*-2-D atomistic treatments. To illustrate such a correspondence we will use a general geometry describing a slab made up of $n_x n_y n_z$ crystallographic cubic unit cells with lattice parameter a_0 and dimensions $L_\alpha = n_\alpha a_0$, $\alpha = x,y,z$. The x-dimension, which should be of zero length in a fully-2-D approximation, will have instead a size equal to twice the potential cut-off radius, to satisfy the minimum-image criterion [19].

The *plane strain* 2-D approximation is meant to represent a *thick* slab in the x-direction for which each x-plane behaves in the same way, *i.e.*, the strain components are constrained to $\varepsilon_{\alpha x} = 0$, for $\alpha = x,y,z$. In order make contact with the continuum-mechanics description of a clamped, thick specimen under plane strain, a homogeneous strain $\varepsilon_{zz} = \varepsilon_0$ is initially imposed along z to the atomistic system. The system is then allowed to contract along y, due to the Poisson effect; *i.e.*, L_z is kept fixed at $L_z = n_z a_0 (1+\varepsilon_0)$ while L_y is allowed to change by means of an appropriately modified Parrinello-Rahman scheme [20]. This condition represents an externally applied, constant uniaxial stress $\sigma_{zz} = \sigma_0$, $\sigma_{yy} = 0$. (A *biaxial* state of stress could also be imposed, by adding a homogeneous deformation ε_{yy} and not relaxing the stress σ_{yy}.) Plane strain is then globally maintained by keeping L_x fixed at $L_x = n_x a_0$. Note that atoms in the x-planes may individually move, although the size L_x is constrained.

The *plane stress* 2-D approximation is meant to represent a *thin* slab in the x-direction for which each x-plane behaves differently, so as to relax the stress components to $\sigma_{\alpha x} = 0$, for $\alpha = x,y,z$. Solutions strictly satisfying such requirements are rather difficult to obtain in continuum mechanics. A more useful approximation in practice is *generalized plane stress*. Considering the integrals of the stress components over the thickness L_x of the slab, generalized plane stress requires the average values of the stresses to be $<\sigma_{\alpha x}> = 0$, for $\alpha = y,z$; the condition $<\sigma_{xx}> = 0$ cannot be rigorously imposed, but it is usually assumed as a reasonable approximation. In an atomistic system this condition can easily be obtained by using a different modification of the Parrinello-Rahman scheme allowing also L_x to adjust its size until $\sigma_{xx} = 0$ so as to match the generalized plane stress approximation.

<u>Crack shape and crack initiation</u>

Two basic types of crack have been usually considered in atomistic fracture simulations: the tip of a semi-infinite crack, or "notch", and a double-edged elliptical crack of finite length.

The concept of semi-infinite crack is established in the LEFM theory as the asymptotic limit of the linear elasticity equations inside a small region around the crack tip [1]; since the resulting theory is crack-length-independent, the actual extent of the region surrounding the crack tip is not a concern. Now, in atomistic simulations the tip of the semi-infinite crack is embedded in a system which necessarily is of finite size; as a consequence, depending on the choice of the initial crack-tip position, the same level of loading could correspond to different amounts of elastic energy stored. The absolute value of the loading level for the semi-infinite crack can only be obtained by measuring the profile of the stress field close to the crack tip and using the asymptotic expression of the stress-intensity factor, a procedure not exempt from ambiguities. An exception is represented by those semi-infinite crack simulations in which the initial crack shape is determined according to a displacement field obtained from a LEFM calculation [21,22]: in that case, the initial loading level is simply prescribed by the prefactor in the continuum-elastic field. However, such a method can only be used in those cases for which an equivalent LEFM problem can be formulated.

By contrast, a finite-length crack has the basic advantage of supplying an intrinsic length scale for the simulation; we will show in the following that the ratio between the typical system size and the crack length is the most natural parameter to study the size-dependence of the results. As a consequence of the presence of such an intrinsic length scale, the Griffith critical load can be unambiguously determined from a very simple equation (see below), and represents the natural energy scale for the externally imposed load.

Concerning crack initiation, for either a semi-infinite or a finite-length crack, the two main choices are: (1) *ad hoc* breaking of some interatomic bonds, thus creating the equivalent of the mathematical singularity in the LEFM continuum-elastic equations, or (2) removing some atoms in the shape of a notch or a slit. Although this second option is often adopted for the sake of simplicity, we believe the first one to have at least two advantages. First, it is completely reversible to a bulk situation: if the crack closes back, *e.g.*, under a sub-critical loading, a perfectly crystalline system of well-known energy and elastic moduli is recovered. Second, the crack configuration which is obtained is much closer to the ideal conditions of LEFM, it is free of defects and thus of additional stresses that might render ambiguous any comparison with LEFM solutions.

Figure 2 describes the procedure we use [16,18] to initiate a crack of finite length by bond-breaking: the external load is initially applied as a homogeneous strain (a tensile, mode-I load in Fig. 2a) (the load is applied in an incremental way in order to remain below the critical load for propagation); a cut of finite length is then introduced into the system by breaking all the bonds across the surfaces delimited by the thick segment (Fig. 2b); the bond-breaking criterion is that atoms in the upper half of the system do not interact with those in the lower half across the crack-defining area, and *vice versa*; the system is then allowed to relax to its new minimum-energy configuration by lattice statics, and the cut opens into an elliptical crack (Fig 2c). All the interactions across the crack surfaces are restored at this point. This is the initial configuration for our crack propagation studies.

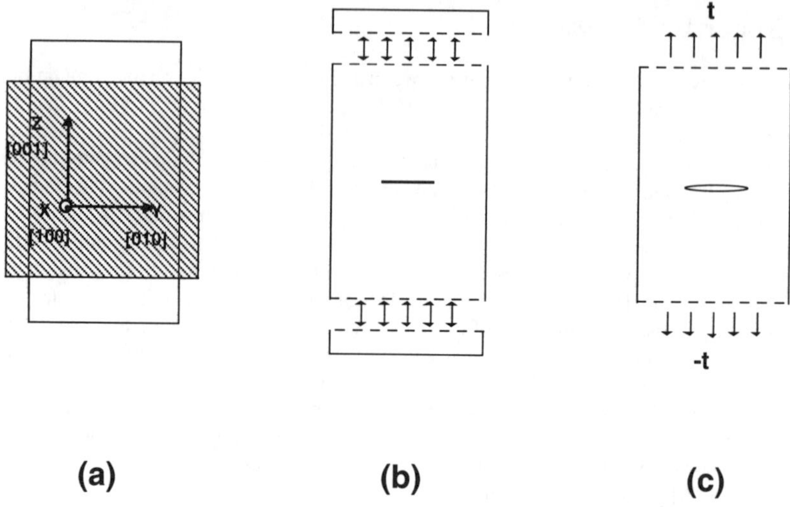

(a)   **(b)**   **(c)**

Fig. 2 - Schematic of the creation of the initial condition for fracture simulations.

<u>Comparison of atomic-level and linear-elastic stress fields</u>

The key requirement for any atomistic description of fracture is that the size of the system surrounding the crack be sufficient for the far stress field to reach the asymptotic value of the applied external load. In simulation methods based on the application of a prescribed displacement field obtained from a known continuum-mechanical solution this requirement is automatically satisfied, at least for the initial loading condition. However, our objective is to set up a self-consistent, fully atomistic method requiring no input from continuum mechanics. Such a method would then be capable of treating without any additional modification crack problems in, *e.g.*, bicrystals, multilayers, polycrystalline and multiphase materials, for which linear-elastic fracture mechanics (LEFM) solutions are difficult to obtain.

We have shown [18] that the most appropriate convergence parameter to study system-size effects is the ratio

$$R \; = \; \frac{L}{2l} \qquad\qquad (4)$$

between the typical system size L and the length $2l$ of a finite-size microcrack. To unambiguously determine a minimum value of R for atomistic simulations of static fracture we compared the stress fields obtained from the anisotropic-elasticity LEFM theory [23] with stresses obtained from atomistic simulations [18] through the virial formula, Eq. (2), for systems with different ratios R. All the results were obtained for a purely tensile, mode-I loading under plane strain condition with a fixed dimension $L_x=3a_0$, necessary to satisfy the minimum-image criterion for our potential in presence of periodicity along x. For the sake of simplicity, we used the (12-6) Lennard-Jones (LJ) interatomic potential with σ=2.315 Å and ε=0.167 eV already

448

used in a number of previous studies [16,18]. The comparison of the behavior of the far stress field can be summarized by stating that a size of about 20 crack-lengths along the direction of applied loading is sufficient for obtaining good convergence to the far stress field. By a detailed comparison at different mode-I loading levels, we verified that the various loading methods (constant-displacement, constant-traction, mixed) give almost identical results at this value of R_z , whereas residual discrepancies persist below such a size. Thus, the same value $R_z{\sim}20$ is the size at which the different loading methods become indeed equivalent.

The comparison of the stress fields close to the crack tip is most easily carried out along the y-direction, because the LEFM solution for the stress along this line is particularly simple and monotonic. We concluded [18] that the minimum system size required for the reproduction of the infinite-plate LEFM solutions with constant displacement or constant tractions border conditions is about $R {\sim}10\text{-}20$, the larger values being necessary in the direction perpendicular to the crack plane. This central finding poses a strict lower limit on the size of the supercell. In fact, for a size of the crack of about $2l{\sim}10a_0$ a slab of about $100\text{x}200\ a_0^2$ is necessary; when this is multiplied by twice the potential cutoff length along the thickness (*i.e.*, $3a_0$ for our 4th-nearest-neighbors LJ potential), and by the number of atoms in each unit cell (4 for the fcc lattice) one obtains a minimum dimension for the system of 240,000 atoms.

A possible objection, sometimes raised against atomistic simulations concerns the actual length of the crack. Indeed, because of the above minimum-size limits, finite-length cracks that can be simulated atomistically cannot exceed a few nm even on big supercomputers. This length is such that the region surrounding the crack tip is of the order of the interatomic distance, and it could be argued that this is too short for the actual $r^{-1/2}$ dominance of the LEFM solutions in the near field, the main character of a true crack in the LEFM context, to be achieved. However, our results show that continuum elasticity compares surprisingly well with atomistic results down to distances of just a few a_0 [18], and that such distances are indeed sufficient for the $r^{-1/2}$ character of the LEFM solutions to be captured by the atomistic simulations, thus confirming previous results obtained with different atomistic techniques. As a consequence, atomic-scale microcracks with lengths in the range of about $10a_0$ can also be safely called "cracks" from the LEFM point of view.

<u>Energy conservation and the Griffith criterion</u>

For purely brittle fracture, the crack-tip behavior is determined by the Griffith criterion [2], *i.e.*, strictly by the competition between the work supplied by the external loading and the energy, 2γ, required to form two free surfaces upon bond breaking. In order to compare the values of the critical stress-intensity factors obtained in our simulations and those based on LEFM theory, the Griffith-predicted value K_{GC} of the critical load for the fcc (001)[010] microcrack configuration was computed using the materials parameters for the LJ potential, Young's modulus Y and Poisson's ratio ν, from the expression:

$$K_{GC} = \left(\frac{2\gamma Y}{1-\nu^2}\right)^{1/2} \tag{5}$$

The stress-intensity factors for our mode-I configurations, to be compared to the above predicted Griffith value, can be estimated at each value of the microcrack half-length l with the infinite-plate LEFM expression [24]

$$K_I = {<}\sigma_{zz}{>}_b \sqrt{\pi l} \tag{6}$$

where $<\sigma_{zz}>_b$ is the average value of the σ_{zz} component of the atomistic stress field σ_{zz} measured along the $y/l=\pm R_y/2$, $z/l=\pm R_z/2$ system borders. For each microcrack configuration the two limiting values of K_I at which opposite crack-tip behavior (closure or propagation) are observed. The critical value of the stress-intensity factor K_{IC} from the atomistic method is estimated as the average between these two extreme values.

Our results show good agreement with the values predicted from the Griffith theory for the critical stress. While for the very small system size corresponding to $R_yR_z=5\times5$ the error is ~3-4%, this decreases to less than 1% when increasing the system size to $R_yR_z=5\times10$. Notably, such a level of agreement is already reached for the smallest crack length $2l=4a_0$. This level of agreement with the Griffith prediction must be considered as very good, since for most of the previous atomistic fracture simulations the error was 5 to 10% or worse.

It is of interest to note that the stress-intensity factor appears to be less sensitive to the system size than is the stress-field behavior. Thus, carrying out a comparison between LEFM theory and the results of atomistic simulations based simply on matching the Griffith criterion is likely to underestimate system-size effects.

CONCLUSIONS

We described the central role of atomistic simulations in providing the necessary level of understanding of the mechanical response of materials, starting from which information can be integrated into mesoscopic and macroscopic descriptions.

Because the mechanical properties of materials, being connected with the presence of extended defects in solids, span a wide range of length scales from nanometer to macroscopic, a major issue is the possibility of linking atomic-scale simulations with mesoscopic and macroscopic models. Several schemes to link atomistic and continuum-mechanics methods into hybrid computational techniques have been reviewed; such approaches, however, seem unlikely to provide the expected big leap in multiscale simulations. Thanks to the development of new algorithms and to the progress in computer hardware, atomistic methods are becoming increasingly powerful and alternative linking schemes, such as the constitutive-relations (CR) approach, can be explored. In the same spirit, fully atomistic methods could supply important parameters to mesoscopic modeling, such as the annihilation distance of a dislocation dipole and the junction strength of two intersecting dislocations, necessary in microplasticity models [25]. However, in order to obtain such information, atomistic simulation methods compatible with continuum mechanics are necessary.

In this work we described how to assess a self-consistent, fully atomistic simulation method which, although requiring no input from continuum mechanics, can make contact with continuum-mechanical problems. In the particular case of fracture simulations we established the range of validity of such a fully atomistic method, through molecular dynamics (MD) simulations of a double-edged, finite-length microcrack embedded in a fully atomistic system with periodic borders in the crack plane. The ratio $R=L/2l$ between the typical system dimensions and the microcrack length was identified as the critical parameter in terms of which system size effects should be assessed. Such size effects involve the convergence of the simulated stress field to the known linear-elastic fracture mechanics (LEFM) far- and near-field solutions, and validation to the Griffith criterion for brittle fracture.

ACKNOWLEDGEMENTS

Over the years I benefited from the constant encouragement and contributions of my friends D. Wolf and S.R. Phillpot (ANL) and S. Yip (MIT). I thank these people also for the wonderful summer times we spent together talking of science in the Argonne forest. This work was supported by the NATO under Collaborative Research Grant CRG-950369.

REFERENCES

1. H. Liebowitz (ed.), "Fracture: An Advanced Treatise. Vol.II: Mathematical Fundamentals", Academic Press, New York, 1968; G.C. Sih, H. Nisitani and T. Ishihara (eds.), "Role of Fracture Mechanics in Modern Technology", Elsevier, Amsterdam, 1987; M.H. Aliabadi and D.P. Rooke, "Numerical Fracture Mechanics", Kluwer, Dordrecht, 1991.
2. A.A. Griffith, Philos. Trans. Royal Soc. (London) A **221** (1920) 163.
3. R.L. Blumberg Selinger, J.J Mecholsky, A. E. Carlsson and E. R. Fuller Jr. (eds.), "Fracture-Instability Dynamics, Scaling and Ductile/Brittle Behavior", MRS vol. 409, Pittsburgh, 1996; A.K. Cheetam et al., J. Comput.-Aided Mater. Design **3** (1996) 1.
4. W.W. Gerberich et al., Acta Metall. Mater. **43** (1995) 1569.
5. W.T. Ashurst and W.G. Hoover, Phys. Rev., B **14** (1976) 1465.
6. P.S. Lomdahl et al., Int. J. Mod. Phys. C**42** (1993) 1075.
7. F.F Abraham et al., Phys. Rev. Lett. **73** (1994) 272.
8. B.L. Holian and R. Ravelo, Phys. Rev. B **51** (1995) 11275.
9. A. Nakano, R.K. Kalia and P. Vashista, Phys. Rev. Lett. **73** (1994) 2336.
10. S.J. Zhou et al., Phys. Rev. Lett. **76** (1996) 2318.
11. D. Mullins and A. Dokainish, Phil. Mag. A **46** (1982) 771.
12. S. Kohloff, P. Gumbsch and H.F. Fischmeister, Phil. Mag. A **64** (1991) 851.
13. E.B. Tadmor, M. Ortiz and R. Phillips, Phil. Mag. A **73** (1996) 1529.
14. R. Miller and R. Phillips, Phil. Mag. A **73** (1996) 803.
15. E. Kaxiras and M.S. Duesbery, Phys. Rev. Lett. **70** (1993) 3752.
16. F. Cleri, S. Yip, D. Wolf and S.R. Phillpot, Phys. Rev. Lett. **79** (1997) 1309; Acta Mater. **45** (1997) 4993.
17. A.E.H. Love, Appendix B in "A Treatise on the Mathematical Theory of Elasticity", 4th ed., Dover, New York, 1944; M. Born and K. Huang, "Dynamical Theory of Crystal Lattices". Oxford University Press, Oxford, 1954.
18. F. Cleri, S.R. Phillpot, D. Wolf and S. Yip, J. Am. Cer. Soc. **81** (1998) 503.
19. M.P. Allen and D.J. Tildesley, "Computer simulation of liquids", Oxford Science Publications, Oxford, 1989.
20. M. Parrinello and A. Rahman, J. Appl. Phys. **52** (1981) 7182.
21. K.S. Cheung and S. Yip, Phys. Rev. Lett. **65** (1990) 2804.
22. R.G. Hoagland, M.S. Daw and J.P. Hirth, J. Mater. Res. **6** (1991) 2565.
23. S.G. Lekhnitskii, pp.153-162 in "Theory of Elasticity of an Anisotropic Elastic Body", Holden-Day, San Francisco, 1963.
24. V.Z. Parton and E.M. Morozov, pp.103-123 in "Mechanics of Elastic-Plastic Fracture", Hemisphere, Washington, 1989.
25. V. Bulatov, F.F. Abraham, L. Kubin, B Devincre and S. Yip, Nature **391** (1998) 669.

NONLINEAR RESPONSE OF SOLIDS AND MOLECULES
IN INTENSE INFRARED RADIATION

Sokrates T. Pantelides,[1,2] Roland Winkler,[1,*] Maurizio Ferconi[1,†] and J. J. Vicente Alvarez[1]

[1]Department of Physics and Astronomy, Vanderbilt University, Nashville, TN 37235

[2]Solid State Division, Oak Ridge National Laboratory, Oak Ridge, TN 37831

ABSTRACT

Infrared radiation, which couples with vibrations in molecules and solids, has long been considered as a promising way to selectively control chemical reactions, materials processing, and biomedical applications. The expectation is that particular frequencies couple to specific atomic motions, in contrast to heat which imparts energy indiscriminately. The promise has yet to be fulfilled, but recent successes and the development of new lasers have rekindled interest. We report calculations of the dynamics of solids and molecules under intense infrared radiation. In solids, using a model calculation, we find that narrowly-defined "windows of opportunity" exist for resonant enhancement of impurity diffusion at moderate intensities. In molecules, using a generalization of Car-Parrinello dynamics, we find that, for very intense fields, energy absorption at normal modes is not very efficient and that selective bond-breaking may occur at nonresonant frequencies by "concerted kicks". Finally, time-dependent density functional theory is used to show that the effect of very intense infrared radiation on atoms and molecules is to produce very high harmonics as found experimentally.

INTRODUCTION

Infrared radiation is characterized by a range of frequencies that overlaps the range of vibrational frequencies in solids and molecules. It has, therefore, been used extensively to probe vibrational states by direct absorption and by other more complex methods such as Raman spectroscopy. It has also presented a promise that, as a source of energy, it can be used for selective chemistry or selective materials modification. The basic idea derives from the fact that vibrations in molecules and solids are in effect harmonic oscillators with characteristic frequencies. When a harmonic oscillator interacts with an external sinusoidal field, it absorbs energy resonantly when the frequency of the external field is equal to that of the oscillator. Since distinct bonds between different species of atoms have distinct "stretch modes" associated with them, one would expect that targeting a particular stretch mode with intense infrared radiation would lead to selective break of that bond or at least an enhanced rate. Similarly, targeting a particular vibrational mode of an impurity might lead to selectively enhanced diffusion of that particular impurity. The potential ramifications for selective chemical reactions or materials modification are enormous. In contrast to thermal processing, which, for a given temperature, drives all possible reactions with activation energies up to a certain maximum, selective processing would allow one to pick and choose which processes to drive by targeting infrared radiation at selected frequencies.

The promise for selective processing has, however, for the most part, remained just that, a promise. Early work with molecules showed that targeting either weak or strong bonds inevitably leads to breaking of the weak bonds.[1-3] The fundamental problem is that vibrational normal modes in molecules and solids are generally not localized on just a few atoms; in addition, different normal modes are coupled by anharmonic terms. The failure to achieve selectivity was, there-

---

[*] Present address: Institut für Technische Physik III, Universität Erlangen-Nürnberg, Staudtstr. 7, D-91058 Erlangen, Germany

[†] Present address: TechHackers, Inc., 5 Hanover Sq., New York, NY 10004

Mat. Res. Soc. Symp. Proc. Vol. 538 © 1999 Materials Research Society

fore, attributed to rapid redistribution of the absorbed energy so that the end result is the same as that produced by heat, which imparts energy indiscriminately to the whole system.[4]

In the 1990's, several experimental successes were reported,[5-7] mostly selective bond breaking by preparing molecules in particular vibrational excited states. These successes led to optimistic commentaries for the future.[8] In addition, the realization of Free Electron Lasers as functional tools for research and new table-top ultrafast infrared lasers have generated new interest. Such lasers deliver high-intensity short pulses that may induce selective reactions on time scales much shorter than any relevant relaxation times. Promising examples are the identification of a particular infrared frequency that produces very sharp, clean incisions in biological tissue[9] and the observation of a reduction in the ablation threshold of hydrogenated diamond films at the C-H stretch vibrational mode.[10] The origin of such selectivity, however, has not been probed at the atomic scale by either experiments or theory.

Theory has been a major player in this field, starting with early studies of energy redistribution within molecules in 1972.[4] In the 1980's theoretical calculations predicted that branching ratios of bimolecular reactions depend strongly on the vibrational state of precursor molecules.[11] The prediction was realized a few years ago [5,7,8]. More recently, model calculations predicted that "chirping" (rapid sweeping of a frequency range), a sequence of several frequencies, or particular pulse structures can significantly increase the probability of breaking specific bonds in small molecules.[12-14] These predictions are yet to be verified experimentally. In the case of solids, a model theoretical calculation led to the conclusion that resonant desorption of atoms on a crystal surface may be possible,[15] but experimental verification is again lacking.

All the above theories[4,11-15] share a common assumption, namely that the dynamics of a molecule under infrared radiation can be described in terms of the potential energy surfaces, the vibrational normal modes, and the dependence of the electrical dipole on interatomic distances <u>in the absence of the radiation field</u>. This assumption is expected to be valid for weak and moderate field intensities, but should break down for higher intensities when the electron density and bonding potentials are altered significantly by the time-dependent radiation field. This regime has remained unexplored and can best be investigated by first-principles, parameter-free approaches. The latter have so far been used to study free atoms or the dynamics of molecules in the visible and uv ranges where the primary effect of radiation is electronic excitations.[16]

In this paper, we summarize our recent theoretical research on the effect of intense infrared radiation on solids and molecules and solids. In the case of molecules we have carried out first-principles density functional calculations at the same level of sophistication and accuracy as is generally practiced for problems without radiation. We find dramatically different behavior at moderate and high intensities. At moderate intensities, resonant absorption occurs at the normal modes and energy is indeed redistributed rapidly so that the weakest bonds are likely to break first. At high intensities, however, the normal modes are no longer relevant and the characteristic resonant absorption does not occur. The electrical dipole is a strong function of the oscillating field. The external oscillating potential is comparable to the bonding potentials so that perturbative analysis is not applicable. Instead we find that strong bonds can be broken by a novel mechanism that we have labeled "concerted kick". In order to deduce macroscopic bond-breaking rates for different frequencies, it would be necessary to repeat the simulations many times with different initial conditions to sample enough of phase space for a meaningful average. Such calculations are not practical at this point. In the case of solids, simulations are even more time consuming so that we carried out a preliminary investigation of selective impurity diffusion using a model crystal described by an empirical classical potential. With this choice we were able to carry out adequate statistics and obtain macroscopic diffusion constants for a range of intensities, frequencies and temperatures. The net result is that a "window of opportunity" for enhanced diffusion exists for the lightest impurities such as hydrogen. This window is defined by a small range of near-resonance frequencies, moderate intensities and relatively low temperatures.

SELECTIVELY ENHANCED IMPURITY DIFFUSION IN MATERIALS

In this work, we considered a model face-centered cubic (fcc) solid described by a Lennard-Jones potential containing a model interstitial impurity, very similar to the system described in Ref. 17. We viewed the impurity as hydrogen, having a very small mass compared with the mass of the host atoms. We also assigned a fixed charge q* to the impurity, allowing it to couple to an external electromagnetic field. The calculations were carried out in a supercell containing 108 host atoms and four impurity atoms. The external field was taken to be $V_{ext} = \sum_l q_l^* \mathbf{E} \cdot \mathbf{R}_l \cos \omega t$, where

E is the amplitude of the external field and q_l^* on each impurity was taken to be one unit. Simulations were carried out using standard procedures (Verlet algorithm, time step of 1 fs).

Before we present the diffusion results, we show in Fig. 1 the power spectrum of the impurity (density of local vibrational modes) in our model system for several temperatures. The broad peak at 16.8 THz is the impurity mode which broadens considerably as a function of temperature. The host spectrum ranges from 0 to 7 THz. On the same plot we show the linear absorption coefficient of the system defined as

$$\alpha(\omega) = \frac{1}{N}\sum_{j=1}^{N}\frac{\langle dT/dt \rangle}{(eE)^2}, \tag{4}$$

where, as in (3), the sum over j corresponds to different initial conditions. The brackets denote a time average taken over the simulation time of each run. Typically 20 runs were needed to produce good statistics. This calculation was most efficiently done by using a microcanonical ensemble where the absorbed energy was allowed to raise the temperature. As it should, the linear absorption coefficient tracks the density of vibrational modes.

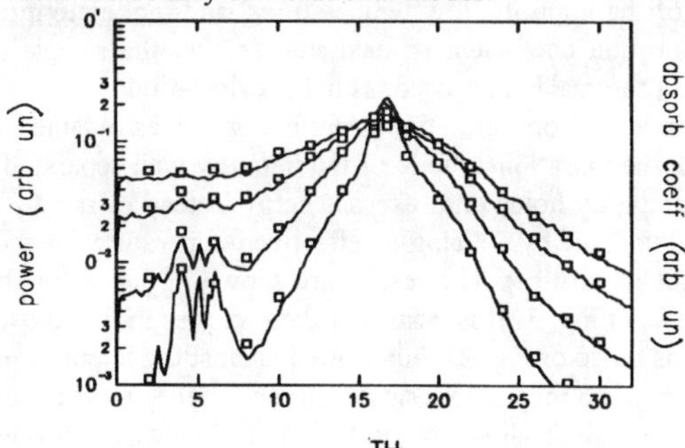

Fig. 1. Power spectrum (solid lines) of the vibrational motion of the impurities and the absorption coefficient (symbols) at different temperatures (500 K, 700 K, 1000 K and 1300 K from bottom to top). For each temperature, the absorption coefficient is rescaled by a constant factor.

Diffusion constants were calculated by using a canonical ensemble and computing the following time average:[17]

$$D = \frac{1}{MN}\sum_{j=1}^{N}\sum_{l=1}^{M}\frac{<|\mathbf{R}_l(t)-\mathbf{R}_l(0)|^2>}{6\Delta t}, \tag{3}$$

where $\mathbf{R}_I$ is the position vector of the Ith impurity and the sum over j corresponds to restarting the simulation with a different set of initial conditions to sample the phase space. Again, 20 runs were needed to produce good statistics. The brackets denote a time average over the simulation time of each run, typically 0.1 ns.

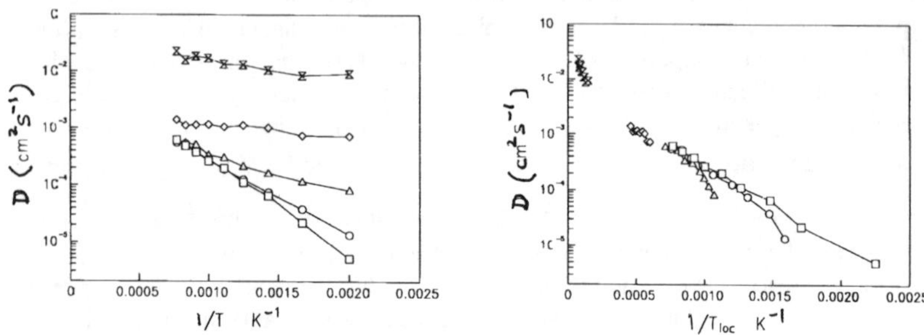

Fig. 2. Arrhenius plot of the diffusion coefficient D (left) versus the inverse temperature and (right) versus the local temperature T_{loc} of the impurities for different values of the electric field.

In Fig. 2a we show the calculated diffusion constant of our model system for ω=16.8 THz in an Arrhenius plot. The curves correspond to increasingly higher intensities, with the lowest curve corresponding to zero field. The activation energy for diffusion decreases with increasing intensity, demonstrating resonant enhanced diffusion. We performed similar calculations for a range of frequencies. In Fig. 3 we show D as a function of frequency for four different temperatures. Again the different curves correspond to increasing intensity. These curves show dramatically the "window of opportunity" for selective enhancement of diffusion. It is a small window, confined around the resonant frequency of the impurity, low temperatures, and <u>moderate intensities</u>. At the resonant frequency, the absorption coefficient is maximum so that the sample's temperature would rise fast unless an efficient thermal bath is used, as in the calculation.

We note that the window of opportunity evident in Fig. 3 was obtained with effectively optimal conditions, i.e., with the vibrational mode of the impurity well separated from the host spectrum, ensuring that the impurity holds onto excess energy instead of rapidly transferring it to the lattice. We investigated this issue by defining an effective temperature for the impurities (average kinetic energy of only the impurities). The results are shown in Fig. 4 for the same sample temperatures and intensities as in Fig. 3. It is clear that these curves mirror those of Fig 3. In Fig. 2b we show the same data as those of Fig. 2a, but plotted against the impurity effective temperature. We see that virtually all the data map onto a single curve with the zero-field activation energy.

The two plots Fig. 2a and 2b allow two distinct but obviously equivalent interpretations of the resonant enhanced diffusion: according to Fig. 2a, the field leads to a decrease in the activation energy. This decrease occurs because the field induces the impurity to vibrate with larger amplitude. In quantum language, the field generates excess local phonons or assists the impurity to partially climb the local phonon ladder. Fig. 2b offers an alternative interpretation: the field induces local heating so that the diffusion constant corresponds to the value appropriate for that local temperature, without change in the activation energy. In the final analysis, at resonance, the impurity absorbs substantial energy from the field and *is capable of retaining a good amount of it in a steady state*. At non-resonant frequencies, significantly less energy is absorbed by the impurity. If the system is in contact with an efficient thermal bath as in the calculation, the impurity retains no significant excess energy to achieve enhanced diffusion.

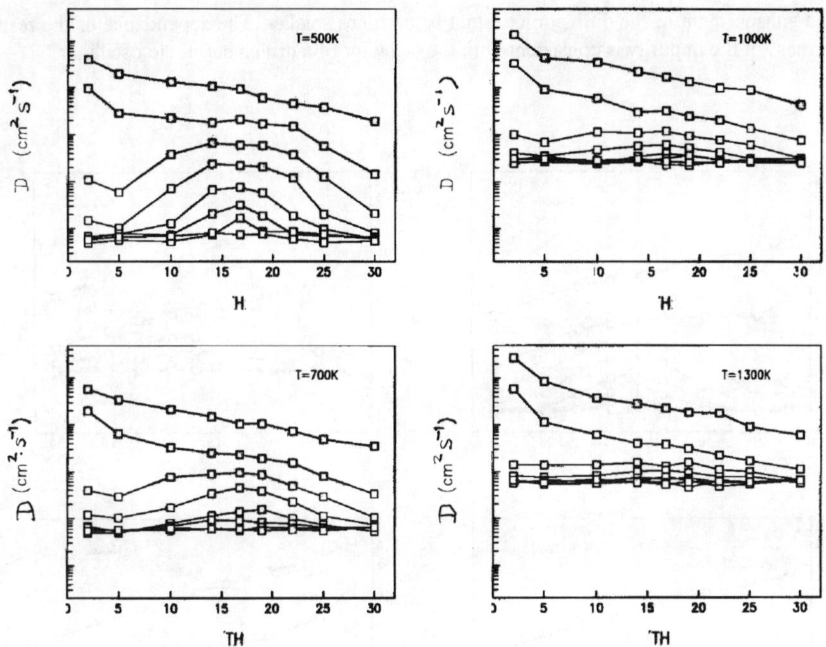

Fig. 3 Diffusion coefficient D versus field frequency for four different values of temperature.

In order to explore the nonlinear aspects of the radiation-induced processes, we computed the rate of energy absorption by the system (dT/dt) as a function of the field intensity, namely E^2. The results are shown for four different temperatures in Fig. 5. If the system was linear, the result would be a straight line parallel to the dashed line. We note a strong nonlinear behavior as the curves of different frequencies cross. The behavior shown here is consistent with the behavior of the diffusion constant as a function of frequency: at low intensities, absorption and hence diffusion occurs primarily at the impurity resonant frequency, but at higher intensities other frequencies finally overtake. The underlying physics is that higher intensities alter the vibrational spectrum significantly (see Fig. 1) and strong absorption occurs over a different range of frequencies.

The calculations could not be performed at lower temperatures because of the slow accumulation of statistics. Below the Debye temperature, quantum phenomena start dominating anyway. The overall conclusion is clear, however. Selective enhancement of diffusion would best be achieved at low temperatures in a set-up where the temperature of the sample is not allowed to increase substantially. The pulsed nature of intense lasers is therefore helpful. In addition, the laser may have to be turned off completely between spurts of pulses in order to allow intermittent cooling of the sample. Equally important, the intensity must be within a range of moderate values. The precise range must be determined by trial and error or by calculations that employ reliable potentials. First-principles calculations at this point for solids are out of the question because of the statistical averaging that must be done.

Finally, the window of opportunity depends strongly on the mass of the impurity. We repeated the calculations for an impurity with a tenfold larger mass and found that the radiation-induced enhancement of the diffusion constant is ten times smaller. The dependence of the results on the mass of the impurity is consistent with the behavior of a driven harmonic oscillator.

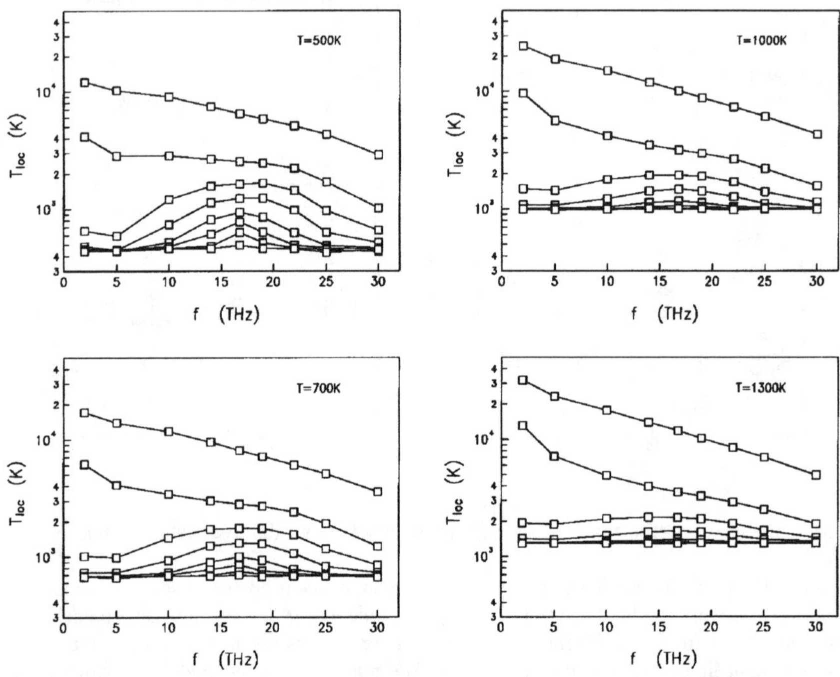

Fig. 4. Local temperature of the impurity T_{loc} versus field frequency for the same values of temperature used in Fig. 6.

It is obvious that our generic conclusions apply to other similar phenomena such as the desorption of atoms or molecules from a surface. In all cases, the biggest problem for enhanced impurity diffusion is likely to be coupling with the host. The optimal conditions correspond to light impurities with a large effective charge and a resonant frequency outside the range of host vibrational frequencies. That is a severe limitation as H looms as the only likely candidate. When the impurity resonant frequency lies within the host's continuum, the phenomenon of enhanced diffusion or desorption would be reduced further or eliminated altogether unless the resonant frequency coincides with a minimum in the density of host vibrational states or by some set of circumstances the coupling with the continuum is small.

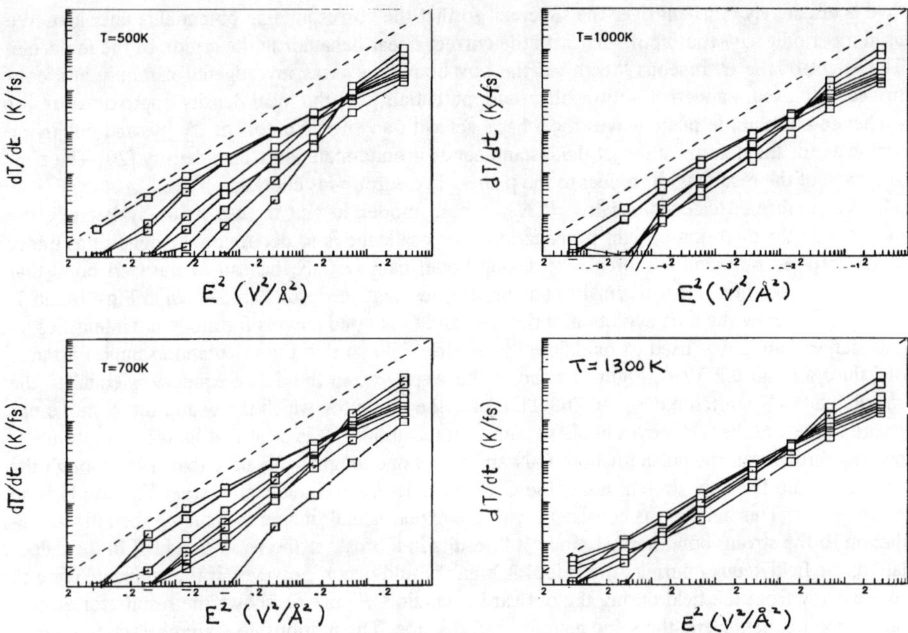

Fig. 5. Rate of temperature increase in Kelvins per femtosecond as a function of the field intensity (i.e. the square of the amplitude of the electric field) for the resonant frequency of the impurity for four different temperatures.

QUANTUM MOLECULAR DYNAMICS IN INTENSE RADIATION – APPLICATION TO MOLECULES

Density functional theory[18] has been widely used for solids and more recently for molecules. The original formulation describes a many-electron system in its ground state for a particular arrangement of the nuclei. In the Car-Parrinello scheme[19,20] for dynamical simulations, the nuclei (or ions in a pseudopotential formalism) are treated as classical particles and the electrons are kept in their instantaneous ground state. The inclusion of a time-dependent external potential in this scheme requires a reformulation of the theory for the description of the electronic system (time-dependent density functional theory[21]). However, for molecules such as HCN, the lowest electronic excitation energy is significantly larger than infrared photon energies, requiring a highly multiphoton process to induce electronic excitations. If such multiphoton processes are negligible, the electrons can be kept at their instantaneous ground state (adiabatic approximation) so that the standard Car-Parrinello scheme can be used. In this section we will describe results obtained with this approximation. In the next section, we will show results of time-dependent density functional theory that demonstrate the validity of this approximation even for the high intensities used in the present work. We note, for comparison, that all semiempirical calculations on molecules such as HCN in the presence of infrared radiation implicitly also assume no electronic excitations.[12,14]

In the adiabatic approximation, the treatment of electrons is identical to Car-Parrinello dynamics.[19,20] We used available codes and included coupling to the electric field. In the long-wavelength dipole approximation, coupling with the magnetic field drops out. We used standard supercells, placing a molecule in a box with periodic boundary conditions. The external electric

459

field is effectively constant over the supercell so that the corresponding potential is liner in **x**. We used a periodic saw-tooth potential with the correct linear behavior in the region of the molecule. The effect of the extraneous "teeth" at the box boundaries was investigated carefully and controlled. We used Vanderbilt's ultrasoft pseudopotentials,[22] the local-density approximation for exchange-correlation, plane waves for a basis set with an energy cutoff of 25 Ry, and conjugate gradients for the determination of the instantaneous ground-state electronic density.[20] The convergence of the results with respect to the plane-wave cutoff was checked.

We performed calculations for HCN for both moderate and intense fields by keeping the molecule in the direction of light propagation. The challenge is to determine the conditions under which infrared radiation can break the strong bond, namely C-N, instead of the C-H bond that normally breaks first under thermal treatment. Representative results are shown in Figs. 6 and 7. In Fig. 6 we show the time evolution of the C-H and C-N bond lengths for moderate intensities. A gaussian envelope was used to modulate the electric field so that the calculation samples a range of values around 0.2 V/ Å. Figure 6a shows the results when the field frequency is equal to the calculated "C-N stretch mode," (64.68 THz.) i.e. the mode for which the C and the N move opposite to one another (H moves in phase with C in this mode). It is clear that harmonic motion persists throughout the pulse for both C-N and C-H bond lengths. We note that, even though the field is driving the C-N stretch mode, the C-H excursions are significantly larger because H is so much lighter. This behavior is consistent with the experimental finding that tuning the infrared radiation to the strong-bond normal mode still results in a break of the weak bond.[2] In the calculation, the field keeps returning the H ion. A break would occur, however, if the molecule were to rotate away from the field during the outward excursion. Figure 6b shows the results for an off-resonance frequency and the same gaussian pulse shape. The motions now are characterized by a superposition of frequencies (the normal modes plus the driving frequency) and the amplitudes are significantly smaller, which is typical of weak off-resonance driving forces.

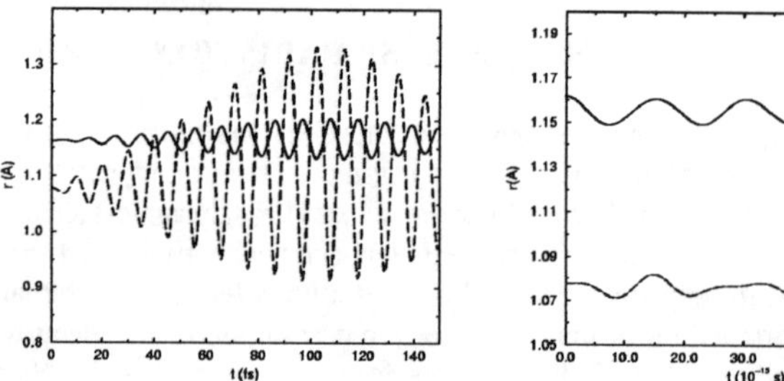

Fig. 6. Time evolution of the bond lengths (solid line for the C-N bond and dashed line for the H-C bond) for a weak field at left) the resonant CN frequency (64.68 THz) and right) at the frequency of 56.76 THz. In both cases a gaussian pulsed was used.

In Fig. 7 we show the response of the molecule to a strong field of constant amplitude E_o=4 V/Å. As in Fig. 6a, Fig. 7a shows the response at the theoretical frequency of the "C-N stretch mode". We note that the molecule is *not driven in the normal mode*, i.e. the phases of the C-N and C-H vibrations are not correlated. Instead, the C-N amplitude rises quickly and remains fairly constant, following a nearly harmonic pattern. In contrast, the C-H motion has a relatively small amplitude, is not harmonic and is out of phase with the C-N motion. It is noteworthy that the C-N amplitude seems to saturate, whereby a C-N break is unlikely. Overall, the system is not absorbing energy monotonically as it does (resonant absorption) in the weak-field case.

The fact that radiation at the frequency of the C-N normal mode is not absorbed in a characteristic "resonant" way suggests that the system is in fact no longer in resonance. The intense radiation modulates the electron density in a way that the "spring constants" vary with time as the amplitude of the field varies. Clearly, this behavior is bound to persist, though modified in detail, if multiphoton electronic excitations were to be included. We conclude that, in the case of very intense fields, one can no longer describe a molecule in terms of resonant absorption and subsequent redistribution of the absorbed energy

Figure 7b shows the response of the molecule to a strong off-resonance field at a frequency of 56.76 THz. It is clear that there is no longer any harmonic pattern for either the C-N or the C-H motion. The molecule again is not absorbing energy monotonically. Nevertheless, a break of the C-N bond occurs after about 140 fs. The break is shown in an expanded scale in Fig. 8.

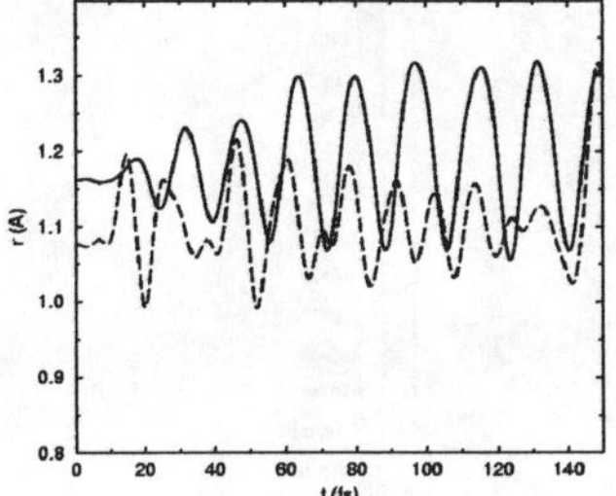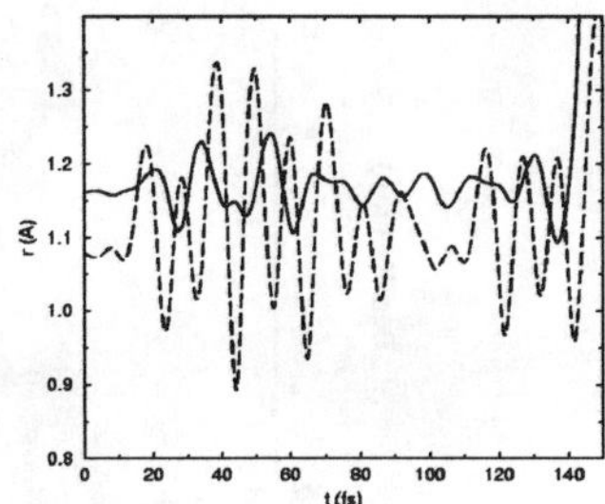

Fig. 7. Time evolution of the bond lengths (same conventions as in Fig. 6) for a strong field at left) the resonant CN frequency (64.68 THz) and right) at the frequency of 56.76 THz.

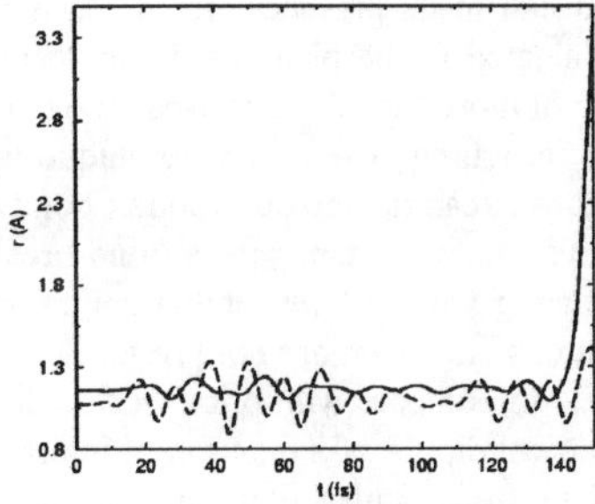

. Fig. 8 Same as Fig. 7 in an expanded form to show the break of the C-N bond.

The intensity used in the simulation that produced the bond break is quite high, namely 2.13×10^{14} Watts/cm^2, corresponding to E_o=4 V/Å. The break at a frequency other than the C-N mode is not surprising for external potentials that are comparable to the depth of bonding potentials in the molecule. The system in fact does not consist of a molecule which responds to the electric field in some perturbative fashion. Instead, the system consists of a molecule and a field and the total potential seen by the nuclei and electrons at any given time has no relationship to the molecule's ground-state potential that determines the normal modes. Instead, insights into the process can be gained by examining the evolution of the C-H and C-N bond lengths. Just before the break, the C-N bond length is at its minimum and begins to expand while the C-H bond length is at its maximum and begins to contract. In other words, the N atom is beginning to accelerate away from the C atom and the H atom is beginning to accelerate toward the C atom from the backside (the center of mass of the molecule is not moving). It seems that the N atom is getting a "concerted kick" at a time of large excursion from C and acquires an "escape velocity." The phenomenon is a direct consequence of the amount of charge around each nucleus which is a strong function of the field intensity. In Fig. 9 we illustrate this point by showing the electron density rearrangement for a particular value of the electric field.

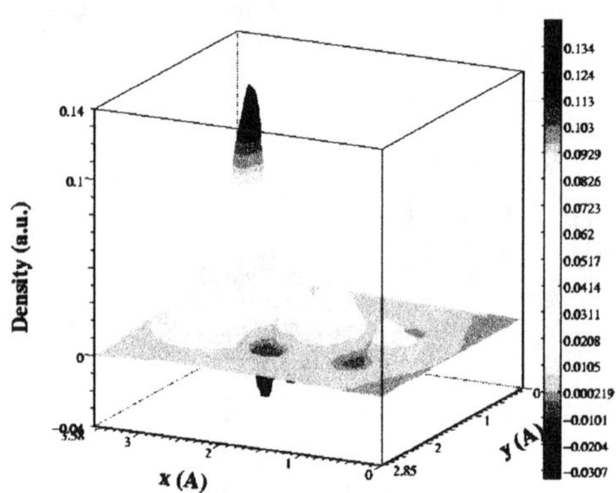

Fig. 9. Difference between the electron density of the HCN molecule (in a.u.) in the absence of the field and in its presence, for a given configuration and an electric field of 4 V/A, evaluated at the plane of the molecule. The highest peak corresponds to a difference of more than 20% between the two densities.

The result described above is intriguing. There may be unique frequencies that lead to such configurations where "concerted kicks" can induce unorthodox bond breaking. In order to ascertain the significance of such events, one must compute a bond-breaking rate by carrying out a phase-space average, i.e. by simulating with different initial conditions for the positions and velocities of the nuclei. Such calculations, however, are not presently practical.

Finally, it would be worthwhile to compare with prior work. All prior work has used empirical models for the molecule's electrical dipole as a function of interatomic spacing, assuming that it is independent of the external field. In particular, these models must make a specific assumption about the asymptotic behavior of the effective charges on the ions at large separations. We investigated such empirical models in the hope that we could fit our first-principles results and then explore a larger portion of phase and calculate rates. We found that the effective "spring constants" and effective charges are strong functions of the radiation field and bond lengths. In particular we found that the dynamics of the molecule is very sensitive to the model one assumes for the effective charges in the asymptotic limit of large bond lengths. This charge determines the magnitude of the force that the radiation field will exert on the ion to arrest or enhance its outward motion. For example, if the model assumes the effective charge goes to zero at some asymptotic rate (see e.g. Ref. [23-24]), the force from the infrared field will also go to zero at the same rate.

NONLINEAR RESPONSE OF ELECTRONS UNDER INTENSE RADIATION

In order to study the effect of intense radiation on the electron system it is necessary to implement time-dependent density functional theory. At each time step, the electron wave functions are determined from a time-dependent one-electron equation. One can thencompute the time-dependent electron density and electrical dipole. Details of the calculation will be published elsewhere.

The time step is now much smaller, making the simulation of combined electron-and-ion dynamics a much more demanding task. So far we have implemented the theory for electron dynamics and have used it to test the adiabatic approximation of the previous section. In Fig. 10 we show the electrical dipole of the electron system of an HCN molecule in the presence of an infrared field of two different intensities at the same frequency as the calculations of Fig. 7b. The ions are kept at their equilibrium positions. In Fig. 10a the intensity corresponds to a field of 2 V/Å and in Fig. 10b to a field of 4 V/Å. The latter is the value at which the C-N bond break was observed in Fig. 8. We see that fine structure is superposed on the overall oscillation of the electrical dipole, which follows the field. For the purposes of this paper, the fine structure represents a small correction, confirming the validity of the adiabatic approximation employed in the calculations (the electron density change corresponding to the fine structure is only a 2-3% correction). Clearly, the fine structure may alter some of the details of the result but will not alter the basic conclusion that the dynamics of the molecule in the presence of intense infrared radiation is not governed by the molecule's normal modes.

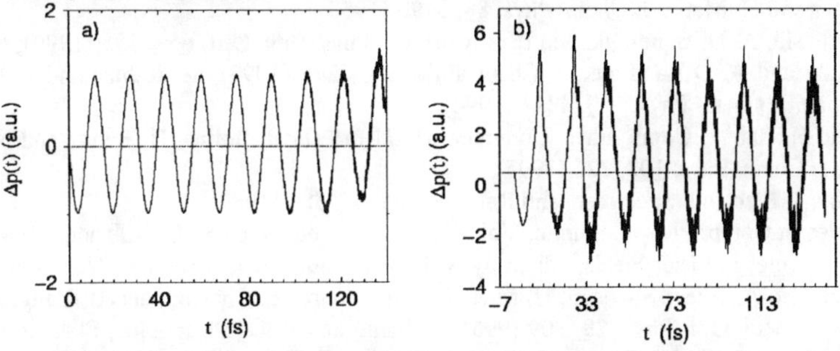

Fig. 10. The electrical dipole of HCN as a function of time for HCN in the presence of infrared radiation of two different intensities (see text). The ions are kept frozen at their equilibrium positions.

The fine structure in Fig. 10 is not numerical noise, as verified by thorough testing of convergence parameters. The numerical accuracy is easier verified in the case of free atoms whose symmetry requires that the nonlinear response is only odd harmonics. We have carried out such calculations and verified the accuracy needed for the present purposes. More details will be published elsewhere, where the nonlinear response will be discussed in more detail. We merely note here that the nonlinear oscillations lead to the emission of light at the corresponding frequencies, a phenomenon that has in fact been seen in rare gas atoms under ultrashort infrared pulses.[25]

Acknowledgment: The work was supported in part by the Office of Naval Research Grant N00014-96-1-1042, by a Grant from the Deutsche Forschungsgemeinschaft and by the William A. and Nancy F. McMinn Endowment at Vanderbilt University. Some calculations were done at the Cornell Theory Center. We thank M. Ramamoorthy and G. Vignale for useful discussions.

REFERENCES

1. For a review of work before 1980, see A. H. Zewail, Physics Today, November 1980, p.2. See also articles in A. H. Zewail, Femtochemistry – Ultrafast Dynamics of the Chemical Bond, (World Scientific, Singapore, 1994);
2. N. Bloembergen and A. H. Zewail, J. Phys. Chem. 88, 5459 (1984).
3. T. J. Chuang and H. Seki, Phys. Rev. Lett. 49, 382 (1982).
4. E.g. the theory by O. K. Rice, H. C. Ramsberger, L. S. Kassel and R. A. Marcus (RRKM) described by P. J. Robinson and K. A. Holbrook in Unimolecular Reactions, (Wiley-Interscience, New York, 1972).
5. A. Sinha, M. C. Hsiao, and F. F. Crim, J. Chem. Phys. 92, 633 (1990); 94, 4928 (1991); 97, 2204 (1993); R. B. Metz, J. D. Thoemke, J. M. Pfeiffer, and F. F. Crim, J. Chem. Phys. 99, 1744 (1993).
6. E. D. Potter, J. L. Herek, S. Pedersen, Q. Liu and A. H. Zewail, Nature 355, 66 (1992).
7. M. J. Bronikowski, W. R. Simpson, and R. N. Zare, J. Phys. Chem,. 97, 2194 (1993).
8. R. N. Zare, Nature 365, 105 (1993); D. J. Tannor, Nature 369, 445 (1994); P. Brumer and M. Shapiro, Scientific American, March 1995, p. 56.
9. G. Edwards et al. Nature 371, 416 (1994).
10. A. Ueda et al. Nucl. Instr. and Meth. B 100, 427 (1994).
11. G. C. Schatz, M. C. Colton, and J. L. Grant, J. Phys. Chem. 88, 2971 (1984).
12. R. Heather and H. Metiu, J. Chem. Phys. 88, 5496 (1988).
13. S. Chelkowski, A. D. Bandrauk, and P. B. Corkum, Phys. Rev. Lett. 65, 2355 (1990); S. Chelkowski and A. D Bandrauk, J. Chem. Phys. 99, 4279 (1993); M. Kaluza and J. T. Muckerman, J. Chem. Phys. 100, 4211 (1994).
14. S. Shi and H. Rabitz, Comp. Phys. Commun. 63, 71 (1991); J. Botina, H. Rabitz and N. Rahman, J. Chem. Phys. 102, 226 (1995).
15. J. C. Tully, J. Electron Spectrosc. and Rel. Phenom. 45, 381 (1987).
16. See articles in Comp. Phys. Commun. Vol. 63, Nos. 1-3, edited by K. C. Kulander (1991 and in Molecules in Laser Fields, edited by A. D. Bandrauk (marcel Dekker, New York, 1994); see also A. Giusti-Suzor, F. H. Mies, L. F. DiMauro, E. Charron, and B. Yang, J. Phys. B: At. Mol. Opt. Phys. 28, 309 (995); J. Zhang and P. Lambropoulos, Phys. Rev. Lett. 77, 2186 (1997); M. Brewczyk, K. Rzazewski, and C. W. Clark, Phys. Rev. Lett. 78, 191 (1997).
17. M. J. Gillan, J. Phys. C: Solid State Phys. 19, 6169 (1986).
18. P. Hohenberg and W. Kohn, Phys. Rev. 136, B864 (1964); W. Kohn and L. J. Sham, Phys. Rev. 140, A1133 (1965).
19. R. Car and M. Parrinello, Phys. Rev. Lett. 55, 2471 (1985).
20. M. C. Payne, M. P. Teter, D. C. Allan, T. A. Arias, and J. D. Joannopoulos, Rev. Mod. Phys. 64, 1045 (1992).
21. E. Runge and E. K. U. Gross, Phys. Rev. Lett. 52, 997 (1984).
22. D. Vanderbilt, Phys. Rev. B 41, 7892 (1990).
23. A. K. Rappé and W. A. Goddard, J. Phys. Chem. 95, 3358 (1991).
24. R. Winkler and S. T. Pantelides, J. Chem. Phys. 106, 7714 (1997).
25. C. Kan, N. H. Burnett, C. E. Capjack, and R. Rankin, Phys. Rev. Lett. 79, 2971 (1997); M. Schnürer, Ch. Spielmann, P. Wobrauschek, C. Streli, N. H. Burnett, C. Kan, K. Ferencz, R. Koppitsch, Z. Cheng, T. Brabec, and F. Krausz, Phys. Rev. Lett. 80, 3236 (1998).

Finite Temperature Quasicontinuum Methods

Vivek Shenoy, Vijay Shenoy and Rob Phillips

Division of Engineering, Brown University, Providence, RI 02912

Abstract

In this paper we extend the quasi-continuum method to study equilibrium properties of defects at finite temperatures. We present a derivation of an effective energy function to perform Monte Carlo simulations in a mixed atomistic and continuum setting. It is shown that the free energy minimization technique can be easily incorporated into the quasi-continuum frame work, permitting a reduction of the full set of atomistic degrees of freedom even in the finite temperature setting. The validity of the proposed methods is demonstrated by computing the thermal expansion and the temperature dependence of the elastic moduli for Cu. We also employ the quasi-continuum free energy minimization method to study the finite temperarure structure of a dislocation core in Al.

Introduction

Defects and their interactions play a crucial role in a wide range of phenomena that determine the mechanical properties of solids. For a long time, theories based on continuum elasticity have been used successfully to describe a number of aspects of defect interactions[1]. With the advent of ever more powerful computers, it has become possible to understand the properties of defects on the basis of atomistic simulations. While simulations involving a million atoms, performed using supercomputers, are now commonplace, alternative classes of models have exploited atomistic insights without abandoning the advantages offered by the continuum approach altogether. For example while studying a solid with many defects, in the immediate vicinity of individual defects it is necessary to have atomic level resolution to capture the complex rearrangement of atoms, while far from defects, a continuum description would be efficient. Such approaches enable us to handle situations where the phenomenon under study inherently depends on physics on multiple length scales, without significantly increasing the demand on computational resources. One such approach, is the quasi-continuum method(QC)[2] for performing lattice statics calculations at $0K$. This method is based on adaptive strategies in which in the course of a calculation, the region of full atomistic resolution can vary in response to evolution of both loading and defects. In this paper we extend the method to study the equilibrium properties of defects at finite temperatures.

A number of approaches have been taken to study the equilibrium properties of defects. Some of the most common methods include molecular dynamics (MD), Monte Carlo (MC)[3] and lattice dynamics using the harmonic[4] and local harmonic approximations[5,6]. While the MD and MC methods can be used to accurately determine defect properties, they require significantly larger computational resources compared to $0K$ lattice statics calculations. The methods that employ harmonic approximations are less accurate, but require computational resources comparable to a $T = 0$ calculation. In this paper we will develop the framework for the quasi-continuum Monte Carlo(QCMC) and quasi-continuum

Mat. Res. Soc. Symp. Proc. Vol. 538 © 1999 Materials Research Society

free energy minimization (QCFEM) methods. The paper is organized as follows. In the next section we present the theoretical formulation leading to an effective energy function to perform the QCMC simulations and a coarse grained free energy function for the QCFEM method. The following section is devoted to the validation of the methods, where we start by applying the QCMC method to a 1-D chain and demonstrate the validity of the effective energy function. We then apply the methods to compute the thermal expansion and the temperature dependence of the elastic moduli in Cu. The temperature dependence of the core structure of the conventional dislocation in Al is then studied using the QCFEM method. We conclude by pointing out the problems that can be tackled in the future using the methods developed here.

Theoretical Formulation

We start by giving a brief review of the QC method at $0K$. At $0K$, the configuration of a solid is determined by a set of atomic displacements $\mathbf{u}_i$ which minimize the potential energy function $H(\{\mathbf{u}_i\}) = \sum_{i=1}^{N} E_i(\mathbf{u}_1, ..., \mathbf{u}_N)$. In the QC method the potential energy function is approximately written in terms of the displacements of a set of representative atoms that form the nodes of a finite element mesh. The number of representative atoms is usually much smaller than the total number of atoms. For later use, we denote the coordinates of the representative atoms and the "slave" atoms with indices r and s respectively. The displacement of a slave atom can be determined in terms of the displacements of the representative atoms by using a finite element interpolation, $\mathbf{u}_i^s = \sum_j \mathbf{u}_j^r N_j(\mathbf{X}_i^s)$, where N_j's are the finite element shape functions. The representative atoms fall into either the atomistic region or the finite element region depending on a criterion to be described below. The former type of representative atoms are essential to capture the atomistic nature of defect cores and interfaces while the latter represents the continuum limit and is used in regions of the solid undergoing nearly homogeneous deformations. The energy of the atoms in the atomistic region is computed by an explicit consideration of all of their neighbors, while the energy of the atoms in the finite element region is computed from the local deformation gradient using the Cauchy-Born Rule. A given atom is treated as local or non-local depending on the magnitude of the differences in deformation gradients that it senses from the different elements around it. This can be characterized by the inequality $\max_{a,b;k} |\lambda_k^a - \lambda_k^b| < \epsilon_L$, where a and b run over all elements that are within some radius r_s of the representative atom of interest, λ_k^e is the kth eigenvalue of the right stretch tensor obtained from the deformation gradient $\mathbf{F}_e$ in element e, and ϵ_L is an empirically selected constant. This criterion implies that the error in the computation of the energy using the local deformation gradients when compared with a fully atomistic calculation of the energy is within a specified tolerance. Thus all atoms for which the above criterion is satisfied are assigned to the finite-element region, while for the remaining atoms the energies are computed atomistically. Using the above considerations the approximate energy function can be written as

$$H_{QC}(\{\mathbf{u}_i^r\}, T = 0) = \underbrace{\sum_{\beta} E_\beta(\{\mathbf{u}_i^r\})}_{\text{Atomistic Region}} + \underbrace{\sum_e \nu_e \mathcal{E}(\mathbf{F}_e)}_{\text{FE Region}} \tag{1}$$

where $\mathcal{E}(\mathbf{F})$ is the energy of a single atom undergoing a homogeneous deformation given by the deformation gradient tensor $\mathbf{F}$, and ν_e is the total number of atoms falling in element *e*. The approximate energy function H_{QC} can now be minimized using standard methods like the conjugate gradient method.

At finite temperatures, the probability of finding a solid in any configuration $\{\mathbf{u}_i\}$, can be expressed as $P(\{\mathbf{u}_i\}) \propto \exp\left[-\beta H(\{\mathbf{u}_i\})\right]$. In a Monte Carlo simulation, the equilibrium properties of the solid can be studied by using the Metropolis algorithm to sample configurations distributed according to $P(\{\mathbf{u_i}\})$. In formulating the QC version of the Monte Carlo algorithm, we need to determine the effective energy function $H_{QC}(\{\mathbf{u}_i^r\}, T)$, so that probability of finding the system in a configuration described by the representative degrees of freedom is expressed as $P(\{\mathbf{u}_i^r\}) \propto \exp\left[-\beta H_{QC}(\{\mathbf{u}_i^r\}, T)\right]$. On first thought it might be tempting to guess that $H_{QC}(\{\mathbf{u}_i^r\}, T) = H_{QC}(\{\mathbf{u}_i^r\}, 0)$ and to perform MC simulations with the effective energy function in Eq. (1). We will now show that this guess turns out to be wrong. The effective energy function for any temperature is *exactly* determined by

$$\exp\left[-\beta H_{QC}(\{\mathbf{u}_i^r\}, T)\right] = \int \exp\left[-\beta H(\{\mathbf{u}_i^r\}, \{\mathbf{u}_i^s\})\right] du^s, \tag{2}$$

where all the slave degrees of freedom are integrated out. In order to perform this integration, let us write $\mathbf{u}_i^s = \sum_j \mathbf{u}_j^r N_j(\mathbf{X}_i^s) + \boldsymbol{\eta}_i$, where we have introduced a fluctuational variable $\boldsymbol{\eta}_i$, which accounts for the fact that at finite temperatures, the co-ordinates of the slave atoms are not simply determined by a simple finite element interpolation. The quantity $\boldsymbol{\eta}_i$ may also be thought of as a means of bringing about random fluctuations in the finite element shape functions at finite temperatures. In terms of the random fluctuations $\boldsymbol{\eta}_i$, the energy function can be expanded as

$$H(\{\mathbf{u}^r\}, \{\boldsymbol{\eta}\}) = H_{QC}(\{\mathbf{u}^r\}, 0) + \sum_j \left.\frac{\partial H}{\partial \mathbf{u}_j^s}\right|_R \boldsymbol{\eta}_j + \frac{1}{2}\sum_{i,j} \left.\frac{\partial^2 H}{\partial \mathbf{u}_i^s \partial \mathbf{u}_j^s}\right|_R \boldsymbol{\eta}_i \boldsymbol{\eta}_j + \cdots, \tag{3}$$

where R is used to indicate that the derivatives are evaluated by setting all η_i's to zero. Since it is very difficult to integrate out the slave degrees of freedom by retaining all the terms in the above expansion, we will attempt to perform the integration by terminating the expansion to second order. Including the fluctuational terms up to quadratic order, and integrating the slave degrees of freedom, requires the knowledge of the normal modes of the slave atoms while the representative atoms are fixed at positions $\{\mathbf{u}^r\}$. This requires the diagonalization of a matrix whose size is proportional to the number of slave atoms. Since doing this is once again computationally prohibitive, the integral in Eq. (2) is approximately evaluated by computing it in each element and then summing up the contributions from all the elements. In each element we can approximate the integral by the product of the free energy of an atom in an infinite crystal undergoing a homogeneous deformation gradient and the number of slave atoms in the element. Using the well known expression for the free energy of an atom in an infinite crystal, we can approximate the effective energy function by

$$H_{QC}(\{\mathbf{u}_i^r\}, T) = H_{QC}(\{\mathbf{u}_i^r\}, 0) + k_B T v_a \sum_e \nu_e^s \int_B d\mathbf{k} \sum_\sigma ln\left(\frac{\hbar\omega(\mathbf{F}_e, \mathbf{k}, \sigma)}{k_B T}\right), \tag{4}$$

where v_a is the volume occupied by an atom in the crystal, ν_e^s is the number of slave atoms in the element e and $\omega(\mathbf{F}_e, \mathbf{k}, \sigma)$ is the frequency of a phonon with wavevector $\mathbf{k}$ and branch index σ in a crystal undergoing a homogeneous deformation gradient $\mathbf{F}_e$ and the integral is over B, the first Brillouin zone of the crystal. At finite temperatures, while the first term accounts for the internal energy of the deformed crystal, the second term can be thought of as arising from the entropic contributions of the slave degrees of freedom that are integrated out.

The procedure that we have suggested for integrating out the slave degrees of freedom in the system can still be computationally demanding. When a node is displaced at a MC step, we have to compute the phonon spectrum of all the elements surrounding the displaced node. To expedite computations, we adopt a method suggested by Lesar et.al[5], for computing the free energy of solids in the local harmonic (LH) approximation. Within this approximation, the effective energy function can be written as

$$H_{QC}(\{\mathbf{u}_i^r\}, T) = H_{QC}(\{\mathbf{u}_i^r\}, 0) + 3k_B T \sum_e \nu_e^s ln \left(\frac{\hbar D^{1/6}(\mathbf{F}_e)}{k_B T} \right), \tag{5}$$

where $D(\mathbf{F}_e)$ is the determinant of the local dynamical matrix of an atom in a crystal undergoing a homogeneous deformation gradient $\mathbf{F}_e$.

So far we have developed the effective energy function to carry out MC simulations within the quasi-continuum frame work. For realistic problems, while this is a significant improvement in the computational resources compared to the conventional MC simulations, the effort required is significantly greater than that of a $0K$ simulation. It is therefore desirable to develop a method that requires computational effort comparable to the $T = 0$ simulation. This can be achieved if the free energy of the solid can be obtained in terms of the coordinates of the representative atoms. As noted earlier, LeSar et.al[5] have shown that the free energy is readily expressed in terms of the atomic configurations, if the Einstein model is used to compute the vibrational free energy of all the atoms in the solid. This method is easily implemented within the quasi-continuum approach, by writing the local harmonic free energy in the form $A_{QC} = \sum_\beta A_\beta(\{\mathbf{u}_i^r\}) + \sum_e \nu_e \mathcal{A}(\mathbf{F}_e)$, where the first term represents the local harmonic free energy in the atomistic region computed by explicitly considering the neighbors, while the second term represents the free energy of an atom in the finite element region computed from the knowledge of the deformation gradients. The equilibrium configuration of the atoms can be then computed by minimizing A_{QC}. While this free energy minimization method is fast, it is not always accurate. In the process of minimizing the energy, if an atom moves to a region where energy landscape has negative curvature, the determinant of the dynamical matrix becomes imaginary and it is unclear how to proceed.

Validation

In this section, we validate the QCMC method by first applying it to study a simple 1D model. We will present a detailed derivation that shows that if the free energy of the slave degrees of freedom is not included, the model behaves as if it were at a lower temperature. Consider a 1D chain of N atoms of unit mass, interacting with near neighbors. We apply periodic boundary conditions, so that the first atom and the Nth atoms are near neighbors. If x is the distance between the atoms, the interaction is taken to be of the form $E(x) = e^{2(x-1)} - 2e^{-(x-1)}$ for $x \geq 0$ and $E(x) = \infty$ for $x < 0$, so that the neighbors cannot cross each other. Since the interactions are only between the near neighbors, it is easy to show that the partition function has the form $Z(N, T) = \left(\int_0^\infty \exp\left[-\beta E(x)\right] dx \right)^N$. From the partition function any equilibrium property of interest can be computed. For example by defining the ensemble average of any quantity $O(x)$ as

$$\langle O \rangle = \frac{\int_0^\infty O(x) \exp\left[-\beta E(x)\right] dx}{\int_0^\infty \exp\left[-\beta E(x)\right] dx}, \tag{6}$$

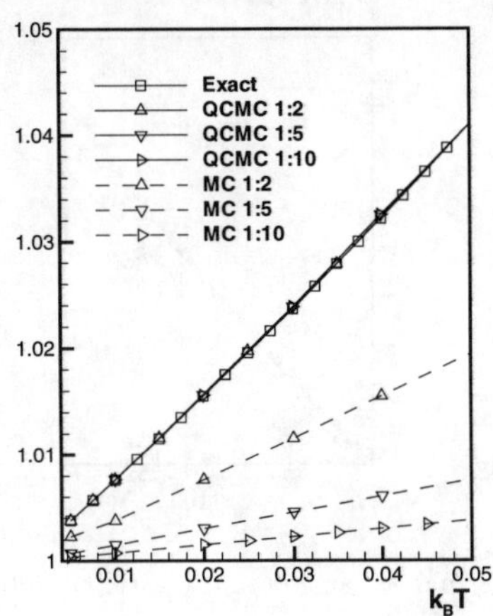

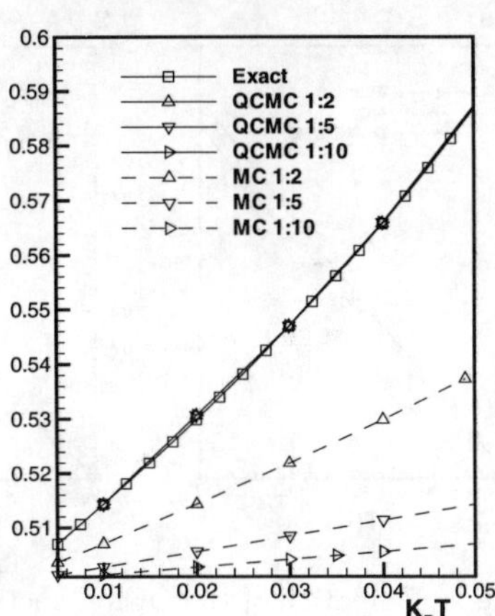

Figure 1: Validation of the QCMC method in 1D. Lattice parameter and thermal expansion for the 1-D near neighbor lattice.

it is easily shown that the lattice parameter is given by $a(T) = \langle x \rangle$ and that the elastic compliance is given by $S(T) = \frac{N}{k_B T}\left(\langle x^2 \rangle - [\langle x \rangle]^2\right)$. Using these formulae the lattice parameter and the elastic compliance at any temperature can be computed simply by e-valuating one dimensional integrals. Let us now scale the system by a factor s, and choose $(N-1)/s + 1$ atoms equally spaced at distance s to be the representative atoms. These atoms are the nodes of the finite element mesh that has $N-1$ elements. We evaluate the mean displacement between the two nodes of an element, by using the effective energy function derived in the last section. If u_{i+1} and u_i are the nodal displacements of the nodes $i+1$ and i respectively, the deformation gradient in the element i is $F_i = (u_{i+1} - u_i)/s$. This implies that each of the s atomic bonds in the element have lengths $b = 1 + F_i$. Within the local harmonic approximation, the effective energy function in the element is given by $H_{QC}(b,T) = sE(b) + k_B T(s-1)\log(\hbar\omega/k_B T)$, where $\omega = \sqrt{E''(b)}$. Let b_{QC} and b_{MC} denote the ensemble average of the bond length computed using the correct effective energy $H_{QC}(b,T)$ and the zero temperature effective energy $H_{QC}(b,0) = sE(b)$ respectively. We find that

$$b_{QC} = \frac{\int_0^\infty b\exp\left[-s\beta E(b)\right]\left[E''(b)\right]^{\frac{1-s}{2}}db}{\int_0^\infty \exp\left[-s\beta E(b)\right]\left[E''(b)\right]^{\frac{1-s}{2}}db} \quad and \quad b_{MC} = \frac{\int_0^\infty b\exp\left[-s\beta E(b)\right]db}{\int_0^\infty \exp\left[-s\beta E(b)\right]db}. \tag{7}$$

Note that when $s = 1$, both b_{QC} and b_{MC} are equal, while they are unequal when $s > 1$. It is also transparent that $b_{MC} = a(T/s)$, i.e, using the zero temperature effective energy, gives us the lattice parameter at a reduced temperature determined by the scale fator s. We have also performed MC simulations on a 1-D chain of 100 atoms using the constant stress ensemble to compute the lattice parameter and the elastic compliance for systems scaled by factors 2, 5 and 10. The results are shown in Fig. 1, along with the curve obtained by computing the one dimensional integrals mentioned earlier. It is clear that

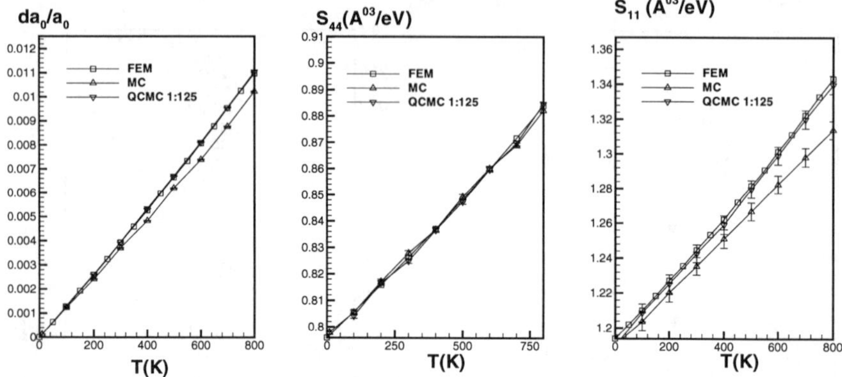

Figure 2: Validation of the QCMC method in 3D. Thermal expansion and elastic compliance for Cu.

the QCMC simulations reproduce the correct results, while using the zero temperature effective energy (denoted by MC in Fig. 1) gives incorrect results. We have verified that the results obtained by using the wrong effective energy correspond to a lower temperature T/s as predicted by Eq. (7). As a part of the validation of our methods in 3D, we have computed the temperature dependence of the lattice parameter and elastic compliances. In Fig. 2, we show the results obtained by QCMC, QCFEM and the conventional MC simulations. The conventional MC simulations were performed with 500 atoms in a periodic cube in the constant stress ensemble. In this method the dimensions of the cube as well the orientation of the cube axes are altered as a part of the MC sampling procedure and the elastic compliances are computed from the fluctuations in the cube dimensions. The QCMC simulations performed on a cube with 64 nodes, each node representing 125 atoms. The results indicate that the magnitude of the difference in the of the various quantities computed using the three methods is quite small. We have also used the QCFEM method to look at the structure of the core of the conventional dislocation in Al (we have used the EAM potentials of Ercolessi and Adams[7]). At 0K, dislocation core region is known to have a width of about 15Å. Our results show that the width decreases as a function of temperature. The widths at 300 K and 600 K were found to be 13.7Å and 12.9Å respectively. We are now trying to understand these results using continuum elasticity arguments.

Summary and Conclusion

In this paper we have presented two approaches that extend the quasicontinuum method to finite temperatures. We derived the effective energy that accounts for the entropic contribution of the slave degrees of freedom. The validity of the QCMC method was demonstrated by applying it to a tractable 1-D model. The method was also applied to compute thermal expansion and temperature dependence of the elastic moduli in Cu. We also presented preliminary results of a study of the core structure of the conventional dislocation in Al using the QCFEM method, although the results obtained in this case raise more questions than they answer.

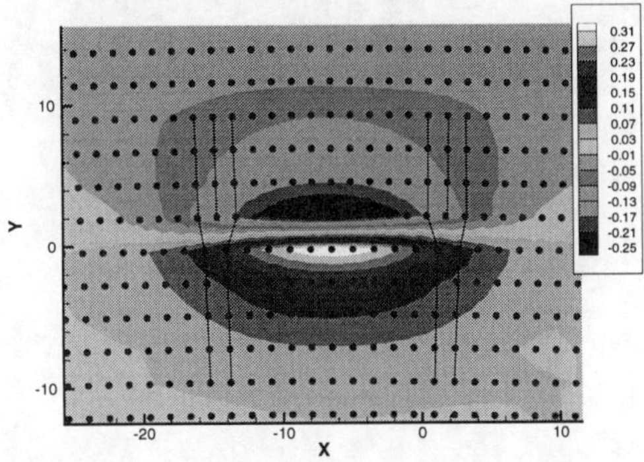

Figure 3: The structure of the core of the conventional dislocation in Al at 300K. The contours indicate the magnitude of the out of plane displacements (in angstroms).

The methods proposed in the paper result in a significant reduction in the full set of atomistic degrees of freedom. In the future we hope to apply these methods to compute temperature dependence of a number of material parameters such as the Peierls stress, dislocation line tension and core energies. It is also possible that the method of this type can also be applied to study other physical problems like domain walls in magnets and disinclinations in liquid crystals.

References

[1] See for example J. D. Eshelby in *Prog. Solid State Physics,* edited by F. Sietz and D. Trunbull, (Academic Press, NewYork, 1956), **3**, pp. 79-114.

[2] E. B. Tadmor *et.al, Phil. Mag. A,* **73**, p. 1529, (1996); V. B. Shenoy *et.al, Phys. Rev. Lett.,* **80**, p. 742, (1998).

[3] *Simulation of Liquids and Solids*, edited by G. Ciccotti, D. Frenkel and I. R. McDonald, (North-Holland, Amsterdam, 1987).

[4] T. H. K. Baron and M. L. Klein, in *Dynamical Properties of Solids,* edited by G. K. Horton and A. A. Maradudin (North-Holland, Amsterdam, 1975), Vol. 1, pp. 391-450.

[5] R.LeSar, R. Najafabadi and D. J. Srolovitz, *Phys. Rev. Lett.,* **63**, p. 624, (1989).

[6] For a critical evaluation of the harmonic methods along with numerous references see S. M. Foiles, *Phys. Rev. B,* **49**, p. 14930, (1994).

[7] F. Ercolessi, J. Adams, *Europhys. Lett.,* **26**, p. 583, (1993).

IMPOSING FIELD BOUNDARY CONDITIONS IN MD SIMULATION OF FLUIDS: OPTIMAL PARTICLE CONTROLLER AND BUFFER ZONE FEEDBACK

Ju Li, Dongyi Liao, Sidney Yip
Department of Nuclear Engineering, MIT, Cambridge, MA 02139 *(http://mmm.mit.edu)*

ABSTRACT

We formulate a method for imposing continuous field boundary conditions on an MD simulation with little or no disturbance to its dynamics. Our approach combines the previously developed thermodynamic field estimator, which extracts macroscopic fields from particle data, with a novel procedure, optimal in the sense of least disturbance, for imposing prescribed continuum field boundary conditions on the atomistic system. By means of feedback control and assuming molecular chaos in fluids, that disturbance can further be eliminated entirely, thus providing an exact solution for general steady-state fluid problems, where the desired fields at the real boundary is achieved by adjusting actions in a region *separated* from the real boundary by a buffer zone.

OPTIMAL PARTICLE CONTROLLER (OPC)

A basic problem in fluid simulation, which was first investigated by O'Connell and Thompson[1], then Hadjiconstantinou and Patera[3], and then by us[4, 5], is the following: we would like the particles in an MD simulation to satisfy the following distribution in their positions and velocities $\{\mathbf{x}_i, \mathbf{v}_i\}$,

$$dP = f\left(\mathbf{x}, \mathbf{v}|\rho(\mathbf{x}), T(\mathbf{x}), \bar{\mathbf{v}}(\mathbf{x})\right) d\mathbf{x}d\mathbf{v} = \frac{\rho(\mathbf{x})d\mathbf{x}}{(2\pi T(\mathbf{x}))^{3/2}} \exp\left(-\frac{|\mathbf{v} - \bar{\mathbf{v}}(\mathbf{x})|^2}{2T(\mathbf{x})}\right) d\mathbf{v} + f^{(2)}d\mathbf{x}d\mathbf{v}, \quad (1)$$

at a prescribed boundary $\partial\mathcal{C}$, where $\rho(\mathbf{x})$, $T(\mathbf{x})$, $\bar{\mathbf{v}}(\mathbf{x})$ are the macroscopic density, velocity and temperature fields. For this to be sensible, $\partial\mathcal{C}$ should be in small-gradient regions where $\rho(\mathbf{x})$ etc. are well-defined; conversely, if that is true, the transition from particle to continuum description like the Navier-Stokes equation, can be shown to be exact by the Chapman-Enskog expansion, where $f^{(2)}$ is the the second-order correction to the local Maxwellian that contains the spatial gradients of these fields. So, both continuum and particle descriptions should be valid when close to $\partial\mathcal{C}$, and (1) is the link that allows us to couple the two.

In a previous paper[4] we have shown how to solve the opposite problem, that is how to *infer* the macroscopic fields in $\mathcal{C}$ from current particle data, using a technique named thermodynamic field estimator (TFE); the upshot is that we can accurately determine the macroscopic fields at an inner boundary $\partial\mathcal{U}$ inside $\mathcal{C}$ in real time even when the particle data contains considerable "thermal noise". And so $\partial\mathcal{U}$ provides the BC for a continuum solver in the classical domain decomposition and alternating Schwarz iteration formalism[2], where $\partial\mathcal{C} - \partial\mathcal{U}$ is the continuum-MD overlap region. Because of space limitation we will not further discuss the TFE here, nor the full Schwarz iteration implementation which we have carried out for Couette flow, but will focus on *just* the second part of the iteration, which is how to impose continuous field BC on $\mathcal{C}$. Interested readers should follow the pioneering paper of Hadjiconstantinou and Patera[3], and much detailed reports of our work in [4, 5]. Suffice it here to note that in order for the whole scheme to work to second order accuracy in (1), we can simply use the first term in (1) for TFE, ignoring $f^{(2)}$.

Now, suppose one has inferred the *current* fields to be $\rho'(\mathbf{x})$, $T'(\mathbf{x})$, $\bar{\mathbf{v}}'(\mathbf{x})$ on $\partial\mathcal{C}$ using TFE, but wants the fields to be $\rho(\mathbf{x})$, $T(\mathbf{x})$, $\bar{\mathbf{v}}(\mathbf{x})$, how should one modify the particle coordinates such that the desired distribution is achieved? That is, say there is a random variable set $\{X_n\}$ satisfying distribution $f(X)$, but we want them to satisfy distribution $g(X)$, so we begin to replace X_n's by Y_n's hoping that $\{Y_n\}$ will satisfy $g(Y)$, what should be the optimal $X_n \to Y_n$ transformation?

One may ignore X_n and use only $g(Y)$ to resample the random variable Y_n. While this would indeed lead to a BC imposing scheme, it is quite conceivable that ignoring the current state will result in a procedure that strongly disturbs the particle dynamics. If decoupling the two distributions, or equivalently, the two sets of particle coordinates, is not a good idea, then one should look for a way to relate them such that the disturbance is minimized. We now introduce a quantitative measure of this disturbance:

$$B = \sum_n |\Delta \mathbf{v}_n|^2 = \sum_n |\mathbf{v}_n^{\text{out}} - \mathbf{v}_n^{\text{in}}|^2, \tag{2}$$

where $\Delta \mathbf{v}_n$ is the change in particle velocity of the nth particle ($\mathbf{v}_n^{\text{in}}$ and $\mathbf{v}_n^{\text{out}}$ are the velocity before and after the transformation). This quantity characterized the artificial disturbance to particle dynamics which should be minimized as much as possible.

As an example let us consider a simple problem of 1D heat conduction, for which we wish to impose a boundary condition of high temperature T_h at $x = 0$ and low temperature T_l at $x = 1$. Intuitively one could imagine doing the following: when a particle crosses either boundary $x = 0$ or 1, give it a random velocity drawn from distribution (1) with parameters $\bar{\mathbf{v}} = 0$ and $T = T_h$ or T_l. However, this procedure does not work. When implemented in practice in the case of homogeneous heating ($T_h = T_l$), results show that the bulk temperature reaches a value of $T_h/2$.

The scheme fails because one is dealing with conditional probability. The speed distribution of atoms which cross the boundary is *different* from the speed distribution of atoms in the bulk. Instead it is weighted by the normal velocity,

$$dP = \frac{v \exp\left(-\frac{v^2}{2T}\right) dv}{\int_0^{+\infty} v \exp\left(-\frac{v^2}{2T}\right) dv} = \frac{v}{T} \exp\left(-\frac{v^2}{2T}\right) dv, \quad 0 < v < +\infty. \tag{3}$$

$\langle v^2 \rangle$ from distribution (3) is $2T$, not T. Thus, if we sample the *boundary crossing atoms* using bulk distribution (1) with parameter $T = T_h = T_l$, the energy can only be balanced in a statistical sense when the bulk temperature reaches $T_h/2$.

A more subtle defect of this scheme is that, for whatever the incoming velocity v_{in} of the particle before hitting the boundary, a new velocity v_{out} is drawn from a given distribution, say $g(v)$, entirely *independent* of v_{in}. Thus if we evaluate the disturbance to particle dynamics using (2), it is always substantial. In the case of *homogeneous heating*, even when the correct distribution (3) is used in drawing v_{out}'s and the system has reached the desired temperature $T = T_h = T_l$, the scheme continues to disturb the particles by giving each boundary crossing atom a new v_{out}. On the other hand, if we just let $v_{\text{out}} = v_{\text{in}}$, i.e., do nothing, the system temperature stays at T_h! A more intelligent particle controller should automatically tune down its influence as the system approaches the desired state, a behavior we may call the coalescence property.

We now formulate the above ideas mathematically. Suppose $\{X_n\}$ conform to distribution function $f(X)$:

$$dP(\eta < X < \eta + d\eta) = f(\eta)d\eta, \tag{4}$$

and we would like the series to conform to a different distribution g. We propose to achieve this by replacing every X with another number Y which is distributed according to g. Thus the goal is to find a transformation $\mathcal{T}$,

$$\mathcal{T} : X_n \to Y_n, \tag{5}$$

with the requirement that if $\{X_n\}$ conforms to distribution $f(X)$, $\{Y_n\}$ will conform to $g(Y)$:

$$dP(\xi < Y < \xi + d\xi) = g(\xi)d\xi. \tag{6}$$

There are many possible $\mathcal{T}$'s. However our previous discussion shows that the following property is desirable: *if $f \equiv g$, then $\mathcal{T}$ gives $Y_n \equiv X_n$.* Generally, to incorporate the idea of *minimally disturbing the dynamics*, we adopt the reasonable criterion that

$$B[\mathcal{T}] = \langle (Y - X)^2 \rangle \tag{7}$$

be minimized among all possible $\mathcal{T}$'s.

We see that if $\mathcal{T}$ is randomly drawing Y from $g(Y)$ without referencing to X, it satisfies the basic requirement (6) but not the coalescence property. We call this $\mathcal{T}_1$ transformation. To incorporate coalescence one may consider $\mathcal{T}_2$,

$$\mathcal{T}_2 : \begin{cases} Y = X : & p \le \frac{g(X)}{Kf(X)} \\ \text{draw } Y \text{ randomly from } g(Y) : & p > \frac{g(X)}{Kf(X)} \end{cases}$$

where p is a random number uniformly distributed over $[0, 1]$, and K is a constant (scheme fails if K does not exist) such that

$$Kf(X) \ge g(X) \quad \text{for} \quad -\infty < X < +\infty. \tag{8}$$

But is $\mathcal{T}_2$ the best one?

We propose the following transformation, in the form of an implicit relation,

$$\mathcal{T}_3(X \to Y) : \int_{-\infty}^{X} f(\xi)d\xi = \int_{-\infty}^{Y} g(\xi)d\xi. \tag{9}$$

It can be checked that Y_n indeed conforms to distribution g if the incoming random variable X_n conforms to f, and it satisfies the coalescence property. We believe that $\mathcal{T}_3$ is the mathematically optimal transformation which minimizes (7). A tentative proof is given in [5]. This assertion should come as no surprise because it is the only one-to-one continuous mapping which satisfies (6) without extra randomness like in $\mathcal{T}_2$. Our experience shows that while $\mathcal{T}_3$ works rather well, it is an implicit algorithm and could be computationally demanding.

One may choose to operate $\mathcal{T}$ in a finite-volume region or just at a certain boundary, which are called bulk or boundary particle controllers respectively. The subtleties of boundary particle controllers are already mentioned in the first example. When the average disturbance to microscopic particle dynamics, defined by some reasonable measure, due to the operation of a family of particle controllers serving certain fixed purpose, is minimized, it is called the optimal particle controller (OPC). We believe that for one- or decoupled multi-dimensional distributions, $\mathcal{T}_3$ is OPC. The explicit formulas for general 3D $T(\mathbf{x})$, $\bar{\mathbf{v}}(\mathbf{x})$ boundary particle controller are given in [5], while $\rho(\mathbf{x})$ coupling involves particle insertion or removal which is more difficult to treat (see [4]). But explicit $\rho(\mathbf{x})$ boundary condition is rare. Also, those derivations ignore $f^{(2)}$, which cause certain problems, and we will derive more accurate versions by factoring $f^{(2)}$ into the distribution function. In all cases the general discussions about OPC hold.

EXTENDED BOUNDARY CONDITION

In last Section a method was proposed to control the field boundary condition of an MD simulation that results in least disturbance to the particle dynamics in the sense of (2). Nevertheless the disturbance still exists for particles in the skin region near the boundary, which can be shown to be proportional to the rate of dissipation in the system. In this Section we will formulate a procedure which eliminates that disturbance entirely at the specified boundary, thus providing an *exact* theoretical solution to the problem. Since any artificial action necessarily alters the particle dynamics in the vicinity where it is imposed, the best one can do is to act some distance away from the intended boundary and cause the macroscopic field at the boundary to be what is prescribed.

This can be done using a three-region approach which we will call the *Extended Boundary Condition (EBC)*, shown in Fig.1, and through a feedback control mechanism. The physical region of interest, $\mathcal{C}$ (core), is surrounded by a buffer zone $\mathcal{B}$. Artificial actions are applied on an outer MD region $\mathcal{A}$ which is sufficiently separated from $\mathcal{C}$, its aim being to induce the prescribed field boundary conditions on the core boundary $\partial\mathcal{C}$. Due to molecular chaos in fluids, short wavelength

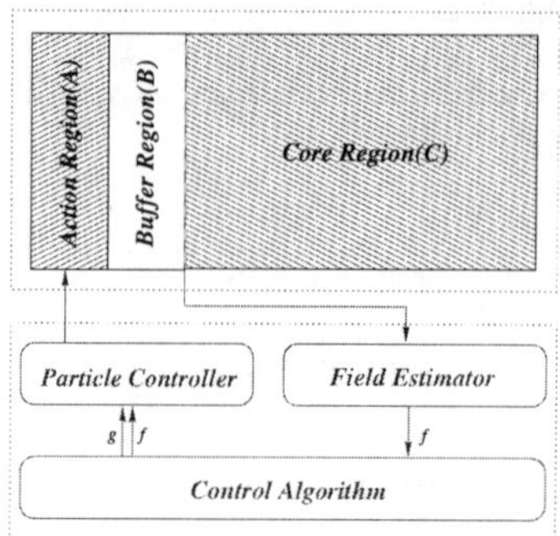

Figure 1: Schematic diagram of the Extended Boundary Condition (EBC), which incorporates a field estimator, a particle controller and a feedback control algorithm.

perturbations on particle dynamics will decay over a distance of a few collisional mean free paths, thus setting a lower limit for $\mathcal{A}$-$\mathcal{C}$ separation. The action in $\mathcal{A}$ is in general a matter of choice, but here we will use the OPC developed in the last section because it minimizes the artificial disturbance, thus $\mathcal{B}$ could be as thin as possible. The thermodynamic field estimator (TFE) serves as the detector of the control loop, inferring current fields on $\partial\mathcal{C}$ based on particle data from the entire $\mathcal{C}$ region. A control algorithm compares the inferred current field with the prescribed field on $\partial\mathcal{C}$, and gives instructions to the particle controller.

Note that although $\mathcal{A}$ and $\mathcal{B}$ are discrete-particle regions governed by molecular dynamics as $\mathcal{C}$, they are not real physical domains that appear in the Schwarz domain decomposition. They are constructs existing in the MD simulation solely to ensure that the physical region of interest, $\mathcal{C}$, has the correct field boundary conditions and evolves according to natural particle dynamics. Properly implemented, it is a powerful tool for studying fluid systems because a fluid atom in $\mathcal{C}$ has no way to tell *any* difference from reality. It in turn can be used as a perfect benchmark for comparing the actual performances of $\mathcal{T}_1$ and $\mathcal{T}_3$, as we will see.

In applying the EBC, the buffer region $\mathcal{B}$ is between $\mathcal{A}$ and $\partial\mathcal{C}$. Actions are applied on $\mathcal{A}$ instead of $\partial\mathcal{C}$ in such a way that the desired field at $\partial\mathcal{C}$ is obtained. The question then becomes, given the desired flow velocity at $\partial\mathcal{C}$, v_C^*, how does one control the desired flow velocity at $\mathcal{A}$ (prescribed value to the OPC), v_A^*? Adopting a first-order feedback control algorithm, we may write,

$$\frac{dv_A^*(t)}{dt} = -\kappa(v_C(t) - v_C^*),\qquad(10)$$

where the rate constant κ should be picked according to the approximate response time at $\partial\mathcal{C}$ due to an action on $\mathcal{A}$, making $v_C(t)$ converge to v_C^* as quickly as possible without oscillation.

Compared to directly applying OPC to $\partial\mathcal{C}$ without a buffer zone, EBC has two advantages: 1. It practically generates no disturbance to the core region. 2. We do not need to worry about $f^{(2)}$ or higher-order terms, which come out automatically on $\partial\mathcal{C}$. The drawback is that for general BC, multi-variable feedback control is needed, which is quite a challenging task.

COUETTE FLOW TEST

In this Section we demonstrate how the above ideas are implemented for Couette flow simulation. The process is somewhat reversed: we first implement the direct BC's using $\mathcal{T}_1$ and OPC,

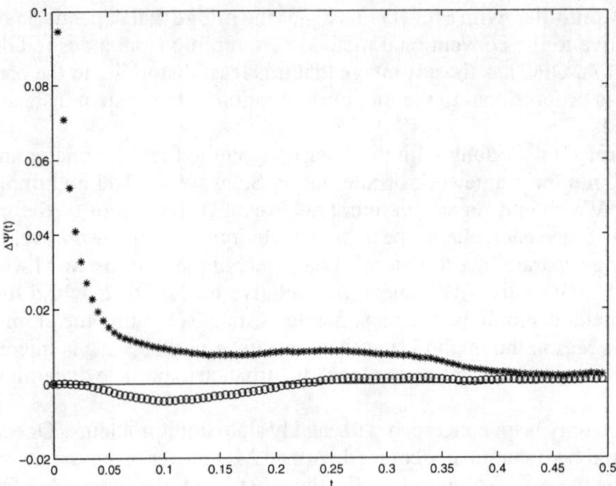

Figure 2: Comparison of the change to the normalized velocity autocorrelation function $\Psi(t)$ due to $\mathcal{T}_1$ and OPC. $\Psi_{\text{EBC}}(t) - \Psi_{\mathcal{T}_1}(t)$ is plotted in stars and $\Psi_{\text{EBC}}(t) - \Psi_{\text{OPC}}(t)$ is plotted in circles.

obtain the actual flow velocity value at $\partial\mathcal{C}$, and then use EBC to reproduce that actual value at $\partial\mathcal{C}$ by acting on $\mathcal{A}$. We then in turn calculate a specially defined particle velocity auto-correlation function with EBC, and compare those of the direct approaches to it, to show that OPC is indeed much more effective than $\mathcal{T}_1$ in preserving the particle dynamics.

The velocity autocorrelation function is a well-used quantity which characterizes microscopic dynamics. We can define its localized and normalized version

$$\Psi(t) = \frac{\langle \mathbf{v}_i(0) \cdot \mathbf{v}_i(t) \rangle}{\langle \mathbf{v}_i(0)^2 \rangle}, \tag{11}$$

for any i^{th} particle that crosses $\partial\mathcal{C}$ and comes into $\mathcal{C}$ region at $t = 0$; its contribution to $\Psi(t)$ is terminated whenever the particle re-crosses $\partial\mathcal{C}$. Since the EBC acts on $\mathcal{A}$ instead of $\partial\mathcal{C}$, $\Psi_{\text{EBC}}(t)$ should be the same as that of nature where no artificial disturbance can be discerned on $\partial\mathcal{C}$. On the other hand, it is not so if we directly act on $\partial\mathcal{C}$. However, $\Psi(t)$ of direct OPC should be better than that of direct $\mathcal{T}_1$ in comparison with $\Psi_{\text{EBC}}(t)$. Fig. 2 shows the difference $\Psi_{\text{EBC}}(t) - \Psi_{\mathcal{T}_1}(t)$ in stars and $\Psi_{\text{EBC}}(t) - \Psi_{\text{OPC}}(t)$ in circles. Notice there is a drastic loss of correlation in the first 0.05 ps, clearly attributable to the reaction of surrounding atoms when its direction is suddenly altered from its original motion. On the other hand, $\Psi_{\text{OPC}}(t)$ is almost indistinguishable from $\Psi_{\text{EBC}}(t)$.

DISCUSSIONS

We have developed a hybrid continuum-atomistic simulation scheme[4, 5] based on the concept of domain decomposition, in the same spirit as Hadjiconstantinou and Patera[3]. Such methods are useful for problems where much of the region of interest can be treated by continuum description, but a small critical, embedded part requires atomistic simulation. While Hadjiconstantinou and Patera implemented their method with attention to both the atomistic and continuum representations, we have focused our efforts on techniques for imposing continuum BC to atomistic simulations. In this work matching of the two different levels of representation, each with its own set of degrees of freedom, takes place in the overlap region through an implicit transformation which we have called

477

the Optimal Particle Controller, with Eqn (1) serving as the bridge that depends on both particle and field variables. Relative to the conventional method of sampling from a desired distribution such as the $\mathcal{T}_1$ transformation, OPC has the advantage that it is least disturbing to the particle dynamics, and that disturbance is proportional to the rate of dissipation in the system instead of the absolute magnitude of the fields.

We have incorporated OPC along with the Thermodynamic Field Estimator and the Extended Boundary Condition into the framework of alternating Schwarz method and implemented it in a study of shear flow. We should regard the numerical results[5] as proof of the principles of the entire hybrid scheme. Since each one of these three techniques has its own novel features, further applications, including separate investigations of each technique, will be worthwhile to bring out their capabilities and limitations. We believe that relative to the other hybrid methods recently proposed [1, 3] our method should be the most gentle as long as treating the atomistic subdomain is concerned. For this reason the method should be most useful when one is interested in delicate or subtle molecular effects where minimizing local disturbance to particle dynamics is a significant concern.

Up to now we have only been concerned with steady state fluid problems. One may wonder if a continuum solver could be dynamically coupled with an MD solver using the above techniques in real time. Nobody has the exact solution yet. On the other hand, the timescale of an macroscopic dynamical event is usually much greater than that of its underlying microscopic mechanism, thus to the critical microscopic region at any given moment of the event, the outside world would seem to be in a steady state flow condition and could be modeled by our techniques.

The case for solids is different. Solids, unlike fluids, have long range order. The assumption of molecular chaos is invalid here, and a buffer zone cannot cover up the disturbed dynamics elsewhere, which is the underpinning of the Extended Boundary Condition. Thus, it is quite difficult to isolate a region in a crystalline material since the phonons have very long mean free paths, and their scattering and reflections determine important macroscopic properties such as the thermal conductivity. Although researchers have developed successful techniques to couple continuum with atomistic regions for static calculations, no one has yet claimed to successfully implement such a scheme to study finite temperature properties.

ACKNOWLEDGMENTS

This work was supported by the Sandia National Laboratory and by AFOSR Grant No. F49620-96-1-0447. We express our appreciation to C.-N. Wong for suggesting this problem as well as interest and support throughout its development. We also acknowledge helpful discussions with N. Hadjiconstantinou and A. Patera.

REFERENCES

[1] S. T. O'Connell and P. A. Thompson, Phys. Rev. E **52**, 5792 (1995).

[2] B. F. Smith, P. E. Bjorstad, W. D. Gropp, *Domain Decomposition: Parallel Multilevel Methods for Elliptic Partial Differential Equations* (Cambridge University Press, 1996).

[3] N. Hadjiconstantinou and A. T. Patera, *Int. J. Modern Phys.* C **8** 967, (1997).

[4] J. Li, D. Liao and S. Yip, Phys. Rev. E **57**, 7259 (1998).

[5] D. Liao, J. Li and S. Yip, submitted to Phys. Rev. E.

ELECTRONIC STRUCTURE BOUNDARY VALUE PROBLEMS WITHOUT ALL OF THE ATOMS

H. T. JOHNSON, R. PHILLIPS, AND L. B. FREUND
Division of Engineering, Brown University, Providence, RI 02912

ABSTRACT

A mixed atomistic/continuum technique, in the spirit of the quasicontinuum method, is formulated and used to solve boundary value problems in strained semiconductor structures in which the mechanical fields and the local electronic structure are fully coupled. The technique is implemented by means of a standard structural mechanics finite element package. Within each element in the mesh, tight binding calculations are made for mechanical properties based on the local electronic structure. The finite element program calculates equilibrium mechanical fields based on this atomistic constitutive information; the underlying electronic properties can then be extracted at the element level. The technique is demonstrated by examining several simple plane strain boundary value problems for coherently strained silicon.

INTRODUCTION

Strain in semiconductor structures arises from the effects of lattice parameter or thermal expansion mismatch or due to the presence of defects. In small structures with complex geometries, like quantum dots and quantum wires, the strain is inherently coupled to the electronic properties and quantum effects which make the devices practically useful. Most studies of coupled strain and electronic properties in semiconductor structures are based on accurate but computationally intensive pure atomistic models, such as the work on pyramidal island quantum dots by Zunger[1]. A few other studies, including those of Grundmann *et al.* [2], Pryor[3], and Johnson *et al.* [4] are based on continuum models which use perturbation theory to calculate strain effects in quantum dots and quantum wires. However, fully atomistic models are computationally limited to the smallest devices, and continuum models are limited by approximate treatments of quantum and elastic effects.

The goal of the technique presented here is to study coupling of strain and electronic properties in semiconductor structures using a boundary value problem approach. The advantages of using the continuum finite element method to solve boundary value problems are combined with the advantages of a quantum mechanical description of the material based on a tight binding treatment of the electronic structure. The mixed formulation exploits these strengths while avoiding the need to explicitly account for each atomic degree of freedom present in the entire boundary value problem. Furthermore, the method takes advantage of the convenience of meshing and postprocessing available in a standard structural mechanics finite element package. In the following section, the method is formulated. Several examples of simple boundary value problems are then presented.

FORMULATION

Mixed continuum/atomistic procedure

The technique is based in principle on the quasicontinuum method of Tadmor *et al.* [5] The finite element method serves as a framework for the analysis, and atomistics provides constitutive information at the element level. Here, tight binding is used to provide the atomistic information. A local formulation is used, so that the finite element mesh size is always considerably larger than the atomic scale, and it is assumed that atoms within a particular element experience a uniform environment. The approach can be considered standard, nonlinear elastic finite element analysis, but with the nonlinear constitutive behavior based on tight binding calculations. These calculations for constitutive information within each element are simplified by recourse to the Cauchy-Born rule, which allows for the fields within elements to be treated as uniform.

The finite element framework of the method is provided by the structural mechanics finite element package Abaqus[6]. The Abaqus elasticity formulation allows for material behavior within any element to be characterized by the user, as a function of the evolving deformation gradient tensor $\mathbf{F}$. Nonlinearity is accomodated by dividing the nonlinear elastic problem into a series

Mat. Res. Soc. Symp. Proc. Vol. 538 © 1999 Materials Research Society

of linear loading steps. At the beginning of each load step, the atomistic subroutine determines material properties based on the incoming deformation gradient tensor $\mathbf{F}$. Abaqus then minimizes the total energy of the deformable body with respect to nodal displacements using a Newton-Raphson type algorithm. The new nodal displacements at the end of each step define $\mathbf{F}$ in each element for the beginning of the next step. The process continues until the entire load has been applied and equilibrium mechanical fields have been found. The atomistic calculations determine cohesive energy, stress and elastic constants in each element based on local electronic structure. Thus, after each load step is complete, it is possible to extract both mechanical and electronic properties from the tight binding calculation.

Tight binding atomistic calculation

The tight binding approach, also referred to as the linear combination of atomic orbitals method (LCAO)[7], is based on the assumption that the total cohesive energy of a crystalline material can be written as the sum of two contributions: one due to the bonding valence electrons associated with each atom, and the other due to repulsive sources including ion-ion interactions. The total cohesive energy per atom for an arrangement of N atoms is given by

$$E_{tot} = \frac{1}{N}\left(\sum_{j=1}^{N} E_j^{elec} + \frac{1}{2}\sum_{i=1}^{N}\sum_{k=1}^{N} \phi^{rep}(r_{ik})\right) \qquad (1)$$

where the second term is the repulsive part of the energy in pair potential form, and the first term is the electronic part of the energy, which is given by

$$E^{elec} = 2\sum_{i=1}^{n} \alpha_i E(k_i) \qquad \forall E(k_i) < E_f \qquad (2)$$

where i ranges over a set of special quadrature points k_i that span the Brillouin zone, α_i are corresponding weights, and E_f is the Fermi energy. States with energies $E(k_i) < E_f$ are occupied. The energies of states at a particular point k_i in the Brillouin zone are found as eigenvalues of the orthogonal Slater-Koster tight binding Hamiltonian, so that $E = \langle \Psi^* | H^{tb} | \Psi \rangle$ where Ψ is the wavefunction, and all terms are implicitly functions of k_i. Because the deformation within each element is assumed to be uniform, the basis for the Hamiltonian for Si includes the electronic orbitals associated with only two atoms. Symmetry is exploited to account for the other lattice sites in the crystal. In this way, the atomistic calculation within each element amounts only to the diagonalization of a matrix with dimension $(2m \times 2m)$, where m is the number of orbitals considered per atom. One s orbital and three p orbitals are sufficient to model Si, so here the calculation in each element consists of the diagonalization of an 8×8 matrix. The tight binding parameterization for Si used in the examples discussed here follows Wang $et\ al.$ [8] This parameterization is based on the fitting of six parameters to the results of first principles calculations; it reproduces bulk mechanical properties fairly reliably, although it deviates from some experimentally observed electronic properties. However, the key idea in the work reported here is the presence of the coupling itself.

Internal relaxation degree of freedom

In order to prescribe atomic positions in a diamond cubic crystalline material like Si, with two atoms per primitive unit cell, it is necessary to consider not only $\mathbf{F}$, but also the possibility of internal relaxation of the atoms in the crystal. This internal degree of freedom, or shuffle vector, can result in a substantial reduction of the total energy of the crystal and must be considered[9]. In the technique presented here, the shuffle vector δ is calculated numerically as a function of the incoming deformation gradient tensor $\mathbf{F}$ at the beginning of each loading step. For a particular $\mathbf{F}$ input by the finite element program, the generalized force $\mathbf{f}(\mathbf{F})$ that induces δ is given by

$$\mathbf{f}(\mathbf{F}) = -\frac{\partial E_{tot}}{\partial \delta} = \frac{-1}{N}\left(\sum_{j=1}^{N} \frac{\partial E_j^{elec}}{\partial \delta} + \frac{1}{2}\sum_{i=1}^{N}\sum_{k=1}^{N} \frac{\partial \phi^{rep}(r_{ik})}{\partial \delta}\right) \qquad (3)$$

where the derivative of the repulsive potential is calculated directly and the derivative of the electronic energy is determined analytically using the Hellman-Feynman theorem[10], and is given by,

$$\frac{\partial E^{elec}}{\partial \delta} = \sum_{i=1}^{n} \alpha_i \langle \Psi^*(k_i) | \frac{\partial H^{tb}(k_i)}{\partial \delta} | \Psi(k_i) \rangle \qquad \forall k_i \;\; s.t. \;\; E(k_i) < E_f \tag{4}$$

The components of $\mathbf{f}$ and the cohesive energy $E_{tot}(\mathbf{F})$ are used to seek the energy minimizing δ by means of a numerical conjugate-gradient algorithm. In plane strain problems like the cases discussed here, δ consists of an out of plane displacement only. The equilibrium δ then gives a new, relaxed $E_{tot}(\mathbf{F})$ which corresponds to the correct electronic and mechanical properties for the given state of deformation.

Electronic properties

The tight binding calculation of the energy in each element is based on quantum mechanics. The dispersion relations $E(\vec{k})$ and the density of states $\rho(E)$ are intrinsic to the calculation of the electronic portion of the energy. In the context of the mixed atomistic/continuum formulation, these properties are measures of the *local* electronic structure in the material subjected to the *local* equilibrium state of deformation. The density of states, which is the property of interest here, is available because the summation of energy eigenvalues calculated in (2) is an approximate representation of the integral given by

$$E^{elec} = 2 \int_{-\infty}^{E_f} E \rho(E) dE \tag{5}$$

where $\rho(E)$ is the density of states for $-\infty < E < \infty$. Thus, a reasonably accurate approximation to $\rho(E)$ can be constructed using the energy eigenvalues $E(k_i)$ and the corresponding weights α_i. A gaussian broadening technique is used to generate a continuous density of states curve based on the set of discrete energies.

Mechanical properties

Mechanical behavior in each element is also characterized atomistically. The Newton-Raphson numerical algorithm in the finite element procedure requires the components of the tangential elastic modulus tensor, which are second derivatives of the total energy of the crystal with respect to various combinations of strain components, and which can be written as

$$C_{ijkl} = \frac{\partial^2 E_{tot}}{\partial \epsilon_{ij} \partial \epsilon_{kl}} \tag{6}$$

In principle, these derivatives can be calculated analytically using Hellman-Feynman theory[10]. However, in the analyses described here, the derivatives are calculated numerically. There are only six nonzero components of the tangential elastic modulus tensor necessary for a plane strain analysis for arbitrary crystal symmetry. Assuming that for an arbitrary state of strain $\tilde{\epsilon}$ the energy surfaces with respect to small perturbations in strain are parabolic, these components can be calculated numerically using a centered difference scheme. The C_{1111} component of the elastic modulus tensor, for example, is given by

$$C_{1111} = \frac{E^{tot}[\tilde{\epsilon} + \delta\epsilon] + E^{tot}[\tilde{\epsilon} - \delta\epsilon] - 2E^{tot}[\tilde{\epsilon}]}{\varepsilon^2} \qquad \text{where} \;\; \delta\epsilon = \varepsilon \vec{e_1} \otimes \vec{e_1} \tag{7}$$

The other five nonzero components are determined similarly. This numerical second derivative calculation requires several energy calculations per elastic modulus tensor component, as well as the introduction of small perturbations in strain which introduces some numerical error.

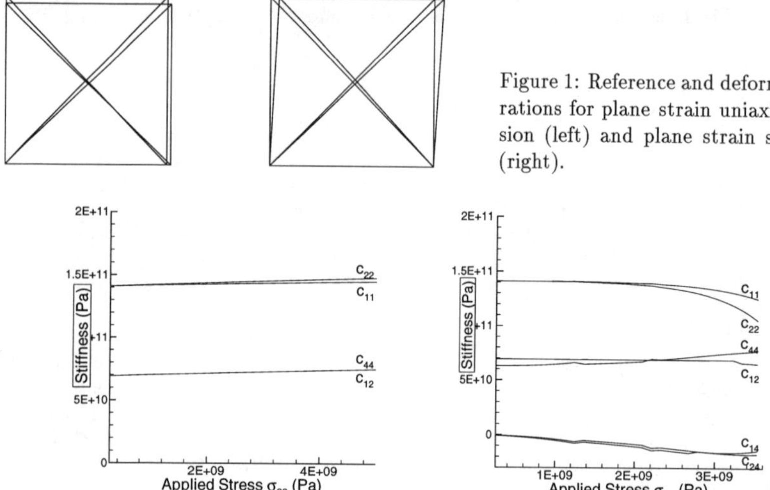

Figure 1: Reference and deformed configurations for plane strain uniaxial compression (left) and plane strain simple shear (right).

Figure 2: Effect of uniaxial compression (left) and simple shear (right) on elastic moduli. There is a stronger effect in the case of simple shear, including the development of nonzero C_{14} and C_{24} components associated with the loss of cubic symmetry in the crystal.

RESULTS

Simple uniform boundary value problems

To illustrate the technique, two simple plane strain Si boundary value problems are briefly considered. Uniaxial compression and simple shear are shown in Figure . Each mesh consists of four constant strain triangle elements and is loaded on the upper surface while the lower surface is simply supported. The effects of the deformation on the tangential elastic moduli during the loading history of the meshes are shown in Figure 2. In both cases, there is considerable material nonlinearity at moderate stresses on the order of 1 GPa. In the simple shear case, nonzero C_{14} and C_{24} components develop as a result of the loss of cubic symmetry in the crystal. Density of states plots corresponding to reference and deformed configurations are shown in Figure 3. In each diagram, the inset shows the density of states near the top of the valence band; there is a noticeable strain effect in both cases, but particularly in the case of uniaxial compression, where there is a shift on the order of 10 meV. In these simple examples, relatively small deformations are shown to affect both the electronic and mechanical properties of the material, which is consistent with continuum analyses of other coupled problems.[2-4]

Nonuniform boundary value problems

Two more complicated boundary value problems, featuring nonuniform deformation that is characteristic of strained quantum devices, are also considered. A wedge shaped mesh, with a symmetry boundary condition on the left surface and constrained to four percent extensional strain along the bottom surface to simulate lattice mismatch is shown in Figure 4 (left). This structure is typical of a strained island quantum dot or wire: it is constrained by the mismatched lattice parameter of the substrate material and it has traction free lateral surfaces. The strain is almost completely relaxed at the top corner of the wedge but is equal to the system mismatch strain along the interface. Figure 4 (right) illustrates the effect of the nonuniform strain on electronic properties in the device; densities of states are plotted for the element at the edge of the base of the island and for the element at the apex of the island. The inset shows a shift in the conduction band on the order

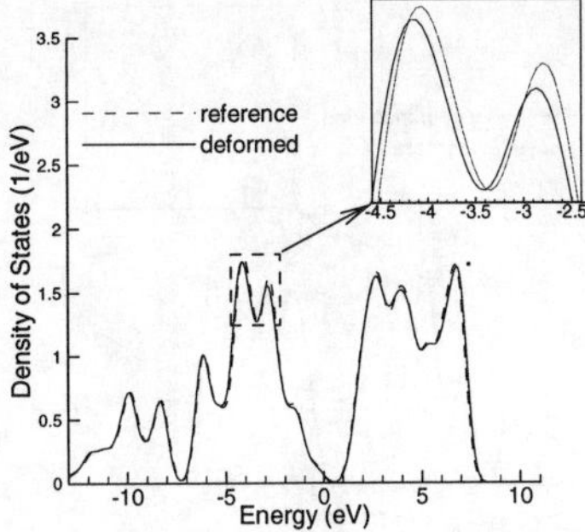
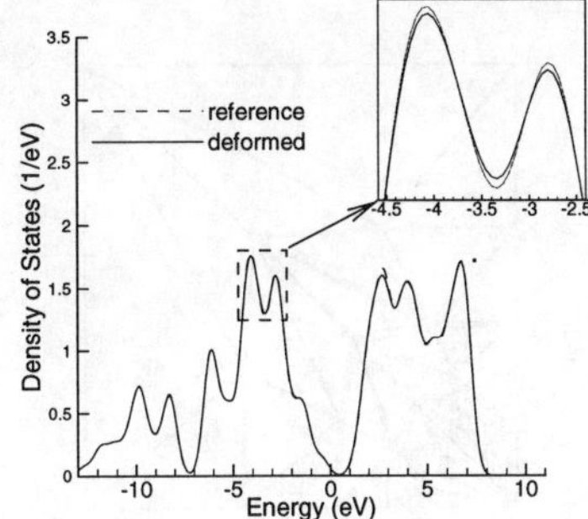

Figure 3: Densities of states for uniaxially compressed (left) and simply sheared (right) material compared to the unstrained reference configuration. The strain effects is most significant near the top of the valence band.

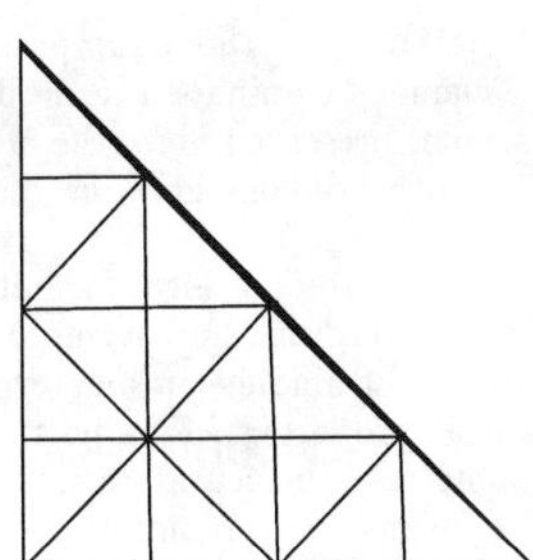
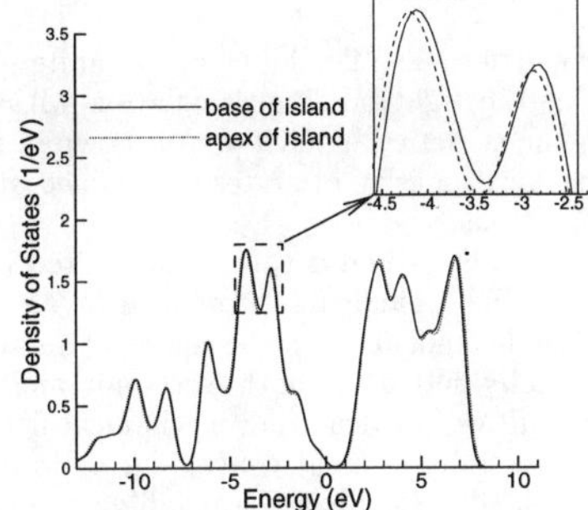

Figure 4: Deformed cross-section of island (left). The mesh for the reference configuration, which corresponds to the case in which the *entire* island is fully strained to the substrate, appears behind the deformed mesh. The density of states at the strained edge of the island base is significantly different than the density of states at the relatively unstrained apex of the island (right).

of 10 meV in the element at the base relative to the element at the apex. There is also a slight decrease in the density of states in the low end of the conduction band in the interface region.

An elastic inclusion problem is shown in Figure 5. The block has symmetry boundary conditions on all edges, and the Si material in the block is subjected to the four percent uniform dilatation of the much stiffer material inside the circular region. This generates a large uniaxial compressive strain near the inclusion, but the strain is small in the remote region. This geometry is representative of problems in which defects induce highly strained regions in Si which can act as quantum wells[11]. The analysis shown here combines the mixed atomistic/continuum technique with standard linear elastic finite element analysis, which is used to treat the rigid inclusion where electronic structure is not important. The density of states, also shown in Figure 5 in the highly strained element near the inclusion, is considerably different than the density of states in the remote region.

CONCLUSIONS

The mixed atomistic/continuum technique presented here is a useful tool for solving boundary value problems in strained semiconductors because it treats the fully coupled nature of mechanical and

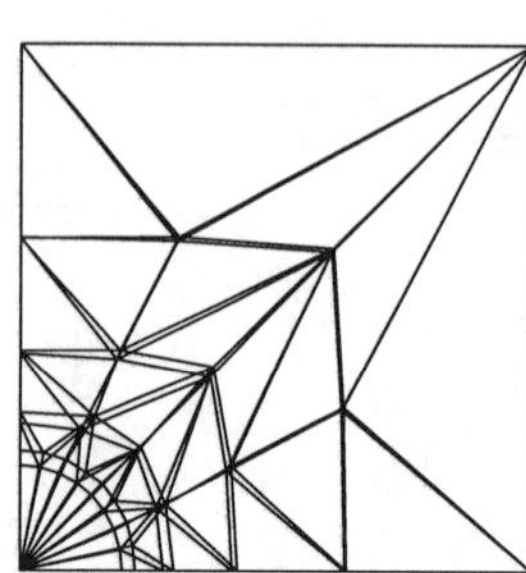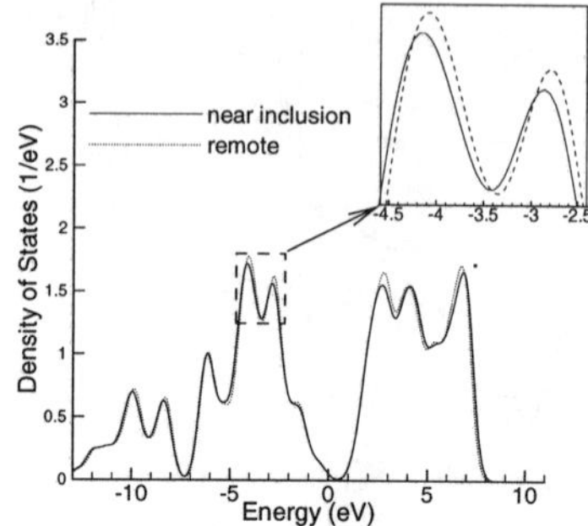

Figure 5: Reference and deformed configurations (with displacements exaggerated) for a Si block with a rigid inclusion (left). The arc shaped rigid inclusion expands uniformly and strains the surrounding material. The density of states in the region closest to the inclusion at the midpoint of the arc differs from the density of states in the remote region (right).

electronic properties. The flexibility of the finite element method, and the assumption of uniform deformation in the tight binding calculation within each element eliminate the need to consider each atom in the structure explicitly. As shown in the examples presented here, the method can be used to probe local density of states in strained structures with nonhomogeneous geometries and boundary conditions.

As it stands, the method is somewhat limited because it is based on a local formulation, so the environment around a particular atom is always assumed to be uniform. Furthermore, the coupled calculations are dependent upon the choice of an appropriate tight binding parameterization which may not be ideally suited for both electronic and mechanical properties. Finally, there are some numerical difficulties in calculating mechanical fields; in some cases inaccurate elastic moduli lead to problems with finite element convergence. However, the means to overcome these obstacles are within reach. The development of a reliable nonlocal formulation will allow for the study of atomic scale defects and bimaterial interface problems. More accurate tight binding formulations have been recently reported, and better methods for calculating elastic moduli are under development.

ACKNOWLEDGEMENTS

This work was supported in part by the MRSEC Program of the National Science Foundation under Award DMR-9632524, and in part by the Office of Naval Research under Contract N00014-95-1-0239. This support is gratefully acknowledged.

REFERENCES

1. A. Zunger, MRS Bull. **23** 15, (1998).
2. M. Grundmann, O. Stier, D. Bimberg, Phys. Rev. B **52**, 11969 (1995).
3. C. Pryor, Phys. Rev. B **57**, 7190 (1998).
4. H. T. Johnson, L. B. Freund, C. D. Akyüz, A. Zaslavsky, J. Appl. Phys. **84**, 3714 (1998).
5. E. B. Tadmor, M. Ortiz, R. Phillips, Phil. Mag. A **73**, 1529 (1996).
6. ABAQUS, Version 5.7, Hibbitt, Karlsson & Sorensen, Inc., Pawtucket, RI 02860, U. S. A. (1997).
7. W. A. Harrison, *Electronic Structure and the Properties of Solids*, 2nd ed. (Dover Publications, New York, 1989) chapter 3.
8. C. Z. Wang, C. T. Chan, K. M. Ho, Phys. Rev. B **39**, 8586 (1989).
9. E. B. Tadmor, N. Bernstein, G. S. Smith, E. Kaxiras, in press, Phys. Rev. B (1998).
10. R. P. Feynman, Phys. Rev. **56**, 340 (1939).
11. H. Weman, B. Monemar, G. S. Oehrlein, S. J. Jeng, Phys. Rev. B **42**, 3109 (1990).

BIAS POTENTIALS FOR HYPERDYNAMICS SIMULATIONS

W. G. RUDD* and A. F. VOTER**

*Department of Computer Science, Oregon State University, Corvallis, Oregon, 97333
**Theoretical Division, Los Alamos National Laboratory, Los Alamos, NM 87545

ABSTRACT

We describe our studies of two methods for generating bias potentials for use in hyperdynamics simulations. In the first method, first reported by Steiner, et al, the bias potential is the additional energy needed to keep the total potential energy above a fixed level. This potential exerts a negligible computational load and is very easy to code. The second technique involves computing the iterative Hessian-based bias potential as usual, but including in those calculations only the atoms in a small "active region" that surrounds the area in which a state transition is expected to occur. This "subspace hyperdynamics" method is less costly than full HD. The extra computational effort required for HD scales as the size of the active region(s), instead of with the size of the entire simulation domain.

We have carried out extensive tests of both methods on two problems, the diffusion of adatoms on silver surfaces and the migration of vacancies in bulk silver. Using the first method, to obtain large boosts the bias potential must be so high that many transitions are prevented from occurring. Our results for subspace hyperdynamics are promising; we obtain the same results as with full hyperdynamics, but with considerably less computational effort.

I. INTRODUCTION

Hyperdynamics (HD) [1,2] is a new technique for accelerating the progress of systems in time in molecular dynamics (MD) simulations. The method is designed for problems in which there are infrequent transitions between equilibrium states of the system. It becomes even more powerful when combined with the parallel replica method [3,4] for running simulations on separate phase space trajectories in parallel. Examples of appropriate problems for the method include surface and bulk diffusion, and dislocation and fracture dynamics at low strain rates.

MD codes that use HD add a bias to the interatomic potential in such a way as to reduce the time spent in regions of configuration space in which transitions are unlikely to occur, while preserving the accuracy of measured transition rates and other phenomena. The success of HD has been well-documented for small systems. One of our long-term objectives is to determine more efficient ways for doing HD so that we can apply it to systems of tens of thousands of atoms and larger.

Here we describe two methods for reducing the overhead in computing bias potentials for HD. In the first, we explore a technique suggested by Steiner, et al, [5] in which we replace the ordinary potential with a constant when the former falls below the constant level. In the second method, we apply the method described in [2]. However, in our new technique, called "subspace hyperdynamics" (SHD), we use only a subspace of the configuration space of the system in this process.

Mat. Res. Soc. Symp. Proc. Vol. 538 ©1999 Materials Research Society

II. HYPERDYNAMICS

HD is an extension of MD that accelerates the calculations when the system is in potential wells far from regions of transitions between states. In the HD approach the potential energy surface is modified with a bias potential that raises the energy in wells so that less computer time is spent simulating the trajectories while the system is in the wells. For a trajectory that has evolved for n time steps, the elapsed hypertime is

$$t_{hyper} = \sum_{i=1}^{n} \Delta t_{MD} \exp[\beta \Delta V_b(\mathbf{r}_i)] \tag{1}$$

Here $\mathbf{r}_i$ is the position on the trajectory in configuration space at step i and $\beta = 1/k_B T$, where k_B is Boltzmann's constant and T is the absolute temperature. $\Delta V_b(\mathbf{r}_i)$ is the bias that is added to the unbiased potential before computing the forces used in the equations of motion. The time integration for the MD step proceeds as usual.

The raw boost, t_{hyper}/n, increases exponentially with ΔV_b. Restrictions on the forms of ΔV_b. are described in [1]. The most important of these is that ΔV_b should be zero near the dividing surfaces between wells so that the trajectory is not blocked from following lowest-energy paths across barriers.

The iterative Hessian-based method generates bias potentials that obey the restrictions. The method involves a numerical determination of the lowest two eigenvalues of the Hessian matrix $\mathbf{H}$ the matrix of second derivatives of the potential V, followed by a projection of the gradient onto the eigenvector that corresponds to the lowest eigenvalue. The technique requires only first derivatives of the potential, which are available from the MD force routine, and it avoids the necessity for computing the full Hessian matrix. Each iteration of the numerical minimizer scales as O(N), where N is the number of atoms.

Using this bias potential one can obtain boosts of two or three orders of magnitude. Some of this gain is negated by the fact that computing ΔV_b ordinarily requires from 30 to 50 times as long per step as it does to carry out an ordinary MD step. In other words, each iteration requires several gradient (force) function calls and several iterations are required to converge to the minimum eigenvalues. Thus the effective boost is reduced by a corresponding factor.

If the bias potential is nonzero on the dividing surface, the observed mean time transitions will be related to the "exact" transition state theory average τ_{TST} via the relation

$$<\tau> = \tau_{TST}/f_{loss} \tag{2}$$

in which

$$f_{loss} = \left\langle \exp[-\beta \Delta V_b(\mathbf{r}_i)] \right\rangle_\delta \tag{3}$$

The average in Equation (3) is taken on the transition boundaries. f_{loss} behaves like a transmission factor. If $\Delta V_b = 0$ on the boundaries and the other restrictions are satisfied, the measured mean transition times should match the transition state theory predictions.

Simple forms for the bias potential are easy to code and fast, but those that offer large boosts tend to be nonzero on the dividing surface. The penalty for using such a bias potential is that we lose part of the reduction in computational time we achieved through the use of HD. This is because the blocked transitions result in longer times between transitions. Furthermore, we no longer have estimates of $\langle\tau\rangle$ that are statistically the same as the true values. We could correct the error in $\langle\tau\rangle$ if we could obtain an estimate of f_{loss} or if f_{loss} is near 1. This is the idea behind the constant biased potential.

III. CONSTANT BIASED POTENTIALS

The constant or "flat" biased potential is defined by

$$\Delta V_b = \begin{cases} V_b - V & \text{if } V < V_b \\ 0 & \text{if } V < V_b \end{cases} \tag{4}$$

for some constant potential level V_b.

Appealing to the Maxwellian distribution of energies appropriate for systems in thermal equilibrium with a heat bath, one can derive the forms of f_{loss} and the raw boost $< \exp[\beta\Delta V] >$ for the constant bias potential. Initially, the raw boost increases exponentially as V_b increases, but then the f_{loss} factor becomes large enough to cancel the raw boost. The net result is that there is a range of values of V_b over which the effective boost tracks the raw boost; within that range $f_{loss} \cong 1$. Therefore, we expect values of diffusion coefficients and transition rates to be reasonably accurate in those ranges.

We conducted an extensive set of experiments to determine the effect of the choice of V_b on the behavior of the HD simulations. The system studied is that described in [2] in which a single Ag adatom moves on the (100) surface of silver. There is, as the theory predicts, a range of values of V_b in which the effective boost does coincide with the raw boost, we get accurate values of $\langle\tau\rangle$, and boosts are on the order of 10. With larger raw boosts we obtain larger speedups, but the effective boost and the accuracy of the measured transition rates decrease simultaneously.

In another set of experiments, we studied self-interstitial migration in silicon, using the Stillinger-Weber [5] potential. Diffusion coefficients measured via HD match closely those obtained via MD in simulations involving 217 atoms over a wide range of temperatures. In those studies, V_b was set to the MD average of V.

A problem with the constant biased method is that good values of V_b depend upon the barrier heights, i.e. upon the system being simulated and its thermodynamic state. Furthermore, we have not determined how one might use the method when there might be more than one transition barrier of differing heights.

IV. SUBSPACE HYPERDYNAMICS

SHD is based upon the idea that the dynamics of transitions depend most strongly on the positions and momenta of relatively small sets of atoms, "active regions," that are close to an existing defect. Using a subspace of configuration space as the basis for computing a bias potential cannot alter the dynamics of transitions so long as the subspace is large enough to encompass all of the coordinates that might be involved in transitions. In the case of point defects in solids, for example, we expect that the relevant subspace need include only those atoms that lie within a few neighbor shells of the defect(s). Regions further away can be expected to have little influence on the transition dynamics.

For the case of point defects in solids, which atoms to include in the active region can be determined either geometrically or by selecting those atoms that have the highest potential energy. How to select active regions for more complicated systems is not as clear. Consideration of relative potential energies alone will not suffice in some cases, such as when defects are near surfaces or when low-energy atoms are involved in transitions.

In SHD, we compute the bias potential using the same numerical method as in ordinary hyperdynamics, but we use the (implied) Hessian matrix for only the coordinates in the active region. In other words, instead of solving the minimization problem for a system of N atoms, we need solve a much smaller problem of size N_{active}, where N_{active} is the number of atoms in the active region. The additional effort to compute the bias potential therefore scales like that of problems of size N_{active} instead of N.

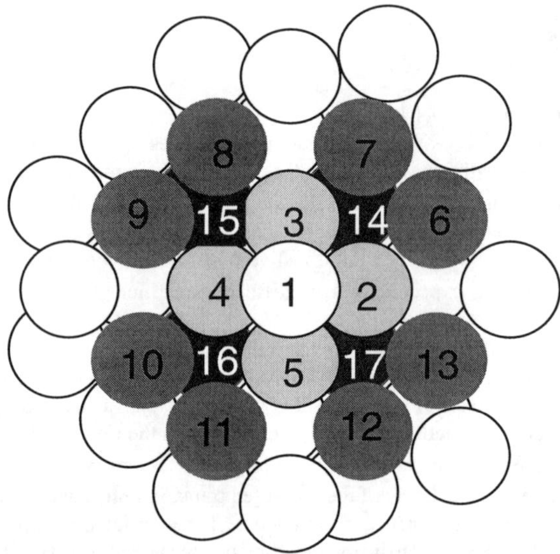

Figure 1. An Ag atom on the Ag(100) surface, showing the 5-atom and 18-atom active regions..

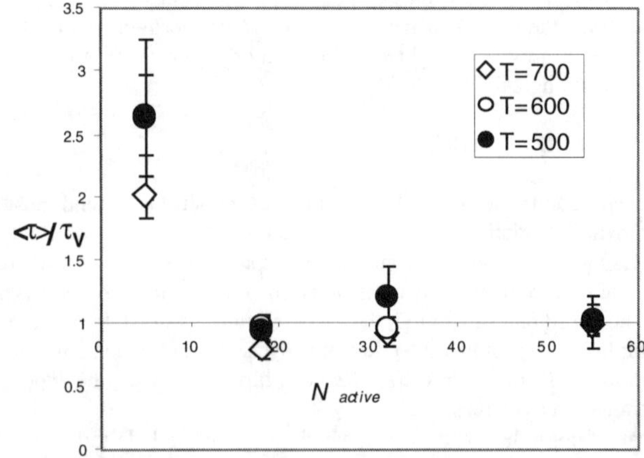

Figure 2. Mean reduced transition time. τ_V is the Vineyard harmonic approximation [7] for $\langle\tau\rangle$.

SHD reduces the HD computational overhead because individual iterations in converging to the eigenvalues require less computational effort. Furthermore, the number of iterations required to converge the eigenvalues sometimes increases at least linearly with the size of the active space, so using a small active region instead of the full space mitigates against this factor.

In our initial tests of SHD, we again considered the diffusion of an Ag adatom on the Ag(100) surface [2]. The potential and forces the adatom feels depend most strongly upon the positions of the atoms closest to it. Because there are 4, 13, and 14 atoms in respective neighbor cells of the adatom (Figure 1), we expect that SHD using a bias potential formed from a sequence of calculations involving 5, 18, and 32 atoms, including the adatom, would converge to the results we obtain from using all the atoms in the system.

We do indeed obtain the predicted effect. Figure 2 shows mean transition times obtained from SHD using the parallel replica method relative to transition times from the Vineyard harmonic approximation [7]. This is for the 91-atom system as described in [2], in which 55 atoms move on a base of 36 stationary atoms. The results for $N_{active} = 18$ and for $N_{active} = 32$ are indistinguishable from the "exact" result $N_{active} = 55$ for the three temperatures treated. This result is despite the fact that in our current implementation the derivatives are not exact for atoms near, but not in, the active regions.

We note that for $N_{active} = 5$ the measured transition times are considerably longer than the Vineyard times. Not shown are the results for $N_{active} = 2$, for which the measured transition times are about a factor of 10 longer. These active regions are too small to include all the atoms involved in the reaction coordinates for hoops or exchanges, so dividing surface blocking occurs. In particular, we observe almost no exchange transitions, which account for roughly half of the transitions during the correct dynamics. Exchange transitions require concerted movement of larger numbers of atoms than these small active regions can describe.

Figure 3 shows the improvements in execution time we obtain through the use of the SHD method. Wall-clock computational times T_W can decrease by more than a factor of 3, depending upon temperature. For $T = 700K$ and $600K$ the additional boost at the lower temperature compensates for the longer transition times, so T_W are roughly the same. However, for $T = 500K$

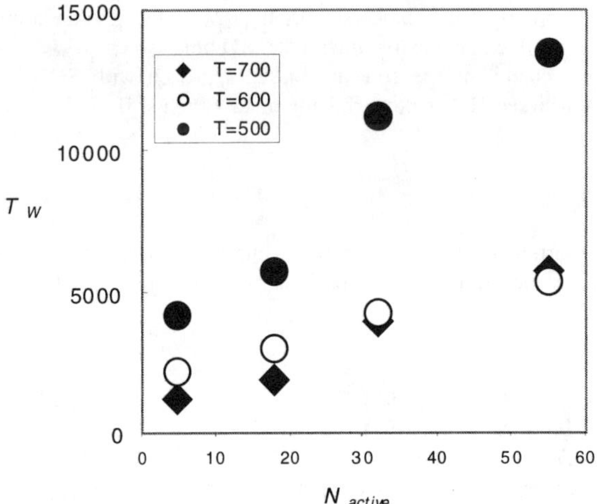

Figure 3. Wall-clock time T_W per transition on seven processors versus sizes of active regions.

the increase in transition times outweighs the increase in boost, so that the net computational time is longer than for the other two temperatures. These two figures show that the best choice for this problem is N_{active}=18, which includes the adatom and its first two neighbor shells.

This result is despite the fact that the code does not yet take full advantage of the potential speedup. With a properly designed code, simulations of large systems (e.g. thousands of atoms) with small active regions should run as fast as direct MD.

Finally, we report the success of the combined SHD/HD/parallel replica method for vacancy migration in silver at T=300K. Using SHD we obtain the same results for $<\tau>$ with N_{active} = 18 as we obtain using ordinary HD for a system of 107 atoms. It is worth noting here that $<\tau> \sim 10^{-4}$ seconds for this problem, which shows that these methods enable us to simulate systems for milliseconds and longer, periods that are inaccessible via ordinary MD.

V. CONCLUSIONS

We have presented results of studies of two forms of bias potential designed to speed up calculations in HD simulations. The constant biased method requires essentially no additional computer time and is simple to implement. But, in order to obtain large boosts with this potential, one must raise the bias to levels above the potential level at which transitions occur. Doing so prevents some transitions from occurring, resulting in erroneous increases in measured transition times. This inaccuracy could be tolerable if the intent of the simulation is to search for new transition mechanisms. To use the method for accurate determinations of transition rates it will be necessary to develop methods for obtaining accurate estimates of f_{loss} and to find efficient ways of determining good values of V_b.

SHD shows great promise. For some systems, the number of iterations required to find the minimum eigenvalues appears to increase with the size of the system. We suspect that the additional effort required for HD could scale as $O(N^2)$ for larger systems. Given such behavior, the computational overhead for SHD would scale as $O(N^2_{active})$, which constitutes a significant reduction in computational complexity. In any case, for large systems with localized defects we can expect N_{active} to be many orders of magnitude smaller than N, so that the additional time for HD using SHD should be negligible in comparison with the MD time.

Finally, we note that combining the constant biased approach with SHD should work even better because constant-biased HD is powerful for small systems[5], and presumably also for small *sub*systems.

ACKNOWLEDGMENTS

This work was supported by the Office of Basic Energy Sciences of the United States Department of Energy, and by the Computer Science Department of Oregon State University.

REFERENCES

[1] A. F. Voter, J. Chem. Phys. **106**, 4665 (1997).
[2] A. F. Voter, Phys. Rev. Lett. **78**, 3908 (1997).
[3] A. F. Voter, Phys. Rev. B, **57**, 13985 (1998).
[4] A. F. Voter and T. C. Germann, Mat. Res. Soc. Symp. Proc. **528**, 221 (1998).
[5] M. M. Steiner, P.-A. Genilloud, and J. W. Wilkins, Phys. Rev. B **57**, 10236 (1998).
[6] Frank H. Stillinger and Thomas A. Weber, Phys Rev. B **31**, 5262(1985).
[7] G. H. Vineyard, J. Phys. Chem. Solids **3**, 121(1957).

PRESSURE WAVES IN MICROSCOPIC SIMULATIONS OF LASER ABLATION

LEONID V. ZHIGILEI and BARBARA J. GARRISON

Department of Chemistry, The Pennsylvania State University, University Park, PA 16802

ABSTRACT

Laser ablation of organic solids is a complex collective phenomenon that includes processes occurring at different length and time scales. A mesoscopic breathing sphere model developed recently for molecular dynamics simulation of laser ablation and damage of organic solids has significantly expanded the length-scale (up to hundreds of nanometers) and the time-scale (up to nanoseconds) of the simulations. The laser induced buildup of a high pressure within the absorbing volume and generation of the pressure waves propagating from the absorption region poses an additional challenge for molecular-level simulation. A new dynamic boundary condition is developed to minimize the effects of the reflection of the wave from the boundary of the computational cell. The boundary condition accounts for the laser induced pressure wave propagation as well as the direct laser energy deposition in the boundary region.

INTRODUCTION

Traditional computational material modeling methods are based either on macroscopic continuum-mechanics theories or atomistic simulation techniques. Recently, substantial progress has been made in development of advanced multiscale simulation techniques aimed at bridging the gap between the atomistic and continuum descriptions.[1,2,3,4,5] These techniques include a direct combination of atomistic and continuum finite element methods to simulate an adequate response of surrounding material to the active localized processes in the atomistic computational cell[2,3,4] and multiscale hierarchic approaches, in which accurate atomic-scale calculations provide data for parametrization of macroscopic or mesoscopic models.[1,5] These techniques, however, are hardly applicable for systems in which the critical events are not localized and the processes occurring at different length scales are inherently coupled and should be considered simultaneously.

An example of such a complex multiscale phenomenon considered in this work is laser ablation of an organic solid.[6] The processes involved in laser ablation include laser excitation of absorbing molecules, energy transfer from the excited molecules into the internal and translational modes of other molecules in the solid, formation of a highly energetic high-temperature and high-pressure region, explosive disintegration and prompt forward ejection of a volume of material, intensive processes in the ejected plume, and propagation of the pressure wave away from the ablation region. The complex character of the involved intertwined processes hinders an adequate analytical formulation for a continuum description of the phenomenon whereas a collective character of the laser ablation occurring at the mesoscopic rather than molecular scale does not permit a direct application of the atomistic simulation approach. An alternative mesoscopic model that has advantage of both addressing the effects of laser irradiation at a submicron resolution and yet incorporating a realistic description of energy relaxation of individual molecules internally excited by photon absorption has been developed recently for molecular dynamics (MD) simulation of laser ablation and damage of organic solids.[7]

In this paper we shortly review the basic features of the model, give a few examples of simulation setups used to study the laser induced processes in practically important applications, and focus on the computational approach used to deal with the laser induced pressure waves.

Mat. Res. Soc. Symp. Proc. Vol. 538 © 1999 Materials Research Society

THE BREATHING SPHERE MODEL

In this section we give only a brief description of the breathing sphere model for molecular dynamics simulations of laser ablation and damage in organic solids. Complete details of the model are given in Ref. 7.

The model assumes that each molecule (or appropriate group of atoms) can be represented by a single particle that has the true translational degrees of freedom but an approximate internal degree of freedom. The internal degree of freedom is attributed to each molecule by allowing the particles to change their sizes. The characteristic frequency of the internal motion, Fig. 1, is controlled by the parameters of an anharmonic potential. The rate of the intermode energy transfer

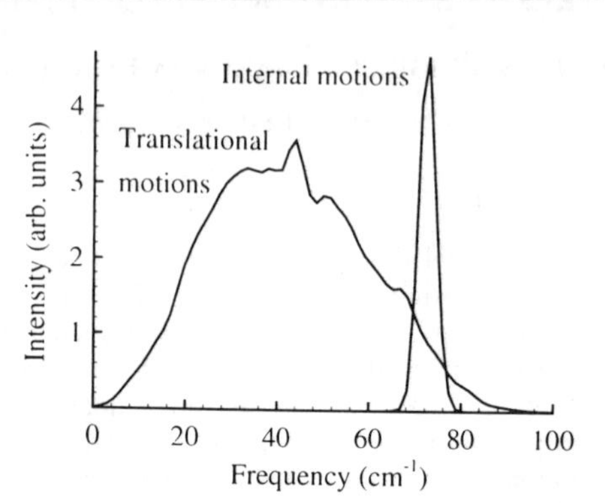

Figure 1. Vibrational spectrum of the breathing sphere model at 300 K. The coupling between internal and translational motions defines the rate of the vibrational relaxation of excited molecules.

is determined by the size of the anharmonicity and frequency mismatch between vibrational modes. Thus, the parameters of the internal potential can be used to affect the coupling between internal and translational molecular motions and to achieve a desired rate of the conversion of internal energy of the molecules excited by the laser to the translational and internal motion of the other molecules.

Because the molecules and not the atoms are the particles of interest, the system size can be significantly larger. Moreover, since explicit atomic vibrations are not followed, the timestep in the numerical integration is longer. One more advantage of the breathing sphere model is the ability to simulate complex multicomponent organic materials. We can easily include bonding interactions, different strengths and absorptions of different components. The rate of energy transfer within an individual component as well as between components can be precisely controlled.

The effect of laser irradiation is simulated by vibrational excitation of random molecules during the time of the laser pulse within the penetration depth appropriate for a given wavelength. Vibrational excitation is modeled by depositing a quantum of energy equal to the photon energy into the kinetic energy of internal vibration of a given molecule. The absorption

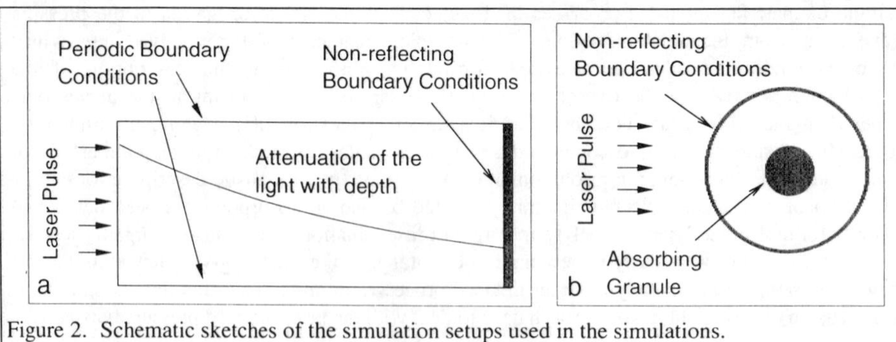

Figure 2. Schematic sketches of the simulation setups used in the simulations.

probability can be modulated by Beer's law to reproduce the exponential attenuation of the laser light with depth or can be restricted to a certain component within a complex material.

Two simulation setups that have been used to study the laser induced processes are shown in Fig. 2. Simulations performed with a setup shown in Fig. 2a are used to study the basic microscopic mechanisms of laser ablation,[7,8] and have yielded predictions on the threshold behavior,[8,9] composition of the ejected plume[7,8,9] and velocity distributions of ejected molecules.[10] The computational setup used to investigate mechanisms of laser damage in the case of spatially localized absorbers (absorbing particles embedded in a transparent medium)[11] is shown in Fig. 2b.

Owing to the approximations adopted in the breathing sphere model, computational cells schematically shown in Fig. 2 can be large enough to reproduce the collective dynamics leading to laser ablation and damage. One effect, however, that can not be directly simulated within the MD model is propagation of the laser induced elastic wave. The origin of the wave and the method used to minimize the effects of the reflection of the wave from the boundary of the computational cell are discussed in the next section.

PRESSURE WAVES AND NON-REFLECTING BOUNDARY CONDITION

The role of the laser induced pressure and associated photomechanical effects in laser ablation is defined by the irradiation parameters.[12,13] When the laser pulse width is shorter than a characteristic time of mechanical relaxation, the laser heating takes place at nearly constant volume conditions and leads to a high thermoelastic pressure buildup within the absorbing volume. For longer laser pulses the laser heating is more gradual, the material has sufficient time to expand during the laser pulse, and material removal is induced by thermal evaporation occurring under ambient pressure.

It is the former case of the short pulse irradiation that poses an additional challenge for molecular-level simulations. The laser induced pressure leads to the formation of the pressure wave that propagates in all directions from the absorption region. Propagation of the stress wave occurs at the length-scales that are beyond the capability of the breathing sphere model. This is illustrated in Fig. 3 where the formation and propagation of a plane pressure wave within the MD computational cell is given for the simulation setup shown in Fig. 2a. Even for a large, 180 nm in the direction of the wave propagation, computational cell, the pressure wave reaches the boundary of the simulation zone in less than 70 ps. This time is significantly shorter than the

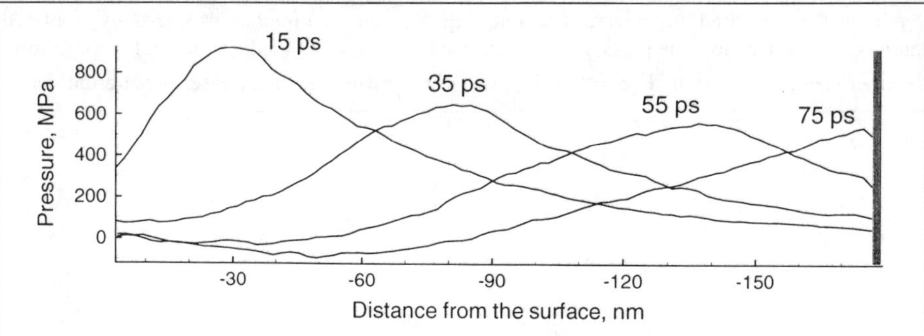

Figure 3. Formation and propagation of the pressure wave for the computational setup shown in Fig. 2a. Laser fluence is 55 J/m^2, penetration depth is 50 nm, and pulse width is 15 ps. Non-reflecting boundary condition applied at the depth of 180 nm is described in the text below.

characteristic time of the processes involved in laser ablation and artificial border effects can interfere with the simulation results. Both rigid and free boundary conditions lead to the complete reflection of the pressure wave and can cause fracturing at a certain depth when the tensile strength is exceeded. Such process of a subsurface fracture is termed spallation and is illustrated in Fig. 4. In this simulation spallation at the back side of the irradiated sample is caused by reflection of the unloading wave propagating from the surface of the irradiated sample. The laser penetration depth in this case is longer than the size of the computational cell leading to nearly even initial energy and pressure distributions within the sample. Interestingly, the spallation at the back is observed at a laser fluence about half of the threshold fluence for the onset of the massive material ejection or ablation[8,9] at the front surface.

Several approaches can be used to minimize the finite size artifacts and simulate the non-reflecting propagation of the stress wave out from the MD computational cell. A number of "reservoir methods" in which a viscous damping applied in an outer region of a

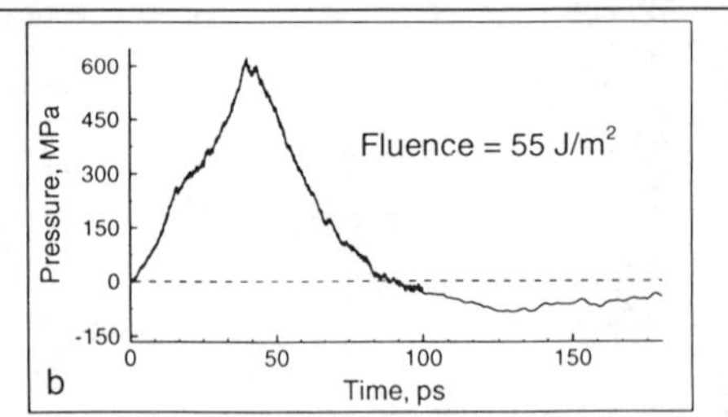

Figure 4. Back spallation due to interaction of the laser induced pressure wave with a rigid boundary of the computational cell.

MD computational cell has been used to adsorb sound waves and dislocations generated in the process of crack propagation[14] or energetic particle bombardment.[15] An alternative approach is to combine the MD model with the continuum finite element method.[2,3,4,16] The advantage of this approach is the ability to study the long-range propagation of the waves and their interaction with each other and inhomogeneities in the media.[16] For the cases, however, when the aim is merely to avoid artifacts due to pressure wave reflection the above approaches appear to be excessively complex and computationally expensive. In this work we adopt a simple and computational efficient non-reflecting boundary condition based on analytical evaluation of the forces acting at the molecules in the boundary region from the outer "infinite medium".

To understand the origin of the forces acting at the boundary in the setup shown in Fig. 2a, we first consider a larger computational cell, Fig. 3, and plot the pressure development at the depth of the proposed boundary, 100 nm, Fig. 5. In the regime of inertially confined thermoelastic expansion, the pressure profile formed by the end of the laser pulse, Fig. 3, follows the laser energy deposition, $P(z) = \Gamma \Phi / L_p \exp\left[-z/L_p\right]$, where L_p is the laser penetration depth,

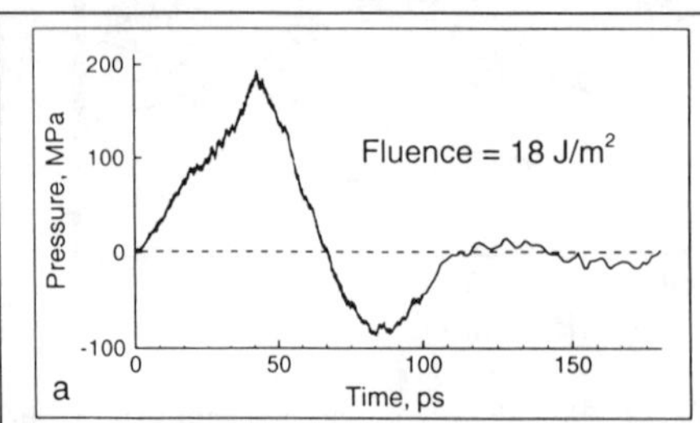

Figure 5. Temporal pressure profiles at 100 nm below the surface for the computational setup shown in Fig. 2a. Laser penetration depth is 50 nm, and pulse width is 15 ps.

Φ is the fluence, and Γ is the Grüneizen coefficient (we neglecting here the temperature dependence of Γ). At the depth of 100 ps this leads to a nearly linear increase of the pressure during the laser pulse, 15 ps, with the final pressure proportional to the laser fluence, Fig. 5. The initial pressure distribution is a source of pressure waves that, due to interaction with the free surface of the sample, form a characteristic bipolar pulse propagating deeper into the sample, Fig. 5a. At laser fluences above the threshold for ablation the tensile phase (negative pressure) is obscured by superposition with the compressive recoil pressure from the ejection of the ablation plume, Fig. 5b.

In our approach the boundary condition is a set of terminating forces F_T that are applied to the atoms in the boundary region. The width of the boundary region is equal to the cut-off distance of the interaction potential used in the MD region. For molecules in the boundary region the parts of the force component in the direction normal to the surface that are due to the interaction with neighbors located closer to the boundary are substituted by terminating forces. In the calculation of the terminating forces that are updated at each integration step, we take into account three effects, namely, the static forces that mimic interaction with molecules beyond the computational cell, F_0, the forces due to the pressure wave propagation through the boundary region, F_W, and the forces due to the direct laser energy absorption in and around the boundary region during the laser pulse, F_L.

The contribution of the pressure wave to the terminating force, F_W, is based on the assumption that the pressure wave propagates as a traveling wave with velocity c. The F_W per molecule then should be proportional to the instantaneous velocity of the boundary, u, as $F_W = S_M Zu$, where $Z=\rho c$ is the acoustic impedance, S_M is the surface area per molecule, and ρ is the density. The force due to the laser energy absorption, F_L, is assumed to rise linearly with time, t, during the laser pulse. In order to account for relaxation of the laser induced pressure we multiply F_L by a damping factor taken in the form proposed in Ref. 17, $D=L_p(1-\exp(tc/L_p))/tc$. As shown in Fig. 6, the resulting terminating force, $F_T(\Phi,t,u)=F_0+F_L(\Phi,t)+F_W(u)$ provides a good approximation of the real forces from the outer part of material obtained from larger scale simulations.

We have used the boundary condition described above in two and three-dimensional simulations with both simulation setups shown in Fig. 2 and were able to completely eliminate simulation artifacts associated with the reflection of the pressure wave from the boundary of the MD computational cell. The boundary condition allows one to restrict area of the MD simulation

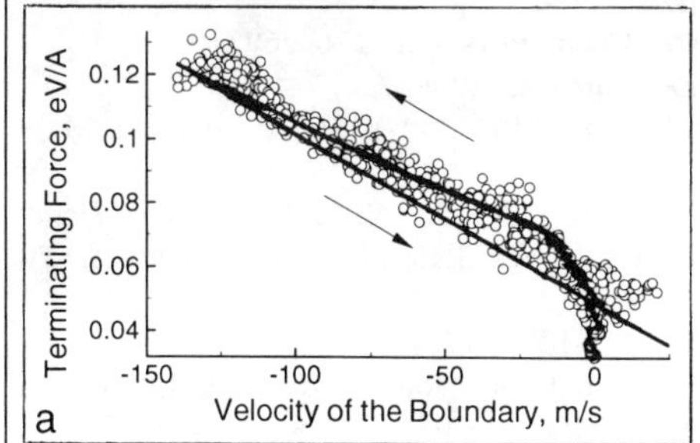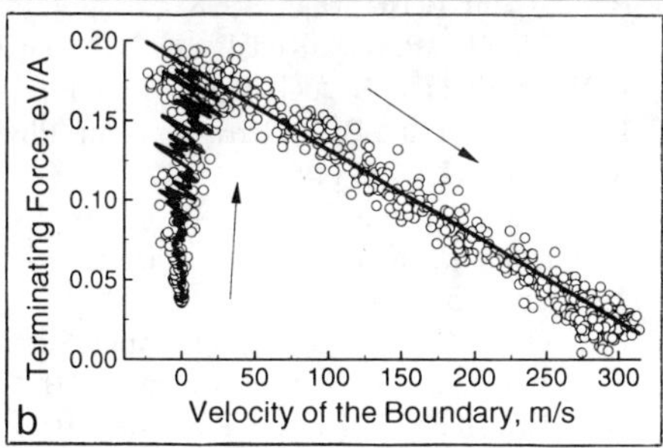

Figure 6. Terminating force per molecule applied as a boundary condition (lines) and calculated directly from a larger scale simulation (data points). The forces for two cases are shown. The laser penetration depth is a) half the depth of the computational cell and b) much longer than the depth of the cell (even energy deposition). Arrows show the direction of time development.

to the region where active processes of laser induced melting, ablation and damage occur. This significantly expands the scope of phenomena that can be addressed and allows to use irradiation parameters comparable to the experimental values.[9] A snapshot from such a large scale simulation is shown in Fig. 7.

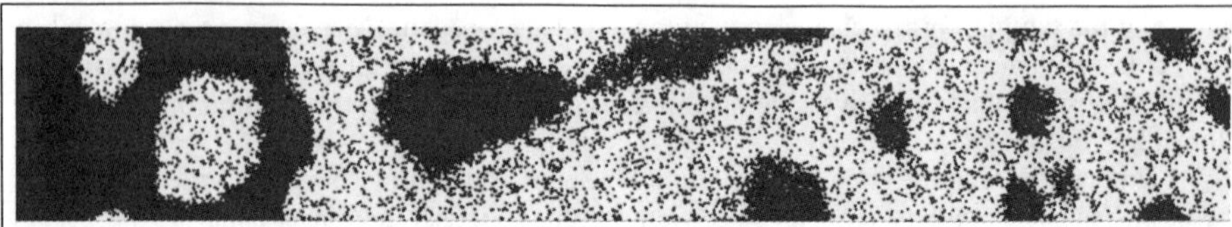

Figure 7. Snapshot from a large-scale three-dimensional molecular dynamics simulation at 400 ps after irradiation with 150 ps laser pulse at a fluence of 55 J/m^2. The laser penetration depth is 50 nm. The initial depth of the computational cell in the direction of incident laser pulse is 100 nm. Non-reflecting boundary condition is applied at the bottom of the computational cell. The simulation is performed with the breathing sphere model.

ACKNOWLEDGMENTS.

This work was supported by NSF and ONR through the MFEL Program. The computational support for this work was provided by the IBM-SUR Program and the Center for Academic Computing at PSU. We thank Victor Sparrow for a stimulating discussion.

REFERENCES

1. F. Cleri, S. R. Phillpot, D. Wolf, and S. Yip, J. Am. Ceram. Soc. **81**, 501 (1998).
2. S. Kohlhoff, P. Gumbsch, and H. F. Fischmeister, Phil. Mag. A **64**, 851 (1991).
3. E. B. Tadmor, M. Ortiz, and R. Phillips, Phil. Mag. A **73**, 1529 (1996).
4. H. Rafii-Tabar, L. Hua, and M. Cross, J. Phys.: Condens. Matter **10**, 2375 (1998).
5. O. Shenderova, D. W. Brenner, A. Nazarov, A. Romanov, and L. Yang, Phys. Rev. B. **57**, R3181 (1998).
6. *Proceedings of the 4th International Conference on Laser Ablation*, edited by R. E. Russo, D. B. Geohegan, R. F. Haglund, Jr., and K. Murakami, Appl. Surf. Sci. **127-129** (1998).
7. L. V. Zhigilei, P. B. S. Kodali, and B. J. Garrison, J. Phys. Chem. B **101**, 2028 (1997); J. Phys. Chem. B **102**, 2845 (1998).
8. L. V. Zhigilei, P. B. S. Kodali, and B. J. Garrison, Chem. Phys. Lett. **276**, 269 (1997).
9. L. V. Zhigilei and B. J. Garrison, Appl. Phys. Lett., in press, 1999.
10. L. V. Zhigilei and B. J. Garrison, Appl. Phys. Lett. **71**, 551, (1997); Rapid Commun. Mass Spectrom., **12**, 1273 (1998).
11. L. V. Zhigilei and B. J. Garrison, Appl. Surf. Sci. **127-129**, 142 (1998); *Laser-Tissue Interaction IX*, edited by S. L. Jacques (SPIE Proceedings Series, Vol. 3254, Washington, 1998), p. 135.
12. D. Kim and C. P. Grigoropoulos, Appl. Surf. Sci. **127-129**, 53 (1998).
13. V. Venugopalan, N. S. Nishioka, and B. B. Mikic, Biophysical Journal. **70**, 2981 (1996).
14. B. L. Holian and R. Ravelo, Phys. Rev. B 51, 11275 (1995).
15. M. Moseler, J. Nordiek, and H. Haberland, Phys. Rev. B **56**, 15439 (1997).
16. J. A. Smirnova, L. V. Zhigilei, and B. J. Garrison, Comput. Phys. Commun., in press, 1999.
17. R. S. Dingus and R. J. Scammon, *Laser-Tissue Interaction IX*, edited by S. L. Jacques (SPIE Proceedings Series, Vol. 1427, Washington, 1991), p.45.

QUANTITATIVE CALCULATION OF DISLOCATION MOBILITY

S. SWAMINARAYAN, D. L. PRESTON
Los Alamos National Laboratory, MS G-755, Los Alamos, NM 87545

ABSTRACT:

We present a new method to calculate the response of dislocations to applied stress. This new method, called the dislocation treadmill, can be used to study the effect of vacancies, interstitials, stresses, strain rate, temperature etc., on the steady state velocity of the dislocation. We demonstrate the use of the method by calculating the response of a dislocation to a constant applied shear stress.

INTRODUCTION:

In recent years there have been a number of simulations of dislocation dynamics [1-4] where the long range interactions between dislocations is used to predict the evolution of the dislocation microstructure of metallic materials subjected to external stresses. The results of these simulations are then used to provide predictions of the mechanical behavior of the material. A critical input to these simulations of dislocation dynamics is the response of individual dislocations to an applied stress. Unfortunately, there do not exist data which describes the velocity dependence on the stress in the system. Consequently these codes use arbitrary relationships to describe this dependence. The most common relationship used is a linear relationship between the velocity and the resolved shear stress on the slip plane. The constant of proportionality is called the mobility of the dislocation and is assumed to be constant over the entire range of velocities encountered during the course of the simulation.

The mobility, i.e. the dependence of velocity on applied stress, can only be obtained through atomistic simulations. All atomistic simulations of dislocation response to stress [5-7] to date have been performed either on a block with free boundary conditions on four faces and periodic boundary conditions along the dislocation line or on a box with periodic boundary conditions in all three directions. The dislocation is placed at one end of the box and the box is stressed. The dislocation moves in response to this applied stress and exits at the other end of the box. However, the size of the box is small enough that surface effects dominate the dislocation response and the dislocation never reaches equilibrium velocity before it exits at the far side. The results of these simulations are qualitative and cannot be used to generate a quantitative data base to predict the mobility of dislocations as a function of applied stress.

In this paper we present a new method to predict the dislocation mobility as a function of applied stress. The method, which we call a *dislocation treadmill*, works by introducing material ahead of the dislocation and removing it from behind the dislocation as it moves. Thus, at steady state, the environment around the dislocation is unchanging and we get a quantitative prediction of the dislocation response to the applied stress for a given material. The advantage of the treadmill method is that it allows the dislocation to reach steady state without boundary conditions playing a strong role in the process. Further, because of the nature of the boundary conditions, it is very easy to study the effect of vacancies, interstitials, stresses, strain rate, temperature etc., on the steady state velocity of the dislocation.

METHOD:

A: Geometry of the dislocation treadmill

We run the simulations in a thin band of material. This material has an edge dislocation located in the center of the cell. The cell has mixed boundary conditions as shown in Figure 1. All the atoms in the cell are displaced according to the positions prescribed by linear elasticity for a straight dislocation gliding with velocity v, and located at the center of the cell [8,9]. The entire cell is then sheared such that the bulk shear stress corresponds to the required value. The atoms at the end of

Mat. Res. Soc. Symp. Proc. Vol. 538 © 1999 Materials Research Society

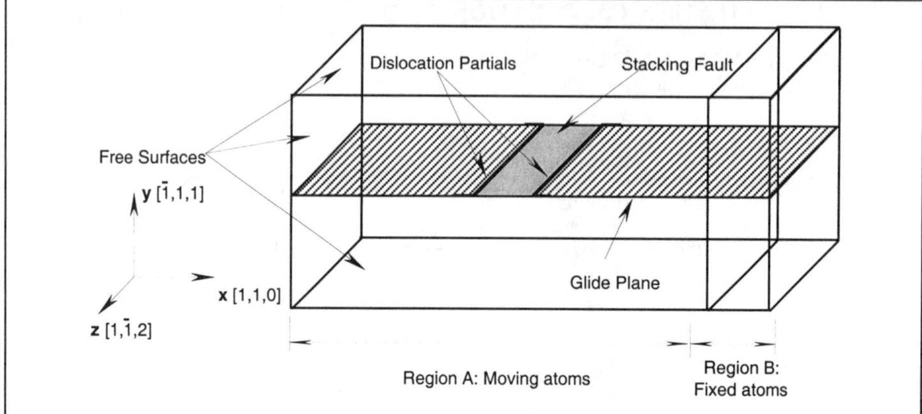

Figure 1: Geometry of the dislocation treadmill. Periodic boundary conditions are used along the dislocation line (z-direction). Free boundary conditions are used along top, bottom and left (-x) face. Atoms in region B are held fixed while those in Region A are allowed to move.

the cell ahead of the moving dislocation (region B in Figure 1) are held fixed at these positions. The rest of the atoms (region A in Figure 1) are free to move. Region B is typically 10 burgers vectors thick, while the rest of the cell is about 200 burgers vectors long. The cell has free boundary conditions in the y-direction and along the negative x-direction (the left hand side face).

The procedure for running the dislocation treadmill is shown pictorially in Figure 2. As the dislocation moves, we track its motion. When the dislocation moves more than some critical distance (typically one burgers vector), we stop the dynamics. We then discard one burgers vector worth of material, henceforth called a frame of material, from the free end of the system (region C in Figure 2b), translate the fixed atoms in region B forward by the frame thickness and add one frame of material (marked as region D in Figure 2c) in the space between the moving atoms (region A) and those in region B. The atoms in regions D and E are then moved to their equilibrium positions using a simple minimization scheme while keeping the rest of the atoms in the system fixed. Thus, as the dislocation moves, we move the simulation cell with it, hence the name dislocation treadmill. This gives us a moving window in the material within which we track the dislocation as it marches along.

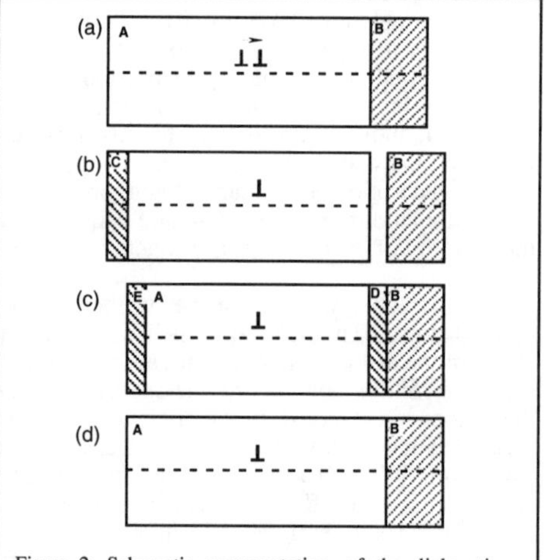

Figure 2: Schematic representation of the dislocation treadmill. As the dislocation moves along, we remove material behind it (C), and add material ahead of it (D).

The minimization of the energy with respect to positions in regions D and E ensure that any elastic waves caused by the addition and removal of material are damped out. Since we are running MD on all the atoms in region A, at equilibrium the displacement field around the dislocation reaches a steady state configuration which by necessity has the correct elastic field built in. Once the system reaches steady state, the local environment around the dislocation does not change. At this stage, the rate at which we add frames directly gives us the dislocation velocity for the applied stress. This dislocation velocity is compared with the initial velocity used to obtain the starting positions of the atoms. If the velocities are off by more than 10%, we reinitialize the cell using the new velocity obtained in the simulation and iterate until the simulation converges

B: Atomic interactions

We employ the embedded atom method (EAM) potential to describe the interactions between the different atoms in the system. We present only a brief overview of the EAM potentials here. The details of the EAM can be found elsewhere [10,11]. Using the EAM method, the potential energy of the system, U, can be written as:

$$U = \frac{1}{2} \sum_{j \neq i} \phi_{ij} + \sum_{i} F(\rho_i) \qquad (1)$$

where the sums over i and j are over all the atoms in the system, ϕ_{ij} is a repulsive pair interaction between atom i and its neighbors j, ρ_i is the electron density at atom i due to the rest of the atoms and $F(\rho_i)$ is the energy released when atom i is embedded in the cloud of density ρ_i and is attractive. Typically, all terms in the potential are cut off (i.e., set to zero) at some prescribed distance. For our simulations, we use the EAM potential for Cu as described by Voter [11] with a cutoff of 4.961Å.

C: Molecular Dynamics:

We used molecular dynamics [12] to calculate the time evolution of the atomic positions within the dislocation treadmill. The initial temperature for all runs was set to 0K. The equations of motion were integrated using a Stoermer finite-difference or leap-frog method, with the time step for all the simulations set to 10^{-15}s.

RESULTS AND DISCUSSIONS:

The primary concern in simulations carried out in small systems is the effect of the finite cell size on the response of the dislocations to the applied stress. To study this effect, we carried out the simulations at two different sizes. Figure 3 shows the velocity of the dislocation as a function of time for two different system sizes: the open circles represent calculations done with a system size of 384.7Å x 233.5Å x 17.7Å and the closed circles represent the results for a system size of 511.7Å x 311.3Å x 22.1Å. Both systems were subjected to a bulk shear strain, ε_{xy}=0.6%. This corresponds to a stress of 486 MPa for the Cu EAM potential used here. Both systems were started at 0K and allowed to run for 3.5 picoseconds (ps) with a time step of 10^{-15}s.

We compared the average dislocation velocities between 1.5 ps and 3.5 ps. The arithmetic means of the two test cases agreed to within 1%. However, we find that the RMS deviation from this average was much larger for the smaller system. The smaller system showed RMS deviations of ±16% with some points off

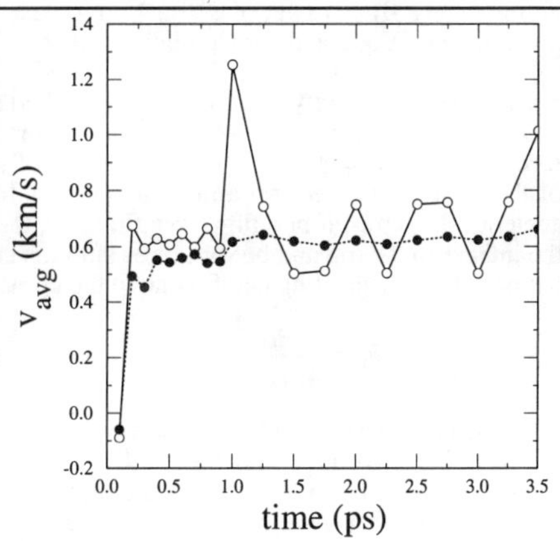

Figure 3: Dislocation velocity as a function of time for system sizes of 384.7Å x 233.5Å x 17.7Å (open circles) and 511.7Å x 311.3Å x 22.1Å (filled circles) for a bulk shear stress of 486 MPa. Although the means of the two curves agree well, the smaller system shows large deviations from this mean value.

the average velocity by as much as 100%. The bigger system, on the other hand, was stable with

RMS deviations about the mean less than ±5%. This indicates that we have achieved a terminal velocity for the dislocation for the applied stress. Since the means of the two runs agree well and the deviations from the mean for the larger system were small, we used the larger system size for all our simulations.

As a demonstration of the applicability of this method to predict dislocation response to applied stress, we present in Figure 4 the result of a series of simulations conducted at strains up to 0.8% for a system size of 511.7Å x 311.3Å x 22.1Å.

Also shown on Fig. 4 are the results predictions from linear elasticity using equation 7-78 from Hirth [9, p. 209] which has been reproduced below as equation 2:

$$v = \frac{50\,C_t}{\mu\, g\!\left(\dfrac{T}{\theta}\right)}\,\sigma \qquad (2)$$

where v is the velocity of the dislocation, σ is the applied stress, μ is the shear modulus (~810 GPa for the Cu potential used here), θ is the debye temperature (315K for Cu) and $g(T/\theta)$ is a function of temperature. Although we find that the average temperature of the solid

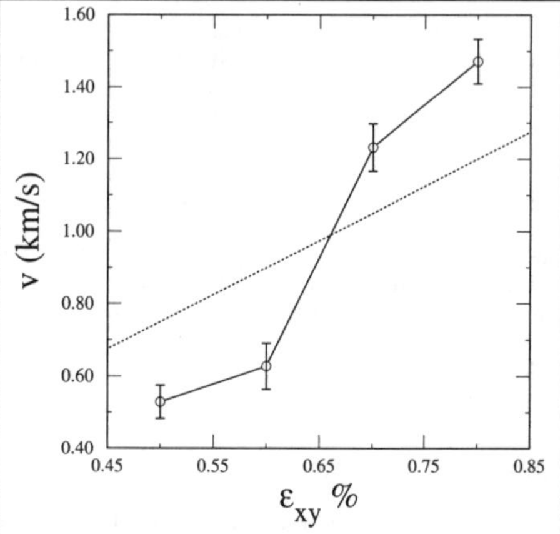

Figure 4: Velocity of dislocations as a function of bulk shear strain obtained using the Dislocation Treadmill method. The dotted line is the prediction of Eq. 2.

in all of the runs is less than 5K, the temperature near the dislocation cores increases to approximately 50k. Therefore, for our comparison with Eq. 2, we used a value of $T/\theta = 0.16$, giving us $g(T/\theta) = 0.1$ [9, p. 210]. Using these values we find reasonable agreement between the predictions of Eq. 2 and the simulation data over the entire range of stress studied.

The plot of velocity as a function of applied stress shows a jump in v between a strain of 0.6% and 0.7%. We believe there the main factors causing this jump are lattice trapping and image forces (or edge effects). To quantify effect of image forces from the top, bottom, and rear walls of the system, we perform an elastic analysis for the image forces on a dislocation at rest in the center of the semi-infinite film. We find that, the forces on the dislocation in the x-direction due to the images in the top and bottom faces sum to zero. The force in the x-direction due to the image in the back face (neglecting cutoffs due to the corners) is:

$$F_x = \frac{-\mu b^2}{4\pi(1-v)l} \qquad (3)$$

where μ is the shear modulus, b the burgers vector, v is the poisson ratio, and l is the distance between the dislocation and the wall. The force on the dislocation due to the image forces as calculated using Eq. 3 is 0.03N, while that due to the applied stress varies from 0.10N at 405 MPa ($\varepsilon_{xy} = 0.5\%$) to 0.17N at 648 MPa ($\varepsilon_{xy} = 0.8\%$). Thus, as long as we apply stresses greater than 550 MPa ($\varepsilon_{xy} = 0.7\%$), the wall effects are negligible compared to the applied stresses.

A dislocation moving in response to an applied stress is analogous to a particle moving in a biased periodic potential [13,14]. For a bias greater than some critical value (the peierls stress), the dislocation moves in the direction of the bias with a velocity which oscillates about some mean value. We deduce from our results that close to the sudden increase in velocity at a stress of 550 MPa, we are close to the critical bias value. The reasons for the sudden increase in velocity lies outside the scope of the present paper which is a proof of principle that the dislocation treadmill works.

One shortcoming of the treadmill method is that dislocations are constrained to be straight during the duration of the simulations. Hence, the formation of kinks which aid the motion of dislocations is suppressed. This can be overcome by increasing the periodic length of the cell along the dislocation line. At higher temperatures, where the effect of kinks is more important, the velocity predicted by the treadmill method will be a lower bound to the actual velocity of a dislocation, and the effect of periodic cell size on the results will become more important.

CONCLUSIONS:

We have presented a new method for calculating the velocity of isolated dislocations in response to an applied stress. The method is stable and can be used to study the effect of temperature, applied stress, and vacancy and impurity concentrations on dislocation motion. Edge effects are small and can be ignored if the stresses are greater than 550 MPa.

ACKNOWLEDGEMENT:

This work was performed at the Los Alamos National Laboratory supported by the U.S. Department of Energy.

REFERENCES:

1. H.Y. Wang and R. LeSar, *Phil. Mag. A*, **71** (1995).
2. D.B. Barts and A.E. Carlsson, *Phys. Rev. E*, **52** (1995).
3. N. Zacharopoulos, D.J. Srolovitz and R. LeSar, *Acta Mater.*, **45**, #9, (1997).
4   Hussein M. Zbib, M. R. Rhee, and, John P. Hirth, *Int. J. Mech. Sci.*, **40**, #2-3, (1998).
5   A. Serra, a. D. J. B. *Acta Metall. et Mat.*, **43**, (1995).
6   P.C. Clapp, M.V. Glazov, and J.A. Rifkin, *J. de Phys. IV*, **3**, (1993).
7   M.I. Baskes, and M.S. Daw, *Sandia National Laboratories Report SAND89-8460*, (1990).
8   J.D. Eshelby, *Proc. Phys. Soc., 62A, 307 (1949).*
9   J.P. Hirth, and J. Lothe, in *"Theory of Dislocations,"* John Wiley & Sons, New York, 1984.
10 S.M. Foiles, M.I. Baskes and M.S. Daw, *Phys. Rev. B* **33.2** (1986).
11 A.F. Voter, *J. Chem. Phys.*, **82**, (1985).
12 M.P. Allen, and D.J. Tildesley, in "*Computer Simulation of Liquids*," Oxford Science Publications, Oxford, (1996).
13 P, Hänggi, Peter Talkner, and Michal Bokovek, *Rev. Mod. Phys.*, **62** #2, (1990).
14 A. Kumar, F.E. Hauser, and J.E. Dorn, *Acta. Met.*, **16**, (1968).

THERMAL CONDUCTIVITY OF SOLID ARGON BY CLASSICAL MOLECULAR DYNAMICS

Hideo Kaburaki, Ju Li[(*)], Sidney Yip[(*)]
Center for Promotion of Computational Science and Engineering, Japan Atomic Energy Research Institute, Tokai,Ibaraki 319-1195, Japan *(kaburaki@sugar.tokai.jaeri.go.jp)*
(*)Department of Nuclear Engineering, MIT, Cambridge, MA 02139 *(http://mmm.mit.edu)*

ABSTRACT

Following the Green-Kubo formalism in linear response theory, the lattice thermal conductivity of solid argon is determined by using classical molecular dynamics simulation to calculate the heat current correlation function. Comparing the absolute conductivities obtained using the Lennard-Jones potential with experiments, we find the predicted results to uniformly underestimate the measurements in magnitude, whereas the calculated temperature dependence corresponds well with the data. The temporal behavior of the heat current autocorrelation function shows that while a single exponential decay description is appropriate at elevated temperatures, below the half of the Debye temperature, the heat current relaxation clearly consists of two stages, an initial rapid decay associated with local dynamics followed by a slower component associated with the dynamics of lattice vibrations (phonons).

INTRODUCTION

Molecular dynamics simulation is by now a well established method for calculating transport properties of liquids, especially when a reliable interatomic potential model is available [1]. For rare-gas systems the Lennard-Jones pair potential has long been in use as a reference model. While generally useful for qualitative purposes, it is known to have limitations associated with the neglect of many-body or quantum effects [2, 3].Thermal conductivity happens to be a property for which the Lennard-Jones potential has given results for liquid argon which are in quite good agreement with experiment [4]. Thus it may be expected that using this model to describe solid argon should meet with similar success.

In contrast to liquid-state studies, the use of molecular dynamics simulation to analyze thermal conduction in solids has been relatively limited [5, 6, 7, 8, 9]. None of these studies to date can claim to provide an unambiguous demonstration of the quantitative accuracy of this method against experimental data. On the other hand, several attempts at phonon scattering calculations in the analysis of thermal conduction in rare-gas crystals have been reported [10, 11, 12], showing varying degrees of agreement with experiments. Because each calculation involves certain approximations, it is at present unclear how one can separate the issue of validity of the potential model from that of the theoretical assumptions.

In this work we report the calculation of thermal conductivity of single crystal argon using classical molecular dynamics in conjunction with the conventional model of Lennard-Jones interatomic pair potential. Besides the goal of establishing a benchmark for the prediction of thermal conductivity of a solid, our motivation also includes understanding further the dynamical details of thermal conduction in a crystalline lattice. We have obtained results at constant pressure (P=0) which span the entire temperature range down to 10 K. Fitting our simulation data to T^{-n}, we find, whereas fitting the experimental data in the same way gives n = 1.5. For the magnitude of the conductivity coefficient, we find the experimental values to be uniformly greater by about a factor of 1.9. As indicated below, we regard this discrepancy as largely associated with the interatomic potential model.

According to the Green-Kubo formalism [5], the thermal conductivity is given by the time interval of the heat current autocorrelation function $< \mathbf{J}(t)\mathbf{J}(0) >$. In this approach, besides the numerical value of the thermal conductivity, one can probe the dynamics of thermal conduction through the temporal decay of the correlation function. Our results show that near the melting point the relaxation of $< \mathbf{J}(t)\mathbf{J}(0) >$ can be reasonably well described as a single exponential decay. As temperature is lowered past the half of the Debye temperature (about 80-90K), a second (slower) relaxation component becomes discernible.

THERMAL CONDUCTIVITY AND HEAT CURRENT AUTOCORRELATION FUNCTION

Our simulations were carried out using a periodic cell containing N argon atoms which interact through the Lennard-Jones 6-12 potential with constants $\epsilon = 119.8[K], \sigma = 3.405[\text{Å}]$. While most runs were made with $N = 256$, a few runs also have been made with $N = 500$. Fifth-order predictor-corrector (Gear) method was used to integrate the equations of motion with a time step size of $10fs$. The thermal conductivity is given by

$$\kappa = \frac{1}{3Vk_BT^2} \int_0^\infty < \mathbf{J}(t) \cdot \mathbf{J}(0) > dt, \tag{1}$$

where the heat current $\mathbf{J}$ is

$$\mathbf{J} = \sum_i E_i\mathbf{v}_i + \frac{1}{2} \sum_{\substack{i>j \\ pairs}} \mathbf{r}_{ij}[\mathbf{F}_{ij} \cdot (\mathbf{v}_i + \mathbf{v}_j)]. \tag{2}$$

The heat current autocorrelation function was obtained by averaging over a number of time origins. Typically a simulation run of 10^6 time steps was performed to determine the thermal conductivity. In the case of $10K$, almost 10^4 time steps were needed for the autocorrelation function to decay to zero value. Also, the point of zero crossing was found to vary appreciably from one run to another . Therefore, a long run of 10^7 steps was made to obtain ten samples which were then averaged to give the final smooth autocorrelation function.

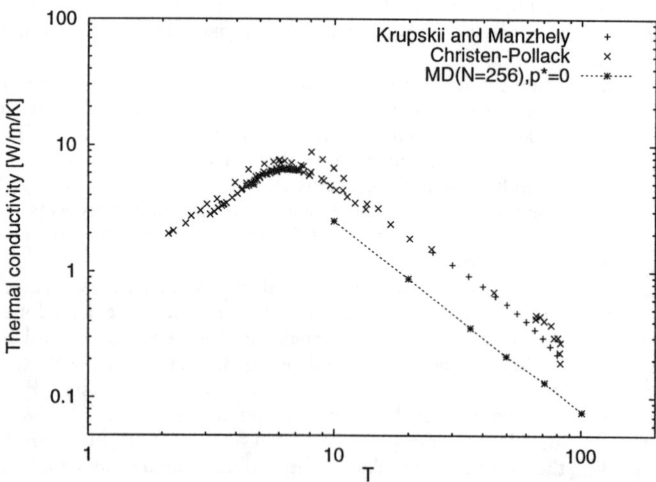

Figure 1: Temperature dependence of thermal conductivity of solid argon under the free standing condition.

Figure 1 shows the calculated thermal conductivity at six temperatures, $10, 20, 35, 50, 70$ and $100K$, under the free standing condition, achieved in each case by varying the density to maintain effectively zero pressure. Errors incurred in determining the absolute values of the conductivity are estimated to be between 5 to 10%.It is found that these data exhibit a temperature dependence varying more rapidly than T^{-1}. Fitting the data to T^{-n} gives $n = 1.516 \pm 0.006$.

Figure 1 also shows two sets of experimental data [13, 14]. We first note that the temperature variation of simulation results is very similar to that observed. A good fit of the experimental results is obtained with $n = 1.5$. Secondly the magnitude of the calculated conductivity is uniformly low compared to the experiments, by a factor of about 0.52.

As a test of system size effects, additional simulations with $N = 500$ were performed for 10^6 and 10^7 steps. As shown in Figure 2, there appears to be little change in going from $N = 256$ to $N = 500$. It should be noted that we have made no attempt to introduce any correction for quantum effects [6, 9]. This is because zero-point motion effects are not significant in the thermal conductivity of argon, even at the lowest temperature studied, $T = 10K$.

The thermal conductivity of argon crystal is known to be sensitive to variations in molar volume. We have performed several constant volume simulations , obtaining the conductivity values shown in Figure 3. From these results we find a T^{-1} temperature variation which is sometimes taken to be a rough estimate of the phonon lifetime at high temperatures. Also shown in Figure 3 are the measurements at constant molar volume [15]. A comparison of the simulation results with these data over the range of molar volume studied at 70K, given in Figure 4, shows the same trend of increasing conductivity with decreasing volume.

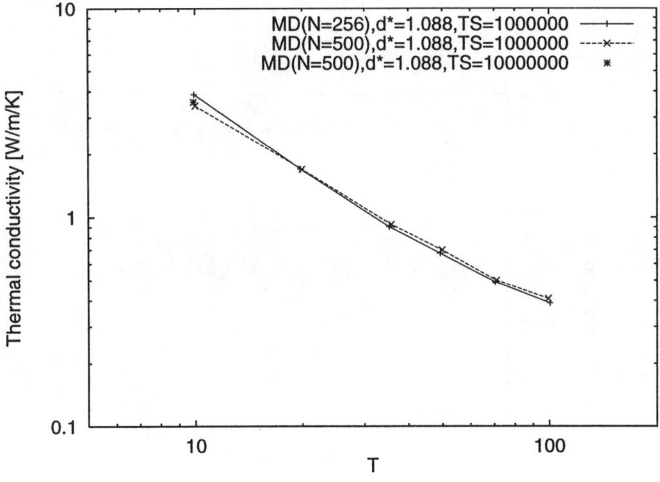

Figure 2: System size effect of thermal conductivity of solid argon.

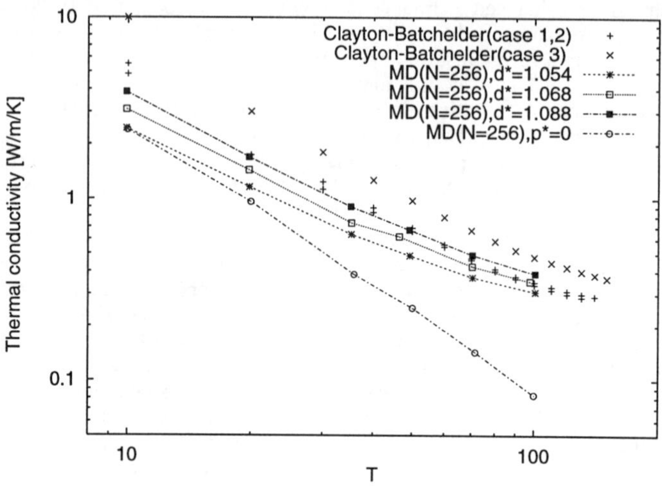

Figure 3: Temperature dependence of thermal conductivity of solid argon under constant molar volume conditions.

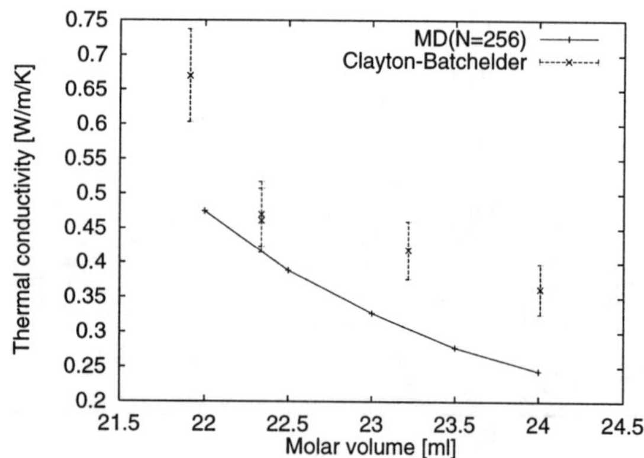

Figure 4: Molar volume dependence of thermal conductivity of solid argon at 70K.

Figure 5 shows the heat current autocorrelation function for the free standing condition at $20K$ and $70K$. The absolute values are rescaled by the maximum $t = 0$ value $< \mathbf{J}(0) \cdot \mathbf{J}(0) >$. The existence of a two-stage decay at the lower temperature can be seen clearly. In the composite of the six temperatures we have studied, Figure 6, the systematic appearance of the second relaxation component becomes increasingly more pronounced as the temperature is lowered. We interpret the initial rapid decay as characteristic of local dynamics, while the slower decay is associated with the dynamics of phonon transport, the collective effects of lattice vibrations.

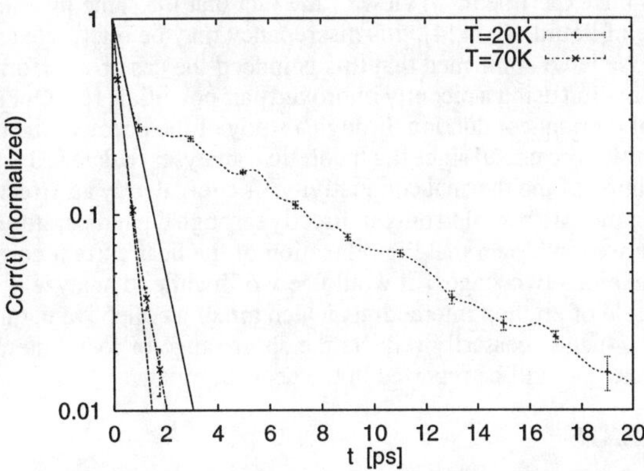

Figure 5: The heat current autocorrelation function for the free standing condition at 20K and 70K.

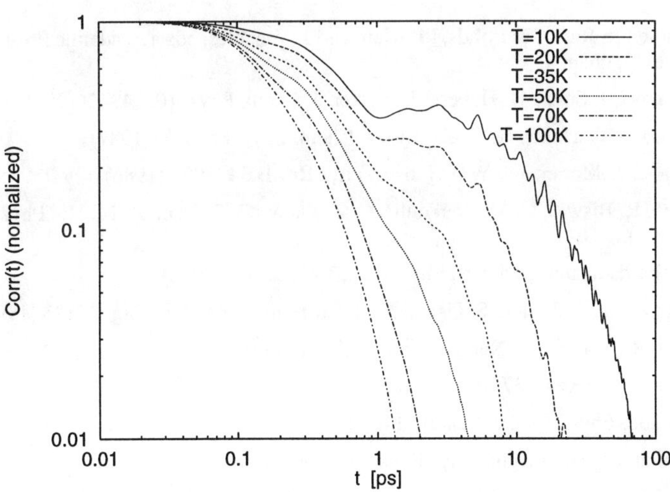

Figure 6: Temperature dependence of the second relaxation component for the heat current autocorrelation function.

DISCUSSION

This work has a two-fold purpose. The first is to establish how well one can predict the thermal conductivity of a simple solid such as argon for which a reasonably well established interatomic potential model is available. We find that the conventional Lennard-Jones potential gives results which are lower than the experiment. In view of the fact that the same potential gives quite good agreement in the case of liquid argon [4], this discrepancy may be interpreted as an inadequacy of the model potential. We have confirmed that this is indeed the case by performing the same thermal conductivity calculation using a recently improved pair potential [16]. Our second purpose is to probe the dynamics of thermal conduction through a study of the decay of the heat current correlation function. This is felt to be useful since the theoretical analyses to date [10, 11, 12] are concerned with the direct calculation of the thermal conductivity, or equivalently an effective relaxation time, in which case the dynamics are revealed only indirectly through the temperature dependence. In the results presented here we have seen that the relaxation of the heat current correlation at moderate to low temperatures exhibits two stages. It would be worthwhile to analyze the simulation results further to clarify the role of phonon interactions which are of a collective nature as opposed to the short-time relaxation which necessarily reflects the short-range or local interatomic interactions. This work is in progress and will be reported in the near future.

ACKNOWLEDGMENTS

Work of J. Li and S. Yip is supported by AFOSR Grant No. F49620-1-0447.

REFERENCES

[1] M. P. Allen and D. J. Tildesley, "Computer Simulation of Liquids",Clarendon Press, Oxford, 1987.

[2] J. A. Barker, in Rare Gas Solids, M. Klein and J. Venables, eds. (Academic Press, New York, 1976), vol. 1, chap. 4.

[3] E. Ermakova, J. Solca, H. Huber, M. Welker, J. Chem. Phys. 102, 4942(1995).

[4] R. Vogelsang, C. Hoheisel, G. Ciccotti, J. Chem. Phys. 86, 6371(1987).

[5] A.J.C.Ladd, B.Moran, and W.G.Hoover, Phys.Rev.B **34**, 5058 (1986).

[6] Y. H. Lee, R. Biswas, C. M. Soukoulis, C. Z. Wang, C. T. Chan, K. M. Ho, Phys. Rev. B 43, 6573 (1991).

[7] C. F. Richardson and P. Clancy, Phys. Rev. B 45, 12260 (1992).

[8] H. Kitagawa, Y. Shibutani, S. Ogata, Modell. Simul. Mater. Sci. Eng. 3, 521 (1995).

[9] J. Li, L. Porter, S. Yip, J. Nuc. Mater. 255, 139 (1998).

[10] C.L.Julian, Phys.Rev. **137**, A128 (1965).

[11] G.Niklasson, Phys.kondens.Materie 14,138(1972).

[12] M.Omini and A.Sparavigna, Phys.Rev.B **35**, 9064(1996-II).

[13] D.K.Christen G.L.Pollack,Phys.Rev.B **12**, 3380(1975).

[14] I.N.Krupskii and Manzhelii, Sov.Phys.JETP **28**,1097(1969).

[15] F.Clayton and D.N.Batchelder, J. Phys. C **6**, 1213(1973).

[16] H. Kaburaki, J. Li, S. Yip, to be published.

DIGITAL MATERIAL: A FRAMEWORK FOR MULTISCALE MODELING OF DEFECTS IN SOLIDS

C. R. MYERS,[1] S.R. ARWADE,[2] E. IESULAURO,[2] P.A. WAWRZYNEK,[2] M. GRIGORIU,[2]
A.R. INGRAFFEA,[2] P.R. DAWSON,[3] M.P. MILLER,[3] AND J.P. SETHNA[4]
[1]Cornell Theory Center
[2]School of Civil and Environmental Engineering
[3]Sibley School of Mechanical and Aerospace Engineering
[4]Laboratory of Atomic and Solid State Physics
Cornell University, Ithaca, NY 14853

ABSTRACT

We are designing and developing a novel software environment – the Digital Material –
to support materials modeling across many length and time scales, in order to develop
improved descriptions of material structure and evolution. Software support is required for
high-performance numerical kernels, lightweight infrastructures for prototyping, steering
and analysis, information transfer across scales, coupling of disparate simulation modules,
and collaboration between a multidisciplinary collection of researchers.

INTRODUCTION

The need for multiscale modeling of materials

There is widespread recognition – fueled to a considerable extent by the continued growth
of available computing power – that improved descriptions of material structure and response will result from the development of models which incorporate information from
smaller length and time scales into material models at larger scales. Efforts typically focus
either on summarizing small scale processes for use in larger scale models, explicitly coupling material models at multiple length scales, or some combination of both approaches.
Extracting meaningful information, however, from complex, multiscale, multidisciplinary
models is a nontrivial task, requiring new types of software support for scientific computing. As part of several coupled scientific and engineering research efforts, we have begun to
design and develop a flexible and expressive software environment for multiscale materials
modeling, dubbed *Digital Material*. In this paper we highlight some of its key design and
implementation features.

Digital Material makes extensive use of a number of software engineering techniques
which have, to date, not seen widespread use in computational science and engineering.
These include, most importantly, the use of interpreted scripting languages to coordinate,
steer and integrate a complex suite of simulation and analysis tools, and the use of object-
oriented design principles and related techniques such as "design patterns" to introduce
needed flexibility into the modeling process.

Mat. Res. Soc. Symp. Proc. Vol. 538 © 1999 Materials Research Society

Some challenges of multiscale modeling of materials

Much of material behavior – especially in engineering systems of interest to us, such as poly-crystalline metallic alloys used in structural applications – is driven by the complex interaction and evolution of defect structures at mesoscales. Therefore, a suitable modeling framework must provide support for simulating and understanding complex, three-dimensional dynamical process of many degrees of freedom. Since models of such phenomena are still not well-developed, there must be support for prototyping, validating and interpreting different sorts of models. Support for rapid prototyping cannot impinge, however, on the ever-present need for computational performance.

In the course of research, scientists and engineers naturally describe material entities and structures at various length scales, including electrons, atoms, vacancies, dislocations, voids, lattices, grains, grain boundaries, twin boundaries, dislocation structures, cracks, surfaces, and precipitates. It is therefore important that there be modeling and programming support to describe those structures. It is equally important that researchers be able to develop simulation codes in which the underlying representation of such structures can be altered; this will support not only the process of validating reduced-order models but also the need for run-time adaptivity, whereby certain material features can be simulated more faithfully by explicit treatment of small scales if such a need arises.

Because the nature of material structures and mathematical models differ considerably from scale to scale, it is important that multiscale descriptions be able to represent material information consistently across scales. We have found it useful to recall that various descriptions of material structures are typically approximations to the fully $3N-$dimensional configuration space described by N atoms in a system, or the $4N-$dimensional configuration spacetime describing the evolution of the system. (We are choosing for now not to explicitly include electronic degrees-of-freedom in such a description, but instead to summarize them those into attributes attached to atomic degrees-of-freedom.) One could presumably, with sufficient computer resources, reconstruct the structure and evolution of all the larger scale features in a material (grains, grain boundaries, precipitates, cracks, etc.) from the underlying atomic configuration. Fortunately, most atomic degrees of freedom are not actually required to achieve an accurate description of material structure and response, because of the strongly localized nature of deformation in materials. Coarse-grained theories (such as linear elasticity) often adequately describe the deformations of many atoms in a material.

Such descriptions involve various *approximations* to the underlying atomic configuration space. A major challenge, therefore, is to develop an *optimal* approximation – one that balances *fidelity* as provided by smaller scale structures and processes with *efficiency* as provided by reduced-order descriptions at larger scales. Since there is no one situation or set of resources to dictate which is *the* optimal tradeoff, one must actually develop a systematic framework for deciding and computing an optimal approximation based on questions of interest and resources at hand. In some respects, we envision something that is a generalization of some of the ideas at the core of the quasicontinuum method.[1] But instead of instantiating only atomic degrees-of-freedom to resolve small scale structure, we envision the need to instantiate a hierarchy of structures (atoms, dislocations, dislocation tangles, voids, cracks, etc.) at various scales in order to optimally approximate the underlying configuration space.

DIGITAL MATERIAL

The Digital Material framework is being built to address these and other challenges. At its heart, Digital Material is intended as a problem-solving environment that will support the development of reduced-order descriptions of material structure and response for use at larger scales, the explicit coupling of material models at multiple scales, and the extraction of information from complex, coupled, dynamical processes in materials.

Digital Material as a Material Representation

The material and defect structures that constitute the focus of materials research must be available to the programmer as computational building blocks, to be accessed, manipulated and combined in arbitary ways. We may have, for example, a geometric model of a grain boundary triple junction (derived, say, from experimental measurements), which we decide to model atomistically. To do so, we would like to be able to compose the description of the junction with a recipe for instantiating atomic degrees-of-freedom that is independent of the details of the junction. Should we decide to represent the junction by a collection of dislocations rather than a collection of atoms, we should be able to replace the atomistic generator with one that creates dislocations.

Developing models of material and defect structures so that they can be treated as computational building blocks is naturally achieved through the use of object-oriented techniques. Constructing flexible couplings between such objects, so that we have a composable suite of simulation modules rather than a single simulation program, requires further design considerations, which fall under the category of "design patterns".[2] Design patterns are a class of software engineering techniques which describe the interactions between collections of objects, in such a way as to encapsulate aspects of a problem which need to be varied. Our vision of multiscale modeling involves varying, potentially at run time, a number of aspects, including: the internal representation of material structures, the assembly of objects which are combined to define a specific material "sample", the identity of collective objects (i.e., whether an object is modeled as a collection of small-scale structures for use at one scale or as a primitive object for use at a larger scale). Perhaps most importantly, design patterns provide natural support for the core of our modeling architecture: the separation of geometric structures from attributes attached to those structures and from the tools, models and other external entities that act to query or modify those structures.

Digital Material as an Interactive Problem-Solving Environment

The field of scientific programming continues to unfold, exploiting new tools and methodologies to allow for more expressive computational modeling. Press and Teukolsky[3] have identified the need for a middle ground between traditional low-level programming languages that are "close to the machine" and more recent "total environments" that purport to provide support for all aspects of scientific computing (but which have dubious performance for large-scale numerical computing). In addition, interactivity in programming and analysis environments is increasingly recognized as a key to unraveling complex phenomena.[4]

Although the analogy is not entirely correct, it would not be inappropriate to describe our Digital Material as something akin to a "Matlab for Materials". Matlab is very successful at providing high-level support for research in linear algebra, allowing computational experimentation by freeing the programmer from low-level details. Unfortunately, we find that Matlab is less well-suited for problems outside the realm of linear algebra, so we have decided on a different core.

The central element of Digital Material which supports high-level interactive programmability as well as "glue" to integrate many different modules is Python, an interpreted, object-oriented programming language.[5] In addition to being a high-level programming language, Python is also an extension language capable of dynamically loading compiled modules written in more traditional programming languages. By allowing the incorporation of low-level extension modules, Python is able to provide high-level language support for rapid prototyping and interactive exploration, while not sacrificing computational performance in time-critical sections of code. Python is freely available and runs on all platforms of interest; for scientific and numerical computing, modules are available to support array manipulations, linear algebra, FFTs, geometric operations, interpolation, polynomial fits, automatic differentiation, plotting, GUI development, 3D visualization, image processing, Internet programming, and database access (to name a few). The process of incorporating one's own simulation or analysis code into the larger Python environment is vastly facilitated by the SWIG package[6], developed by David Beazley originally in support of the SPaSM materials modeling project at LANL.[7] SWIG is a nonintrusive "interface and wrapper generator" which automatically produces the "glue code" that enables Python to communicate with C or C++. SWIG also ingeniously allows for the development of object-oriented structure on top of legacy codes which do not possess such structure.

With tools such as these, we will be able to build a flexible modeling, simulation and analysis environment, by mixing and matching components as needed. Rapid prototyping can be done very effectively within Python, with time-critical pieces migrated down to a compiled language as performance becomes an issue. Ultimately, we will be able to use heuristic insights gained through interactive exploration to automate highly complex processes, such as adaptive creation of more refined small-scale models in response to particular types of material deformation.

INITIAL APPLICATIONS

Because our assembled group is diverse in its background and focus, research is ongoing on a number of problems at a number of different length scales. We are in the process now of developing new simulation capabilities, and integrating existing capabilities into a more unified framework. Initial applications include: determining the structure, energetics and mobility of elementary defects in atomic lattices in two dimensions (with J. Tomasi and X. Chen), integrating components for three-dimensional finite-element simulations of crack propagation (as part of a related project with K. Pingali, S. Vavasis, P. Chew, D. Schneider, P. Stoghill and N. Chrisochoides), and investigating the nature of texture and misorientation and their roles in microstructural failure processes such as intergranular fracture. We will breifly describe below some of the ongoing research in this last area.

Texture, misorientation, dislocation structures and fracture in polycrystals

The large-scale plastic deformation of polycrystalline metals can involve the formation of intricate dislocation structures and, at larger scales, the development of preferred crystalline orientations, or texture. Such structures and patterns are intellectually fascinating in their own right, and can also have a profound impact on material properties, e.g., on the fracture toughness of a metal prone to intergranular cracking. We are working to couple simulations of large-scale plasticity and texture evolution[8] to codes focused on predicting crack growth.[9] Increasingly sophisticated experimental tools – such as Electron Backscatter Diffraction Patterns (EBSP) – are also available to probe lattice orientations and misorientations on mesoscales. An example of EBSP orientation data for an aluminum alloy are shown in Figure 1. Data such as these will enable us to characterize the statistical properties of orientation distributions, generate synthetic material samples for simulation, and test hypotheses concerning the detailed form of dislocation structures that form to delineate grain and subgrain boundaries.

Figure 1: Experimental lattice orientation data on the surface of a polycrystalline aluminum alloy, as determined by Electron Backscatter Diffraction. Local lattice orientation is measured on a regular triangular grid of side length 5 microns; a section of a full scan of size $540\mu m \times 540\mu m$ is shown. (Line directions and grayscales represent orientation axis and angle, respectively, in a Rodrigues axis-angle parameterization.) Misorientation between scan locations indicates the presence of grain or subgrain boundaries. By incorporating such data into the Digital Material environment, we can validate models of dislocation structure formation and texture evolution, and develop statistical models to generate synthetic material samples for simulation.

CONCLUSIONS

The peculiar demands of multiscale modeling of materials require new strategies for simulation, analysis, interrogation and collaboration. The complex, dynamic adaptivity of materials needs to be addressed with computational tools that are equally complex, dynamic and adaptive. Multiscale modeling environments must provide systematic programming support for information transfer across scales. Lightweight scripting and extension environments like Python can effectively balance the conflicting requirements of high performance and rapid prototyping. All of these elements are conspiring to bring life to Digital Material.

ACKNOWLEDGMENTS

This work is supported by the NSF and the AFOSR under grants NSF 9873214 and AFOSR F49620-98-1-0401.

References

[1] E.B. Tadmor, M. Ortiz, and R. Phillips, Phil. Mag. A **73**, no.6, 1529 (1996).

[2] E. Gamma, R. Helm, R. Johnson, and J. Vlissides, *Design Patterns: Elements of Reusable Object-Oriented Software*, Addison Wesley Longman, Reading, MA (1995).

[3] W.H. Press and S.A. Teukolsky, Computers in Physics **11** (5), 416 (1997).

[4] P.F. Dubois, Computers in Physics **8** (1), 70 (1994).

[5] The Python Home Page, http://www.python.org.

[6] SWIG: Simplified Wrapper and Interface Generator, http://www.swig.org.

[7] D.M. Beazley and P.S. Lomdahl, Proceedings of Supercomputing '96, Pittsburgh, PA (1996).

[8] A. Kumar and P.R. Dawson, Comp. Methods Appl. Mech. Engr., in press.

[9] Gray, L., D. Potyondy, E. Lutz, P. Wawrzynek, L. Martha, and A. Ingraffea, Mathematical Models and Methods in Applied Sciences **4(2)**, 179 (1994).

Modeling of inelastic deformation of f.c.c. single- and polycrystalline materials with low stacking fault energies

A. Staroselsky* and L. Anand**

* UTC Research Center, 411 Silver Lane, MS 129-73 East Hartford, CT 06108 USA
** Department of Mechanical Engineering MIT, Cambridge, MA 02139, USA

ABSTRACT

A new rate-independent constitutive model for plastic deformation of crystalline materials deforming by slip and twinning has been formulated, and implemented in a finite-element program. We have simulated *three different structural levels* by choosing representative volume elements (RVEs) as *(i)* a small part of a single crystal for the analysis of the heterogeneity of plastic deformation in single crystals, *(ii)* a whole single crystal for polycrystal simulations, and *(iii)* a group of crystals for a Taylor-type model of polycrystals. We show that the predictions for the texture and stress-strain response from the model are in reasonably good agreement with experiments in plane-strain compression for a different single crystal and polycrystalline f.c.c. materials.

INTRODUCTION

The two main mechanisms for the plastic deformation of metals are crystallographic slip and deformation twinning. Twinning plays an important role in the plastic deformation of high symmetry materials with low stacking fault energy. The analytical modeling and computational accounting for twinning as a mechanism of plastic deformation is in its nascent stages. Review of the early considerations of twinning in crystal plasticity models is given by Chin (1975). In these early models the shear due to twinning was treated as pseudo-slip, but the rotation due to twinning was not accounted for, and hence these models were not capable of predicting crystallographic texture evolution. Evolution of texture due to slip and twinning has been considered by Van Houtte (1978), and more recently by Lebensohn and Tome (1994). However, their considerations have been targeted for the rigid-plastic, non-hardening case, and have been limited to polycrystalline materials.

We focus our attention on developing an elastic-plastic constitutive model and attendant computational procedure to simulate the stress-strain response and texture evolution in f.c.c. metals which deform by slip on twelve $\{111\} < 110 >$ slip systems and by twinning on twelve $\{111\} < 11\bar{2} >$ twin systems[1].

The overall plastic deformation of a crystal is always inhomogeneous at length scales associated with slip and twinning, and should be defined as an average over a volume element that must contain enough dislocation loops and twins to result in an acceptably smooth process at the continuum level of interest here. The smallest such volume element above which the plastic response can be considered smooth, is labeled as a representative-volume element. In our model we shall take a *small part of a crystal* as a representative-volume element (RVE) for single crystal analysis and a *whole crystal* for polycrystal simulations.

---

[1]Note that unlike slip, which can occur in either the positive or negative $< 110 >$ slip direction, twinning, because the underlying atomic arrangement is polar in nature, can occur in only one $< 11\bar{2} >$ type direction on a $\{111\}$ plane.

Mat. Res. Soc. Symp. Proc. Vol. 538 © 1999 Materials Research Society

In this paper we report results of finite element modeling to predict the operative twinning-dominated deformations and crystallographic texture evolution during plane-strain compression of f.c.c. (*i*) *single crystals* which have been experimentally studied by Chin *et al.* (1969), and (*ii*) *polycrystalline structure.* We also report our results on Taylor-type simulation of polycrystalline α-brass. We show that our model is able to reproduce well the experimentally measured pole figures and stress levels. Thus, the crystal plasticity based elastic-plastic model is able to predict material behavior on different structural levels: from single crystal inhomogeneity to overall polycrystalline response.

CONSTITUTIVE EQUATIONS

The two major kinematic issues in modeling twinning are: (*i*) accounting for the shear associated with twinning; and (*ii*) accounting for the reorientation of the crystal lattice due to twinning. Here we follow Van Houtte (1978) idea that the shear due to twinning is first accumulated as a pseudo-slip, that is, the shearing rate on the twin systems is taken to be given by $\dot{\gamma}^i = \dot{f}^i \gamma_0$, where γ_0 is the amount of twinning shear, and the crystal lattice is given the twinning-related orientation only if a probabilistic criterion, based on the relative volume fractions of the twinned and non-twinned parts of a crystal, is met.

The governing variables in the constitutive model are taken as: (*i*) The Cauchy stress, $\mathbf{T}$. (*ii*) The deformation gradient, $\mathbf{F}$. (*iii*) Crystal slip and twin systems labeled by integers i. Each system is specified by a unit normal $\mathbf{n}_0^i$ to the slip/twin plane, and a unit vector $\mathbf{m}_0^i$ denoting the slip/twin direction. The slip and twin systems $(\mathbf{m}_0^i, \mathbf{n}_0^i)$ are assumed to be known in the reference configuration. The amount of shear, γ_0, and the lattice rotation accompanying twinning, $\mathbf{R}^{tw}$, are also assumed to be known. (*iv*) A plastic deformation gradient, $\mathbf{F}^p$, with $\det\mathbf{F}^p = 1$. This represents the cumulative effect of dislocation motion and shear due to twinning on the active slip and twin systems in the crystal. (*v*) The slip and twin system deformation resistances $s^i > 0$, with units of stress. (*vi*) The twin volume fractions $f^i \geq 0$, with $\sum_i f^i \leq 1$ and $\dot{f}^i \geq 0$, which means the absence of detwinning events.

The elastic deformation gradient is defined by $\mathbf{F}^e \equiv \mathbf{F}\,\mathbf{F}^{p-1}$ with $\det\mathbf{F}^e > 0$, and it describes the elastic distortion of the lattice; it is this distortion that gives rise to the stress $\mathbf{T}$. Then, $\mathbf{E}^e \equiv (1/2)\left\{\mathbf{F}^{eT}\mathbf{F}^e - \mathbf{1}\right\}$ and $\mathbf{T}^* \equiv (\det\mathbf{F}^e)\,\mathbf{F}^{e-1}\mathbf{T}\mathbf{F}^{e-T}$ the Green elastic strain measure and the symmetric second Piola-Kirchoff stress tensor relative to the relaxed configuration.

Elastic stretches in metallic single crystals are generally small. Accordingly, the constitutive equation for the stress in a metallic single crystal is taken as the linear relation

$$\mathbf{T}^* = \mathcal{C}\left[\mathbf{E}^e\right], \tag{1}$$

where $\mathcal{C}$ is a fourth-order anisotropic elasticity tensor, where $\mathbf{E}^e$ and $\mathbf{T}^*$ are the strain and stress measures defined above.

Let $\mathbf{S}_0^i = \mathbf{m}_0^i \otimes \mathbf{n}_0^i$ denote the Schmid tensors, and let $\tau^i = (\mathbf{C}^e\,\mathbf{T}^*)\cdot\mathbf{S}_0^i$ denote the resolved shear stress on the ith slip/twin system. Then, the conditions for slip and twinning are taken as

$$\phi^i = \left|\tau^i\right| - s^i \leq 0. \tag{2}$$

During plastic flow the following consistency conditions must be satisfied: $\dot{\gamma}^i \dot{\phi}^i = 0$ if $\phi^i = 0$. The consistency conditions serve to determine the shearing rates $\dot{\gamma}^i \geq 0$ on the slip and twin systems. The shearing rates also are restricted by Kuhn-Tacker complementarity conditions $\dot{\gamma}^i \phi^i = 0$.

The evolution of the plastic deformation gradient is

$$\dot{\mathbf{F}}^p = \mathbf{L}^p\,\mathbf{F}^p, \quad \text{with } \mathbf{L}^p = \sum_i \dot{\gamma}^i\,\text{sign}(\tau^i)\,\mathbf{S}_0^i. \tag{3}$$

Here $\mathbf{L}^p$ given by the sum of the shearing rates on all the slip and twin systems.

The evolution equations for slip and twin resistances may be generically taken as $\dot{s}^i = \sum_j h^{ij}\,\dot{\gamma}^j$, where h^{ij} are the hardening moduli.

For the twin systems, evolution equations for twin volume fractions is given as $\dot{f}^i = \dot{\gamma}^i/\gamma_0 \geq 0$, where γ_0 is the twinning shear. The twinning shear corresponding to a twin system may be written as $\mathbf{S} = \mathbf{1} + \gamma_0\,\mathbf{m}_0 \otimes \mathbf{n}_0$, $\quad \mathbf{m}_0 \cdot \mathbf{n}_0 = 0$, $\quad \gamma_0 = 1/\sqrt{2}$, where $\mathbf{m}_0$ is a unit vector in a $< 11\bar{2} >$ direction, and $\mathbf{n}_0$ the unit normal to the associated $\{111\}$ plane. The corresponding twinning rotation is given by $\mathbf{R}^{tw} = 2\,\mathbf{n}_0 \otimes \mathbf{n}_0 - \mathbf{1}$.

During the "pseudo-slip" phase, the slip and twin system deformation resistances will be taken as constant. Leffers and co-workers (e.g., 1991, 1993) have reported that during plane-strain compression of brass, twins form thin lamellae which cluster to form bundles in grains, and that subsequent slip is restricted to planes which are parallel to these twin bundles. We have modeled this important kinematic restriction on the activity of the slip systems as follows. When the fraction $f = \max\{f^\alpha\}$, the maximum value of f^α taken over all twin systems, reaches a value $\lambda \approx 0.05$, slipping and twinning in systems whose slip/twin planes are not parallel to the plane of the twin system with maximum f^α are restricted by choosing appropriate values of slip and twin resistances.

The lattice reorientation condition is that if $f > \xi$, with $\xi \in [0.2; 1]$ a random number, then the orientation of the RVE be replaced by the twin related orientation. That is, if $\{\mathbf{e}_i^c | i = 1, 2, 3\}$ denotes a local orthonormal basis associated with the crystal lattice in the old relaxed configuration, then once this criterion is met the crystal basis in the new relaxed configuration of the crystal be taken as $\mathbf{e}_i^{c\,*} = \mathbf{R}^{tw}\mathbf{e}_i^c$.

Further, if the lattice reorientation condition is met, then the f^i are reinitialized to zero, and subsequent twinning in the RVE is suppressed by setting the twin system deformation resistances to a large value. However, subsequent slip in this RVE is allowed, and the values of all slip system resistances are set equal to the value of the resistance for the slip system(s) parallel to the twin system with the maximum f^i prior to the reorientation.

The constitutive equations and the time-integration procedure have been implemented in the finite-element program ABAQUS/Explicit (1995) by writing a "user material" subroutine.

PLANE-STRAIN DEFORMATION of F.C.C. Co-8% Fe SINGLE CRYSTAL

We shall simulate plane-strain compression of $(110)[\bar{3}34]^2$ specially oriented single crystals Co-8%Fe tested by Chin et al. (1969). The single-crystal calculations shown in this section were carried out by modeling a part of a single crystal as a single finite element.
For the numerical simulation of plane-strain compression, 900 two-dimensional ABAQUS-CPE4R elements (continuum, plane-strain, 4-noded, reduced integration) were assigned to the single crystal. It is important to note that our 2-D finite element simulations incorporate the full 3-D slip and twin system structure. Plane-strain compression was modeled by constraining the top and bottom boundaries of the mesh to remain straight, with the bottom boundary subjected to zero displacement in the Y-direction, and top boundary subjected to

---

[2]These initial crystal orientations may alternatively be described in terms of the Euler angle notation $\{\theta, \phi, \omega\}$.

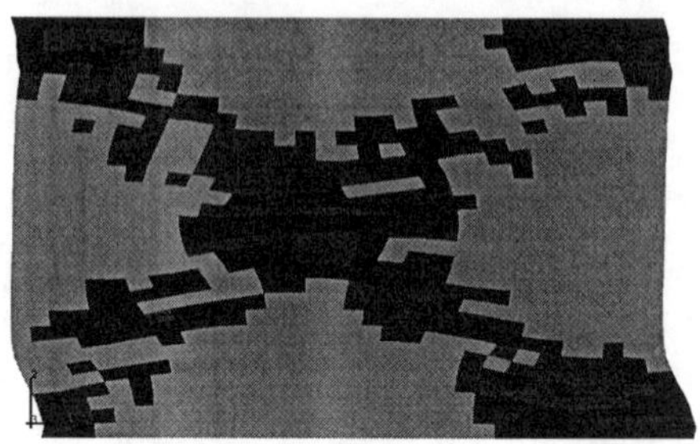

Figure 1: (a) Micrograph of the $(110)[\bar{3}34]$ oriented crystal showing twin bands (from Chin *et al.*, 1969). (b) The deformed FEM mesh with calculated twin bands.

a negative displacement in the Y-direction for a total compressive strain of 20%. For the vertical boundaries we imposed periodic boundary conditions, defined as follows. Let $\mathbf{u}_l$ and $\mathbf{u}_r$ respectively denote the displacements for a node on the left boundary and another on the right boundary which is at the same horizontal level in the initial mesh, and let $\mathbf{d} = \mathbf{u}_r - \mathbf{u}_l$ denote the relative displacement for this pair of corresponding nodes on the left and right boundaries. Then, partially periodic boundary conditions are specified by requiring that $\mathbf{U}_{right} - \mathbf{U}_{left} = \mathbf{d}$, where $\mathbf{U}_{left}$ and where $\mathbf{U}_{right}$ are the vectors of displacements of all the nodes for the left and right boundaries, respectively.

In the numerical simulations reported below, we have used the following values for the initial slip and twin resistances: $s_0^{twin} = 70\,\text{MPa}$; $s_0^{slip} = 40\,\text{MPa}$. A micrograph of the deformed specimen taken from Chin *et al.* (1969) is shown in Fig. 1a. This picture shows that a mixture of slip and twinning has occurred on the (111) and $(11\bar{1})$ planes. Our corresponding finite element simulation is shown in Fig.1b.

Our calculations also predict that slip occurs on the $(111)[\bar{1}01]$ and $(11\bar{1})[0\bar{1}\bar{1}]$ systems, and twinning occurs on the $(111)[1\bar{2}1]$ and $(\bar{1}\bar{1}1)[\bar{1}21]$ systems. The elements twinned by the $(111)[1\bar{2}1]$ system are shaded black, and those twinned by the $(\bar{1}\bar{1}1)[\bar{1}21]$ system are shaded dark grey. The calculated twin bands intersect at an angle close to that observed in the experiments (Fig. 1a).

Texture evolution is one of the most important characteristics of slip/twin systems activity. Fig.2(a) presents the {111} stereographic pole figure of the crystal in its initial orientation. The pole figure predicted by the finite element calculations is shown in Fig.2(b), together with the experimentally measured (Chin et al., 1969) pole figure in Fig. 2(c). The agreement between numerically predicted and experimentally measured texture is very good.

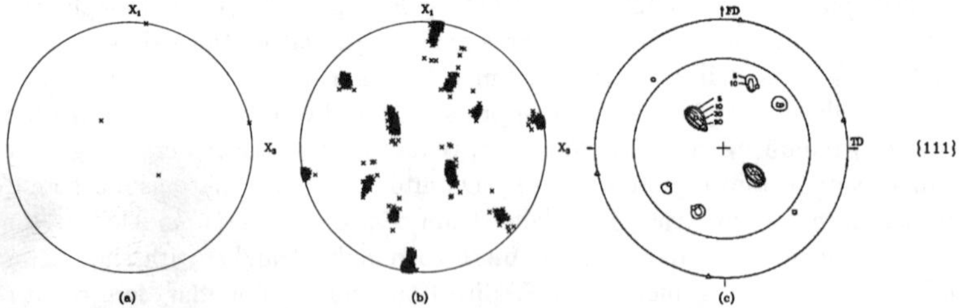

Figure 2: {111} pole figures for $(110)[\bar{3}34]$ oriented crystal. Initial (a), FEM calculated after plane-strain compression to 20% (b), and experimentally measured (Chin et al., 1969) (c).

PLANE-STRAIN DEFORMATION of F.C.C. α-BRASS POLYCRYSTAL

Finite-element calculations for plane strain compression of an aggregate of 225 initially randomly-oriented grains were carried out with various values of the material parameters. Each grain was considered as RVE and modeled as plane strain finite element. The results of plane strain finite element calculations are very close to the full 3D plane strain simulations with 343 (cube $7 \times 7 \times 7$) ABAQUS brick elements (for details of this see Staroselsky and Anand 1998). The process of curve-fitting the plane strain compression stress-strain data to obtain the value of the hardening parameters yields $s_0^{slip} = 25$ MPa, $s_0^{twin} = 155$ MPa,

The quality of the curve-fit is shown in Figure 3. One can see that the calculated stress-strain response is very close to the experimentally observed one. The jumps on the numerically calculated curve at strains greater than ≈ 0.5 are due to the crystal lattice rotations of the grains during twinning.

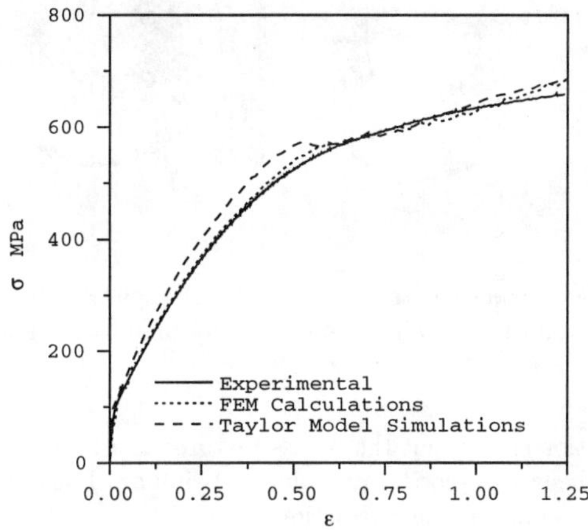

Figure 3: Comparison of the experimentally-measured stress-strain curves for plane strain compression of α-brass against a finite element model as well as against a Taylor model.

The numerical calculations show that because of the different values of slip and twin resistances, in the initial stages of deformation the grains deform by crystallographic slip only. The slip deformation resistances increase due to strain-hardening, and at a level of macroscopic strain of approximately 10%-15%, the first threshold, for the twin fraction used to restrict slip, is reached in some grains. At this stage the slip resistances increase very fast for all systems other than those which are co-planar with the dominant twin system; dominant crystallographic shearing occurs along the chosen crystallographic plane in a given crystal. When the second threshold for the twin fraction $f = \max\{f^\alpha\}$ is reached, $f > \xi$, with ξ a random number distributed over the interval $[0.2; 1]$, the crystal lattice is replaced with a twin-related one, and as discussed previously, all twin system deformation resistances and all slip system resistances are set to the values corresponding to those for the active twin system. With our choice of material parameters, the crystal lattice rotation due to twinning starts at about 45% macroscopic strain. Before this level of deformation, the calculated pole figures are very close to those typical for f.c.c. materials deforming by slip alone. Experimentally measured and numerically predicted equal area {111} and {100} pole figures for plane strain compression after 100% of compression are shown in Fig. 4.

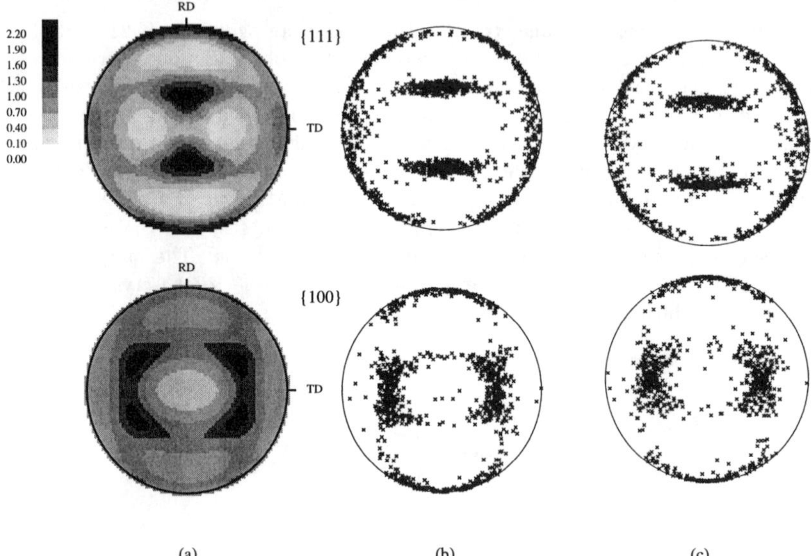

(a)   (b)   (c)

Figure 4: Comparison of the (a) experimentally-measured pole figures after 100% plane strain compression of α-brass against (b) finite element model predictions and (c) Taylor model predictions.

Our numerical experimentation shows that the "brass-texture" is a result of both mechanisms: constrained slip, *and* the lattice reorientation due to twinning. Each of these mechanisms alone is insufficient for accurate model predictions.

A very large body of literature exists on application of the Taylor model for texture prediction. The main assumption of this model for a polycrystal is that the deformation gradient in each grain is homogeneous and equal to the macroscopic one at a material point. If, in addition, we assume that all grains have equal volume, then the average Cauchy stress at each macroscopic continuum point is simply the number-averaged stress (Asaro and Needleman, 1985). With the material parameters calibrated for the finite-element model of the polycrystal, our Taylor model simulations slightly overpredict the stress-strain responses in both plane strain compression and simple compression. The texture predictions from the two modeling schemes are very similar. Thus, the Taylor model may be used for obtaining computationally inexpensive and reasonably accurate predictions of both the stress-strain curve and the crystallographic texture of f.c.c. materials deforming by combined slip and twinning.

CONCLUDING REMARKS AND FUTURE DIRECTIONS

Our calculations clearly demonstrate the ability of our constitutive model and computational procedure to capture the major features of plastic deformation of crystalline f.c.c. material due to slip and twinning at different scales: from a part of a single crystal to a large polycrystalline aggregate.

However, as formulated, the constitutive model has a number of limitations: (*i*) The

520

important role of twin-boundary energy is neglected in the model. (*ii*) There is no length scale in the model. Much work needs to be done to improve our understanding of the slip and twin hardening and hardening interactions, and their mathematical representation. For this and for the analysis of the width of the twin bands, a suitable length scale associated with non-local effects of twinning needs to be introduced into the constitutive model.

ACKNOWLEDGEMENTS

The support for this work was provided by the U.S. Army Research Office under Grant DAAH04-94-G-0060, and the National Science Foundation under Grant CMS-9610130. The ABAQUS finite element software was made available under an academic license from HKS, Inc., Pawtucket, R.I. One of the authors (A. S.) thanks Pratt & Whitney for the support of his participation in MRS meeting.

REFERENCES

ABAQUS (1995) *Reference Manuals*. Hibbitt, Karlsson, & Sorensen Inc., Pawtucket, R.I.

Asaro, R., and Needleman, A. (1985) Texture Development and Strain Hardening in Rate Dependent Polycrystals. *Acta Metallurgica*, **33**, 923–953.

Chin, G. Y., Hosford, W. F., and Mendorf, D. R. (1969), Accommodation of Constrained Deformation in f.c.c. Metals by Slip and Twinning. *Proceedings of The Royal Society A*, **309**, 433–456.

Chin, G. Y. (1975) Development of Deformation Textures, in *Constitutive Equations in Plasticity*, Argon, A. S., ed., 431 – 447.

Lebensohn, R. A., and Tome, C. N. (1994) A Self-Consistent Viscoplastic Model: Prediction of Rolling Textures of Anisotropic Polycrystals. *Materials Science and Engineering*, **A175**, 71 – 82.

Leffers, T.(1991) Microstructures, Textures, and Deformation Patterns at Large Strain. In *MECAMAT'91*, Balkema, Rotterdam, 73-86.

Leffers, T. and Jensen, D. J. (1993) in *Textures and Microstructures*, **14-18**, 933-952.

Staroselsky, A., and Anand, L. (1998) Inelastic Deformation of Polycrystalline Face Centered Cubic Materials by Slip and Twinning *Journal of The Mechanics and Physics of Solids*, **46**, 671-696.

Tome, C. N., Lebensohn, R. A., and Kocks, U. F. (1991) A Model for Texture Development Dominated by Deformation Twinning: Application to Zirconium Alloys. *Acta Metallurgica et Materialia*, **39**, 2667 – 2680.

Van Houtte, P. (1978) Simulation of the Rolling and Shear Texture of Brass by the Taylor Theory adapted for Mechanical Twinning. *Acta Metallurgica*, **26**, 591 – 604.

AB INITIO STUDY OF CHANGES IN THE MAGNETISM OF IRON DURING THE BCC-HCP PHASE TRANSFORMATION

M. Šob[*], M. Friák[*†], L.G. Wang[*1], and V. Vitek[‡]
[*]Institute of Physics of Materials, Academy of Sciences of the Czech Republic, Žižkova 22, CZ-616 62 Brno, Czech Republic, mojmir@ipm.cz
[†]Faculty of Science, Masaryk University, Kotlářská 2, CZ-611 37 Brno, Czech Republic
[‡]Department of Materials Science and Engineering, University of Pennsylvania, 3231 Walnut St., Philadelphia, PA 19104-6272, U. S. A.

ABSTRACT

The pressure-induced bcc-hcp transformation in iron is studied ab initio by following a constant-volume one-parameter transformation path. The calculations are performed in the generalized gradient approximation (GGA). It is shown that the analysis of both spin-polarized and non-spin-polarized states is essential in order to correctly obtain the energetics of the phase transformation. Our calculations make it possible to locate, with high precision, the transition configuration at which the bcc ferromagnetic structure transforms into the hcp nonmagnetic structure.

1. INTRODUCTION

Although the structure of a newly discovered high-temperature and high-pressure β-phase of iron has been discussed extensively in recent years [1], the bcc–hcp phase transformation that occurs at much lower pressure (about 13 GPa) [2] is not properly understood. In a recent experimental study [3], a structurally distorted hcp phase with an anomalously large c/a ratio and a bcc phase with an anomalously large lattice parameter were found to occur during this phase transition. Ekman et al. [4] performed an extensive *ab initio* study of the bcc-hcp transformation in iron, but have not analyzed the relation of the spin-polarized and non-spin-polarized phases in the neighborhood of the phase boundary.

The purpose of this paper is to investigate, from first principles, the structural and magnetic behavior of iron along a one-parameter transformation path that connects the bcc and hcp structures. The important aspect of this study is that it considers the role of non-spin-polarized states in the transition region. In fact it demonstrates that, in order to understand the energetics of the phase transformation correctly, both ferromagnetic (FM) and nonmagnetic (NM) states have to be analyzed.

2. ONE-PARAMETER BCC-HCP TRANSFORMATION PATHS

One possible path leading to a bcc-hcp transformation may be constructed as a combination of a homogeneous deformation and a shuffling of alternate close-packed atomic planes in opposite directions [5, 6, 3]. These two degrees of freedom may be coupled such as to obtain a one-parameter transformation path. On this path, described here by a parameter Δ, we consider the bcc and hcp structures, as well as the structures in between, as orthorombic

---

[1]On leave from the Central Iron and Steel Research Institute, Beijing, China. Present address: Fritz-Haber Institut der Max-Planck Gesellschaft, Faradayweg 4-6, D-14195 Berlin-Dahlem, Germany.

523

Mat. Res. Soc. Symp. Proc. Vol. 538 © 1999 Materials Research Society

with four-atoms in the unit cell. The lattice contants a, b and c are then given by the relations

$$a = a_0 \sqrt{2} / (V/V_0)^{1/3}, \tag{1}$$
$$b = a\,[\Delta\,(2\sqrt{3} - 3\sqrt{2})/6 + \sqrt{2}/2], \tag{2}$$
$$c = a\,[\Delta\,(2\sqrt{2} - 3)/3 + 1], \tag{3}$$

where a_0 is the lattice constant of the bcc structure and

$$V/V_0 = \sqrt{2}\,[\Delta\,(2\sqrt{3} - 3\sqrt{2})/6 + \sqrt{2}/2]\,[\Delta\,(2\sqrt{2} - 3)/3 + 1] \tag{4}$$

is the factor chosen such that the volume per atom is conserved. The positions of atoms are, in the units of lattice constants a, b and c, $(0,0,0)$, $(1/2, 1/2, 0)$, $(-\Delta/6, 1/2, 1/2)$, $(1/2-\Delta/6, 0, 1/2)$. For $\Delta = 0$ we obtain the bcc structure and $\Delta = 1$ corresponds to the hcp structure with the ideal c/a ratio $\sqrt{8/3}$. This path is very similar to the recently proposed path [7] which corresponds to linearly increasing strain in two crystallographic directions, while preserving the volume and shuffling adjacent close-packed planes; the shuffling is linearly coupled to the magnitude of the strain. A different one-parameter bcc-hcp path has been proposed in [8]. These one-parameter paths connect the bcc and hcp structures more directly. They avoid high energy configurations that are encountered if only a shuffling or only a lattice deformation is applied, and are close to the minimum energy paths at constant volume.

Craievich et al. [9] have shown that some energy extrema on constant-volume transformation paths are dictated by the symmetry. Namely, most of the structures encountered along the transformation path have a symmetry that is lower than cubic or hexagonal, while for some values of the parameter Δ the symmetry of the structure is higher. The derivative of the total energy with respect to this parameter must then be zero at these points. In our case this occurs for $\Delta = 0$ (bcc structure) and $\Delta = 1$ (hcp structure). These are the so-called symmetry-dictated extrema. However, other extrema may occur that are not dictated by symmetry and reflect properties of the specific material. Such extrema are not found in the present study. Configurations corresponding to energy minima at the transformation paths represent stable or metastable structures and may mimic atomic arrangements that could be encountered when investigating extended defects such as interfaces and dislocations [10, 7].

3. DETAILS OF THE CALCULATIONS

In order to obtain reliable *ab initio* total energies for structures encountered along the transformation path studied, the methods using a shape approximation of the crystal potential (for example spheroidization, as in the LMTO-ASA or in standard KKR and APW approaches) are not adequate [11, 10]. Instead, full-potential treatments must be employed. In this study we utilized the full-potential linearized augmented plane waves (FLAPW) code described in detail in [12]. The electronic structure calculations were performed self-consistently within the generalized gradient approximation (GGA) [13]. This is crucial in iron as the local-density spin approximation (LSDA) does not yield the correct bcc ferromagnetic ground state. We performed both spin-polarized (ferromagnetic, FM) and non-spin-polarized (nonmagnetic, NM) calculations.

When following the transformation path, crystal lattices are severely distorted and some atoms may move very close together. Therefore, the muffin-tin radii must be sufficiently small to guarantee non-overlapping of the muffin-tin spheres at every point along the path. We used the muffin-tin radius of 2.1455 au for iron atoms and kept it constant in all the calculations presented here.

4. RESULTS AND DISCUSSION

Fig. 1a displays the variation of total energies calculated for the FM and NM states along the studied transformation path for the experimental lattice volume corresponding to the bcc lattice constant $a_{bcc} = 5.408$ au. It is seen that the energy for FM states exhibits a minimum at $\Delta = 0$ (bcc structure) and a maximum at $\Delta = 1$ (hcp structure). The energy of the NM states has a maximum at the bcc structure and a minimum at the hcp structure. All NM states have a higher energy than the FM states. The magnetic moment (Fig. 1c) does not show any dramatic variations.

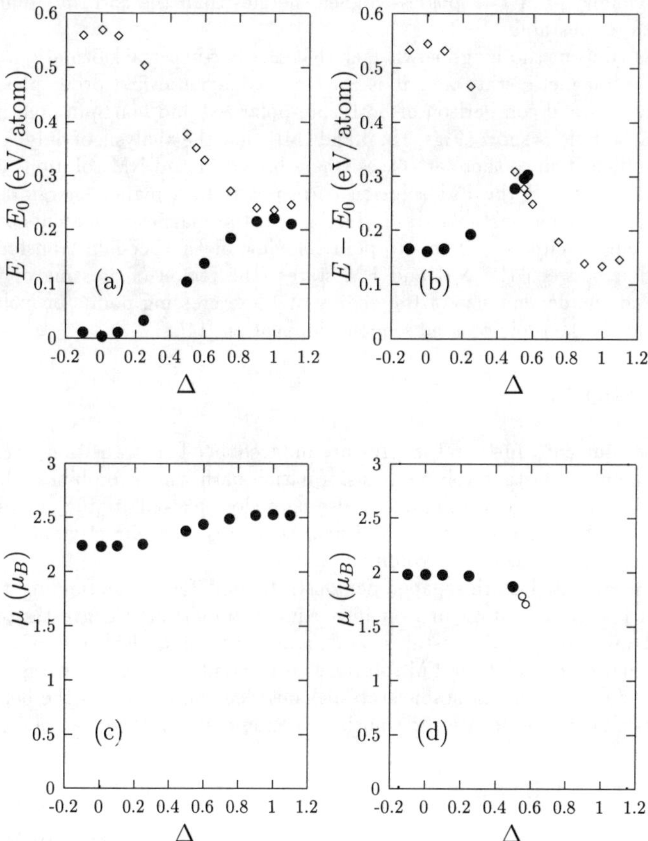

Fig. 1. Variations of total energies and magnetic moments along the constant-volume bcc-hcp transformation path. Here E_0 is the ground state energy of bcc FM iron and Δ is the parameter of the path defined in Sec. 2. Full circles represent results of FM calculations and diamonds of NM calculations. Figures (a) and (c) correspond to the experimental lattice volume per atom (a_{bcc}=5.408 au), figures (b) and (d) to the equilibrium volume of the hcp structure (the lattice parameter of the bcc lattice with the same volume per atom is a_{bcc}=5.1644 au). Empty circles in Fig. (d) show magnetic moments of FM states which have energy higher than the corresponding NM states at the same value of Δ.

The situation is completely different at the equilibrium volume of the NM hcp structure corresponding to the bcc lattice parameter of 5.1644 au (Fig. 1b). The energy of the FM bcc structure increased while the energy of the NM hcp structure decreased so much that it is lower than the energy of the FM bcc structure. However, most interesting is the region of Δ between 0.5 and 0.6. The FM and NM curves cross there, and at Δ close to 0.6 the FM states have a higher energy than the corresponding NM states so that they are not stable any more. For this volume, we were not able to find any FM solution of Kohn-Sham equations for $\Delta > 0.6$. Notwithstanding, for slightly higher volumes (corresponding to $a_{bcc} \geq 5.30$ au), the whole FM curves (up to $\Delta = 1$) could be generated [14]. Up to $a_{bcc} = 5.395$ au, the FM states in the vicinity of $\Delta = 1$ possess higher energies than the corresponding NM states and are, therefore, unstable.

Our results confirm the long-known fact that the bcc-hcp transformation also involves a change of the magnetic state and it is, in fact, a magnetic first-order phase transition [4]. Therefore, a careful comparison of both spin-polarized and non-spin-polarized states is critical. Our calculations show (Figs. 1b, d and [14]) that the analysis of only spin-polarized states is not sufficient since there are cases where both FM and NM solutions of the Kohn-Sham equations exist. If the FM solution corresponds to a higher energy than the NM solution, it represents an unstable state. In Fig. 1d, the magnetic moments of such states are shown as empty circles. From the point of view of the bcc-hcp transformation, the transition region between the NM and FM states (the region of crossing of the FM and NM curves) and the dependence of the energy at these crossing points on volume is most interesting. This is the topic of a subsequent publication [14].

5. CONCLUSIONS

In order to elucidate further the pressure-induced bcc-hcp transition, we have introduced one-parameter constant-volume transformation path which combines a homogeneous deformation of the lattice with shuffling of alternate close-packed atomic planes in opposite directions. This path connects the bcc and hcp structures more directly and is close to the minimum-energy path for constant volume.

The results presented in this paper demonstrate that for understanding the pressure-induced bcc-hcp transformation in iron it is equally important to investigate both spin-polarized and non-spin-polarized states. In many cases, both FM and NM solutions of Kohn-Sham equation exist. If the FM solutions correspond to a higher energy than the NM solutions, they do not represent stable states. A detailed comparison of the behavior of FM and NM states allows us to locate the transition region between these two states accurately.

ACKNOWLEDGEMENTS

This research was supported by the Ministry of Education of the Czech Republic (Project No. ME-264) and by the National Science Foundation - International Programs (Grant No. INT-96-05232). A part of this study has been performed in the framework of the COST Action P3 (Project No. OC P3.10). The use of the computer facility at the Boston University SCVC within the grant NSF30266 and at Masaryk University, Brno, is acknowledged.

REFERENCES

1. L. Dubrovinsky, S.K. Saxena, P. Lazor, and H.P. Weber, Science **281**, 11a (1998); D. Andrault, G. Fiquet, M. Kunz, F. Visocekas, and D. Haüserman, Science **281**, 11a (1998).
2. D. Bancroft, E.L. Peterson, and S. Minshall, J. Appl. Phys. **27**, 291 (1956).
3. F.M. Wang and R. Ingalls, Phys. Rev. B **57**, 5647 (1998).
4. M. Ekman, B. Sadigh, K. Einarsdotter, and P. Blaha, Phys. Rev. B **58**, 5296 (1998).
5. W.G. Burgers, Physica **1**,561 (1934).
6. Y. Chen, K.M. Ho, and B.N. Harmon, Phys. Rev. B **37**, 283 (1988).
7. V. Paidar, L.G. Wang, M. Šob, and V. Vitek, Modelling and Simulation in Mat. Sci. Eng., to be published.
8. P. Craievich, J.M. Sanchez, R.E. Watson, and M. Weinert, Phys. Rev. B **55**, 787 (1997).
9. P. Craievich, M. Weinert, J.M. Sanchez, and R.E. Watson, Phys. Rev. Lett. **72**, 3076 (1994).
10. M. Šob, L.G. Wang, and V. Vitek, Comp. Mat. Sci. **8**, 100 (1997).
11. O.K. Andersen, M. Methfessel, C.O. Rodriguez, P. Blöchl, and H.M. Polatoglou, in *Atomistic Simulations of Materials: Beyond Pair Potentials*, edited by V. Vitek and D.J. Srolovitz (Plenum, New York-London, 1989), p. 1.
12. P. Blaha, K. Schwarz, P. Dufek, and R. Augustyn, WIEN95, Technical University of Vienna 1995 (improved and updated Unix version of the original copyrighted WIEN-code, which was published by P. Blaha, K. Schwarz, P. Sorantin, and S.B. Trickey, Comput. Phys. Commun. **59**, 399 (1990).
13. J.P. Perdew, J.A. Chevary, S.H. Vosko, K.A. Jackson, M.R. Pederson, D.J. Singh, and C. Fiolhais, Phys. Rev. B **46**, 6671 (1992).
14. M. Šob, M. Friák, L.G. Wang, and V. Vitek, to be published.

BOND-ORDER POTENTIALS FOR MOLYBDENUM AND NIOBIUM: AN ASSESSMENT OF THEIR QUALITY

M. MROVEC[1], V. VITEK[1], D. NGUYEN-MANH[2],
D. G. PETTIFOR[2], L. G. WANG[3] and M. SOB[3]

[1]Department of Materials Science and Engineering, University of Pennsylvania, Philadelphia, PA 19104-6272, U. S. A.; [2]Department of Materials, University of Oxford, Parks Road, Oxford OX1 3PH, U. K.; [3]Institute of Physics of Materials, Academy of Sciences of the Czech Republic, Zizkova 22, Brno, Czech Republic.

ABSTRACT

The bond-order potentials (BOP) have been constructed for Mo and Nb. These potentials are based on the real-space parametrized tight-binding method in which diagonalization of the Hamiltonian is avoided by direct calculation of the bond-order. In this scheme the energy consists of three parts: The bond part that comprises contributions of d electrons and introduces into the scheme the covalent character of bonding, the central-force many-body part that reflects the environmental dependence of sp overlap repulsion and a pair-wise contribution. The potentials were tested by calculation of energy differences between the bcc and several alternate structures and by investigating the trigonal deformation path. These calculations have been made in parallel using BOP and the full-potential linearized augmented plane-wave method. The central-force many-body Finnis-Sinclair type potentials have also been included into the study of the deformation path. This evaluation of BOP reveals that the potentials reproduce very closely the ab initio results and are, therefore, very suitable for atomistic studies of extended defects in the transition metals.

INTRODUCTION

Extended crystal defects, in particular dislocations, grain boundaries and other interfaces, control the mechanical properties of materials. It has been established since the early seventies that the plastic deformation of body-centered-cubic (bcc) metals, which include a number of transition elements, is dominated by the special features of the cores of 1/2<111> screw dislocations, namely their non-planar spreading (see reviews [1-5]). Clearly, the structure of dislocation cores is an atomic level phenomenon and cannot be analyzed by standard methods of the dislocation theory [6]. For this reason theoretical investigations of the dislocation cores have been performed using atomistic computer modeling (see, for example [3, 7]). The essential precursor of such studies is a description of atomic interactions. Over the period of the last thirty years calculations were made using pair-potentials [8, 9], many-body central-force potentials of the embedded atom type [5] and, recently, interatomic potentials derived from first-principles generalized pseudopotential theory (MGPT potentials) [10-12]. While the spreading of the core into three {110} planes was found in all these studies, features distinguishing specific materials depend sensitively on the description of atomic interactions. For example, the core of the screw dislocation in tantalum was found to possess six-fold symmetry when using the central-force potentials [5] but three-fold symmetry when using MGPT potentials [12]. On the other hand the three-fold symmetry was always found in the case of molybdenum [10, 5].

Similarly, in a study of the $\Sigma 5$ (310) grain boundary in niobium [13] the calculations employing central-force many body potentials predict different relative displacement of grains than potentials based on the generalized pseudopotential theory [14]. The latter are in a better agreement with electron microscope observations. At the same time, in a recent ab initio study the relative displacement of the grains for the same boundary in molybdenum [15] was found to be similar, though not identical, to that detected in calculations employing central-force potentials.

An important distinction of both the ab initio calculations and MGPT potentials is that they include possible covalent character of bonding that is neglected in the central-force schemes. It is, of course, well known that in transition metals the relative stability of alternate crystal structures is controlled by the level of the filling of the d-band [16, 17] and the bonding mediated by the d-electrons has a covalent character. Consequently, this covalent component of bonding may also play a significant role in determination of low energy structures of extended lattice defects.

In this paper we present potentials for two transition metals, niobium and molybdenum,

529

which include the covalent character of d-electron bonding and can be relatively easily used in studies of extended defects since all the calculations are done in the real space. They are the Bond-order Potentials (BOP) [18-23] based on a reformulation of the tight-binding method with the two-center d-d hopping integrals. The quality of these potentials is assessed by calculating structural energies of alternate crystal structures and energies of distorted structures obtained on a trigonal transformation path that connects bcc, face-centered-cubic (fcc) and simple cubic (sc) lattices. These energies are compared with those calculated ab initio using a full-potential linearized augmented plane wave (FLAPW) code described in detail in [24, 25]. At the same time, the transformation path was also studied using the central-force many body potentials of the Finnis-Sinclair type [26], constructed in [27]. The calculations employing BOP are in very good agreement with ab initio results and in the case of the trigonal path reproduce some subtle features that the central-force potentials cannot capture.

CONSTRUCTION OF BOND-ORDER POTENTIALS

Bond-order potentials are based on the parametrized tight-binding method in real space. The computationally demanding diagonalization of the Hamiltonian matrix is avoided by direct calculation of the density matrix and/or bond-order and the local charge neutrality is imposed as an ersatz for self-consistency. In this scheme the cohesive energy is composed of three terms as follows:

$$E_{coh} = E_{bond} + E_{pair} + E_{env} \tag{1}$$

The bond energy $E_{bond} = 2 \sum_{i\alpha \neq j\beta} \Theta_{j\beta,i\alpha} H_{i\alpha,j\beta}$ with $\Theta_{j\beta,i\alpha}$ and $H_{i\alpha,j\beta}$ the bond-order and the hopping integrals, respectively, each associated with atoms i and j and orbitals α and β. The details of evaluation of the bond-order via Green's functions represented as continued fractions, can be found in Refs. [21, 22, 28]. Since d-d bonding provides the main contribution to cohesion in bcc transition metals, only d electrons are included explicitly into E_{bond}. The spherical part of the hopping integrals has been determined by Slater and Koster [30] and the radial part in this study is taken as an inverse power

$$dd\alpha(R) = dd\alpha_0 \left(\frac{R_0}{R}\right)^n \tag{2}$$

for $R < R_1$ and for $R_1 < R < R_{cut}$ as a polynomial spline smoothly approaching zero at R_{cut}, close to the third nearest neighbor spacing, similarly as in [28]; R_0 is the nearest neighbor spacing and $\alpha = \sigma, \pi$ or δ. The ratios $dd\sigma : dd\pi : dd\delta$ have been fixed by the screened LMTO band theory [29] which provides the correct shape for the density of states and they are -17.926:8.249:-1.0 for Mo and -18.295:8.399:-1.0 for Nb, respectively. Other parameters entering the bond energy are the number of electrons in the d-band (4.4 for Mo and 3.3 for Nb), the number of recursion levels in the continued fractions used to evaluate the Green's functions (4 in this study) and the effective electronic temperature introduced in order to dampen down the very long-range Friedel oscillations that can lead to slow convergence of the real-space bond-order expansions [21, 22, 28].

The environmental part of the energy, E_{env}, has been introduced only recently in the context of possible negative Cauchy pressures [31]. It represents the overlap repulsion that arises from the valence sp electrons. It is described by a central-force many body potential so that

$$E_{env} = \frac{1}{2} \sum_{i \neq j} B_{ij} \frac{\exp[-\frac{1}{2}(\lambda_i + \lambda_j)(r_{ij} - 2r_c)]}{r_{ij}} \tag{3a}$$

where

$$\lambda_i = \lambda_\infty + \left[\sum_{k \neq i} C_{ik} \exp(-v_{ik} r_{ik})\right]^{1/\gamma} \tag{3b}$$

Here r_{ij} is the separation of atoms i and j and λ_∞, B_{ij}, C_{ik}, r_c v_{ik} and γ are all adjustable parameters; in monoatomic solids, such as those analyzed here, B_{ij}, C_{ik} and v_{ik} are constants independent of the subscripts.

The third part of the energy is given as a sum of pair potentials

$$E_{pair} = \frac{1}{2} \sum_{i \neq j} \Phi(r_{ij})$$ (4)

where $\Phi(r_{ij})$ is represented functionally as in [28] and contains four adjustable parameters.

It is important to emphasize that the parameters entering the bond, environmental and pairwise parts were determined sequentially: First the parameters entering the bond and environmental parts of the energy were established using a knowledge of the electronic structure, by fitting the Cauchy pressure $c_{12} - c_{44}$ and assuring stability of the phonon corresponding to the N point in the Brillouin zone. The pair potential was then constructed to reproduce exactly the equilibrium lattice parameter, the cohesive energy and the two remaining elastic moduli. The details of the fitting and the values of the corresponding parameters will be presented in detail elsewhere and they can be obtained from the first author at request.

TESTING OF BOND-ORDER POTENTIALS

The first and most important requirement is that the fitted equilibrium bcc structure is stable relative to possible alternate crystal structures. This has been tested by calculating the energies for the fcc, hexagonal-close packed (hcp) and A15 with the same density as the bcc lattice. The energy differences between these structures and the equilibrium bcc structure are summarized in the table. Furthermore, these energy differences were also evaluated using ab initio FLAPW code [24, 25]. Since all the energy differences are positive the bcc lattice is most stable, as required. Moreover, the agreement between the BOP and ab initio calculated energy differences is very good. An exception is the A15 structure in the case of Nb. Its energy evaluated in the BOP scheme is about twice as high as the ab initio value. This is caused by too strong contribution from the environmental term but this discrepancy could be improved by further adjustments of the fitting parameters in this term.

	STRUCTURE	BOP (eV/atom)	FLAPW(eV/atom)
Mo	A15-BCC	0.183	0.218
	FCC-BCC	0.401	0.409
	HCP-BCC	0.455	0.543
Nb	A15-BCC	0.301	0.164
	FCC-BCC	0.352	0.336
	HCP-BCC	0.357	0.354

Comparison of energy differences between the equilibrium bcc and alternate structures

Since the potentials are intended for atomistic modeling of extended defects they have to be applicable when the atomic environment is considerably different from that in the ideal lattice. While this can never be fully tested since the variety of environments encountered in such defects as interfaces, dislocations etc. is very rich, an assessment can be made by investigating the highly distorted structures encountered along certain deformation paths (see, for example, [32-34]). The most common path of this type is the Bain deformation path between the bcc and fcc lattices corresponding to the continuous variation of the c/a ratio, where a is the lattice parameter in the [100] and [010] directions and c in the [001] direction. However, in the present study we investigate a more complex trigonal path which is also a deformation path between the bcc and fcc lattices but passing through the sc structure.

Starting from the bcc lattice, the trigonal path concurs with the homogeneous deformation corresponding to the extension along the [111] axis while keeping the atomic volume constant. In the coordinate system with the x, y and z axes parallel to [1$\bar{1}$0], [11$\bar{2}$] and [111] directions it is described by the Lagrange strain tensor for large deformations

$$\varepsilon_{11} = \varepsilon_{22} = (p^{-2/3} - 1)/2$$

$$\varepsilon_{33} = (p^{4/3} - 1)/2$$ (5)

$$\varepsilon_{12} = \varepsilon_{13} = \varepsilon_{23} = 0$$

where p is a parameter characterizing the deformation; it varies from 1 for bcc, through 2 for sc to 4 for fcc lattices. The magnitude of the vectors parallel to $[1\bar{1}0]$ and $[11\bar{2}]$ directions is proportional to $p^{-1/3}$ while the magnitude of the vectors parallel to the [111] direction is proportional to $p^{2/3}$.

Mo

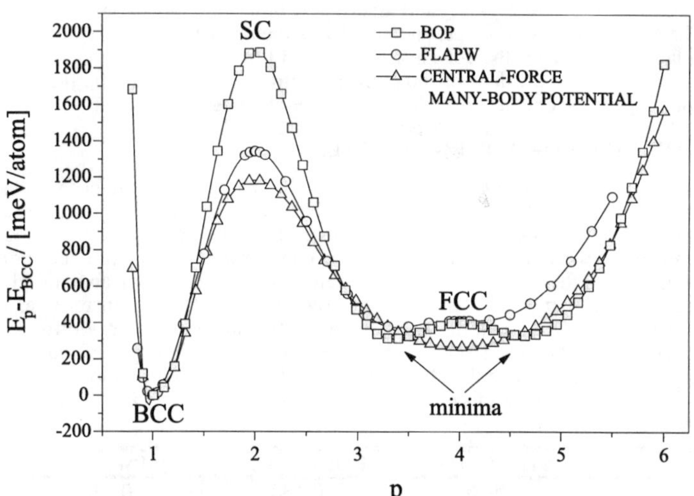

Fig. 1. Trigonal deformation path for molybdenum.

Nb

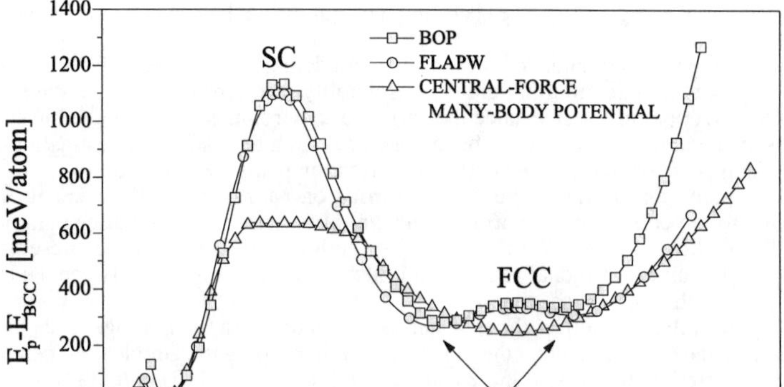

Fig. 2. Trigonal deformation path for niobium.

The dependence of the energy per atom, relative to the energy for the bcc lattice, on the parameter p is shown in Figs. 1 and 2 for Mo and Nb, respectively. These calculations have been made using the constructed BOP potentials, ab initio FLAPW code [24, 25] and also Finnis-Sinclair type central-force many body potentials constructed in [27]. It should be noted that the extrema for p = 1, 2 and 4, corresponding to different cubic structures, are dictated by the symmetry. However, auxiliary minima found in the vicinity of p = 4 are specific to the material and depend on the interatomic forces. These minima, predicted by ab initio calculations, are well reproduced by BOP but not by the central-force potentials. In fact, the latter lead to a local minimum for the fcc structure while both ab initio and BOP calculations suggest that the fcc structure corresponds to a local maximum with subsidiary minima on each side of this maximum. Clearly, this feature is rather subtle since only small energy differences are involved and whether it originates in the case of BOP in directional bonding or many-body environmental term merits further investigation.

In other parts of the calculated energy vs p dependence the agreement between all three descriptions of atomic interactions is very good for molybdenum although the energy of the sc structure is somewhat overestimated by BOP. In the case of niobium BOP and ab initio calculation also agree very closely but calculations using central-force potentials display significant deviations in the vicinity of p = 2 and for p > 5. This suggests that central-force potentials are less reliable for Nb than Mo which acquiesces with the above mentioned studies of dislocations and grain boundaries.

CONCLUSIONS

The bond-order potentials for Mo and Nb discussed in this study are eminently suitable for atomistic studies of extended crystal defects. First, they reproduces exactly the equilibrium lattice parameters, cohesive energies, and elastic moduli and guarantee the stability of the bcc lattice with respect to most likely alternate structures, A15, fcc and hcp. Moreover, the differences in the cohesive energies between the bcc and the alternate structures are in an excellent agreement with ab initio calculations. While these aspects of the atomic interactions can often be attained in the framework of central-force potentials, the bond-order potentials reflect correctly the important feature of cohesion in bcc transition metals, the covalent character of bonding, arising due to partially filled d-band, as well as the many-body character of the overlap repulsion resulting from the valence sp electrons. The importance of these features of bonding is reflected in the ability of bond-order potentials to reproduce not only salient but also rather subtle features of the variation of the energy along the trigonal deformation path when the deviations of the structure away from the ideal bcc lattice are very substantial.

ACKNOWLEDGMENTS

This research was supported in part by the Advanced Strategic Computing Initiative of the U.S. Department of Energy through LLNL, grant no. B331542 (MM and VV) and by the NSF - International Programs Grant no. INT-96-05232 and the Grant Agency of the Academy of Sciences of the Czech Republic, project no. A1010817 (MS). Screened LMTO calculations were performed in Oxford on the HP EXEMPLER computer that was funded by Higher Education Funding Council of England and by Hewlett-Packard.

REFERENCES

1. V. Vitek, *Crystal Lattice Defects* **5**, 1 (1975).
2. M. S. Duesbery, *Dislocations in Solids*, edited by F. R. N. Nabarro (Amsterdam, North Holland), Vol. 8, p. 67 (1989).
3. M. S. Duesbery and G. Y. Richardson, *CRC Critical Reviews in Solid State and Materials Science* **17**, 1 (1991).
4. V. Vitek, *Prog. Mater. Sci.* **36**, 1 (1992).
5. M. S. Duesbery and V. Vitek, *Acta Mater.* **46**, 1481 (1998).
6. J. P. Hirth and J. Lothe, *Theory of Dislocations*, (Wiley-Interscience, New York, 1982).
7. V. Vitek, *Stability of Materials: NATO Advanced Science Institute*, edited by A. Gonis, P. E. A. Turchi and J. Kudrnovsky (New York, Plenum Press), p. 53 (1996).

8. V. Vitek, R. C. Perrin and D. K. Bowen, *Philos. Mag. A* **21**, 1049 (1970).
9. Z. S. Basinski, M. S. Duesberry and R. Taylor, *Can. J. Phys.* **49**, 2160 (1971).
10. W. Xu and J. A. Moriarty, *Phys. Rev. B* **54**, 6941 (1996).
11. W. Xu and J. A. Moriarty, *Comp. Mat. Sci.* **9**, 348 (1998).
12. L. H. Yang, W. Xu and J. A. Moriarty, *Multiscale Modeling of Materials*, edited by V. Bulatov, N. Ghoniem, T. Diaz de la Rubia and T. Kaxiras (Pittsburgh, Materials Research Society), this volume (1999).
13. G. H. Campbell, W. E. King, S. M. Foiles, P. Gumbsch and M. Ruhle, *Structure and Properties of Interfaces in Materials*, edited by W. A. T. Clark, U. Dahmen and C. L. Briant (Pittsburgh, Materials Research Society), Vol. 238, p. 163 (1992).
14. J. A. Moriarty, *Many-Atom Interactions in Solids*, edited by R. M. Nieminen, M. J. Puska and M. J. Manninen (Berlin, Springer), Vol. 48, p. 158 (1990).
15. T. Ochs, O. Beck, C. Elsaesser and B. Meyer, *Philos. Mag. A,* to be published (1999).
16. D. G. Pettifor, *Physical Metallurgy*, edited by R. W. Cahn and P. Haasen (Amsterdam, Elsevier), p. 147 (1983).
17. D. G. Pettifor, *Bonding and Structure of Molecules and Solids*, (Oxford University Press, Oxford, 1995).
18. D. G. Pettifor, *Phys. Rev. Lett.* **63**, 2480 (1989).
19. M. Aoki, *Phys. Rev. Lett.* **71**, 3842 (1993).
20. M. Aoki and D. G. Pettifor, *Mat. Sci. Eng. A* **176**, 19 (1994).
21. A. P. Horsfield, A. M. Bratkovsky, M. Fearn, D. G. Pettifor and M. Aoki, *Phys. Rev. B* **53**, 12694 (1996).
22. A. P. Horsfield, A. M. Bratkovsky, D. G. Pettifor and M. Aoki, *Phys. Rev. B* **53**, 1656 (1996).
23. D. R. Bowler, M. Aoki, C. M. Goringe, A. P. Horsfield and D. G. Pettifor, *Modelling and Simulation in Mat. Sci. Eng.* **5**, 199 (1997).
24. P. Blaha, K. Schwartz, P. Sorantin and S. B. Trickey, *Comp. Phys. Commun.* **59**, 399 (1990).
25. P. Blaha, K. Schwartz, P. Dufek and R. Augustyn, *Wien 95*, (Technical University of Vienna, 1995).
26. M. W. Finnis and J. E. Sinclair, *Philos. Mag. A* **50**, 45 (1984).
27. G. J. Ackland and R. Thetford, *Philos. Mag. A* **56**, 15 (1987).
28. A. Girshick, A. M. Bratkovsky, D. G. Pettifor and V. Vitek, *Philos. Mag. A* **77**, 981 (1998).
29. O. K. Andersen, O. Jepsen and D. Glötzel, *Highlights of Condensed Matter Theory*, edited by F. Bassani, F. Fumi and M. P. Tosi (Amsterdam, North Holland), p. 59 (1985).
30. J. C. Slater and G. F. Koster, *Physical Review* **94**, 1498 (1954).
31. D. Nguyen-Manh, D. G. Pettifor, S. Znam and V. Vitek, *Tight-Binding Approach to Computational Materials Science*, edited by P. E. A. Turchi, A. Gonis and L. Colombo (Pittsburgh, Materials Research Society), Vol. 491, p. 353 (1998).
32. F. Milstein, H. E. Fang and J. Marschall, *Philos. Mag. A* **70**, 621 (1994).
33. M. Sob, L. G. Wang and V. Vitek, *Comp. Mat. Sci.* **8**, 100 (1997).
34. M. Sob, I. Turek and V. Vitek, *Tight-Binding Approach to Computational Materials Science*, edited by P. E. A. Turchi, A. Gonis and L. Colombo (Pittsburgh, Materials Research Society), Vol. 491, p. 79 (1998).

INTERATOMIC POTENTIALS FOR Al AND Ni FROM EXPERIMENTAL DATA AND AB INITIO CALCULATIONS

Y. MISHIN*, D. FARKAS*, M. J. MEHL** and D. A. PAPACONSTANTOPOULOS**
*Department of Materials Science and Engineering, Virginia Polytechnic Institute and State University, Blacksburg, VA 24061-0237; ** Complex Systems Theory Branch, Naval Research Laboratory, Washington, DC 20375-5345

ABSTRACT

New embedded-atom potentials for Al and Ni have been developed by fitting to both experimental data and the results of *ab initio* calculations. The *ab initio* data were obtained in the form of energies of different alternative computer-generated crystalline structures of these metals. The potentials accurately reproduce basic equilibrium properties of Al and Ni such as the elastic constants, phonon dispersion curves, vacancy formation and migration energies, stacking fault energies, and surface energies. The equilibrium energies of various alternative structures not included in the fitting database are calculated with these potentials. The results are compared with predictions of total-energy tight-binding calculations for the same structures. The embedded-atom potentials correctly reproduce the structural stability trends, which suggests that they are transferable to different local environments encountered in atomistic simulations of lattice defects.

INTRODUCTION

In our previous paper [1] we proposed a procedure for developing interatomic potentials by fitting to a large database including both experimental data and structural energies generated by *ab initio* calculations. The underlying idea is that the incorporation of *ab initio* structural energies can improve the accuracy and transferability of potentials due to sampling regions of configuration space that are not accessible experimentally. Such potentials can be more reliable in representing the energies of different atomic configurations probed in atomistic simulations of lattice defects, plastic deformation, fracture or atomic diffusion. This approach may help to bridge the gap between *ab initio* and empirical methods in materials simulations.

The embedded-atom method (EAM) potentials for Al and Ni developed in Ref. [1] were based on cubic splines. In this work we apply the same approach, but develop EAM potentials for these metals using analytical functions. The new potential functions are, thus, smooth together with their higher derivatives, which makes the potentials more suitable for quasi-harmonic calculations of thermal expansion, thermodynamic functions, attempt frequencies of diffusive jumps and other characteristics. Moreover, the smoothness of the second derivatives of potential functions (which is not satisfied by cubic splines) is essential in some advanced simulation techniques, such as hyperdynamics [2]. For Al, we have a additional opportunity of testing the potential by comparing the EAM-predicted structural energies with the results of total-energy tight-binding (TB) calculations using the recently proposed TB parameterization [3]. It should be mentioned that EAM potentials for Al and Ni were proposed by many previous authors (see Baskes [4] for an overview; see also Ercolessi and Adams [5] (Al), Rohrer [6] (Al) and Baskes *et al.* [7](Ni) for more recent potentials).

PARAMETERIZATION AND FITTING

In the EAM formalism[8], the total energy of a monoatomic system is represented as

$$E_{\text{tot}} = \frac{1}{2} \sum_{ij} V(r_{ij}) + \sum_{i} F(\overline{\rho}_i). \tag{1}$$

Mat. Res. Soc. Symp. Proc. Vol. 538 © 1999 Materials Research Society

Here $V(r_{ij})$ is a pair potential as a function of the distance r_{ij} between atoms i and j, and F is the embedding energy as a function of the host electron density $\bar{\rho}_i$ induced at site i by all other atoms in the system. The latter is given by

$$\bar{\rho}_i = \sum_{j \neq i} \rho(r_{ij}), \tag{2}$$

$\rho(r)$ being the electton density function. In this work, the potentials were generated directly in the effective pair format [9]. The pair interaction function was represented as

$$V(r) = \left[E_1 M\left(r, r_0^{(1)}, \alpha_1\right) + E_2 M\left(r, r_0^{(2)}, \alpha_2\right) + \delta \right] \times \psi\left(\frac{r - r_c}{h}\right), \tag{3}$$

where

$$M(r, r_0, \alpha) = \exp\left(-2\alpha(r - r_0)\right) - 2\exp\left(-\alpha(r - r_0)\right) \tag{4}$$

is a Morse function, $\psi(x) = 0$ if $x \geq 0$ and $\psi(x) = x^4/(1 + x^4)$ if $x < 0$. The electron density function was fitted in the form

$$\rho(r) = \left[a \exp\left(-\beta_1(r - r_0^{(3)})^2\right) + \exp\left(-\beta_2(r - r_0^{(4)})\right) \right] \times \psi\left(\frac{r - r_c}{h}\right). \tag{5}$$

The cutoff function $\psi(x)$ guarantees that functions $V(r)$, $\rho(r)$ and their derivatives up the third one turn to zero at the cutoff distance r_c. Function $\rho(r)$ was normalized to $\bar{\rho} = 1$ in the equilibrium FCC crystal,

$$\bar{\rho} = \sum_m N_m \rho_m = 1. \tag{6}$$

Also, function $V(r)$ was required to satisfy the mechanical equilibrium condition at the experimental lattice period a_0,

$$\sum_m N_m R_m V_m' = 0. \tag{7}$$

In Eqs. (6) and (7), $V_m \equiv V(R_m)$ and $\rho_m \equiv \rho(R_m)$, where R_m and N_m are the redius and the number of atoms at mth coordination shell. In Eqs. (6) and (7) can be easily solved for a and E_1, respectively, which reduces the number of free fitting parameters by two. The embedding function was represented by a polynomial

$$F(\bar{\rho}) = F^{(0)} + \frac{1}{2} F^{(2)} (\bar{\rho} - 1)^2 + \sum_{n=1}^{k} q_n (\bar{\rho} - 1)^{n+2}. \tag{8}$$

The coefficients $F^{(0)}$ and $F^{(2)}$ can be expressed in terms of the experimental values of the cohesive energy E_0 and bulk modulus B [1],

$$F^{(0)} = E_0 - \frac{1}{\sum_m} N_m V_m, \tag{9}$$

$$\frac{1}{2} \sum_m N_m V_m'' R_m^2 + F^{(2)} \left(\sum_m N_m \rho_m' R_m\right)^2 = 9B\Omega_0, \tag{10}$$

Ω_0 being the equilibrium atomic volume. Furthermore, one of the coefficients q_i, say q_1, can be determined from the condition $F(0) = 0$. Then, only coefficients q_i with $i \geq 2$ can be used as free fitting parameters. Overall, this parameterization includes $11 + k$ free parameters. Note that this scheme guarantees an exact fit to a_0, E_0 and B.

Table 1: Properties of Al and Ni predicted by the present potentials (EAM) and the spline-fit potentials (SF-EAM) [1] in comparison with experimental data. The properties marked by an asterisk were fitted to when generating the present potentials, all other properties are listed for comparison. The literature references to the experimental data can be found in Ref. [1]. Notations: a_0, equilibrium lattice period; E_0, cohesive energy; B, bulk modulus; c_{ij}, elastic constants; $\nu_{L,T}(P)$, longitudinal (L) or transversal (T) phonon frequency at a high symmetry point P; E_v^f, vacancy formation energy; E_v^m, vacancy migration energy; γ_{SF}, intrinsic stacking fault energy; γ_{us}, unstable stacking fault energy; γ_T, symmetrical twin energy; $\gamma_s(hkl)$, surface energy for (hkl) orientation.

	Al			Ni		
	Experiment	SF-EAM	EAM	Experiment	SF-EAM	EAM
Lattice properties:						
a_0 (Å)*	4.05	4.05	4.05	3.52	3.52	3.52
E_0 (eV/atom)*	−3.36	−3.36	−3.36	−4.45	−4.45	−4.45
B (10^{11} Pa)*	0.79	0.79	0.79	1.81	1.81	1.81
c_{11} (10^{11} Pa)*	1.14	1.14	1.13	2.47	2.47	2.47
c_{12} (10^{11} Pa)*	0.619	0.616	0.620	1.47	1.48	1.48
c_{44} (10^{11} Pa)*	0.316	0.316	0.314	1.25	1.25	1.26
Phonon frequencies:						
$\nu_L(X)$ (THz)*	9.69	9.31	9.48	8.55	8.71	8.71
$\nu_T(X)$ (THz)*	5.80	5.98	5.84	6.27	6.38	6.05
$\nu_L(L)$ (THz)	9.69	9.64	9.20	8.88	8.53	8.60
$\nu_T(L)$ (THz)	4.19	4.30	3.90	4.24	4.31	4.03
$\nu_L(K)$ (THz)	7.59	7.30	7.63	7.30	6.98	7.53
$\nu_{T_1}(K)$ (THz)	5.64	5.42	5.45	5.78	5.68	5.62
$\nu_{T_2}(K)$ (THz)	8.65	8.28	8.71	7.93	8.04	8.22
Vacancy:						
E_v^f (eV)*	0.68	0.68	0.69	1.60	1.60	1.59
E_v^m (eV)*	0.65	0.64	0.61	1.30	1.29	1.16
Planar faults:						
γ_{SF} (mJ/m^2)*	120-166	146	145	125	125	124
γ_{us} (mJ/m^2)		168	152		366	176
γ_T (mJ/m^2)	75	76	77	43	63	65
Surfaces:						
$\gamma_s(110)$ (mJ/m^2)	980[a]	1006	820	2280[a]	2049	2212
$\gamma_s(100)$ (mJ/m^2)	980[a]	943	689	2280[a]	1878	2002
$\gamma_s(111)$ (mJ/m^2)	980[a]	870	586	2280[a]	1629	1771

[a]For average orientation, Ref. [10, 11].

The experimental properties included in the fitting database are marked by an asterisk in Table 1. Additionally, the crystal energy was required to follow the universal equation of state of Rose *et al.* [12] around the equilibrium as accurately as possible. The fitting database also included *ab initio* energies of HCP, BCC and simple cubic (SC) structures relative to the FCC energy. Such energies were obtained by the Linearized Augmented Plane Wave (LAPW) method, including all electrons and allowing for a general potential (see [1] for more details). Following the scaling procedure introduced in Ref. [1], the structures calculated with the EAM potentials were taken with the experimental first neighbour distance in the FCC phase, while the structures calculated by the LAPW method were taken with the equilibrium first neighbour distance in the FCC phase determined by LAPW calculations. The LAPW energies involved in the fitting process were reported in Ref. [1] and will not be repeated here.

RESULTS AND DISCUSSION

The best parameterization was found with $k = 2$ for Al and $k = 3$ for Ni. The optimized fitting parameters are listed in Table 2. Although we used less fitting parameters and the parameterization scheme was not as flexible as that based on cubic splines [1], the new potentials reproduce the basic experimental properties with almost the same accuracy (Table 1). However, the vacancy migration energies are slightly underestimated. The phonon dispersion curves are reproduced quite accurately. The phonon spectra, calculated in the harmonic approximation, are also in good agreement with experimental spectra [13]. The linear thermal expansion was calculated by zero-pressure Monte Carlo simulations using these potentials. The results are in reasonable agreement with experimental data (see Figure below).

In Table 3 we summarize the equilibrium cohesive energies of different crystalline structures of Al and Ni. The TB data for (non-magnetic) Ni were published previously [14], while the TB data for Al were calculated in this work using the TB parameters of Ref. [3]. The non-cubic structures were also relaxed with respect to the c/a ratio. To facilitate the comparison, all TB energies were shifted in a way that the equilibrium TB energy of the FCC structure would coincide with its experimental cohesive energy. The structures considered here include A5 (β-Sn), A9 (graphite), A12 (α-Mn), A13 (β-Mn), A15 (β-W), C32 (ideal omega phase), L1$_2$ (FCC phase with 1 vacancy per cubic unit cell) and D0$_3$ (Fe$_3$Al where Fe sites are occupied by Ni while Al sites are vacant). The agreement between the EAM and TB energies is very good for more compact structures, but becomes poorer as we go to more open structures. The EAM-predictions are the least accurate for strongly anisotropic (A9) and vacancy-containing (L1$_2$ and D0$_3$) structures. The discrepancies are probably due to the intrinsic limitations of the EAM model, especially in comparison with the TB scheme. However, both sets of EAM potentials correctly represent the relatively high stability of the "exotic" structures A12, A13 and A15. The energy of the ideal omega phase in Al is also reproduced very accurately.

Acknowledgements. This work was supported by the National Science Foundation (grant number DMR-9753243), Office of Naval Research and U.S. Department of Defense.

References

[1] Y. Mishin, D. Farkas, M. J. Mehl and D. A. Papaconstantopoulos, *Phys. Rev.* B (1999), in press.

[2] A. F. Voter, *J. Chem. Phys.* **106**, 4665 (1997).

Table 2: Parameters describing the EAM potential functions for Al and Ni.

Parameter	Al	Ni
r_c (Å)	6.78040	5.31116
h (Å)	1.41584	0.50000
E_1 (eV)	2.65219×10^2	0.17505
E_2 (eV)	7.67238×10^{-3}	4.99515×10^{-5}
$r_0^{(1)}$ (Å)	1.06880	2.47780
$r_0^{(2)}$ (Å)	6.45771	5.08677
α_1 (Å^{-1})	2.09043	2.67763
α_1 (Å^{-1})	1.03062	0.75002
δ (Å)	0.10432	2.15076×10^{-2}
a	5.47491×10^{-2}	7.84142×10^{-2}
$r_0^{(3)}$ (Å)	2.74668	2.52195
$r_0^{(4)}$ (Å)	-6.92647	-31.11899
β_1 (Å^{-2})	1.92116	2.06662
β_2 (Å^{-1})	0.42589	24.12636
$F^{(0)}$ (eV)	-2.81468	-3.778129
$F^{(2)}$ (eV)	5.57686	34.45341
q_1 (eV)	-6.24655	68.19974
q_2 (eV)	-21.52522	1.49629×10^2
q_3 (eV)	-15.30493	1.65259×10^2
q_4 (eV)	0.00000	0.70382×10^2

Table 3: Cohesive energies (in eV) of different crystalline structures of Al and Ni as predicted by the present potentials (EAM), the splite-fit potentials (SF-EAM) [1], and by total-energy tight-binding calculations (TB).

Al				Ni			
Structure	TB	SF-EAM	EAM	Structure	TB	SF-EAM	EAM
FCC	-3.36	-3.36	-3.36	FCC	-4.45	-4.45	-4.45
HCP	-3.31	-3.33	-3.32	HCP	-4.41	-4.43	-4.42
BCC	-3.25	-3.25	-3.26	BCC	-4.34	-4.34	-4.36
A15	-3.27	-3.35	-3.29	A12	-4.35	-4.31	-4.32
C32	-3.24	-3.24	-3.28	A13	-4.39	-4.34	-4.37
A5	-3.05	-3.04	-3.14	A15	-4.25	-4.30	-4.33
A9	-2.80	-2.64	-2.42	L1$_2$	-3.79	-3.80	-3.63
L1$_2$	-3.21	-3.06	-2.94	D0$_3$	-3.67	-3.74	-3.54
SC	-2.96	-2.96	-2.94	SC	-3.42	-3.62	-3.57
Diamond	-2.54	-2.47	-2.10	Diamond	-2.51	-2.56	-2.47

Figure 1: Thermal linear expansion of Al and Ni calculated by the Monte Carlo method using the present EAM potential in comparison with experimental data [15]

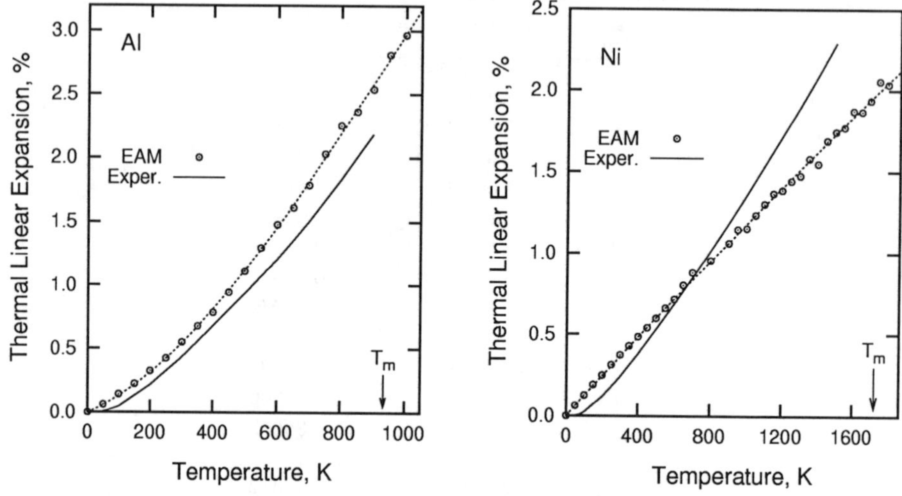

[3] S. H. Yang, M. J. Mehl and D. A. Papaconstantopoulos, *Phys. Rev.* B **57**, R2013 (1998).

[4] M. I. Baskes, *Acta Metall. Sinica* **8**, 287 (1995).

[5] F. Ercolessi and J. B. Adams, *Europhys. Lett.* **26**, 583 (1994).

[6] C. L. Rohrer, *Modelling Simul. Mater. Sci. Eng.* **2**, 119 (1994).

[7] M. I. Baskes, X. Sha, J. E. Angelo and N. R. Moody, *Modelling Simul. Mater. Sci. Eng.* **5**, 651 (1997).

[8] M. S. Daw and M. I. Baskes, *Phys. Rev.* B **29**, 6443 (1984).

[9] R. A. Johnson, *Many Atom Interactions*, volume 48 of *Proceedings in Physics*, Springer Verlag, Berlin (1990), p. 85.

[10] L. E. Murr, *Interfacial Phenomena in Metals and Alloys*, Addison-Wesley, Reading, MA (1975).

[11] W. R. Tyson and W. R. Miller, *Surf. Sci.* **62**, 267 (1977).

[12] J. H. Rose, J. R. Smith, F. Guinea and J. Ferrante, *Phys. Rev.* B **29**, 2963 (1984).

[13] *Landolt-Bornstein, New Series, Group 13a*, volume 12, Springer, Berlin (1981).

[14] M. J. Mehl and D. A. Papaconstantopoulos, *Phys. Rev.* B **54**, 4519 (1996).

[15] *Thermal Expansion*, volume 12 of *Thermophysical Properties of Matter*, Plenum, NY-Wash. (1975).

SELF CONSISTENT-CHARGE DENSITY-FUNCTIONAL TIGHT-BINDING METHOD FOR SIMULATIONS OF BIOLOGICAL MOLECULES

M. Elstner[1,2], D. Porezag[1] , G. Seifert[1] , Th. Frauenheim[1] S. Suhai[2]
[1]Universität-GH Paderborn, Fachbereich Physik, Theoretische Physik, D - 33098 Paderborn, Germany
[2] German Cancer Research Center, Department of Molecular Biophysics, D-69120 Heidelberg

Abstract

We apply a self-consistent charge tight-binding scheme to biomolecules. This method has been shown to give a reliable description of reaction energies, geometries and vibrational frequencies of small organic molecules. We discuss the performance of this method for model peptides and non-bonding interactions in biologically relevant molecular complexes. A comparison with semi-empirical methods and ab initio calculations will be given for DNA base pair H-bonding and stacking interactions.

1 Introduction

Recently, we have presented the development of a self-consistent charge tight binding scheme (SCC-TB) [1, 2]. This method is derived from density functional theory (DFT) by a second order expansion of the DFT total energy functional with respect to the charge density fluctuations $\Delta\rho$ at a given reference density ρ_0. The second order terms in the density fluctuations are efficiently approximated by a simple distribution of atom centered point charges $\Delta q_\alpha = q_\alpha - q_\alpha^0$, estimated by a Mulliken charge analysis, while all other terms maintain the standard tight-binding energy functional form. Hence, this method can be applied in a straightforeward manner to any tight binding scheme, in which the total energy is written in terms of the band structure energy $E_{bs} = \sum_i^{occ} \sum_{\mu\nu} c_\mu^i c_\nu^i H_{\mu\nu}[\rho_0]$ and a short ranged repulsive pair potential, E_{rep}, both determined at the reference desity ρ_0. The approximate DFT energy functional transforms into:

$$E_{tot} = \sum_i^{occ} \sum_{\mu\nu} c_\mu^i c_\nu^i H_{\mu\nu}[\rho_0] + E_{rep}[\rho_0] + \frac{1}{2} \sum_{\alpha\beta} \Delta q_\alpha \Delta q_\beta \gamma_{\alpha\beta}. \tag{1}$$

The third term represents the longe-range Coulomb interactions between point charges at different sites and includes the self-interaction contributions of the single atoms[2]. Making this approximate Kohn-Sham functional subject to a variational principle within a LCAO representation for the single-particle electronic states, we derive a Kohn-Sham equation for minimizing the second order energy functional which is self-consistent in the Mulliken charge distribution through the modification of the Hamitonian matrix [2]. We implemented this self-consistent charge extension into our density-functional based tight-binding method [3], where the $H_{\mu\nu}[\rho_0]$ are calculated within DFT-GGA in a two-center approximation using a minimal basis of atomic-like wavefunctions ϕ_μ.

The results for reaction energies, geometries and frequencies for small organic molecules have been presented recently [1, 2], yielding mean average deviations from experimental values comparable to full DFT calculations. Here we focus on benchmark calculations for biologically relevant molecules, H-bonding complexes, small peptides and DNA H-bonding and stacking interactions. We compare our results from the SCC-TB scheme to those obtained from semi-empirical methods like AM1 [4], PM3 [5], MNDO/M [6] and high level ab initio calculations on the DFT and MP2(MP4) level.

2 Results

2.1 H-bond complexes

A large variety of hydrogen bonding complexes have been intensively studied at different levels of theoretical models, including empirical force-field and semiempirical methods, DFT, HF and post HF methods. Here we present some of the most frequently studied complexes and focus only on their lowest energy conformations. The lowest energy conformers at the ab initio level have always been found to be also the lowest energy conformations within the SCC-TB method. All geometrical parameters of these conformations are in good agreement with the reported ab initio values.

Recently, Hadzi and Koller [7] have reviewed the performance of semiempirical methods like AM1, PM3 and MNDO/M for H-bonding compounds in comparison with ab initio results. The H-binding energies of some complexes, calculated with the present SCC-TB method are shown in Table 2.1 in comparison with PM3, MNDO/M and MP2 (MP4) results form Ref. [7] (see also references therein). We do not show the results for the AM1 method, since its performance is slightly worse than that of the PM3 method. Furthermore, AM1 often does not predict the right ground state structure, e.g. bifurcated H-bonds are favored against linear ones [7]. Consider as an example $H_3O^+ - H_2O$ and $OH^- - H_2O$, which are symmetric ionic complexes, with the proton centered between the oxygens. While SCC-TB predicts the correct structures and energies, both, AM1 and PM3 fail to reproduce these global minima, yielding assymetric structures. The performance of the MNDO/M method is very good for H-bonding interactions. However, it was parametrized with the special emphasis to model H-bonds properly, five empirically evaluated parameters have been introduced for this purpose.

For weak H-bonds, the SCC-TB underestimates H-bonding by about 2 kcal/mole. This also holds for other weak H-bonded complexes not presented here. Although the interaction energies are underestimated, the geometries and general trends predicted by the SCC-TB turn out to be very reliable.

complex	SCC-TB	PM3	MNDO/M	AI	
$H_2O - H_2O$	3.3	(3.5)	(5.5)	5.4	(3.2)
$NH_3 - NH_3$	1.8	(0.6)	(4.0)		(3.1)
$NH_3 - H_2O$	4.2	(3.0)	(4.5)		(4.2)
$HCOOH - HCOOH$	15.5	(8.4)	(14.1)	15.9	(12)
$H_3O^+ - H_2O$	31.3	$(23.6)^+$	(30.6)	33.9	
$NH_4^+ - NH_3$	23.0	(17.3)	(26.2)		(24.4))
$OH^- - H_2O$	34.6	$(26.5)^{++}$			(27-28.5)
$NH_4^+ - OH_2$	20.9	(13.5)	(20.7)	20.6	

Table 1: H-bonding energies of H-bond complexes. The interaction energies are in kcal/mole. Heats of association are given in brackets. AI denotes MP2/MP4 ab initio calculations.
[+] assymetric structure, proton is off center.
[++] not linear, most stable structure is bifurcated.

2.2 Peptides

Small model peptides, like the N-acetyl-L-alanine-N'-methylamide (NALANMA) molecule have been intensely studied at the HF, MP2 and DFT levels of theory in order to understand basic energetical and geometrical properties of polypeptides. These ab initio data also serve for benchmarking more approximate methods, like empirical force fields and semi-empirical methods.

NALANMA has six stable conformers on the B3LYP/6-31G* and MP2/6-31G* potential energy surfaces (PES) [8], which have been taken as starting structures for further geometry optimizations with the SCC-TB, AM1 and PM3 methods. The relative energies of these

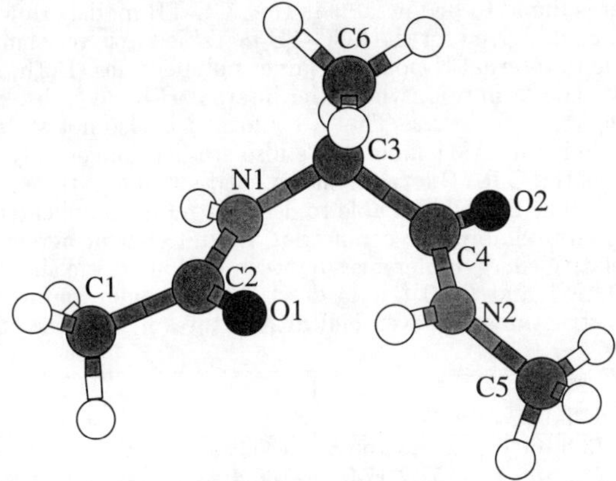

Figure 1: The C_7^{ax} conformer of NALANMA, see text.

conformers for different methods are shown in Table 2.2. The three lowest energy conformers form internal H-bonds, whereas the higher energy conformers do not. The C_7^{ax} structure is shown in Fig. 1. The C_7^{ax} and C_7^{eq} configurations differ in the orientation of the methyl group attached to the central C-atom. In the C_7^{ax} conformer, this methyl group is perpendicular to the seven membered ring, which forms the H-bond, whereas in the C_7^{eq} this methyl group is in plane. In the case of the C_7^{ax} conformer, the steric interaction of this methyl group with the hydrogen bonded ring seems to be underestimated in the SCC-TB, AM1 and PM3 methods, which results in an underestimation of its relative energy. Also the α helical structures are

conf.	DFT-GGA	MP2	HF	SCC-TB	PM3	AM1
C_7^{eq}	0.00	0.00	0.00	0.00	0.00	0.00
C_5^{ext}	1.43	1.76	0.41	1.15	-1.55*	1.72*
C_7^{ax}	2.58	2.61	2.82	1.05	0.87	0.72
β_2	3.18	3.37	2.58	2.30+	-	-
α_L	5.82	4.60	4.72	3.90	3.53	-
α_P	6.85	6.34	5.74	4.96	1.10	3.29

Table 2: Relative energies of the different conformers of NALANMA for different methods as described in the text. DFT-GGA refers to the hybrid method B3LYP. The geometries at the DFT, MP2 and HF level have been determinsd with the 6-31G* basis set.
+ The β_2 conformer is not stable within the SCC-TB, but the maximum force at the B3LYP geometry is very small. The energy is given for the geometry, when the forces are smaller than 0.00065 a.u.
* distorted structure.

predicted to be too low in energy. Since these two structures form no internal hydrogen bonds the smaller energy difference with respect to the ground state could be due to the fact that the energy of the H-bonds is underestimated in the SCC-TB method. Therefore, the stabilization of the C_7 and C_5 structures is underestimated compared to the α helical structures.

The most important geometrical parameters are the two torsion angles Φ and Ψ. The Φ angle is defined by rotations about the N1-C3 bond, $\Phi(C2 - N1 - C3 - C4)$, whereas the Ψ angle is defined by rotations of the C3-C4 bond , $\Psi(N1 - C3 - C4 - N2)$, see Fig. 1. Values of Φ and Ψ for the different conformers at the various levels of theory are given in Table 2.2.

The β_2 conformer is found to be unstable in the SCC-TB model. However, the maximum force at the dihedral angles given for the SCC-TB in Table 2.2 is very small. This conformer may be stabilized due to internal H-bonds in larger polypeptides. Both, AM1 and PM3 find a very distorted C_5^{ext} conformation, where the internal H-bond is broken, as can be seen from the dihedral angles in Table 2.2. The β_2 conformer is also not stable in both methods and the α_L is unstable in the AM1 model. We also studied longer polypeptide chains with up to eight residues at the B3LYP level of theory, and compared the results with the semi-empirical methods [9]. The SCC-TB is able to describe more complicated conformations for these longer chains quite reliably, the geometries are in excellent agreement with the DFT data, whereas the relative energy differences between conformers are slightly underestimated. In contrast to this, AM1 and PM3 fail to describe important conformations, like β-turn structures, extended structures and even helical structures in the case of PM3 [9].

	C_7^{eq} $\Phi\ \Psi$	C_7^{ax} $\Phi\ \Psi$	C_5^{ext} $\Phi\ \Psi$	β_2 $\Phi\ \Psi$	α_L $\Phi\ \Psi$	α_P $\Phi\ \Psi$
DFT	-81.9 72.3	73.8 -60.0	-157.3 165.3	-135.9 23.4	68.5 24.5	-169.4 -37.8
SCC	-83.3 68.3	74.5 -66.4	-152.2 174.6	-136.4 24.5[+]	65.1 13.1	-173.6 -51.6
AM1	-84.4 68.5	76.6 -64.0	-117.7 141.5	- -	- -	-115.5 -55.2
PM3	-71.4 77.7	68.8 -67.9	-93.9 147.9	- -	62.3 39.6	-137.6 -60.5

Table 3: Dihedral angles of the NA-LA-NMA conformers for different methods as described in the text
[+]: The β_2 conformer is not stable within the SCC-TB method. The $\Phi\ \Psi$ values refer to a configuration, where the maximum force os lower than 0.00065 a.u.

2.3 DNA base pairs: H-bonding and stacking energies

The three-dimensional structure of DNA is affected by stacking and hydrogen bonding of the DNA bases. Recently, a systematic study of the DNA base pair stacking and hydrogen bonding has been carried out at the HF and MP2 level of theory [10, 11, 12, 13, 14]. These calculations provide a useful database for benchmarking more approximate methods such as the SCC-TB method, before attempting to simulate such complex structures like the DNA double helix. Both H-bonding and stacking are rather challenging tests for approximate methods like the SCC-TB, since the contribution of the correlation energy and basis set effects appear to be very important. Hobza et al. [13] investigated the H-bonding energies of 26 base pair conformations of the bases adenine (a), cytosine (c), guanine (g) and thymine (t). Fig. 2 shows the configurations of the adenine cytosine H-bond and the guanine uracil (u) stacking complex. The geometries of all complexes were optimized at the HF level of theory, the interaction energies were calculated at the MP2/6-31/G*(0.25) level of theory with subsequent correction for the basis set superposition error (BSSE) [12]. The interaction energies for the different pairs are between 10 and 25.4 kcal/mole. A set of empirical force fields have been tested as well as the semi-empirical AM1 and PM3 methods. While most of the force field methods show a very small average error compared to the MP2 reference interaction energies of the 26 base pairs, leading to mean average errors of 0.9 to 2.4 kcal/mole, the semi-empircal methods AM1 and PM3 show much larger deviations. The mean average errors are 7.3 kcal/mole and 6.3 kcal/mole for the AM1 and PM3 methods respectively. However, the MNDO/M is in good agreement with the MP2 results with a mean average error of only 2.5 kcal/mole. In contrast to that, the interaction energies at the SCFD (ab initio HF plus dispersion energy) were found to be overestimated, leading to

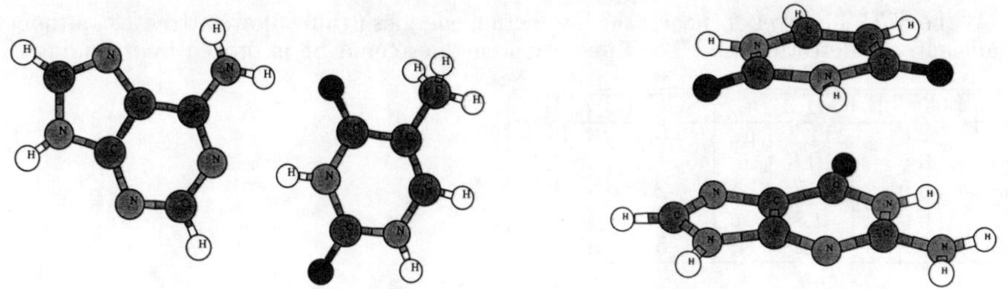

Figure 2: The H-bonded ac base pair and the gu stacked base pair

	SCC-TB	MP2	PM3		SCC-TB	MP2	PM3
gcwc	21.0	25.4	14.9	atwc	9.7	12.4	6.9
gg1	22.5	24.0	11.9	atrwc	9.5	12.4	7.1
cc	13.6	18.8	13.2	aa1	8.7	11.5	6.2
gg3	12.5	17.1	9.2	ga4	8.4	11.1	6.2
ga1	12.3	15.7	8.4	tc2	9.1	11.8	6.1
gt1	14.4	14.7	6.2	tc1	8.8	11.6	5.4
gt2	13.9	14.3	7.2	aa2	7.7	11.0	5.9
ac1	10.5	14.3	8.8	tt2	10.4	10.6	4.9
gc1	10.3	13.9	9.8	tt1	10.3	10.6	5.2
ac2	9.5	14.1	9.2	tt3	10.2	10.5	5.4
ath	9.5	13.3	6.0	ga2	7.5	10.4	5.6
ga3	10.4	15.2	7.8	gg4	6.9	10.3	5.2
atrh	9.5	13.2	7.5	aa3	6.8	10.0	4.4

Table 4: Interaction energies (kcal/mole) of H-bonding base pairs, as described in the text.

a mean average error of 5.6 kcal/mole. In the SCC-TB model, the interaction energies are underestimated, as could be expected from the results for the H-bonding energies discussed in the last section. Only those base pairs which form O..H-N bonds, i.e. the tt and gt pairs, have interaction energies which are very close to those at the MP2 level of theory. This is consistent with the H-bonding energies of the water and ammonia dimer, where the binding energy of the ammonia dimer is understimated by a larger amount than that of the water dimer. This underestimation might be due to the minimal basis, employed in the SCC-TB method. However, the relative stabilities of the base pairs are reproduced quite well, as can be seen from Table 4 (only the MP2 and PM3 results from Ref. [13] and the SCC-TB results are shown), we find a mean average error of 2.8 kcal/mole.

Base pair stacking is an even more challenging test than H-bonding, since correlation is responsible for the stacking stabilization energies to a large extent and very diffuse polarization fuctions (exponent of 0.25) are of primary importance for a proper description. Further, HF has been shown to reproduce stacking energies quite poorly and MP2 seems to overestimate correlation contributions by 15-30 % [14]. Stacking energies have been evaluated at the MP2/6-31G*(0.25) level of theory with a posteriori correction for BSSE at geometries resulting from empirical force field optimizations [11]. Empirical force fields are capable of describing the stacking interactions quite satisfactorily, although there are discreancies between the force fields themselves and with respect to the MP2 data of up to 100 % in the interaction energies [13]. The AM1 and PM3, as well as the MNDO/M method, have been shown to be unable to reproduce the attractive stacking interactions, which are erroneously found to be repulsive (2-10 kcal/mole), leading to a destabilization of the stacked base pairs [13]. This might lead to a destablization of the DNA helix and makes the use of these methods in this system questionable.

At the SCC-TB level of theory, the interaction energies (Table 5) are attractive, although significantly underestimated. The interaction energies could be improved by including the

	SCC-TB	MP2		SCC-TB	MP2
ga	2.6	11.2	gg	5.6	11.3
gu	6.5	10.6	aa	1.5	8.8
ac	3.0	9.5	cc	2.6	8.3
gc	6.0	9.3	uu	3.3	6.5
au	3.8	9.1	cu	5.3	8.5

Table 5: Attractive interaction energies (kcal/mole) for stacked base pairs at the SCC-TB and MP2 level of theory, as described in the text.

disperion energies within an empirical treatment of the van der Waals interactions on the basis of the Slater-Kirkwood approximation [15] as has been implemented by Lewis and Sankey [16].

3 Conclusion

We have presented results of the self-consistent charge density functional based tight-binding method for peptides and non-bonding interactions relevant in biological systems. Although we found an underestimation of the interaction energies, the geometrical parameters of the systems considered are described very well. Compared to the semi-empirical methods like AM1 and PM3 which exhibit qualitative failors, the SCC-TB method leads to a reliable overall description of biological systems.

References

[1] M. Elstner, D. Porezag, G. Jungnickel, T. Frauenheim, S.Suhai, G. Seifert in Tight-binding approach to Computational Materials Science, Edited by P. Turchi, A.Gonis and L. Colombo, MRS Symp. Proc. No. **491** (Materials Research Society, Pittsburgh 1998) 131

[2] M. Elstner, D. Porezag, G. Jungnickel, J. Elsner, M. Haugk, T. Frauenheim, S.Suhai, G. Seifert, Phys. Rev. **B 58** (1998) 7260

[3] D. Porezag, T. Frauenheim, T. Köhler, G. Seifert, R. Kaschner, Phys. Rev. **B 51** (1995) 12947

[4] J. S. Dewar, E. Zoebisch, E. F. Healy, J. J. P. Stewart, J. Am. Chem. Soc. **107**, 3902 (1985).

[5] J. J. P. Stewart, J. Comp. Chem. **10** (1989) 209,221

[6] A. A. Voityuk, A. A. Blizniuk, Theor. CHim. Acta **72** (1987) 223

[7] D. Hadzi, J. Koller, Hydrogen Bonding by semi-empirical Molecular Orbital Methods, in: D. Hadzi (ed.), Theoretical treatment of Hydrogen Bonding, John Wiley and Sons 1997

[8] K. J. Jalkanen, S. Suhai, Chem. Phys. **208** (1996) 81

[9] M. Elstner, K. Jalkanen, S.Suhai, F.Herrman, T. Frauenheim to be published

[10] J. Sponer, J. Leszczynski, P.Hobza, J. Comp. Chem. **17** No. 7 (1996) 841

[11] J. Sponer, J. Leszczynski, P.Hobza, J. Phys. Chem. **100** No.13 (1996) 5590

[12] J. Sponer, J. Leszczynski, P.Hobza, J. Phys. Chem. **100** (1996) 1965

[13] P. Hobza et al., J. Comp. Chem. **18** No. 9 (1997) 1136

[14] J. Sponer, P. Hobza, Chem. Phys. Lett. **267** (1997) 263

[15] T. A. Halgren, J. Am. Chem. Soc. **114** (19992) 7827

[16] J. P. Lewis, O. F. Sankey, Biophysical Journal **69** (1995) 1068

FIRST ATTEMPT TO DEVELOP A MATERIAL MODEL DESCRIBING SINGLE CRYSTAL CREEP BEHAVIOR

Holger Brehm and Uwe Glatzel
Technisches Institut, Metallische Werkstoffe, Friedrich-Schiller-Universität Jena, Löbdergraben 32, D-07743 Jena, Germany

ABSTRACT

A material model is proposed which describes single crystal creep behavior by evolution equations for dislocation densities on individual slip systems. An interaction matrix determines the influence from one glide system to the other. Assuming a face centered cubic crystal, allowing deformation on octahedral glide planes and cube glide planes with a Burgers vector of the type $a/2 <110>$, nine independent parameters of the interaction matrix can be distinguished. A parameter check of the nine independent parameters has been carried out, showing the influence of parameters on specific orientations of the load axis. If one assumes dislocation interaction of a glide system only with itself a smooth behavior is predicted with a maximum creep rate for [001] orientation, followed by [011] and [111]. If a strong interaction is assumed, the orientation dependent creep behavior is not at all smooth, instead it shows a sharp drop in creep rates mainly in symmetric positions of the standard orientation triangle. The orientations with highest creep rates are in this case those which favor single glide. Highly symmetric orientations, such as [001], [011] and [111] have strongly decreased stationary creep rates.

INTRODUCTION

Several material models have been proposed in order to explain the mechanical behavior of single crystals. Schmid and Boas [1] derive the resolved shear stress based on single glide deformation mode. The tensile yield point as function of orientation of single crystals can be described successfully by this theory (see for example Burke and Hibbard [2]).

Many phenomenological models have been developed describing stress-strain behavior of single crystals, for example Chaboche [3], Cailletaud [4], Toth [5] and Franciosi [6], [7]. In the model of Franciosi [6] one takes into account single slip and multislip hardening. He also considers the contribution of active and passive glide systems and their mutual interaction to the creep behavior [7]. The dislocation distribution-dependence of the hardening is given by an interaction matrix. Toth's model [5] considers the hardening of a system as dependent on the dislocation density and the dislocation movement. He also describes the interaction between distinct types of dislocations using an interaction matrix but deals only with 12 different slip systems for the f.c.c. lattice. In the model of Cailletaud [4], describing stress-strain behavior during cyclic loading, micromechanical characteristics such as glide planes are considered.

Despite the very good agreement with experimental observations these models have in common that many material parameters, up to 30, are needed. Not all of these parameters have a physical meaning.

Mat. Res. Soc. Symp. Proc. Vol. 538 ©1999 Materials Research Society

A heavily studied system with respect to orientation dependent mechanical behavior are nickel-based superalloys (for example: Leverant, Kear and Oblak [8]). This is due to the fact that these alloys are used as material for turbine blades in aircraft engines in single crystalline form. In these studies it was shown that [001] oriented samples have superior creep resistance than other orientations. Fortunately, this orientation coincides with the fastest crystal growth direction.

Several models have been proposed explaining the orientation dependent creep behavior of superalloys (for example: MacKay and Maier [9] and Sass, Glatzel and Feller-Kniepmeier [10]). Most of them are based on the combination of the number of active systems, including stacking fault systems with Burgers vector a/3 <112> on {111} glide planes, and the phase boundary orientations of these two phase alloys. In superalloys, many parameters such as γ' particle size, γ' volume fraction, dislocation pile up at phase boundaries, different operating glide systems in different phases, coherency stresses and so on have to be considered. This complexity would appear to make it impossible to develop a successful model describing the orientation dependent superalloy behavior. In this stage of model development we have restricted ourselves to single phase, single crystal alloys.

EXPERIMENT

Single crystals of three different alloys with various orientations have been creep tested. Details are published by Siebörger and Glatzel [11].

Fig. 1 shows a selection of creep curves of the superalloy CMSX-4 (trademark of Cannon Muskegon Corporation) as function of orientation at 1123 K and a stress of 500 MPa. The orientations with the lowest creep rates are close to [001], but do show strong differences in creep response even for fairly small deviations, see sample A1 and A2.

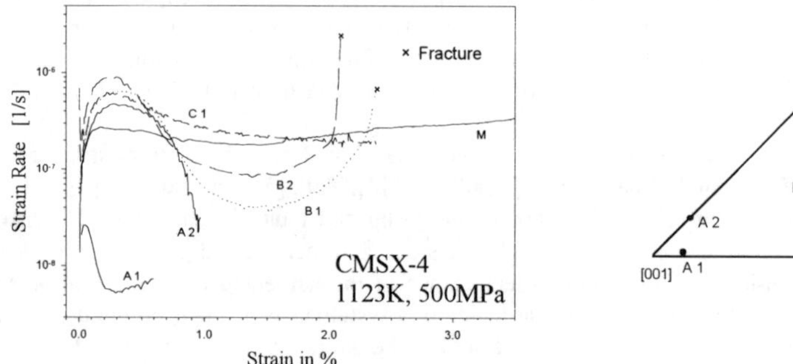

Fig. 1: Creep curves of CMSX-4 samples with different orientations indicated in the standard unit triangle.

Fig. 2 shows the creep behavior of single phase γ' material at 350 MPa. This alloy shows steadily increasing strain rates up to a strains of about 1%, known as inverse primary creep. In contradiction to the superalloy behavior orientations close to [001] and [011] have highest strain rates. Minimum creep rates are obtained for [111]. The shape of the creep curves as well as the orientation dependence varies significantly between the single phase alloy and the two phase superalloy.

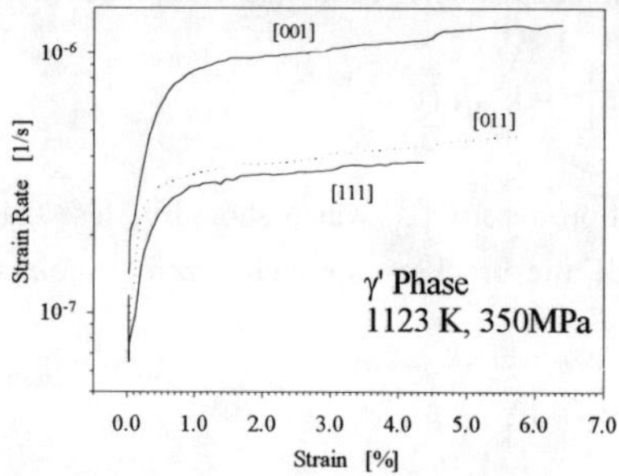

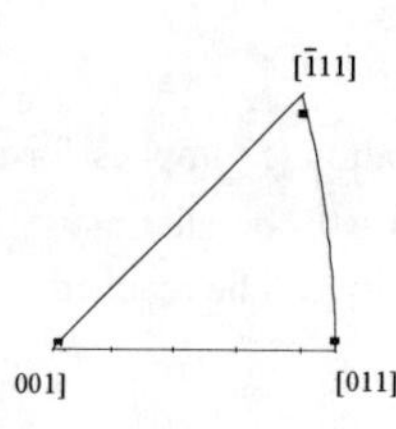

Fig. 2: Creep curves of the γ' phase single crystals as function of orientation.

Results of transmission electron microscope (TEM) observations for the γ' phase single crystals are described in detail by Knobloch and Glatzel [12]. It could be shown that in sample A5, which was only very weakly deformed, two octahedral glide systems <110>{111} govern the deformation process. After long creep exposure (sample A6), the distribution of the observed glide systems is very homogeneous on all slip systems with high Schmid factors.

RESULTS

Following the TEM observations of dislocation densities on individual slip systems, there is a need to develop a dislocation density based material model describing single crystal creep behavior. The major input should be the orientation of the load axis with respect to the single crystal lattice. The proposed model is based on Alexander & Haasen [13] and Haasen [14] starting out with the Orowan equation:

$$\dot{\varepsilon}(t) = \rho(t)\,\left|\vec{b}\right|\,v(t) \qquad (1)$$

using a constant Burgers vector $\vec{b}$ and the time dependent creep rate $\dot{\varepsilon}(t)$, dislocations density $\rho(t)$ and dislocations velocity $v(t)$. Based on TEM observations an energy considerations for fcc crystals the Burgers vector $\vec{b}$ is of the type a/2 <011> and the magnitude $\left|\vec{b}\right|$ is kept constant. The evolution of the dislocation density is given by:

$$\frac{\partial \rho}{\partial t} \equiv \dot{\rho}(t) = \rho(t)\,v(t)\,\delta(t) \qquad (2)$$

The multiplication rate δ and the dislocation velocity v(t) are mainly controlled by an effective stress σ_{eff}. The effective stress is given by the external stress σ, which is reduced by a back stress $k_1\sqrt{\rho}$.

$$\delta(t) = k_2\,\sigma_{eff}^{n} = k_2\left[\sigma - k_1\sqrt{\rho(t)}\right]^{n} \qquad (3)$$

$$v(t) = v_0\,\sigma_{eff}^{m} = v_0\left[\sigma - k_1\sqrt{\rho(t)}\right]^{m} \qquad (4)$$

leading to a differential equation for the time dependence of the dislocation density:

$$\dot{\rho}(t) = \rho(t)\, v_0\, k_2 \left[\sigma - k_1 \sqrt{\rho(t)} \right]^{m+n} \qquad (5)$$

Equation (5) implies for an initial dislocation density ρ_0, which should be less than $(\sigma/k_1)^2$ that $\dot{\rho}(t) \geq 0$ for all times. If the value inside the bracket approaches zero a constant dislocation density will be reached.

$$\rho_{ss} = \rho(t \to \infty) = \left(\frac{\sigma}{k_1} \right)^2 \qquad (6)$$

Since $v(t \to \infty) = 0$ the steady state creep rate in the Haasen-Alexander model approaches in general zero for $t \to \infty$ and this leads to a specific final plastic strain.
The Haasen Alexander model was extended in order to describe the orientation dependent creep behavior of single crystals. A face centered cubic crystal was chosen with Burgers vector $\vec{b}$ of the type a/2 <110>. Main glide planes $\vec{n}$ are octahedral glide planes of the type {111} and cube glide planes of the type {001}.

We have extended equation (5) to 18 coupled differential equations for the above mentioned slip systems.

$$\dot{\rho}_i(t) = \rho_i(t)\, k_2\, \overline{v}_0 \left[(m_s)_i\, \sigma - k_1 \sqrt{\sum_{j=1}^{18} c_{ij}\, \rho_j(t)} \right]^{m+n} , \quad (i = 1,\dots,18) \qquad (7)$$

with an average initial dislocation velocity $\overline{v}_0$, and correspondingly:

$$v_i(t) = (v_0)_i \left[(m_s)_i\, \sigma - k_1 \sqrt{\sum_{j=1}^{18} c_{ij}\, \rho_j(t)} \right]^{m} , \quad (i = 1,\dots,18) \qquad (8)$$

The components c_{ij} of the 18x18 interaction matrix determine how strong a high dislocation density in system j slows down the multiplication rate of the dislocation density in system i. The numerical solution of the system of 18 coupled differential equations (7), allows us to determine the dislocation density of each glide system at every moment. Solving the system of coupled differential equations we receive 18 individual, time dependent shear tensor rates.

$$\dot{\hat{\gamma}}_i(t) = \rho_i(t)\, |\vec{b}_i|\, v_i \left(\hat{\vec{b}}_i \otimes \hat{\vec{n}}_i \right) \qquad (9)$$

The integration of the $\dot{\hat{\gamma}}_i$ leads to the time dependent strain rate.

$$\dot{\varepsilon}(t) = \frac{\partial}{\partial t} \left| \vec{\sigma}_{new} \right| = \frac{\partial}{\partial t} \left[\left(\int_{t'=0}^{t} \sum_{i=1}^{18} \dot{\hat{\gamma}}_i\, dt' \right) \hat{\vec{\sigma}}_0 + \hat{\vec{\sigma}}_0 \right] \approx \sum_{i=1}^{18} \dot{\hat{\gamma}}_i\, \hat{\vec{\sigma}}_0 \qquad (10)$$

We have chosen to plot the calculated steady state creep rate as three dimensional topographical plot (with the logarithm of the creep rate plotted in z-direction) and as contour lines for constant creep rates in a standard orientation triangle in the x-y plane, for different sets of the interaction matrix (Table I).

Table I: Steady state creep rate for different sets of interaction parameter. Note the different scale for the 3D plot in row 1.

3 D Plot of the logarithm of the steady state creep rates in units of s^{-1}.	Plot of constant steady state creep rates together with the standard orientation triangle.	Interaction assumptions and $\dot{\varepsilon}_{ss}^{[001]} : \dot{\varepsilon}_{ss}^{[011]} : \dot{\varepsilon}_{ss}^{[\bar{1}11]}$ in units of 10^{-9} s^{-1}
1e-6 6e-7 4e-7		Each slip system develops individually, see equation (5), i. e. **no interaction** 895 : 593 : 457
1e-6 1e-7 1e-8		**Strong interaction** between all glide syst. (cube + octah.) glide orientations are indicated (**+** octah., **x** cube). 25 : 25 : 23

If stress axis orientations are negligible, e.g. small strains, the creep behavior for no interactions can be approximated by:

$$\dot{\varepsilon}_{ss} \approx \left(\frac{\sigma}{k_1}\right)^2 b \sum_{i=1}^{18} v_i \left[(m_s)_i\right]^3 \qquad (11)$$

In this case of no interaction the steady state strain rate is governed by the sum up of the cube of the schmid factors of all glide systems $\sum_{i=1}^{18} \left[(m_s)_i\right]^3$.

Assuming high values for all parameters of the interaction matrix leads to a steady state creep rate behavior as shown in row 2 of Table I. The creep rates are minimal for orientations close to the edges of the unit triangle and overall strongly reduced as compared to the no interaction case. The maximum creep rates are now given for single glide orientations (marked with an **x** for single cube glide and a **+** for single octahedral glide.

CONCLUSIONS

The presented model tries to explain the orientation-dependent creep behavior of single crystals. Some interesting conclusions can be drawn at present. If dislocation interactions are not operative, a smooth orientation dependence should be observed, only determined by the Schmid factors of the allowed glide systems. This smooth behavior given by theory was experimentally observed for the two phase material (see figure 2). If strong dislocation interactions are active, sharply decreased creep rates should be detected for orientations only a few degrees deviated from each other in special regions of the unit triangle. This very sensitive behavior given by theory was experimentally observed for the one phase material (see figure 1).

REFERENCES

1. Schmid, E. and Boas, W. (1935) Kristallplastizität; Springer, Berlin

2. Burke, E.C. and Hibbard, W.R.Jr. (1952), *Trans. AIME* **194**, 295

3. Chaboche, J.-L. (1977), Bulletin de L'Academie des Siences, Series des Techniques **18**, 33

4. Cailletaud, C. (1992), *Int. J. Plasticity* **8**, 55

5. Zhou, Y., Neale, K. W. and Tóth, L. S. (1993), *Int. J. Plasticity* **9**, 961-978

6. Franciosi, P. (1983) *Acta met.* **31**, 1331-1342

7. Franciosi, P. (1985) *Acta met.* **33**, 1601-1612

8. Leverant, G.R., Kear, B.H. and Oblak, J.M. (1973) *Metall. Trans.* **4**, 355

9. MacKay, R.A. and Maier, R.D. (1982), *Metall. Trans.* **13A**, 1747

10. Sass, V. Glatzel, U. and Feller-Kniepmeier, M. (1996b) in *Superalloys 1996* (eds Kissinger *et al.*), 283

11. Siebörger, D. and Glatzel, U. (1998) Orientation Dependent Creep Behavior and Microstructure of Nickel Solid Solution Single Crystals, submitted to *Acta Materialia*

12. Knobloch, C., Toloraia, V.N. and Glatzel, U. (1997) Anisotropic Creep in $Ni_3(AlTiTa)$. *Scripta Materialia* **37**, 1491-1498

13. Alexander, H. and Haasen, P. (1968) Dislocations and Plastic Flow in the Diamond Structure. *Solid State Physics* **22**, 27-158

14. Haasen P. (1968) Dislocation Dynamics in the Diamond Structure. *Dislocation Dynamics*, eds.: Rosenfield, Hahn, Bement, Jaffee, 701-722

Part V

Non-Crystalline and Nanocrystalline Materials

HYDROGENATED AMORPHOUS SILICON NITRIDE: STRUCTURAL AND ELECTRONIC PROPERTIES

J. F. JUSTO[1], F. DE BRITO MOTA[1,2], and A. FAZZIO[1]

(1) Instituto de Física da USP, CP 66318, CEP 05315-970, São Paulo - SP, Brazil
(2) Instituto de Física da UFBa,
   Campus Universitario de Ondina, CEP 40210-340, Salvador - BA, Brazil

ABSTRACT

We combined empirical and *ab initio* methods to study structural and electronic properties of amorphous silicon nitride. For such study, we developed an interatomic potential to describe the interactions between silicon, nitrogen, and hydrogen atoms. Using this potential, we performed Monte Carlo simulations in a simulated annealing scheme to study structural properties of amorphous silicon nitride. Then this potential was used to generate relevant structures of a-SiN$_x$:H$_y$ which were input configurations to *ab initio* calculations. We investigated the electronic and structural role played by hydrogen incorporation in amorphous silicon nitride.

INTRODUCTION

Computer simulation of materials at atomistic level has long faced a remarkable challenge. Although *ab initio* methods have provided considerable information about the structural and electronic properties of materials [1], they are computationally too expensive and their use has been limited to systems involving up to a few hundred atoms. On the other hand, interatomic potentials have been developed as alternatives in studying structural properties of materials, and can handle systems involving million atoms. The gain in computational cost comes upon a poorer description of the atomic interactions in the systems. Since the calculation of thermodynamical properties are not feasible within the first principles calculations, combining *ab initio* and empirical methods may be the best way to access these type of properties. This is a compromise between precision and computational efficiency.

Here we combine empirical and *ab initio* methods to study structural and electronic properties of hydrogenated silicon nitride. Silicon nitride is a material of great technological interest because of its mechanical and electronic properties which make it suitable to a number of applications [2]. Due to its superior mechanical properties (high density and high melting temperature), it is a prototypical material for engine components and cutting tools. Due to its high dielectric constant and large electronic gap, it has been widely used in electronic devices.

In this paper we investigated properties of unhydrogenated and hydrogenated silicon nitride combining empirical and *ab initio* descriptions. We used an interatomic potential, recently developed by us, to generate relevant amorphous structures, which were input configurations to electronic structure calculations. We showed that hydrogen has an important role in stabilizing the amorphous phase, and is very effective in removing energy levels which appear in the gap of the material.

555

STRUCTURAL AND ELECTRONIC PROPERTIES OF a-SiN

We have recently developed an interatomic potential to describe silicon nitride systems [3] using the Tersoff functional form [4]. The model provided a considerably good description of crystalline and amorphous SiN_x in a wide range of nitrogen contents ($0 < x < 1.5$) as compared to available theoretical [5, 6] and experimental data [7]. The details of the functional form, the strategy to find the best set of fitting parameters, and tests of the empirical model are presented elsewhere [3, 4]. Here we extend that model to include hydrogen interactions. The fitting parameters for all those interactions were found using a database which included *ab initio* or experimental results of the crystalline phase and molecules involving these interactions. Table 1 gives the best set of parameters for all interactions according to the functional form described elsewhere [3].

We used this model to investigate structural properties of hydrogenated amorphous silicon nitride. We modeled the amorphous material using Monte Carlo simulations, with periodic boundary conditions and up to 3000 atoms. Each simulation started with the material in a very high a temperature (6000 K), when the material was liquid. The temperature was slowly reduced during the simulation in a simulated annealing schedule until it reached 300 K. Then the statistical properties were computed over several configurations. The internal stress was released by allowing volume relaxation (NPT ensemble) [8] during the simulations.

Table 1: Best set of the parameters for the Tersoff interatomic potential which describe the Si-Si, Si-N, Si-H, N-H, and H-H interactions. The parameters were obtained from a simulated annealing fit to a database described in the text.

	N	Si	H
$A(eV)$	6.36814×10^3	1.8308×10^3	86.7120
$B(eV)$	5.11760×10^2	4.7118×10^2	43.5310
$\lambda(\mathring{A}^{-1})$	5.43673	2.4799	3.7879
$\mu(\mathring{A}^{-1})$	2.70000	1.7322	1.9800
β	5.29380×10^{-3}	1.1000×10^{-6}	4.0000
n	1.33041	7.8734×10^{-1}	1.00
c	2.03120×10^4	1.0039×10^5	0.00
d	2.55103×10^1	1.6217×10^1	1.00
h	-5.62390×10^{-1}	-5.9825×10^{-1}	1.00
$R(\mathring{A})$	1.80	2.70	0.80
$S(\mathring{A})$	2.10	3.00	1.00
χ			
N	0.00	0.65	0.76
Si	0.65	1.00	0.78
H	0.76	0.78	1.00

Figure 1 shows the average coordination numbers at Si and N atoms in a-SiN_x ($0 < x < 1.5$) as compared to experimental results [7]. The figure shows the coordination $n_A(B)$, the average number of neighbors A around atom type B. Figures 1a and 1b show the case of unhydrogenated and hydrogenated SiN, respectively. For unhydrogenated SiN, at

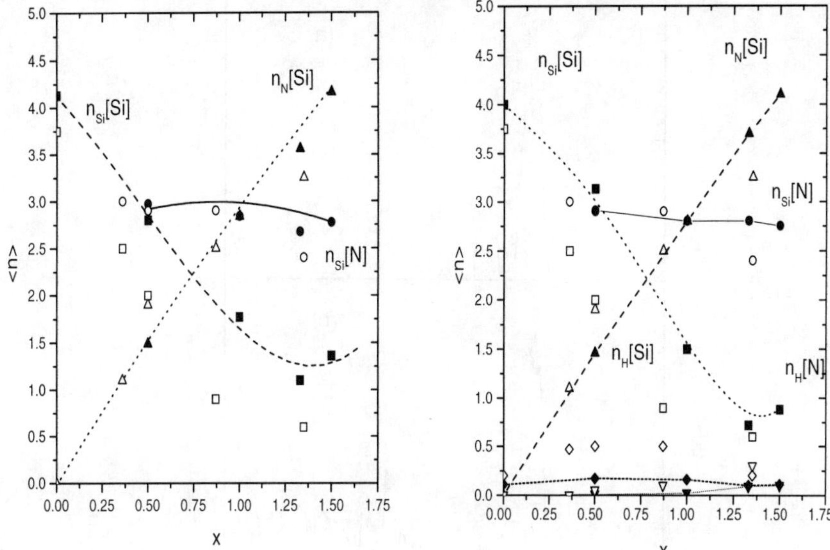

Figure 1: The average coordination number on Si and N atoms as function of the nitrogen content x. The simulation results (full symbols) are compared to experimental data (open symbols). The figure shows $n_{Si}(Si)$ ($\square$), $n_N(Si)$ ($\triangle$), and $n_{Si}(N)$ ($\circ$), in (a) unhydrogenated and (b) hydrogenated silicon nitride. Fig. (b) also shows $n_H(N)$ ($\triangledown$) and $n_H(Si)$ ($\diamond$).

low x ($x < 0.3$), the average coordination $n_{Si}(Si)$ is around four. As x increases, the N atoms start to compete with Si atoms to form bonds, such that $n_{Si}(Si)$ drops from 4.1 to about 0.9 at $x = 1.5$. On the other hand, the coordination $n_N(Si)$ grows linearly from zero to four. The coordination $n_{Si}(N)$ is around 3.0 for any x. Although the figure 1a shows the correct trends for the average coordinations, all the theoretical coordinations are an overestimation of the experimental results [7]. This is because the samples, in which the experiments were performed, had considerably high hydrogen concentrations. Since hydrogen competes with the other atoms to form bonds, a more realistic comparison should also includes hydrogen. Fig. 1b presents the results for hydrogenated silicon nitride. Hydrogen incorporation does not change the overall trends of the average coordinations. However, there is a decrease in the $n_{Si}(N)$, $n_N(Si)$, and $n_{Si}(Si)$ coordinations. Fig. 1b also shows the competition between Si-H and N-H bonds. For low nitrogen contents ($x < 1.25$), hydrogen has a chemical preference to bind to silicon rather than to nitrogen. On the other hand, hydrogen binds prefferentially to nitrogen for $x > 1.25$. This trend is in good agreement with the experimental data [7].

Now we present calculations of the electronic properties of amorphous silicon nitride and investigate the role of hydrogen in eliminating dangling bond levels which appear in the amorphous matrix. We used our empirical model to generate relevant amorhous structures to be input configurations in the *ab initio* calculations. To generate these configurations

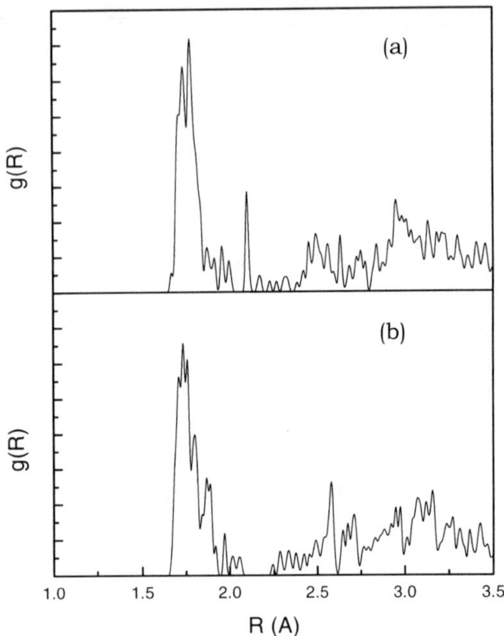

Figure 2: Total radial distribution function $g(r)$ for amorphous silicon nitride for a system of 56 atoms. Figure (a) shows the distribution function resulting from the Monte Carlo simulation using the interatomic potential. Figure (b) shows the distribution function resulting from the relaxed *ab initio* calculations.

with the empirical model, we used the same procedure as described earlier in this paper. However, the large unit cell of all the systems studied had less than 100 atoms. Using theses configurations, we computed the electronic properties of silicon nitride within the density functional theory [1, 9]. The Kohn-Sham equations were solved using the Car-Parrinello scheme [10] with Troullier-Martins type pseudopotentials [11] in the Kleinmann-Bylander form [12]. The basis-set was expanded in plane-waves, with kinetic energy up to 60 Ry. Full atomic relaxation was allowed during the *ab initio* calculations, with the final forces being smaller than 10^{-4} eV/Å.

Figure 2 shows the total radial distribution function $g(r)$ for a-SiN$_{1.33}$ consisting of 56 atoms (24 Si and 32 N atoms). Figure 2(a) shows the $g(r)$ for the system described by the interatomic potential. In the figure, the first peak around $1.7\mathring{A}$ corresponds to the Si-N bonds, while the second peak around $2.5\mathring{A}$ corresponds to the Si-Si bonds. These values are in excellent agreement with experimental results obtained by neutron scattering [13]. That structure was used as input to the *ab initio* calculation. Figure 2(b) shows $g(r)$ after full atomic relaxation is allowed in the *ab initio* calculation. The $g(r)$ before and after relaxation by the *ab initio* method are very similar, except some spurious peaks between $2.0\mathring{A}$ and

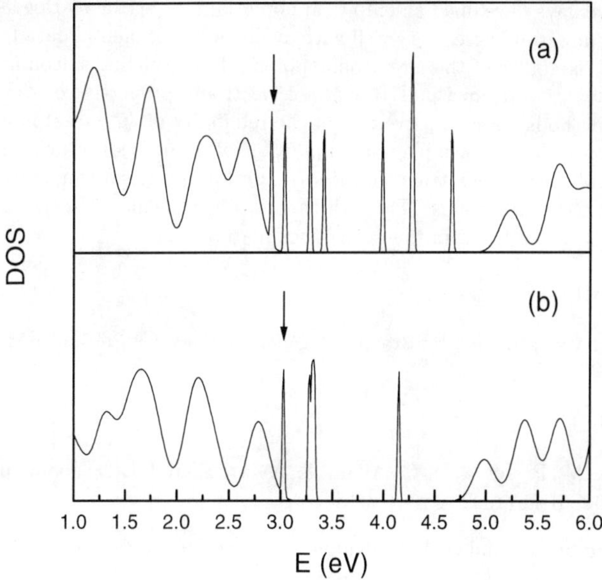

Figure 3: Density of states for the amorphous silicon nitride $SiN_{1.33}$ obtained by *ab initio* calculations. The figures shows: (a) the DOS of the unhydrogenated SiN for the structure after *ab initio* relaxation was allowed. (b) the DOS of the hydrogenated SiN after *ab initio* relaxation was allowed. The figures only show the DOS close to the gap of the material. The top of the valence bands are indicated by arrows.

2.4Å. Besides, Hellmann-Feynman forces computed on the structure before relaxation were on average 10^{-2} eV/Å. Atomic relaxation by the *ab initio* method reduced the total energy of the system by 0.4 eV/atom. All these results show that the configurations generated by the empirical model are already close to a realistic configuration according to a method based on quantum mechanics. These results corroborate the reliability of the empirical model in describing the amorphous silicon nitride.

Now we investigate the role of hydrogen in the electronic properties of amorphous silicon nitride. We used the empirical model combined with Monte Carlo simulations to generate the $a\text{-}SiN_{1.33}\text{:}H_y$ configuration with 66 atoms (32 N, 24 Si, and 10 H atoms). The final structures were used as input in the *ab initio* calculations. Figure 3 shows the density of states (DOS) for the electronic levels close to the gap of the material. The top of the valence band is at 3.0 eV and the bottom of the conduction band is around 5.0 eV. The figure shows the (a) unhydrogenated SiN and (b) hydrogenated SiN. In figure (a) the silicon nitride has several energy levels in the gap. These levels are generally related to Si and N dangling bonds which appear as result of the random network of the amorphous matrix. Hydrogenation of the silicon nitride is very effective in removing the gap levels, as can be seen in the fig. 3b.

CONCLUSIONS

In summary, we have developed an empirical potential to describe hydrogenated silicon nitride systems. The results agree very well with available experimental data for the amorphous phase. The reliability of the interatomic potential in modeling silicon nitride allows to combine it with *ab initio* methods to explore electronic properties of the amorphous phase. *Ab initio* methods can also serve to test the reliability of an interatomic potential.

Our results show that hydrogen plays an important role in both structural and electronic properties of amorphous silicon nitride. In the structural properties it competes with silicon and nitrogen atoms to form bonds. Through the *ab initio* calculations, we observed that hydrogen removes the gap levels in the amorphous material.

ACKNOWLEDGMENTS

Partial support was provided by Brazilian agencies Capes, CNPq, and FAPESP.

References

[1] M. C. Payne, M. P. Teter, D. C. Allan, T. A. Arias, and J. D. Joannopoulos, Rev. Mod. Phys. **64**, 1045 (1992).

[2] F. H. P. M. Habraken and A. E. T. Kuiper, Mat. Science and Eng. R **12**, 123 (1994).

[3] F. de Brito Mota, J. F. Justo, A. Fazzio, Phys. Rev. B **58**, 8323 (1998).

[4] J. Tersoff, Phys. Rev. B **39**, 5566 (1989).

[5] S.-Y. Ren and W. Y. Ching, Phys. Rev. B **23**, 5454 (1981).

[6] A. M. Liu and M. L. Cohen, Phys. Rev. B **41**, 10727 (1990).

[7] M. M. Guraya, H. Ascolani, G. Zampieri, J. I. Cisneros, J. H. Dias da Silva, and M. P. Cantão, Phys. Rev. B **42**, 5677 (1990).

[8] M. P. Allen and D. J. Tidesley, *Computer Simulation of Liquids* (Oxford Univ. Press, Oxford, 1987).

[9] P. Hohenberg and W. Kohn, Phys. Rev. **136**, 864B (1964); W. Kohn and L. J. Sham, Phys. Rev. **140**, 1133A (1965).

[10] R. Car and M. Parrinello, Phys. Rev. Lett. **55**, 2471 (1985).

[11] N. Troullier and J. L. Martins, Phys. Rev. B **43**, 1993 (1991).

[12] L. Kleinmann and D. M. Bylander, Phys. Rev. Lett. **48**, 1425 (1982).

[13] M. Misawa, T. Fukinga, K. Nihara, T. Hirai, and K. Suzuki, J. Non-Cryst. Solids **34**, 313 (1979).

A MODEL OF STRAIN DISTRIBUTION IN NANOCRYSTALLINE SiC AND DIAMOND AT VERY HIGH PRESSURES; IN-SITU X-RAY DIFFRACTION STUDY AND COMPUTER MODELLING

R. PIELASZEK, B. PALOSZ, S. GIERLOTKA, S. STEL'MAKH & U.BISMAYER*

High Pressure Research Center UNIPRESS, ul.Sokolowska 29, PO Box 65, 01 142 Warsaw, Poland

* Mineralogisch-Petrografisches Institut, Uni Hamburg, Grindelallee 48, 20146 Hamburg, Germany

ABSTRACT

A modeling of nanoparticles and *ab initio* simulation of the scattered intensity from the Debye functions is used as a tool for an examination of the strain induced under high pressure in nanocrystalline silicon carbide and diamond. The analysis of the experimental intensity profiles includes a determination of the atomic structure and microstructure of the materials. The advantages of modeling over conventional methods of the analysis of powder diffraction data are discussed. Examples of using the modeling for determination of the shape and size and of one dimensional disordering in very small particles (2-4 nm), and development of internal strains in 10 nm SiC nanocrystals subjected to high pressures are given.

INTRODUCTION

In a standard analysis of diffraction patterns of "ordinary" microcrystalline powders the positions and intensities of Bragg reflections are used to determine the crystallographic structure and microstructure of the material. For nanocrystals of several nm in size, a similar procedure does not work because the Bragg lines loose their unique meaning: an experimental diffraction pattern of a nanosize powder is a convolution of (i) the pattern of the basic crystallographic structure of the interior of the grain with (ii) the scattering from the atoms located at the surface and having the atomic surrounding (and, thus, symmetry) different from the bulk atoms. The surface diffraction effects can be neglected for ordinary (micrometer size) powder particles. For nanocrystalline materials with particles of several nm in size the number of atoms in the interior of the particle is comparable to the number of atoms located at the surface. Those contributions of the bulk and surface atoms to the total scattering are comparable. Therefore, the analysis of diffraction patterns obtained for nanocrystalline materials requires a different approach - a modeling of the atomic structure of nanoparticles and *ab initio* calculation of the corresponding diffraction effects [1,2]. Modeling of very small particles gives a unique opportunity to define the structure of the individual particles atom by atom. Because the procedure of computing the diffraction patterns from the first principles using the Debye functions is time consuming, we limited the size of the particles for which we calculate the diffraction patterns *ab initio* to 20 nm in diameter. In this work we discuss the advantages of the method of modeling and *ab initio* simulation of diffraction by silicon carbide and diamond polycrystals which have a one dimensionally disordered cubic structure.

The analysis of strains in conventional polycrystalline materials (i.e. those with micrometer size grains) can be done using powder diffraction techniques, and is usually based on the analysis of the shape of the Bragg reflections. The analysis of the strain in nanocrystalline silicon carbide and diamond, which is the subject of our work, cannot be based on the shape of individual Bragg reflections alone and must include the full intensity profile of the diffraction pattern. To obtain a

Mat. Res. Soc. Symp. Proc. Vol. 538 © 1999 Materials Research Society

unique image of the strain, what requires a model accounting for the gradient of lattice parameters in individual powder particles, the structure of the starting material has to be determined. Below we present some examples of the diffraction analysis of SiC nanocrystals with 2 to 4 nm size particles [3,4] to demonstrate the sensitivity of our method of modeling to different structure parameters. The analysis is based on selecting such theoretical intensity profile that fits a given experimental pattern best. Nanocrystals with very small particles (2 to 4 nm) have a narrow distribution of sizes and, for the sake of the diffraction analysis, are well approximated by a one size particle material. Determination of strains in SiC nanocrystals of an average grain size of 10 nm [5] is based on the determination of the grain size distribution as a function of pressure.

RESULTS AND DISCUSSION

Computer simulation of the diffraction patterns of small (2 - 4 nm) nanocrystals

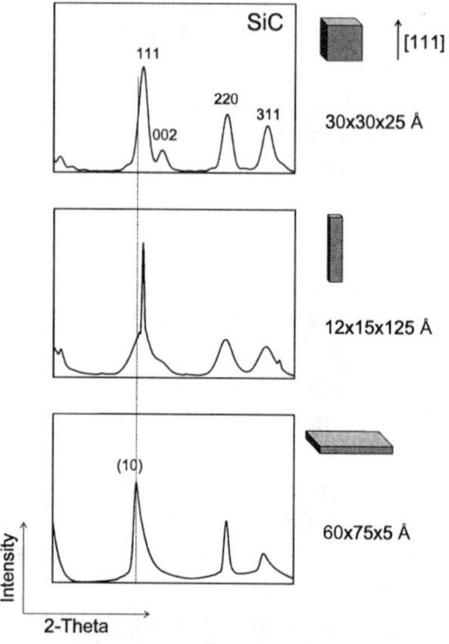

Fig.1   Effect of shape of very small SiC nanoparticle on diffraction pattern

A.   Grain size and grain size distribution function (GSD)

Usually, the size of the grains in a powder is deduced directly from FWHM (Full-Width at Half-Maximum) measured for a Bragg reflection and compared with FWHMs calculated for particles of different sizes. This procedure gives a unique result only when the experimental and theoretical peaks have exactly the same shape. A theoretical Bragg line calculated for a spherical crystallite has a pure Gaussian shape. Any deviation from the spherical shape and/or a combination of particles with more than one characteristic size leads to a non-Gaussian shape of the Bragg lines. In practice, one has always to consider that particles have non-spherical shape and that there is always a certain distribution of sizes of the powder particles. Therefore, determination of the particle size from FWHM is always an approximation. The term "average grain size", which is often used in the literature, does not have a unique meaning and should not be used for very small, several nm in size particles. The reason is that, in such cases, the shape of the Bragg line is a strong convolution of the shape with size effects. It is demonstrated in Fig.1 which shows three diffraction patterns calculated for grains with the same crystallographic lattice and the same number of atoms but having different shapes (a cube, a plate and a needle). Here a conventional description of the patterns using the peak positions, FWHM of the Bragg lines, and the relative intensities of Bragg reflecions is not applicable because the anisotropy of the shape has a strong effect not only on the FWHM and the diffuse scattering, but on the peak positions as well. For instance, the smallest and the largest distances in a cube, the edge and the diagonal, differ by a factor of √3 only, while for the plate the length and thickness differ by a factor of 15. For a cube all Bragg reflections of the cubic structure are visible and have similar profiles. The

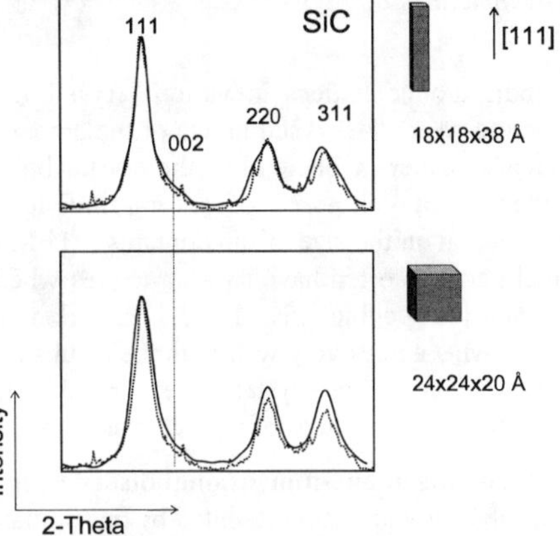

Fig.2 Experimental and theoretical diffraction patterns of SiC nanocrystals patterns calculated for two different models: the pattern of the particle alongated in [111] direction fits the experimental pattern best.

pattern of a plate does not have all cubic reflections visible, and the calculated intensity profile looks like that resulting from a two-dimensional scattering. For the plate, the strongest intensity maximum, which is (111) Bragg reflection for large grain polycrystals, is replaced by a diffuse intensity band characteristic for two-dimensional scattering. Maximum of this intensity band is at the position of (110) Bragg reflection of a trigonal structure of SiC [6]. The strong effect of the shape of the particles which dominates the effect of size in the intensity profiles of very small particles, can usually be ignored for large particles: the effect of size diminishes with inrease of the particle dimensions: it is relatively weak already for grains with dimensions larger than 20-30 nm, and nearly disappears for the grain dimensions larger than 200-300 nm. The anisotropy of shape of the particles with the largest dimension of the order of 20-30 nm and the smallest of several nm, has no effect on the positions of Bragg reflections, and no effect on the diffuse scattering but affects the relative intensities of the reflections. Sensitivity of the calculated intensity profiles to the shape of very small powder particles is demonstrated in Fig. 2 where the experimental pattern of SiC nanocrystals is compared to two theoretical patterns of two grains with different shapes. As seen in the fugure, the change of the grain shape changes the width of the reflections and, simultaneously, affects the relative intensities of the Bragg lines. The pattern calculated for the needle-like shape particle fits much better the experimental intensity profile than that for the more regularly shaped crystallite.

The diffraction method discerns the domains as two single crystallites. One has to remember that the shape derived from the diffraction data does not necessarily correlate with the physical shape of the powder particle: the diffraction gives the information on the lengths of the

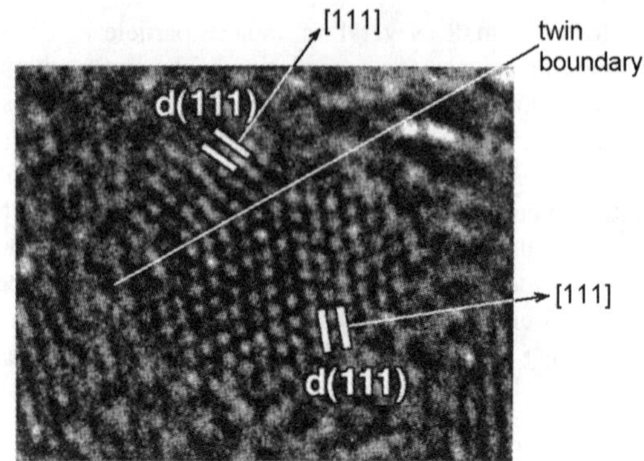

Fig.3 HREM image of SiC particle with the (011) plane perpendicular to the electron beam. Two domains with different orientation of cubic [111] axis are visible.
(HREM image provided by V.Buchmann et al. [2])

periodic atomic lattices in various crystallographic directions. A twin is an example of the structure where the crystal lattice of the domains is terminated by twin boundaries. Because such domains scatter incoherently, the crystal built of several domains having different orientations scatters as if composed of several individual crystallites. Diffraction experiments provide information on the size of the domains. TEM images of nanocrystalline SiC show [2], that very small particles often have twin structures where the neighbouring twin domains have differently oriented [111] cubic axis, Fig.3. Such domains have elongated shape in the stacking direction [111], what agrees very well with the results of our diffraction analysis, Fig.2. This result shows that modeling of nanocrystals gives a reliable information not only about the atomic structure, but also about the real shape of the nanoparticles.

B. Modeling of one-dimensional disordering in nanoparticles

The cubic structure is represented by one domain consisting of only cubic layers "c". A disordered structure consists of small cubic domains where every domain is terminated by a single hexagonal-type layer "h" [6]. Fig. 4 shows a typical HREM image of a SiC nanoparticle where individual layers can be identified and the stacking sequence determined in the crystallite.

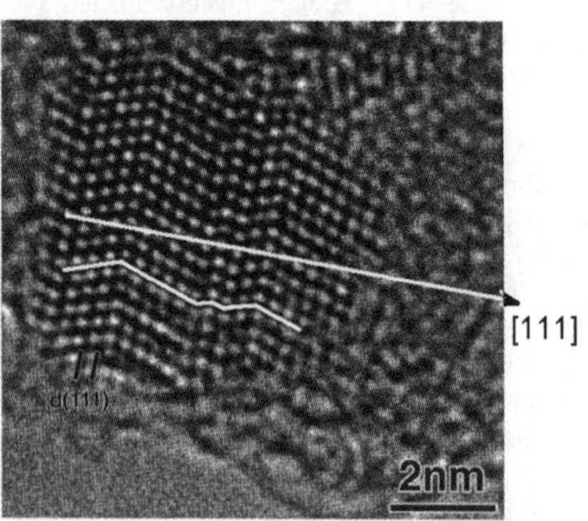

Fig.4 HREM image of one dimensionally disordered SiC particle with the (011) plane perpendicular to the electron beam.
(HREM image provided by V.Buchmann et al. [2])

However, electron microscope images give a very useful and important information on the structure of single particles only. The "general image" of the material has to be obtained from the diffraction measurements.

For very small size nanoparticles modeling the sequence of the layers and simulation of the stacking faults can be done. The stacking of the layers seen on the HREM images can also be obtained by powder diffraction methods combined with the modeling providing an "average structure" of the material. Our modeling allows us to determine the average size of the cubic domains (each terminated by the stacking fault – hexagonal layer h) with individual particles having different structure. Fig.5 shows an experimental diffraction pattern of nanocrystalline SiC and two intensity profiles calculated for two models of one dimensionally disordered cubic structure. Sensitivity of the calculated patterns to the layer stacking, drawn schematically in the figure, is quite obvious.

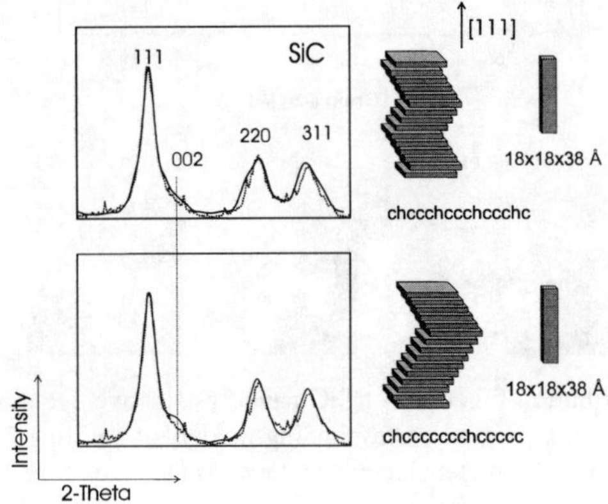

Fig.5 Experimental and theoretical diffraction patterns of   30x30x42 Å SiC nanoparticle calculated for two different layer sequences. A schematic image of the layer stacking together with the corresponding hc notations are given in the figure.

C. Distribution of strain under high pressures

There is no direct way of extracting the information on the compressibility and strain present in the material from the diffraction data obtained under high external pressure. We used the method of modeling to interprete diffraction patterns of silicon carbide and diamond nanocrystals measured under high pressure, as discussd below. We assumed that, as a result of the stress induced by external pressure, a gradient of the lattice constant develops along the radius of the particle, the a-parameter varying from 4.2 to 4.35 A [7]. A shift of the outer (closeto the surface) parts relative to the bulk of the particle makes the grain behave like being split into several smaller particles having different lattice constants. Based on our modeling technique, we determined the change of the grain size distribution function from the diffraction patterns measured under high pressure. Fig. 6 shows an example of changes of the grain size distribution function determined by modeling for 10 nm average grain size SiC nanocrystals. Although at the present stage of our studies the model provides only a rough approximation of the real structure, it gives some information about changes to the lattice under high stress and provides a semi-quantitative description of the related changes to the microstructure of strongly strained very hard materials like silicon carbide and diamond.

SUMMARY - CONCLUSIONS

For very small particles (a few nanometers) the intensity maxima of the diffraction patterns

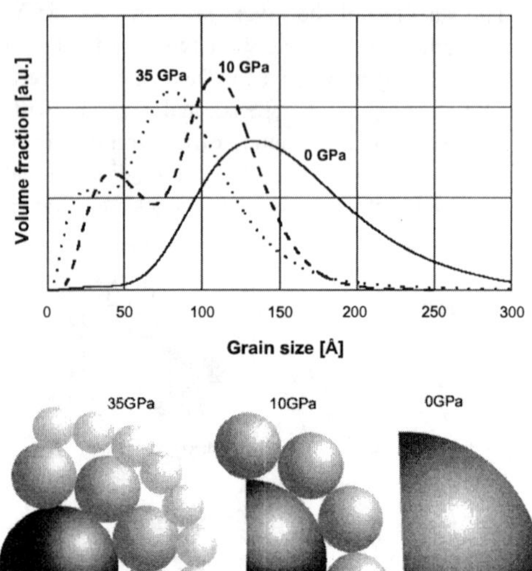

Fig.6   Grain size distribution functions of SiC nanocrystals (average grain size 10 nm) under the pressure of 0, 10 and 30 Gpa, and corresponding models of splitting 10 nm grains into smaller crystallites with the lattice parameters increasing towards the core of the particle.

lose their unique meaning as the Bragg reflections. In such cases, instead of examining individual Bragg reflections the analysis of the full intensity diffraction profile is necessary. Such analysis allows for a direct and simultaneous modeling of the shape and size of the particles, including one-dimensional disordering in close packed structures like silicon carbide and diamond. The method allows for the assessment of strain in the materials and a quantitative description of changes of their microstructure under high external pressures.

ACKNOWLEDGEMENTS

This work was supported by the State Committee for Scientific Research, grant P40708906 and 1076/T08/96/11. The experiments were performed in HASYLAB at DESY in Hamburg, Project II-97-25, with financial support from DLR, Germany, Project POL-153-96.

REFERENCES

[1]   B.Bondars, S.Gierlotka, B.Palosz & S.Smekhnov, Mat. Sc. Forum, **166-169** , 737 (1994).
[2]   R.Pielaszek, M.Aloshina, B.Palosz, S.Gierlotka, S.Stel'makh and U.Bismayer MRS Symp Proc. **501**, 305-310 (1998).
[3]   V.Buschmann, S. Klein, H. Fuess and H. Hahn, J.Cryst. Growth **193** , 335 (1998)
[4]   H.-P.Martin, E.Müller, R.Richter, G.Roever & E.Brendler, J.Mater.Science **32** , 381 (1997).
[5]   D. G. Keil, H. F. Calcote & R. J. Hill, Mat. Res. Soc. Symp. Proc.**410**, 167 (1996)
[6]   S.Stel'makh, PhD.Thesis, Warsaw, (1997).
[7]   B. Palosz, S. Gierlotka, S. Stelmakh, R. Pielaszek, P. Zinn, M.Winzenick, U. Bismayer and H. Boysen, J.Alloys and Compounds, in press

MOLECULAR ORBITAL CALCULATIONS ON ATOMIC STRUCTURES OF Si-BASED COVALENT AMORPHOUS CERAMICS

K. MATSUNAGA and H. MATSUBARA
Fine Ceramics Research Association, Synergy Ceramics Laboratory
2-4-1 Mutsuno Atsuta-ku Nagoya, 456-8587 JAPAN

ABSTRACT

We have performed *ab-initio* Hartree-Fock molecular orbital calculations of local atomic structures and chemical bonding states in Si-N covalent amorphous ceramics. Solute elements such as boron, carbon and oxygen were considered in the Si-N network, and the bonding characteristics around the solute elements were analyzed. When a nitrogen atom is substituted by a carbon atom, it was found that Si-C bonds reinforce the Si-N network due to strong covalency.

INTRODUCTION

Silicon nitride based amorphous ceramics (Si-N) have excellent high-temperature stability, and are considered as promising materials for high temperature applications. The materials can be synthesized by pyrolysis of organic precursors such as polysilazane at around 1000°C. Riedel *et al.* [1] reported that Si-N amorphous materials thus obtained preserve their amorphous state even at temperatures above 1000°C. The high temperature stability of the Si-N amorphous state also depends on incorporation of additional elements such as B, C and O. Amorphous Si-N materials are able to accommodate C atoms (amorphous Si-C-N). It was reported that the amorphous state is maintained up to 1450°C [1]. Thus, the presence of additional C atoms promotes the stability of amorphous Si-N.

The distinctive high-temperature stability of Si-N and Si-C-N amorphous ceramics may have a close relationship with their atomic structure and chemical bonding in the amorphous state. The atomic structures of these amorphous materials are considered to consist of covalent Si-N bonds. Seitz *et al.* [2] suggested that the local atomic structure of Si atoms in Si-C-N can be represented by SiN_xC_y (x+y=4) tetrahedra. C atoms are present in the random network of Si-C-N, forming Si-C bonds. The additional C atoms are homogeneously distributed throughout the material. Thus, the substitution of N atoms by C atoms is expected to play an important role in the stability of Si-N network. In order to clarify the effect of the additional elements in amorphous Si-N, it is important to understand the chemical bonding state around the constituent atoms.

In the present study, we have applied molecular orbital (MO) calculations to amorphous Si-N in order to examine chemical bonding states of the amorphous network. MO calculations are a powerful means of obtaining detailed information on the chemical environment around constituent atoms. In particular, we have paid attention to effects of substituted elements (B, C and O) for N on the stability of the Si-N network, from the viewpoint of chemical bonding. At the same time, we have also carried out molecular dynamics (MD) simulations of amorphous Si-N to investigate characteristics of local atomic structures in the amorphous Si-N network.

COMPUTATIONAL METHODS

For MO calculations, the *ab-initio* Hartree-Fock method was employed as implemented in a general computer program, *gaussian 94* [3]. Local atomic structures of amorphous Si-N are represented by model clusters, and molecular orbitals in clusters are calculated from first principles, giving 6-31G* type basis functions to atoms in clusters.

MD simulations were carried out for amorphous $SiN_{1.33}$ (stoichiometric) with a density of 2.8 g/cm^3. We employed the canonical *NTV* ensemble, and performed MD simulations for 1120 atoms (Si:480, N:640) in a rectangular cell ($29.8 \times 29.8 \times 14.9 \text{Å}^3$). The starting structure for MD

Mat. Res. Soc. Symp. Proc. Vol. 538 © 1999 Materials Research Society

simulations consisted of these Si and N atoms arranged in a β-Si$_3$N$_4$ framework. In order to generate an amorphous system, the initial temperature was set at 4000 K for 4.0 ps to agitate the constituent atoms. Secondly, the temperature was reduced to 1273 K for 3.0 ps, and finally the system was kept at 1273K for 6.0 ps. The equations of motion were integrated using a time interval of 0.2 fs. As interatomic potentials, the three-body Tersoff potential was used [4]. This type of interatomic potential is applicable to MD simulations of covalent systems. The potential parameters for Si and N determined by Kroll [5] were employed because it was demonstrated that crystal structures and properties of silicon nitrides were well reproduced using this potential.

RESULTS AND DISCUSSION

<u>MD Simulation of Amorphous SiN$_{1.33}$</u>

Figure 1 shows the time-averaged atomic positions of amorphous SiN$_{1.33}$ at 1273 K, obtained from the present MD simulation. In crystalline Si$_3$N$_4$, Si atoms are tetrahedrally coordinated by four N atoms, while N atoms are surrounded by three Si atoms in a planar configuration. In amorphous Si$_3$N$_4$, the three dimensional periodicity of the atomic arrangement is lost and a number of Si and N atoms with deficient coordination numbers are expected to be present. As is seen in Fig. 1, the atomic arrangement generated by our MD simulation is random and there are a number of undercoordinated Si and N atoms, although Si and N atoms with coordination numbers of 4 and 3, respectively, can also be found.

Fig. 1 The time-averaged atomic structures generated by the present MD simulations.

Table I lists the average coordination numbers of N atoms ($N_{\text{N-j}}$, where j=Si, N). These coordination numbers were obtained by calculating pair correlation functions during the MD simulation using a cutoff radius of 2.3 Å. $N_{\text{N-Si}}$ is found to be smaller than that in the crystalline state ($N_{\text{N-Si}}$=3.0). A small value of $N_{\text{N-N}}$ can be observed. The average interatomic distance of nearest neighboring N-Si pairs was calculated to be 1.75 Å, which is in good agreement with the experimental value (1.729Å) of chemically vapor deposited (CVD) amorphous Si$_3$N$_4$ with an averaged density of 2.8 g/cm^3 determined by Misawa et $al.$ [6].

The calculated $N_{\text{N-Si}}$ exhibits a smaller value, as compared to the experimental value of $N_{\text{N-Si}}$ (2.78). This may be due to the existence of N-N bonds ($N_{\text{N-N}}$=0.46). In the present MD simulation, we used the Tersoff potential, where attractive interaction of N-N is included. As a result, in a random atomic structure of amorphous Si_3N_4, N-N bonding pairs are partially found in the amorphous state.

As can be seen in Fig. 1, the bond angles of $\angle$Si-N-Si at the two-fold coordinated N sites have a tendency to having larger values than 120°. The $\angle$Si-N-Si angles were found to vary from 110° to 178°, from the present simulation (the average angle : 139.7°).

Table I The average coordination numbers of nitrogen.

$N_{\text{i-j}}$	This work	Experiment [6]
N-Si	2.30	2.78
N-N	0.46	—

MO Calculation of Solute Atoms in Amorphous Si-N Network

As shown above, amorphous Si-N contains a considerable amount of N atoms with deficient coordination numbers, and there are also N atoms surrounded by three Si atoms, in a similar manner to the crystalline state. When additional elements such as C are included in the material, it is possible to substitute N atoms at the two-fold and three-fold coordinated sites. We have investigated chemical bonding states of additional elements in the Si-N network, using ab-$initio$ MO calculations.

Figure 2 displays the model clusters for local atomic structures in the Si-N network that were used in the present MO calculations. Two types of model clusters were considered: (a) the two-fold coordinated model, ($Si_3N_7H_{12}$), (b) the three-fold coordinated model, ($Si_3N_{10}H_{18}$). In these model clusters, for simplicity, Si atoms are assumed to be tetrahedrally coordinated by four N atoms, although a number of Si atoms may have deficient coordination numbers in the amorphous Si-N network. For calculations involving solute elements such as B, C and O, a central N atom in these model clusters (N* indicated by the arrows in Fig. 2) was replaced by the additional elements.

The distance of $r_{\text{X*-Si}}$ and the bond angle of $\angle$Si-X*-Si, θ, (X = substitutional elements) were optimized in the MO calculations. For example, the Si-C*-Si and Si-N*-Si linkages of the two-fold coordinated models were found to have linear configurations (θ=180.0°). As stated above, our MD simulations of amorphous $SiN_{1.33}$ in the previous section also showed that the two-fold coordinated N sites have a tendency to exhibiting much opener bond angles of $\angle$Si-N*-Si more than 120°. In contrast, the Si-C*-Si and Si-N*-Si linkages exhibited planar configurations with the $\angle$Si-X*-Si of 120.0° in the three-fold coordinated models.

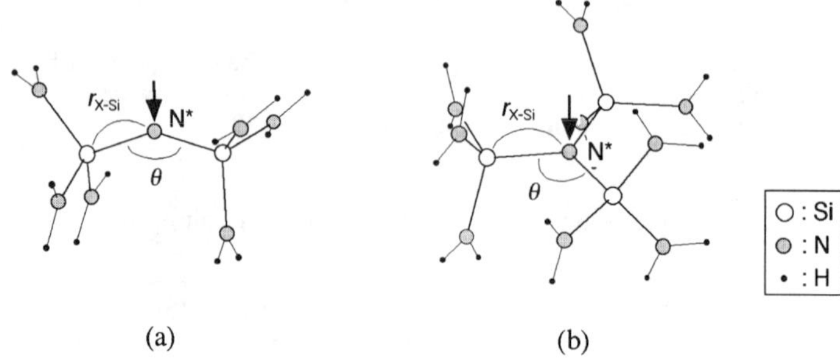

(a)   (b)

Fig. 2 Model clusters for local atomic structures in amorphous Si-N.

To analyze bonding properties of the solute elements (X*) in the Si-N network, we have studied the electron-density difference between the calculated models with equilibrium geometry and the clusters of isolated Si, N and X* atoms superposed in the same configurations as the geometry-optimized models. The electron-density difference indicates electron densities contributing to formation of chemical bonds of Si-X*. When the electron-density difference shows large positive amplitudes at around the midpoint between two nuclei, two nuclei are bonded to each other with strong covalency.

Among the solute elements considered in the present study, it was found that the C atom substituted for a N atom showed the highest covalency in the Si-N network. Figure 3 indicates the contour maps of the electron-density differences of the two-fold coordinated models containing C and N atoms at the X* site. As can be seen in Fig. 3(a), the Si-C* pairs exhibit a higher concentration of electron density between C* and Si, as compared to the case of Si-N pairs. Therefore, in the presence of a C atom in the Si-N network, a C atom locally reinforces the amorphous Si-N network due to strong directional bonding.

In amorphous Si-N materials, atomic rearrangement in the random network is expected to occur at high temperatures. This is a primary step of crystallization of the amorphous state. It was found that Si-C bonds showed the highest covalency in the amorphous network, from the present results. Such covalently bonded Si-C linkages become rigid nodes in the Si-N network, and may inhibit atomic rearrangement around C atoms when C atoms are randomly distributed throughout the Si-N network by substituting N atoms.

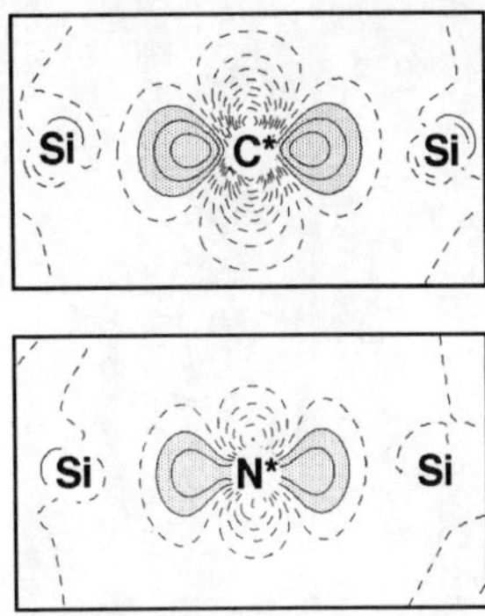

Fig. 3   Contour maps of electron-density differences around C and N atoms in the two-fold coordinated model clusters. Shaded regions indicate positive values of the electron-density differences. A contour spacing is 0.02 electrons/a.u.^{3}.

CONCLUSIONS

We have performed MD simulations of amorphous $SiN_{1.33}$ to examine atomic coordinations in the amorphous state. *Ab-initio* Hartree-Fock MO calculations have also been carried out to investigate effects of solute elements (B, C, O) on the stability of the Si-N amorphous network from the viewpoint of chemical bonding states. In the presence of a C atom, it was found that Si-C bonds exhibited the highest covalency. This contributes to the reinforcement of the Si-N network.

ACKNOWLEDGMENTS

This work has been entrusted by NEDO as part of the Synergy Ceramics Project under the Industrial Science and Technology Frontier (ISTF) Program promoted by AIST, MITI, Japan.

REFERENCES

1. R. Riedel and W. Dressler, Ceramics International **22**, 233 (1996).

2. J. Seitz, J. Bill, N. Egger and F. Aldinger, J. Euro. Ceram. Soc. **16**, 885 (1996).

3. C. C. Roothaan, Rev. Mod. Phys. **23**, 69 (1951).

4. J. Tersoff, Phys. Rev. B **39**, 5566 (1989).

5. P. M. Kroll, PhD thesis, Technische Hochschule Darmstadt, 1996.

6. M. Misawa, T. Fukunaga, K. Niihara, T. Hirai and K. Suzuki, J. Non-Cryst. Solid **34**, 313 (1979).

STRUCTURAL CHARACTERIZATION OF AlON BY ^{27}Al MAS NMR AND QUANTUM CHEMISTRY METHOD

Ying Dai, Xin-Min Min, Ce-Wen Nan, Xin-Mei Pei, Hai-Lan Ren
National Key Laboratory of Advanced Technology for Materials Synthesis and Processing, Wuhan University of Technology, Wuhan, Hubei 430070, P. R. China

ABSTRACT

Aluminum oxynitride (AlON) is a nitrogen-stabilized defective spinel phase of alumina (γ-Al_2O_3). Its structure has not been well understood so far. Solid-state ^{27}Al magic-angle spinning nuclear magnetic resonance and quantum chemistry calculation are used to characterize γ-Al_2O_3 and AlON ceramics in the present study. The resonance spectra clearly show different units (e.g. [AlN_4], [AlO_4] and [AlO_6]) and vary with composition. The calculation shows that the vacancy located at octahedral site is more stable than that located at the tetrahedral site, and nitrogen atoms preferentially replace oxygen atoms in the tetrahedral site, which is in agreement with the experimental analysis by nuclear magnetic resonance.

INTRODUCTION

Aluminum oxynitride spinel (AlON) is one of the compounds in the AlN-Al_2O_3 system and has many diversified applications. In particular, it is an isotropic material that can be sintered into a fully dense ceramic having properties similar to γ-Al_2O_3[1]. Unlike Al_2O_3, however, the cubic AlON structure allows for production of optically transparent polycrystalline material and can eliminate other problems associated with anisotropy. Initially, it was believed that reducing the aluminum species from Al3+ to Al2+ stabilized this phase. In 1959 Yamaguchi[2] reported that the spinel phase was stabilized not by Al^{3+} but by nitrogen. AlON was thought of a nitrogen-stabilized defective spinel phase of γ-Al_2O_3. Since that time various efforts have been made to understand the processing, structure and properties of AlON[3]. Gourmet's studies[4] by neutron diffraction have demonstrated the spinel structure of AlON but have been able to say little about the local coordination changes which occur across this series.

High-resolution nuclear magnetic resonance (NMR) in the solid state is increasingly used as an aid to structural characterization since the NMR frequency is influenced by small changes in the electronic environment of the nucleus. The technique of magic-resolution NMR spectra, which readily distinguish different local aluminum coordination units, has been particularly useful where techniques such as X-ray diffraction (XRD) provide little information about the aluminum distribution due either to the similarity in scattering factors of the constituent elements or to the lack of long-range order. In this paper, structural characterization of AlON phase is assessed by the use of a self-consistent-field discrete variation X_α method (SCF-DV-X_α), one of calculation methods of molecular orbital in quantum chemistry, and ^{27}Al MAS NMR. The calculated results are discussed and compared with the experimental data. XRD analysis was also used to monitor the synthesis of AlON and its phase composition.

EXPERIMENT

Commercially available AlN powder and Al_2O_3 powder, with average particle sizes of $0.6\mu m$ and $1\mu m$, respectively, were used. The AlN powder contained 0.8wt% oxygen. Based on the Al_2O_3-AlN phase diagram[3], three ranges are divided. The single AlON formation range is 27-40 mol% AlN. Samples X_1, X_2, X_3 and X_4 with 25.0, 30.0, 35.7 and 62.4mol% AlN, respectively, lie

Mat. Res. Soc. Symp. Proc. Vol. 538 © 1999 Materials Research Society

in the three typical ranges of Al_2O_3-AlN equilibrium diagram. The composition points of sample X_2 and X_3 lie in the single-phase AlON field. These raw materials were ball-milled in ethanol for two hours, and then dried in vacuum. The powders were isostatically pressed at 400 MPa before sintering. Pressureless sintering was performed at 1923K for 3 hours under a flowing oxygen-free nitrogen atmosphere. The samples were readily identified by X-ray diffraction. The MAS-NMR was performed on a Bruker MSL 400 spectrometer equipped with a 9.4-T magnet, operating at 104.3 MHz. The aluminum spectrum was acquired on a 3-mm DB MAS probe at a rotation rate of 8KHz,with 2 s (π/4 tip angle) pulses, for 500 scans. All spectra were referenced to an external standard of $[Al (H_2O)_6]^{3+}$ in dilute aqueous aluminum nitrate. No spectral smoothing for Al was required. A standard spectra of original mixture(AlN+γ-Al_2O_3) was tested under the same conditions for compare with sintering samples.

CALCALATION MODELS

The AlON phase is a result of the stabilization of the defective spinel structure of γ-Al_2O_3 at high temperature by the introduction of nitrogen. In the typical spinel structure of $MgAl_2O_4$, there are 32 anions (O^{2-}) within the cubic cell to form 8 fcc sub-cells, 8 cations (Mg^{2+}) at the tetrahedral sites to form 4 A-blocks, and 16 cations (Al^{3+}) at the octahedral sites to form 4 B-blocks. In γ-Al_2O_3, 2.667 vacancy sites are randomly chosen either among the 8 tetrahedral sites in A-blocks, or the 16 octahedral sites in B-blocks, denoted by $Al_{21.333}\square_{2.667}O_{32}$, where "$\square$" represents the vacancies. For illustrative and qualitative purposes to AlON, seven classes of various structural unit models are constructed, as listed in Table 1. Each model consists of one A- and one B-block. Several models are constructed in each class by altering sites of N atoms and vacancies. There is one vacancy from models of class I to V. In fact, the calculated models of class I to V should contain 2/3 vacancy, so the models without vacancy should also be discussed. Therefore, the models of class VI without vacancy are selected. Models of class VII with two vacancies are selected, too, for comparison.

RESULTS AND DISCUSSION

^{27}Al MAS NMR

The AlN-Al_2O_3 system is known to contain three crystalline phases, AlN, AlON and Al_2O_3. The crystal structures and hence aluminum coordination of AlN and Al_2O_3 are well known. The monoclinic AlN contains $[AlN_4]$, and Al_2O_3 is believed to have two forms, a high-temperature α-Al_2O_3 phase with $[AlO_6]$ and a low-temperature γ-Al_2O_3 form (<1373K) containing $[AlO_4]$ and $[AlO_6]$. The XRD patterns and the ^{27}Al NMR spectra of our sintering samples are shown in Fig.1 and Fig.2, respectively. The spectra show three distinct peaks at 114, 65 and 12 ppm which are comparable to the peak position for AlN (113 ppm), γ-Al_2O_3 (66 and 8 ppm) and α-Al_2O_3 (11.5 ppm). These peaks are tentatively assigned to $[AlN_4]$, $[AlO_4]$ and $[AlO_6]$. XRD pattern of the sample X_1 shows AlON as the major phase with some α-Al_2O_3. The peaks at 114 ppm for $[AlN_4]$ and 65 ppm for $[AlO_4]$ are assigned to AlON, and the peak at 12 ppm for $[AlO_6]$ must arise from AlON and α-Al_2O_3, i.e. there are $[AlN_4]$ and $[AlO_4]$ coordination in AlON. XRD pattern of the sample X_4 shows AlON as the major phase with some AlN, The peaks at 65 ppm for $[AlO_4]$ and 12 ppm for $[AlO_6]$ must arise from AlON, and the peak at 114 ppm for $[AlN_4]$ should arise from both AlN and AlON, i.e. there are $[AlO_4]$ and $[AlO_6]$ coordinations in AlON. It can be concluded that the aluminum will have a range of local tetrahedral and octahedral environments: $[AlN_4]$, $[AlO_4]$ and $[AlO_6]$ in AlON.

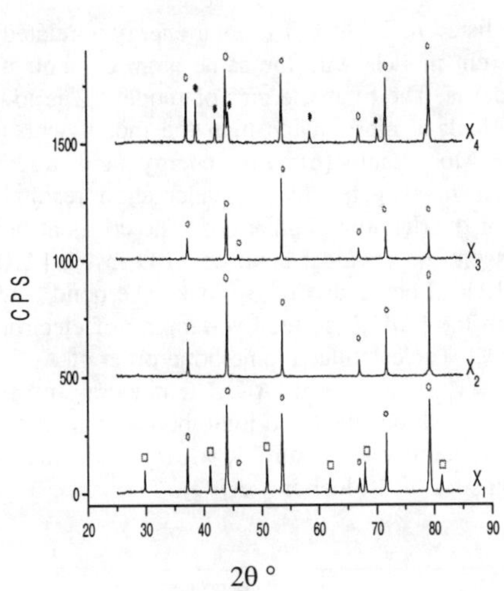

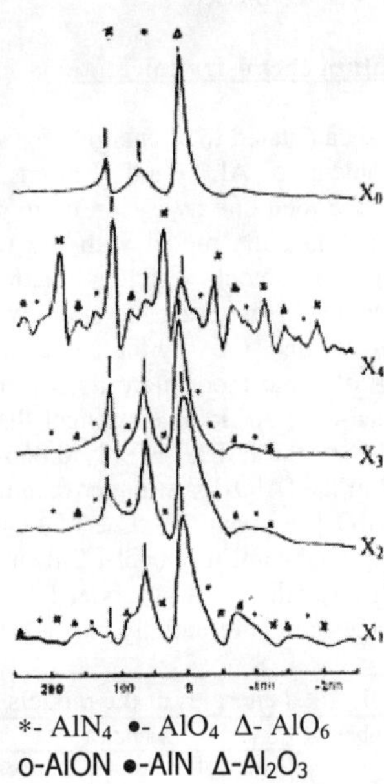

Fig.1 XRD patterns of the samples

Fig.2 ^{27}Al MAS NMR spectra of the samples with different content of AlN

The samples X_2 and X_3 are confirmed to be single-phase by XRD. The similarity of the XRD (Fig.1) makes unequivocal understanding of atomic ordering in the structure difficult. The ability of NMR to distinguish local coordination makes it an ideal complement to diffraction studies for the elucidation of such structures. As seen in Fig.2, the ^{27}Al NMR spectra clearly show the three aluminum coordinations at 114, 65 and 12 ppm. They arise from AlON. It is in agreement with the discussion of the samples X_1 and X_4 above. But the strength of peaks in the samples X_2 and X_3 reveal some difference, i. e., the [AlN₄] peak becomes stronger, the [AlO₄] and [AlO₆] signal become weaker with increasing AlN mol%. This means that more N atoms can substitute O atoms with increase the AlN mol% in single-phase AlON.

Since γ-Al_2O_3 and AlON are isostructural, it is quite likely that the peak positions for [AlO₄] and [AlO₆] may be very similar in these two materials. A comparison is made between X_0 and X_2. The observed peak position in X_0 is at 113, 66 and 8ppm. These peaks have to be due to the AlN and γ-Al_2O_3. The observed peak positions in X_2 are at 114, 66 and 12ppm. They must arise from AlON. There is little difference for the peak positions, but the intensity of the peaks changes. It means that the structure of AlON is similar to γ-Al_2O_3, but reveals different microstructure as N substitutes O at high temperature. The AlON has the typical peaks at 114, 66 and 12ppm, corresponding to [AlN₄], [AlO₄] and [AlO₆], respectively.

Work on β'-sialons has shown that there are mixed anion coordinations in β'-sialons. Mixed aluminum local coordinations $AlO_xN_{4-x}(x=1, 2, \text{ or } 3)$ were observed by Smith [5] for the first time, who employed a high-speed magic angle spinning (MAS)(up to 16 KHz) and high applied magnetic fields(up to 11.7T). In our study, no mixed coordinations are observed in AlON. The signal may be missed.

Quantum chemistry calculations

The calculated total energies for all models are listed in Table 1. The total energy is related to the contents of Al, O and N atoms. For the different models with the same atom contents, the lower the total energy is, the more stable the model is. The total energies of model I-2 is lower than I-1, i.e., the model with the vacancy in B-block is more stable than the model with the vacancy in A-block, which is in agreement with the Mo's results [6] by the energy band structure calculation. This order of stability can be further investigated by the calculation results of chemical bond, i. e., by the covalent bond orders or overlapping populations. The covalent bond orders of some models are listed in Table 2. As seen, the covalent bond of Al-O in the $[AlO_4]$ tetrahedral of A-block is stronger than that in the $[AlO_6]$ octahedral of B-block. The bond length of Al-O in the $[AlO_4]$ is only 0.866 time of that in the $[AlO_6]$, so the overlapping of electronic cloud in the $[AlO_4]$ is stronger than that in the $[AlO_6]$. The calculated ionic bond order of Al-O in the $[AlO_4]$ is about 1.5 time of that in the $[AlO_6]$. There are more $[AlO_4]$ tetrahedral and less $[AlO_6]$ octahedral in model I-2 than in model I-1, so both covalent and ionic bonds in model I-2 are stronger than that in model I-1, and model I-2 is more stable than I-1, which means that the vacancies are preferentially located at the octahedral sites of B-block in γ-Al_2O_3.

Table 1. Total energies of the models

number		vacancies			N atoms			total energies (a.u.)
		total	A-block	B-block	total	A-block	B-block	
I	1	1	1	0	0	0	0	-99073.74
	2	1	0	1	0	0	0	-99149.14
II	1	1	1	0	1	1	0	-98647.04
	2	1	0	1	1	1	0	-98756.15
	3	1	1	0	1	0	1	-98665.05
	4	1	0	1	1	0	1	-98703.36
III	1	1	1	0	2	2	0	-98301.31
	2	1	0	1	2	2	0	-98355.59
	3	1	1	0	2	0	2	-98292.25
	4	1	0	1	2	0	2	-98307.48
IV	1	1	1	0	3	3	0	-97775.62
	2	1	0	1	3	3	0	-97858.68
	3	1	1	0	3	0	3	-97814.99
	4	1	0	1	3	0	3	-97837.68
V	1	1	1	0	4	0	4	-97407.01
	2	1	0	1	4	0	4	-97402.31
	3	1	1	0	4	1	3	-97402.73
	4	1	0	1	4	1	3	-97405.78
	5	1	1	0	4	2	2	-97396.55
	6	1	0	1	4	2	2	-97411.35
	7	1	1	0	4	3	1	-97393.46
	8	1	0	1	4	3	1	-97416.27
	9	1	1	0	4	4	0	-97390.65
	10	1	0	1	4	4	0	-97420.69
VI	1	0	0	0	0	0	0	-103372.21
	2	0	0	0	1	1	0	-102966.30
VII	1	2	1	1	0	0	0	-94886.50
	2	2	1	1	1	1	0	-94457.34

Table 2. Covalent bond orders of some models

number		A block		B block	
		Al-O	Al-N	Al-O	Al-N
I	1	0.4835		0.3639	
	2	0.4960			
II	1	0.4741	0.5433	0.3614	
	2	0.4859	0.5550		
	3	0.4802		0.3644	0.3686
	4	0.4994			
VI	1	0.4785		0.3589	
	2	0.4661	0.5360	0.3620	
VII	1	0.4501			
	2	0.4414	0.5151		

For AlON, the total energies show that the stability order is II-2>II-1 and II-4>II-3, which is consistent with the conclusion above that the model with the vacancy in B-block is more stable than the model with the vacancy in A-block. As the valence electrons of N atom are $2s^2 2p^3$ and that of O atom is $2s^2 2p^4$, there are emptier valence orbits in N atom than in O atom. The overlapping of valence orbit between Al and N atoms is stronger than that between Al and O atoms, so the covalent bond of Al-N is stronger than that of Al-O. As seen in Table 2, the covalent bond of Al-N is stronger than that of Al-O in $[AlO_4]$ or $[AlO_6]$, and Al-N in $[AlO_3N]$ is stronger than in $[AlO_5N]$. The vacancy is in the B-block and there is $[AlO_3N]$ tetrahedral in the model II-2, so the model II-2 is the most stable in the models of class II. The same conclusion can be obtained from models of classes III and IV, i.e., model 2 is more stable than model 1, and model 4 is more stable than model 3, and model 2 is the most stable of four models in each class.

As ^{27}Al MAS NMR spectra of AlON are characterized by three resonances corresponding to $[AlN_4]$, $[AlO_4]$ and $[AlO_6]$, models of class V, with four N atoms in each models, are studied in more details. It contains 10 models (Table 1). From our calculations, we can give the following results. (1) Model V-4, 6, 8, or 10 is more stable than model V-3, 5, 7 or 9, respectively, which also means that the model with the vacancy in B-block is more stable than the model with the vacancy in A-block. Model V-1 is more stable than V-2, since there is no vacancy but 4 N atoms in the B-block of the model V-1. (2) The stability order of models, V-9 < 7 < 5 < 3 < 1, with no vacancy in B-block in each model, is corresponding to the increase in the content of the N atoms in the B-block. (3) The stability order of models, V-2 < 4 < 6 < 8 < 10, with no vacancy in A-block in each model, is corresponding to the increase in the content of the N atoms in the A-block. (4) Model V-10, with no vacancy but four N atoms in A-block, is the most stable, which also shows that the vacancy located in B-block is more stable than that located in A-block. Nitrogen atoms preferentially replace oxygen atoms in $[AlO_4]$ tetrahedral, and the model with $[AlN_4]$ is more stable than that with $[AlO_{4-x}N_x]$ (x=1, 2 and 3) and $[AlO_{6-x}N_x]$ (x=1, 2, 3 and 4). This seems consistent with the ^{27}Al MAS NMR spectra of AlON characterized by three resonances for $[AlN_4]$, $[AlO_4]$ and $[AlO_6]$ discussed above.

As nitrogens preferentially replace oxygens in $[AlO_4]$ tetrahedral, the models $[AlO_3N]$ with one N atom replacing are included in the classes VI and VII. As discussed above, The order of total energies is (VI-2) + (VII-2) < 2(II-10) or (II-1) + (II-2), but < 2(II-2). Model II-2 is the most stable in these 4 models. This also shows that the vacancies are preferentially located at the octahedral sites of B-block in AlON, and nitrogen atoms preferentially replace oxygen atom in $[AlO_4]$ tetrahedral. The same results also can be obtained from the models with more N atoms replacing.

CONCLUSION

^{27}Al MAS NMR and quantum chemistry calculation have been used to study the structure of γ-Al_2O_3 and AlON. ^{27}Al MAS NMR spectra of AlON are characterized by three resonances at 114, 66 and 12 ppm corresponding to $[AlN_4]$, $[AlO_4]$ and $[AlO_6]$ respectively, and vary with composition (AlN mol%). The calculations have shown that the vacancy located at octahedral site in B-block is more stable than that located at tetrahedral site in A-block; Nitrogen atoms preferentially replace oxygen atoms in $[AlO_4]$ tetrahedral, and the model with $[AlN_4]$ is more stable than that with $[AlO_{4-x}N_x]$ (x=1, 2 and 3) and $[AlO_{6-x}N_x]$(x=1, 2, 3 and 4), which is in agreement with the experimental analysis by ^{27}Al MAS NMR.

ACKNOWLEDGEMENT

This work was supported by the National Science Foundation of China through Outstanding Young Scientist Foundation under No.59825102.

REFERENCES

1.T. M. Hartnett, E. A. Maguire, Ceram. Eng. Sci. Proc, 3, 67 (1982).
2.G.Yamaguchi, Chem. Soc. Japan Bull. 32, 1264 (1959).
3.N. D. Corbin, J. Eur. Ceram. Soc. 5, 143 (1989).
4.P. Goursat, M.Billey, Mat. Chem. 6, 81 (1981).
5.M E. Smith, J. Phys. Chem. 96, 1444 (1992).
6.Shang.Di Mo,Yong-Nian Xu, J. Am. Ceram. Soc, 80, 1193 (1997).

STRUCTURAL CHARACTERISTICS AND QUANTUM CHEMISTRY CALCULATION OF Si-DOPED BORON CARBIDES

Min Xinmin, Nan Cewen and <u>Cai Kefeng</u>
National Key Laboratory of Advanced Technology for Materials Synthesis and Processing, Wuhan University of Technology, Wuhan, Hubei 430070, P. R. China

ABSTRACT

Structural characteristics, chemical bonds and thermoelectric properties of Si-doped boron carbides are studied through calculations of various structural unit models by using a self-consistent-field discrete variation X_α method. The calculations show that Si atom doped in boron carbide is in preference to substituting B or C atoms on the end of boron carbide chain, and then may occupy interstitial sites, but it is difficult for Si to substitute B or C atom in the centers of chain or in the icosahedra. A representative structural unit containing a Si atom is $[C\text{-}B\text{-}Si]^{\varepsilon+}\text{-}[B_{11}C]^{\varepsilon-}$, while the structural unit without Si is $[C\text{-}B\text{-}B(C)]^{\delta-}\text{-}[B_{11}C]^{\delta+}$, and the coexistence of these two different structural units makes the electrical conductivity increases. As the covalent bond of Si-B or Si-C is weaker than that of B-B or B-C, the thermal conductivity decreases when Si is added into boron carbides. With the electrical conductivity increases and the thermal conductivity decreases, Si doping has significant effect on thermoelectric properties of boron carbides.

INTRODUCTION

Boron carbide has attracted much attention in recent years[1-3], because of its large Seebeck coefficient, low thermal conductivity and high-temperature stability. Besides of many other uses, Boron carbide is an important potential candidate for applications in high-temperature thermoelectric conversion. Most work on the boron carbides has been focused on the effect of carbon concentration on the thermoelectric properties and structure of boron carbides of B_xC[4]. However, work on doping effect in the boron carbides is limited[3,5-6]. Wood[5] reported a p-type sample of B_9C doped with 0.5% Mg with value of thermoelectric figure-of-merit (Z) as large as 0.5×10^{-3} at 1000 °C; Aselage[3] studied P-doped boron carbide. Our experiment (it will be reported elsewhere in detail) shows that Si doping has significant effect on thermoelectric properties of boron carbides.

In order to understand the structure and behaviour of boron carbides, some band-structure theoretical calculations of boron carbides have been done[7-9]. However, there have been a few controversies in those existing results. For example, calculations of electronic band structure of $B_{13}C_2$ and $B_{12}C_3$ by Armstrong *et al*[8] shown that inter-icosahedral bonds are much stronger than intra-icosahedral bonds. By contrast, Bylabder *et al*'s results[9] shown that the intra-icosahedral bonds are much stronger than inter-icosahedral bonds. On the other hand, Aselage[3] considered that the chain is positively charged and icosahedron is negatively charged, i.e., $[CBC]^+\text{-}[B_{11}C]^-$, but calculations by Kirfel *et al*[7] did not indicate a deviation of the icosahedron from electrical neutrality, while calculations by Armstrong *et al*[8] illustrated that the icosahedron is positively charged. By using self-consistent-field discrete variation X_α (i.e., SCF-DV-X_α) method[10], our calculations[11] for nine different structure models illustrated that the total strength of five covalent bonds between an atom and its five co-ordinating atoms in the icosahedron is slightly larger than the three times of the strength of a single bond such as

Mat. Res. Soc. Symp. Proc. Vol. 538 © 1999 Materials Research Society

one inter-icosahedral bond. Therefore, by comparing the strengths between one intra-icosahedral bond and one inter-icosahedral bond, the inter-icosahedral bond is much stronger than intra-icosahedral bond, which is consistent with that the inter-icosahedral bond is shorter than the intra-icosahedral bond. But by comparing the bond strengths between the intra-icosahedral and inter-icosahedral bonds on the whole, the intra-icosahedral bonds are stronger than inter-icosahedral bonds. Our calculations also indicated that the icosahedron is slightly positively charged and approaches electric neutrality, i.e., $[C\text{-}B\text{-}B(C)]^{\delta-}\text{-}[B_{11}C]^{\delta+}$ (δ being close to zero), which is consistent with the analysis of electronegativity and similar to the calculations by Kirfel[7] and Armstrong et al[8].

To our knowledge, the calculations of the doped boron carbides is still missing. In this paper, also by using SCF-DV-X_α method, we calculated the structural characteristics of Si-doped boron carbides. The correlation between the composition, structure, chemical bond and thermoelectric properties of Si-doped boron carbides is discussed.

STRUCRURAL MODELS FOR CALCULATIONS

As boron carbides contain different B_{12} and $B_{11}C$ icosahedra, and C-B-C and C-B-B chains, this complicated structure makes it difficult to study the correlation between the structure, chemical bond and properties by using only single model. For illustrative and qualitative purposes, various structural unit models of 5 classes are constructed, as shown in table 1. As an example, model I-4 of $B_{13}C_2$, or C-B-C-B_{12} (i.e. C-B-C chains and B_{12} icosahedra as shown in Fig. 1, without Si(i)), boron atoms in the B_{12} icosahedra can be divided into two types. One is six boron atoms in the equatorial puckered hexagon, which is called as B1. The other is six boron atoms in the top and bottom triangles, symbolised as B2. The boron atoms in the center of the chain are denoted as B3. The C-B-C chain lies along the three-fold axis and bonds to the B_{12} icosahedra through C-B1. On the other hand, the icosahedra bond to one another directly along the B2-B2 bond. This model contains 54 atoms with the operation of the three-fold axis. These 54 atoms can be divided into 16 classes according to their positions of symmetry. The representative one of each class of atoms is symbolised by the numbers in the brackets (shown in Fig.1), and these numbers are different from other numbers in B1, B2 and B3 which are not in the brackets. For convenience, this model is simply expressed as C-B-C-B_{12} and denoted as model I-4. Based on model I-4, other models can be constructed by changing the atoms in the chain or icosahedron. Six different models of class I(table 1) are chosen. The structure of each model is the same as that in Fig.1, but with some different atoms. For example, model I-5, C-B-C-$B_{11}C(9)$, represents that B1(9) in model I-4 is substituted with C(9); and for model I-6, C-B-C-$B_{11}C(5)$, B2(5) in model I-4 is substituted with C(5). Both models I-5 and I-6 contain C-B-C chains and $B_{11}C$ icosahedra, but the sites of C atoms in the $B_{11}C$ icosahedra are different in this two models.

The models of class II represent that Si atoms substitute for the atoms on the end of chain of class I. In the models of class III, Si atoms occupy the interstitial sites between the chain and icosahedra(see from Fig.1 with Si(i)). In the models of class IV and V, Si atoms substitute for the center atoms in the chains and the atoms in the icosahedra of class I, respectively.

RESULTS AND DISCUSSION

Table 1 shows the calculated total energies of all models, which are related to the contents of B, C and Si atoms. For the different models with the same B, C and Si contents, the lower the

nearest to the Si(13) at the chain end. Around the Si(i) in the models of class III, there are more first-nearest atoms of B or C with larger electronegativity than Si, which leads to higher net charges of Si in the models of class III than in class II(table 1). This is an important difference between models of classes III and II, and is also the reason that models of class III is less stable than models of class II. The representative structural unit is $[C\text{-}B\text{-}Si]^{\varepsilon+}\text{-}[B_{11}C]^{\varepsilon-}$ for the models of class II, but $[C\text{-}B\text{-}B(C)]^{\varepsilon-}\text{-}[B_{11}C]^{\varepsilon'-}\text{-}Si(i)^{(\varepsilon+\varepsilon')+}$ for class III, which is another important difference between models of classes III and II.

The net charges of Si(14) atoms in models of class IV are 0.7 $\sim$ 0.8 and higher than those of Si(13) in class II, which is one reason why models of class IV is less stable than models of class II. On the other hand, the net charges of B3(14) atoms in models of class II are 0.0625, -0.0670 and 0.0720, respectively. Compared with B3(14) in class II, the Si(14) atoms in class IV have very high positive charges, and less electrical cloud on the Si(14) than B3(14) in class II. Correspondingly, the calculated covalent bond orders between Si(14) and its coordinating atoms in class IV are less than those between B3(14) and its coordinating atoms in class II. In other words, the covalent bonds between Si(14) and its coordinating atoms in class IV is weaker than those between B3(14) and its coordinating atoms in class II. For example, the calculated covalent bond order of C(13)-Si(14) is 0.6954 in model IV-3, but the bond order of Si(13)-B3(14) is 0.8450 in model II-3. This is another reason that models of class IV is less stable than models of class II.

In the models of class V, as Si substitutes for the atoms in icosahedra, the bond lengths in the icosahedra become larger. Correspondingly, their covalent bond orders, or the strengths of the covalent bond decrease obviously. For example, the calculated covalent bond order of Si(9)-B1(7) is 0.2955 in model V-3, but that of B1(9)-B1(7) is 0.5295 in model II-3. Therefore, the models of class V is less stable than the models of class II.

Finally, let us discuss the relation between the structural calculations above and electronic transport, or thermoelectric properties. The electronic transport measurements (electrical conductivity, Seebeck coefficient and Hall mobility)[12] have been shown to be polaronic hopping and that the carriers are holes for all carbon concentrations. It has been suggested that since the carriers are associated with the icosahedra, the bipolaron formation may occur by the disproportionation reaction[3] of $2[B_{11}C]^0 \rightarrow [B_{11}C]^- + [B_{11}C]^+$, where $[B_{11}C]^+$ represents a bipolaronic hole. Then the electrical transport is through polaronic hopping of $[B_{11}C]^+ + [B_{11}C]^- \rightarrow [B_{11}C]^- + [B_{11}C]^+$. If Si occupies only the end of chain, for example, in the model II-3, C-B-Si-B_{11}C, the content of Si is 15.5wt%. In our experiment, when the content of Si is about 4.0wt%, the electrical conductivity has the maximum and the sample is with the best improvement of properties. For the 4.0wt% content of Si-doped boron carbides, the larger part of the structural units is still C-B-B-B_{11}C and C-B-C-B_{11}C, or $[C\text{-}B\text{-}B]^{\delta-}$, $[C\text{-}B\text{-}C]^{\delta-}$ and $[B_{11}C]^{\delta+}$. The disproportionation reaction is still through the reaction of B_{11}C linking the chain without Si. Compared with boron carbides without Si, because of existing both $[C\text{-}B\text{-}B]^{\delta-}$ and $[C\text{-}B\text{-}Si]^{\varepsilon+}$ chains linking to the icosahedra in Si-doped boron carbides, $[C\text{-}B\text{-}B]^{\delta-}$ and $[C\text{-}B\text{-}Si]^{\varepsilon+}$ both may combine to the products of $[B_{11}C]^+$ and $[B_{11}C]^-$ of the disproportionation reaction, respectively. The energy of the disproportionation reaction decreases and the reaction becomes more easily, which is favorable for the formation of $[B_{11}C]^+$ and $[B_{11}C]^-$. Therefore, the concentration of the carriers increase. On the other hand, as $[B_{11}C]^+$ may get electrons from both of $[B_{11}C]^-$ and $[B_{11}C]^{\delta-}$, the polaronic hopping of $[B_{11}C]^+ + [B_{11}C]^- \rightarrow [B_{11}C]^- + [B_{11}C]^+$ can transfer through more paths. Because of the above two reasons, i.e., the effect of the coexistence of the two different structural units of $[C\text{-}B\text{-}B(C)]^{\delta-}\text{-}[B_{11}C]^{\delta+}$ and $[C\text{-}B\text{-}Si]^{\varepsilon+}\text{-}[B_{11}C]^{\varepsilon-}$, the electrical conductivity increases when Si is doped into boron carbide. If Si occupies the interstitial site, the representative structural unit is $[C\text{-}B\text{-}B(C)])^{\varepsilon^-}\text{-}[B_{11}C]^{\varepsilon'^-}\text{-}Si(i)^{(\varepsilon+\varepsilon')+}$. There is also the effect of the coexistence of the different structural units and the same result can be obtained.

Compared with B and C atoms, as Si atom has larger positive charge and less electronic cloud, the covalent bond of Si-B or Si-C is weaker than that of B-B or B-C. For example, the covalent bond order of Si(13)-B3(14) is 0.8207 in model II-1, and that of B(13)-B3(14) is 1.1808 in model I-1. It is consistent with our experimental result that the thermal conductivity decreases when Si is added into boron carbides. With the electrical conductivity increases and the thermal conductivity decreases, Si doping has significant effect on thermoelectric properties of boron carbides.

IV. CONCLUSION

By using SCF-DV-X_α method, five classes of models, four classes of them with Si atoms occupying the different sites, are calculated, and the results show that the doped Si atoms are in preference to substitute B or C atoms on the ends of chain, and then may occupy interstitial sites, but it is difficult for Si to substitute B or C atom in the centers of chain or in the icosahedra. A representative structural unit containing a Si atom is $[C\text{-}B\text{-}Si]^{\varepsilon^+}\text{-}[B_{11}C]^{\varepsilon^-}$, while the structural unit without Si is $[C\text{-}B\text{-}B(C)]^{\delta^-}\text{-}[B_{11}C]^{\delta^+}$. Because of the effect of the coexistence of these different structural units in Si doped boron carbide, $[C\text{-}B\text{-}B]^{\delta^-}$ and $[C\text{-}B\text{-}Si]^{\varepsilon^+}$ both may combine to the products of $[B_{11}C]^+$ and $[B_{11}C]^-$ of the disproportionation reaction, respectively, so the disproportionation reaction becomes more easily. On the other hand, as $[B_{11}C]^+$ may get electrons from both of $[B_{11}C]^-$ and $[B_{11}C]^{\delta^-}$, the polaronic hopping of $[B_{11}C]^+ + [B_{11}C]^- \rightarrow [B_{11}C]^- + [B_{11}C]^+$ can transfer through more paths. Therefore, the electrical conductivity increases when Si dopes into boron carbide. As the covalent bond of Si-B or Si-C is weaker than that of B-B or B-C, the thermal conductivity decreases when Si is added into boron carbides. With the electrical conductivity increases and the thermal conductivity decreases, Si doping has significant effect on thermoelectric properties of boron carbides.

ACKNOWLEDGEMENT

This work was supported by the National Science Foundation of China through Outstanding Young Scientist Foundation under No.59825102.

REFERENCES

1. H.Werheit, *J. Mater. Sci. Eng. B* **29**, p.228(1995).
2. F.Thevenot, *J. Europ. Ceramic Soc.* **6**, p.205(1990).
3. T.L.Aselage, *Mat. Res. Soc. Symp. Proc.* **234**, p.145(1991).
4. M.Bouchacourt and F. Thevenot, *J. Mat. Sci.* **20**, p.1237(1985).
5. C.Wood, *Boron-Rich Solids*, **140**, p.362(1986).
6. R.Telle, *The Physics and chemistry of carbides: Nitrides, and Borides*, **185**, p.249(1990).
7. A.Kirfel, A.Gupta and G.Will, *Acta Cryst. B* **35**, p.1052(1979).
8. D.R.Armstrong, J.Bolland and P.G.Perkins, *Acta Cryst. B* **39**, p.324(1983).
9. D.M.Bylander, L.Kleinman and S.Lee, *Phys. Rev. B* **42**, p.1394(1990).
10. D.E.Ellis and G.S.Painter, *Phys. Rev. B* **2**, p.2887(1970).
11. Min Xinmin, Nan Cewen and Cai Kefeng, The other paper in this Meeting.
12. C.Wood and D.Emin, *Phys. Rev. B* **29**, p.4582(1984).

AUTHOR INDEX

SUBJECT INDEX

stress, 353, 497
structural unit model, 383
structure, 573, 579
subspace hyperdynamics, 27
substitution behavior, 377
superalloys, 547
surface
    chemistry, 275
    diffusion, 263
    morphology, 323
    strain mapping, 179

temperature effects, 93
texture clustering, 179
thermal expansion, 535
thermoelectric property, 579
thermomechanical stress, 263

tight-binding, 275, 353, 413, 529
trigonal path, 529
transient enhanced diffusion, 291
transition state, 51
treadmill, 497
twinning(-), 515
    anti-twinning asymmetry, 87

vacancy(ies), 395, 389
    cluster, 223
    diffusion, 359

weak beam microscopy, 329

yield
    point, 3
    stress anomaly, 329